Optical Engineering

The Design of Optical Systems

Warren J. Smith

Chief Scientist, Consultant
Rockwell Collins Optronics
Carlsbad, California

Fourth Edition

Mc
Graw
Hill

New York   Chicago   San Francisco   Lisbon   London   Madrid
Mexico City   Milan   New Delhi   San Juan   Seoul
Singapore   Sydney   Toronto

1 2 3 4 5 6 7 8 9 0   DOC/DOC   0 1 3 2 1 0 9 8 7

ISBN 9781265902650
MHID 1265902658

Sponsoring Editor
Taisuke Soda

Production Supervisor
Richard C. Ruzycka

Editorial Supervisor
Stephen M. Smith

Project Manager
Sam RC, International
Typesetting and Composition

Copy Editor
Priyanka Sinha, International
Typesetting and Composition

Proofreader
Benodini Banerjee

Indexer
Robert Swanson,
ARC Films, Inc

Art Director, Cover
Jeff Weeks

Composition
International Typesetting
and Composition

Affectionately dedicated to my wife Dung My Luong

ABOUT THE AUTHOR

Warren J. Smith is semi-retired at Rockwell Collins Optronics in Carlsbad, California, after more than six decades of practice in optical engineering, design, and fabrication. He is a consultant, writer, lecturer, occasional expert witness, and one of the most widely known writers and educators in the field of optical design. Mr. Smith is a Fellow and Past National President of both SPIE and the Optical Society of America. He is also the author of *Modern Lens Design*, now in its Second Edition, and *Practical Optical System Layout*, both published by McGraw-Hill. He resides with his wife in Vista, California.

Contents

Preface

Welcome to the fourth edition of *Modern Optical Engineering*, known fondly to many by its acronym, MOE. Many of the revisions from the third edition should be quite apparent—it is a bigger book, with reorganized content and six more chapters. There are dozens of smaller changes and additions which are embedded in the new MOE—too many to list here.

MOE now has 21 chapters, two appendices, and a glossary, compared with 15 chapters in the third edition, and 14 chapters in the first and second editions. MOE has grown steadily over the years, beginning with 476 pages in the first edition, then 524 in the second, 617 in the third edition, and now 768 pages in this one. The earlier editions of MOE competently filled a niche in the field, and I hope this edition will do the same. The preface to the first edition began with this paragraph:

This book is dedicated to the practicing engineer or scientist who requires effective practical technical information on optical systems and their design. The recent increase in the utilization of optical devices in such fields as alignment, metrology, automation, and space/defense applications has brought about a need for technical people conversant with the optical field. Thus, many individuals whose basic training is in electronics, mechanics, physics, or mathematics, find themselves in positions requiring a relatively advanced competence in optical engineering. It is the author's hope that this volume will enable them to undertake their practice of optics soundly and with confidence.

Over the years the number of people practicing optics has increased many-fold, and the field of optical engineering has broadened to include many new, unforeseen applications as well, but the need cited above continues. The wide acceptance of MOE as a "must have" book has been gratifying, and I hope that this new edition will be equally well received.

Among the more apparent changes in this edition are: A modest revision of the organization of the entire book. The old second chapter has been divided into three chapters, gaussian optics, paraxial calculations, and the optics of combinations. The chapter on aberrations has been enhanced by adding a chapter on third-order aberration theory and computation. Most of the old chapter on raytracing has been banished to App. A. An entirely new chapter of first-order system layout case studies, consisting of worked-out numerical examples, has been added, including an example of zoom lens layout and an example of an infrared cooled detector system. The two chapters of the earlier editions which dealt with lens design are now three (The Basics of Lens Design, Lens Design for Eyepieces, Microscopes, Cameras, etc., and Design of Mirror and Catadioptric Systems). The previous Chap. 14 which had "Some 44 More Lens Designs" has been expanded to some 62 designs, and the presentation of the aberrations has been improved to include not only the ray intercept plots and longitudinal spherical (now in three wavelengths), monochromatic distortion and field curvature, but also longitudinal color and lateral color (in three wavelengths). This same presentation style has been used throughout the book, replacing the old-fashioned monochromatic longitudinal plots of spherical and field curvature. An entirely new chapter on the use of "stock" lenses has been added following the chapter on the practical side of the practice of optics. The numerical exercises which follow the early chapters are now completely worked out, instead of simply providing the correct answers. There is a new Glossary, which I have attempted to make the best, most complete, most authoritative, and most accurate glossary in optical system design literature. The two new appendices deal with raytracing and standard dimensions. A worked-out demonstration of designing a tolerance budget has been added to the optical practice chapter (Chap. 20). A series of entertaining do-it-yourself visual experiments conclude the chapter on the eye.

In my opinion no single individual or small group can write "the complete book" on optical engineering and optical system design. Any book tends to deal primarily with its author's personal experience (and none of us has done everything). So I take this opportunity to strongly suggest that the reader who finds MOE interesting and worthwhile should also read Kingslake, *Optical System Design*, Academic, 1983, and also Fischer and Tadic, *Optical System Design*, McGraw-Hill, 2000, in order to avail themselves of additional (and competent) viewpoints on the subject.

In closing, I want to express my gratitude to the many friends and colleagues who have shared their knowledge with me during six decades of enjoying my work in optics—with their help I learn something every day.

In particular, I want to acknowledge, with thanks, Jerry Carollo and Greg Newbold of Rockwell Collins Optronics for my office in Carlsbad, and Leo Gardner of Lambda Research for providing OSLO, the optical design program which I use daily.

I wish you well.

Warren J. Smith

1

Optics Overview

1.1 The Electromagnetic Spectrum

This book deals with certain phenomena associated with a relatively narrow slice of the electromagnetic spectrum. Optics is often defined as being concerned with radiation visible to the human eye; however, in view of the expansion of optical applications in the regions of the spectrum on either side of the visible region, it seems not only prudent, but necessary, to include certain aspects of the infrared and ultraviolet regions in our discussions.

The known electromagnetic spectrum is diagramed in Fig. 1.1 and ranges from cosmic rays to radio waves. All the electromagnetic radiations transport energy and all have a common velocity in vacuum of $c = 2.998 \times 10^{10}$ cm/s. In other respects, however, the nature of the radiation varies widely, as might be expected from the tremendous range of wavelengths represented. At the short end of the spectrum, we find gamma radiation with wavelengths extending below a billionth of a micron (one micron or micrometer = 1 μm = 10^{-6} m) and at the long end, radio waves with wavelengths measurable in miles. At the short end of the spectrum, electromagnetic radiation tends to be quite particle-like in its behavior, whereas toward the long wavelength end the behavior is mostly wavelike. Since the optical portion of the spectrum occupies an intermediate position, it is not surprising that optical radiation exhibits both wave and particle behavior.

The visible portion of this spectrum (Fig. 1.2) takes up less than one octave, ranging from violet light with a wavelength of 0.4 μm to red light with a wavelength of 0.76 μm. Beyond the red end of the spectrum lies the infrared region, which blends into the microwave region at a

WAVELENGTH
IN MICROMETERS
(MICRONS)

FREQUENCY
IN CYCLES PER SECOND (Hz)

Frequency (Hz)	Region	Wavelength	Unit
10^{23}	COSMIC RAYS	10^{-9}	
10^{22}		10^{-8}	
10^{21}	GAMMA RAYS	10^{-7}	X-UNIT
10^{20}		10^{-6}	
10^{19}		10^{-5}	
10^{18}	X-RAYS	10^{-4}	ANGSTROM UNIT (Å)
10^{17}		10^{-3}	NANOMETER MILLIMICRON
10^{16}		10^{-2}	
10^{15}	ULTRAVIOLET	10^{-1}	
10^{14}	VISIBLE	1	MICROMETER (μm) (MICRON)
10^{13}	INFRARED	10	
10^{12}		10^{2}	
10^{11} EHF		10^{3}	MILLIMETER (mm)
10^{10} SHF	MICROWAVE	10^{4}	CENTIMETER (cm)
10^{9} UHF		10^{5}	
10^{8} VHF	F.M. RADIO	10^{6}	METER (m)
10^{7} HF	TELEVISION	10^{7}	
10^{6} MF	A.M. RADIO	10^{8}	
10^{5} LF		10^{9}	KILOMETER (km)
10^{4}		10^{10}	
10^{3}		10^{11}	
10^{2}		10^{12}	
10		10^{13}	
1		10^{14}	

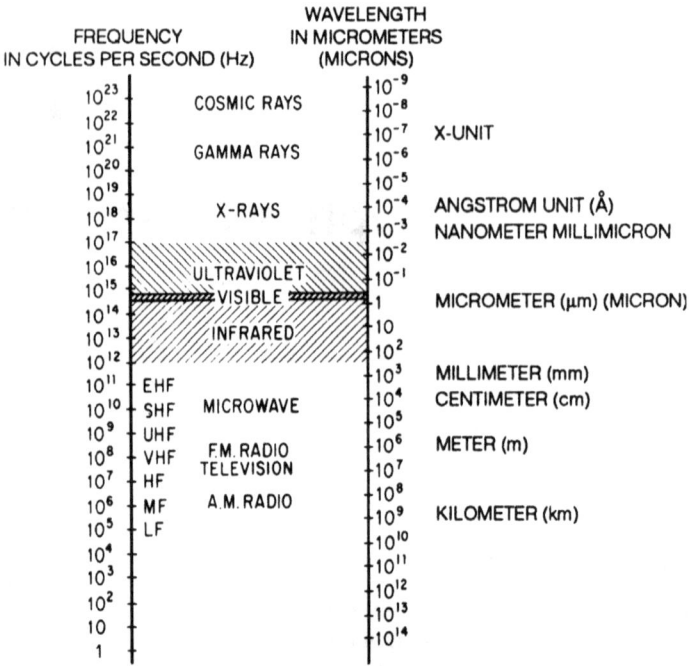

Figure 1.1 The electromagnetic spectrum.

wavelength of about 1 mm. The ultraviolet region extends from the lower end of the visible spectrum to a wavelength of about 0.01 μm at the beginning of the x-ray region. The wavelengths associated with the colors seen by the eye are indicated in Fig. 1.2.

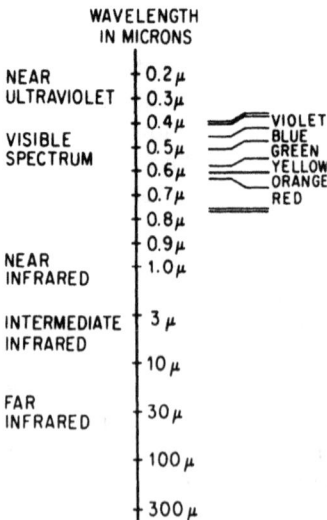

WAVELENGTH
IN MICRONS

Region	Wavelength	Color
NEAR ULTRAVIOLET	0.2 μ	
	0.3 μ	
	0.4 μ	VIOLET
VISIBLE SPECTRUM	0.5 μ	BLUE / GREEN
	0.6 μ	YELLOW / ORANGE
	0.7 μ	RED
	0.8 μ	
	0.9 μ	
NEAR INFRARED	1.0 μ	
INTERMEDIATE INFRARED	3 μ	
	10 μ	
FAR INFRARED	30 μ	
	100 μ	
	300 μ	

Figure 1.2 The "optical" portion of the electromagnetic spectrum.

TABLE 1.1 Commonly Used Wavelength Units

Centimeter	=	10^{-2} meter		
Millimeter	=	10^{-3} meter		
Micrometer	=	10^{-6} meter	=	10^{-3} millimeter
Micron	=	10^{-6} meter	=	10^{-3} millimeter
Millimicron	=	10^{-3} micron	=	1.0 nanometer
	=	10^{-6} millimeter		
	=	10^{-9} meter		
Nanometer	=	10^{-9} meter	=	1.0 millimicron
Angstrom	=	10^{-10} meter	=	0.1 nanometer

The ordinary units of wavelength measure in the optical region are the angstrom (Å); the millimicron (mμ), or nanometer (nm); and the micrometer (μm), or micron (μ). One micron is a millionth of a meter, a millimicron is a thousandth of a micron, and an angstrom is one ten-thousandth of a micron (see Table 1.1). Thus, 1.0 Å = 0.1 nm = 10^{-4} μm. The frequency equals the velocity c divided by the wavelength, and the wavenumber is the reciprocal of the wavelength, with the usual dimension of cm^{-1}.

1.2 Light Wave Propagation

If we consider light waves radiating from a point source in a vacuum as shown in Fig. 1.3, it is apparent that at a given instant each wave front is spherical in shape, with the curvature (reciprocal of the radius) decreasing as the wave front travels away from the point source. At a sufficient distance from the source, the radius of the wave front may be regarded as infinite. Such a wave front is called a plane wave.

The distance between successive waves is of course the wavelength of the radiation. The velocity of propagation of light waves in vacuum is approximately 3×10^{10} cm/s. In other media the velocity is less than in vacuum. In ordinary glass, for example, the velocity is about two-thirds of the velocity in free space. The ratio of the velocity in vacuum to the velocity in a medium is called the index of refraction of that medium, denoted by the letter n.

$$\text{Index of refraction } n = \frac{\text{velocity in vacuum}}{\text{velocity in medium}}$$
$$= \frac{\text{wavelength in vacuum}}{\text{wavelength in medium}} \quad (1.1)$$

Both wavelength and velocity are reduced by a factor of the index; the frequency remains constant.

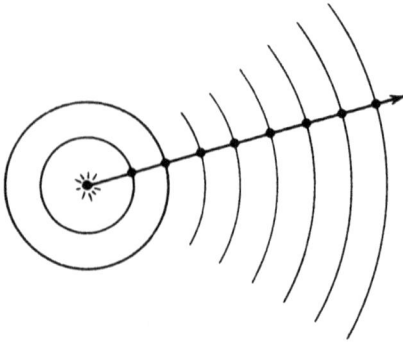

Figure 1.3 Light waves radiating from a point source in an isotropic medium take a spherical form; the radius of curvature of the wave front is equal to the distance from the point source. The path of a point on the wave front is called a light ray, and in an isotropic medium is a straight line. Note also that the ray is normal to the wave front.

Ordinary air has an index of refraction of about 1.000277, and since almost all optical work (including measurement of the index of refraction) is carried out in a normal atmosphere, it is a highly convenient convention to express the index of a material relative to that of air (rather than vacuum), which is then assumed to have an index of exactly 1.0.

The actual index of refraction for air at 15°C is given by

$$(n-1) \times 10^8 = 8342.1 + \frac{2{,}406{,}030}{(130-v^2)} + \frac{15{,}996}{(38.9-v^2)}$$

where $v = 1/\lambda$ (λ = wavelength, in μm). At other temperatures the index may be calculated from

$$(n_t - 1) = \frac{1.0549 \, (n_{15°} - 1)}{(1 + 0.00366t)}$$

The change in index with pressure is 0.0003 per 15 lb/in², or 0.00002/psi.

If we trace the path of a hypothetical point on the surface of a wave front as it moves through space, we see that the point progresses as a straight line. The path of the point is thus what is called a ray of light. Such a light ray is an extremely convenient fiction, of great utility in understanding and analyzing the action of optical systems, and we shall devote the greater portion of this volume to the study of light rays. Note well that the ray is normal to the wave front, and vice versa.

The preceding discussion of wave fronts has assumed that the light waves were in a vacuum, and of course that the vacuum was isotropic, i.e., of uniform index in all directions. Several optical crystals are anisotropic; in such media wave fronts as sketched in Fig. 1.3 are not spherical. The waves travel at different velocities in different directions, and thus at a given instant a wave in one direction will be further from the source than will a wave traveling in a direction for which the media has a larger index of refraction.

Although most optical materials may be assumed to be isotropic, with a completely homogeneous index of refraction, there are some significant exceptions. The earth's atmosphere at any given elevation is quite uniform in index, but when considered over a large range of altitudes, the index varies from about 1.0003 at sea level to 1.0 at extreme altitudes. Therefore, light rays passing through the atmosphere do not travel in exactly straight lines; they are refracted to curve toward the earth, i.e., toward the higher index. Gradient index optical glasses are deliberately fabricated to bend light rays in controlled curved paths. In this text we shall assume homogeneous media unless specifically stated otherwise. We also assume that lens elements are immersed in air.

1.3 Snell's Law of Refraction

Let us now consider a plane wave front incident upon a plane surface separating two media, as shown in Fig. 1.4. The light is progressing from the top of the figure downward and approaches the boundary surface at an angle. The parallel lines represent the positions of a wave front at regular intervals of time. We shall call the index of the upper medium n_1 and that of the lower n_2. From Eq. 1.1, we find that the velocity in the upper medium is given by $v_1 = c/n_1$ (where c is the velocity in vacuum $\approx 3 \times 10^{10}$ cm/s) and in the lower by $v_2 = c/n_2$. Thus, the velocity in the upper medium is n_2/n_1 times the velocity in the lower, and the distance which the wave front travels in a given interval of time in the upper medium will also be n_2/n_1 times that in the lower. In Fig. 1.4 the index of the lower medium is assumed to be larger so that the velocity in the lower medium is less than that in the upper medium.

At time t_0 our wave front intersects the boundary at point A; at time $t_1 = t_0 + \Delta t$ it intersects the boundary at B. During this time it has moved a distance

$$d_1 = v_1 \Delta t = \frac{c}{n_1} \Delta t \qquad (1.2a)$$

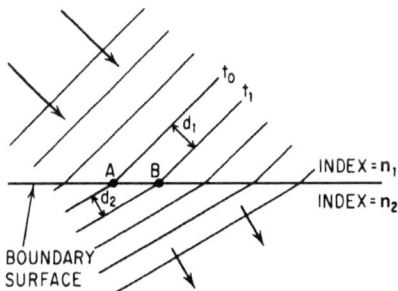

Figure 1.4 A plane wave front passing through the boundary between two media of different indices of refraction ($n_2 > n_1$).

in the upper medium, and a distance

$$d_2 = v_2 \, \Delta t = \frac{c}{n_2} \, \Delta t \qquad (1.2b)$$

in the lower medium.

In Fig. 1.5 we have added a ray to the wave diagram; this ray is the path of the point on the wave front which passes through point B on the surface and is normal to the wave front. If the lines represent the positions of the wave at equal intervals of time, AB and BC, the distances between intersections, must be equal. The angle between the wave front and the surface (I_1 or I_2) is equal to the angle between the ray (which is normal to the wave) and the normal to the surface XX'. Thus we have from Fig. 1.5

$$AB = \frac{d_1}{\sin I_1} = BC = \frac{d_2}{\sin I_2}$$

and if we substitute the values of d_1 and d_2 from Eqs. 1.2a and 1.2b, we get

$$\frac{c \, \Delta t}{n_1 \sin I_1} = \frac{c \, \Delta t}{n_2 \sin I_2}$$

which, after canceling and rearranging, yields

$$n_1 \sin I_1 = n_2 \sin I_2 \qquad (1.3)$$

This expression is the basic relationship by which the passage of light rays is traced through optical systems. It is called *Snell's law* after one of its discoverers.

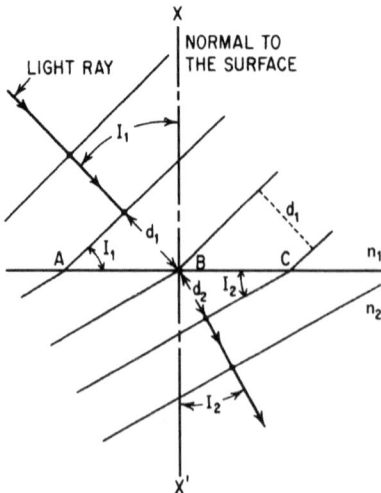

Figure 1.5 Showing the geometry of refraction at a surface bounding two different media, and illustrating Snell's law (Eqn. 1.3).

Since Snell's law relates the sines of the angles between a light ray and the normal to the surface, it is readily applicable to surfaces other than the plane which we used in the example above; the path of a light ray may be calculated through any surface for which we can determine the point of intersection of the ray and the normal to the surface at that point.

The angle I_1 between the incident ray and surface normal is customarily referred to as the angle of incidence; the angle I_2 is called the angle of refraction.

For all optical media the index of refraction varies with the wavelength of light. In general the index is higher for short wavelengths than for long wavelengths. In the preceding discussion it has tacitly been assumed that the light incident on the refracting surface was monochromatic, i.e., composed of only one wavelength of light. Figure 1.6 shows a ray of white light broken into its various component wavelengths by refraction at a surface. Notice that the blue light ray is bent, or refracted, through a greater angle than is the ray of red light. This is because n_2 for blue light is larger than n_2 for red. Since $n_2 \sin I_2 = n_1 \sin I_1 =$ a constant in this case, it is apparent that if n_2 is larger for blue light than red, then I_2 must be smaller for blue than red. This variation in index with wavelength is called dispersion; when used as a differential it is written dn, otherwise dispersion is given by $\Delta n = n_{\lambda 1} - n_{\lambda 2}$, where λ_1 and λ_2 are the wavelengths of the two colors of light for which the dispersion is given. *Relative* dispersion is given by $\Delta n/(n - 1)$ and, in effect, expresses the "spread" of the colors of light as a fraction of the amount that light of a median wavelength is bent.

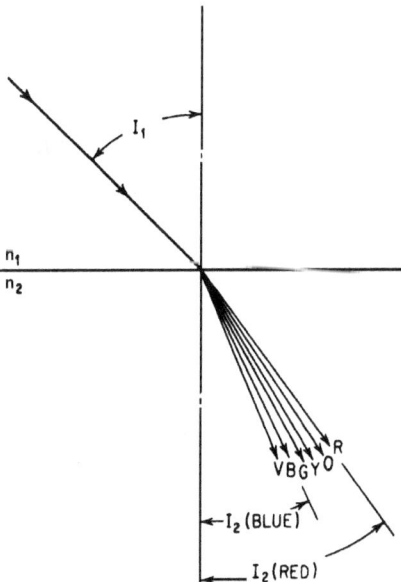

Figure 1.6 The dispersion of white light into its constituent colors by refraction (exaggerated for clarity).

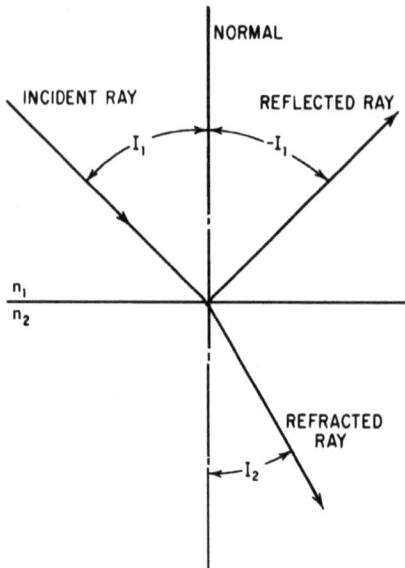

Figure 1.7 Relationship between a ray incident on a plane surface and the resultant reflected and refracted rays.

All of the light incident upon a boundary surface is not transmitted through the surface; some portion is reflected back into the incident medium. A construction similar to that used in Fig. 1.5 can be used to demonstrate that the angle between the surface normal and the reflected ray (the angle of reflection) is equal to the angle of incidence, and that the reflected ray is on the opposite side of the normal from the incident ray (as is the refracted ray). Thus, for reflection, Snell's law takes on the form

$$I_{\text{incident}} = -I_{\text{reflected}} \tag{1.4}$$

Figure 1.7 shows the relationship between a ray incident on a plane surface and the resultant reflected and refracted rays.

At this point it should be emphasized that the incident ray, the normal, the reflected ray, and the refracted ray all lie in a common plane, called the plane of incidence, which in Fig. 1.7 is the plane of the paper.

1.4   The Action of Simple Lenses and Prisms on Wave Fronts

In Fig. 1.8 a point source P is emitting light; as before, the arcs centered about P represent the successive positions of a wave front at regular intervals of time. The wave front is incident on a biconvex lens consisting of two surfaces of rotation bounding a medium of (in this instance) higher index of refraction than the medium in which the source is

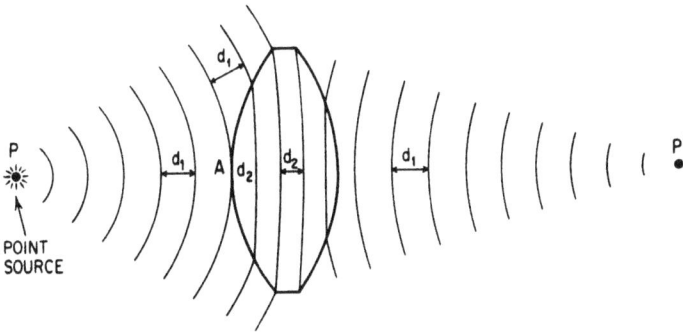

Figure 1.8 The passage of a wave front through a converging, or positive, lens element.

located. In each interval of time the wave front may be assumed to travel a distance d_1 in the medium of the source; it will travel a lesser distance d_2 in the medium of the lens. (As in the preceding discussion, these distances are related by $n_1 d_1 = n_2 d_2$.) At some instant, the vertex of the wave front will just contact the vertex of the lens surface at point A. In the succeeding interval, the portion of the wave front inside the lens will move a distance d_2, while the portion of the same wave front still outside the lens will have moved d_1. As the wave front passes through the lens, this effect is repeated in reverse at the second surface. It can be seen that the wave front has been retarded by the medium of the lens and that this retardation has been greater in the thicker central portion of the lens, causing the curvature of the wave front to be reversed. At the left of the lens the light from P was diverging, and to the right of the lens the light is now converging in the general direction of point P'. If a screen or sheet of paper were placed at P', a concentration of light could be observed at this point. The lens is said to have formed an image of P at P'. A lens of this type is called a converging, or positive, lens. The object and image are said to be *conjugates*.

Figure 1.8 diagrams the action of a convex lens—that is, a lens which is thicker at its center than at its edges. A convex lens with an index higher than that of the surrounding medium is a converging lens, in that it will increase the convergence (or reduce the divergence) of a wave front passing through it.

In Fig. 1.9 the action of a concave lens is sketched. In this case the lens is thicker at the edge and thus retards the wave front more at the edge than at the center and increases the divergence. After passing through the lens, the wave front appears to have originated from the neighborhood of point P', which is the image of point P formed by the lens. In this case, however, it would be futile to place a screen at P' and

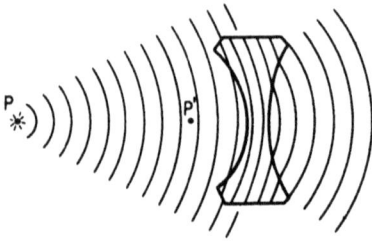

Figure 1.9 The passage of a wave front through a diverging, or negative, lens element.

expect to find a concentration of light; all that would be observed would be the general illumination produced by the light emanating from *P.* This type of image is called a *virtual* image to distinguish it from the type of image diagramed in Fig. 1.8, which is called a *real* image. Thus a virtual image may be observed directly or may serve as a source to be reimaged by a subsequent lens system, but it cannot be produced on a screen. The terms "real" and "virtual" also may be applied to rays, where "virtual" applies to the extended part of a real ray.

The path of a *ray* of light through the lenses of Figs. 1.8 and 1.9 is the path traced by a point on the wave front. In Fig. 1.10 several ray paths have been drawn for the case of a converging lens. Note that the rays originate at point P and proceed in straight lines (since the media involved are isotropic) to the surface of the lens where they are refracted according to Snell's law (Eq. 1.3). After refraction at the second surface the rays converge at the image P'. (In practice the rays will converge exactly at P' only if the lens surfaces are suitably chosen surfaces of rotation, usually nonspherical, the axes of which are coincident and pass through P.) This would lead one to expect that the concentration of light at P' would be a perfect point. However, the wave nature of light causes it to be diffracted in passing through the limiting aperture of the lens so that the image, even for a "perfect" lens, is spread out into a small disc of light surrounded by faint rings as discussed in Chap. 9.

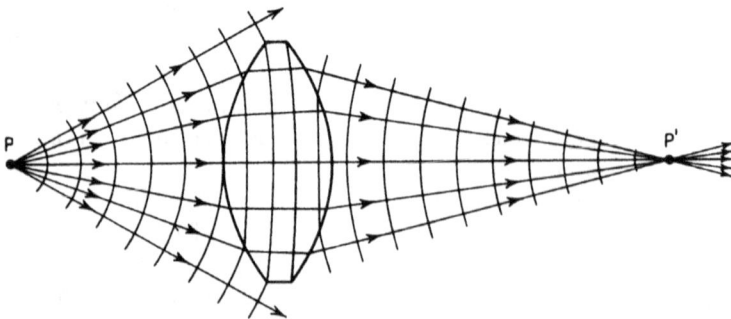

Figure 1.10 The relationship between light rays and the wave front in passing through a positive lens element.

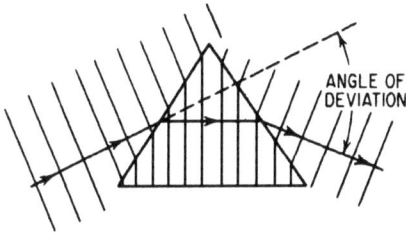

Figure 1.11 The passage of a plane wave front through a refracting prism.

In Fig. 1.11 a wave front from a source so far distant that the curvature of the wave front is negligible is shown approaching a prism, which has two flat polished faces. As it passes through each face of the prism, the light is refracted downward so that the direction of propagation is deviated. The angle of deviation of the prism is the angle between the incident ray and the emergent ray. Note that the wave front remains plane as it passes through the prism.

If the radiation incident on the prism consisted of more than one wavelength, the shorter-wavelength radiation would be slowed down more by the medium composing the prism and thus deviated through a greater angle. This is one of the methods used to separate different wavelengths of light and is, of course, the basis for Isaac Newton's classic demonstration of the spectrum.

1.5   Interference and Diffraction

If a stone is dropped into still water, a series of concentric ripples, or waves, is generated and spreads outward over the surface of the water. If two stones are dropped some distance apart, a careful observer will notice that where the waves from the two sources meet there are areas with waves twice as large as the original waves and also areas which are almost free of waves. This is because the waves can reinforce or cancel out the action of each other. Thus if the crests (or troughs) of two waves arrive simultaneously at the same point, the crest (or trough) generated is the sum of the two wave actions. However, if the crest of one wave arrives at the same instant as the trough of the other, the result is a cancellation. A more spectacular display of wave reinforcement can often be seen along a sea wall where an ocean wave which has struck the wall and been reflected back out to sea will combine with the next incoming wave to produce an eruption where they meet.

Similar phenomena occur when light waves are made to interfere. In general, light from the same point on the source must be made to travel two separate paths and then be recombined, in order to produce optical interference. The familiar colors seen in soap bubbles or in oil films on wet pavements are produced by interference.

Young's experiment, which is diagrammed schematically in Fig. 1.12, illustrates both diffraction and interference. Light from a source to the left of the figure is caused to pass through a slit or pinhole s in an opaque screen. According to *Huygens' principle,* the propagation of a wave front can be constructed by considering each point on the wave front as a source of new spherical wavelets; the envelope of these new wavelets indicates the new position of the wave front. Thus s may be considered as the center of a new spherical or cylindrical wave (depending on whether s is a pinhole or a slit), provided that the size of s is sufficiently small. These diffracted wave fronts from s travel to a second opaque screen which has two slits (or pinholes), A and B, from which new wave fronts originate. The wave fronts again spread out by diffraction and fall on an observing screen some distance away.

Now, considering a specific point P on the screen, if the wave fronts arrive simultaneously (or in phase), they will reinforce each other and P will be illuminated. However, if the distances AP and BP are such that the waves arrive exactly out of phase, destructive interference will occur and P will be dark.

If we assume that s, A, and B are so arranged that a wave front from s arrives simultaneously at A and B (that is, distance sA exactly equals distance sB), then new wavelets will start out simultaneously from A and B toward the screen. Now if distance AP exactly equals distance BP, or if AP differs from BP by exactly an integral number of wavelengths, the wave fronts will arrive at P in phase and will reinforce. If AP and BP differ by one-half wavelength, then the wave actions from the two sources will cancel each other.

If the illuminating source is monochromatic, i.e., emits but a single wavelength of light, the result will be a series of alternating light and dark bands of gradually changing intensity on the screen (assuming that s, A, and B are slits), and by careful measurement of the geometry of the slits and the separation of the bands, the wavelength of the radiation may be computed. (The distance AB should be less than a millimeter and the distance from the slits to the screen should be to the order of a meter to conduct this experiment.)

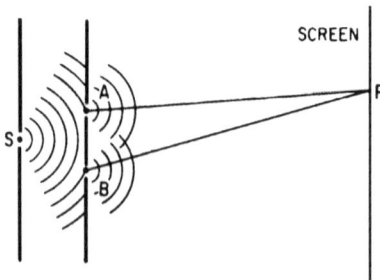

Figure 1.12 Young's diffraction experiment.

With reference to Fig. 1.13, it can be seen that, to a first approximation, the path difference between *AP* and *BP*, which we shall represent
by Δ, is given by

$$\Delta = \frac{AB \cdot OP}{D}$$

Rearranging this expression, we get

$$OP = \frac{\Delta \cdot D}{AB} \tag{1.5}$$

Now as Fig. 1.13 is drawn, it is obvious that the optical paths *AO* and
BO are identical, so the waves will reinforce at *O* and produce a bright
band. If we set Δ in Eq. 1.5 equal to (plus or minus) one-half wavelength,
we shall then get the value of *OP* for the first dark band

$$OP \text{ (1st dark)} = \frac{\pm \lambda D}{2AB} \tag{1.6}$$

and if we assume that the distance from slits to screen *D* is one meter,
that the slit separation *AB* is one-tenth of a millimeter, and that the
illumination is red light of a wavelength of 0.64 μm, we get the following
by substitution of these values in Eq. 1.6:

$$OP \text{ (1st dark)} = \frac{\pm \lambda 10^3}{2 \cdot 10^{-1}} = \frac{\pm 10^4 \lambda}{2} = \frac{\pm 10^4 \cdot 0.64 \cdot 10^{-3}}{2} = \pm 3.2 \text{ mm}$$

Thus the first dark band occurs 3.2 mm above and below the axis.
Similarly the location of the next light band can be found at 6.4 mm by
setting Δ equal to one wavelength, and so on.

If blue light of wavelength 0.4 μm were used in the experiment, we
would find that the first dark band occurs at ±2 mm and the next
bright band at ±4 mm.

Now if the light source, instead of being monochromatic, is white and
consists of all wavelengths, it can be seen that each wavelength will
produce its own array of light and dark bands of its own particular
spacing. Under these conditions the center of the screen will be illuminated by all wavelengths and will be white. As we proceed from the
center, the first effect perceptible to the eye will be the dark band for

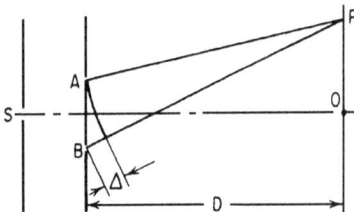

Figure 1.13 Geometry of Young's
experiment.

blue light which will occur at a point where the other wavelengths are still illuminating the screen. Similarly, the dark band for red light will occur where blue and other wavelengths are illuminating the screen. Thus a series of colored bands is produced, starting with white on the axis and progressing through red, blue, green, orange, red, violet, green, and violet, as the path difference increases. Further from the axis, however, the various light and dark bands from all the visible wavelengths become so "scrambled" that the band structures blend together and disappear.

Newton's rings are produced by the interference of the light reflected from two surfaces which are close together. Figure 1.14 shows a beam of parallel light incident on a pair of partially reflecting surfaces. At some instant a wave front AA' strikes the first surface at A. The point on the wave front at A travels through the space between the two surfaces and strikes the second surface at B where it is partially reflected; the reflected wave then travels upward to pass through the first surface again at C. Meanwhile the point on the wave front at A' has been reflected at point C and the two paths recombine at this point.

Now if the waves arrive at C in phase, they will reinforce; if they arrive one-half wavelength out of phase, they will cancel. In determining the phase relationship at C we must take into account the index of the material through which the light has traveled and also the phase change which occurs on reflection. This phase change occurs when light traveling through a low-index medium is reflected from the surface of a high-index medium; the phase is then abruptly changed by 180°, or one-half wavelength. No phase change occurs when the indices are encountered in reverse order. Thus with the relative indices as indicated in Fig. 1.14, there is a phase change at C for the light following the $A'CD$ path, but no phase change at B for the light reflected from the lower surface.

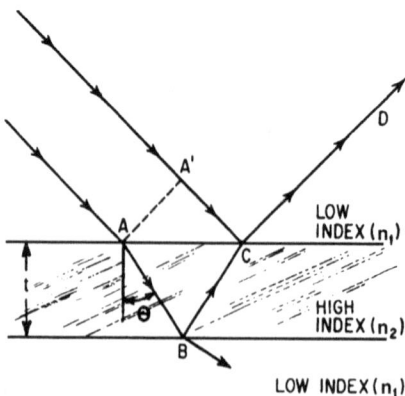

Figure 1.14   Relative indices.

As in the case of Young's experiment described above, the difference between the optical paths ABC and $A'C$ determines the phase relationship. Since the index of refraction is inversely related to the velocity of light in a medium, it is apparent that the length of time a wave front takes to travel through a thickness d of a material of index n is given by $t = nd/c$ (where $c \approx 3 \cdot 10^{10}$ cm/s = velocity of light). The constant frequency of electromagnetic radiation is given by c/λ, so that the number of cycles which take place during the time $t = nd/c$ is given by $(c/\lambda) \cdot (nd/c)$ or nd/λ. Thus, if the number of cycles is the same, or differs by an integral number of cycles, over the two paths of light traversed, the two beams of light will arrive at the same phase.

In Fig. 1.14, the number of cycles for the path $A'C$ is given by $\frac{1}{2} + n_1 A'C/\lambda$ (the one-half cycle is for the reflection phase change) and for the path ABC by $n_2 ABC/\lambda$; if these numbers differ by an integer, the waves will reinforce; if they differ by an integer plus one-half, they will cancel.

The use of cycles in this type of application is often inconvenient, and it is customary to work in *optical path length*, which is the physical distance times the index and is a measure of the "travel time" for light. It is obvious that if we consider the difference between the two path lengths (arrived at by multiplying the above number of cycles by the wavelength λ), exactly equivalent results are obtained when the difference is an integral number of wavelengths (for reinforcement) or an integral number plus one-half wavelength (for cancellation). Thus, for Fig. 1.14, the *optical path difference* (OPD) is given by

$$\text{OPD} = \frac{\lambda}{2} + n_1 A'C - n_2 ABC \qquad (1.7)$$

or

$$\text{OPD} = \frac{\lambda}{2} + 2n_2 t \cos \theta$$

when the phase change is taken into account by the $\lambda/2$ term.

The term "Newton's rings" usually refers to the ring pattern of interference bands formed when two spherical surfaces are placed in intimate contact. Figure 1.15 shows the convex surface of a lens resting on a plane surface. At the point of contact the difference in the optical paths reflected from the upper and lower surfaces is patently zero. The phase change on reflection from the lower surface causes the beams to rejoin exactly out of phase, resulting in complete cancellation and the appearance of the central "Newton's black spot." At some distance from the center the surfaces will be separated by exactly one-quarter wavelength, and this path difference of one-half wavelength plus the phase change results in reinforcement, producing

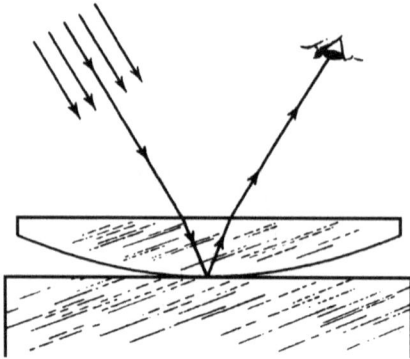

Figure 1.15 Newton's rings are formed by the interference between the light reflected by two closely spaced surfaces. If the two surfaces are spherical, the interference pattern is a series of alternating light and dark rings.

a bright ring. A little further from the center, the separation is one-half wavelength, resulting in a dark ring, and so on.

Just as in Young's experiment, the dark and bright bands for different wavelengths will occur at different distances from the center, resulting in colored circles near the point of contact which fade away toward the edge.

A setup similar to Fig. 1.15 can obviously be used to measure the wavelength of light if the radius of curvature of the lens is known and a careful measurement of the diameters of the light and dark fringes is made. The spacing between the surfaces is the sagittal height (SH) of the radius (R), given by

$$\text{SH} = R - (R^2 - Y^2)^{1/2} \approx \frac{Y^2}{2R} \tag{1.8}$$

where Y is the semidiameter of the ring measured. SH is equal to $\lambda/4$ for the first bright ring, $\lambda/2$ for the first dark ring, $3\lambda/4$ for the second bright ring, and so on.

Two other useful forms of Eq. 1.8 are:

$$R = \frac{(Y^2 + \text{SH}^2)}{2 \cdot \text{SH}} \tag{1.8a}$$

$$Y = \sqrt{2R \cdot \text{SH} - \text{SH}^2} \tag{1.8b}$$

1.6   The Photoelectric Effect

In the preceding section, the discussion was based upon the assumption that light was wavelike in nature. This assumption provides reasonable explanations for reflection, refraction, interference, diffraction, and dispersion, as well as other effects. The photoelectric effect, however, seems to require for its explanation that light behaves as if it consisted of particles.

In brief, when short-wavelength light strikes a photoelectric material, it can knock electrons out of the material. As stated, this effect could be explained by the energy of the light waves exciting an electron sufficiently for it to break loose. However, when the nature of the incident radiation is modified, the characteristics of the emitted electrons change in an unexpected way. As the intensity of the light is increased, the number of electrons is increased just as might be expected. If the wavelength is increased, however, the maximum velocity of the electrons emitted is reduced; if the wavelength is increased beyond a certain value (this value is characteristic of the particular photoelectric material used), the maximum velocity drops to zero and no electrons are emitted, regardless of the intensity. The energy of a photon in electron volts is given by 1.24 divided by the wavelength in micrometers (microns).

Thus the energy necessary to break loose an electron is not stored up until enough is available (as one would expect of the wavelike behavior of light). The situation here is more analogous to a shower of particles, some of which have enough energy to break an electron loose from the forces which bind it in place. Thus the particles of shorter wavelength have sufficient energy to release an electron. If the intensity of light is increased, the number of electrons released is increased and their velocity remains unchanged. The longer-wavelength particles do not have enough energy to knock electrons loose, and when the intensity of the long-wavelength light is increased, the effect is to increase the number of particles striking the surface, but each particle is still insufficiently powerful to release an electron from its bonds.

The apparent contradiction between the wave and particle behavior of light can be resolved by assuming that every "particle" has a wavelength associated with it which is inversely proportional to its momentum. This has proved true experimentally for electrons, protons, ions, atoms, and molecules; for example, an electron accelerated by an electric field of a few hundred volts has a wavelength of a few angstroms (10^{-4} μm) associated with it. Reference to Fig. 1.1 indicates that this wavelength is characteristic of x-rays, and indeed, electrons of this wavelength are diffracted in the same patterns (by crystal lattices) as are x-rays.

Bibliography

Born, M., and E. Wolf, *Principles of Optics,* Cambridge, England, Cambridge University Press, 1997.
Brown, E., *Modern Optics,* New York, Reinhold, 1965.
Ditchburn, R., *Light,* New York, Wiley-Interscience, 1963.
Drude, P., *Theory of Optics,* New York, Dover, 1959.

Greivenkamp, J. E., "Interference," in *Handbook of Optics,* Vol. 1, New York, McGraw-Hill, 1995, Chap. 2.

Hardy, A., and P. Perrin, *The Principles of Optics,* New York, McGraw-Hill, 1932.

Hecht, E., and A. Zajac, *Optics,* Reading, MA, Addison-Wesley, 1974.

Jacobs, D. *Fundamentals of Optical Engineering,* New York, McGraw-Hill, 1943.

Jenkins, F., and H. White, *Fundamentals of Optics,* New York, McGraw-Hill, 1976.

Kingslake, R., *Optical System Design,* New York, Academic, 1983.

Levi, L., *Applied Optics,* New York, Wiley, 1968.

Marathay, A. S., "Diffraction," in *Handbook of Optics,* Vol. 1, New York, McGraw-Hill, 1995, Chap. 3.

Strong, J., *Concepts of Classical Optics,* New York, Freeman, 1958.

Walker, B. H., *Optical Engineering Fundamentals,* New York, McGraw-Hill, 1995.

Wood, R., *Physical Optics,* New York, Macmillan, 1934.

Exercises

1 What is the index of a medium in which light has a velocity of 2×10^{10} cm/s?

ANSWER: Eq. 1.1 n = (velocity in vacuum)/(velocity in medium)
$$= 3 \cdot 10^{10}/2 \cdot 10^{10}$$
$$= 1.5$$

2 What is the velocity of light in water ($n = 1.33$)?

ANSWER: Eq. 1.1 $1.33 = 3 \cdot 10^{10}/$(velocity in water)
Velocity in water $= 3 \cdot 10^{10}/1.33$
$$= 2.26 \cdot 10^{10} \text{ cm/s}$$

3 A ray of light makes an angle of 30° with the normal to a surface. Find the angle to the normal after refraction if the ray is in:

   (a)   air and the other material is $n = 1.5$.
   (b)   water, $n = 1.33$ and the other material is air.
   (c)   water and the other material is $n = 1.5$.

ANSWER: Eq. 1.3 $n_1 \sin I_1 = n_2 \sin I_2$
$$I_2 = \arcsin [(n_1/n_2) \sin I_1]$$

   (a) $I_2 = \arcsin [(1.0/1.5) \cdot 0.5] = 19.47°$
   (b) $I_2 = \arcsin [(1.33/1.0) \cdot 0.5] = 41.68°$
   (c) $I_2 = \arcsin [(1.33/1.5) \cdot 0.5] = 26.32°$

4 Two 6-in-diameter optical flats are in contact at one edge and separated by a piece of paper (0.003-in thick) at the opposite edge. When illuminated by light of 0.000020-in wavelength, how many fringes will be seen? Assume normal incidence.

ANSWER: The airgap is 0.003 in, or 0.003/0.000020 = 150 wavelengths. At one fringe per half wavelength, there will be 300 fringes between the contact point and the paper (or about 50 fringes per inch).

5 In Exercise 4, if the space between the flats is filled with water ($n = 1.333$), how many fringes will be seen?

ANSWER: The optical path ($= n \cdot d$) is $1.333 \cdot 0.003$ in $= 0.004$ in, or 200 wavelengths. At one fringe per half wavelength, there will be 400 fringes.

6 The convex surface of a lens is in contact with a flat plate of glass. If the radius of the lens surface is 20 in, at what diameter will the first dark interference band/ring be seen? The second? The third? What are the ring diameters if the radius is 200 in?

ANSWER: There will be a dark spot at the center contact point ("Newton's black spot") because of the reflection phase change at the lower surface. The first dark ring will occur where the airspace is one-half wavelength, or 0.000010 in. This is of course simply the sagittal height of the lens surface. Per Eq. 1.8 this is:

$$\text{SH} = R - (R^2 - Y^2)^{1/2} = 0.000010 \text{ in} = 20 - (400 - Y^2)^{1/2}$$

Squaring, we get: $Y^2 = 400 - 19.99999^2 = 0.0004$, and $Y = 0.02$ in; the ring diameter is 0.040 in.

The second ring is located where the airspace equals one wavelength (0.000020 in); its diameter is 0.05657 in. The third ring is at 1.5 wavelength spacing, and its diameter is 0.06928 in. Note that the diameters are related by approximately the square root of the ring number; ring #2 diameter equals ring #1 diameter times $\sqrt{2}$, and ring #3 diameter equals #1 diameter times $\sqrt{3}$.

When the radius is 200 in, the diameters are 0.1265, 0.1789, and 0.2191 in. These are less than for the 20-in radius surface by a factor of about 3.162, which is about $\sqrt{10}$, or $\sqrt{200/20}$. This ratio is exact only for small diameters.

2

Gaussian Optics: The Cardinal Points

2.1 Introduction

The action of a lens on a wave front was briefly discussed in Sec. 1.4. Figures 1.8 and 1.9 showed how a lens can modify a wave front to form an image. A wave front is difficult to manipulate mathematically, and for most purposes the concept of a light ray (which is the path described by a point on a wave front) is much more convenient. In an isotropic medium, light rays are straight lines normal to the wave front, and the image of a point source is formed where the rays converge (or appear to converge) to a concentration or focus. In a "perfect" lens the rays converge to a point at the image.

For purposes of calculation, an extended object may be regarded as an array of point sources. The location and size of the image formed by a given optical system can be determined by locating the respective images of the sources making up the object. This can be accomplished by calculating the paths of a large number of rays from each object point through the optical system, applying Snell's law (Eq. 1.3) at each ray-surface intersection in turn. However, it is possible to locate optical images with considerably less effort by means of simple equations derived from the limiting case of the trigonometrically traced ray (as the angles involved approach zero). These expressions yield image positions and sizes which would be produced by a perfect optical system; they are **paraxial** or **first-order**.

The term "first-order" refers to a power series expansion equation which can be derived to define the intersection point of a ray in the image plane as a function of h, the position of the ray in the object plane, and y, the position of the ray in the aperture of the optical system. If the system is symmetrical about an axis (called the *optical axis*), the power series expansion has only odd power terms (in which the sum of the exponents of h and y add up to 1, 3, 5, etc.) The first-order terms of this expansion effectively describe the position and size of the image. (See Eqs. 5.1 and 5.2.)

First-order (or gaussian or paraxial) optics is often referred to as the optics of perfect optical systems. The first-order equations can be derived by reducing the exact trigonometrical expressions for ray paths to the limit when the angles and ray heights involved approach zero. As indicated in Chap. 3, these equations are completely accurate for an infinitesimal threadlike region about the optical axis, known as the *paraxial* region. The value of first-order expressions lies in the fact that a well-corrected optical system will follow the first-order expressions almost exactly, and also that the first-order image positions and sizes provide a convenient reference from which to measure departures from perfection. In addition, the paraxial expressions are linear and are much easier to use than the trigonometrical equations.

We shall begin this topic by considering the manner in which a "perfect" optical system forms an image, and we will discuss the expressions which allow the location and size of the image to be found when the basic characteristics of the optical system are known. In a subsequent chapter we will take up the determination of these basic characteristics from the constructional parameters of an optical system. And finally, methods of image calculation by paraxial ray-tracing as well as compound optical systems will be discussed.

2.2 Cardinal Points of an Optical System

The Mathematician Gauss discovered that the imagery of an optical system (i.e., the location of an image, the size of the image, and the orientation of the image) could easily be calculated if one knew the location of a few points on the optical axis of the system. Henceforth, in addition to assuming isotropic media, we will also assume that an optical system is one of axial symmetry, in that all surfaces are figures of rotation about a common axis, called the *optical axis*.

A well-corrected optical system can be treated as a "black box" the characteristics of which are defined by its cardinal points, which are its first and second *focal points,* its first and second *principal points,* and its first and second *nodal points.* The focal points are those points at which light rays (from an infinitely distant axial object point) parallel to the optical axis* are brought to a common focus on the axis. If the rays entering the system and those emerging from the system are extended until they intersect, the points of intersection will define a surface, usually referred to as the principal plane. In a well-corrected optical system the principal surfaces are spheres, centered on the object and image. In the paraxial region where the distances from the axis are infinitesimal, the surfaces can be treated as if they were planes, hence the name, principal "planes." The intersection of this surface with the axis is the principal point. The "second" focal point and the "second" principal point are those defined by rays approaching the system from the left. The "first" points are those defined by rays from the right.

The *effective focal length* (efl) of a system is the distance from the principal point to the focal point. The *back focal length* (bfl), or back focus, is the distance from the vertex of the last surface of the system to the second focal point. The *front focal length* (ffl) is the distance from the front surface to the first focal point. These are illustrated in Fig. 2.1.

The *nodal points* are two axial points such that a ray directed toward the first nodal point appears (after passing through the system) to emerge from the second nodal point parallel to its original direction. The nodal points of an optical system are illustrated in Fig. 2.2 for an ordinary thick lens element. When an optical system is bounded on both sides by air (as is true in the great majority of applications), the nodal points coincide with the principal points.

Unless otherwise indicated, we will assume that our optical systems are axially symmetrical and are bounded by air. Equations 2.11 through 2.15 cover the case where the surrounding medium is not air.

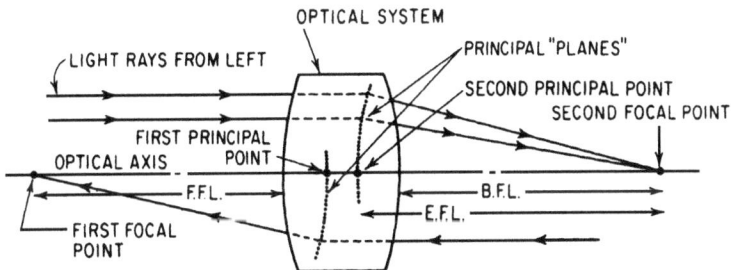

Figure 2.1 The location of the focal points and principal points of a generalized optical system.

---

*The optical axis is a line through the centers of curvature of the surfaces which make up the optical system. It is the common axis of rotation for an axially symmetrical optical system. Note that in real life, systems of more than two surfaces do not have a unique axis, because three or more real points are rarely exactly aligned on a straight line.

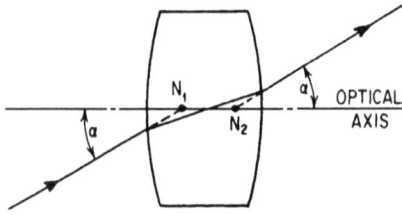

Figure 2.2 A ray directed toward the first nodal point (N_1) of an optical system emerges from the system without angular deviation and appears to come from the second nodal point (N_2).

The *power* of a lens or an optical system is the reciprocal of its effective focal length; power is usually symbolized by the Greek letter phi (ϕ). If the focal length is given in meters, the power (in reciprocal meters) is measured in *diopters*. The dimension of power is reciprocal distance, e.g., in^{-1}, mm^{-1}, cm^{-1}, etc.

2.3   Image Position and Size

When the cardinal points of an optical system are known, the location and size of the image formed by the optical system can be readily determined. In Fig. 2.3, the focal points F_1 and F_2 and the principal points P_1 and P_2 of an optical system are shown; the object which the system is to image is shown as the arrow AO. Ray OB, parallel to the system axis, will pass through the second focal point F_2; the refraction will appear to have occurred at the second principal plane. The ray OF_1C passing through the first focal point F_1 will emerge from the system parallel to the axis. (Since the path of light rays is reversible, this is equivalent to starting a ray from the right at O' parallel to the axis; the ray is then refracted through F_1 in accordance with the definition of the first focal point in Sec. 2.2.)

The intersection of these two rays at point O' locates the image of point O. A similar construction for other points on the object would locate additional image points, which would lie along the indicated

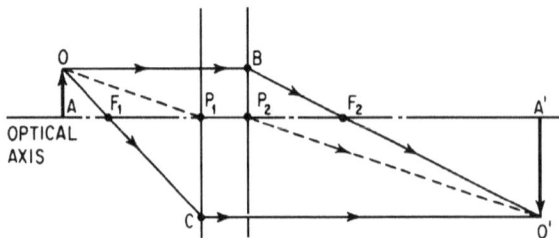

Figure 2.3 Showing the ray paths through the focal points and principal points.

arrow $O'A'$. A plane object normal to the axis is imaged as a plane, also normal to the axis. See Sec. 4.5 for a tilted object.

A third ray could be constructed from O to the first nodal point; this ray would appear to emerge from the second nodal point and would be parallel to the entering ray. If the object and image are both in air, the nodal points coincide with the principal points, and such a ray is drawn from O to P_1 and from P_2 to O', as indicated by the dashed line in Fig. 2.3.

At this point in our discussion, it is necessary to adopt a convention for the algebraic signs given to the various distances involved. The following conventions are used by most workers in the field of optics. There is nothing sacrosanct about these conventions, and many optical workers adopt their own, but the use of some consistent sign conventions is a practical necessity.

1. Heights above the optical axis are positive (e.g., OA and P_2B). Heights below the axis are negative (P_1C and $A'O'$).

2. Distances measured to the left of a reference point are negative; to the right, positive. Thus P_1A is negative and P_2A' is positive.

3. The focal length of a converging lens is positive and the focal length of a diverging lens is negative.

Image position

Figure 2.4 is identical to Fig. 2.3 except that the distances have been given single letters; the heights of the object and image are labeled h and h', the focal lengths are f and f', the object and image distances (from the principal planes) are s and s', and the distances from focal point to object and image are x and x', respectively. According to our sign convention, h, f, f', x', and s' are positive as shown, and $x, s,$ and h' are negative. Note that the primed symbols refer to dimensions associated with the image and the unprimed symbols to those associated with the object.

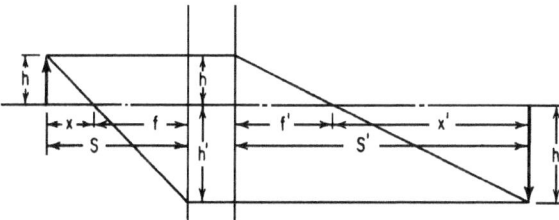

Figure 2.4 The ray sketch of Fig. 2.3 with the distances labeled as focal length or object and image distances.

From similar triangles we can write

$$\frac{h}{(-h')} = \frac{(-x)}{f} \quad \text{and} \quad \frac{h}{(-h')} = \frac{f'}{x'} \tag{2.1}$$

Setting the right-hand members of each equation equal and clearing fractions, we get

$$ff' = -xx' \tag{2.2}$$

If we assume the optical system to be in air, then f will be equal to f' and

$$x' = \frac{-f^2}{x} \tag{2.3}$$

This is the "newtonian" form of the image equation and is very useful for calculations where the locations of the focal points are known.

If we substitute $x = s + f$ and $x' = s' - f$ in Eq. 2.3, we can derive another expression for the location of the image, the "gaussian" form.

$$f^2 = -xx' = -(s + f)(s' - f)$$

$$= -ss' + sf - s'f + f^2$$

Canceling out the f^2 terms and dividing through by $ss'f$, we get

$$\frac{1}{s'} = \frac{1}{f} + \frac{1}{s} \tag{2.4}$$

or alternatively,

$$s' = \frac{sf}{(s + f)} \quad \text{or} \quad f = \frac{ss'}{(s - s')} \tag{2.5}$$

Image size

The *lateral* (or *transverse*) *magnification* of an optical system is given by the ratio of image size to object size, h'/h. By rearranging Eq. 2.1, we get for the magnification m,

$$m = \frac{h'}{h} = \frac{f}{x} = \frac{-x'}{f} \tag{2.6}$$

Substituting $x = s + f$ in this expression to get

$$m = \frac{h'}{h} = \frac{f}{(s + f)}$$

and noting from Eq. 2.5 that $f/(s + f)$ is equal to s'/s, we find that

$$m = \frac{h'}{h} = \frac{s'}{s} \tag{2.7a}$$

Other useful relations are

$$s' = f(1 - m) \tag{2.7b}$$

$$s = f\left(\frac{1}{m} - 1\right) \tag{2.7c}$$

Note that Eqs. 2.3 through 2.7 assume that both object and image are in air, and also that Figs. 2.3 and 2.4 show a *negative* magnification.

 Longitudinal magnification is the magnification *along* the optical axis, i.e., the magnification of the longitudinal *thickness* of the object or the magnification of a longitudinal *motion* along the axis. If s_1 and s_2 denote the distances to the front and back edges of the object and s'_1 and s'_2 denote the distances to the corresponding edges of the image, then the longitudinal magnification $\overline{m}$ is, by definition,

$$\overline{m} = \frac{s'_2 - s'_1}{s_2 - s_1}$$

 Substituting Eqs. 2.7b and 2.7c for the primed distances and manipulating, we get

$$\overline{m} = \frac{s'_1}{s_1} \cdot \frac{s'_2}{s_2} = m_1 \cdot m_2 \tag{2.8}$$

noting that $m = s'/s$. As $(s'_2 - s'_1)$ and $(s_2 - s_1)$ approach zero, then m_1 approaches m_2, and

$$\overline{m} = m^2 \tag{2.9}$$

This indicates that longitudinal magnification is positive, and that object and image always move in the same direction.

Example 2.1

Given an optical system with a positive focal length of 10 in, find the position and size of the image formed of an object 5-in high which is located 40 in to the left of the first focal point of the system. See Fig 2.5a.

 Using the newtonian equation, we get, by substituting in Eq. 2.3,

$$x' = \frac{-f^2}{x} = \frac{-10^2}{-40} = +2.5 \text{ in}$$

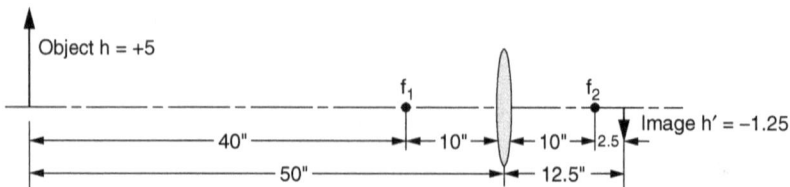

Figure 2.5a The formation of a real image. See Example 2.1.

Therefore the image is located 2.5 in to the right of the second focal point. To find the image height, we use Eq. 2.6.

$$m = \frac{h'}{h} = \frac{f}{x} = \frac{10}{-40} = -0.25$$

$$h' = mh = (-0.25)(5) = -1.25 \text{ in}$$

Thus if the base of the object were on the optical axis and the top of the object 5 in above it, the base of the image would also lie on the axis and the image of the top would lie 1.25 in below the axis.

The gaussian equations can be used for this calculation by noting that the distance from the first principal plane to the object is given by $s = x - f = -40 - 10 = -50$; then, by Eq. 2.4,

$$\frac{1}{s'} = \frac{1}{f} + \frac{1}{s} = \frac{1}{10} + \frac{1}{(-50)} = 0.1 - 0.02 = 0.08$$

$$s' = \frac{1}{0.08} = 12.5 \text{ in}$$

and the image is found to lie 12.5 in to the right of the second principal plane (or 2.5 in to the right of the second focal point, in agreement with the previous solution).

The height of the image can now be determined from Eq. 2.7a.

$$m = \frac{h'}{h} = \frac{s'}{s} = +\frac{12.5}{-50} = -0.25$$

$$h' = mh = (-0.25)(5) = -1.25 \text{ in}$$

Example 2.2

If the object of Example 2.1 is located 2 in to the *right* of the first focal point, as shown in Fig. 2.5b, where is the image and what is its height? Using Eq. 2.3,

$$x' = \frac{-f^2}{x} = \frac{-10^2}{+2} = -50 \text{ in}$$

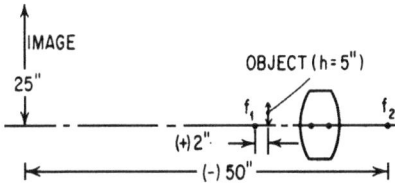

Figure 2.5b The formation of a virtual image. See Example 2.2.

Notice that the image is formed to the *left* of the second focal point; in fact, if the optical system is of moderate thickness, the image is to the left of the optical system and also to the left of the object. From Eq. 2.6 we get the magnification

$$m = \frac{h'}{h} = \frac{f}{x} = \frac{10}{2} = +5$$

$$h' = mh = (5)\,(5) = +25 \text{ in}$$

The magnification and image height are both positive. In this case the image is a *virtual* image. A screen placed at the image position will not have an image formed on it, but the image may be observed by viewing through the lens from the right. A positive sign for the lateral magnification of a simple lens indicates that the image formed is virtual; a negative sign for the magnification of a simple lens indicates a real image. Figure 2.5 shows the relationships in these examples.

Example 2.3

If the object of Example 2.2 is 0.1-in thick, what is the apparent thickness of the image? Since the lateral magnification was found to be 5 times in Example 2.2, the longitudinal magnification, by Eq. 2.9, is approximately 5^2, or 25. Thus the apparent image thickness is approximately 25 times (0.1 in), or 2.5 in. If an exact value for the apparent thickness is required, the image position for each surface of the object must be calculated. Assuming that the front of the object was given in Example 2.2 as 2 in to the right of the first focal point, then its rear surface must lie 1.9 in to the right of f_1. Its image is located at

$$x' = \frac{-f^2}{x} = \frac{-100}{1.9} = -52.63 \text{ in}$$

to the left of the second focal point. Thus the distance between the image positions for the front and rear surfaces is 2.63 in, in reasonable agreement with the approximate result of 2.5 in. Had we computed the thickness for the case where the front and back surfaces of the object were 1.95 and 2.05 in from the focal point, the results from the exact and approximate calculations would have been in even better agreement, yielding an image thickness of 2.502 in.

2.4   A Collection of Imagery Equations

These equations are derived from the newtonian and gaussian equations in this chapter. See Fig. 2.6.

Newtonian:

$$x' = f^2/x \qquad x = -f^2/x' \qquad f = \sqrt{-xx'}$$
$$x' = -mf \qquad x = f/m \qquad m = f/x = -x'/f$$

Gaussian:

$$(1/s') = (1/f) + (1/s)$$
$$s' = sf/(s + f) \qquad s = s' \cdot f/(f - s') \qquad f = s \cdot s'/(s - s')$$
$$s' = f(1 - m) \qquad s = f(m - 1)/m$$

$$T \equiv s' - s \qquad\qquad T = -f(m - 1)^2/m$$
$$f = -Tm/(m - 1)^2 \qquad s = [-T \pm \sqrt{(T^2 - 4fT)}]/2$$
$$f/\# = -1/[2(u - u')] \qquad\quad = 1/[2\,(m - 1)\,NA']$$
$$\qquad\qquad\qquad\qquad\quad = m/[2(m - 1)\,NA]$$
$$u' = 1/[2\,(f/\#)\,(m - 1)] \qquad = u/m \qquad\qquad\qquad m = u/u'$$
$$u = m/[2\,(f/\#)\,(m - 1) \qquad = mu'$$
$$NA = m/[2(m - 1)\,(f/\#)] \quad NA' = 1/[2(m - 1)(f/\#)] \quad NA' = NA/m$$

Where f is the focal length (EFL)

s and s' are the object and image distances from the principal points

x and x' are the object and image distances from the focal points

$T = (s' - s) =$ track length (object to image distance)

$f/\# =$ relative aperature $= f$/diameter

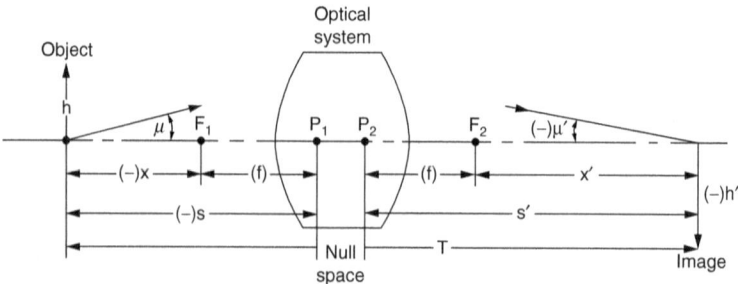

Figure 2.6   The meaning of the symbols of Sec. 2.4.

$m = h'/h = $ transverse magnification
u and u' are the ray slopes at the object, and image
NA and NA' are the numerical apertures at object and image
$(= u$ and $u')$

2.5   Optical Systems *Not* Immersed in Air

If the object and image are not in air, as assumed in the preceding paragraphs, the following equations should be used instead of the standard expressions of Eqs. 2.2 through 2.9.

Assume an optical system with an object-side medium of index n, and an image-side medium of index n'. The first and second effective focal lengths, f and f', respectively, may differ; they are related by

$$\frac{f}{n} = \frac{f'}{n'}$$
(2.10)

Note that
$$\phi = \frac{nu}{y'} = \frac{n'u'}{y}$$
(2.10a)

The focal lengths can be determined by a ray-tracing calculation, just as with an air-immersed system. For example, $f' = -y_1/u'_k$ (see Eq. 3.19).

Object and image distances

$$\frac{n'}{s'} = \frac{n}{s} + \frac{n}{f} = \frac{n}{s} + \frac{n'}{f'}$$
(2.11)

$$x' = \frac{-ff'}{x}$$
(2.12)

Magnifications

$$m = \frac{h'}{h} = \frac{ns'}{n's} = \frac{f}{x} = \frac{-x'}{f'}$$
(2.13)

for an object at infinity,

$$h' = fu_p = f'u_p n/n'$$
(2.14)

$$\overline{m} = \frac{\Delta s'}{\Delta s} = \frac{ff'}{x^2} \quad \text{(note that } \overline{m} \neq m^2)$$
(2.15)

Focal point to nodal point distance equals the *other* focal length.

Bibliography

Bass, M., *Handbook of Optics,* Vol. 1, New York, McGraw-Hill, 1995.
Fischer, R. E., and B. Tadic-Galeb, *Optical System Design,* New York, McGraw-Hill, 2000.
Greivenkamp, J., *Field Guide to Geometrical Optics,* Bellingham, WA, SPIE, 2004.
Kingslake, R., *Optical System Design,* Orlando, Academic, 1983.
Smith, W. J., *Modern Lens Design,* New York, McGraw-Hill, 2002.
Smith, W. J., *Practical Optical System Layout,* New York, McGraw-Hill, 1997.

Exercises

1   A 10 in focal length lens forms an image of a telephone pole which is 200 ft away (from its first principal point). Where is the image located (a) with respect to the second focal point of the lens, and (b) with respect to the second principal point?

ANSWER: (a) Using Eq. 2.3 $x' = -f^2/x = -10^2/(-200 \cdot 12 + 10)$
$$= -100/(-2390)$$
$$= +0.041841 \text{ in}$$

(b) Using Eq. 2.5 $s' = sf/(s + f) = -2400 \cdot 10/(-2400 + 10)$
$$= -24000/-2390$$
$$= +10.041841 \text{ in}$$

2   (a) How big is the image (in Exercise 1) if the telephone pole is 50-ft high? and (b) what is the magnification?

ANSWER: (a) Solving Eq. 2.6 for h', we get
$$h' = -h \cdot x'/f = -50 \cdot 12 \cdot 0.041841/10$$
$$= -2.510460 \text{ in}$$

(b) Again using Eq. 2.6, we can get
$$m = h'/h = -2.510460/12 \cdot 50$$
$$= -0.00418\times$$

or

(a) Solving Eq. 2.7a for $m = s'/s = 10.041841/(-200 \cdot 12)$
$$= -0.004184\times$$
and $h' = m \cdot h = -0.004184 \cdot 12 \cdot 50$
$$= -2.510460 \text{ in}$$

3   A 1-in cube is 20 in away from the first principal point of a negative lens of 5 in focal length. Where is the image, and what are its dimensions (height, width, thickness)?

ANSWER: Using Eq. 2.5 (or 2.4) we get $s' = s \cdot f/(s + f)$
$$s' = -20(-5)/(-20 - 5) = -4.0 \text{ in}$$

The lateral magnification is thus $m = s'/s = -4/(-20) = +0.2\times$, and the longi-tudinal magnification is approximately (by Eq. 2.9) $\overline{m} = m^2 = +0.04\times$. Thus, the height and width of the image are both 0.20 in, and the thickness is ≈ 0.04 in.

Note that a more exact value for the image thickness and the longitudinal magnification can be obtained by calculating s' for two object distances, 1 in apart. If we use object distances of -20 in and -21 in, we get an image distance difference (image thickness) of 0.038462 in. Using -19 in and -20 in, we get 0.041667 in. Using -19.5 in and -20.5 ft, the result is an image thickness of 0.040016 in.

4 The first principal point of a 2 in focal length lens is 1 in from the object. Where is the image, and what is the magnification?

ANSWER: Using Eq. 2.5, we find the image at
$$s' = s \cdot f/(s + f)$$
$$s' = -1 \cdot 2/(-1 + 2) = -2/1 = -2 \text{ in}$$

The image is 2 in to the left of the lens, and the magnification $m = s'/s = -2/(-1) = +2\times$.

3

Paraxial Optics and Calculations

3.1 Refraction of a Light Ray at a Single Surface

As mentioned in Chap. 1, the path of a meridional light ray through an optical system can be calculated from Snell's law (Eq. 1.3) by the application of a modest amount of geometry and trigonometry. Figure 3.1 shows a light ray (GQP) incident on a spherical surface at point Q. The ray is directed toward point P where it would intersect the optical axis at a distance L from the surface if the ray were extended. At Q the ray is refracted by the surface and intersects the axis at P', a distance L' from the surface. The surface has a radius R with center of curvature at C and separates two media of index n on the left and index n' on the right. The light ray makes an angle U with the axis before refraction, U' after refraction; angle I is the angle between the incident ray and the normal to the surface (HQC) at point Q, and angle I' is the angle between the refracted ray and the normal. Notice that plain or unprimed symbols are used for quantities before refraction at the surface; after refraction, the symbols are primed.

The sign conventions which we shall observe are as follows:

1. A radius is positive if the center of curvature lies to the right of the surface.

2. As before, distances to the right of the surface are positive; to the left, negative.

3. The angles of incidence and refraction (I and I') are positive if the ray is rotated clockwise to reach the normal.

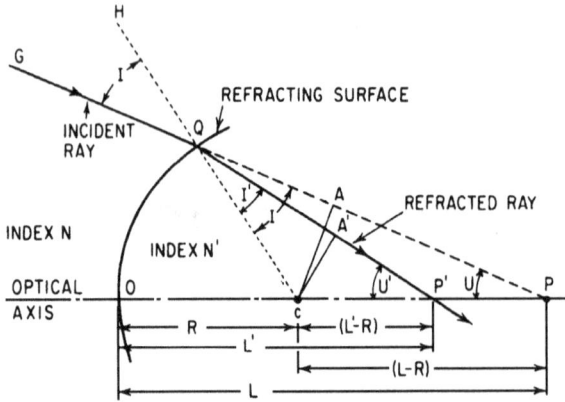

Figure 3.1   Refraction of a ray at a spherical surface.

4. The slope angles (U and U') are positive if the ray is rotated clockwise to reach the axis. (*Historical Note:* Until the latter part of the Twentieth Century, the accepted convention for the sign of the slope in optics was the reverse of the current one, and Fig. 3.1 was an "all-positive diagram.")

5. The light travels from left to right.

(In Fig. 3.1 all quantities are positive except U and U', which are negative.)

A set of equations which will allow us to trace the path for the ray may be derived as follows. From right triangle PAC,

$$CA = (R - L) \sin U \tag{3.1}$$

and from right triangle QAC,

$$\sin I = \frac{CA}{R} \tag{3.2}$$

Applying Snell's law (Eq. 1.3), we get the sine of the angle of refraction,

$$\sin I' = \frac{n}{n'} \sin I \tag{3.3}$$

The exterior angle QCO of triangle PQC is equal to $-U + I$, and, as the exterior angle of triangle $P'QC$, it is also equal to $-U' + I'$. Thus, $-U + I = -U' + I'$, and

$$U' = U - I + I' \tag{3.4}$$

From right triangle $QA'C$ we get

$$\sin I' = \frac{CA'}{R} \tag{3.5}$$

and substituting Eqs. 3.2 and 3.5 into Eq. 3.3 gives us

$$CA' = \frac{n}{n'} \, CA \tag{3.6}$$

Finally, the location of P' is found by rearranging $CA' = (R - L') \sin U'$ from right triangle $P'A'C$ into

$$L' = R - \frac{CA'}{\sin U'} \tag{3.7}$$

Thus, beginning with a ray defined by its slope angle U and its intersection with the axis L, we can determine the corresponding data, U' and L', for the ray after refraction by the surface. Obviously, this process could be applied surface by surface to trace the path of a ray through an optical system.

3.2   The Paraxial Region

The paraxial region of an optical system is a thin threadlike region about the optical axis which is so small that all the angles made by the rays (i.e., the slope angles and the angles of incidence and refraction) may be set equal to their sines and tangents. At first glance this concept seems utterly useless, since the region is obviously infinitesimal and seemingly of value only as a limiting case. However, calculations of the performance of an optical system based on paraxial relationships are of tremendous utility. Their simplicity makes calculation and manipulation quick and easy. Since most optical systems of practical value form good images, it is apparent that most of the light rays originating at an object point must pass at least reasonably close to the paraxial image point. The paraxial relationships are the limiting relationships (as the angles approach zero) of the exact trigonometric relationships derived in the preceding section, and thus give locations for image points which serve as an excellent approximation for the imagery of a well-corrected optical system.

Paradoxically, the paraxial equations are frequently used with relatively large angles and ray heights. This extension of the paraxial region is useful in estimating the necessary diameters of optical elements and in approximating the aberrations of the image formed by a lens system, as we shall demonstrate in later chapters. This works because the paraxial equations are linear, not trigonometric, and can be scaled.

Although paraxial calculations are often used in rough preliminary work on optical systems and in approximate calculations (indeed, the term "paraxial approximation" is often used), the reader should bear in mind that the paraxial equations are perfectly exact for the paraxial region and that, as an exact limiting case, they are used in aberration determination as a basis of comparison to indicate how far a trigono-metrically computed ray departs from its ideal location.

The simplest way of deriving a set of equations for the paraxial region is to substitute the angle itself for its sine in the equations derived in the preceding section. Thus we get

from Eq. 3.1 $$ca = (R - l)u \qquad (3.8)$$

from Eq. 3.2 $$i = ca/R \qquad (3.9)$$

from Eq. 3.3 $$i' = ni/n' \qquad (3.10)$$

from Eq. 3.4 $$u' = u - i + i' \qquad (3.11)$$

from Eq. 3.6 $$ca' = n\,ca/n' \qquad (3.12)$$

from Eq. 3.7 $$l' = R - ca'/u' \qquad (3.13)$$

Notice that the paraxial equations are distinguished from the trigono-metric equations by the use of lowercase letters for the paraxial values. This is a widespread convention and will be observed throughout this text. Note also that the angles are in radian measure, not degrees.

Equations 3.8 through 3.13 may be materially simplified. Indeed, since they apply exactly only to a region in which angles and heights are infinitesimal, we can totally eliminate i, u, and ca from the expressions without any loss of validity. Thus, if we substitute into Eq. 3.13, Eq. 3.12 for ca' and Eq. 3.11 for u', and continue the substi-tution with Eqs. 3.8, 3.9, and 3.10, the following simple expression for l' is found:

$$l' = \frac{ln'R}{(n'-n)l + nR} \left[= \frac{n'R}{(n'-n)} \quad \text{if } l = \infty \right] \qquad (3.14)$$

By rearranging we can get an expression which bears a marked simi-larity to Eq. 2.4 and Eq. 2.11 (relating the object and image distances for a complete lens system):

$$\frac{n'}{l'} = \frac{(n'-n)}{R} + \frac{n}{l} \qquad (3.15a)$$

These two equations are useful when the quantity of interest is the distance l'. If the object and image are at the axial intersection distances l and l', the magnification is given by

$$m = \frac{h'}{h} = \frac{nl'}{n'l} \tag{3.15b}$$

In Sec. 2.2 we noted that the power of an optical system was the reciprocal of its effective focal length. In Eq. 3.15a the term $(n' - n)/R$ is the power of the surface. A surface with positive power will bend (converge) a ray toward the axis; a negative-power surface will bend (diverge) a ray away from the axis. If R is in meters, the power is in diopters.

3.3 Paraxial Raytracing through Several Surfaces

The *ynu* raytrace

Another form of the paraxial equations is more convenient for use when calculations are to be continued through more than one surface. Figure 3.2 shows a paraxial ray incident on a surface at a height y from the axis, with the ray-axis intersection distances l and l' before and after refraction. The height y in this case is a fictitious extension of the paraxial region, since, as noted, the paraxial region is an infinitesimal one about the axis. However, since all heights and angles cancel out of the paraxial expressions for the intercept distances (as indicated above), the use of finite heights and angles does not affect the accuracy of the expressions. For systems of modest aperture these fictitious heights and angles are a reasonable approximation to the corresponding values obtained by exact trigonometrical calculation.

In the paraxial region, every surface approaches a flat plane surface, just as all angles approach their sines and tangents. Thus we can express the slope angles shown in Fig. 3.2 by $u = -y/l$ and $u' = -y/l'$, or $l = -y/u$ and $l' = -y/u'$. If we substitute these latter values for l and l' into Eq. 3.15a, we get

$$\frac{n'u'}{y} = \frac{-(n'-n)}{R} + \frac{nu}{y}$$

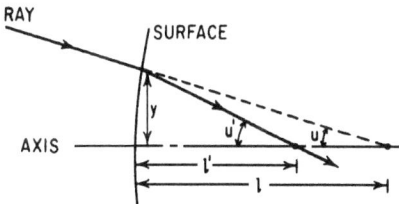

Figure 3.2 The relationship $y = -lu = -l'u'$ for paraxial rays.

and multiplying through by y, we find the slope after refraction.

$$n'u' = nu - y \frac{(n' - n)}{R} \tag{3.16}$$

It is frequently convenient to express the curvature of a surface as the reciprocal of its radius, $C = 1/R$; making this substitution, we have

$$n'u' = nu - y (n' - n) C \tag{3.16a}$$

To continue the calculation to the next surface of the system, we require a set of transfer equations. Figure 3.3 shows two surfaces of an optical system separated by an axial distance t. The ray is shown after refraction by surface #1; its slope is the angle u'_1. The intersection heights of the ray at the surfaces are y_1 and y_2, respectively, and since this is a paraxial calculation, the difference between the two heights can be given by tu'_1. Thus, it is apparent that

$$y_2 = y_1 + tu'_1 = y_1 + t \frac{n'_1 u'_1}{n'_1} \tag{3.17}$$

And if we note that the slope of the ray incident on surface #2 is the same as the slope after refraction by #1, we get the second transfer equation

$$u_2 = u'_1 \quad \text{or} \quad n_2 u_2 = n'_1 u'_1 \tag{3.18}$$

These equations can now be used to determine the position and size of the image formed by a complete optical system, as illustrated by the following example. Note that the paraxial ray heights and ray slopes are scalable (i.e., they may be multiplied by the same factor). The result of scaling is the data of another ray (which has the same axial intersection).

Example 3.1

Figure 3.4 shows a typical problem. The optical system consists of three surfaces, making a "doublet" lens the radii, thicknesses, and indices of

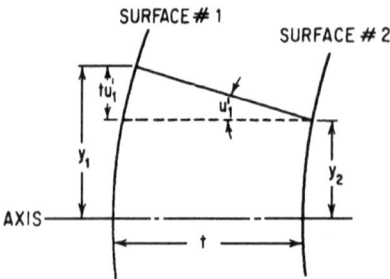

Figure 3.3 The transfer of a paraxial ray from surface to surface by $y_2 = y_1 + tu'_1$. Note that although the surfaces are drawn as curved in the figure, mathematically they are treated as planes. Thus the ray is assumed to travel the axial spacing t in going from surface #1 to surface #2.

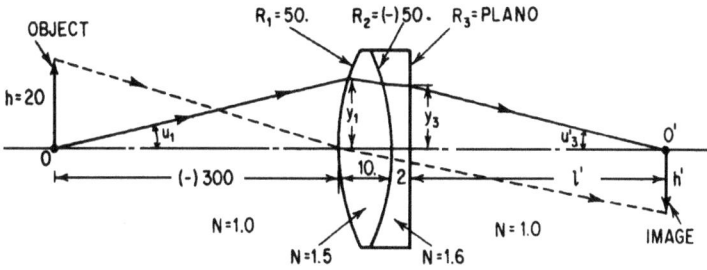

Figure 3.4 The rays traced in Example 3.1.

which are indicated in the figure. The object is located 300 mm to the left of the first surface and extends a height of 20 mm above the axis. The lens is immersed in air, so that object and image are in a medium of index $n = 1.0$.

The first step is to tabulate the parameters of the problem with the proper signs associated. Following the sign convention given above, we have the following:

$h = +20$ mm

$l_1 = -300$ mm $n_1 = 1.0$

$R_1 = +50$ mm $C_1 = +0.02$ $t_1 = 10$ mm $n'_1 = n_2 = 1.5$

$R_2 = -50$ mm $C_2 = -0.02$ $t_2 = 2$ mm $n'_2 = n_3 = 1.6$

$R_3 = $ plano $C_3 = 0$ $n'_3 = 1.0$

The location of the image can be found by tracing a ray from the point where the object intersects the axis (O in the figure); the image will then be located where the ray recrosses the axis at O'. We can use any reasonable value for the starting data of this ray. Let us trace the path of the ray starting at O and striking the first surface at a height of 10 mm above the axis. Thus $y_1 = +10$ and we get the initial slope angle by

$$u_1 = \frac{-y_1}{l_1} = \frac{-10}{-300} = +0.0333$$

and since $n_1 = 1.0$, $n_1 u_1 = +0.0333$. The slope angle after refraction is obtained from Eq. 3.16a.

$$n'_1 u'_1 = -y_1 (n'_1 - n_1) C_1 + n_1 u_1$$

$$= -10 (1.5 - 1.0) (+0.02) + 0.0333$$

$$= -0.1 + 0.0333$$

$$n'_1 u'_1 = -0.0666$$

The ray height at surface #2 is found by Eq. 3.17.

$$y_2 = y_1 + \frac{t_1\,(n'_1 u'_1)}{n'_1}$$

$$= 10 + \frac{10\,(-0.0666)}{1.5}$$

$$= 10 - 0.444$$

$$y_2 = 9.555$$

Noting that $n_2 u_2 = n'_1 u'_1$, the refraction at the second surface is carried through by

$$n'_2 u'_2 = -y_2\,(n'_2 - n_2)\,C_2 + n_2 u_2$$

$$= -9.555\,(1.6 - 1.5)\,(-0.02) - 0.0666$$

$$= +0.019111 - 0.0666$$

$$= -0.047555$$

and the ray height at the third surface is calculated by

$$y_3 = y_2 + \frac{t_2\,(n'_2 u'_2)}{n'_2} = 9.555 + \frac{2\,(-0.04755)}{1.6}$$

$$= 9.555 - 0.059444 = 9.496111$$

Since the last surface of the system is plane, i.e., of infinite radius, its curvature is zero and the product nu is unchanged at this surface:

$$n'_3 u'_3 = -y_3\,(n'_3 - n_3)\,C_3 + n_3 u_3$$

$$= -9.496111\,(1.0 - 1.6)\,(0) - 0.047555 = -0.047555$$

and

$$u'_3 = \frac{n'_3 u'_3}{n'_3} = -0.047555$$

Now the location of the image is given by the final intercept length l', which is determined by

$$l'_3 = \frac{-y_3}{u'_3} = \frac{-9.496111}{-0.047555}$$

$$= +199.6846$$

Note that the choice of y_1 or u_1 may be an arbitrary one. We can scale y and u, but l and and l' remain the same.

The execution of a long chain of calculations such as the preceding is much simplified if the calculation is arranged in a convenient table form. By ruling the paper in squares, a simple arrangement of the constructional parameters at the top of the sheet and the ray data below helps to speed the calculation and eliminate errors. The following table (Fig. 3.5) sets forth the curvatures, thicknesses, and indices of the lens in the first three rows; the next two rows contain the ray heights and index-slope angle products of the calculation worked out above.

The image height can now be found by tracing a ray from the top of the object and determining the intersection of this ray with the image plane we have just computed. Such a ray is shown by the dashed line in Fig. 3.4. If we elect to trace the ray which strikes the vertex of the first surface, then y_1 will be zero and the initial slope angle will be given by

$$u_1 = \frac{-(y_1 - h)}{l_1} = \frac{-(0 - 20)}{-300} = -0.0666$$

The calculation of this ray is indicated in the sixth and seventh rows of Fig. 3.5 and yields $y_3 = -0.52888...$ and $n'_3 u'_3 = -0.067555$.

The height of the image, h' in Fig. 3.4, can be seen to equal the sum of the ray height at surface #3 plus the amount the ray climbs or drops in traveling to the image plane.

$$h' = y_3 + l'_3 \frac{n'_3 u'_3}{n'_3} = -0.52888 + 199.6846 \frac{-0.067555}{1.0}$$

$$= -14.0187$$

Notice that the expression used to compute h' is analogous to Eq. 3.17; if we regard the image plane as surface #4 and the image distance

		Surface #1		Surface #2		Surface #3	
Curvature		+0.02		-0.02		0.0	
thickness			10.		2.		
index	1.0		1.5		1.6		1.0
Ray height (y)		10.		9.555		9.496111	
Nu	+0.0333		-0.0666		-0.047555		-0.047555
y		0.0		-0.444		-0.52888	
Nu	-0.0666		-0.0666		-0.067555		-0.067555

Figure 3.5 An orderly layout of raytracing calculation.

$l'_3 = 199.6846$ as the spacing between surfaces #3 and #4, Eq. 3.17 can be used to calculate y_4, which is h'.

Similarly, Eq. 3.17 can be used to determine the initial slope angle u_1 by regarding the object plane as surface zero and rearranging the equation to solve for $u'_0 = u_1$ as shown below:

$$y_1 = y_0 + t_0 \frac{n'_0 u'_0}{n'_0}$$

$$u'_0 = u_1 = \frac{y_1 - y_0}{t_0} = \frac{h - y_1}{l_1}$$

Note that all paraxial rays from a given object point will intersect the paraxial focus image plane at exactly the same point.

3.4 Calculation of the Focal Points and Principal Points

In general, the focal lengths of an optical system can easily be calculated by tracing a ray parallel to the optical axis (i.e., with an initial slope angle u equal to zero) completely through the optical system. Then the effective focal length (efl) is minus the ray height at the first surface divided by the ray slope angle u'_k after the ray emerges from the last surface. Similarly, the back focal length (bfl) is minus the ray height at the last surface divided by u'_k. Using the customary convention that the data of the last surface of the system are identified by the subscript k, we can write

$$\text{efl} = \frac{-y_1}{u'_k} \tag{3.19}$$

$$\text{bfl} = \frac{-y_k}{u'_k} \tag{3.20}$$

The cardinal points of a single lens element can be readily determined by use of the raytracing formulas given in the preceding section. The focal point is the point where the rays from an infinitely distant axial object cross the optical axis at a common focus. As indicated, this point can be located by tracing a ray with an initial slope (u_1) of zero through the lens and determining the axial intercept.

The reader may wish to test his or her understanding and skill at raytracing by calculating the focal lengths of the doublet lens of Fig. 3.4. The results should be:

$$\text{efl} = +122.950820$$
$$\text{bfl} = +113.504098$$
$$\text{ffl} = -124.590164$$

The efl from the (right to left) calculation of ffl should be exactly the same as the efl from the (left to right) calculation for bfl.

Figure 3.6 shows the path of such a ray through a single lens element. The principal plane (p_2) is located by the intersection of the extensions of the incident and emergent rays. The effective focal length (efl) or focal length (usually symbolized by f), is the distance from p_2 to f_2 and, for the paraxial region, is given by

$$\text{efl} = f = \frac{-y_1}{u'_2}$$

The back focal length (bfl) can be found from

$$\text{bfl} = \frac{-y_2}{u'_2}$$

Because of the frequency with which these quantities are used, it is worthwhile to work up a single equation for each of them. If the lens has an index of refraction n and is surrounded by air of index 1.0, then $n_1 = n'_2 = 1.0$ and $n'_1 = n_2 = n$. The surface radii are R_1 and R_2, and the surface curvatures are c_1 and c_2. The thickness is t. At the first surface, using Eq. 3.16a,

$$n'_1 u'_1 = n_1 u_1 - (n'_1 - n_1) y_1 c_1 = 0 - (n-1) y_1 c_1$$

The height at the second surface is found from Eq. 3.17:

$$y_2 = y_1 + \frac{t n'_1 u'_1}{n'_1} = y_1 - \frac{t(n-1) y_1 c_1}{n} = y_1 \left[1 - \frac{(n-1)}{n} t c_1 \right]$$

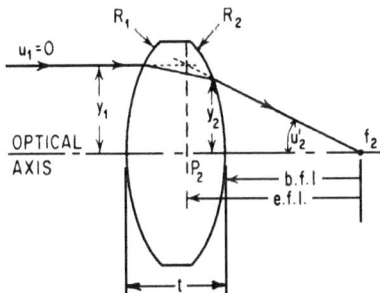

Figure 3.6 A ray parallel to the axis is traced through an element to determine the effective focal length and back focal length.

And the final slope is found by Eq. 2.31a:

$$n'_2 u'_2 = n'_1 u'_1 - y_2(n'_2 - n_2) c_2$$

$$= - (n - 1) y_1 c_1 - y_1 \left[1 - \frac{(n - 1)}{n} tc_1 \right] (1 - n) c_2$$

$$(1.0) u'_2 = u'_2 = - y_1(n - 1) \left[c_1 - c_2 + tc_1 c_2 \frac{(n - 1)}{n} \right]$$

Thus the power ϕ (or reciprocal focal length) of the element is expressed as

$$\phi = \frac{1}{f} = \frac{-u'_2}{y_1} = (n - 1) \left[c_1 - c_2 + tc_1 c_2 \frac{(n - 1)}{n} \right] \qquad (3.21)$$

or, if we substitute $c = 1/R$,

$$\phi = \frac{1}{f} = (n - 1) \left[\frac{1}{R_1} - \frac{1}{R_2} + \frac{t (n - 1)}{R_1 R_2 n} \right] \qquad (3.21a)$$

The back focal length can be found by dividing y_2 by u'_2 to get

$$\text{bfl} = \frac{-y_2}{u'_2} = f - \frac{ft (n - 1)}{n R_1} \qquad (3.22)$$

The distance from the second surface to the second principal point is just the difference between the back focal length and the effective focal length (see Fig. 3.6); this is obviously the last term of Eq. 3.22.

The above procedure has located the second principal point and second focal point of the lens. The "first" points are found simply by substituting R_1 for R_2 and vice versa.

The focal points and principal points for several shapes of elements are diagramed in Fig. 3.7. Notice that the principal points of an equiconvex or equiconcave element are approximately evenly spaced within the element. In the plano forms, one principal point is always at the curved surface and the other is about one-third of the way into the lens. In the meniscus forms, one of the principal points is completely outside the lens; in extreme meniscus shapes, both the principal points may lie outside the lens and their order may be reversed from that shown. **Note well** that the focal points of the negative elements are in reversed order compared to a positive element.

If the lens element is not immersed in air, we can derive a similar expression for it. Assuming that the object medium has an index of n_1, the lens index is n_2, and the image medium has an index of n_3, then the

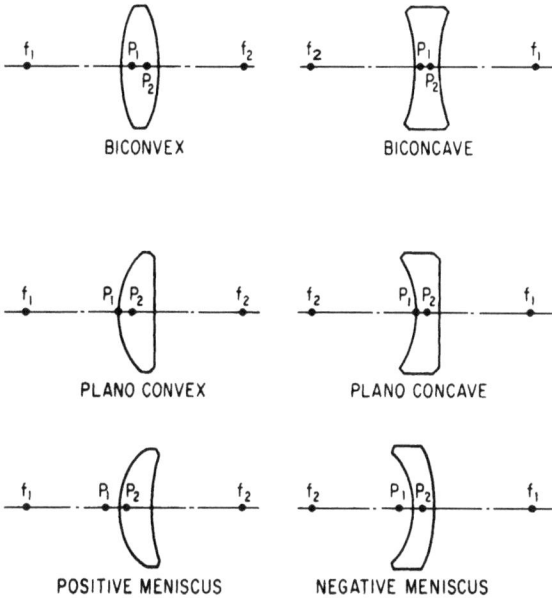

Figure 3.7 The location of the focal points and principal points for several shapes of converging and diverging elements.

two effective focal lengths and the back focal length can be calculated from

$$\frac{n_1}{f} = \frac{n_3}{f'} = \frac{(n_2 - n_1)}{R_1} - \frac{(n_2 - n_3)}{R_2} + \frac{(n_2 - n_3)(n_2 - n_1)t}{n_2 R_1 R_2} \quad (3.23)$$

$$bfl = f' - \frac{f't(n_2 - n_1)}{n_2 R_1} \quad (3.24)$$

Note that if n_1 and n_3 are equal to 1.0 (i.e., the index of air), these expressions reduce to Eqs. 3.21 and 3.22.

3.5 The "Thin Lens"

If the thickness of a lens element is small enough so that its effect on the accuracy of the calculation may be neglected, the element is called a thin lens. The "thin lens" concept is an extremely useful one for the purposes of quick preliminary calculations and analysis and as a design tool.

The focal length of a thin lens can be derived from Eq. 3.21 by setting the thickness equal to zero.

$$\frac{1}{f} = (n - 1)\,(c_1 - c_2) \tag{3.25}$$

$$\frac{1}{f} = (n - 1)\left(\frac{1}{R_1} - \frac{1}{R_2}\right) \tag{3.25a}$$

Since the lens thickness is assumed to be zero, the principal points of a "thin lens" are coincident with the location of the lens. Thus, in computing object and image positions, the distances s and s' of Eqs. 2.4, 2.5, 2.7, etc., are measured from the lens itself. The term $(c_1 - c_2)$ is often called the *total curvature*, or simply the curvature of the element.

Note that if the lens index is 1.5, the radii of an equiconvex or equiconcave element equals the focal length ($R = \pm f$) and the radius of a planoconvex or planoconcave is one-half the focal length ($R = \pm f/2$).

Example 3.2

An object 10-mm high is to be imaged 50-mm high on a screen that is 120-mm distant. What are the radii of an equiconvex lens of index 1.5 which will produce an image of the proper size and location?

The first step in the calculation is the determination of the focal length of the lens. Since the image is a real one, the magnification will have a negative sign, and by Eq. 2.7a we have

$$m = \frac{h'}{h} = (-)\,\frac{50}{10} = \frac{s'}{s} \qquad \text{or} \qquad s' = -5s$$

For the object and image to be 120 mm apart,

$$120 = -s + s' = -s - 5s = -6s$$

$$s = -20 \text{ mm}$$

and
$$s' = -5s = +100 \text{ mm}$$

Substituting into Eq. 2.4 and solving for f, we get

$$\frac{1}{100} = \frac{1}{f} + \frac{1}{-20}$$

$$f = 16.67 \text{ mm}$$

Noting that for an equiconvex lens $R_1 = -R_2$, we use Eq. 3.25a to solve for the radii

$$\frac{1}{f} = +0.06 = (n-1)\left(\frac{1}{R_1} - \frac{1}{R_2}\right) = 0.5\frac{2}{R_1}$$

$$R_1 = \frac{1}{0.06} = 16.67 \text{ mm}$$

$$R_2 = -R_1 = -16.67 \text{ mm}$$

3.6 Mirrors

A curved mirror surface has a focal length and is capable of forming images just as a lens does. The equations for paraxial raytracing (Eqs. 2.31 and 2.32) can be applied to reflecting surfaces by taking into account two additional sign conventions. The index of refraction of a material was defined in the first chapter as the ratio of the velocity of light in vacuum to that in the material. Since the direction of propagation of light is reversed upon reflection, it is logical that the sign of the velocity should be considered reversed, and the sign of the index reversed as well. Thus the conventions are as follows:

1. The signs of the indices following a reflection are reversed, so the index is negative when light travels right to left.

2. The signs of the spacings following a reflection are reversed if the following surface is to the left.

Obviously if there are two reflecting surfaces in a system, the signs of the indices and spacings are changed twice and, after the second change, revert to the original positive signs, since the direction of propagation is again left to right.

Figure 3.8 shows the locations of the focal and principal points of concave and convex mirrors. The ray from the infinitely distant source which defines the focal point can be traced as follows, setting $n = 1.0$ and $n' = -1.0$:

$$nu = 0 \quad \text{(since the ray is parallel to the axis)}$$

$$n'u' = nu - y\frac{(n'-n)}{R} = 0 - y\frac{(-1-1)}{R} = \frac{2y}{R}$$

thus

$$u' = \frac{n'u'}{n'} = \frac{n'u'}{-1} = \frac{-2y}{R}$$

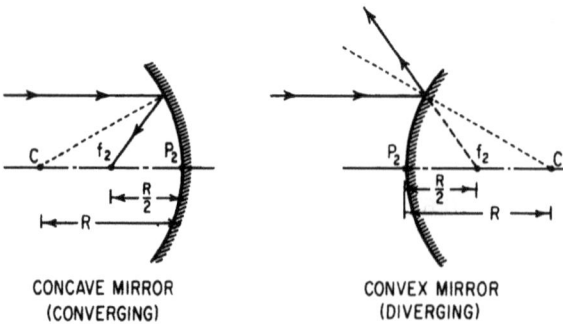

CONCAVE MIRROR
(CONVERGING)

CONVEX MIRROR
(DIVERGING)

Figure 3.8 The location of the focal points for reflecting surfaces.

The final intercept length is

$$l' = \frac{-y}{u'} = \frac{yR}{2y} = \frac{R}{2}$$

and we find that the focal point lies halfway between the mirror and its center of curvature.

The concave mirror is the equivalent of a positive converging lens and forms a real image of distant objects. The convex mirror forms a virtual image and is equivalent to a negative element. Because of the index sign reversal on reflection, the sign of the focal length is also reversed and the focal length of a simple mirror is given by

$$f = -\frac{R}{2}$$

so that the sign conforms to the convention of positive for converging elements and negative for diverging elements.

Example 3.3

Calculate the focal length of the Cassegrain mirror system shown in Fig. 3.9 if the radius of the primary mirror is 200 mm, the radius of the secondary mirror is 50 mm, and the mirrors are separated by 80 mm.

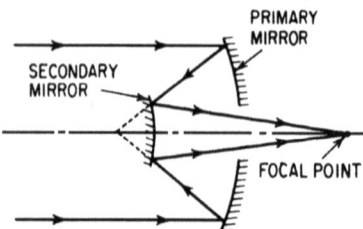

Figure 3.9 Cassegrain mirror system. The image formed by the primary mirror is the virtual object for the secondary mirror.

Radius (R)		-200		-50	
Thickness (t)			-80		
Index (N)	+1.0		-1.0		+1.0
Ray height (y)		1.0		+0.2	
Ray slope × index (Nu)	0		-0.01		-0.002

Figure 3.10 Raytrace through a two mirror system.

Following our sign convention, the radii are both negative and the distance from primary to secondary mirror is also considered negative, since the light traverses this distance right to left. The index of the air is taken as +1.0 before the primary and after the secondary; between the two, the index is −1.0. Thus the optical data of the problem and the computation are set up and carried through as shown in Fig. 3.10. Careful attention to signs is necessary in this calculation to avoid mistakes.

The focal length of the system is given by $-y_1/u'_2 = -1.0/-0.002 = 500$ mm. The final intercept distance (from R_2 to the focus) is equal to $-y_2/u'_2 = -0.2/-0.002 = 100$ mm, and the focal point lies 20 mm to the right of the primary mirror. Notice that the (second) principal plane is completely outside the system, 400 mm to the left of the secondary mirror, and that this type of system provides a long focal length and a large image in a small, compact system.

Bibliography

Bass, M., *Handbook of Optics,* Vol. 1, New York, McGraw-Hill, 1995.
Fischer, R. E., and B. Tadic-Galeb, *Optical System Design,* New York, McGraw-Hill, 2000.
Greivenkamp, J., *Field Guide to Geometrical Optics,* Bellingham, WA, SPIE, 2004.
Kingslake, R., *Optical System Design,* Orlando, Academic, 1983.
Smith, W. J., *Modern Lens Design,* New York, McGraw-Hill, 2002.
Smith, W. J., *Practical Optical System Layout,* New York, McGraw-Hill, 1997.

Exercises

1 A detector, 1 mm on a side, is "immersed" on the plano surface of a plano convex lens with an index of 1.50 and a radius of 10.0 mm. When viewed through the convex surface, where is the image, and what is the image size if the immersion lens is:

   (a) 7.0-mm thick
   (b) 10.0-mm thick
   (c) 16.666 . . . mm thick

ANSWER: (a) Using Eq. 3.16, trace a ray from the axial intercept of the plane surface through the curved surface, at a height of 1 mm on that surface. The ray slope in the glass is $1/7 = +0.142857$ and the slope-index product nu is thus $+0.214286$. The slope after refraction (in air $n = 1.0$) will be given by

$n'u' = u' = nu - y(n' - n)/R = +0.214286 - 1.0 \,(-0.5)/(-10) = +0.164286$, and the image location can be found from $l' = -y/u' = -1/0.164286 = -6.086961$. The magnification can be found from an oblique raytrace from the top of a 1 mm object, or from Eq. 3.15b, $m = h'/h = nl'/n'l$. Thus $m = 1.5 \cdot (-6.086961)/1.0 \cdot 7 = -1.304349\times$ and the image size is 1.304349 mm.

Tabulating the results for the three thicknesses, we have:

(a) th = 7.0 $u' = +0.164286$ $l' = -6.08961$ $h' = 1.304349$ mm

(b) th = 10.0 $= +0.100000$ $= -10.00000$ $= 1.500000$ mm

(c) th = 16.66 . . . $= +0.090000$ $= -11.11111$ $= 1.000000$ mm

Note that the image size can also be found from Eq. 4.16, $m = nu/n'u'$.

2 Given an equiconvex lens, radii = ±100, thickness = 10, and index = 1.50, trace a ray (parallel to the axis) through the lens, beginning at a ray height of (a) 1.0, and (b) 10.0, and determine the axial intersection of the ray.

ANSWER:

R	+100.	−100.	
t		10.	
n	1.0	1.5	1.0

(a) y   1.0   0.9666 . . . $l' = -y/u' = +98.3051$

      nu   0.0   −0.005   −0.0098333 . . .

(b) y   10.0   9.666 . . . $l' = -y/u' = +98.3051$

      nu   0.0   −0.050   −0.098333 . . .

3 Determine the effective and back focal lengths of the lens in Exercise 2., (a) from the raytrace data, and (b) using the thick lens equations.

ANSWER:

(a)   Eq. 3.19 $\text{efl} = -y_1/u'_k = -1.0/(-0.0098333) = +101.6949$

       Eq. 3.20 $\text{bfl} = -y_K/u'_k = -.9666/(-0.0098333) = +98.3051$

(b)   Eq. 3.21 $\phi = (1/f) = (n - 1)\,[c_1 - c_2 + tc_1c_2\,(n - 1)/n]$

                       $= 0.5[0.01 - (-0.01) + 10 \cdot 0.01 \cdot (-0.01) \cdot 0.5/1.5]$

                       $= 0.5[0.02 - 0.000333] = 0.0098333 . . .$

                 $\text{efl} = 101.6949$

       Eq. 3.22 $\text{bfl} = f - ft\,(n - 1)c_1/n]$

                     $= 101.6949 - 101.6949 \cdot 10 \cdot 0.5 \cdot 0.01/1.5$

                     $= 101.6949 - 3.3898$

                     $= 98.3051$

4 What is the focal length of the lens in Exercise 3 if it is treated as a thin lens?

ANSWER: Eq. 3.25 $\phi = (1/f) = (n - 1)[c_1 - c_2]$

                      $= 0.5[0.01 - (-0.01)] = 0.5 \cdot 0.02 = 0.01$

             $\text{efl} = 1/\phi = 100.0$

4

Optical System Considerations

4.1 Systems of Separated Components

An *optical element* is a single irreducible optical lens or mirror.

An *optical component* may be an element, or several elements which are treated as a unit.

An *optical member* is one of two parts of a system, separated by a diaphragm: the front member and the back member.

An *optical system* is a complete set of optics which produces an image of the desired size, in the desired location, and with the desired orientation.

In order to simplify and organize the design of optical systems, it is convenient (especially in preliminary work) to treat an optical system as an arrangement of components, each with zero thickness. In preliminary work we can deal with components as simple, unified pieces. A component may consist of several elements, but in creating a system layout scheme, we simply specify the power (or focal length) of a component and its location. When the initial layout is done, the zero thickness components are replaced by real, physically possible components.

This avoids having to handle the system by means of surface-by-surface calculation. To this end we can introduce the paraxial ray height y into the equations of Sec. 2.3, just as we did in Sec. 3.3.

An optical component (which may be made up of a number of elements) is shown in Fig. 4.1 with its object a distance s from the first principal plane and its image a distance s' from the second principal plane. The principal planes are planes of unit magnification, in that the incident and emergent ray paths appear to strike (and emerge from) the same height on both the first and second principal planes. Thus, in Fig. 4.1

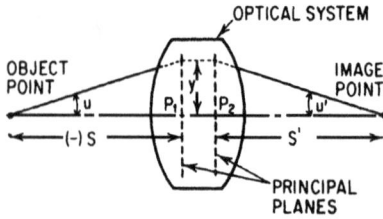

Figure 4.1 The principal planes are planes of unit magnification, so a ray appears to leave the second principal plane at the same height (y) that it appears to strike the first principal plane.

a ray from the object point, which would (if extended) strike the first principal plane at a distance y from the axis, emerges from the last surface of the system as if it were coming from the same height y on the second principal plane. For this reason we can write the following delightfully simple relationships:

$$u = \frac{-y}{s} \quad \text{and} \quad u' = \frac{-y}{s'}$$

and substitute $s = -y/u$ and $s' = -y/u'$ into Eq. 2.4:

$$\frac{1}{s'} = \frac{1}{s} + \frac{1}{f}$$

$$\frac{-u'}{y} = \frac{-u}{y} + \frac{1}{f}$$

$$u' = u - \frac{y}{f}$$

If we now replace the reciprocal focal length ($1/f$) with the component power ϕ, we get the first equation of our ray:

$$u' = u - y\phi \tag{4.1}$$

The transfer equations to the next component in the system are the same as those used in the paraxial surface-by-surface raytrace of Sec. 3.3:

$$y_2 = y_1 + du'_1 \tag{4.2}$$

$$u'_1 = u_2 \tag{4.3}$$

where y_1 and y_2 are the ray heights at the principal planes of components #1 and #2, u'_1 is the slope angle after passing through component #1, and d is the axial distance from the second principal plane of component #1 to the first principal plane of component #2.

Note that these equations are equally applicable to systems composed of either thick or "thin" lenses. Obviously, when applied to thin lenses, d becomes the spacing between elements, since the element and its principal planes are coincident.

Focal lengths of two-component systems

The preceding equations may be used to derive compact expressions for the effective focal length and back focal length of a system comprised of two separated components. Let us assume that we have two components of powers ϕ_a and ϕ_b separated by a distance d (if the lenses are "thin"; if they are thick, d is the separation of their principal points). The system is sketched in Fig. 4.2.

Beginning with a ray parallel to the axis which strikes lens a at y_a, we have

$$u_a = 0$$

$$u'_a = 0 - y_a\phi_a \quad \text{by Eq. 4.1}$$

$$y_b = y_a - dy_a\phi_a = y_a(1 - d\phi_a) \quad \text{by Eq. 4.2}$$

$$u'_b = -y_a\phi_a - y_a(1 - d\phi_a)\,\phi_b \quad \text{by Eq. 4.1}$$

$$= -y_a(\phi_a + \phi_b - d\phi_a\phi_b)$$

The power (reciprocal focal length) of the system is given by

$$\phi_{ab} = \frac{-u'_b}{y_a} = \frac{1}{f_{ab}} = \phi_a + \phi_b - d\phi_a\phi_b$$

$$= \frac{1}{f_a} + \frac{1}{f_b} - \frac{d}{f_a f_b} \tag{4.4}$$

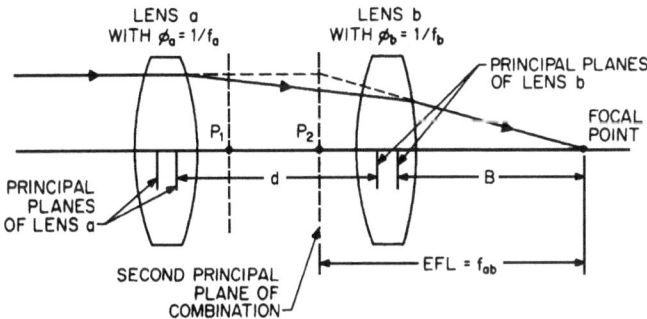

Figure 4.2 Raytrace through two separated components to determine the focal length and back focus distance of the combination.

and thus

$$f_{ab} = \frac{f_a f_b}{f_a + f_b - d} \tag{4.5}$$

The back focus distance (measured from the second principal point of b) is given by

$$B = \frac{-y_b}{u'_b} = \frac{y_a (1 - d\phi_a)}{y_a (\phi_a + \phi_b - d\phi_a\phi_b)} \tag{4.6}$$

$$= \frac{(1 - d/f_a)}{1/f_a + 1/f_b - d/f_a f_b} = \frac{f_b (f_a - d)}{f_a + f_b - d}$$

By substituting f_{ab}/f_a from Eq. 4.5, we get

$$B = \frac{f_{ab} (f_a - d)}{f_a} \tag{4.6a}$$

The front focus distance (ffd) for the system is found by reversing the raytrace (i.e., trace from right to left) or more simply by substituting f_b for f_a to get

$$(-)\text{ffd} = \frac{f_{ab} (f_b - d)}{f_b} \tag{4.6b}$$

The inverse solution

Frequently it is useful to be able to solve for the focal lengths of the components when the focal length, back focus distance, and spacing are given for the system. Manipulation of Eqs. 4.5 and 4.6a will yield

$$f_a = \frac{df_{ab}}{f_{ab} - B} \tag{4.7}$$

$$f_b = \frac{-dB}{f_{ab} - B - d} \tag{4.8}$$

Equations 4.7 and 4.8 are probably the most widely used equations in optical system layout work.

General equations for two-component finite conjugate systems

Using the same technique, we can derive expressions which give us the solution to all two-component optical problems. There are two types of problems which occur. With reference to Fig. 4.3, the first type occurs

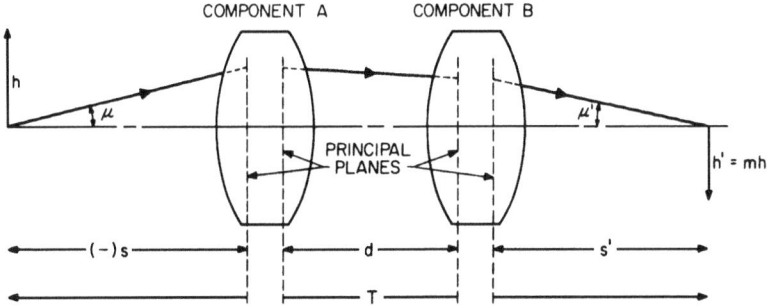

Figure 4.3   A two-component system operating at finite conjugates.

when we are given the required system magnification, the positions of the two components, and the object-to-image distance (neglecting the spaces between the principal planes of the components). Thus, knowing s, s', d, and the magnification m, we wish to determine the powers (or focal lengths) of the two components, which are given by

$$\phi_A = \frac{(ms - md - s')}{msd} \qquad (4.9)$$

$$\phi_B = \frac{(d - ms + s')}{ds'} \qquad (4.10)$$

In the second type of problem we are faced with the inverse case, in that we know the component powers, the desired object-to-image distance, and the magnification; we must determine the locations for the two components. Under these circumstances the mathematics result in a quadratic relationship, and thus there may be two solutions, one solution, or no solution (i.e., an imaginary solution). The following quadratic equation (Eq. 4.11) in d (the spacing) is first solved for d [using the standard equation $x = (-b \pm \sqrt{b^2 - 4ac})/2a$ to solve $0 = ax^2 + bx + c$].

$$0 = d^2 - dT + T(f_A + f_B) + \frac{(m - 1)^2 f_A f_B}{m} \qquad (4.11)$$

Then s and s' are easily determined from

$$s = \frac{(m - 1)d + T}{(m - 1) - md\phi_A} \qquad (4.12)$$

$$s' = T + s - d \qquad (4.13)$$

Thus Eqs. 4.4 through 4.13 constitute a set of expressions which can be used to solve *any* problem involving two components. Since two-component systems constitute the vast majority of optical systems, these are extremely useful equations. Note that a change of the sign of the magnification *m* from plus to minus will result in two completely different optical systems. They will produce the same enlargement (or reduction) of the image. One will have an erect, and the other an inverted, image, but one system may be significantly more suitable than the other for the intended application.

4.2 The Optical Invariant

The optical invariant, or Lagrange invariant, is a constant for a given optical system, and it is a very useful one. Its numerical value may be calculated in any of several ways, and the invariant may then be used to arrive at the value of other quantities without the necessity of certain intermediate operations or raytrace calculations which would otherwise be required.

Let us consider the application of Eq. 3.16a to the tracing of two paraxial rays through an optical system. One ray (the "axial" ray) is traced from the foot, or axial intercept, of the object; the other ray (the "oblique" ray) is traced from an off-axis point on the object. Figure 4.4 shows these two rays passing through a generalized system.

At *any* surface in the system, we can write out Eq. 3.16a for each ray, using the subscript *p* to denote the data of the oblique ray.

For the axial ray

$$n'u' = nu - y\,(n' - n)\,c$$

For the oblique ray

$$n'u'_p = nu_p - y_p\,(n' - n)\,c$$

We now extract the common term $(n' - n)\,c$ from each equation and equate the two expressions:

$$(n' - n)\,c = \frac{nu - n'u'}{y} = \frac{nu_p - n'u'_p}{y_p}$$

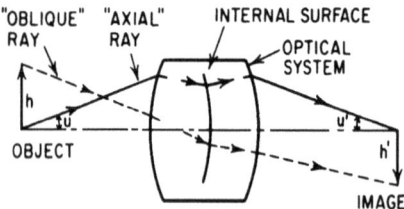

Figure 4.4 The axial and oblique rays used to define the optical invariant, $hnu = h'n'u'$.

Multiplying by yy_p and rearranging, we get

$$y_p nu - ynu_p = y_p n'u' - yn'u'_p$$

Note that on the left side of the equation the angles and indices are for the left side of the surface (that is, before refraction) and that on the right side of the equation the terms refer to the same quantities after refraction. Thus $y_p nu - ynu_p$ is a constant which is invariant across any surface.

By a similar series of operations based on Eq. 3.17, we can show that $(y_p nu - ynu_p)$ for a given surface is equal to $(y_p nu - ynu_p)$ for the next surface. Thus this term is not only invariant across the surface but also across the space between the surfaces; it is therefore invariant throughout the entire optical system or any continuous part of the system.

$$\text{Invariant} \qquad \text{Inv} = y_p nu - ynu_p = n\,(y_p u - yu_p) \tag{4.14}$$

The invariant and magnification

As an example of its application, we now write the invariant for the object plane and image plane of Fig. 4.4. In an object plane $y_p = h$, $n = n$, $y = 0$, and we get

$$\text{Inv} = hnu - (0)\,nu_p = hnu$$

In the corresponding image plane $y_p = h'$, $n = n'$, $y = 0$, and we get

$$\text{Inv} = h'n'u' - (0)\,n'u'_p = h'n'u'$$

Equating the two expressions gives

$$hnu = h'n'u' \tag{4.15}$$

which can be rearranged to give a very generalized expression for the magnification of an optical system

$$m = \frac{h'}{h} = \frac{nu}{n'u'} \tag{4.16}$$

Equation 4.16 is, of course, valid only for the extended paraxial region; this relationship is sometimes applied to trigonometric calculations, where it takes the form of Eq. 4.17 for the magnification at a zone of the aperture.

$$hn \sin u = h'n' \sin u' \tag{4.17}$$

$$\text{or } m = \frac{n \sin u}{n' \sin u'}$$

Etendue

Note that *etendue* or *throughput*, used in radiometry and radiative transfer considerations, is the pupil aperture area times the solid angle field of view, or the object/image area times the solid angle of the acceptance/imaging cone, and is thus related to the square of the optical invariant.

Example 4.1

We can apply the invariant to the calculation made in Example 3.1 by assuming that only the axial ray has been traced. The axial ray slope at the object was +0.0333 . . . and the corresponding computed slope at the image was found to be −0.047555. . . . Since the object and image were both in air of index 1.0, we can find the image height from Eq. 4.16,

$$m = \frac{h'}{h} = \frac{h'}{20} = \frac{nu}{n'u'} = \frac{1.0\,(+0.0333\ldots)}{1.0\,(-0.047555\ldots)}$$

$$h' = \frac{20\,(+0.0333)}{-0.047555)}$$

$$h' = -14.0187$$

This value agrees with the height found in Example 3.1 by tracing a ray from the tip of the object to the tip of the image. The saving of time by the elimination of the calculation of this extra ray indicates the usefulness of the invariant.

Image height for object at infinity

Another useful expression is derived when we consider the case of a lens with its object at infinity. At the first surface the invariant is

$$\text{Inv} = y_p n\,(0) - y_1 n u_p = -y_1 n u_p$$

since the "axial" ray from an infinitely distant object has a slope angle u of zero. At the image plane y_p is the image height h', and y for the "axial" ray is zero; thus

$$\text{Inv} = h'n'u' - (0)\,n'u'_p = h'n'u'$$

Equating the two expressions for Inv, we get

$$h'n'u' = -y_1 n u_p \tag{4.18}$$

$$h' = -u_p \frac{n y_1}{n'u'}$$

which is useful for systems where the object and image are not in air. If both object and image are in air, we set $n = n' = 1.0$, and recalling that $f = -y_1/u'$, we find

$$h' = u_p f \qquad (4.19)$$

$$= \tan U_P \cdot f \quad \text{(for nonparaxial rays)}$$

Telescopic magnification

If we evaluate the invariant at the entrance and exit pupils of a system, y_p is (by definition) equal to zero, and the invariant becomes

$$\mathrm{Inv} = -ynu_p = -y'n'u'_p$$

where y is the pupil semidiameter, and u_p is the angular half field of view. For an afocal system we can equate the invariant at the entrance and exit pupils and then solve for the afocal (or telescopic) angular magnification to get

$$MP = \frac{u'_p}{u_p} = \frac{yn}{y'n'}$$

which indicates that the telescopic magnification is equal to the ratio of entrance pupil diameter to exit pupil diameter (assuming that $n = n'$). This is discussed further in Chap. 13.

Data of a third ray from two traced rays

As one might suspect from the preceding, an optical system is actually completely defined by the paraxial raytrace data of any two unrelated rays, i.e., rays with different axial intersections.

Paraxial raytrace data may be scaled. In other words the ray heights and slopes can be multiplied or divided by a scaling constant; the result is a new raytrace. The new ray will still intersect the axis at the same point(s) as the old, but the ray heights and slopes will be different.

If we treat raytrace data as just a set of equations, or equalities, it is apparent that since one may add or subtract equalities and thereby obtain another equality, one can add the scaled data (ray height or slope) of two rays and get the data for a third ray. If A and B are scaling constants, we can express the data of the third ray as the sum of the scaled data of rays 1 and 2.

$$y_3 = Ay_1 + By_2 \qquad (4.20)$$

$$u_3 = Au_1 + Bu_2 \qquad (4.21)$$

At a location in the system where we know the data of all three rays, we can determine the scaling constants by a simultaneous solution of Eqs. 4.20 and 4.21, and get

$$A = (y_3 u_2 - u_3 y_2)/(u_2 y_1 - y_2 u_1) \tag{4.22}$$

$$B = (u_3 y_1 - y_3 u_1)/(u_2 y_1 - y_2 u_1) \tag{4.23}$$

Typically, the axial and principal rays are chosen as rays 1 and 2. The third ray is often defined in object space, although it may be defined anywhere in the system where the data of all three rays are known. The scaling factors A and B are calculated from Eqs. 4.22 and 4.23. Then, after putting these values for A and B into Eqs. 4.20 and 4.21, the data of ray 3 in image space can be found by inserting the image space data of rays 1 and 2 into the resulting equations. (Assuming that the data of ray 3 in image space is what is desired.)

Focal length determination

For example, if we have the raytrace data for the axial ray #1 and oblique ray #2 and define ray 3 as having $u_3 = 0$ and $y_3 = 1$ in object space, Eqs. 4.20 and 4.21 can give us the final ray height and slope in image space for ray 3. Then, with the primed data (y' and u') indicating the values in image space, we get:

$$efl = -1/u_3' = -(y_1 u_2 - u_1 y_2)/(u_1 u_2' - u_2 u_1') \tag{4.24}$$

$$bfl = -y_3'/u_3' = -(u_2 y_1' - u_1 y_2')/(u_1 u_2' - u_1 u_1') \tag{4.25}$$

If we reverse the whole process and set $u_3' = 0$ and $y_3' = 1$, we can get the (normally negative) front focal length

$$ffl = -y_3/u_3 = -(-u_2'y_1)/(u_1 u_2' - u_2 u_1') \tag{4.26}$$

Formula for two specific rays

The above formulas are perfectly general. If we select certain rays to trace, a simplified expression can often be derived. For example, the OSLO reference manual gives the following variation on this scheme: If we start ray #1 at the foot of the object, and have it strike the first surface at $y_1 \equiv y_3$, and if we start ray #2 at height h on the object and send it through the center of the first surface, then the ray coordinates in object space (i.e., at the object) are:

for ray #1: $y_1 = 0$ and $u_1 = y_3/s$,
for ray #2: $y_2 = h$ and $u_2 = -h/s$

where s is the distance from the object to the first surface. (Note that this surface may be the entrance pupil if desired.)

Now, substituting into Eq. 4.24 for the focal length, we get:

$$\text{efl} = -y_3/u'_3 = -y_3 h/(y_3 u'_2 + h u'_1)$$

Where the primed data are in image space.

Most optical computer programs make use of Eqs. 4.24 to 4.26 (or the equation immediately above) to calculate the focal lengths, because such programs usually put a nominally infinitely distant object at a large, but finite, distance, and thus cannot use the axial ray to exactly calculate the focal length directly by $f = -y_1/u'_k$.

Aperture stop and entrance pupil

Another optical software application of this principle involves the determination of the entrance pupil location when the location of the aperture stop is specified. Again, assuming that an axial ray (y and u) and a principal ray (y_p and u_p) have been traced, we determine the constant B for use in Eqs. 4.20 and 4.21 which will shift the principal ray so that its height at the desired stop surface is zero. This yields

$$B = -y_p/y$$

where y_p and y are taken at the stop surface. Then the new principal ray data at the **first** surface are

$$\text{New } y_p = \text{old } y_p + By$$
$$\text{New } u_p = \text{old } u_p + Bu$$

The entrance pupil location corresponding to this stop position is then $L_p = -y_p/u_p$, and a principal ray aimed at the center of the pupil will pass through the center of the stop.

4.3   Matrix Optics

The general form of the paraxial raytracing equations (Eqs. 3.16 and 3.17 or Eqs. 4.1 and 4.2) is $A = B + CD$. Using Eqs. 4.1 and 4.2, for example, and adding two obvious identities, we have

$$u' = u - y\phi \qquad (\text{plus } y = y)$$
$$y_2 = y_1 + du'_1 \qquad (\text{plus } u_2 = u'_1)$$

We can write the first set in matrix notation as

$$\begin{bmatrix} u' \\ y \end{bmatrix} = \begin{bmatrix} 1 & -\phi \\ 0 & 1 \end{bmatrix} \begin{bmatrix} u \\ y \end{bmatrix} \tag{4.27}$$

The second set becomes

$$\begin{bmatrix} u_2 \\ y_2 \end{bmatrix} = \begin{bmatrix} 1 & 0 \\ d & 1 \end{bmatrix} \begin{bmatrix} u'_1 \\ y_1 \end{bmatrix} \tag{4.28}$$

Substituting the left side of Eq. 4.27 into Eq. 4.28 and multiplying the two inner matrices, we get

$$\begin{bmatrix} u_2 \\ y_2 \end{bmatrix} = \begin{bmatrix} 1 & -\phi \\ d & 1 - d\phi \end{bmatrix} \begin{bmatrix} u_1 \\ y_1 \end{bmatrix}$$

which is the matrix form of Eqs. 4.1 and 4.2.

This process can be chained to encompass an entire optical system if desired, and the final product of all the inner matrices can be interpreted to yield the cardinal points, focal lengths, etc., of the system.

Note well that there is absolutely no magic in this process. The amount of computation involved is exactly the same as in the corresponding paraxial raytrace. To this author it seems far more informative to trace the ray paths and to have the added benefit of a knowledge of the paraxial ray heights and slopes. However, for those to whom matrix manipulation is second nature, this formulation has a definite appeal, although no advantage.

4.4 The y-ybar Diagram

The y-ybar diagram is a plot of the ray height y of an axial ray versus the ray height, ybar, of an oblique (i.e., principal or chief) ray. Thus each point on the plot represents a component (or surface) of the system.

Figure 4.5a shows an erecting telescope and Fig. 4.5b shows the corresponding y-ybar diagram. Note that point A in the y-ybar diagram corresponds to component A, etc. An experienced practitioner can quickly sketch up a system in y-ybar form in the same way that a system can be sketched using elements and rays.

The reduction of either a y-ybar diagram or a sketch with rays to a set of numerical values for the component powers and spacings involves the same amount of computation in either case. Although the y-ybar diagram is simpler to draw than a ray sketch, there is obviously more information in the ray sketch, and an experienced practitioner can easily draw a ray sketch accurately enough to allow conclusions to be drawn as to its practicality, size, etc. which the y-ybar diagram does not readily provide.

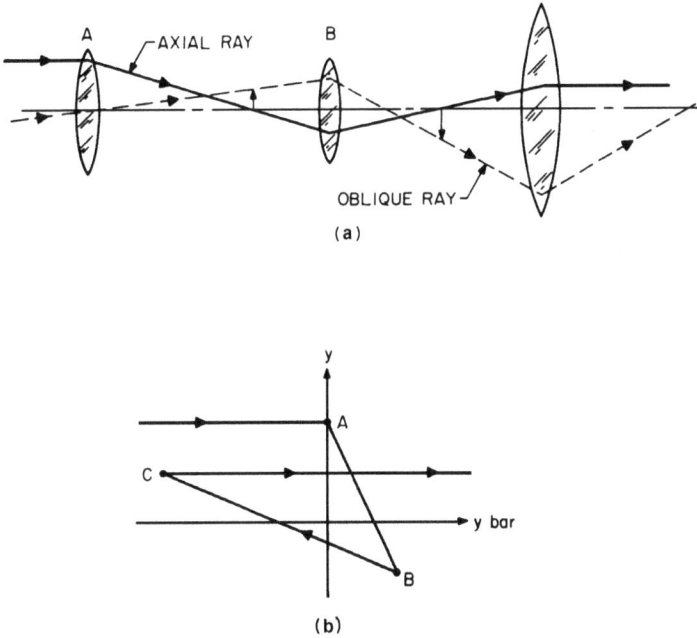

Figure 4.5 (a) Schematic of an optical system and (b) the corresponding y-ybar diagram.

4.5 The Scheimpflug Condition

To this point we have assumed that the object is defined by a plane surface which is normal to the optical axis. However, if the object plane is tilted with respect to the vertical, then the image plane is also tilted. The Scheimpflug condition is illustrated in Fig. 4.6a, which shows the tilted object and image planes intersecting at the plane of the lens. Or, stated more precisely for a thick lens, the extended object and image planes intersect their respective principal planes at the same height.

For small tilt angles in the paraxial region, it is apparent from Fig. 4.6a that the object and image tilts are related by

$$\theta' = \theta \, \frac{s'}{s} = m\theta \tag{4.29}$$

where m is the magnification. For finite (real) angles

$$\tan \theta' = \frac{s'}{s} \tan \theta = m \tan \theta \tag{4.30}$$

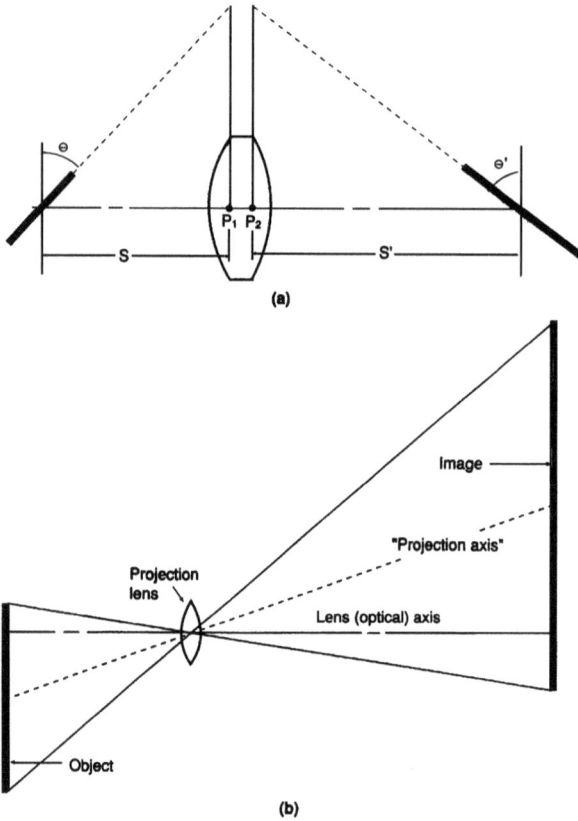

Figure 4.6 (a) The Scheimpflug condition can be used to determine the tilt of the image surface when the object surface is tilted away from the normal to the optical axis. The magnification under these conditions will vary across the field, producing "keystone" distortion. As diagramed here, the magnification of the top of the object is larger than that of the bottom. (Compare the ratio of image distance to object distance for the rays from the top and bottom of the object.) (b) Keystoning can be avoided if the object and image planes are parallel. The figure shows how the "projection axis" can be tilted upward without producing keystone distortion.

Note that in general a tilted object or image plane will cause what is called *keystone distortion,* because the magnification varies across the field. This results from the variation of object and image distances from top to bottom of the field. This distortion is often seen in overhead projectors when the top mirror is tilted to raise the image projected on the screen. This is equivalent to tilting the screen. As shown in Fig. 4.6b, keystone distortion can be prevented by keeping the plane of the object effectively parallel to the plane of the image. In a projector this means that the field of view of the projection lens must be increased on one side of the axis by the amount that the beam is tilted above the horizontal.

Note that a curved surface is imaged as a curved surface with the same curvature and sign as the object surface. A sphere of radius $(-R)$, for example, is imaged as a sphere of radius $(-R)$. This is because the height y of a point on a surface is imaged as my and the sag z of a point on the surface is imaged as m^2z.

4.6   Summary of Sign Conventions

1. Light normally travels from left to right.

2. Focal length is positive for converging lenses.

3. Heights above the axis are positive.

4. Distances to the right of a reference point are positive.

5. A radius or curvature is positive if the center of curvature is to the right of the surface.

6. Angles are positive if the ray is rotated clockwise to reach the normal or the axis.

7. After a reflection (when light direction is reversed), the signs of subsequent indices and spacings are reversed; i.e., if light travels from right to left, the index is negative; if the next surface is to the left, the space is negative.

It may be noted that, although the discussions of this chapter have centered about spherical surfaces, and the equations derived have utilized the radii and curvatures of spherical surfaces, the paraxial expressions are equally valid for all continuous surfaces of rotation centered on the optical axis when the osculating radius (i.e., the radius of the surface at the axis) of the surface is used. This includes both conic sections and generalized aspheric surfaces.

Bibliography

Bass, M., *Handbook of Optics,* Vol. 1, New York, McGraw-Hill, 1995.
Fischer, R. E. and Tadic-Galeb, B., *Optical System Design,* New York, McGraw-Hill, 2000.
Greivenkamp, J., *Field Guide to Geometrical Optics,* Bellingham, WA, SPIE, 2004.
Kingslake, R., *Optical System Design,* Orlando, Academic, 1983.
Smith, W. J., *Modern Lens Design,* New York, McGraw-Hill, 2002.
Smith, W. J., *Practical Optical System Layout,* New York, McGraw-Hill, 1997.

Exercises

1 A Gregorian telescope objective is composed of a concave primary mirror with a radius of 200 and a concave secondary mirror with a radius of 50. The separation of the mirrors is 130. Find the effective focal length and locate the image. Figure 18.3 shows a Gregorian objective.

ANSWER: There are two ways to approach this: one is by raytracing; the other using the two-component equations.

By raytracing:

(Note well the sign conventions for the radii, the index, and the spacing between the mirrors.)

R		-200		$+50$	
t			-130		
n	1.0		-1.0		$+1.0$
y		1.0		-0.30	efl $= 1/(-.002) = -500.$
nu	0		$+0.01$	-0.002	bfl $= -0.3/(-.002) = +150.$

The focus is $150 - 130 = 20$ behind (to the right of) the primary.

By separated component equations: The focal length of a mirror is $R/2$. Concave mirrors act as positive focal length elements. So we use $f_a = +100$ and $f_b = +25$. In raytracing the 130 mirror spacing is regarded as a negative distance; here we use the optical distance $d \cdot n = -130 \, (-1.0) = +130$, and the sign of the spacing is positive.

Eq. 4.5 $f_{ab} = f_a f_b/(f_a + f_b - d)$

$\qquad = 100 \cdot 25/(100 + 25 - 130)$

$\qquad = -500$

Eq. 4.6a $B = f_{ab}(f_a - d)/f_a$

$\qquad = -500(100 - 130)/100$

$\qquad = +150$

2 Find the effective, back, and front focal lengths of a system the front component of which has a $+10''$ focal length and the rear component of which has a $-10''$ focal length. The separation between them is $5''$.

ANSWER:

Eq. 4.5 $f_{ab} = f_a f_b/(f_a + f_b - d)$

$\qquad = 10(-10)/(10 - 10 - 5)$

$\qquad = +20$

Eq. 4.6a $B = f_{ab}(f_a - d)/f_a$

$\qquad = 20(10 - 5)/10$

$\qquad = +10.$

Eq. 4.6b $(-ffd) = f_{ab}(f_b - d)/f_b$

$$= 20(-10 - 5)/(-10)$$

$$= +30$$

3 What powers are necessary in a two-component system if one requires a 20″ focal length, a 10″ back focus, and a 5″ air space?

ANSWER: Using Eqs. 4.7 and 4.8,

$$f_a = df_{ab}/(f_{ab} - B)$$

$$= 5 \cdot 20/(20 - 10)$$

$$= +10$$

$$f_b = -d \cdot B/(f_{ab} - B - d)$$

$$= -5 \cdot 10/(20 - 10 - 5)$$

$$= -10$$

5

The Primary Aberrations

5.1 Introduction

In the preceding chapters we discussed the image-forming character-istics of optical systems, but we limited our consideration to an infinite-simal thread-like region about the optical axis called the paraxial region. In this chapter we will consider, in general terms, the behavior of lenses with *finite* apertures and fields of view. It has been pointed out that well-corrected optical systems behave nearly according to the rules of paraxial imagery. This is another way of stating that a lens without aberrations forms an image of the size and in the location given by the equations for the paraxial or first-order region. We shall measure the aberrations by the amount by which rays miss the paraxial image point.

It can be seen that aberrations may be determined by calculating the location of the paraxial image of an object point and then tracing a large number of rays (by the exact trigonometrical ray-tracing equations of App. A) to determine the amounts by which the rays depart from the paraxial image point. Stated this baldly, the mathematical determina-tion of the aberrations of a lens which covered any reasonable field at a real aperture would seem a formidable task, involving an almost infinite amount of calculation. However, by classifying the various types of image faults and by understanding the behavior of each type, the work of determining the aberrations of a lens system can be simplified greatly, since only a few rays need be traced to evaluate each aberra-tion; thus the problem assumes more manageable proportions.

Seidel investigated and codified the primary aberrations and derived analytical expressions for their determination. For this reason, the primary image defects are usually referred to as the *Seidel aberrations.*

5.2 The Aberration Polynomial and the Seidel Aberrations

With reference to Fig. 5.1, we assume an optical system with symmetry about the optical axis, so that every surface is a figure of rotation about the optical axis. Because of this symmetry, we can, without any loss of generality, define the object point as lying on the y axis; its distance from the optical axis is $y = h$. We define a ray starting from the object point and passing through the system aperture at a point described by its polar coordinates (s, θ). The ray intersects the image plane at the point x', y'.

We wish to know the *form* of the equation which will describe the image plane intersection coordinates y' and x' as a function of h, s, and θ; the equation will be a power series expansion. While it is impractical to derive an exact expression for other than very simple systems or for more than a few terms of the power series, it *is* possible to determine the general form of the equation. This is simply because we have assumed an axially symmetrical system. For example, a ray which

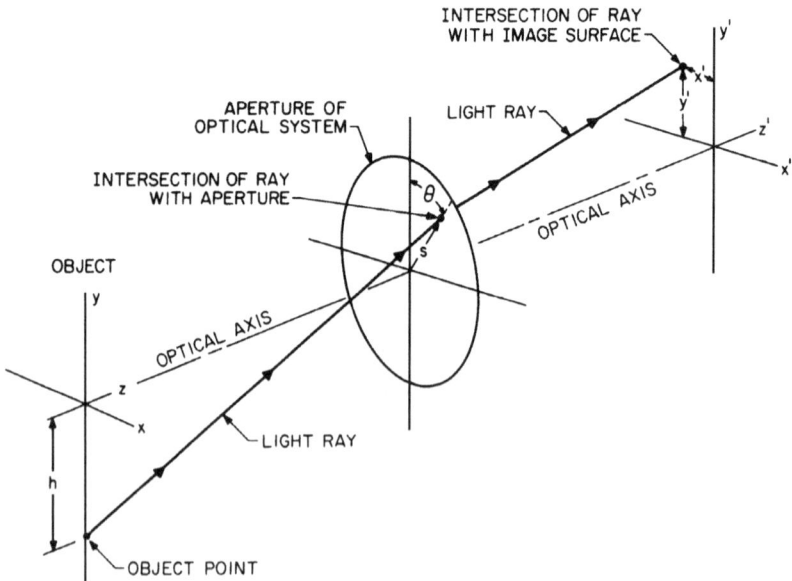

Figure 5.1 A ray from the point $y = h$, $(x = 0)$ in the object passes through the optical system aperture at a point defined by its polar coordinates, (s, θ), and intersects the image surface at x', y'.

intersects the axis in object space must also intersect it in image space. Every ray passing through the same axial point in object space and also passing through the same annular zone in the aperture (i.e., with the same value of s) must pass through the same axial point in image space. A ray in front of the meridional (y, z) plane has a mirror-image ray behind the meridional plane which is identical except for the (reversed) signs of x' and θ. Similarly, rays originating from $\pm h$ in the object and passing through corresponding upper and lower aperture points must have identical x' intersections and oppositely signed y' values. With this sort of logic one can derive equations such as the following:

$$y' = A_1 s \cos \theta + A_2 h$$
$$+ B_1 s^3 \cos \theta + B_2 s^2 h(2 + \cos 2\theta) + (3B_3 + B_4)sh^2 \cos \theta + B_5 h^3$$
$$+ C_1 s^5 \cos \theta + (C_2 + C_3 \cos 2\theta)s^4 h + (C_4 + C_6 \cos^2 \theta)s^3 h^2 \cos \theta$$
$$+ (C_7 + C_8 \cos 2\theta)s^2 h^3 + C_{10} sh^4 \cos \theta + C_{12} h^5 + D_1 s^7 \cos \theta + \cdots \quad (5.1)$$

$$x' = A_1 s \sin \theta$$
$$+ B_1 s^3 \sin \theta + B_2 s^2 h \sin 2\theta + (B_3 + B_4)sh^2 \sin \theta$$
$$+ C_1 s^5 \sin \theta + C_3 s^4 h \sin 2\theta + (C_5 + C_6 \cos^2 \theta)s^3 h^2 \sin \theta$$
$$+ C_9 s^2 h^3 \sin 2\theta + C_{11} sh^4 \sin \theta + D_1 s^7 \sin \theta + \cdots \quad (5.2)$$

where A_N, B_N, etc., are constants, and h, s, and θ have been defined above and in Fig. 5.1.

Notice that in the A terms, the exponents of s and h are unity. In the B terms the exponents total 3, as in s^3, $s^2 h$, sh^2, and h^3. In the C terms the exponents total 5, and in the D terms, 7. These are referred to as the first-order, third-order, and fifth-order terms, etc. There are 2 first-order terms, 5 third-order, 9 fifth-order, 14 seventh-order, 20 ninth-order, and

$$\frac{(n + 3)\,(n + 5)}{8} - 1$$

nth-order terms. In an axially symmetrical system there are no even-order terms; only odd-order terms may exist (unless we depart from symmetry as, for example, by tilting a surface or introducing a toroidal or other nonsymmetrical surface).

It is apparent that the A terms relate to the paraxial (or first-order) imagery discussed in the preceding chapters. A_2 is simply the magnification (h'/h), and A_1 is a transverse measure of the distance from the paraxial focus to our "image plane." All the other terms in Eqs. 5.1 and 5.2

are called *transverse aberrations*. They represent the distance by which the ray misses the ideal image point as described by the paraxial imaging equations.

The B terms are called the third-order, or Seidel aberrations. B_1 is spherical aberration, B_2 is coma, B_3 is astigmatism, B_4 is Petzval, and B_5 is distortion. Similarly, the C terms are called the fifth-order aberrations. C_1 is fifth-order spherical aberration; C_2 and C_3 are linear coma; C_4, C_5, and C_6 are oblique spherical aberrations; C_7, C_8, and C_9 are elliptical coma; C_{10} and C_{11} are Petzval and astigmatism; and C_{12} is distortion.

The 14 terms in D are the seventh-order aberrations; D_1 is the seventh-order spherical aberration. A similar expression for OPD, the wave front deformation, is given in Chap. 15.

As noted above, the Seidel aberrations of a system in monochromatic light are called spherical aberration, coma, astigmatism, Petzval curvature, and distortion. In this section we will define each aberration and discuss its characteristics, its representation, and its effect on the appearance of the image. Each aberration will be discussed as if it alone were present; obviously in practice one is far more likely to encounter aberrations in combination than singly. The third-order aberrations can be calculated using the methods given in Chap. 6.

Spherical aberration

Spherical aberration can be defined as the variation of focus with aperture. Figure 5.2 is a somewhat exaggerated sketch of a simple lens forming an "image" of an axial object point a great distance away. Notice that the rays close to the optical axis come to a focus (intersect the axis) very near the paraxial focus position. As the ray height at the

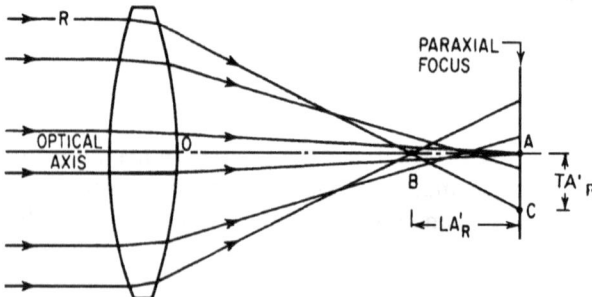

Figure 5.2 A simple converging lens with undercorrected spherical aberration. The rays farther from the axis are brought to a focus nearer the lens.

lens increases, the position of the ray intersection with the optical axis moves farther and farther from the paraxial focus. The distance from the paraxial focus to the axial intersection of the ray is called longitudinal spherical aberration. Transverse, or lateral, spherical aberration is the name given to the aberration when it is measured in the "vertical" direction. Thus, in Fig. 5.2 AB is the longitudinal, and AC the transverse spherical aberration of ray R.

Since the magnitude of the aberration obviously depends on the height of the ray, it is convenient to specify the particular ray with which a certain amount of aberration is associated. For example, marginal spherical aberration refers to the aberration of the ray through the edge or margin of the lens aperture. It is often written as LA_m or TA_m.

Spherical aberration is determined by tracing a paraxial ray and a trigonometric ray from the same axial object point and determining their final intercept distances l' and L'. In Fig. 5.2, l' is distance OA and L' (for ray R) is distance OB. The longitudinal spherical aberration of the image point is abbreviated LA' and

$$LA' = L' - l' \tag{5.3}$$

Transverse spherical aberration is related to LA' by the expression

$$TA'_R = -LA' \tan U'_R = -(L' - l') \tan U'_R \tag{5.4}$$

where U'_R is the angle the ray R makes with the axis. Using this sign convention, spherical aberration with a negative sign is called *undercorrected spherical*, since it is usually associated with simple uncorrected positive elements. Similarly, positive spherical is called *overcorrected* and is generally associated with diverging elements.

The spherical aberration of a system is usually represented graphically. Longitudinal spherical is plotted against the ray height at the lens, as shown in Fig. 5.3a, and transverse spherical is plotted against the final slope of the ray, as shown in Fig. 5.3b. Figure 5.3b is called a *ray intercept curve*. It is conventional to plot the ray through the top of the lens on the right in a ray intercept plot, regardless of the sign convention used for ray slope angles.

For a given aperture and focal length, the amount of spherical aberration in a simple lens is a function of object position and the shape, or bending, of the lens. For example, a thin glass lens with its object at infinity has a minimum amount of spherical at a nearly plano-convex shape, with the convex surface toward the object. A meniscus shape, either convex-concave or concave-convex has much more spherical aberration. If the object and image are of equal size (each being two focal lengths from the lens), then the element shape which gives the minimum spherical is equiconvex. Usually, a uniform distribution of

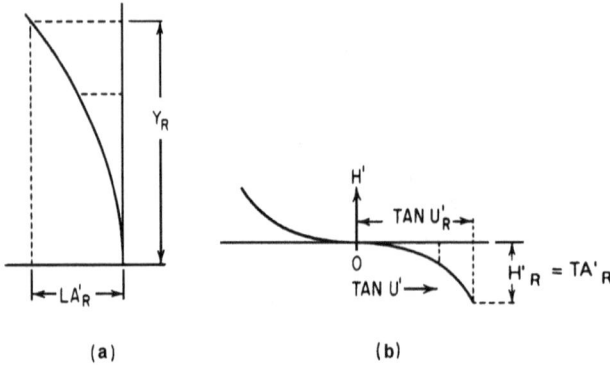

Figure 5.3 Graphical representation of spherical aberration. (a) As a longitudinal aberration, in which the longitudinal spherical aberration (*LA'*) is plotted against ray height (*Y*). (b) As a transverse aberration, in which the ray intercept height (*H'*) at the paraxial reference plane is plotted against the final ray slope (tan *U'*).

the amount that a ray is "bent" or deviated at each surface will minimize the spherical.

The image of a point formed by a lens with spherical aberration is usually a bright dot surrounded by a halo of light; the effect of spherical on an extended image is to soften the contrast of the image and to blur its details.

In general, a positive, converging lens or surface will contribute undercorrected spherical aberration to a system, and a negative lens or a divergent surface, the reverse (although there are certain exceptions to this).

Figure 5.3 illustrated two ways to present spherical aberration, as either a longitudinal or a transverse aberration. Equation 5.4 showed the relation between the two. The same relationship is also appropriate for astigmatism and field curvature and axial chromatic (Sec. 5.3). Note that coma, distortion, and lateral chromatic do not have a longitudinal measure. All of the aberrations can also be expressed as angular aberrations. The angular aberration is simply the angle subtended from the second nodal (or in air, principal) point by the transverse aberration. Thus

$$AA = \frac{TA}{s'} \tag{5.5}$$

Yet a fourth way to measure an aberration is by OPD (Optical Path Difference), the departure of the actual wave front from a perfect reference sphere centered on the ideal image point, as discussed in Sec. 5.6 and Chap. 15.

The transverse measure of an aberration is directly related to the size of the image blur. Graphing it as a ray intercept plot (e.g., Fig. 5.3b and Fig. 5.24) allows the viewer to identify the various types of aberration afflicting the optical system. This is of great value to the lens designer, and the ray intercept plot of the transverse aberrations is an almost universally used presentation of the aberrations. As discussed later (in Chap. 15), the OPD, or wave-front deformation, is the most useful measure of image quality for well-corrected systems, and a statement of the amount of the OPD is usually accepted as definitive in this regard. The longitudinal presentation of the aberrations is most useful in understanding field curvature and axial chromatic (especially secondary spectrum).

Coma

Coma can be defined as the variation of magnification with aperture. Thus, when a bundle of oblique rays is incident on a lens with coma, the rays passing through the edge portions of the lens may be imaged at a different height than those passing through the center portion. In Fig. 5.4, the upper and lower rim rays A and B, respectively, intersect the image plane above the ray P which passes through the center of the lens. The distance from P to the intersection of A and B is called the tangential coma of the lens, and is given by

$$\text{Coma}_T = H'_{AB} - H'_P \qquad (5.6)$$

where H'_{AB} is the height from the optical axis to the intersection of the upper and lower rim rays, and H'_P is the height from the axis to the intersection of the ray P with the plane perpendicular to the axis and passing through the intersection of A and B. The appearance of a

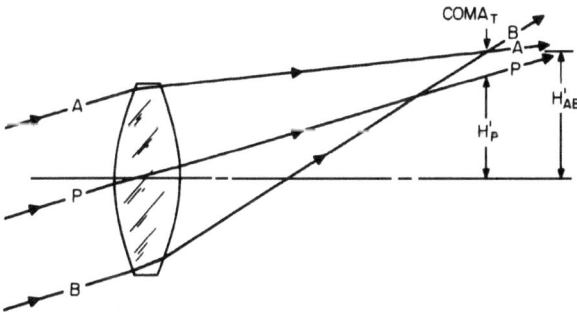

Figure 5.4 In the presence of coma, the rays through the outer portions of the lens focus at a different height than the rays through the center of the lens.

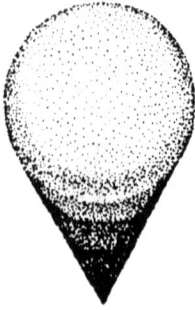

Figure 5.5 The coma patch. The image of a point source is spread out into a comet-shaped flare.

point image formed by a comatic lens is indicated in Fig. 5.5. Obviously the aberration is named after the comet shape of the figure.

Figure 5.6 indicates the relationship between the position at which the ray passes through the lens aperture and the location which it occupies in the coma patch. Figure 5.6a represents a head-on view of the lens aperture, with ray positions indicated by the letters A through H and A' through D', with the primed rays in the inner circle. The resultant coma patch is shown in Fig. 5.6b with the ray locations marked with corresponding letters. Notice that the rays which formed a circle on the aperture also form a circle in the coma patch, but as the rays go around the aperture circle once, they go around the image

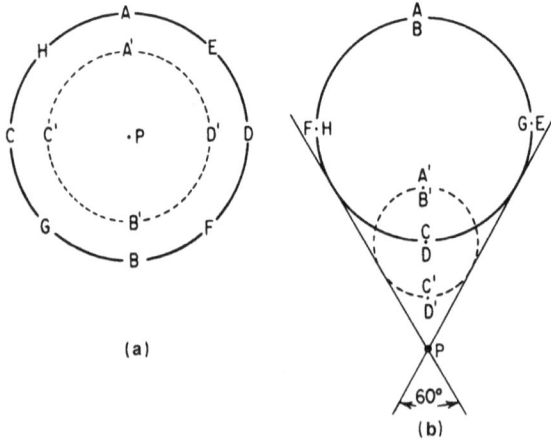

(a)

(b)

Figure 5.6 The relationship between the position of a ray in the lens aperture and its position in the coma patch. (a) View of the lens aperture with rays indicated by letters. (b) The letters indicate the positions of the corresponding rays in the image figure. Note that the diameters of the circles in the image are proportional to the *square* of the diameters in the aperture.

circle twice in accord with the B_2 terms in Eqs. 5.1 and 5.2. The primed rays of the smaller circle in the aperture also form a correspondingly smaller circle in the image, and the central ray P is at the point of the figure. Thus the comatic image can be viewed as being made up of a series of different-sized circles arranged tangent to a 60° angle. The size of the image circle is proportional to the square of the diameter of the aperture circle.

In Fig. 5.6b the distance from P to AB is the tangential coma of Eq. 5.6. The distance from P to CD is called the sagittal coma and is one-third as large as the tangential coma. About half of all the energy in the coma patch is concentrated in the small triangular area between P and CD; thus the sagittal coma is a somewhat better indication of the *effective* size of the image blur than is the tangential coma.

Coma is a particularly disturbing aberration since its flare is non-symmetrical. Its presence is very detrimental to accurate determination of the image position since it is much more difficult to locate the "center of gravity" of a coma patch than for a circular blur such as that produced by spherical aberration.

Coma varies with the shape of the lens element and also with the position of any apertures or diaphragms which limit the bundle of rays forming the image. In an axially symmetrical system there is no coma on the optical axis. The size of the coma patch varies linearly with its distance from the axis. The offense against the Abbe sine condition (OSC) is discussed in Chap. 6. Coma is zero on the axis.

Astigmatism and field curvature

In the preceding section on coma, we introduced the terms "tangential" and "sagittal"; a fuller discussion of these terms is appropriate at this point. If a lens system is represented by a drawing of its axial section, rays which lie in the plane of the drawing are called *meridional* or *tangential* rays. Thus rays A, P, and B of Fig. 5.6 are tangential rays. Similarly, the plane through the axis is referred to as the *meridional* or *tangential* plane, as may *any* plane which includes the axis.

Rays which do not lie in a meridional plane are called *skew* rays. The oblique meridional ray through the center of the aperture stop of a lens system is called the *principal,* or *chief, ray.* If we imagine a plane passing through the chief ray and perpendicular to the meridional plane, then the (skew) rays from the object which lie in this *sagittal* plane are sagittal rays. Thus in Fig. 5.6 all the rays except A, A', P, B', and B are skew rays, and the sagittal rays are C, C', D', and D.

As shown in Fig. 5.7, the image of a point source formed by oblique fans of rays in the tangential plane will be a line image; this line, called the *tangential image,* is perpendicular to the tangential plane; i.e., it lies

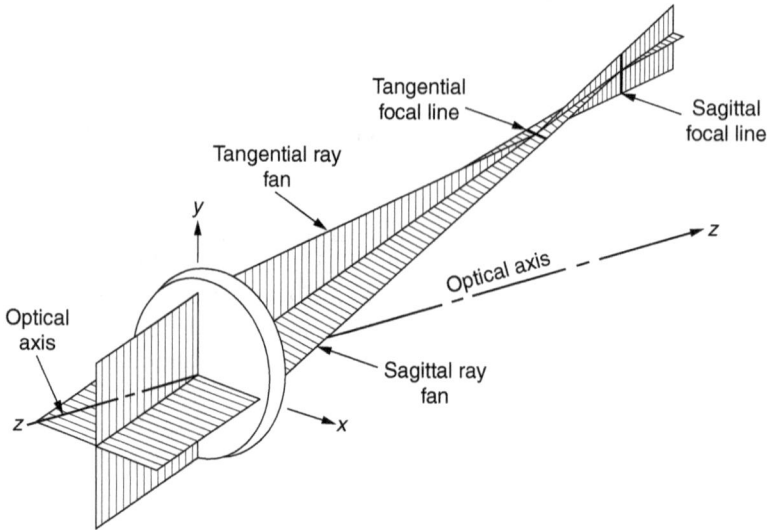

Figure 5.7 Astigmatism (negative or inward-curving).

in the sagittal plane. Conversely, the image formed by the rays of the sagittal fans is a line which lies in the tangential plane.

Astigmatism occurs when the tangential and sagittal (sometimes called radial) images do not coincide. In the presence of astigmatism, the image of a point source is not a point, but takes the form of two separate lines as shown in Fig. 5.7. Between the astigmatic foci the image is an elliptical or circular blur. (Note that if diffraction effects are significant, this blur may take on a square or diamond characteristic because the line images are effectively acting as slit apertures.)

Unless a lens is poorly made, there is no astigmatism (or coma) when an *axial* point is imaged. As the imaged point moves further from the axis, the amount of astigmatism gradually increases. Off-axis images seldom lie exactly in a true plane; when there is primary astigmatism in a lens system, the images lie on curved surfaces which are paraboloid in shape. The shape of these image surfaces is indicated for a simple lens in Fig. 5.8. This drawing is to scale; it is not exaggerated or distorted.

The amount of astigmatism in a lens is a function of the power and shape of the lens and its distance from the aperture or diaphragm which limits the size of the bundle of rays passing through the lens. In the case of a simple lens or mirror, where its own diameter limits the size of the ray bundle, the astigmatism is equal to the square of the distance from the axis to the image (i.e., the image height) divided by the focal length of the element, i.e., $-h^2/f$.

Every optical system has associated with it a sort of basic field curvature, called the Petzval curvature, which is a function of the

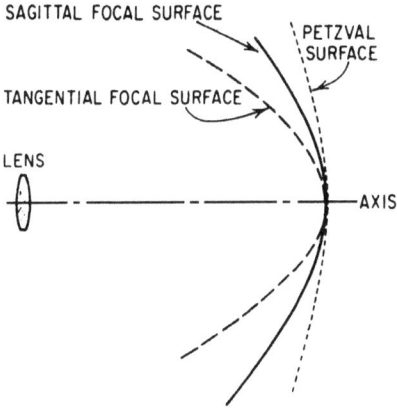

Figure 5.8 The primary astigmatism of a simple lens. The tangential image is three times as far from the Petzval surface as the sagittal image. Note that the figure is to scale.

index of refraction of the lens elements and their surface curvatures. When there is no astigmatism, the sagittal and tangential image surfaces coincide with each other and lie on the Petzval surface. When there is primary astigmatism present, the tangential image surface lies three times as far from the Petzval surface as the sagittal image; note that both image surfaces are on the same side of the Petzval surface, as indicated in Fig. 5.8.

When the tangential image is to the left of the sagittal image (and both are to the left of the Petzval surface) the astigmatism is called negative, undercorrected, or inward-(toward the lens)curving. When the order is reversed, the astigmatism is overcorrected, or backward-curving. In Fig. 5.8, the astigmatism is undercorrected and all three surfaces are inward-curving. It is possible to have overcorrected (backward curving) Petzval and undercorrected (inward) astigmatism, or vice versa.

Positive lenses introduce inward curvature of the Petzval surface to a system, and negative lenses introduce backward curvature. The Petzval curvature (i.e., the *longitudinal* departure of the Petzval surface from the ideal flat image surface) of a thin simple element is equal to one-half the square of the image height divided by the focal length and index of the element, $-h^2/2nf$. Note that "field curvature" means the *longitudinal* departure of the focal surfaces from the ideal image surface (which is usually flat) and not the reciprocal of the radius of the image surface. The radius of the Petzval surface (at the axis) for a simple element is given by $\rho = -nf$.

Distortion

When the image of an off-axis point is formed farther from the axis or closer to the axis than the image height given by the paraxial expressions, the image of an extended object is said to be distorted. The amount of distortion is the displacement of the image from the paraxial position, and can be expressed either directly or as a percentage of the ideal image height, which, for an infinitely distant object, is equal to $h' = f \tan \theta$.

The amount of distortion ordinarily increases as the image size increases; the distortion itself usually increases as the cube of the image height (percentage distortion increases as the square). Thus, if a centered rectilinear object is imaged by a system afflicted with distortion, it can be seen that the images of the corners will be displaced more (in proportion) than the images of the points making up the sides. Figure 5.9 shows the appearance of a square figure imaged by a lens system with distortion. In Fig. 5.9a the distortion is such that the images are displaced outward from the correct position, resulting in a flaring or pointing of the corners. This is overcorrected, or pincushion, distortion. In Fig. 5.9b the distortion is of the opposite type and the corners of the square are pulled inward more than the sides; this is negative, or barrel, distortion.

A little study of the matter will show that a system which produces distortion of one sign will produce distortion of the opposite sign when

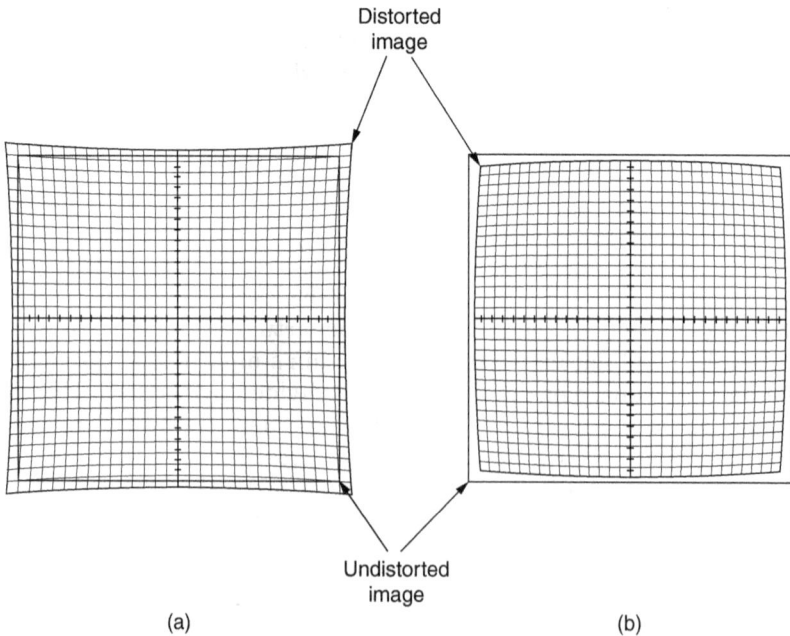

Distorted image

Undistorted image

(a)   (b)

Figure 5.9 Distortion. (a) 8% positive, or pincushion, distortion. (b) 6% negative, or barrel, distortion. The sides of the image are curved because the amount of distortion varies as the cube of the distance from the axis. Thus, in the case of a square, the corners are distorted $2\sqrt{2}$ as much as the center of the sides. 1% or less distortion is considered "good." 2% or 3% is often ok.

object and image are interchanged. Thus a camera lens with barrel distortion will have pincushion distortion if used as a projection lens (i.e., when the film is replaced by a slide). Obviously if the same lens is used both to photograph and to project the slide, the projected image will be rectilinear (free of distortion) since the distortion in the slide will be canceled out upon projection.

5.3 Chromatic Aberrations

Because of the fact that the index of refraction varies as a function of the wavelength of light, the properties of optical elements also vary with wavelength. Axial chromatic aberration is the longitudinal variation of focus (or image position) with wavelength. In general, the index of refraction of optical materials is higher for short wavelengths than for long wavelengths; this causes the short wavelengths to be more strongly refracted at each surface of a lens so that in a simple positive lens, for example, the blue light rays are brought to a focus closer to the lens than the red rays. The distance along the axis between the two focus points is the longitudinal axial chromatic aberration. Figure 5.10 shows the chromatic aberration of a simple positive element. When the short-wavelength rays are brought to a focus to the left of the long-wavelength rays, the chromatic is termed undercorrected, or negative.

The image of an axial point in the presence of chromatic aberration is a central bright dot surrounded by a halo. The rays of light which are in focus, and those which are nearly in focus, form the bright dot. The out-of-focus rays form the halo. Thus, in an undercorrected visual instrument, the image would have a yellowish dot (formed by the orange, yellow, and green rays) and a purplish halo (due to the red and blue rays). If the screen on which the image is formed is moved toward the lens, the central dot will become blue; if it is moved away, the central dot will become red.

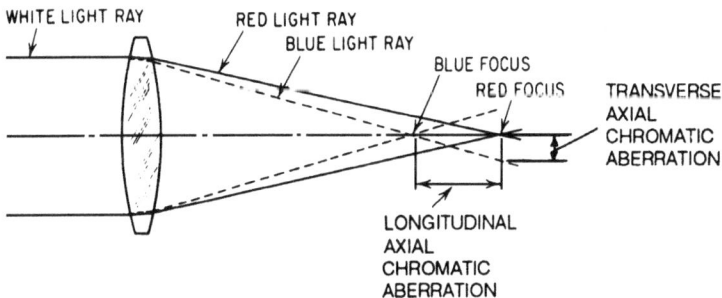

Figure 5.10   The undercorrected longitudinal chromatic aberration of a simple lens is due to the blue rays undergoing a greater refraction than the red rays.

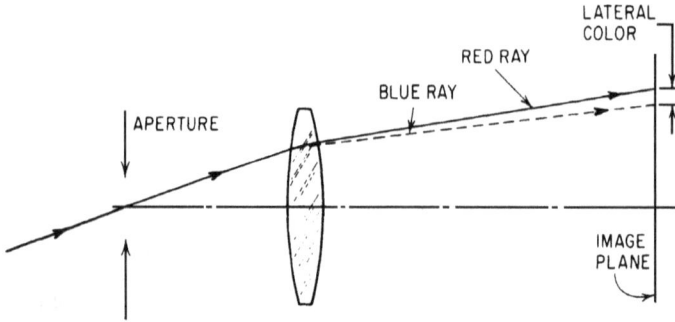

Figure 5.11 Lateral color, or chromatic difference of magnification, results in different-sized images for different wavelengths.

When a lens system forms images of different sizes for different wavelengths, or spreads the image of an off-axis point into a rainbow, the difference between the image heights for different colors is called *lateral color,* or *chromatic difference of magnification.* In Fig. 5.11 a simple lens with a displaced diaphragm is shown forming an image of an off-axis point. Since the diaphragm limits the rays which reach the lens, the ray bundle from the off-axis point strikes the lens above the axis and is bent downward as well as being brought to a focus. The blue rays are bent downward more than the red and thus form their image nearer the axis. If the stop is to the right of the lens the blue rays will be imaged further from the axis than the red ray.

The chromatic variation of index also produces a variation of the monochromatic aberrations discussed in Sec. 5.2. Since each aberration results from the manner in which the rays are bent at the surfaces of the optical system, it is to be expected that, since rays of different color are bent differently, the aberrations will be somewhat different for each color. In general this proves to be the case, and these effects are of practical importance when the basic aberrations are well corrected.

5.4   The Effect of Lens Shape and Stop Position on the Aberrations

A consideration of either the thick-lens focal length equation

$$\frac{1}{f} = (n-1)\left(\frac{1}{R_1} - \frac{1}{R_2} + \frac{n-1}{n}\frac{t}{R_1 R_2}\right)$$

or the thin-lens focal length equation

$$\frac{1}{f} = (n-1)\left(\frac{1}{R_1} - \frac{1}{R_2}\right) = (n-1)(C_1 - C_2)$$

reveals that for a given index and thickness, there is an infinite number of combinations of R_1 and R_2 which will produce a given focal length. Thus a lens of some desired power may take on any number of different shapes or "bendings." The aberrations of the lens are changed markedly as the shape is changed; this effect is the basic tool of optical design.

As an illustrative example, we will consider the aberrations of a thin positive lens made of borosilicate crown glass with a focal length of 100 mm and a clear aperture of 10 mm (a speed of $f/10$) which is to image an infinitely distant object over a field of view of $\pm 17°$. A typical borosilicate crown is 517:642, which has an index of 1.517 for the helium d line ($\lambda = 5876$ Å), an index of 1.51432 for C light ($\lambda = 6563$ Å), and an index of 1.52238 for F light ($\lambda = 4861$ Å).

(The aberration data presented in the following paragraphs were calculated by means of the thin-lens third-order aberration equations of Chap. 6.)

If we first assume that the stop or limiting aperture is in coincidence with the lens, we find that several aberrations do *not* vary as the lens shape is varied. Axial chromatic aberration is constant at a value of -1.55 mm (undercorrected); thus the blue focus (F light) is 1.55 mm nearer the lens than the red focus (C light). The astigmatism and field curvature are also constant. At the edge of the field (30 mm from the axis) the sagittal focus is 7.5 mm closer to the lens than the paraxial focus, and the tangential focus is 16.5 mm inside the paraxial focus. Two aberrations, distortion and lateral color, are zero when the stop is at the lens.

Spherical aberration and coma, however, vary greatly as the lens shape is changed. Figure 5.12 shows the amount of these two aberrations plotted against the curvature of the first surface of the lens. Notice that coma varies linearly with lens shape, taking a large positive value when the lens is a meniscus with both surfaces concave toward the object. As the lens is bent through plano-convex, convex-plano, and convex meniscus shapes, the amount of coma becomes more negative, assuming a zero value near the convex-plano form.

The spherical aberration of this lens is always undercorrected; its plot has the shape of a parabola with a vertical axis. Notice that the spherical aberration reaches a minimum (or more accurately, a maximum) value at approximately the same shape for which the coma is zero. This, then, is the shape that one would select if the lens were to be used as a telescope objective to cover a rather small field of view. Note that if both object and image are "real" (i.e., not virtual), the spherical aberration of a positive lens is always negative (undercorrected).

Let us now select a particular shape for the lens, say, $C_1 = -0.02$ and investigate the effect of placing the stop away from the lens, as

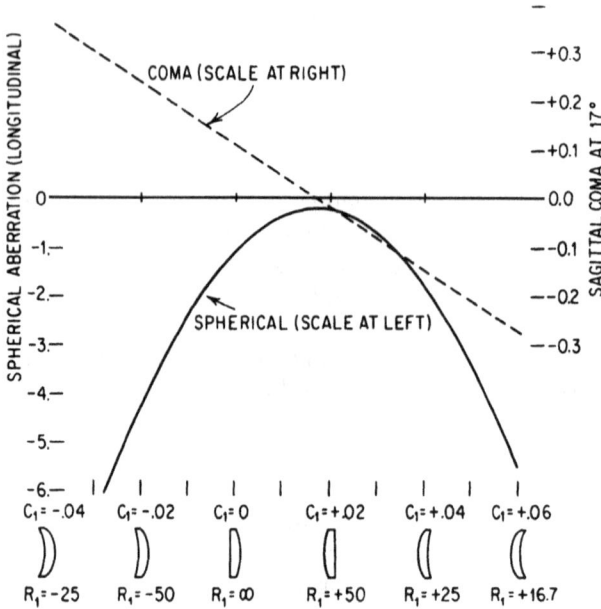

Figure 5.12 Spherical aberration and coma as a function of lens shape. Data plotted are for a 100-mm focal length lens, index = 1,517, (with the stop at the lens) at $f/10$ covering $\pm 17°$ field.

shown in Fig. 5.13. The spherical and axial chromatic aberrations are completely unchanged by shifting the stop, since the axial rays strike the lens in exactly the same manner regardless of where the stop is located. The lateral color and distortion, however, take on positive values when the stop is behind the lens and negative when it is before the lens. Figure 5.14 shows a plot of lateral color, distortion, coma, and

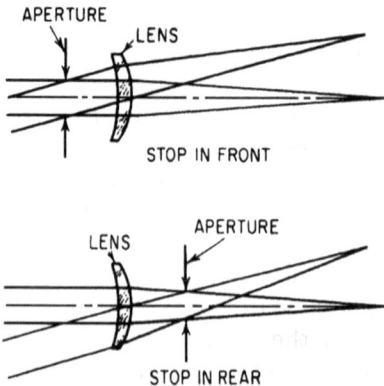

Figure 5.13 The aperture stop away from the lens. Notice that the oblique ray bundle passes through an entirely different part of the lens when the stop is in front of the lens than when it is behind the lens.

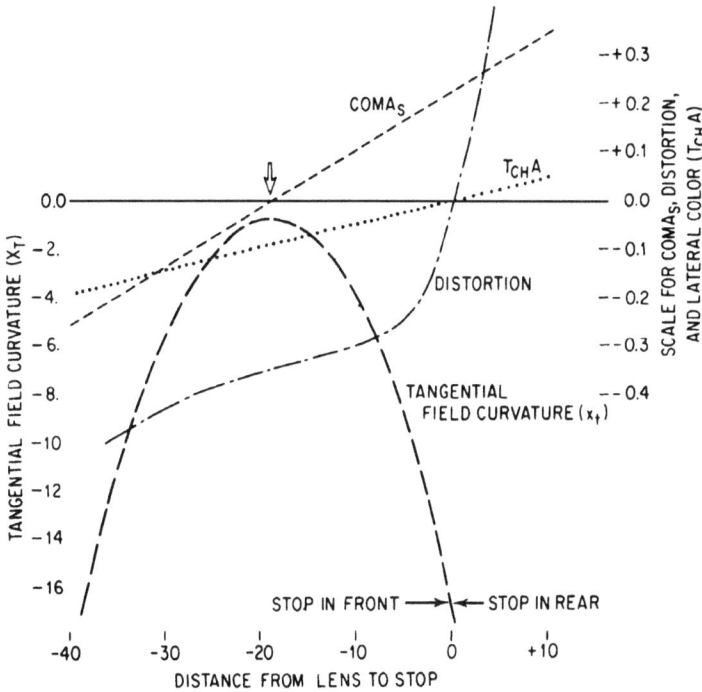

Figure 5.14 Effect of shifting the stop position on the aberrations of a simple lens. The arrow indicates the "natural" stop position where coma is zero. (efl = 100, C_1 = −0.02, speed = $f/10$, field = ±17°.)

tangential field curvature as a function of the stop position. The most pronounced effects of moving the stop are found in the variations of coma and astigmatism. As the stop is moved toward the object, the coma decreases linearly with the stop position, and has a zero value when the stop is about 18.5 mm in front of the lens. The astigmatism becomes less negative so that the position of the tangential image approaches the paraxial focal plane. Since astigmatism is a quadratic function of the stop position, the tangential field curvature (x_t) plots as a parabola. Notice that the parabola has a maximum at the same stop position for which the coma is zero. This is called the *natural* position of the stop, and for all lenses with undercorrected primary spherical aberration, the natural, or coma-free, stop position produces a more backward curving (or less inward curving) field than any other stop position.

Figure 5.12 showed the effect of lens shape with the stop fixed in contact with the lens, and Fig. 5.14 showed the effect of the stop position with the lens shape held constant. There is a "natural" stop position for each shape of the simple lens we are considering. In Fig. 5.15, the

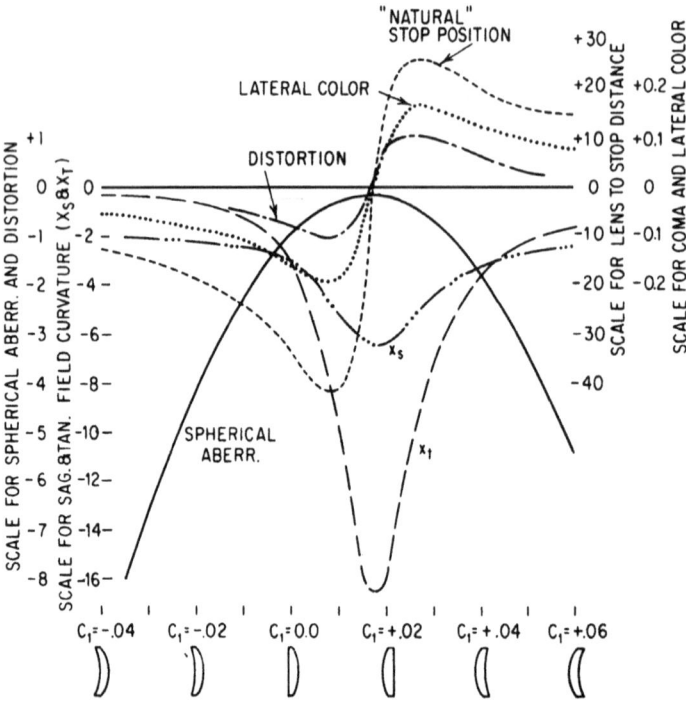

Figure 5.15 The variation of the aberrations with lens shape when the stop is located in the "natural" (coma free) position for each shape. Data are for 100-mm $f/10$ lens covering $\pm 17°$ field, made from BSC-2 glass (517:645).

aberrations of the lens have again been plotted against the lens shape; however, in this figure, the aberration values are those which occur when the stop is in the natural position. Thus, for each bending the coma has been removed by choosing this stop position, and the field is as far backward curving as possible.

Notice that the shape which produces minimum spherical aberration also produces the maximum field curvature, so that this shape, which gives the best image near the axis, is not suitable for wide field coverage. The meniscus shapes at either side of the figure represent a much better choice for a wide field, for although the spherical aberration is much larger at these bendings, the field is much more nearly flat. This is the type of lens used in inexpensive cameras at speeds of $f/11$ or $f/16$.

5.5 Aberration Variation with Aperture and Field

In the preceding section, we considered the effect of lens shape and aperture position on the aberrations of a simple lens, and in that discussion we assumed that the lens operated at a fixed aperture of

$f/10$ (stop diameter of 10 mm) and covered a fixed field of $\pm 17°$ (field diameter of 60 mm). It is often useful to know how the aberrations of such a lens vary when the size of the aperture or field is changed.

Figure 5.16 lists the relationships between the primary aberrations and the semi-aperture y (in column one) and the image height (or field angle) h (in column two). To illustrate the use of this table, let us assume that we have a lens the aberrations of which are known; we wish to determine the size of the aberrations if the aperture diameter is increased by 50 percent and the field coverage reduced by 50 percent. The new y will be 1.5 times the original, and the new h will be 0.5 times the original.

Since longitudinal spherical aberration is shown to vary with y^2, the 1.5 times increase in aperture will cause the spherical to be $(1.5)^2$, or 2.25, times as large. Similarly transverse spherical, which varies as y^3, will be $(1.5)^3$, or 3.375, times larger (as will the image blur due to spherical).

Coma varies as y^2 and h; thus, the coma will be $(1.5)^2 \times 0.5$, or 1.125, times as large. The Petzval curvature and astigmatism, which vary with h^2, will be reduced to $(0.5)^2$, or 0.25, of their previous value, while the blurs due to astigmatism or field curvature will be $1.5(0.5)^2$, or 0.375, of their original size.

The aberrations of a lens also depend on the position of the object and image. A lens which is well corrected for an infinitely distant object, for example, may be very poorly corrected if used to image a nearby object. This is because the ray paths and incidence angles change as the object position changes.

It should be obvious that if *all* the dimensions of an optical system are scaled up or down, the *linear* aberrations are also scaled in exactly

Aberration	vs. Aperture	vs. Field Size or Angle
Spherical (longitudinal)	y^2	—
Spherical (transverse)	y^3	—
Coma	y^2	h
Petzval curvature (longitudinal)	—	h^2
Petzval curvature (transverse)	y	h^2
Astigmatism and field curvature (longitudinal)	—	h^2
Astigmatism and field curvature (transverse)	y	h^2
Distortion (linear)	—	h^3
Distortion (in percent)	—	h^2
Axial chromatic (longitudinal)	—	—
Axial chromatic (transverse)	y	—
Lateral chromatic	—	h
Lateral chromatic (CDM)	—	—

Figure 5.16 The variation of the primary aberrations with aperture and field.

the same proportion. Thus if the simple lens used as the example in Sec. 5.4 were increased in focal length to 200 mm, its aperture increased to 20 mm, and the field coverage increased to 120 mm, then the aberrations would all be doubled. Note, however, that the speed, or f/number, would remain at f/10 and the angular coverage would remain at $\pm 17°$. The *percentage* distortion would not be changed, nor would the chromatic difference of magnification (CDM).

Aberrations are occasionally expressed as angular aberrations. For example, the transverse spherical aberration of a system subtends an angle from the second principal point of the system; this angle is the angular spherical aberration. Note that the angular aberrations are not changed by scaling the size of the optical system.

5.6 Optical Path Difference (Wave Front Aberration)

Aberrations can also be described in terms of the wave nature of light. In Chap. 1, it was pointed out that the light waves converging to form a "perfect" image would be spherical in shape. Thus when aberrations are present in a lens system, the waves converging on an image point are deformed from the ideal shape (which is a sphere centered on the image point). For example, in the presence of undercorrected spherical aberration the wave front is curled inward at the edges, as shown in Fig. 5.17. This can be understood if we remember that a ray is the path of a point on the wave front and that the ray is also normal to the wave front. Thus, if the ray is to intersect the axis to the left of the paraxial focus, the section of the wave front associated with the ray must be

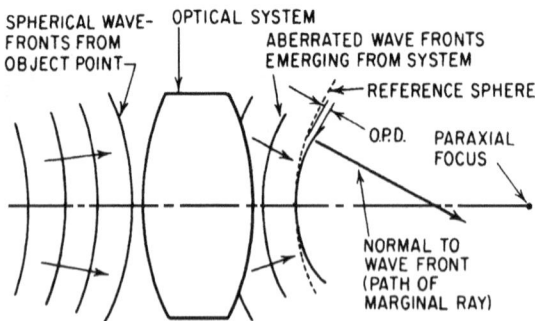

Figure 5.17 The optical path difference (OPD) is the distance between the emerging wave front and a reference sphere (centered in the image plane) which coincides with the wave front at the axis. The OPD is thus the difference between the marginal and axial paths through the system for an axial point.

curled inward. The wave front shown is "ahead" of the reference sphere; the distance by which it is ahead is called the *optical path difference,* or OPD, and is customarily expressed in units of wavelengths. The wave fronts associated with axial aberrations are symmetrical figures of rotation, in contrast to the off-axis aberrations such as coma and astigmatism. For example, the wave front for astigmatism would be a section of a torus (the outer surface of a doughnut) with different radii in the prime meridians. For off-axis imagery, the reference sphere is chosen to pass through the center of the exit pupil (in some calculations, the reference sphere has an infinite radius, for convenience in computing).

5.7   Aberration Correction and Residuals

Section 5.4 indicated two methods which are used to control aberrations in simple optical systems, namely lens shape and stop position. For many applications a higher level of correction is needed, and it is then necessary to combine optical elements with aberrations of opposite signs so that the aberrations contributed to the system by one element are cancelled out, or corrected, by the others. A typical example is the achromatic doublet used for telescope objectives, shown in Fig. 5.18. A single positive element would be afflicted with both undercorrected spherical aberration and undercorrected chromatic aberration. In a negative element, in the other hand, both aberrations are overcorrected. In the doublet a positive element is combined with a less powerful negative element in such a way that the aberrations of each balance out. The positive lens is made of a (crown) glass with a low chromatic dispersion, and the negative element of a (flint) glass with a high dispersion. Thus, the negative element has a greater amount of chromatic aberration *per unit of power,* by virtue of its greater dispersion, than the crown element. The relative powers of the elements are chosen so that the chromatic exactly cancels while the focusing power of the crown element dominates.

The situation with regard to spherical aberration is quite analogous except that element power, shape, and index of refraction are involved instead of power and dispersion as in chromatic. If the index of the

POSITIVE ELEMENT
NEGATIVE ELEMENT

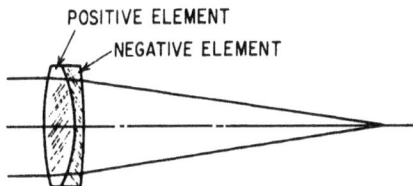

Figure 5.18 Achromatic doublet telescope objective. The powers and shapes of the two elements are so arranged that each cancels the aberrations of the other.

negative element is higher than the positive, the cemented inner surface is divergent, and will contribute overcorrected spherical to balance the undercorrection of the outer surfaces.

Aberration correction usually is exact for only one zone of the aperture of a lens or for one angle of obliquity, because the aberrations of the individual elements do not balance out exactly for all zones and angles. Thus, while the spherical aberration of a lens may be corrected to zero for the rays through the edge of the aperture, the rays through the other zones of the aperture usually do not come to a focus at the paraxial image point. A typical longitudinal spherical aberration plot for a "corrected" lens is shown in Fig. 5.19. Notice that the rays through only one zone of the lens intersect the paraxial focus. Rays through the smaller zones focus nearer the lens system and have undercorrected spherical; rays above the corrected zone show overcorrected spherical. The undercorrected aberration is called residual, or zonal, aberration; Fig. 5.19 would be said to show an undercorrected zonal aberration. This is the usual state of affairs for most optical systems. Occasionally a system is designed with an overcorrected spherical zone, but this is unusual.

Chromatic aberration has residuals which take two different forms. The correction of chromatic aberration is accomplished by making the foci of two different wavelengths coincide. However, due to the nature of the great majority of optical materials, the nonlinear dispersion characteristics of the positive and negative elements used in an achromat do not "match up," so that the focal points of other wavelengths do not coincide with the common focal point of the two selected colors. This difference in focal distance is called secondary spectrum. Figure 5.20 shows a plot of back focal distance versus wavelength for a typical achromatic lens, in which the rays for C light (red) and F light (blue) are brought to a common focus. The yellow rays come to a focus about 1/2400th of the focal length ahead of the C-F focal point.

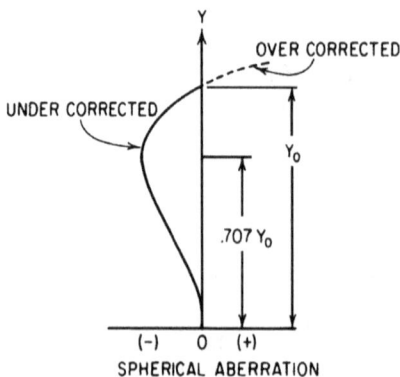

Figure 5.19 Plot of longitudinal spherical aberration versus ray height for a "corrected" lens. For most lenses, the maximum undercorrection occurs for the ray the height of which is 0.707 that of the ray with zero spherical.

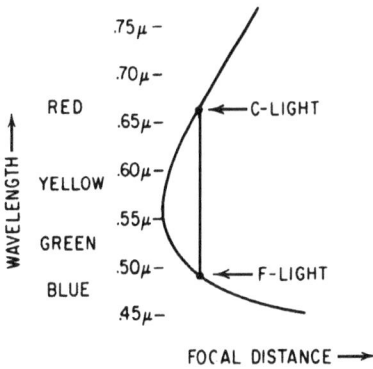

Figure 5.20 The secondary spectrum of a typical doublet achromat, corrected so that C and F light are joined at a common focus. The distance from the common focus of C and F to the minimum of the curve (in the yellow green at about 0.55 μ) is called the *secondary spectrum.*

The second major chromatic residual may be regarded as a variation of chromatic aberration with ray height, or as a variation of spherical aberration with wavelength, and is called *spherochromatism.* In ordinary spherochromatism, the spherical aberration in blue light is overcorrected and the spherical in red light is undercorrected (when the spherical aberration for the yellow light is corrected). Figure 5.21 is a longitudinal spherical aberration plot in three wavelengths for a typical achromatic doublet of large aperture. The correction has been adjusted

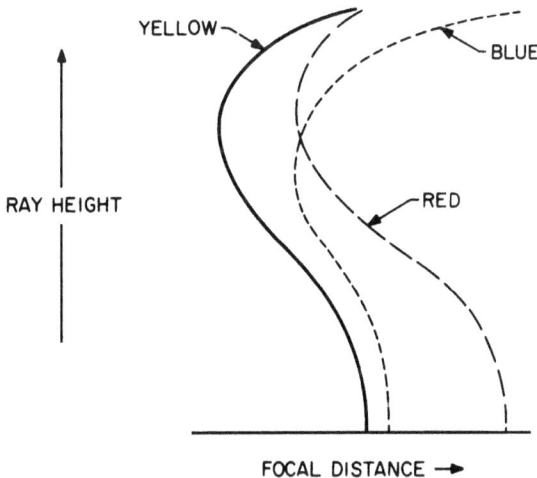

Figure 5.21 Spherochromatism. The longitudinal aberration of a "corrected" lens is shown for three wavelengths. The marginal spherical for yellow light is corrected but is overcorrected for blue light and undercorrected for red. The chromatic aberration is corrected at the 0.707 zone but is overcorrected above it and undercorrected below. A transverse plot of these aberrations is shown in Fig. 5.24k.

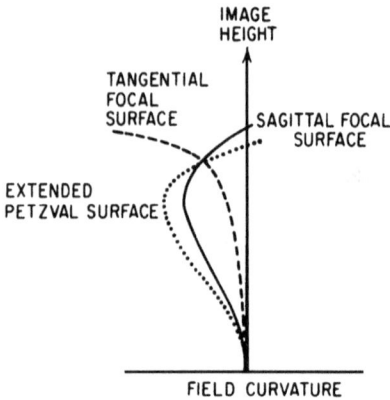

Figure 5.22 Field curvature of a photographic anastigmat. The astigmatism has been corrected for one zone of the field but is overcorrected inside this zone and undercorrected beyond it.

so that the red and blue rays striking the lens at a height of 0.707 of the marginal ray height are brought to a common focus. The distance between the yellow focus and the combined red-blue focus at this height is, of course, the secondary spectrum discussed above. Notice that above this 0.707 zone the chromatic is overcorrected and below it is undercorrected so that one half of the area of the lens aperture is overcorrected and one half undercorrected.

The other aberrations have similar residuals. Coma may be completely corrected for a certain field angle, but will often be overcorrected above this obliquity and undercorrected below it. Coma and color may also undergo a change of sign with aperture, with the central part of the aperture overcorrected and the outer zone undercorrected.

Astigmatism usually varies markedly with field angle. Figure 5.22 shows a plot of the sagittal and tangential field curvatures for a typical photographic anastigmat, in which the astigmatism is zero for one zone of the field. This point is called the *node*, and typically the two focal surfaces separate quite rapidly beyond the node. Astigmatism may also vary with wavelength.

5.8   Ray Intercept Curves and the "Orders" of Aberrations

When the image plane intersection heights of a fan of meridional rays are plotted against the slope of the rays as they emerge from the lens, the resultant curve is called a *ray intercept curve* or an H'–tan U' curve. The shape of the intercept curve not only indicates the amount of spreading or blurring of the image directly, but also can serve as a diagnostic to indicate which aberrations are present. Figure 5.3b, for example, shows simple undercorrected spherical aberration.

In Fig. 5.23, an oblique fan of rays from a distant object point is brought to a perfect focus at point P. If the reference plane passes through P, it is apparent that the H'–tan U' curve will be a straight horizontal line. However, if the reference plane is behind P (as shown) then the ray intercept curve becomes a tilted straight line since the height, H', decreases as tan U' decreases. Thus it is apparent that shifting the reference plane (or focusing the system) is equivalent to a rotation of the H'–tan U' coordinates. A valuable feature of this type of aberration representation is that one can immediately assess the effects of refocusing the optical system by a simple rotation of the abscissa of the figure. Notice that the slope of the line ($\Delta H'/\Delta$ tan U') is exactly equal to the distance (δ) from the reference plane to the point of focus, so that for an oblique ray fan the tangential field curvature is equal to the slope of the ray intercept curve.

The accepted convention for plotting the ray intercepts is that (1) they are plotted for positive image heights (i.e., above the axis) and (2) that the ray through the top of the lens is plotted at the right end of the plot. For compound systems, where the image is relayed by a second component, the ray plotted to the right is the one with the most negative slope, i.e., the one through the bottom of the first component. The result of this is that the sign of the aberrations shown in the ray intercept plot can be instantly recognized. For example, the plot for an undercorrected spherical always curves down at the right end and up at the left, and a line connecting the ends of a plot showing positive coma always passes above the point representing the principal ray.

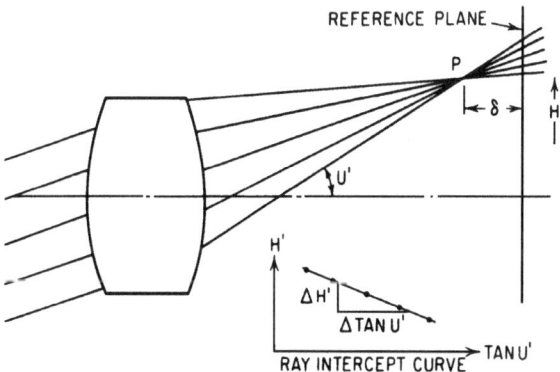

Figure 5.23 The ray intercept curve (H' – tan U') of an image point which does not lie in the reference plane is a tilted straight line. The slope of the line ($\Delta H'/\Delta$ tan U') is mathematically identical to δ, the distance from the reference plane to the point of focus P. Note that δ is equal to X_T, the tangential field curvature, when the paraxial focal plane is chosen as the reference plane.

Note that in an H'–tan U' plot, this plotting convention violates the convention for the sign of the ray slope. This seeming contradiction is the result of the change from the historical optical ray slope sign convention which occurred several decades ago.

Figure 5.24 shows a number of intercept curves, each labeled with the aberration represented. The generation of these curves can be readily understood by sketching the ray paths for each aberration and then plotting the intersection height and slope angle for each ray as a point of the curve. Distortion is not shown in Fig. 5.24; it would be represented as a vertical displacement of the curve from the paraxial image height h'. Lateral color would be represented by curves for two colors which were vertically displaced from each other. The ray intercept curves of Fig. 5.24 are generated by tracing a fan of meridional or tangential rays from an object point and plotting their intersection heights versus their slopes. The imagery in the other meridian can be examined by tracing a fan of rays in the sagittal plane (normal to the meridional plane) and plotting their x-coordinate intersection points against their slopes in the sagittal plane (i.e., the slope relative to the principal ray lying in the meridional plane). Note that Fig. 5.24k is for the same lens as the longitudinal plot in Fig. 5.21.

It is apparent that the ray intercept curves which are "odd" functions, that is, the curves which have a rotational or point symmetry about the origin, can be represented mathematically by an equation of the form

$$y = a + bx + cx^3 + dx^5 + \cdots$$

or

$$H' = a + b \tan U' + c \tan^3 U' + d \tan^5 U' + \cdots \qquad (5.7)$$

All the ray intercept curves for *axial* image points are of this type. Since the curve for an axial image must have $H' = 0$ when $\tan U' = 0$, it is apparent that the constant a must be a zero. It is also apparent that the constant b for this case represents the amount the reference plane is displaced from the paraxial image plane. Thus the curve for lateral spherical aberration plotted with respect to the paraxial focus can be expressed by the equation

$$TA' = c \tan^3 U' + d \tan^5 U' + e \tan^7 U' + \cdots \qquad (5.8)$$

It is, of course, possible to represent the curve by a power series expansion in terms of the final angle U', or sin U', or the ray height at the lens (Y), or even the initial slope of the ray at the object (U_0) instead of tan U'. The constants will, of course, be different for each.

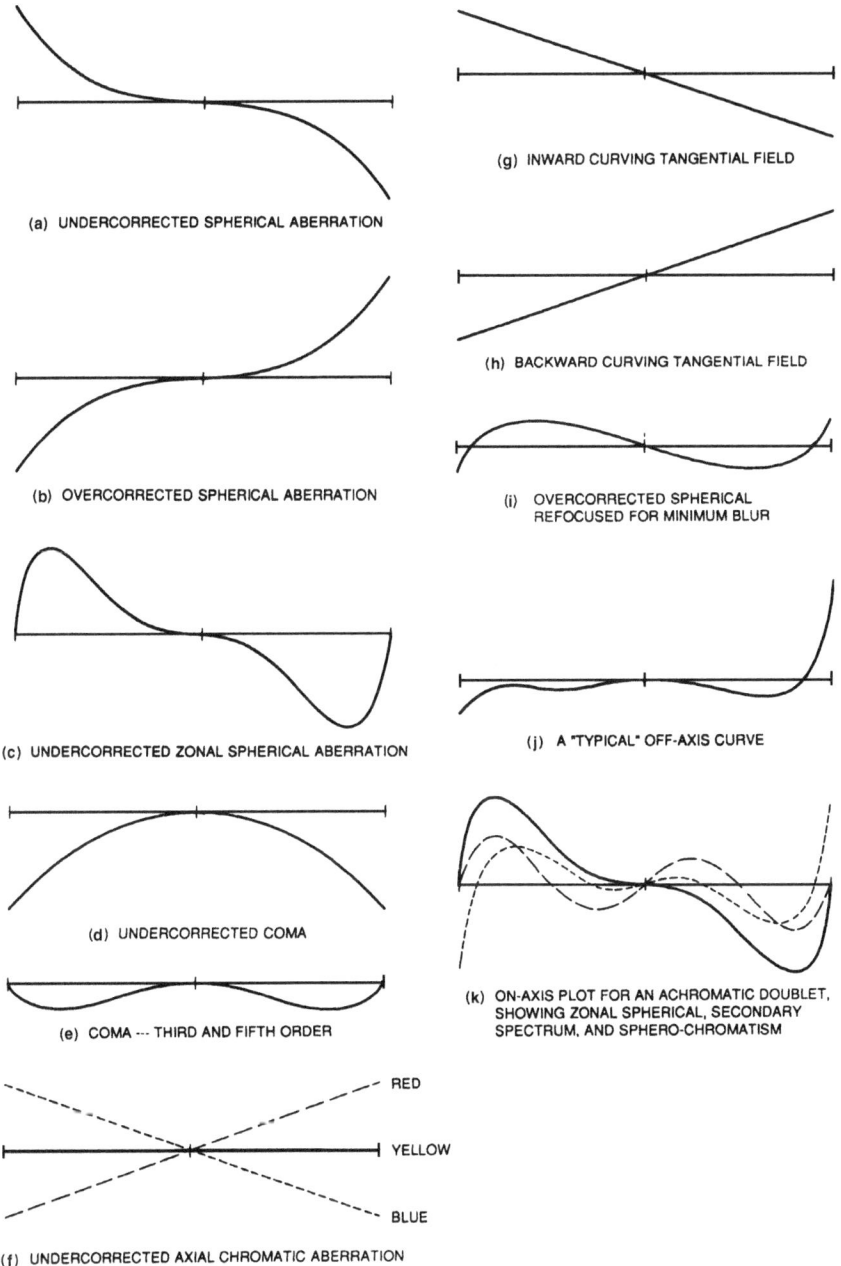

(a) UNDERCORRECTED SPHERICAL ABERRATION

(b) OVERCORRECTED SPHERICAL ABERRATION

(c) UNDERCORRECTED ZONAL SPHERICAL ABERRATION

(d) UNDERCORRECTED COMA

(e) COMA --- THIRD AND FIFTH ORDER

(f) UNDERCORRECTED AXIAL CHROMATIC ABERRATION

(g) INWARD CURVING TANGENTIAL FIELD

(h) BACKWARD CURVING TANGENTIAL FIELD

(i) OVERCORRECTED SPHERICAL REFOCUSED FOR MINIMUM BLUR

(j) A "TYPICAL" OFF-AXIS CURVE

(k) ON-AXIS PLOT FOR AN ACHROMATIC DOUBLET, SHOWING ZONAL SPHERICAL, SECONDARY SPECTRUM, AND SPHERO-CHROMATISM

Figure 5.24 The ray intercept plots for various aberrations. The ordinate for each curve is H, the height at which the ray intersects the (paraxial) image plane. The abscissa is $\tan U$, the final slope of the ray with respect to the optical axis. Note that it is conventional to plot the ray through the top of the lens at the right of the figure, and that curves for image points above the axis are customarily shown.

For simple uncorrected lenses the first term of Eq. 5.8 is usually adequate to describe the aberration. For the great majority of "corrected" lenses the first two terms are dominant; in a few cases three terms (and rarely four) are necessary to satisfactorily represent the aberration. As examples, Figs. 5.3, 5.24a, and 5.24b can be represented by $TA' = c \tan^3 U'$, and this type of aberration is called third-order spherical. Figure 5.24c, however, would require two terms of the expansion to represent it adequately; thus $TA' = c \tan^3 U' + d \tan^5 U'$. The amount of aberration represented by the second term is called the fifth-order aberration. Similarly, the aberration represented by the third term of Eq. 5.8 is called the seventh-order aberration. The fifth-, seventh-, ninth-, etc., order aberrations are collectively referred to as higher-order aberrations.

As will be shown in Chap. 6, it is possible to calculate the amount of the primary, or third-order, aberrations without trigonometric ray-tracing, that is, by means of data from a paraxial raytrace. This type of aberration analysis is called *third-order theory*. The name "first-order optics" given to that part of geometrical optics devoted to locating the paraxial image is also derived from this power series expansion, since the first-order term of the expansion results purely from a longitudinal displacement of the reference plane from the paraxial focus.

Notes on the interpretation of ray intercept plots

The ray intercept plot is subject to a number of interesting interpretations. It is immediately apparent that the top-to-bottom extent of the plot gives the size of the image blur. Also, a rotation of the horizontal (abscissa) lines of the graph is equivalent to a refocusing of the image and can be used to determine the effect of refocusing on the size of the blur.

Figure 5.23 shows that the ray intercept plot for a defocused image is a sloping line. If we consider the slope of the curve at any point on an H–tan U ray intercept plot, the slope is equal to the defocus of a small-diameter bundle of rays centered about the ray represented by that point. In other words, this would represent the focus of the rays passing through a pinhole aperture which was so positioned as to pass the rays at that part of the H–tan U plot. Similarly, since shifting an aperture stop along the axis is, for an oblique bundle of rays, the equivalent of selecting one part or another of the ray intercept plot, one can understand why shifting the stop can change the field curvature and coma, as discussed in Sec. 5.4.

The OPD (optical path difference) or wave-front aberration can be derived from an H–tan U ray intercept plot. The area under the curve between two points is equal to the OPD between the two rays which correspond to the two points. Ordinarily, the reference ray for OPD is either the optical axis or the principal ray (for an oblique bundle). Thus the OPD for a given ray is usually the area under the ray intercept plot between the center point and the ray.

Mathematically speaking, then, the OPD is the integral of the H–tan U plot and the defocus is the slope or first derivative. The coma is related to the curvature or second derivative of the plot, as a glance at Fig. 5.24d will show.

It should be apparent that a more general ray intercept plot for a given object point can be considered as a power series expansion of the form

$$H' = h + a + bx + cx^2 + dx^3 + ex^4 + fx^5 + \cdots \qquad (5.9)$$

where h is the paraxial image height, a is the distortion, and x is the aperture variable (e.g., tan U'). Then the art of interpreting a ray intercept plot becomes analogous to decomposing the plot into its various terms. For example, cx^2 and ex^4 represent third- and fifth-order coma, while dx^3 and fx^5 are the third- and fifth-order spherical. The bx term is due to a defocusing from the paraxial focus and could be due to curvature of field. Note that the constants a, b, c, etc., will be different for points of differing distances from the axis. For the primary aberrations, the constants will vary according to the table of Fig. 5.16, and in general per Eqs. 5.1 and 5.2.

5.9 The Relationships between Longitudinal Aberration, Transverse Aberration, Wave-Front Aberration (OPD), and Angular Aberration

The various ways of assigning a numerical value to an aberration are very simply related. Given the value of one measure of an aberration, the corresponding values for the other measures of that aberration can readily be found. Figure 5.25 illustrates the case of spherical aberration, which can be specified in four different ways:

1. as a longitudinal aberration
2. as a transverse aberration
3. as a wave-front aberration and
4. as an angular aberration

Defocusing, spherical, astigmatism, Petzval curvature, and axial chromatic aberrations can all be expressed in any of the four measures. Since coma, distortion, and lateral chromatic aberrations are

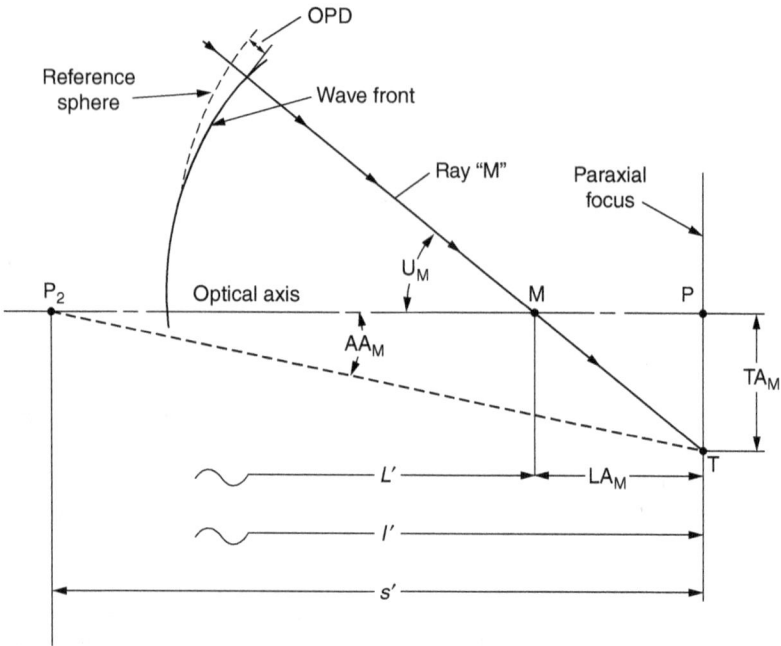

Figure 5.25   Spherical aberration.

not measurable in longitudinal terms, the transverse, angular, and wave-front measures are used for them.

In Fig. 5.25 the marginal ray M intersects the axis at point M, a distance $LA_M = L' - l'$ to the left of the paraxial focus at P. This distance is the longitudinal spherical aberration. Ray M strikes the paraxial focal plane at point T, a distance TA_M below the axis; this is the transverse spherical aberration. The angular spherical aberration is the angle subtended by the transverse aberration from the second nodal or principal point of the optical system, and in air is thus simply the transverse aberration divided by s', the distance from the second principal point to the paraxial focus. The wave-front aberration (or OPD, for Optical Path Difference) is the distance along the ray between the wave front and a reference sphere centered at a reference point (or an "ideal" image point). In Fig. 5.25 the reference sphere is shown centered on the paraxial focus at P. The relationship between OPD and the various aberrations is discussed at greater length in Chap. 15.

OPD is the integral of the angular aberration over the aperture, and is also the integral of the H–tan U plot between rays. The sagittal and tangential field curvatures, x_s and x_t, are the slopes of their H–tan U plots at the principal ray.

The aberrations are related as follows:
The longitudinal aberrations are defined as position differences along the optical (or z) axis.

For spherical $\qquad\qquad LA = L' - l'$

For chromatic $\qquad\qquad LA_{ch} = l'_F - l'_C$

For field aberrations

$$x_s = \text{(sagittal focus distance*)} - l'$$

$$x_t = \text{(tangential focus distance*)} - l'$$

$$x_p = (3x_s - x_t)/2$$

$$\text{astigmatism} = x_t - x_s$$

*measured parallel to the z-axis, from the vertex of the last surface to the image focus (found along the principal ray)

The transverse versions of these aberrations are simply the product of the longitudinal aberration and (the negative of) the slope of the marginal ray. For aberrations such as field curvature and paraxial chromatic in general, the slope of the axial marginal ray (either paraxial or trignometric) is used. For aberrations associated with a specific ray (e.g., spherical or the transverse field curvature of a specific ray in the presence of vignetting), the slope of that ray is used. Thus for marginal spherical,

$$TA_M = LA_M \cdot \tan U_M$$

and for zonal spherical

$$TA_Z = LA_Z \cdot \tan U_Z$$

For astigmatism, either

$$T_{\text{astig}} = (x_t - x_s) \cdot \tan U_M$$

or

$$T_{\text{astig}} = (x_t - x_s) \cdot U_M$$

is commonly used.

Field curvature along any ray in the aperture is equal to the slope of the H–$\tan U$ plot at that ray. In the meridional ray-intercept plot this slope equals x_t; in the sagittal ray-intercept plot it is x_s. These field curvatures are effectively a measure of the imagery of the system if a pinhole aperture were appropriately placed at the aperture stop.

Bibliography

Bass, M., *Handbook of Optics,* Vol. 1, New York, McGraw-Hill, 1995.
Fischer, R. E. and B. Tadic-Galeb, *Optical System Design,* New York, McGraw-Hill, 2000.
Greivenkamp, J., *Field Guide to Geometrical Optics,* Bellingham, WA, SPIE, 2004.
Kingslake, R., *Optical System Design,* Orlando, Academic, 1983.
Smith, W. J., *Modern Lens Design,* New York, McGraw-Hill, 2002.
Smith, W. J., *Practical Optical System Layout,* New York, McGraw-Hill, 1997.
Welford, W., *Aberrattions of Optical Systems,* London, Hilger, 1986.

Exercises

1 The longitudinal spherical aberration of two axial rays which have been traced through a system is -1.0 and -0.5; the ray slopes (tan U) are -0.5 and -0.35 respectively. What are the transverse aberrations (a) in the paraxial focal plane, and (b) in a plane 0.2 before the paraxial focal plane?

ANSWER: (a) In the paraxial focal plane the transverse aberration is the longitudinal aberration times (minus) the ray slope

$$TA = -LA \times \tan U = -(-1) \times (-0.5) = -0.5 \text{ for ray \#1 and}$$
$$= -(-0.5) \times (-0.35) = -0.175 \text{ for ray \#2.}$$

(b) In a plane which is 0.2 before the paraxial focus the transverse spherical is

$$TA = -[LA - (-0.2)] = -(-1 + 0.2) \times (-0.5) = -0.4 \text{ for ray \#1, and}$$
$$= -(-0.5 + 0.2) \times (-0.35) = -0.105 \text{ for ray \#2.}$$

2 A lens has coma$_T$ = +1.0. Plot the focal plane intercepts of the rays which pass through: (a) the marginal zone, (b) the 0.707 zone, and (c) the 0.5 zone. See Fig. 5.6.

ANSWER: Rays from an annular zone of the aperture intersect the focal plane in a circle, the diameter of which varies as the square of the size of the zone of the aperture. For primary coma the circles are tangent to two lines making an angle of 60°. The distance from the apex of the two lines to the top of the circle is the tangential coma. From the Fig. 5.26 it can be seen that the radius R of the circle is the short leg of a 30°–60° right triangle whose hypotenuse equals $2R$. This hypotenuse equals (coma$_T$ − R), so $2R$ = (coma$_T$ − R), and since the coma is equal to 1.0, R must equal 1/3. The center of the circle is 2/3 from the apex of the two lines.

The diagram for the other zones can be similarly derived. The coma varies as the square of the aperture, so the coma for the 0.707 zone is 0.5, and for the 0.5 zone it is 0.25.

3 A certain type of lens has the following primary aberrations at a focal length of 100, an aperture of 10, and a field of ±5°: Longitudinal spherical = +1.0; Coma = +1.0, and X_T = +1.0. What are the aberrations of this type of lens when:

(a) f = 200, aperture = 10, field = ±2.5°?
(b) f = 50, aperture = 10, field = ±10°?

ANSWER: (a) At a focal length of 200, all the dimensions are doubled, and all the aberrations become +2.0. When the aperture is reduced from 20 to 10, the

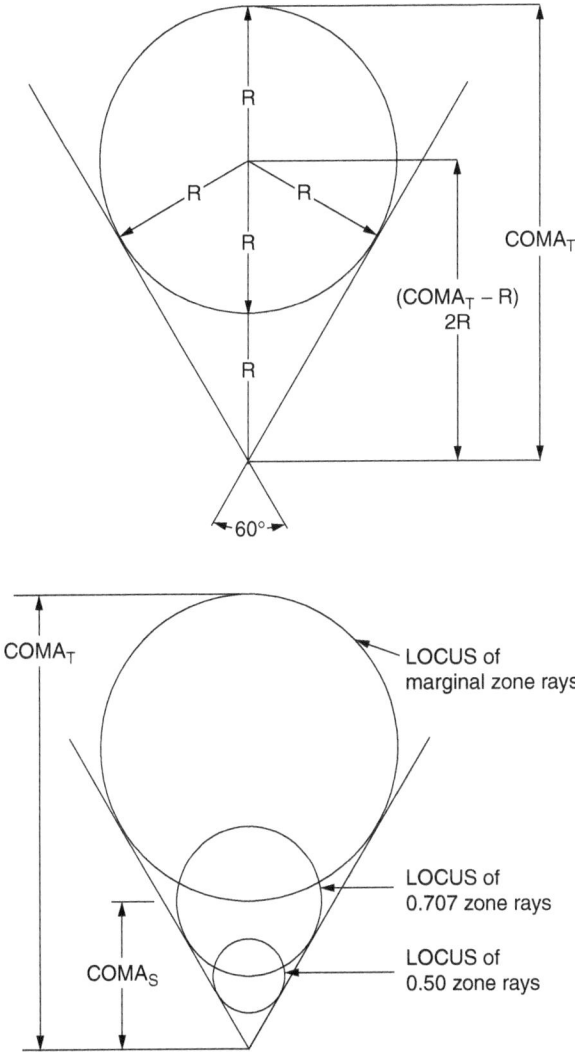

Figure 5.26 Details of ray placement geometry in a drawing of the ray pattern for third order coma.

longitudinal spherical is reduced by $(10/20)^2$ from $+2.0$ to $+0.5$; the coma is reduced by $(10/20)^2$ from $+2.0$ to $+0.5$, and X_T is unchanged at $+2.0$. When the field is reduced to $\pm 2.5°$, the spherical is unchanged at 0.5; the coma is reduced by $(2.5/5)$ from 0.5 to 0.25; X_T is reduced by $(2.5/5)^2$ from 2.0 to 0.5.

Thus we have:

	spherical	coma	tan field
original	$+1.0$	$+1.0$	$+1.0$
(a)	$+0.5$	$+0.25$	$+0.5$
(b)	$+2.0$	$+4.0$	$+2.0$

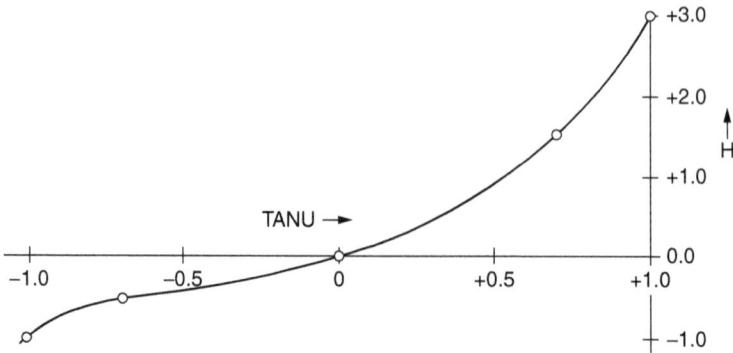

Figure 5.27 Ray intercept plot for a system with transverse spherical equal to +1.0, coma$_T$ equal to +1.0 and transverse X_T equal to +1.0.

4 Plot the ray intercept curve for a lens with transverse spherical, transverse coma, and transverse X_T each equal to 1.0. Assume third-order aberrations and tan $U'_M = 1.0$.

ANSWER: We can write an equation for the total transverse aberration from our knowledge of how the aberrations vary with aperture (as given in Fig. 5.16 and as shown in Fig. 5.24).

$$\Sigma TA = 1.0 \times \tan^3 U + 1.0 \times \tan^2 U + 1.0 \times \tan U$$

We can now simply plot the value of ΣTA for values of tan U from -1.0 to $+1.0$.
    As an alternate, we can evaluate the curves for each aberration and add them up as follows:

tan U	-1.0	-0.707	0.0	$+0.707$	$+1.0$
transverse spherical	-1.0	-0.35	0.0	$+0.35$	$+1.0$
coma	$+1.0$	$+0.35$	0.0	$+0.35$	$+1.0$
transverse X_T	-1.0	-0.707	0.0	$+0.707$	$+1.0$
Summation	-1.0	-0.557	0.0	$+1.557$	$+3.0$

Note that the slope of the plot at the origin is equal to the longitudinal $X_T = +1.0$.

6

Third-Order Aberration Theory and Calculation

6.1 Introduction

The previous chapter described the various aberrations and, in Eqs. 5.1 and 5.2, indicated the manner in which the various "orders" of aberrations varied with the aperture and the field angle of the optical system. For an axially symmetrical optical system only "odd" orders (1st, 3rd, 5th, 7th . . .) may exist. The aberrations of the first-order turn out to be those which are eliminated by locating the reference point at the paraxial image. The first-order aberrations are thus defects of focus or of image size (or height) which vary linearly with aperture or obliquity, such as simple defocusing or the paraxial chromatic aberrations (e.g., transverse axial chromatic or lateral color).

And so we come to the first "real" aberrations, the third-order aberrations wherein the exponents of y (the aperture) and h (the field angle) add up to three. And then we find the fifth-order aberrations, followed by the seventh-, the ninth-, and so on. Limiting our attention to the third- and fifth-order, we have the five third order aberrations, the five corresponding fifth-order aberrations, plus two new fifth-order aberrations, oblique spherical and elliptical coma. The manner in which the aberrations vary with aperture and field are tabulated on the next page:

Third-order aberrations		Fifth-order aberrations	
Exponents	Name	Exponents	Name
y^3	spherical	y^5	5th-order spherical
y^2h	coma	y^4h	linear coma
yh^2	astigmatism	yh^4	5th-order astigmatism
yh^2	Petzval	yh^4	5th-order Petzval
h^3	distortion	h^5	5th-order distortion
		y^3h^2	oblique spherical
		y^2h^3	elliptical coma

An examination of the "*C*" terms in Eqs. 5.1 and 5.2 will indicate the complexity of the fifth-order aberrations, since unlike the third-order terms, for several of the aberrations there is more than one coefficient, and the shape of the aberration blur can vary significantly because of this.

Happily, it turns out that the third-order aberration surface contributions can be calculated from the raytrace data of two paraxial rays, the axial ray and the principal, or chief ray. The fifth-order aberrations can also be calculated from the same data, but the fifth-order contributions from a surface are not determined solely by the ray data at the surface in question, but also from the ray data or aberration contributions from the other surfaces of the system. (See the Buchdahl reference.) An illustration of this effect can be found in the discussion of the design of telescope objectives in Chap.16. Most full-service optical software programs can calculate both the third- and fifth-order aberration contributions.

The chief value of the contribution equations is that they not only allow the calculation of the amount of the aberrations, but they calculate the contribution of each individual surface to the final aberration, which is simply the sum of all the surface contributions. The third-order contributions not only indicate where the third-order aberrations originate, but they are a fair indicator of the source of the higher order aberrations. This relationship is not a simple one, but if a system is cursed with a fifth-order spherical aberration for example, it is quite likely that the problem originates mostly at those surfaces which have outstandingly large third-order spherical contributions.

Another value of the third-order contribution equations lies in the fact that while it is relatively easy to change the third-order aberrations by changing a constructional parameter such as a curvature, a spacing, or an index, the higher order aberrations are relatively stable and difficult to change. Thus if one changes a parameter and finds that a certain amount of change in the third-order aberration is produced, it is likely that a very similar amount of change will be found in the calculation of the aberration by an exact trigonometric raytracing.

6.2   Paraxial Raytracing

Although the paraxial raytracing equations were presented in Chap. 3, they are repeated here for completeness (in slightly modified form).

Opening: 1. Given y and u at the first surface

$$\text{or 2.} \qquad y = -lu \qquad\qquad\qquad (6.1a)$$

$$\text{or 3.} \qquad y = h - su \qquad\qquad\qquad (6.1b)$$

Refraction:

$$u' = \frac{nu}{n'} + \frac{-cy(n' - n)}{n'} \qquad\qquad (6.1c)$$

Transfer to the next surface:

$$y_{j+1} = y_j + tu'_j \qquad\qquad\qquad (6.1d)$$

$$u_{j+1} = u'_j \qquad\qquad\qquad\qquad (6.1e)$$

Closing:

$$l'_k = \frac{-y_k}{u'_k} \qquad\qquad\qquad\qquad (6.1f)$$

or

$$h' = y_k + s'_k u'_k \qquad\qquad\qquad (6.1g)$$

The symbols have the following meanings:

---

y	The height at which a ray strikes the surface; positive above the axis, negative below.
$u(u')$	The slope of the ray before (after) refraction.
$h(h')$	The height of the ray in the object (image) plane; positive above the axis, negative below.
$l(l')$	The intersection distance from the surface before (after) refraction; positive (negative) if the intercept point is to the right (left) of the surface.
$s(s')$	The distance from the first (last) surface to the object (image) plane; positive (negative) if the plane is to the right (left) of the surface.
c	The curvature (reciprocal radius) of the surface, equal to $1/R$; positive if the center of curvature is to the right of the surface, negative if to the left.
$n(n')$	The index of refraction preceding (following) the surface; positive if the ray travels from left to right, negative if it travels right to left.
t_j	The vertex spacing between surfaces (j) and ($j + 1$); positive if surface ($j + 1$) is to the right of surface (j).
k	A subscript indicating the last surface of the system.

Figure 6.1 illustrates the meaning of the symbols.

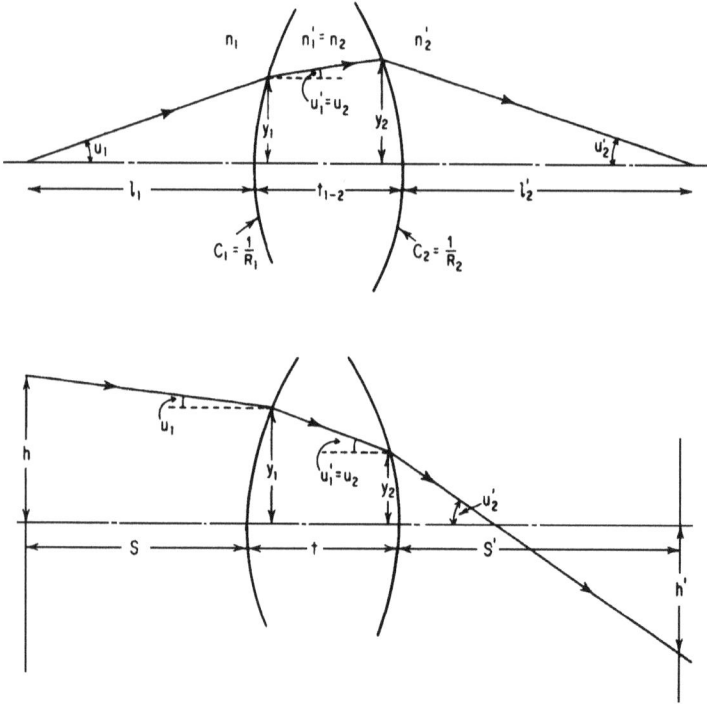

Figure 6.1 Diagrams to illustrate the symbols used in the paraxial ray-tracing equations (6.1a through 6.1g).

6.3 Third-Order Aberrations: Surface Contributions[*]

The third-order aberration surface contributions are readily calculated from the data of two paraxial rays; an axial ray (starting at the axial intercept of the object and passing through the rim of the entrance pupil) and a (paraxial) principal ray (from an off-axis object point through the center of the entrance pupil). These rays are traced by Eqs. 6.1a through 6.1g. In the following, the data of the axial ray will be symbolized by unsubscripted letters (y, u, i, etc.) and that of the paraxial principal ray by letters with the subscript "p" (y_p, u_p, i_p, etc.).

The *optical invariant Inv*, is determined from the data of the two rays at the first surface (or at any convenient surface).

$$Inv = y_p nu - ynu_p = h'n'_k u'_k \tag{6.2a}$$

where the subscript "k" indicates the last surface of the system.

---

[*]D. Feder, "Optical Calculations with Automatic Computing Machines," *J. Opt. Soc. Am.*, vol. 41, pp. 630–636 (1951).

The final image height h' (i.e., the intersection point of the paraxial principal ray with the image plane), is determined from the principal ray, or by

$$h' = Inv/n'_k u'_k \qquad (6.2b)$$

where n'_k and u'_k are the index and slope of the axial ray after passing through the last surface of the system.

Then the following are evaluated for each surface of the system:

$$i = cy + u \qquad (6.2c)$$

$$i_p = cy_p + u_p \qquad (6.2d)$$

$$B = \frac{n(n' - n)}{2n'\,Inv}\,y(u' + i) \qquad (6.2e)$$

$$B_p = \frac{n(n' - n)}{2n'Inv}\,y_p\,(u'_p + i_p) \qquad (6.2f)$$

$$TSC = Bi^2 h \qquad (6.2g)$$

$$CC = Bii_p h \qquad (6.2h)$$

$$TAC = Bi_p^2 h \qquad (6.2i)$$

$$TPC = \frac{-(n - n')\,chInv}{2nn'} \qquad (6.2j)$$

$$DC = h\,[B_p ii_p + \tfrac{1}{2}(u'^2_p - u^2_p)] \qquad (6.2k)$$

$$TAchC = \frac{-yi}{n'_k u'_k}\left(\Delta n - \frac{n}{n'}\Delta n'\right) \qquad (6.2l)$$

$$TchC = \frac{-yi_p}{n'_k u'_k}\left(\Delta n - \frac{n}{n'}\Delta n'\right) \qquad (6.2m)$$

As previously, primed symbols refer to quantities after refraction at a surface. Most of the symbols (y, n, u, c) are defined in Sec. 6.2, or immediately above. Those which have not been previously defined are:

B and B_p	Intermediate steps in the calculation.
i	The paraxial angle of incidence.
Δn	The dispersion of the medium, equal to the difference between the index of refraction for the short wavelength and long wavelength. For visual work, $\Delta n = n_F - n_C$, or $\Delta n = (n - 1)/V$.
Inv	The optical invariant $= hnu = h'n'u'$.

The third-order aberration contributions of the individual surfaces are given by Eqs. 6.2g through 6.2m, where

TSC	is the transverse third-order spherical aberration contribution.
CC	is the sagittal third-order coma contribution.
3CC	is the tangential third-order coma contribution.
TAC	is the transverse third-order astigmatism contribution.
TPC	is the transverse third-order Petzval contribution.
DC	is the third-order distortion.
TAchC	is the paraxial transverse axial chromatic aberration contribution.
TchC	is the paraxial lateral chromatic aberration contribution.

Note that TAchC and TchC are first-order aberrations; since they are conveniently computed at the same time as the third-order aberrations, the equations are presented here.

The longitudinal values of the contributions may be obtained by dividing the transverse values by $-u'_k$, the final slope of the axial ray, thus

$$SC = \frac{-TSC}{u'_k}$$

$$AC = \frac{-TAC}{u'_k}$$

$$PC = \frac{-TPC}{u'_k}$$ (6.2n)

$$LAchC = \frac{-TAchC}{u'_k}$$

The *Seidel coefficients* can be obtained by multiplying the transverse third-order contributions or sums by $(-2n'_k u'_k)$. Thus

$$S1 = -TSC\,(2n'_k u'_k)$$

$$S2 = -CC\,(2n'_k u'_k)$$

$$S3 = -TAC\,(2n'_k u'_k)$$

$$S4 = -TPC\,(2n'_k u'_k)$$

$$S5 = -DC\,(2n'_k u'_k)$$

The third-order aberrations at the final image are obtained by adding together the contributions of all the surfaces to get ΣTSC, ΣCC, ΣTAC, etc. These contribution sums are as follows:

ΣTSC	is the third-order transverse spherical aberration.
ΣSC	is the third-order longitudinal spherical aberration.
ΣCC	is the third-order sagittal coma.
3ΣCC	is the third-order tangential coma.
ΣTAC	is the third-order transverse astigmatism.
ΣAC	is the third-order longitudinal astigmatism.
ΣTPC	is the third-order transverse Petzval sum.
ΣPC	is the third-order longitudinal Petzval sum.
ΣDC	is the third-order distortion.
ΣTAchC	is the first-order transverse axial color.
ΣLchC	is the first-order longitudinal axial color.
ΣTchC	is the first-order lateral color.

To the extent that the first- and third-order aberrations approximate the complete aberration expansions, the following relationships are valid:

$$\Sigma SC \approx L' - l' \qquad \text{(spherical)}$$

$$3\Sigma CC \approx \tfrac{1}{2}(H'_A + H'_B) - H'_p \qquad \text{(tangential coma)}$$

$$z_s \approx \Sigma PC + \Sigma AC \qquad \text{(sag. curvature of field, } x_s\text{)}$$

$$z_t \approx \Sigma PC + 3\Sigma AC \qquad \text{(tan. curvature of field, } x_t\text{)}$$

$$\rho = \frac{h^2}{2\Sigma PC} \qquad \text{(Petzval radius of curvature)}$$

$$= \frac{h^2 u'_p}{2\Sigma TPC}$$

$$\frac{100\Sigma DC}{h} \approx \text{percentage distortion}$$

$$\Sigma LAchC \approx l'_F - l'_C \qquad \text{(axial color)}$$

$$\Sigma TchC \approx h'_F - h'_C \qquad \text{(lateral color)}$$

Contributions from aspheric surfaces

For raytracing purposes, an aspheric surface of rotation is conveniently represented by an equation of the form

$$z = f(x, y) = \frac{cs^2}{[1 + \sqrt{1 - c^2 s^2}]} + A_2 s^2 + A_4 s^4 + \cdots + A_j s^j \qquad (6.2o)$$

where z is the longitudinal coordinate (abscissa) of a point on the surface which is a distance s from the z axis and

$$s^2 = y^2 + x^2$$

For the purposes of computing the third-order contributions, we can assume that the aspheric surface is represented by a power series in s^2

$$z = {}^1\!/_2 C_e s^2 + ({}^1\!/_8 C_e^3 + K)\, s^4 + \cdots \qquad (6.2\text{p})$$

in which the terms in s^6 and higher may be neglected. For aspheric surfaces given in the form of Eq. 6.2o, the equivalent curvature C_e and equivalent fourth-order deformation constant K may be determined from

$$C_e = c + 2A_2 \qquad (6.2\text{q})$$

$$K = A_4 - \frac{A_2}{4}\,(4A_2{}^2 + 6cA_2 + 3c^2) \qquad (6.2\text{r})$$

where c, A_2, and A_4 are the curvature and second- and fourth-order deformation terms, respectively, of Eq. 6.2o. Note that if A_2 is zero, $C_e = c$ and $K = A_4$; see Chap. 18 for conics, where $A_4 = \kappa/8R^3$.

The aspheric surface contributions are determined by first computing the contributions for the equivalent spherical surface C_e using Eqs. 6.2g through 6.2m. Then the contributions due to the equivalent fourth-order deformation constant K are computed by the following equations and added to those of the equivalent spherical surface to obtain the total third-order aberration contribution of the aspheric surface.

$$W = \frac{4K\,(n' - n)}{Inv} \qquad (6.2\text{s})$$

$$TSC_a = Wy^4 h \qquad (6.2\text{t})$$

$$CC_a = Wy^3 y_p h \qquad (6.2\text{u})$$

$$TAC_a = Wy^2 y_p{}^2 h \qquad (6.2\text{v})$$

$$TPC_a = 0 \qquad (6.2\text{w})$$

$$DC_a = Wy y_p{}^3 h \qquad (6.2\text{x})$$

$$TAchC_a = 0 \qquad (6.2\text{y})$$

$$TchC_a = 0 \qquad (6.2\text{z})$$

It is worth noting that if the aspheric surface is located at the aperture stop (or at a pupil), then $y_p = 0$, and the only third-order aberration that is affected by the aspheric term is spherical aberration. The Schmidt camera makes use of this by placing its aspheric corrector plate at the stop so that only the spherical aberration of the spherical mirror is affected by the plate. Conversely, if an aspheric is expected to affect coma, astigmatism, or distortion, it must be located a significant distance from the stop.

6.4 Third-Order Aberrations: Thin Lenses; Stop Shift Equations

When the elements of an optical system are relatively thin, it is frequently convenient to assume that their thickness is zero. As we have previously noted, this assumption results in simplified approximate expressions for element focal lengths, which are nonetheless quite useful for rough preliminary calculations. This approximation can be applied to third-order aberration calculations; the results form a very useful tool for preliminary analytical optical system design. The following equations may be derived by application of the surface contribution equations of the preceding section to a lens element of zero thickness.

The thin-lens third-order aberrations are found by tracing an axial and a principal ray through the system of thin lenses, in the manner outlined in Chap. 4. The equations used are

$$u' = u - y\phi \tag{6.3a}$$

$$y_2 = y_1 + du'_1 \tag{6.3b}$$

where u and u' are the ray slopes before and after refraction by the element, ϕ is the element power (reciprocal focal length), y is the height at which the ray strikes the element, and d is the spacing between adjacent elements.

From Sec. 3.5 we also recall that the power of a thin element is given by

$$\phi = 1/f$$
$$= (n - 1)(c_1 - c_2) \tag{6.3c}$$
$$= (n - 1)c$$

where $c = c_1 - c_2$ and c_1 and c_2 are the curvatures (reciprocal radii) of the first and second surfaces of the element.

After tracing the axial and "principal" rays through the system, the following are computed for each element

$$v = \frac{u}{y} \left(\text{or } v' = \frac{u'}{y} \right) \tag{6.3d}$$

$$Q = \frac{y_p}{y} \tag{6.3e}$$

where u and y are taken from the data of the axial ray and y_p is from the principal ray data.

Then the aberration contributions may be determined using the *stop shift equations:*

$$\text{TSC*} = \text{TSC} \tag{6.3f}$$

$$\text{CC*} = \text{CC} + Q \cdot \text{TSC} \tag{6.3g}$$

$$\text{TAC*} = \text{TAC} + 2Q \cdot \text{CC} + Q^2\text{TSC} \tag{6.3h}$$

$$\text{TPC*} = \text{TPC} \tag{6.3i}$$

$$\text{DC*} = \text{DC} + Q(\text{TPC} + 3\text{TAC}) + 3Q^2\text{CC} + Q^3\text{TSC} \tag{6.3j}$$

$$\text{TAchC*} = \text{TAchC} \tag{6.3k}$$

$$\text{TchC*} = \text{TchC} + Q \cdot \text{TAchC} \tag{6.3l}$$

The starred terms are the contributions from an element which is not at the stop—that is, one for which $y_p \neq 0$. The unstarred terms are the contributions from the element when it is in contact with the stop (and $y_p = 0$) and are given by the following equations:

$$\text{TSC} = \frac{y^4}{u'_k} (G_1 c^3 - G_2 c^2 c_1 - G_3 c^2 v + G_4 c c_1{}^2 + G_5 c c_1 v + G_6 c v^2)$$

$$= \frac{y^4}{u'_k} (G_1 c^3 + G_2 c^2 c_2 + G_3 c^2 v' + G_4 c c_2{}^2 + G_5 c c_2 v' + G_6 c v'^2) \tag{6.3m}$$

$$\text{CC} = -hy^2(0.25 G_5 c c_1 + G_7 c v - G_8 c^2)$$

$$= -hy^2 (0.25 G_5 c c_2 + G_7 c v' + G_8 c^2) \tag{6.3n}$$

$$\text{TAC} = \frac{h^2 \phi u'_k}{2} \tag{6.3o}$$

$$\text{TPC} = \frac{h^2 \phi u'_k}{2n} = \frac{\text{TAC}}{n} \tag{6.3p}$$

$$\text{DC} = 0 \tag{6.3q}$$

$$\text{TAchC} = \frac{y^2 \phi}{V u'_k} \tag{6.3r}$$

$$\text{TchC} = 0 \tag{6.3s}$$

$$\text{TSchC} = \frac{y^2 \phi P}{V u'_k} \tag{6.3t}$$

The symbols in the preceding have the following meanings:

---

u'_k is the final slope of the axial ray (at the image).

h is the image height (the intersection of the "principal" ray with the image plane).

V is the Abbe V-number of the lens material, equal to $(n_d - 1)/(n_F - n_C)$.

P is the partial dispersion of the lens material, equal to $(n_d - n_C)/(n_F - n_C)$.

G_1 through G_8 are functions of the lens material index, listed in Eq. 6.3u.

TSC, CC, TAC, DC, TPC, TAchC, and TchC have the same meanings as in Sec. 6.3.

TSchC is the transverse secondary spectrum contribution, equal to $(l'_d - l'_C)(-u'_k)$.

The transverse aberrations may be converted to longitudinal measure by dividing by $(-u'_k)$ per Eq. 6.2n, as follows:

$$\text{SC} = \frac{-\text{TSC}}{u'_k}$$

$$\text{AC} = \frac{-\text{TAC}}{u'_k}$$

$$\text{PC} = \frac{-\text{TPC}}{u'_k}$$

$$\text{LchC} = \frac{-\text{TAchC}}{u'_k}$$

$$\text{SchC} = \frac{-\text{TSchC}}{u'_k}$$

The relations between the thin-lens contributions and the various measures of the aberrations are the same as indicated in Sec. 6.3.

$$G_1 = \frac{n^2 (n - 1)}{2} \qquad\qquad G_5 = \frac{2 (n + 1)(n - 1)}{n}$$

$$G_2 = \frac{(2n + 1)(n - 1)}{2} \qquad\qquad G_6 = \frac{(3n + 2)(n - 1)}{2n}$$

$$G_3 = \frac{(3n + 1)(n - 1)}{2} \qquad\qquad G_7 = \frac{(2n + 1)(n - 1)}{2n} \qquad (6.3\mathrm{u})$$

$$G_4 = \frac{(n + 2)(n - 1)}{2n} \qquad\qquad G_8 = \frac{n (n - 1)}{2}$$

The contributions, TSC*, CC*, etc., are determined for each element in the system. The individual contributions are then added to get ΣTSC*, ΣCC*, etc., and, to the extent that (1) the thin-lens fiction is valid, and (2) the third-order aberrations adequately represent the total aberration of the system,

$$\Sigma SC \approx L' - l'$$

$$\Sigma CC^* \approx \mathrm{coma}_S \qquad \approx {}^1\!/_3 \mathrm{coma}_T$$

$$\Sigma PC^* + \Sigma AC^* \approx x_s \qquad \text{(sagittal field curvature)}$$

$$\Sigma PC^* + 3\Sigma AC^* \approx x_t \qquad \text{(tangential field curvature)}$$

$$\frac{1}{\Sigma\dfrac{\phi}{n}} = -\rho = \text{Petzval radius}$$

$$\frac{100 \, \Sigma DC^*}{h} \approx \text{percentage distortion}$$

$$\Sigma LchC = l'_F - l'_C$$

$$\Sigma TchC^* = h_F - h_C$$

$$\Sigma SchC = l'_d - l'_C$$

The thin-lens third-order aberration expressions (which are frequently called G-sums) can be used with the specific data of an optical system to determine the (approximate) aberration values. Another usage is in design work where the curvatures and/or spacings and powers of the elements are to be determined in such a way that the aberration values are equal to some desired set of values, as will be evident in Chap. 16. For aspheric surfaced lenses, the contributions from the asphericity are calculated (by Eqs. 6.2r through 6.2y) and added to the contributions calculated for spherical surfaced lenses.

Equations 6.3f to 6.3l are called *stop shift equations.* They may also be applied to the *surface* contributions (from Eq. 6.2) to determine the third-order aberrations for a new, or changed, stop position by setting

$$Q = \frac{(y^*_p - y_p)}{y}$$

where y^*_p is the ray height of the "new" principal ray (i.e., after the stop is shifted) and y_p and y are as indicated in Sec. 6.3. Note that Q is an invariant; thus the values for y^*_p, y_p, and y may be taken at *any* convenient surface. When the equations are used this way the unstarred terms (SC, CC, etc.) refer to the aberrations with the stop in the original position, while the starred terms (SC*, CC*, etc.) refer to the aberrations with the stop in the new position. Another consequence of the invariant nature of this definition of Q is the fact that the stop shift may be applied to either the individual surface contributions or to the contribution sums of the entire system or any portion thereof.

The implications of the stop shift equations (Eqs. 6.3f through 6.3l) are worthy of note. If *all* the third-order aberrations are corrected for a given stop position, then moving the stop will not change them. Similarly, if there is no spherical, the coma is not affected by a stop shift. This is the case with the paraboloid mirror which, because it has no spherical aberration, has the same amount of coma regardless of where the stop is placed. But because it has coma, the astigmatism is a function of the stop position.

6.5 Sample Calculations

Since it is highly unlikely that a reader interested in the material of this chapter will want to carry out any of the aberration calculations "by hand," we will abjure our usual set of exercises in favor of a demonstration of computer software calculation. Our subject will be a quite ordinary Cooke triplet anastigmat at a focal length of 101 mm, a speed of $f/3.5$, and a total field of 23.8°. The reader is invited to duplicate (either by hand or by computer) the calculations to validate his computations.

A fairly complete raytrace analysis of the lens is shown in Fig. 6.2, which is available as a "one click" feature in the software program that I use (OSLO from Lambda Research Corp.).

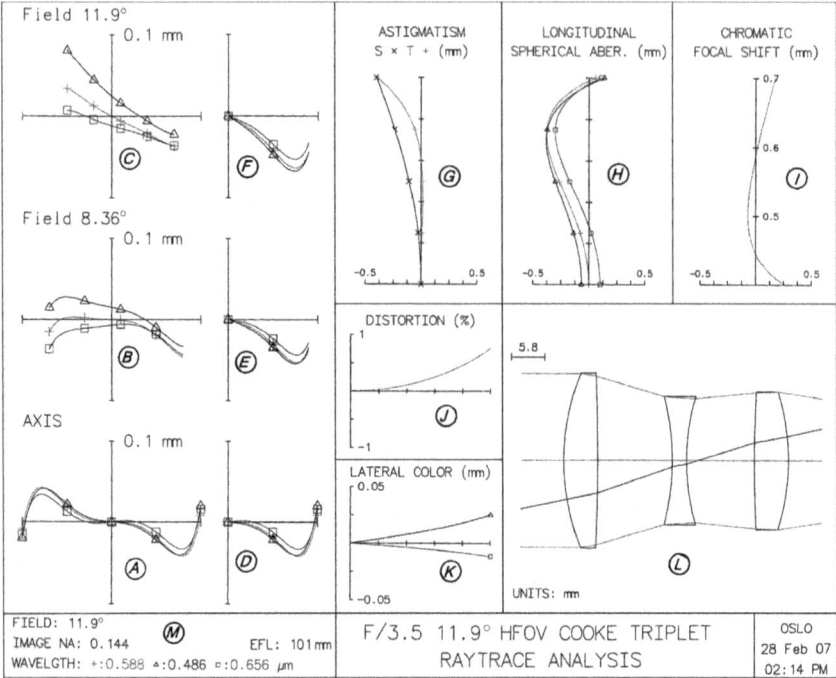

Figure 6.2   Ray aberration plots for the lens described on page 119.

■ Plots *A*, *B*, and *C* are ray intercept plots for meridional ray fans (on axis, at 0.7 field, and at full field, respectively).

■ Plots *D*, *E*, and *F* are for the sagittal ray fans (since the sagittal curves are point symmetric about the origin, only half of the plot needs to be shown).

■ Plot *G* shows the sagittal and tangential field curvatures, x_S and x_T.

■ Plot *H* shows the longitudinal spherical aberration for the three wavelengths. (Note that this is the same data as in plot *A* which shows the transverse aberrations.)

■ Plot *I* shows the paraxial longitudinal chromatic aberration.

■ Plot *J* shows the percent distortion vs. field.

■ Plot *K* is the lateral chromatic, shown separately as the (F-D) and the (C-D) lateral color. The full (F-C) lateral color is the distance between the two curves.

- Plot L is a across-section drawing of the lens, including marginal axial and principal rays.

- The box M contains the half field angle, the numerical aperture, the focal length, and the wavelengths.

The prescription for this lens is as follows:

Radius	Spacing	Glass	Semidiameter				
+37.40	5.90	SK4	(613586) 14.7	EFL	=	101.181	
−341.48	12.93	air	14.7	BFL	=	77.405	
−42.65	2.50	SF2	(648338) 10.8	NA	=	0.1443 (f/3.47)	
+36.40	2.00	air	10.8	GIH	=	21.248 (±11.860°)	
stop	9.85	air	10.3	PTZ/F	=	−2.935	
+204.52	5.90	SK4	(613586) 11.6	VL	=	39.08	
−37.05	77.405		11.6	OD	=	infinity (1.0 e+8)	

Table 6.1 is the paraxial raytrace data for each surface of the lens, including the object and image surfaces and the aperture stop (#5). The column headed PY is the axial ray height (y) at the surface, PU is the ray slope after passing through the surface (u'), and PI is the paraxial angle of incidence (i) at the surface. Columns PYC, PUC, and PIC are the corresponding data for the principal ray.

Table 6.2 lists the paraxial chromatic surface contributions and their sums. Column PAC is the paraxial transverse axial (primary) chromatic (F-C), and SAC is the transverse secondary spectrum (F-d). Column PLC and SLC are the corresponding values for the lateral chromatic.

Table 6.3 gives the third-order (Seidel) aberration surface contributions. Column SA3 is the transverse third-order spherical aberration, CMA3 is the sagittal coma (1/3 of tangential coma), AST3 is the

TABLE 6.1   The Paraxial Ray Trace Data of the Lens Above

SRF	PY	PU	PI	PYC	PUC	PIC
0	—	1.4600e-07	1.4600e-07	−2.1000e+07	0.210000	0.210000
1	14.600000	−0.148315	0.390374	−6.411174	0.195343	0.038578
2	13.724943	−0.263817	−0.188507	−5.258650	0.324469	0.210743
3	10.313791	−0.065055	−0.505641	−1.063264	0.187124	0.349399
4	10.151154	0.073436	0.213823	−0.595454	0.297727	0.170765
5	10.298026	0.073436	0.073436	−3.3307e-15	0.297727	0.297727
6	11.021371	0.025062	0.127325	2.932611	0.179164	0.312066
7	11.169234	−0.144296	−0.276402	3.989677	0.222961	0.071480
8	—	−0.144296	−0.144296	21.248022	0.222961	0.222961

TABLE 6.2   The Paraxial Surface Chromatic Contributions of the Lens Described on p 119

SRF	PAC	SAC	PLC	SLC
1	−0.255960	−0.178181	−0.025295	−0.017608
2	−0.187385	−0.130444	0.209488	0.145831
3	0.419729	0.297051	−0.290034	−0.205263
4	0.287842	0.203711	0.229879	0.162690
5	—	—	—	—
6	−0.063021	−0.043871	−0.154461	−0.107525
7	−0.223595	−0.155651	0.057824	0.040253
SUM	−0.022389	−0.007385	0.027401	0.018377

transverse astigmatism ($0.5NA[x_T - x_S]$), PTZ3 is the transverse Petzval field curvature, and DIS3 is the third-order transverse distortion. These correspond to the coefficients B_1 through B_5 in Eqs. 5.1 and 5.2.

Although this chapter has dealt with the calculation of the first- and third-order aberration surface contributions, the following fifth-order calculations are typically available in optical software programs, so we include sample calculations at this point. For convenience, we repeat the aberration power series expansion equations (Eqs. 5.1 and 5.2) from the previous chapter. The terms h and s, θ are the fractional object height ($0<h<1$) and the fractional pupil coordinates ($0<s<1$).

$$y' = A_1 s \cos \theta + A_2 h$$
$$+ B_1 s^3 \cos \theta + B_2 s^2 h(2 + \cos 2\theta) + (3B_3 + B_4)sh^2 \cos \theta + B_5 h^3$$
$$+ C_1 s^5 \cos \theta + (C_2 + C_3 \cos 2\theta)s^4 h + (C_4 + C_6 \cos^2 \theta)s^3 h^2 \cos \theta$$
$$+ (C_7 + C_8 \cos 2\theta)s^2 h^3 + C_{10} sh^4 \cos \theta + C_{12} h^5 + D_1 s^7 \cos \theta + \cdots \quad (5.1)$$

TABLE 6.3   The Third Order Surface Aberration Contributions

SRF	SA3	CMA3	AST3	PTZ3	DIS3
1	−0.709019	−0.070068	−0.006924	−0.330897	−0.033385
2	−0.755360	0.844458	−0.944066	−0.036241	1.095939
3	2.049816	−1.416428	0.978756	0.300215	−0.883772
4	0.493011	0.393733	0.314447	0.351763	0.532055
5	—	—	—	—	—
6	−0.035845	−0.087854	−0.215325	−0.060510	−0.676055
7	−1.229178	0.317877	−0.082206	−0.334022	0.107641
SUM	−0.186575	−0.018282	0.044681	−0.109691	0.142422

TABLE 6.4   The Fifth-Order Surface Contribution for the Lens on p 119

SRF	SA5	CMA5	AST5	PTZ5	DIS5	SA7
1	−0.061946	−0.001202	0.000323	−0.004476	0.000474	−0.006382
2	−0.154199	0.189546	−0.060252	−0.007701	0.056437	−0.028360
3	0.394409	−0.354457	0.033863	0.007252	−0.026296	0.073525
4	0.110416	0.055693	0.021076	0.016469	0.052491	0.033928
5	—	—	—	—	—	—
6	−0.015016	−0.022169	−0.023359	−0.004057	−0.053149	−0.006118
7	−0.148238	0.092127	0.011081	0.008735	−0.014342	−0.017923
SUM	0.125426	−0.040462	−0.017268	0.025173	0.015616	0.048670

$$x' = A_1 s \sin \theta$$
$$+ B_1 s^3 \sin \theta + B_2 s^2 h \sin 2\theta + (B_3 + B_4)sh^2 \sin \theta$$
$$+ C_1 s^5 \sin \theta + C_3 s^4 h \sin 2\theta + (C_5 + C_6 \cos^2 \theta)s^3 h^2 \sin \theta$$
$$+ C_9 s^2 h^3 \sin 2\theta + C_{11} sh^4 \sin \theta + D_1 s^7 \sin \theta + \cdots \quad (5.2)$$

Table 6.4 lists the fifth-order analogs of the five Seidel aberrations listed in Table 6.3, plus the seventh-order spherical aberration term. These correspond to the coefficients C_1, C_3, $(C_{10} - C_{11})/4$, $(5C_{11} - C_{10})/4$, C_{12}, and D_1 in Eqs. 5.1 and 5.2.

Table 6.5 gives the Buchdahl fifth-order coefficient values (C_1, C_2 . . .) for the Eqs. 5.1 and 5.2.

TABLE 6.5   The Fifth Order Buchdahl Aberrations Surfaces Contributions of the Lens on p 119

SRF	C1/2	C3/4	C5/6	C7/8	C9/10	C11/12
1	−0.061946	−0.001202	−0.018301	0.006375	0.006542	0.004799
	−0.007324	−0.018538	0.033213	0.008183	0.006092	0.000474
2	−0.154199	0.189546	−0.169776	0.361953	0.134015	−0.067954
	0.314507	−0.468290	−0.243766	0.248751	−0.308963	0.056437
3	0.394409	−0.354457	0.226035	−0.313680	−0.109803	0.041115
	−0.574065	0.624135	0.280642	−0.193903	0.176565	−0.026296
4	0.110416	0.055693	0.049073	0.017082	0.025061	0.037545
	0.108491	0.094590	−0.066253	0.009050	0.121850	0.052491
5	—	—	—	—	—	—
	—	—	—	—	—	—
6	−0.015016	−0.022169	−0.034561	0.026508	−0.008375	−0.027415
	−0.045089	−0.083140	0.024848	0.002542	−0.120850	−0.053149
7	−0.148238	0.092127	−0.008556	−0.076892	−0.043850	0.019815
	0.141897	−0.039780	0.047409	−0.055757	0.064138	−0.014342
SUM	0.125426	−0.040462	0.043914	0.021346	0.003592	0.007905
	−0.061583	0.108976	0.076092	0.018867	−0.061167	0.015616

Bibliography

Buchdahl, H., *Optical Aberration Coefficients*, London, Oxford, 1954.

Conrady, A., *Applied Optics and Optical Design,* New York, Oxford, 1929. (This and Vol. 2 also were published by Dover, New York.)

Herzberger, M., *Modern Geometrical Optics,* New York, Interscience, 1958.

Kingslake, R., *Optical System Design,* San Diego, Academic, 1983.

Rimmer, M., *Optical Aberration Coefficients*, Mastek Thesis, University of Rochester, NY, 1963.

Smith, W., *Modern Lens Design*, New York, McGraw-Hill, 2002.

Welford, W., *Aberrations of Optical Systems,* London, Hilger, 1986.

7

Prism and Mirror Systems

7.1 Introduction

In most optical systems, prisms serve one of two major functions. In spectral instruments (spectroscopes, spectrographs, spectrophotometers, etc.) their function is to disperse the light or radiation; that is, to separate the different wavelengths. In other applications, prisms are used to displace, deviate, or reorient a beam of light or an image. In this type of use, the prism is carefully arranged so that it will *not* separate the different colors.

7.2 Dispersing Prisms

In a typical dispersing prism, as shown in Fig. 7.1, a light ray strikes the first surface at an angle of incidence I_1 and is refracted downward, making an angle of refraction I'_1 with the normal to the surface. The ray is thus deviated through an angle of $(I_1 - I'_1)$ at this surface. At the second surface the ray is deviated through an angle $(I'_2 - I_2)$, so the total deviation of the ray is given by

$$D = (I_1 - I'_1) + (I'_2 - I_2) \qquad (7.1)$$

From the geometry of the figure it can be seen that angle I_2 is equal to $(A - I'_1)$, where A is the vertex angle of the prism; making this substitution in Eq. 7.1, we get

$$D = I_1 + I'_2 - A \qquad (7.2)$$

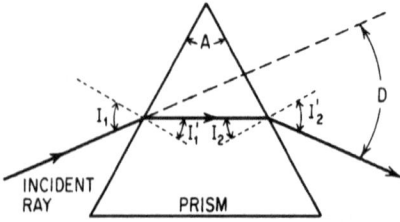

Figure 7.1 The deviation of a ray by a refracting prism.

To compute the deviation produced by the prism we can readily determine the angles in Eq. 7.2 by Snell's law (Eq. 1.3) as follows (where n is the prism index):

$$\sin I'_1 = \frac{1}{n} \sin I_1 \tag{7.3}$$

$$I_2 = A - I'_1 \tag{7.4}$$

$$\sin I'_2 = n \sin I_2 \tag{7.5}$$

While it is ordinarily much more convenient to calculate the deviation step by step, using the equations above, it is possible to combine them into a single expression for D, in terms of I_1, A, and n as follows:

$$D = I_1 - A + \arcsin \left[(n^2 - \sin^2 I_1)^{1/2} \sin A - \cos A \sin I_1 \right] \tag{7.6}$$

It is apparent that the deviation is a function of the prism index and that the deviation will be increased as the index is raised. For optical materials, the index of refraction is higher for short wavelengths (blue light) than for long wavelengths (red light). Therefore, the deviation angle will be greater for blue light than red, as indicated in Fig. 7.2. This variation of the deviation angle with wavelength is called the dispersion of the prism. An expression for the dispersion can be found by

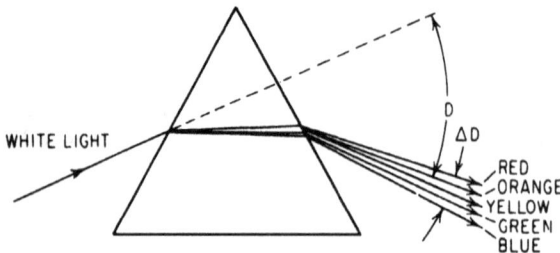

Figure 7.2 The dispersion of white light into its component wavelengths by a refracting prism (highly exaggerated).

differentiating the preceding equations with respect to the index n, assuming that I_1 is constant, yielding,

$$dD = \frac{\cos I_2 \tan I'_1 + \sin I_2}{\cos I'_2} dn \qquad (7.7)$$

The angular dispersion with respect to wavelength is simply $dD/d\lambda$ and is obtained by dividing both sides of Eq. 7.7 by $d\lambda$. The resulting $dn/d\lambda$ term on the right is the index dispersion of the prism material.

7.3 The "Thin" Prism

If all the angles involved in the prism are very small, we can, as in the paraxial case for lenses, substitute the angle itself for its sine. This case occurs when the prism angle A is small and when the ray is almost at normal incidence to the prism faces. Under these conditions, we can write

$$i'_1 = \frac{i_1}{n}$$

$$i_2 = A - i'_1 = A - \frac{i_1}{n}$$

$$i'_2 = ni_2 = nA - i_1$$

$$D = i_1 + i'_2 - A = i_1 + nA - i_1 - A$$

and finally

$$D = A(n - 1) \qquad (7.8a)$$

If the prism angle A is small but the angle of incidence I is *not* small, we get the following approximate expression for D (which neglects powers of I larger than 3).

$$D = A(n - 1)\left[1 + \frac{I^2(n + 1)}{2n} + \cdots\right] \qquad (7.8b)$$

These expressions are of great utility in evaluating the effects of a small prismatic error in the construction of an optical system since it allows the resultant deviation of the light beam to be determined quite readily.

The dispersion of a "thin" prism is obtained by differentiating Eq. 7.8a with respect to n, which gives $dD = Adn$. If we substitute A from Eq. 7.8a, we get

$$dD = D \frac{dn}{(n - 1)} \qquad (7.9)$$

Now the fraction $(n - 1)/\Delta n$ is one of the basic numbers used to characterize optical materials. It is called the reciprocal relative dispersion, Abbe V number, or V-value. Ordinarily n is taken as the index for the helium d line (0.5876 μm) and Δn is the index difference between the hydrogen F(0.4861 μm) and C(0.6563 μm) lines, and the V-value is given by

$$V = \frac{n_d - 1}{n_F - n_C} \tag{7.10}$$

Making the substitution of $1/V$ for $dn/(n - 1)$ in Eq. 7.9, we get

$$dD = \frac{D}{V} \tag{7.11}$$

which allows us to immediately evaluate the chromatic dispersion produced by a thin prism.

7.4 Minimum Deviation

The deviation of a prism is a function of the initial angle of incidence I_1. It can be shown that the deviation is at a minimum when the ray passes symmetrically through the prism. In this case $I_1 = I'_2 = \frac{1}{2}(A + D)$ and $I'_1 = I_2 = A/2$, so that if we know the prism angle A and the minimum deviation angle D_0 it is a simple matter to compute the index of the prism from

$$n = \frac{\sin I_1}{\sin I'_1} = \frac{\sin \frac{1}{2}(A + D_0)}{\sin \frac{1}{2}A} \tag{7.12}$$

This is a widely used method for the precise measurement of index, since the minimum deviation position is readily determined on a spectrometer. This position for the prism is also approximated in most spectral instruments because it allows the largest diameter beam to pass through a given prism and also produces the smallest amount of loss due to surface reflections.

7.5 The Achromatic Prism and the Direct Vision Prism

It is occasionally useful to produce an angular deviation of a light beam without introducing any chromatic dispersion. This can be done by combining two prisms, one of high-dispersion glass and the other of low-dispersion glass. We desire the sum of their deviations to equal $D_{1,2}$ and the sum of their dispersions to equal zero. Using the equations for "thin" prisms (Eqs. 7.8 and 7.11), we can express these requirements as follows:

Deviation $D_{1,2} = D_1 + D_2 = A_1 (n_1 - 1) + A_2 (n_2 - 1)$

Dispersion $dD_{1,2} = dD_1 + dD_2 = 0 = \dfrac{D_1}{V_1} + \dfrac{D_2}{V_2}$

$$= \dfrac{A_1\,(n_1 - 1)}{V_1} + \dfrac{A_2\,(n_2 - 1)}{V_2}$$

A simultaneous solution for the angles of the two prisms gives

$$A_1 = \dfrac{D_{1,2}V_1}{(n_1 - 1)\,(V_1 - V_2)}$$

$$A_2 = \dfrac{D_{1,2}V_2}{(n_2 - 1)\,(V_2 - V_1)}$$

(7.13)

It is apparent that the prism angles will have opposite signs and that the prism with the larger V-value (smaller relative dispersion) will have the larger angle. A sketch of an achromatic prism is shown in Fig. 7.3. Note that the emerging rays are not coincident but are parallel, indicating an identical angular deviation.

In the *direct vision prism* it is desired to produce a dispersion without deviating the ray. By setting the deviation $D_{1,2}$ equal to zero and preserving the dispersion term $dD_{1,2}$ in the preceding equations we can solve for the angles of two prisms which will produce the desired result. The solution is

$$A_1 = \dfrac{dD_{1,2}V_1V_2}{(n_1 - 1)\,(V_2 - V_1)}$$

$$A_2 = \dfrac{dD_{1,2}V_1V_2}{(n_2 - 1)\,(V_1 - V_2)}$$

(7.14)

A two-element direct vision prism is shown in Fig. 7.4a. In order to obtain a large enough dispersion for practical purposes it is often necessary to use more than two prisms. Figure 7.4b shows the application of such a prism to a hand spectroscope.

Since Eqs. 7.13 and 7.14 were derived using the equations for thin prisms, it is obvious that the values of the component prism angles

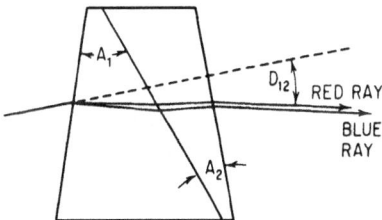

Figure 7.3 An achromatic prism. The red and blue rays emerge parallel to each other; no chromatic dispersion is introduced by the deviation.

(a)

(b)

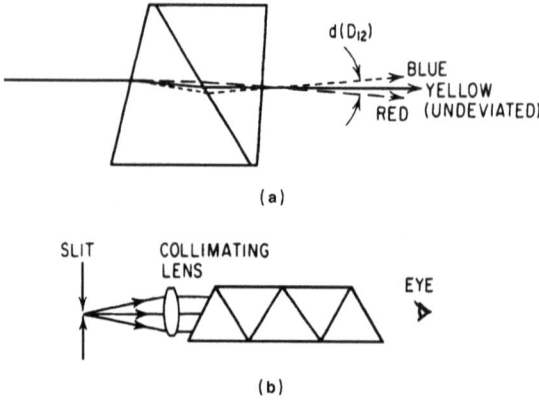

Figure 7.4 (a) A direct vision prism disperses the light into its spectral components without deviation of the beam. (b) Hand spectroscope. The collimating lens produces a magnified image of the slit at infinity for easy viewing. The prism then disperses the light into a spectrum without deviation of the yellow ray.

which they give will be approximations to the exact values when the prisms are other than thin. For exact work, these approximate values must be adjusted by exact ray tracing based on Snell's law.

7.6 Total Internal Reflection

When a light ray passes from a higher index medium to one with a lower index, the ray is refracted away from the normal to the surface as shown in Fig. 7.5a. As the angle of incidence is increased, the angle of refraction increases at a greater rate, in accordance with Snell's law $(n > n')$:

$$\sin I' = \frac{n}{n'} \sin I$$

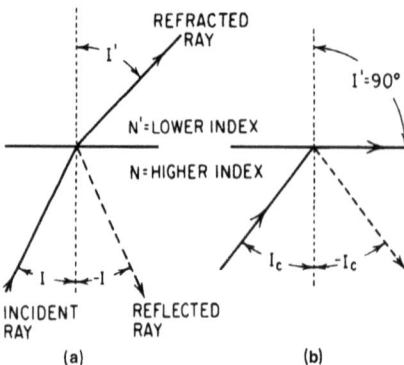

(a)

(b)

Figure 7.5 Total internal reflection occurs when a ray, passing from a higher to a lower index of refraction, has an angle of incidence whose sine equals or exceeds n'/n.

When the angle of incidence reaches a value such that $\sin I = n'/n$, then $\sin I' = 1.0$ and $I' = 90°$. At this point none of the light is transmitted through the surface; the ray is totally reflected back into the denser medium, as is any ray which makes a greater angle to the normal. The angle

$$I_c = \arcsin \frac{n'}{n} \qquad (7.15)$$

is called the *critical angle* and for an ordinary air-glass surface has a value of about 42° if the index of the glass is 1.5; for an index of 1.7, the critical angle is near 36°; for an index of 2.0, 30°; for an index of 4.0, 14.5°.

For practical purposes, if the boundary surface is smooth and clean, 100 percent of the energy is redirected along the totally reflected ray. However, it should be noted that the electromagnetic field associated with the light actually does penetrate the surface for a relatively short (to the order of a wavelength) distance. If there is anything near the other side of the boundary surface, the total internal reflection can be "frustrated" to some extent and a portion of the energy will be transmitted. Since the distance of effective penetration is only to the order of the wavelength of the light involved, this phenomenon has been used as the basis of a light valve, or modulator. In the German "Licht-Sprecher," an external piece of glass was placed in contact with the reflecting face of a prism to frustrate the reflection, and then moved an extremely short distance away (e.g., a few micrometers) to reinstate the reflection.

It should also be noted that the reflection of a totally reflecting surface is *decreased* by aluminizing or silvering the surface. When this is done, the reflectance drops from 100 percent to the reflectance of the coating applied to the surface.

7.7   Reflection from a Plane Surface

Since the prism systems which are discussed in the balance of this chapter are primarily reflecting prisms (the majority of which can be replaced by a system of plane mirrors), we shall first discuss the imaging properties of a plane reflecting surface. Rays originating at an object are reflected according to the law of reflection, which states that both the incident and reflected rays lie in the plane of incidence and that both rays make equal angles with the normal to the surface. The normal to the surface is the perpendicular at the point where the ray strikes the surface, and the plane of incidence is that plane containing the incident ray and the normal.

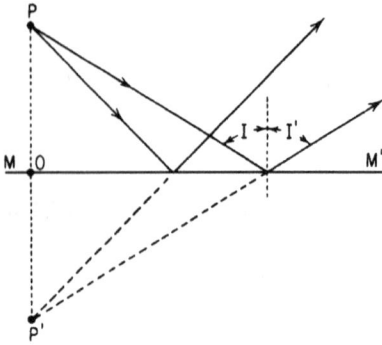

Figure 7.6 A plane reflecting surface forms a virtual image of an object point. Object and image are equidistant from the reflecting surface, and both lie on the same normal to the surface.

In Fig. 7.6, the plane of the page is the plane of incidence. Two rays from point *P* are shown reflected from the surface *MM′*. By extending the rays backward, it can be seen that after reflection they appear to be coming from point *P′*, which is a virtual image of point *P*. Both *P* and *P′* lie on the same normal to the surface (*POP′*), and the distance *OP* is exactly the same as the distance *OP′*.

If we now consider an extended object such as the arrow *AB* in Fig. 7.7, we can readily locate the position of its image by using the principles of the preceding paragraph to locate the images of points *A* and *B*. An observer at *E* looking *directly* at the arrow would see the arrowhead *A* at the top of the arrow. However, in the reflected image, the arrowhead (*A′*) is at the bottom of the arrow. The image of the arrow has been reoriented (or inverted) by the reflection.

If we add a crosspiece *CD* to the arrow, the image is formed as shown in Fig. 7.8, and although the image of the arrow has been inverted, the image of the crosspiece has the same left-to-right orientation as the object.

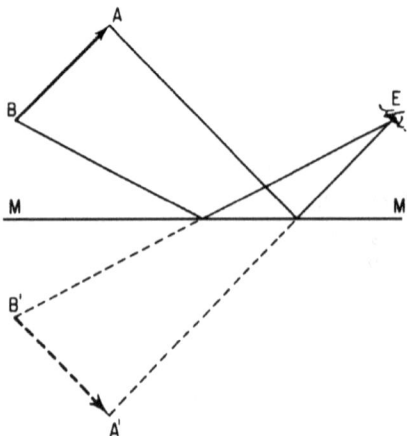

Figure 7.7 The reflected image *A′B′* of the arrow *AB* appears inverted to an observer at *E*.

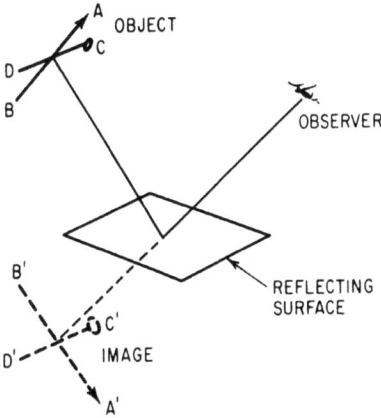

Figure 7.8 The reflected image is inverted top to bottom, but not left to right.

The preceding discussion has treated reflection from the standpoint of an observer viewing a reflected image. Since the path of light rays is completely reversible, we can equally well consider point P' in Fig. 7.6 to be an image formed by a lens at the right. Then P would be the reflected image of P'. Similarly in Figs. 7.7 and 7.8, we may replace the eye with a lens the image of which is the primed figure ($A'B'$ or $A'B'C'D'$) and view the unprimed figures as their reflected images.

A point worth noting is that reflection constitutes a sort of "folding" of the ray paths. In Fig. 7.9, the lens images the arrow at AB. If we now insert reflecting surface MM', the reflected image is at $A'B'$. Notice that if the page were folded along MM', the arrow AB and the solid line rays would exactly coincide with the arrow $A'B'$ and the reflected (dashed) rays. It is frequently convenient to "unfold" a complex reflecting system; one advantage of this device is that an accurate drawing of the ray paths becomes a simple matter of straight lines.

A useful technique to determine the image orientation after passage through a system of reflectors is to imagine that the image is a transverse arrow, or pencil, which is bounced off the reflecting surface, much as a thrown stick would be bounced off a wall. Figure 7.10 illustrates the technique. The first illustration shows the pencil approaching and striking the reflecting surface, the second shows the point bouncing off the reflector

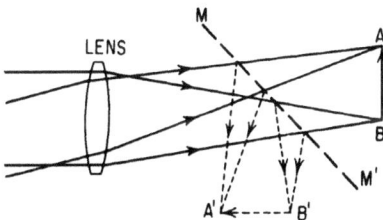

Figure 7.9 The reflecting surface MM' folds the optical system. Note that if the page is folded along MM', the rays and images coincide.

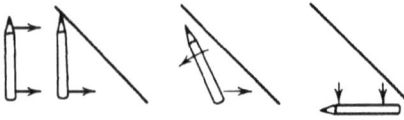

Figure 7.10 A useful technique in determining the orientation of a reflected image is to visualize the image as a pencil "bouncing" off a solid wall as it moves along the system axis.

and the blunt end continuing in the original direction, and the third shows the pencil in the new orientation after the reflection. If the process is repeated with the pencil perpendicular to the plane of the paper, the orientation of the other meridian of the image can be determined. The procedure can then be repeated through each reflection in the system.

A card marked with the arrow and crossbar of Fig. 7.11 is also useful for this purpose. The reader's attention is directed to the fact that the initial orientation of the pencil, or pattern, is chosen so that one meridian of the pattern coincides with the plane of incidence. In the majority of reflecting systems, one or the other of the meridians will be in the plane of incidence throughout the system, and the application of this technique is straightforward. Where this is not the case, the card can be marked with a second set of meridians so that the second set is aligned with the plane of incidence. This second set can then be carried through the reflection as before; the orientation of the final image is of course given by the original set of markings. Figure 7.20b exemplifies this method.

7.8   Plane Parallel Plates

As will become apparent, most prism systems are the equivalent of a thick block of glass. Thus we continue with a discussion of the effects produced by a plane-parallel plate of glass. Figure 7.12 shows a lens which, *in air*, would form an image at *P*. The insertion of the plane parallel plate between the lens and *P* displaces the image to *P'*. If we trace the path of the light rays through the plate, we first notice that the ray emerging from the plate has exactly the same slope angle that it had before passing through the plate, since by Snell's law,

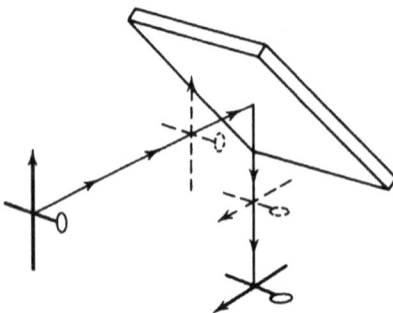

Figure 7.11 Image orientation after reflection.

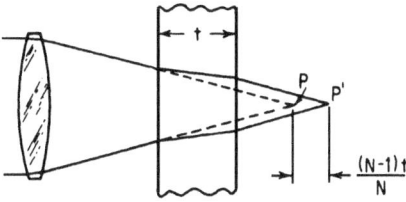

Figure 7.12 The longitudinal displacement of an image by a plane parallel glass plate.

$\sin I'_1 = (1/n) \sin I_1$, and $I_2 = I'_1$ (since the surfaces are parallel). Thus, $\sin I_2 = \sin I'_1 = (1/n) \sin I_1 = (1/n) \sin I'_2$, and $I_1 = I'_2$. Therefore, the effective focal length of the lens system, and the size of the image, are unchanged by the insertion of the plate.

The amount of longitudinal displacement of the image is readily determined by application of the paraxial raytracing formulas of Chap. 3, and is equal to $(n-1)t/n$. The effective thickness of the plate compared to air (the equivalent air thickness) is less than the actual thickness t by the amount of this shift. The *equivalent air thickness* is thus found by subtracting the displacement from the thickness and is equal to t/n. The concept of equivalent thickness is useful when one wishes to determine whether a certain size prism can be fitted into the available air space of an optical system, and also in prism system design.

If the plate is rotated through an angle I as shown in Fig. 7.13, it can be seen that the "axis ray" is laterally displaced by an amount D, which is given by

$$D = t \cos I (\tan I - \tan I') = t \, \frac{\sin (I - I')}{\cos I'}$$

or

$$D = t \sin I \left(1 - \frac{\cos I}{n \cos I'}\right)$$

or

$$D = t \sin I \left[1 - \sqrt{\frac{1 - \sin^2 I}{n^2 - \sin^2 I}}\right]$$

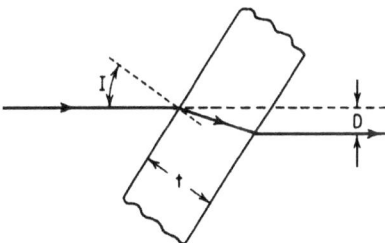

Figure 7.13 The lateral displacement of a ray by a tilted plane parallel plate.

A power series expansion yields the following expression:

$$D = \frac{tI\,(n-1)}{n} \left[1 + \frac{I^2\,(-n^2 + 3n + 3)}{6n^2} \right.$$

$$\left. + \frac{I^4\,(n^4 - 15n^3 - 15n^2 + 45n + 45)}{120n^4} + \cdots \right]$$

For small angles, we can make the usual substitution of the angle for its sine or tangent, or simply use the first term of the expansion to get

$$d = \frac{ti\,(n-1)}{n}$$

This lateral displacement by a tilted plate is used in high-speed cameras (where the rotating plate displaces the image an amount approximately equal to the travel of the continuously moving film) and in optical micrometers. The optical micrometer is usually placed in front of a telescope and used to displace the line of sight. The amount of displacement is read off a calibrated drum connected to the mechanism which tilts the plate.

When used in parallel light, a plane parallel plate is completely free of aberrations (since the rays enter and leave at the same angles). However, if the plate is inserted in a convergent or divergent beam, it does introduce aberrations. The longitudinal image displacement $(n-1)t/n$ is greater for short wavelength light (higher index) than for long, so that overcorrected chromatic aberration is introduced. The amount of displacement is also greater for rays making large angles with the axis; this is, of course, overcorrected spherical aberration. When the plate is tilted, the image formed by the meridional rays is shifted backward while the image formed by the sagittal rays (in a plane perpendicular to the page in the figures) is shifted by a lesser amount, so that astigmatism is introduced.

The amount of aberration introduced by a plane parallel plate can be computed by the formulas below. Reference to Fig. 7.14 will indicate the meanings of the symbols

---

U and u—slope angle of the ray to the axis

U_p and u_p—the tilt of the plate

t—thickness of the plate

n—index of the plate

V—Abbe V number $(n_d - 1)/(n_F - n_C)$

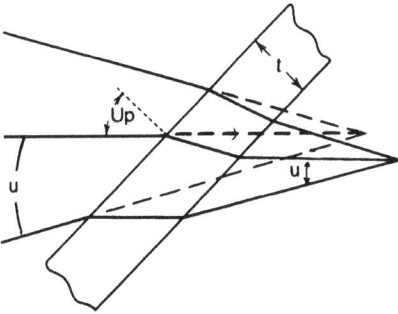

Figure 7.14 Aberration introduced by a plane parallel plate.

Chromatic aberration $= l'_F - l'_C = \dfrac{t(n-1)}{n^2 V}$

Spherical aberration $= L' - l' = \dfrac{t}{n}\left[1 - \dfrac{n\cos U}{\sqrt{n^2 - \sin^2 U}}\right]$ (exact)

$$= \dfrac{tu^2(n^2-1)}{2n^3} \quad \text{(third order)}$$

Astigmatism $= (l'_s - l'_t) = \dfrac{t}{\sqrt{n^2 - \sin^2 U_p}}$

$$\times \left[\dfrac{n^2 \cos^2 U_p}{(n^2 - \sin^2 U_p)} - 1\right] \quad \text{(exact)}$$

$$= \dfrac{-tu_p^2(n^2-1)}{n^3} \quad \text{(third order)}$$

Sagittal coma $= \dfrac{tu^2 u_p(n^2-1)}{2n^3}$ (third order)

Lateral chromatic $= \dfrac{tu_p(n-1)}{n^2 V}$ (third order)

These expressions are extremely useful in estimating the effect that the introduction (or removal) of a plate or a prism system will have on the state of correction of an optical system.

A common use for a glass plate is as a beam splitter, tilted at an angle of 45°. In this orientation the astigmatism is approximately a quarter of the thickness of the plate. Since this can severely degrade the image, such plate beam splitters are not recommended in convergent or divergent beams (i.e., where u in Fig. 7.14 is nonzero). Note that the astigmatism can be nullified by inserting another identical plate which is tilted in a meridian 90° to the original plate, by introducing either a weak cylinder or a tilted spherical surface, or by wedging the plate.

7.9   The Right-Angle Prism

The right-angle prism, with angles of 45°–90°–45°, is the building block of most nondispersing prism systems. Figure 7.15 shows a parallel bundle of rays passing through such a prism, entering through one face, reflecting from the hypotenuse face, and leaving through the second face. If the rays are normally incident on the face of the prism, they are deviated through an angle of 90°. At the hypotenuse face, the rays have an angle of incidence of 45° so that they are subject to total internal reflection. If the entrance and exit faces are low-reflection-coated, this makes the prism a highly efficient reflector for visual usage since the only losses are the absorption of the material and the reflection losses at the faces which total a few percent or less. (In the ultraviolet and infrared portions of the spectrum, the absorption of a prism may be quite objectionable.) It can be seen that the total internal reflection is limited to rays which have angles of incidence greater than the critical angle, and many prism systems are made of high-index glass to permit total reflection over larger angles.

By *unfolding* the prism, as indicated by the dashed lines in Fig. 7.16, it is apparent that the prism is the equivalent of a glass block with parallel faces, with a thickness equal to the length of the entrance or exit faces. The equivalent air thickness of the block is, of course, this thickness divided by the index of the prism.

If the 45°–90°–45° prism is used with the light beam incident on the hypotenuse face as shown in Fig. 7.17, the light is totally reflected twice and the rays emerge in the opposite direction, having been deviated through 180°. Figure 7.17 also indicates the unfolded prism path and the image orientation of this prism. Notice that the image has been inverted, top to bottom, but not left to right. The unfolded prism

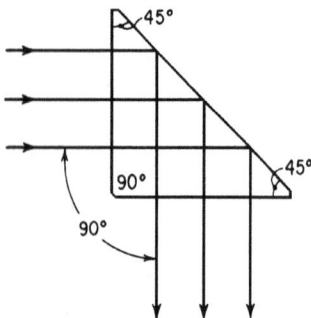

Figure 7.15   Right-angle prism.

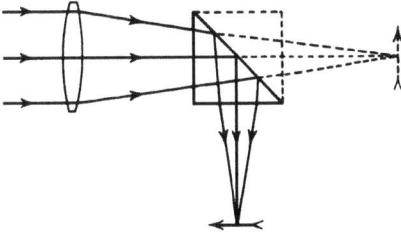

Figure 7.16 "Unfolding" a 90° prism to show that it is equivalent to a block of glass.

path is called a *tunnel diagram*. Such a diagram can be used to determine the angular field of the prism as well as the size of the beam which will pass through the prism.

Used in this way, this prism is a *constant-deviation prism*. Regardless of the angle at which a ray enters the prism, the emergent ray will be parallel, as shown in Fig. 7.18a. This characteristic is a property of the two reflecting surfaces of the prism. A system which directs the light ray back on itself is called a retrodirector; this prism is a retrodirector in one meridian only. (Another of the many constant-deviation systems possible with two reflectors is the 90° deviation arrangement shown in Fig. 7.18b, where the reflecting surfaces are at 45° to each other.) The constant-deviation angle is just twice the angle between the two mirrors.

A prism made by cutting off one corner of a cube, so that there are three mutually perpendicular reflecting surfaces, is retrodirective in both meridians. The corner cube (or cube corner) reflector will return all the light rays striking it back toward their source, although the rays will be displaced laterally.

A third orientation of the 45°–90°–45° prism is shown in Fig. 7.19, in which the bundle of rays arrives parallel to the hypotenuse face of the prism. After being refracted downward at the entrance face, the rays are reflected upward from the hypotenuse and emerge after a second refraction at the exit face. The unfolded path of the rays

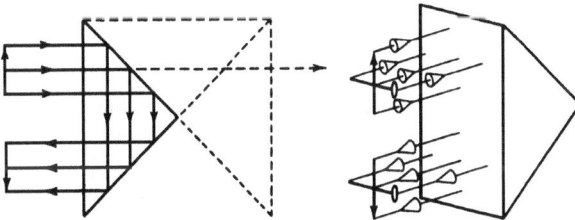

Figure 7.17 Right-angle prism used with hypotenuse as entrance and exit face.

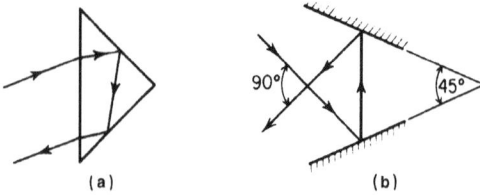

Figure 7.18 (a) The right-angle prism used in the manner shown is a constant-deviation prism, in that each ray is reflected through exactly 180°. The entering and emergent paths are parallel, regardless of the initial angle the ray makes with the prism. (b) A pair of constant-deviation mirrors. In this case, the deviation produced by the two reflections is always exactly 90°.

(shown in dashed lines) indicates that this prism is the equivalent of a plane parallel plate which is tilted with respect to the axis of the bundle, whereas in the preceding examples the prism faces have been normal to the axis. If this prism is used in a convergent light beam, it will introduce a substantial amount of astigmatism (roughly equal to one-quarter of its thickness). For this reason, this prism, which is known as a *Dove prism,* is used almost exclusively in parallel light. Since the apex of the prism is not used by the light beam, the prism is usually truncated at AA'.

The Dove prism has a very interesting effect on the orientation of the image. In Fig. 7.20a, the arrow and crossbar pattern is shown to be inverted from top to bottom but not left to right. If the prism is rotated 45°, as in Fig. 7.20b, the image is rotated through 90°; if the prism is rotated 90° as in Fig. 7.20c, the pattern is rotated 180°. Thus, the image is rotated twice as fast as the prism. (The analysis of the image orientation in Fig. 7.20b is an example of the use of an auxiliary pattern as described in Sec. 7.7. The auxiliary pattern is shown in dotted lines in Fig. 7.20b.)

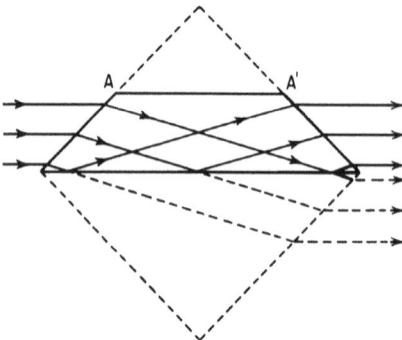

Figure 7.19 The Dove prism. The dashed lines show that the Dove prism is equivalent to a tilted plate and will introduce astigmatism when used in convergent or divergent beams.

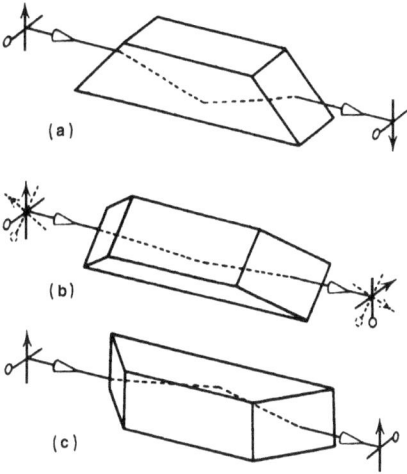

(a)

(b)

(c)

Figure 7.20 The orientation of an image by a Dove prism. (a) Original position. (b) Prism rotated 45°; image is rotated 90°. (c) Prism rotated 90°; image is rotated 180°. Note that the dotted arrow and crossbar in (b) is oriented so that the dotted arrow is in the plane of incidence to simplify the analysis of the image orientation.

The length of the Dove prism is four to five times the diameter of the bundle of rays which it will transmit. If two Dove prisms are cemented hypotenuse to hypotenuse (after silvering or aluminizing these faces), the aperture is thereby doubled with no increase in length. The double Dove prism is used in parallel light as is the Dove. It must be precisely fabricated to avoid producing two slightly separated images. When the double Dove is rotated, or tipped, about its center, it can be used as a scanner to change the direction of sight of a telescope or periscope.

7.10 The Roof Prism

If the hypotenuse face of a right-angle prism is replaced by a "roof," i.e., two surfaces at 90° the intersection of which lies in the hypotenuse, the prism is called a *roof,* or *Amici, prism.* Face and side views of a roof prism are shown in Fig. 7.21. The addition of the roof to the prism serves to introduce an extra inversion to the image, as can be seen by comparing the final orientation of the crossbar in Fig. 7.11 with that

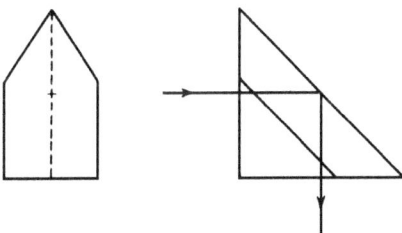

Figure 7.21 Roof, or Amici, prism.

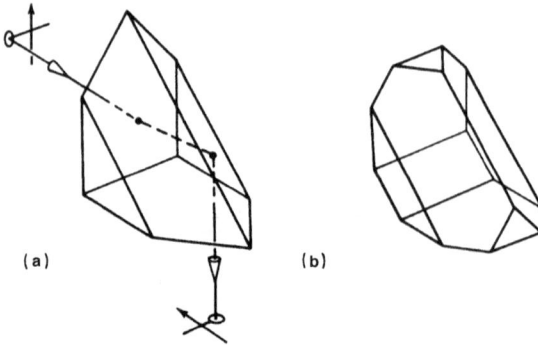

Figure 7.22 Amici prism (a) showing a single ray path through the prism and indicating the image orientation, (b) with truncated corners to reduce weight without sacrifice of useful aperture.

in Fig. 7.22a. This can be understood by tracing the path of the dashed ray in Fig. 7.22a which connects the circles in the arrow and crossbar figures before and after passing through the prism.

The angle of incidence (at the roof surface) of the ray shown in Fig. 7.22a is about 60° instead of the 45° it would be for the same ray in the right-angle prism. Even a ray perpendicular to the roof edge has an angle of incidence of 45°. The result is that a roof surface allows total internal reflection for beam angles which would be transmitted through the hypotenuse face of a right-angle prism.

In practice, the Amici prism is usually fabricated with the corners cut off, as shown in Fig. 7.22b, in order to reduce the size and weight of the prism. The 90° roof angle must be made to a high order of accuracy. If there is an error in the roof angle, the beam is split into two beams which diverge at an angle which is six times the error. Thus, to avoid any apparent doubling of the image, the roof angle is usually made accurate to one or two seconds of arc.

The introduction of a roof degrades the diffraction-limited resolution by a factor approaching 2 in the direction perpendicular to the roof edge (due to a polarization/phase shift on reflection) no matter how perfectly the prism is made*. Multilayer coatings have been developed which will reduce this effect.

---

*A. Mahan, "Focal plane anomalies in roof prisms," *J. Opt. Soc. Am.*, vol. 35, 1945, p. 623.

A. Mahan, "Further studies of focal plane anomalies in roof prisms," *J. Opt. Soc. Am.*, vol. 36, 1946, p. 715A.

A. Mahan, "Focal plane diffraction anomalies in telescopic systems," *J. Opt. Soc. Am.*, vol. 37, 1947, p. 852.

A. Mahan and E. Price, "Diffraction pattern deterioration by roof prisms," *J. Opt. Soc. Am.*, vol. 40, 1950, p. 664.

7.11   Erecting Prism Systems

In an ordinary telescope, the objective lens forms an inverted image of the object, which is then viewed through the eyepiece. The image seen by the eye is upside down and reversed from left to right, as indicated in Fig. 7.23. To eliminate the inconvenience of viewing an inverted image, an erecting system is often provided to re-invert the image to its proper orientation. This may be a lens system or a prism system.

Porro prism of the first type

The most commonly used prism-erecting system is the Porro prism of the first type, illustrated in Fig. 7.24. The Porro system consists of two right-angle prisms oriented at 90° to each other. The first prism inverts the image from top to bottom and the second prism reverses it from left to right. The optical axis is displaced laterally, but is not deviated. One can see that if this system is inserted into the telescope of Fig. 7.23, the final image will have the same orientation as the object. Although the prism system is ordinarily inserted between the objective and eyepiece (to minimize its size), it will erect the image regardless of where it is placed in the system.

The Porro prism (first type) owes its popularity to the fact that the 45°–90°–45° prisms are relatively easy and inexpensive to manufacture, with no critical tolerances. However, if the prisms are not mounted so that their roof edges are exactly at 90° to each other, the final image will be rotated through twice the angular mounting error. This is of special importance in binocular systems where the image presented to one eye must be identical to that presented to the other.

A shallow ground slot is often cut across the center of the hypotenuse face of each prism to prevent unwanted grazing angle reflections from this face which originate from outside the field of view. See also Fig. 7.39.

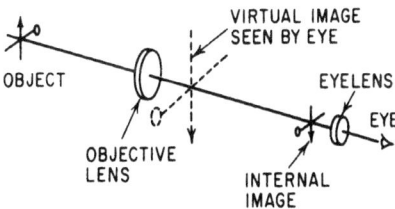

Figure 7.23 In a simple telescope, the objective lens forms a real, inverted internal image of the object, which is reimaged by the eyelens. The image seen by the eye is a virtual inverted image of the object.

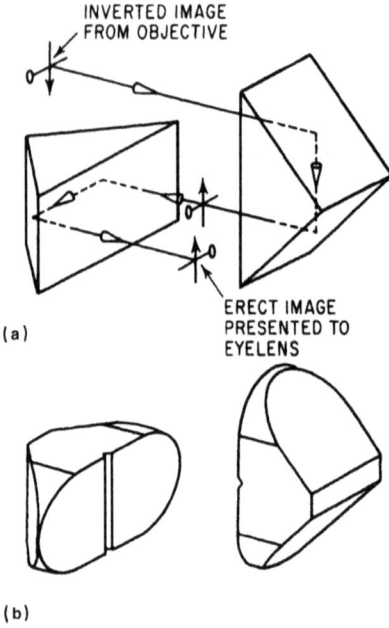

INVERTED IMAGE
FROM OBJECTIVE

ERECT IMAGE
PRESENTED TO
EYELENS

(a)

(b)

Figure 7.24 Porro prism system (first type) (a) indicating the way the Porro system erects an inverted image. (b) Porro prisms are usually fabricated with rounded ends to save space and weight. Note that the spacing between the prisms has been shown increased for clarity.

Porro prism of the second type

The Porro prism of the second type is shown in Fig. 7.25, and serves the same purpose as the Porro #1 system. Both Porro systems function by total internal reflection so that no silvering is required. It is common to round off the ends of the prisms to conserve space and weight.

The second Porro is somewhat more difficult to fabricate than the first type, but in some applications its compactness, and the fact that the prisms can be readily cemented together, offer compensating advantages. The Porro #2 may also be made in three pieces, by

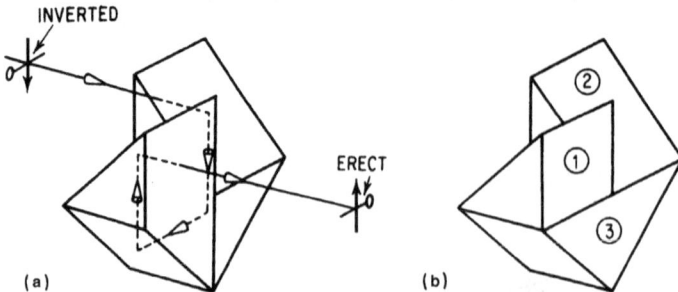

INVERTED

ERECT

(a)

(b)

Figure 7.25 Porro prism system (second type) (a) indicating the erection of an inverted image. This system is shown made from two prisms in (a) and from three prisms in (b).

cementing two small right-angle prisms on the hypotenuse of a large right-angle prism as indicated in Fig. 7.25b. The lateral displacement of the axis is less than that for the Porro #1 system.

Abbe prism

The Abbe (or Koenig, or Brashear-Hastings) prism (Fig. 7.26) is an erecting prism which can be used when it is desired to erect the image without displacing the axis as the Porro prisms do. The roof is necessary to provide the left-to-right reversal of the image; the roof angle must be made accurately to avoid image doubling.

If this prism is made without the roof, it will invert the image in one meridian only, just as the Dove prism. However, since its entrance and exit faces are normal to the system axis, it may be used in a converging beam without introducing astigmatism.

Other erecting prisms

Among the many prisms designed to erect an image are those sketched in Fig. 7.27. The fact that the image is inverted and reversed left to right after passing through these prisms may be verified by the methods outlined in Sec. 7.7. Notice that each prism (except Fig. 7.27f) has been arranged so that the axial ray enters and leaves the prism normal to the prism faces and that all reflections are total internal reflections. In the Leman and Goerz prisms, the axis is displaced but not deviated. In the Schmidt and modified Amici prisms, the axis is deviated through a definite angle, which can be selected by the designer (within the limits allowed by total internal reflection). Note also that the roof surface is used at the location where the angle of incidence is small and where there would be light leakage through an ordinary surface.

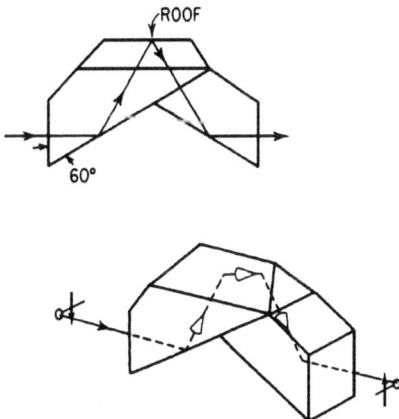

Figure 7.26 Abbe prism. Used as an in-line erecting system, it does not displace the axis as the Porro systems do, nor does it materially displace the image longitudinally.

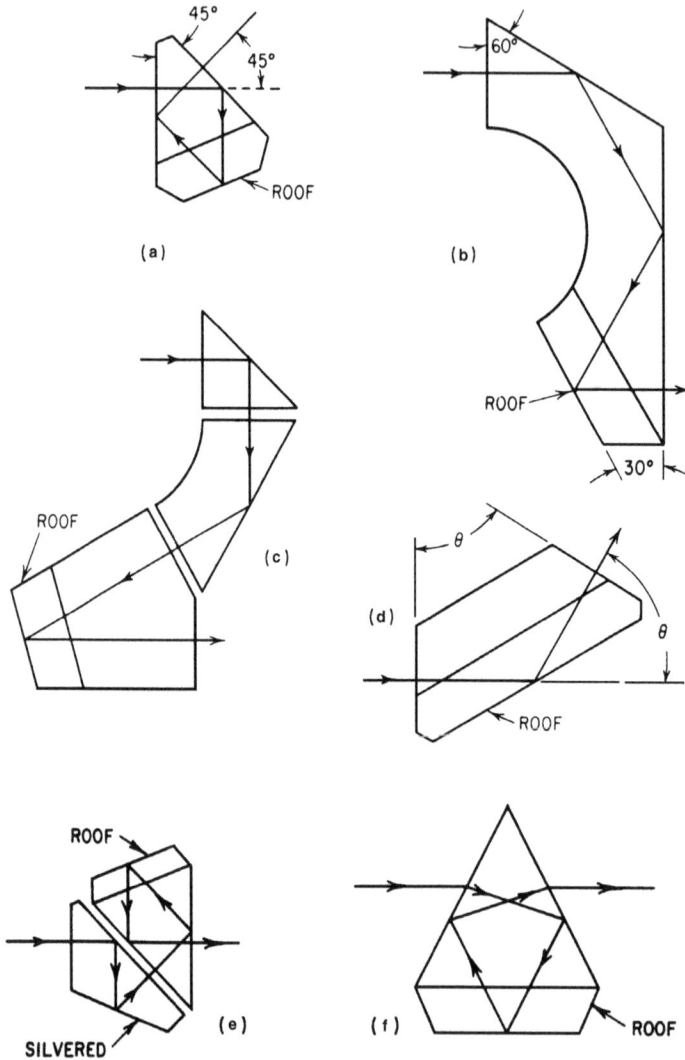

Figure 7.27 Erecting prisms: (a) Schmidt prism; (b) Leman (or Sprenger) prism; (c) Goerz prism; (d) modified Amici prism; (e) roofed Pechan prism; (f) roofed delta prism.

7.12   Inversion Prisms

The Dove prism (Figs. 7.19 and 7.20) and the roofless Abbe prism mentioned in Sec. 7.11 are examples of prisms which invert the image in one meridian but not the other. The plane mirror and the right-angle prism (Figs. 7.11 and 7.16) are also simple inversion systems. Figure 7.28 shows the above prisms plus the Pechan prism, which is a

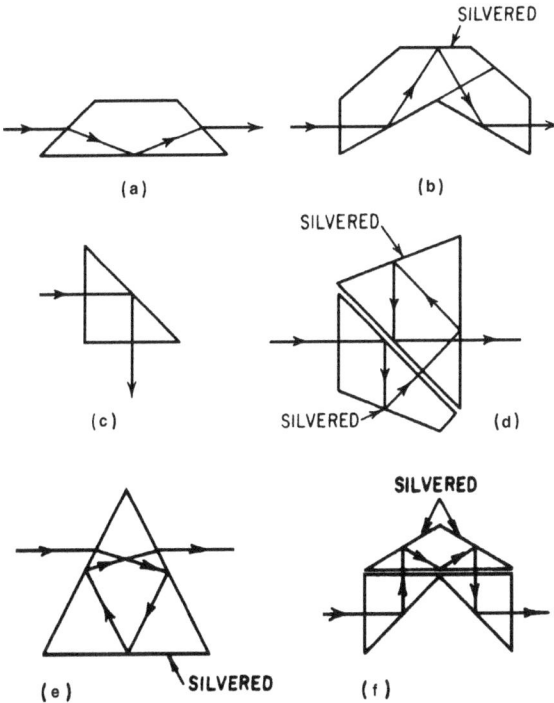

Figure 7.28 Inversion (or derotation) prisms: (a) Dove prism; (b) reversion prism; (c) right-angle prism; (d) Pechan prism; (e) delta, or Taylor, prism; (f) compact prism.

relatively compact prism for this purpose. Notice that the addition of a "roof" to any of these prisms will convert it to an erecting system.

An inversion prism is also known as a *derotation prism,* since all inversion prisms rotate the image in the same manner as the Dove prism, as shown in Fig. 7.20.

The mirror version of Fig. 7.28b is called a *k-mirror* and is useful in infrared and ultraviolet applications where material for a solid prism system is impractical.

7.13 The Penta Prism

The Penta prism (Fig. 7.29a) will neither invert nor reverse the image. Its function is to deviate the line of sight by 90°. It has the valuable property of being a constant-deviation prism, in that it deviates the line of sight through the same angle regardless of its orientation to the line of sight.

Most of the prism systems described in this chapter could be replaced by a series of plane mirrors, and this is sometimes done for reasons of

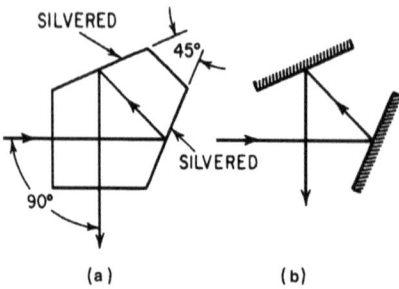

Figure 7.29 The Penta prism (a) and its equivalent mirror system (b).

weight and/or economy. However, a prism, as a monolithic glass block, is a very stable system and is not as subject to environmental variation of angles as is an assemblage of mirrors on a metal support block.

The Penta prism is used where it is desirable to produce an exact 90° deviation without having to orient the prism precisely. The end reflectors of rangefinders are often of this type, and in optical tooling and precise alignment work, the Penta prism is useful to establish an exact 90° angle. In large rangefinders, however, the prism is replaced by two mirrors (Fig. 7.29b), securely cemented to a block in order to avoid the weight, absorption, and cost of a large block of solid glass. The deviation of two mirrors as shown in Fig. 7.29 is equal to twice the angle between the mirrors.

Occasionally a roof is substituted for one of the reflecting faces of the Penta prism to invert the image in one meridian.

7.14 Rhomboids and Beamsplitters

The rhomboid prism is a simple means of displacing the line of sight without affecting the orientation of the image or deviating the line of sight. The rhomboid prism and its mirror system equivalent are shown in Fig. 7.30.

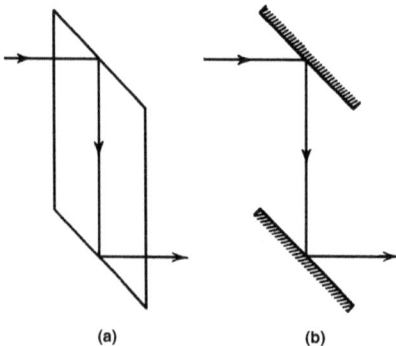

Figure 7.30 (a) Rhomboid prism. (b) An equivalent mirror system. Both systems displace the optical axis without deviation or reorientation of the image.

A beamsplitter is frequently useful for the purpose of combining two beams (or images) into one, or for separating one beam into two. A thin plate of glass with one surface coated with a semireflecting coating, as shown in Fig. 7.31a, can be used for this purpose, but it suffers from two drawbacks. First, if used in a convergent or divergent beam, it would introduce astigmatism, and second, the reflection from the second surface, although faint, would produce a ghost image displaced from the primary image. (Note that in parallel light neither of these objections is valid, provided the surfaces of the plate are accurately parallel.) The astigmatism can be controlled by placing a weak convex surface on the second side of the plate. The beamsplitter cube (Fig. 7.31b) avoids these difficulties. It is composed of two right-angle prisms cemented together. The hypotenuse of one prism is coated with a semireflecting coating before cementing.

Where the weight or absorption of the cube cannot be tolerated, a *pellicle* is often used as a semireflector. A pellicle is a thin (2- to 10-μm) membrane (usually a plastic such as nitrocellulose) stretched over a frame; by virtue of its extreme thinness, both the astigmatism and ghost displacement are reduced to acceptable values.

Obviously, the shape of the pellicle surface is determined by the shape of the frame over which it is stretched, and an accurately plane support is necessary. There are two less obvious features of the pellicle which may be disadvantageous: (1) Interference between light reflected from the two surfaces of the extremely thin pellicle can result in a transmission that varies in a rippled way as a function of wavelength, and (2) the pellicle can act as if it were the diaphragm of a microphone, and any atmospheric vibrations can change the shape of the reflecting surface, introducing significant changes in the imagery of the system. This is the basis for one "talk-on-a-beam-of-light" toy.

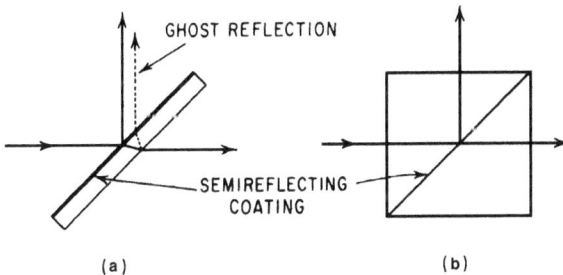

Figure 7.31 Beamsplitters. (a) A thin parallel plate is convenient but may be objectionable because of ghosting and astigmatism, unless used in parallel light. (b) Beamsplitting cube has a semireflecting coating supplied to one of the diagonal faces before cementing.

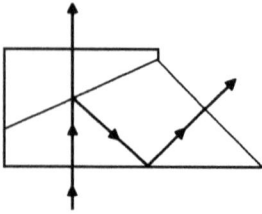

Figure 7.32 45° tilting eyepiece prism.

Figure 7.32 shows a prism which is often used in microscope eyepieces to change the direction of the line of sight from vertical to a more-convenient-to-use 45°. As shown, the prism can be used as a beamsplitter either to provide for coaxial illumination or to allow a second eyepiece; without the beamsplitting feature, it simply redirects the line of sight.

In Fig. 7.33, two binocular eyepiece prism systems are sketched. Both serve the same function, namely splitting the light beam from an objective lens into two parts. The two beams are displaced sufficiently so that they can be presented to two eyepieces and both eyes may simultaneously view the same subject. Notice that in both systems, extra glass has been added to the left-hand path so that the amount of glass in each path is identical; in this way the aberrations introduced by the glass are the same for each path. Most of the glass in these systems could be dispensed with if desired, since each of them is equivalent to a beamsplitting cube plus three reflectors. In the system shown in Fig. 7.33b, the two halves can be rotated about the objective axis to vary the spacing between the eyepieces as shown in Fig. 7.33c. Notice that the image is not rotated by this procedure but

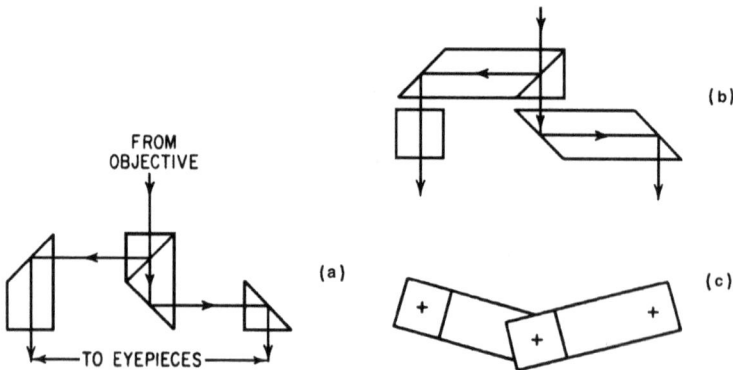

Figure 7.33 Prism systems for binocular eyepiece instruments. System (a) can be adjusted to match the user's eye separation by sliding both outer prisms in or out; this defocuses the instrument. Sketch (c) shows how the halves of (b) can be rotated about the objective axis to make this adjustment.

retains its original orientation, because the reflecting surfaces are in the form of a rhomboid prism.

Often two Porro systems are used in a similar rotatable configuration which allows a change in the eye separation.

7.15   Plane Mirrors

In the preceding discussions we have indicated several times that reflecting prisms may be replaced by mirrors. For most applications, it is necessary that the mirrors be first-surface mirrors, as opposed to ordinary second-surface mirrors. The two types are sketched in Fig. 7.34. The first-surface mirror is usually preferable because it does not produce a ghost image as does the second-surface mirror. In addition, the second-surface mirror requires the processing of an extra surface in its fabrication. It also requires the light to pass through a thickness of glass which may introduce aberrations and which will absorb energy in ultraviolet and infrared applications. The second-surface mirror can be made more durable, however, since its reflecting coating can be protected from the elements by electrodeposited copper and painted coverings. First-surface mirrors are usually made with vacuum-deposited aluminum films protected by a thin transparent overcoating of silicon monoxide or magnesium fluoride.

7.16   The Design of Prism and Reflector Systems

Ordinarily it is required of a prism (or reflector) system that it produce an image with a certain orientation and with the emergent beam of light redirected in a given manner. The design effort is usually best begun by establishing the minimum number of reflectors which will produce the desired result. This is most simply (and perhaps best)

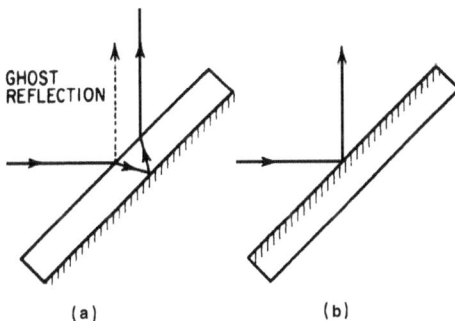

Figure 7.34 (a) Second-surface mirror. (b) First-surface mirror.

(a)   (b)

accomplished by straightforward trial and error. A rough perspective sketch is made to indicate the reflections necessary to locate the image in its desired position. The orientation of the image is then checked by the technique of Sec. 7.7; reflectors are added in various orientations until the image orientation is correct. Usually several roughly equivalent schemes are possible, and a selection can be made based on the requirements of the application.

When the reflection system is completed, the optical system is unfolded, i.e., sketched with the optical axis as a straight line. The object, image, and lens apertures are added to the sketch and the necessary sizes for the reflectors are determined in both meridians. If the system is to be composed of prisms, the unfolded layout is repeated with the axial distances adjusted to the "equivalent air thickness" (t/n) for that portion of the system which is glass so that the ray paths can be drawn as straight lines.

As an example of reflector system design, let us consider the problem presented by Fig. 7.35. The object at A is to be projected by an ordinary lens B onto a screen at S. The plane of S is parallel to the original projection axis and its center is above the axis by some amount Y. The required orientations of object and image are shown in the sketch.

We begin by noting that the image formed by the projection lens will be inverted in both meridians with respect to the object, as shown at C in Fig. 7.35. Now, passing to Fig. 7.36, let us consider the effect of a mirror placed at D. Of the four directions shown as possible reflections at D, the upward reflection labeled D_1 seems the most promising since it sends the light in a direction that it must eventually take, so we elect to pursue this line. Using similar reasoning at E, we should be inclined to select E_2; however, the image at E_2 is rotated 90° from our desired orientation. Selecting E_1 on the basis that its image orientation is closest to the desideratum, we consider a reflection at F. Again, F_3 is in the proper direction, but the image is reversed from left to right. Case F_1

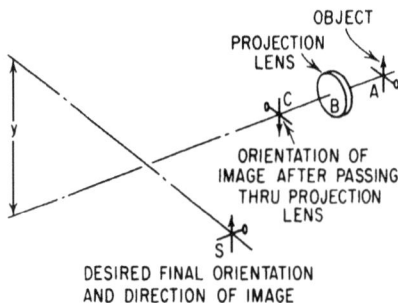

OBJECT
PROJECTION
LENS

ORIENTATION OF
IMAGE AFTER PASSING
THRU PROJECTION
LENS

DESIRED FINAL ORIENTATION
AND DIRECTION OF IMAGE

Figure 7.35 Inversion of image formed by projection lens.

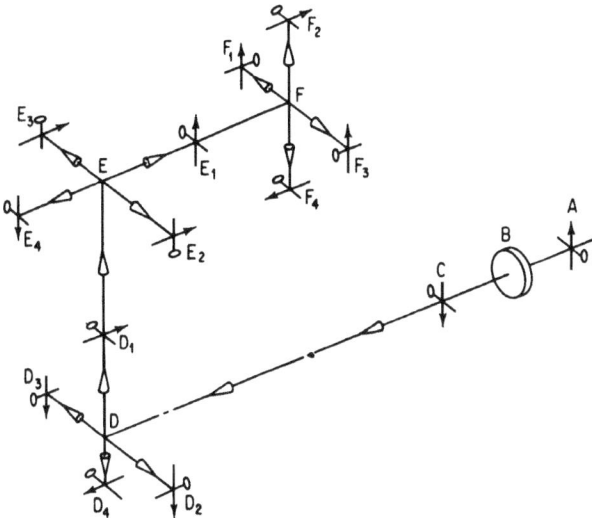

Figure 7.36 Possible orientation paths for the reflecting system exercise in text.

has the proper orientation, but the light is traveling away from the screen. If we add a mirror to reverse the direction of propagation, we will have both orientation and direction as required. To accomplish this without directing the light back through F, we must resort to a figure 4 arrangement as shown in Fig. 7.37, which diagrams the entire system.

It is quite apparent that Fig. 7.37 represents only one of the many possible arrangements of mirrors which could be utilized to accomplish this same end result. The reader may also have noticed that the discussion has been limited to reflections for which the plane of incidence lay in one of the cartesian reference planes, and also that first consideration was given to reflections which deviated the axis by 90°. For the novice, these restrictions have much to recommend them; one is well

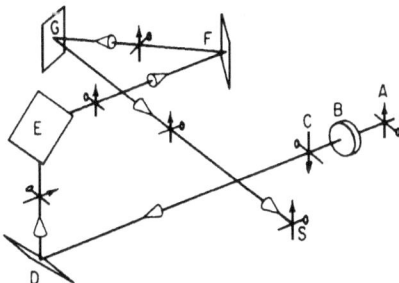

Figure 7.37 One possible solution to the exercise in the text.

advised to keep first trials of this type as simple and uncomplicated as possible. Further, the reduction of the system to practice is much simplified if compound angles are avoided. If our problem had required that the final image be rotated 45°, then we would necessarily have had to depart from the cartesian planes to achieve the desired result.

The Porro erecting prism (Fig. 7.38a) will serve as an illustrative example of the "unfolding" technique used in the design of prism systems. The prisms have been unfolded in Fig. 7.38b (for clarity, the second prism is shown rotated 90° about the axis). Each prism can be seen to be the equivalent of a glass block the thickness of which is twice the size of its end face. Notice that the rays from the lens are refracted at each air-glass surface of the system and that the image has been displaced to the right by the prisms.

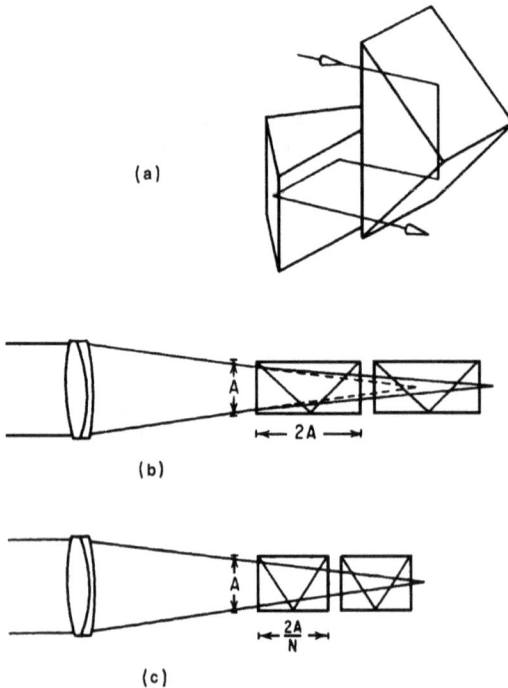

Figure 7.38 (a) Porro prism system (first type). (b) Unfolded prisms. Dashed lines indicate path rays would take without prisms. Solid line shows the displacement of the focal point by the prisms. (c) The prisms are drawn to their equivalent air thickness so that the rays can be drawn as straight lines.

In Fig. 7.38c, the prisms are drawn with their "equivalent air thickness" as discussed in Sec. 7.8. This allows us to draw the (paraxial) light rays through the prism as straight lines, simplifying the construction considerably.

Now let us suppose that we are to design the minimum size Porro system for a 7 × 50 binocular. The objective lens has a focal length of 7 in, an aperture of 2 in, and is to cover a $^5/_8$-in-diameter field, as sketched in Fig. 7.39a. We first note that the proportions of face width to "equivalent air thickness" for each prism (Fig. 7.39a) are $A{:}2A/n = 1{:}2/n$, or, if we assume an index of 1.50, 3:4. We begin the design from the image and work toward the objective. Placing the exit face of the prism $^1/_2$ in from the image (to allow for clearance and to keep the glass surface well out of the focal plane), we construct the dashed line shown in Fig. 7.39a with a slope of 3:8 (one-half the face-to-equivalent-thickness ratio) starting from the axial intercept of the exit face. This line is, of course, the locus of the corners of a family of prisms of various sizes, and the point where it intersects the extreme clearance ray defines the minimum size prism which will transmit the entire cone of light from the objective. For practical purposes, the prism should be made slightly larger than this to allow for bevels and mounting shoulders.

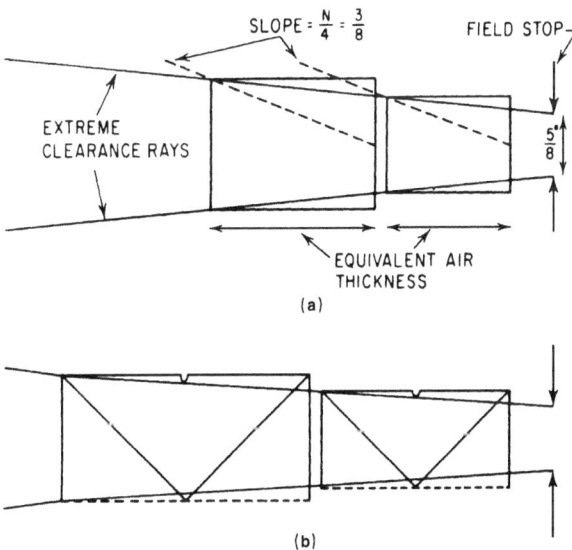

Figure 7.39 The layout of a minimum-size prism system is shown in (a). The extreme clearance rays connect the rim of the objective with the edge of the field of view. The intersection of the dashed lines (see text) with these rays locates the corner of the smallest prism which will pass the full image cone. In (b) the prisms are drawn to scale, showing their true thickness.

The procedure is now repeated for the other prism; an air space is left between the two to allow for the mounting plate to which both prisms are to be fastened. In Fig. 7.39b, the system is drawn to scale, with the prism blocks expanded to their true length. The reason for the ground slot usually cut into the hypotenuse faces of Porro prisms can be understood from an examination of the unfolded drawings. Light rays from outside the desired field of view can be reflected (by total internal reflection) from these faces back into the field where they are quite annoying; the slot intercepts these rays as they graze along the hypotenuse.

7.17 Analysis of Fabrication Errors

The effects produced by errors in prism angles (due to manufacturing tolerances) are readily analyzed. Such angular errors can be treated as equivalent to the rotation of a reflecting surface from its nominal position, and/or the addition of a thin refracting prism to the system.

As an example, consider the right-angle prism shown in Fig. 7.40 and assume that the upper 45° angle is too large by ϵ and that the lower 45° angle is too small by ϵ. A ray normal to the entrance face will make an angle of incidence of 45° + ϵ at the hypotenuse; the angle of reflection will then be 45° + ϵ and the ray will be reflected through an angle of 90° + 2ϵ. Thus, rotating the reflecting face through ϵ has introduced an error of 2ϵ in the direction of the ray.

At the exit face, the ray has an angle of incidence of 2ϵ and, if the prism index is 1.5, an angle of refraction of 3ϵ. Thus, the total deviation of the ray from its nominal direction is 3ϵ. Also, since the ray has been deviated through an angle ϵ by refraction at this surface, the ray will be dispersed and spread out into a spectrum subtending an angle of ϵ/V according to Eq. 7.11.

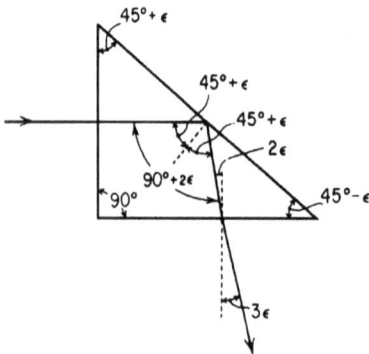

Figure 7.40 The passage of a ray through a right-angle prism the hypotenuse face of which is tilted from its proper position by a small angle ϵ. After reflection, the ray is deviated by 2ϵ; this is increased to 3ϵ (or 2$n\epsilon$) by refraction at the exit face.

Bibliography

Greivenkamp, J. E., *Field Guide to Geometrical Optics*, Bellingham, WA, SPIE, 2004.

Hopkins, R. E., in Kingslake (ed.), *Applied Optics and Optical Engineering*, Vol. 3, New York, Academic, 1966.

Hopkins, R. E., *Handbook of Optical Design* (MIL-HDBK-141), Washington, U.S. Government Printing Office, 1962.

Kingslake, R., *Applied Optics and Optical Engineering*, Vol. 5, New York, Academic, 1969.

Kingslake, R., *Optical System Design,* New York, Academic, 1983.

Pegis, R., and M. Rao, "Mirror Systems," *Applied Optics,* Vol. 2, 1963, Optical Society of America, Washington, D.C., pp. 1271–1274.

Smith, W., "Image Formation: Geometrical and Physical Optics," in W. Driscoll (ed.), *Handbook of Optics,* New York, McGraw-Hill, 1978.

Southall, J., *Mirrors, Prisms, and Lenses,* New York, Dover, 1964.

Walles, S., and R. Hopkins, "Image Orientation" in *Applied Optics,* Vol. 3, Optical Society of America, Washington, D.C., 1964, pp. 1447–1452.

Wolfe, W. L., "Nondispersive Prisms," in *Handbook of Optics,* Vol. 2, New York, McGraw-Hill, 1995, Chap. 4.

Zissis, G. J., "Dispersive Prisms and Gratings," in *Handbook of Optics,* Vol. 2, New York, McGraw-Hill, 1995, Chap. 5.

Characteristics of the Human Eye

8.1 Introduction

A knowledge of the characteristics of the human eye is important to the practice of optical engineering because the majority of optical systems utilize the eye as the final element of the system in one way or another. Thus, it is vital that the designer of an optical system understand what the eye can and cannot accomplish. For example, if a visual optical system is required to recognize a certain size target or to measure to a certain degree of accuracy, the magnification of the image presented to the eye must be sufficient to allow the eye to detect the necessary details. On the other hand, it would be wasteful to design a system with a perfection of image rendition which the eye could not utilize.

The human eye is a living optical system and its characteristics vary widely from individual to individual. For a given individual, the characteristics may vary from day to day, indeed from hour to hour. Therefore, the data presented in this chapter must be considered as central values in a range of values; in fact, some data are useful only as an indication of the order of magnitude of a certain characteristic. The conditions under which the eye is used play a large role in determining the behavior of the eye and must *always* be taken into account.

In physiological optics, the unit of measure for the power of a lens or optical system is the *diopter,* the abbreviation for which is *D*. The diopter power of a lens is simply the reciprocal of its effective focal length, when the focal length is expressed in meters. For example, a lens with a 1-m focal length has a power of 1 diopter; a $\frac{1}{2}$-m focal length, 2 diopters; and a lens of 1-in focal length has a power of 40 diopters (or more exactly, 39.37 D). For a single surface, the dioptric power is given by $(n' - n)/R$, with R the radius in meters. A *1-diopter prism* produces a deviation of 1 cm in a 1-m distance, i.e., a deviation of 0.01 radians, or about 0.57 degrees.

8.2 The Structure of the Eye

The eyeball is a tough plasticlike shell filled largely with a jellylike substance under sufficient pressure to maintain its shape. It rides in a bony socket of the skull on pads of flesh and fat. It is held in place and rotated by six muscles.

Figure 8.1 is a horizontal section of the right eye; the nose is to the left of the figure. The outer shell (sclera) is white and opaque except for the cornea, which is clear. The *cornea* supplies most (about two-thirds) of the refractive power of the eye. Behind the cornea is the *aqueous humor,* which (as its name implies) is a watery fluid. The *iris,* which gives the eye its color, is capable of expanding or contracting to control the amount of light admitted to the eye. The pupil formed by the iris can range in diameter from 8 mm in very dim light to less than 2 mm under very bright conditions. The *lens* of the eye is a flexible capsule suspended by a multitude of fibers, or ligaments, around its periphery. The eye is focused by changing the shape of the lens. When the sphincter muscles to which the suspensory ligaments are connected are relaxed, the lens has its flattest shape and the normal eye is focused at infinity. When these muscles contract, the lens bulges, so that its radii are shorter and the eye is focused for nearby objects. This process is called *accommodation.*

Behind the lens is the *vitreous humor,* a material with the consistency of thin jelly. All of the optical elements of the eye are largely water; in

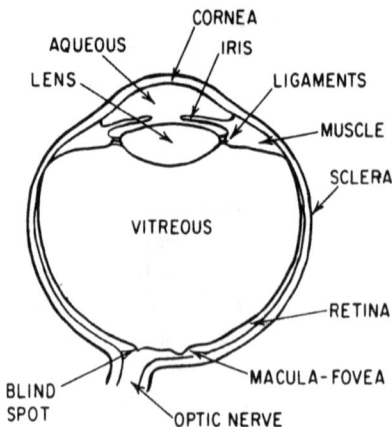

Figure 8.1 Schematic horizontal section of right eyeball (from above).

fact, a reasonable simulation of the optics of the eye can be made by considering the eye as a single refracting surface of water ($n_D = 1.333$, $V = 55$).

The following table lists typical values for the radii, thicknesses, and indices of the optical surfaces of the eye. These, of course, vary from individual to individual.

R_1 (air to cornea) + 7.8 mm	t_1 (cornea) 0.6	n_1 1.376	v_1 57
R_2 (cornea to aqueous) + 6.4 mm	t_2 (aqueous) 3.0	n_2 1.336	v_2 61
R_3 (aqueous to lens) + 10.1 mm	t_3 (lens) 4.0	n_3 1.386–1.406	v_3 48
R_4 (lens to vitreous) −6.1 mm	t_4 (vitreous) 16.9	n_4 1.337	v_4 61
R_5 (vitreous to retina) – 13.4 mm			

The principal points are located 1.5 and 1.8 mm behind the cornea, and the nodal points are 7.1 and 7.4 mm behind the cornea. The first focal point is 15.6 mm outside the eye; the second is, of course, at the retina. The distance from the second nodal point to the retina is 17.1 mm; thus the retinal size of an image can be found by multiplying the angular subtense of the object (from the first nodal point) by this distance. When the eye accommodates (focuses), the lens becomes nearly equiconvex with radii of about 5.3 mm, and the nodal points move a few millimeters toward the retina. The center of rotation of the eyeball is 13 to 16 mm behind the cornea.

An often overlooked fact is that the commonly accepted eye data tabulated above do not give an adequate picture of the quality of the visual system. First, the surfaces of the eye are not spherical. Some surfaces, especially those of the lens, depart significantly from true spheres. In general, the surface curvature tends to be weaker toward the margin of the surface. Second, the index of the lens is not uniform, but is higher in the central part of the lens. This sort of index gradient produces convergent refracting power in and of itself; it also reduces the surface refracting power at the margin of the lens. Note that both the gradient index and the surface asphericities introduce overcorrected spherical aberration, which offsets the undercorrected spherical of the outer surface of the cornea.

The *retina* contains blood vessels, nerve fibers, the light-sensitive rod and cone cells, and a pigment layer, in that order in the direction that the light travels. The optic nerve and the associated blind spot are located where the nerve fibers leave the eyeball and proceed to the brain. Slightly (about 5°) to the temporal (outer) side of the optical axis of the eye is the macula; the center of the macula is the fovea. At the fovea, the structure of the retina thins out and, in the central 0.3-mm diameter, only cones are present. The fovea is the center of sharp vision. Outside this area rods begin to appear; further away only rods are present.

There are about 7 million cones in the retina, about 125 million rods, and only about 1 million nerve fibers. The cones of the fovea are 1 to 1.5 μm in diameter and are about 2 to 2.5 μm apart. The rods are about 2 μm in diameter. In the outer portions of the retina, the sensitive cells are more widely spaced and are multiply connected to nerve fibers (several hundred to a fiber), accounting for the less distinct vision in this area of the retina. In the fovea, however, for some cones there is one cone cell per fiber; but there are 7 million cones and only one million nerve fibers.

The field of vision of an eye approximates an ellipse about 130° high by about 160° wide. The binocular field of vision, seen by both eyes simultaneously, is approximately circular and about 120° in diameter.

8.3 Characteristics of the Eye

Visual acuity

The characteristic of the eye which is probably of greatest interest to the optical engineer is its ability to recognize small, fine details. Visual acuity (VA) is defined and measured in terms of the angular size of the smallest character that can be recognized. The characters most frequently used to test VA are uppercase letters or a heavy ring with a break in the outline. Many uppercase letters can be considered as made up of five elements; e.g., the letter E has three bars and two spaces. Visual acuity is the reciprocal of the angular size (in minutes of arc) of one of the elements of the letter. "Normal" VA is considered to be 1.0, i.e., when the smallest recognizable letter subtends an angular height of 5 minutes from the eye and each element of the letter subtends 1 minute. Acuity is frequently expressed as the ratio between the distance to the target (usually 20 ft) and the distance at which the target element would subtend 1 minute. Thus, a VA of one-half, or 20/40, indicates that the minimum recognizable letter subtends 10 minutes and its elements 2 minutes. In the Landolt broken ring test, the width of the ring and the width of the break correspond to the letter element size, and recognition consists of determining the orientation of the break. Visual acuity may reach 2 (or 3 in unusual individuals) under ideal conditions.

As indicated above, the "normal" visual acuity is 1 minute, and this is also the value for the angular resolution of the eye which is conventionally assumed in connection with the design of optical instruments. Note however that a resolution of one line *pair* (or one cycle) per minute of arc actually corresponds to a VA of 2, or 20/10. However, this is the value of VA under what might be termed "normal conditions," and it is the value *only* for that part of the field of view which corresponds to the fovea of the retina. Outside the fovea, the acuity drops rapidly, as indicated in Fig. 8.2, which is a logarithmic plot of visual acuity (relative to

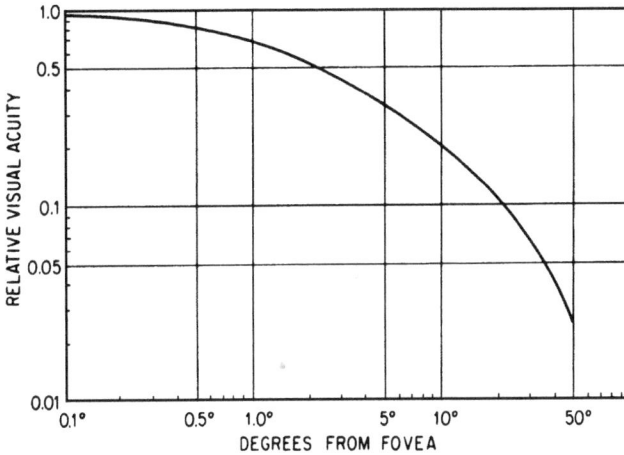

Figure 8.2 The variation of visual acuity (relative to the fovea) with the retinal position of the image. Note that because of the logarithmic scales of the figure, the falloff in visual acuity is far more rapid than the shape of the curve might indicate.

that at the fovea, which is arbitrarily set at unity) versus the angular position of the test target in the field of view. Also note that the vertical VA is 5 to 10 percent higher than horizontal and that the horizontal and vertical VA are about 30 percent higher than oblique (45°) VA.

As the brightness of a scene is diminished, the iris opens wider and the rods take over from the cones. At low illuminations, the eye is color blind and the fovea becomes a blind spot, since the cones lack the necessary sensitivity to respond to low levels of illumination. One result of this process is that the visual acuity drops as the illumination drops. This relationship is plotted in Fig. 8.3, which also indicates the normal pupil size. Note that the brightness of the area surrounding the test target affects the acuity. A uniform illumination seems to maximize the acuity. Figure 8.4 shows that, as might be expected, reducing the contrast of the target will also reduce the acuity.

Because the eye has about 0.75 D of chromatic aberration (C-light to F-light; it is about 3 D from 380 nm to 780 nm), VA is affected by the wavelength of light illuminating the target. Normally, VA is given for white light. In monochromatic light, the acuity is very slightly higher for the yellow and yellow-green wavelengths and slightly lower for red wavelengths. In blue (or far red) light, VA may be 10 to 20 percent lower, and in violet light the reduction in VA is 20 to 30 percent. The chromatic of the eye can be corrected or doubled (by external lenses) without detection; a quadrupling is noticeable. The effect of the chromatic aberration on the acuity of the eye is less than one might expect because the slightly yellow lens blocks out the ultraviolet, and the macula lutea (which is

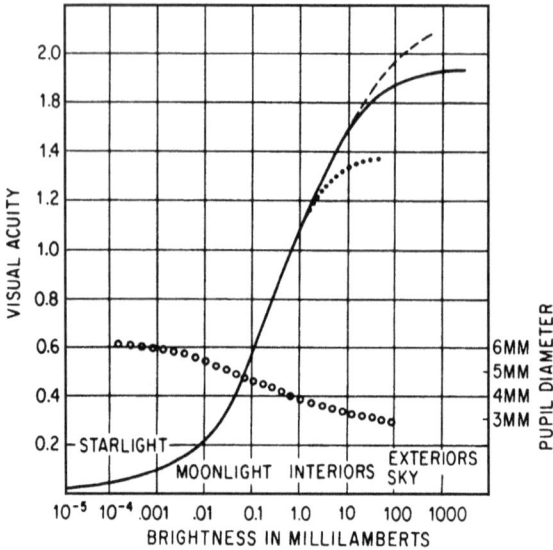

Figure 8.3 Visual acuity as a function of object brightness. Visual acuity in reciprocal minutes. The dashed and dotted lines show the effect of increased and decreased (respectively) surround brightness (1 millilambert is approximately the brightness of a perfect diffuser illuminated by 1 footcandle). The open circle curve indicates the diameter of the pupil; pupil diameters are larger in the young and smaller in the old, especially at lower brightnesses.

Latin for yellow spot) filters out the blue and violet light; the spectral response function of the eye is as shown in Figs. 8.8, 8.9 and 8.10.

Other types of acuity

Vernier acuity is the ability of the eye to align two objects, such as two straight lines, a line and a cross hair, or a line between two parallel lines. In making settings of this type, the eye is extremely capable. In instrument design, it can be safely assumed that the average person

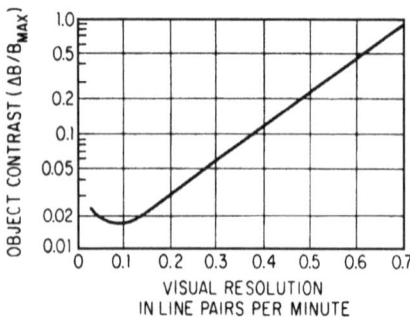

Figure 8.4 The object contrast ($\Delta B/B_{max}$) necessary for the eye to resolve a pattern of alternating bright and dark bars of equal width. Note that this curve shifts upward in reduced light levels and drops as the light level is increased. For this plot the bright bars had a brightness of $B_{max} = 23$ footlamberts.

can repeat vernier settings to better than 5 seconds of arc and that he or she will be accurate to about 10 seconds of arc. Exceptional individuals may do as well as 1 or 2 seconds. Thus, the vernier acuity is 5 or 10 times the visual acuity. Vernier acuity is best when setting one line abutting between two, next best setting a line on cross hairs or aligning two butting lines, and less effective in superimposing two lines.

The narrowest black line on a bright field that the eye can detect subtends an angle of from $\frac{1}{2}$ to 1 second of arc. In conditions of reversed contrast, i.e., a bright line or bright spot, the size of the line is not as important as its brightness. The governing factor is the amount of energy which reaches and triggers the retinal cell into responding. The minimum level seems to be 50 to 100 quanta incident on the cornea (only a few percent of the energy incident on the cornea actually reaches the cell).

The eye is capable of detecting angular motion to the order of 10 seconds of arc. The slowest motion that the eye will detect is 1 or 2 minutes of arc per second of time. At the other extreme, a point moving faster than 200° per second will blur into a streak.

The eyes judge distance from a number of clues. Accommodation, convergence (the turning in of the eyes to view a near object), haze, perspective, experience, etc., each play a part. Three-dimensional, or stereo, vision results from the separation of the two eyes, which causes each eye to see a slightly different picture of an object. The amount of stereo parallax which can be detected is as small as 2 to 4 seconds. In a clueless surround, a test subject can adjust two rods to be equidistant to within about 1 in when the rods are 20 ft away. The detectable ΔD in millimeters is approximately the square of the distance in meters (D^2).

The ability of the eye to detect flicker is a function of the brightness of the scene. The critical flicker fusion frequency is the frequency at which the flicker ceases to be detected. At a low scene brightness (e.g., 0.001 to 0.01 cd/m^2) the flicker fusion frequency (*FFF*) is about 10 or 11 cycles per second. At a high brightness (e.g., 10 cd/m^2) the *FFF* rises to 40 to 50 cycles per second. The *FFF* is lower in the outer portions of the field of view. The *FFF* is what determines the minimum acceptable shutter frequency in movie projection and the refresh rate for television.

Sensitivity

The lowest level of brightness which can be seen or detected is determined by the light level to which the eye has become accustomed. When the illumination level is reduced, the pupil of the eye expands, admitting more light, and the retina becomes more sensitive (by switching from cone vision to rod vision and also by an electrochemical mechanism involving rhodopsin, the visual purple pigment). This process is called dark adaptation. Figure 8.5 illustrates the adaptation process as a function of the length of time that the eye is in darkness. The "fovea only" curve indicates that after 5 or 10 minutes, the level of brightness detectable by the portion of the retina used for distinct vision is as low as it will ever get. At lower levels of illumination, only the outer portions of the retina are useful; the fovea becomes a blind spot. Figure 8.5 is for a target which subtends about 2°; the threshold brightness is lower for larger targets and higher for smaller targets. As indicated by the dashed lines, the conditions of the test have a great bearing on the threshold of vision, and the data of Fig. 8.5 should be regarded as indicating only an order of magnitude for the threshold.

The eye is a poor photometer; it is very inaccurate at judging the absolute level of brightness. However, it is an excellent instrument for comparison purposes, and can be used to match the brightness or color of two adjacent areas with a high degree of precision. Figure 8.6 indicates the brightness difference that the eye can detect as a function of the absolute brightness of the test areas. At ordinary brightness levels, a brightness difference of about 1 or 2 percent is detectable. (Note that in comparison photometry, in which the eye is called upon to match two areas, the precision of setting is increased by making a series of readings. In half the readings, the brightness of the variable area is raised until an apparent match is obtained; in the other half of the readings, the brightness is lowered to obtain the apparent match. The average is then much more accurate than either set.) Contrast sensitivity is best when there is no visible dividing line between the two areas under comparison. When the areas are separated, or if the demarcation between areas is not distinct, contrast sensitivity drops markedly.

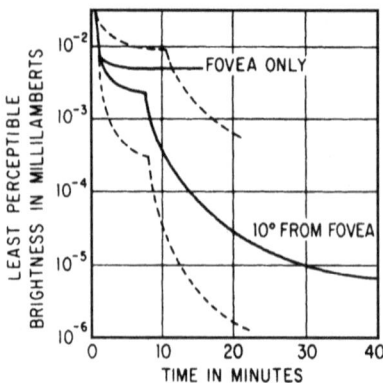

Figure 8.5 The threshold of vision. The minimum brightness perceptible drops sharply with time as the eye adapts itself to darkness. The upper and lower dashed curves show the effect of high and low illumination levels (respectively) before adaptation begins. For areas subtending more than 5° the threshold is almost constant, but rises rapidly as target size is reduced. Curves shown are for a target subtending about 2°.

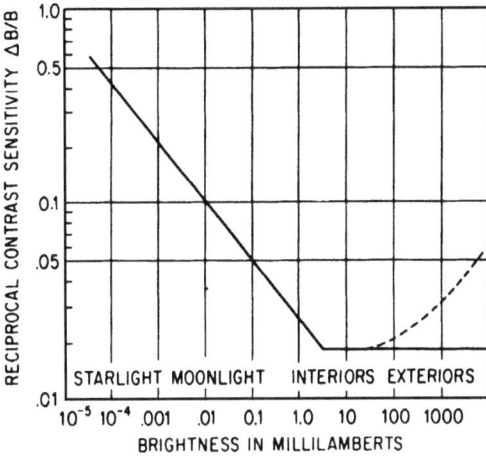

Figure 8.6   The contrast sensitivity of the eye as a function of field brightness. The smallest perceptible difference in brightness between two adjacent fields (ΔB) as a fraction of the field brightness B remains quite constant for brightnesses above 1 millilambert if the field is large. The dashed line indicates the contrast sensitivity for a dark surrounded field. (1 millilambert is approximately the brightness of a perfect diffuser illuminated by 1 footcandle, i.e., 1 foot-lambert.)

Figure 8.7 indicates the capability of the normal eye as a comparison colorimeter. Again, the eye is poor at determining the absolute wavelength of a color but quite good at determining a color match; wavelength differences of a few nanometers are detectable under suitable conditions. The comments of the preceding paragraph regarding dividing lines between test areas apply to color sensitivity as well.

The sensitivity of the eye to light is a function of the wavelength of the light. Under normal conditions of illumination, the eye is most

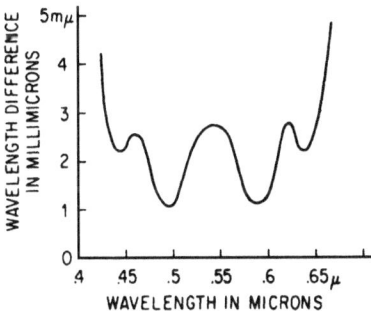

Figure 8.7   Sensitivity of the eye to color differences. The amount by which two colors must differ for the difference to be detectable in a side-by-side comparison is plotted as a function of the wavelength. Some data indicates a more uniform sensitivity of about twice that shown here.

sensitive to yellow-green light at a wavelength of 0.55 μm, and its sensitivity drops off on either side of this peak. For most purposes the sensitivity of the eye may be considered to extend from 0.4 to 0.7 μm. (or 0.38 to 0.78) Thus, in designing an optical instrument for visual use, the monochromatic aberrations are corrected for a wavelength of 0.55 or 0.59 μm and chromatic aberration is corrected by bringing the red and blue wavelengths to a common focus. The wavelengths usually chosen are either e(0.5461 μm) or d(0.5876 μm) for the yellow, C(0.6563 μm) for the red, and F(0.4861 μm) for the blue.

Figure 8.8 shows the sensitivity of the eye as a function of wavelength for normal levels of illumination and also for the dark-adapted eye. The photopic curve applies for brightness levels of 3 cd/m² or more, and the scotopic curve applies for brightness levels of 0.003 cd/m² or less. Between these levels, the term "mesopic" is used. Notice that the peak sensitivity for the dark-adapted eye shifts toward the blue end of the spectrum, from 0.55 μm to a value near 0.51 μm. This "Purkinje shift" is due to the differing chromatic sensitivities of the rods and cones of the retina, as shown in Fig. 8.8. Figure 8.9 is a tabulation of the values used in plotting Fig. 8.8.

The sensitivity of the eye has been measured in the infrared out to 1100 nm, where it is less than 10^{-11} of the sensitivity at the peak response (at 555 nm). Figure 8.10 is a plot of both the photopic and scotopic spectral sensitivity which also has the integral of both plots, showing the fraction of the total response due to wavelengths shorter than the indicated one. This curve allows one to determine the relative response between two wavelengths or to determine a suitable weighting for the wavelengths used in the analysis of an optival system's performance.

The *Troland* is a unit of retinal illumination and equals:

$$\text{Troland} = (\text{Object luminance in cd/m}^2)/(\text{Pupil area in mm}^2)$$

$$= 278 \ \tau \ (\text{illuminance in lm/m}^2)$$

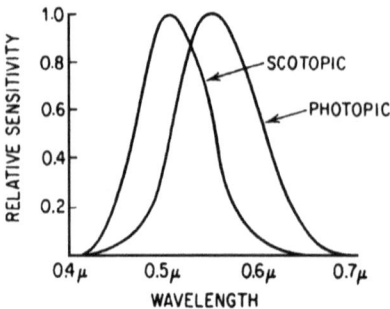

Figure 8.8 The relative sensitivity of the eye to different wavelengths for normal levels of illumination (photopic vision) and under conditions of dark adaptation (scotopic vision).

Wavelength, μm	Photopic	Scotopic	Wavelength, μm	Photopic	Scotopic
0.39	0.0001	0.0022	0.59	0.7570	0.0655
0.40	0.0004	0.0093	0.60	0.6310	0.0332
0.41	0.0012	0.0348	0.61	0.5030	0.0159
0.42	0.0040	0.0966	0.62	0.3810	0.0074
0.43	0.0116	0.1998	0.63	0.2650	0.0033
0.44	0.0230	0.3281	0.64	0.1750	0.0015
0.45	0.0380	0.4550	0.65	0.1070	0.0007
0.46	0.0600	0.5672	0.66	0.0610	0.0003
0.47	0.0910	0.6756	0.67	0.0320	0.0001
0.48	0.1390	0.7930	0.68	0.0170	0.0001
0.49	0.2080	0.9043	0.69	0.0082	0.0000
0.50	0.3230	0.9817	0.70	0.0041	
0.51	0.5030	0.9966	0.71	0.0021	
0.52	0.7100	0.9352	0.72	0.0010	
0.53	0.8620	0.8110	0.73	0.0005	
0.54	0.9540	0.6497	0.74	0.0003	
0.55	0.9950	0.4808	0.75	0.0001	
0.56	0.9950	0.3288	0.76	0.0001	
0.57	0.9520	0.2076	0.77	0.0000	
0.58	0.8700	0.1212			

Figure 8.9   The standard relative luminosity factors (relative sensitivity or response) for photopic and scotopic conditions.

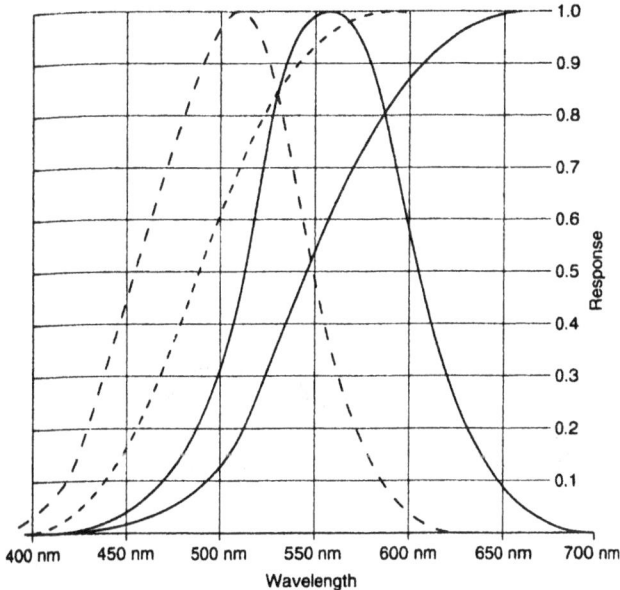

Figure 8.10   The spectral response of the human eye. The solid lines are the photopic (daylight) response, and the dashed lines are the scotopic (dark-adapted) response. There are two curves for each response: one is the relative response at the indicated wavelength; the other is the (integrated) fraction of the response for wavelengths shorter than that indicated.

8.4   Defects of the Eye

Nearsightedness (*myopia*) is a defect of focus resulting from too much power in the lens and cornea and/or too long an eyeball. The result is that the image of a distant object falls ahead of the retina and cannot be focused sharply. Since myopia results from an excessive amount of positive power, it can be corrected by placing a negative lens before the eye. The power of the negative lens is chosen so that its image is formed at the most distant point on which the myopic eye can focus. For example, a person with 2 diopters of myopia cannot see clearly beyond $\frac{1}{2}$ m (20 in), and a -2 diopter lens (focal length $= -\frac{1}{2}$ m or -20 in) is used to correct for this amount of myopia. The onset of myopia frequently coincides with adolescence, when growth is most rapid.

Instrument myopia occurs when an observer (especially an untrained observer) focuses an optical instrument such as a microscope or telescope. There is a tendency to focus the instrument so that the image appears to be about 20 in (2 diopters) away. This may be due to the observer's perception that the image is inside the instrument and therefore should be nearby. Most experienced observers will focus an instrument much nearer to an infinity setting. They do this by moving the microscope toward the object to focus, so that the image is behind the viewer's eye (and thus well out of focus) until it is in focus. Instrument myopia may be related to *night myopia,* where, in the dark and with no stimulus, the eye apparently also focuses at a near distance (60 to 80 in).

Farsightedness (*hyperopia*) is the reverse of myopia and results from too short an eye and/or too little power in the refracting elements of the eye. The image of a distant object is formed (when the eye is relaxed) behind the retina. Hyperopia can be corrected by the use of a positive spectacle lens. Obviously farsighted individuals can, to the extent that their power of accommodation will allow, refocus their eyes to bring the image onto the retina. If prolonged, this may cause headaches.

Astigmatism is a difference in the power of the eye from meridian to meridian and usually results from an imperfectly formed cornea, which has a stronger radius in one direction than in the other. Astigmatism of the eye is corrected by the use of toroidal surfaces on the spectacle lenses.

Astigmatism "with the rule" has a stronger cornea radius in the vertical meridian than in the horizental meridian.

A *contact lens,* placed in contact with the surface of the cornea, effectively changes the curvature of the outer surface of the eye (where most of the visual refractive power occurs). A rigid contact lens can easily correct astigmatism by replacing the toroidal surface of the cornea with its own spherical surface. Obviously, a soft (flexible) contact lens requires an orientation mechanism to align its toroidal power with that of the eye. Myopia and hyperopia can be corrected with contact lenses which effectively flatten or strengthen the curvature of the outer surface of the visual optical system.

Radial keratotomy is a surgical technique where radial cuts are made in the cornea (through most of its thickness). This weakens the cornea, and the internal pressure of the eye causes it to bulge in the region of the cuts, thus changing the shape and the power of the cornea. Two obvious drawbacks to this procedure are light scattering from the corneal scars left by the cuts, and the fact that the power of the eye tends to change as one ages, so that the correction may not be permanent. Another technique (PRK) involves a change in corneal shape by sculpting using laser ablation. LASIK slices a thin flap of the cornea off and then ablates the cornea to change its shape; the flap is then replaced.

The chromatic aberration of the eye was discussed in Sec. 8.3; many eyes have some undercorrected spherical aberration as well. The lens of the eye has aspheric surfaces and a higher index of refraction in the central core of the lens than in the outer portions; both of these factors reduce the power of the system at the margin of the lens and tend to correct the heavy undercorrected spherical from the cornea. A few persons have overcorrected spherical. In most people, the spherical tends toward overcorrection with accommodation, since the lens bulges more at the center than at the edge when the eye focuses on a near point. As much as ±2 diopters of spherical have been measured; however, like chromatic aberration, spherical seems to have little effect on the resolution of the eye.

Presbyopia is the inability to accommodate (focus) and results from the hardening of the material of the lens which comes with age. Figure 8.11 indicates the (typical) relationship between age and the power of accommodation. When the eye can no longer accommodate to reading

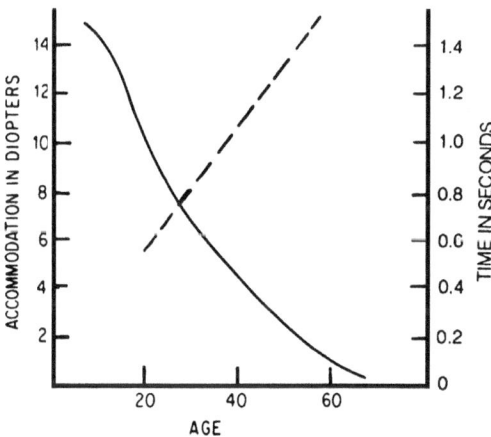

Figure 8.11 The variation of accommodation power with age (solid line). The dashed line indicates the time in seconds to accommodate to 1.3 diopters.

distance (2 or 3 diopters), it is necessary to wear positive lenses to read comfortably.

Keratoconus is a conically shaped cornea and can be corrected by contact lenses which effectively overlay a new spherical surface on the cornea.

An opaque or cloudy lens (*cataract*) is frequently removed surgically to restore vision. The resultant loss of power can be made up by an extremely strong positive spectacle lens. But better solutions are a contact lens or by surgically implanting a plastic intraocular lens near the iris. Such an aphakic eye, lacking a lens, cannot accommodate. Also, the change in retinal image size due to the shift in refractive power from inside to outside the eye (if due to the strong spectacle lens) will preclude binocular vision if only one eye is lensless.

Aniseikonia is the name given to a disparity in retinal image size from one eye to the other, occurring in otherwise normal eyes, and results in lack of binocular vision if the disparity is larger than a few percent. Aniseikonia can be corrected by special thick meniscus lenses or airspaced doublets which are effectively low-power telescopes the magnifications of which balance out the difference in retinal image size.

In instrument design, a number of additional factors should be taken into consideration, especially for binocular instruments. An adjustment must be provided for the variation in interpupillary distance, so that both sides of the instrument can be aligned with the pupils of the eyes. This distance is typically about $2\frac{1}{2}$ in, but it ranges from 2 to 3 in. Both halves of a binocular instrument must have the same magnification (within $\frac{1}{2}$ to 2 percent, depending on the individual's tolerance) and both halves must have their axes parallel (to within $\frac{1}{4}$ prism diopter vertically, $\frac{1}{2}$ diopter divergence, and 1 diopter convergence). Each side must be independently focusable to allow for variations in focus between the two eyes. A focus adjustment of ± 4 diopters will take care of the requirements of all but a few percent of the population; ± 2 diopters will satisfy about 85 percent. The depth of field of the eye (the distance on either side of the point of best focus through which vision is distinct) is about $\pm \frac{1}{4}$ diopter. The Rayleigh quarter wave (see Chap. 11) depth of focus is $\pm 1.1/(\text{pupil diameter})^2$ diopters, which for a 3-mm pupil works out to $\pm \frac{1}{8}$ diopter. For biocular devices, such as head-up displays (HUDs), the angular disparity between the eyes should be less than 0.001 radians.

Bibliography

Adler, F., *Physiology of the Eye—Clinical Applications,* St. Louis, Mosby, 1959.

Alpern, M., "The Eyes and Vision," in W. Driscoll (ed.), *Handbook of Optics,* New York, McGraw-Hill, 1978.

Blaker, W., "Ophthalmic Optics," in Shannon and Wyant (eds.), *Applied Optics and Optical Design,* Vol. 9, New York, Academic, 1983.

Charman, W. N., "Optics of the Eye," *Handbook of Optics,* Vol. 1, New York, McGraw-Hill, 1995, Chap. 24.

Davson, H., *The Physiology of the Eye,* London, Blakiston, 1950.

Dudley, L., "Stereoscopy," in Kingslake (ed.), *Applied Optics and Optical Engineering,* Vol. 2, New York, Academic, 1965.

Fry, G., "The Eye and Vision," in Kingslake (ed.), *Applied Optics and Optical Design,* Vol. 2, New York, Academic, 1965.

Geisler, W. S., and M. S. Banks, "Visual Performance," *Handbook of Optics,* Vol. 1, New York, McGraw-Hill, 1995, Chap. 25.

Hartridge, H., *Recent Advances in the Physiology of Vision,* London, Blakiston, 1950.

Kingslake, R., *Optical System Design,* New York, Academic, 1983.

Lueck, I., "Spectacle Lenses," in Kingslake (ed.), *Applied Optics and Optical Engineering,* Vol. 3, New York, Academic, 1966.

Mouroulis, P., *Visual Instrumentation,* New York, McGraw-Hill, 1999.

Richards, "Visual Optics," in MIL-HDBK-141, *Optical Design,* Washington, Defense Supply Agency, 1962.

Schwiegerling, J., *Field Guide to Visual and Ophthalmic Opitcs,* SPIE, 2004.

Westheimer, J., JOSA, v62, p-1502 (Dec 1952).

Zoethout, W., *Physiological Optics,* Professional Press, 1939.

Experiments

Here are a few experiments which can demonstrate some of the characteristics of your eyes.

1. Usually one eye is dominant. For most (but not all) of us it's the right eye. Extend your arm and point your finger at an object, so that the finger is aligned with the object. Cover or close one eye at a time and note which eye shows the alignment, and which eye shows the finger misaligned. The aligned eye is the dominant one.

2. You can demonstrate the "blind spot" in the retina where the nerve fibers and vessels exit the eye. Cover or close your left eye. Stare at the **O** below. Adjust the distance between the page and your eye until the **X** diappears. The **X** is now on the blind spot. If your left eye is dominant, stare at the **X**. Although it's not easy, you can demonstrate this without the **X** and **O**. Stare straight ahead with your left eye covered and move your finger off to the right. When you find the right location, your finger tip will disappear.

O   **X**

3. Measure the space between the **O** and the **X** and the distance from your eye to the page. Calculate the distance at the retina between the blind spot and the fovea.

4. This one is difficult. Cover one eye and stare at a single letter in a printed page. Without allowing your gaze to wander, consider the definition of the letters around the one at which you are staring. This illustrates that the field over which you see most distinctly is

very small. (Actually, for the very best vision, it's about a half a degree.) You might also try it with the numbers below. Stare at the **x**.

> 9123456789123456789123456789123456789
>
> 9123456789123456789123456789123456789
>
> 9123456789123456789123456789123456789
>
> 123456789123456789**X**123456789123456789
>
> 9123456789123456789123456789123456789
>
> 9123456789123456789123456789123456789
>
> 9123456789123456789123456789123456789

5. In part, vision is a learned skill. You probably feel that you see your entire field of vision quite distinctly. Try this:
   (a) Stare straight ahead and extend your arm out to the side with your fingers spread out. **Q:** How many fingers do you see? **A:** The proper answer is "What fingers?" If you wiggle them or wave you will see motion, buy you can't resolve the fingers.
   (b) Now extend your arm at about 45° to your line of sight. Now you can detect fingers but you can't tell how many.
   (c) Extend your arm straight ahead and, with one eye, stare at your thumb. You are now quite sure there are several fingers, but you still can't tell how many.

Exercises

1 What power telescope is necessary to enable a person with "normal" vision to read letters 1-mm high at a distance of 300 ft? (Note: 1 minute of arc is 0.0003 radians.)

ANSWER: One millimeter at 300 ft is 0.04 in at 3600 in, which subtends 0.0000111 radians. For "normal" vision we need the letter to subtend 5 minutes, or 0.0015 radians. Thus the magnification power must be 0.0015/0.0000111, or 135×.

2 What power corrective lens would be prescribed for a nearsighted person who could not focus clearly on an object more than 5 in away from his eye?

ANSWER: A lens with a focal length of -5 in will produce a virtual image of a distant object at 5 in from the lens. Five inches is about 1/8th of a meter, and such a lens has a power of $1/f = 1/(-1/8) = -8$ diopters. However, since the lens will not be in contact with the eye, the lens focal length plus the lens-to-eye distance must equal 5 in. A spacing of 1.5 in would require a -3.5 in focal length, or -88.9 mm, or -0.0889 m, and its power is -11.2 diopters.

3 Assuming a depth of focus of $\pm^{1}/_{4}$ diopter, over what range of distance is vision clear when the eye is focused at 10 in?

ANSWER: A distance of 10 in is about $^{1}/_{4}$ m, and may be expressed as 4 diopters, so the depth of focus, expressed in diopters is 3.75 to 4.25 diopters. The corresponding distances are 1/3.75 = 0.2667 m and 1/4.25 = 0.2353 m. Thus the depth of focus is 0.2667 − 0.2353 = 0.0314 m = 31.4 mm = 1.235 in.

4 It is desired to set an optical vernier to a precision of 0.0001 in. Assuming that the vernier projects an image of a ruled scale on to a screen which is viewed from a distance of 10 in, and that the setting is made by aligning a scale line with a cross hair on the screen, what magnification must the projection lens of the optical vernier have? Use 10 seconds of arc for the vernier acuity of the eye. (Note: 1 second equals 0.000005 radians.)

ANSWER: At a distance of 10 in, 10 seconds of arc equals (10 × 0.000005) × 10 in = 0.0005 in. For this to be equivalent to 0.0001 in at the scale, the scale must be magnified by 0.0005/0.0001 = 5.0 ×.

5 A convex reflector with a radius of curvature = 10 in is mounted on a spindle and rotated. (a) What is the largest amount that its center of curvature can be displaced from the axis of rotation without the motion of the reflected image being detected by the naked eye? (Assume that the reflected image is viewed from a distance of 10 in.) (b) What are the fastest and slowest speeds of rotation at which the motion caused by a decentration of 0.02 in can be detected?

ANSWER: (a) If the displacement is e, then the total up and down is $2e$, and viewed from a distance of 10 in the motion subtends an angle of 0.2e radians. If the eye can detect a motion of 10 seconds, or 0.000050 radians, then 0.2e equals 0.000050, and e equals 0.000050/0.2 = 0.000250 in. (b) If the displacement is 0.02 in and the image rotates in a circle, the image travels π×2×0.02 = 0.12 in per revolution of 360°. At a viewing distance of 10 in this is 0.012 radians per revolution. If the slowest detectable motion is one (or 2) minutes of arc per second (0.0003 rad/sec), then the rate of rotation is (0.0003 rad/sec)/(0.012 rad/rev), or 0.024 rev/sec = 1.4 rev/min (or 2.9 rpm if we use 2 min). If the fastest visible is 200°/sec (= 3.5 rad/sec), then the rate of rotation is 3.5/0.012 = 280 rps.

6 (a) If a plane parallel plate is specified to have zero ±10 millidiopters power, what is the shortest tolerable focal length it may have? (b) Assuming that one surface is truly flat, what is the strongest (shortest) acceptable radius for the other surface, if the index is 1.6? (c) If the piece has a diameter of 20 mm, how many Newton's rings will be visible when this surface is tested against a true flat? (Assume a wavelength of 555 nm. One fringe occurs for each half wavelength change in the thickness of the airspace.)

ANSWER: (a) 10 millidiopters = 0.010 diopter. The focal length is equal to
1/0.010 = 100 m. (b) Using Eq. 3.25, $\varphi = (n - 1)(c_1 - c_2)$ assuming that c_2 is
zero, solving for c_1 and inverting to get R_1, we get

$$R_1 = f(n - 1) = 100 \times 0.6 = 60 \text{ m}.$$

(c) The sagittal height of a sphere is approximately SH = $y^2/2R$ = $10^2/120{,}000$ =
0.000833 mm. This is equal to 0.000833/0.000550 = 1.515 wavelengths. Since
we get one fringe for each half wavelength, there will be 3.03 fringes.

9

Stops, Apertures, Pupils and Diffraction

9.1 Introduction

In every optical system, there are apertures (or stops) which limit the passage of energy through the system. These apertures are the clear diameters of the lenses and diaphragms in the system. One of these apertures will determine the diameter of the cone of energy which the system will accept from an axial point on the object. This is termed the *aperture stop,* and its size determines the illumination (irradiance) at the image. Another stop may limit the size or angular extent of the object which the system will image. This is called the *field stop.* The importance of these stops to the photometry (radiometry) and performance of the system cannot be overemphasized.

The elements of an inexpensive camera system are sketched in Fig. 9.1 and illustrate both aperture and field stops in their most basic forms. The diaphragm in front of the lens limits the diameter of the bundle of rays that the system can accept and is thus the aperture stop. The mask adjacent to the film determines the angular field coverage of the system and is quite apparently the field stop of the camera.

Not all systems are as obvious as this, however, and we will now consider more complex arrangements. Because the theory of stops is readily explained by the use of a concrete example, the following discussions will be with reference to Fig. 9.2, which is a highly exaggerated sketch of a telescopic system focused on an object at a finite distance. The system shown consists of an objective lens, erector lens, eyelens, and two internal diaphragms. The objective forms an inverted image

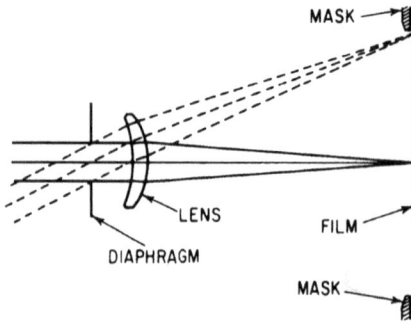

Figure 9.1 The elements of a simple box camera illustrate the functions of elementary aperture and field stops (the diaphragm and mask, respectively).

of the object. This image is then reimaged at the first focal point of the eyelens by the erector lens, so that the eyelens forms the final image of the object at infinity, where it can comfortably be viewed by the eye.

9.2   The Aperture Stop and Pupils

By following the path of the axial rays (designated by solid lines) in Fig. 9.2, it can be seen that diaphragm #1 is the aperture of the system which limits the size of the axial cone of energy from the object. All of the other elements of the system are large enough to accept a bigger cone. Thus, diaphragm #1 is the aperture stop of the system.

The oblique ray through the center of the aperture stop is called the *principal,* or *chief, ray,* and is shown in the figure as a dashed line. The *entrance* and *exit pupils* of the system are the images of the aperture stop in object and image space, respectively. That is, the entrance pupil is the image of the aperture stop as it would be seen if viewed from the axial point on the object; the exit pupil is the aperture stop image as it would be seen if viewed from the final image plane (in this case, at an infinite distance). In the system of Fig. 9.2, the entrance pupil lies near the objective lens and the exit pupil lies to the right of the eyelens.

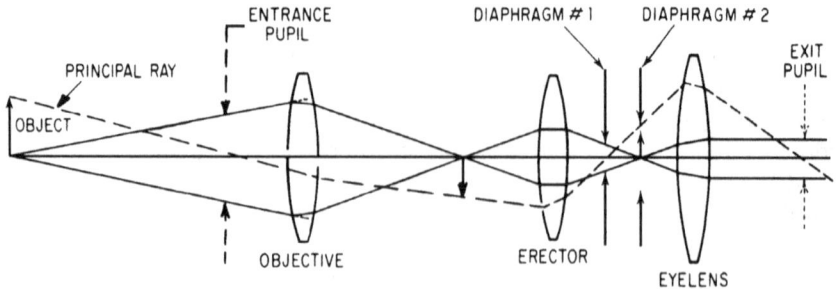

Figure 9.2   Schematic sketch of an optical system to illustrate the relationships between pupils, stops, and fields.

Notice that the initial and final intersections of the dashed principal ray with the axis locate the pupils, and that the diameter of the axial cone of rays at the pupils indicates the pupil diameters. It can be seen that, for any point on the object, the amount of radiation accepted by, and emitted from, the system is determined by the size and location of the pupils. Note that any image of the aperture stop is a pupil. In Fig. 9.2 there is a pupil at diaphragm #1.

9.3   The Field Stop

By following the path of the principal ray in Fig. 9.2, it can be seen that another principal ray starting from a point in the object which is farther from the axis would be prevented from passing through the system by diaphragm #2. Thus, diaphragm #2 is the field stop of this system. The images of the field stop in object and image space are called the *entrance* and *exit windows,* respectively. In the system of Fig. 9.2, the entrance window is coincident with the object and the exit window is at infinity (which is coincident with the image). Note that the windows of a system do not coincide with the object and image unless the field stop lies in the plane of a real image formed by the system.

The angular field of view is determined by the size of the field stop, and is the angle which the entrance or exit window subtends from the entrance or exit pupil, respectively. The angular field in object space is frequently different from that in image space. (Alternate definition: the angular field of view is the angle subtended by the object or image from the first or second nodal point of the system, respectively. Thus, for nontelescopic systems in air, object and image field angles are equal according to this definition. Note that this definition cannot be applied to an afocal system, which has no nodal or principal points.)

9.4   Vignetting

The optical system of Fig. 9.2 was deliberately chosen as an ideal case in which the roles played by the various elements of the system are definite and clear-cut. This is not usually the situation in real optical systems, since the diaphragms and lens apertures often play dual roles.

Consider the system shown in Fig. 9.3, consisting of two positive lenses, A and B. For the axial bundle of rays, the situation is clear; the aperture stop is the clear aperture of lens A, the entrance pupil is at A, and the exit pupil is the image, formed by lens B, of the diameter of lens A.

Some distance off the axis, however, the situation is markedly different. The cone of energy accepted from point D is limited on its lower edge by the lower rim of lens A and on its upper edge by the upper rim

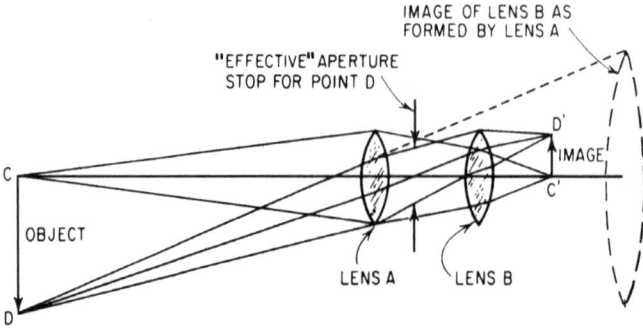

Figure 9.3 Vignetting in a system of separated components. The cone of rays from point D is limited by the lower rim of lens A and the upper rim of B, and is smaller than the cone accepted from point C. Note that the upper ray from D just passes through the image of lens B which is formed by lens A.

of lens B. The size of the accepted cone of energy from point D is less than it would be if the diameter of lens A were the only limiting agency. This effect is called *vignetting,* and it causes a reduction in the illumination at the image point D'. It is apparent that for some object point still farther from the axis than point D, no energy at all would pass through the system; thus there is no field stop per se in this system as shown.

The appearance of the system when viewed from point D is shown in Fig. 9.4. The entrance pupil has become the common area of two circles, one the clear diameter of lens A, and the other the diameter of lens B as imaged by lens A. The dashed lines in Fig. 9.3 indicate the location and size of this image of B, and the arrows indicate the "effective" aperture stop which has a size, shape, and position completely different than that for the axial case.

In a photographic lens with an adjustable iris diaphragm, its location should be such that when stepped down to a small diameter, its clear aperture is centered in the vignetted oblique beam.

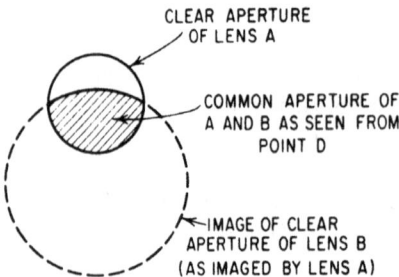

Figure 9.4 The apertures of the optical system of Fig. 9.3 as they are seen from point D.

The determination of vignetting in an optical system can be accomplished by tracing just two paraxial rays. Begin by collecting the powers, spaces, and clear apertures of the system. Then trace an axial paraxial ray from the foot of the object with a height of 1.0 at the first element. Calculate y/ca for each element and aperture; the diameter with the largest y/ca is the aperture stop. Multiply the raytrace data (y and u) by ca/y to get the raytrace data of the marginal ray. Trace an oblique ray through the center of the aperture stop at a convenient slope, say 0.1. Calculate y_p/ca for each element and aperture. The one with the largest value is the field stop. Scale the ray data by ca/y_p to get the data of the principal ray. The intersection of this ray with the object and image planes gives the size of the field. The two rays can be combined as described in Sec. 4.2 to obtain the data of any third ray, without executing another raytrace. Of course the upper and lower rim ray data is simply ($y_p \pm y$) and ($u_p \pm u$). If the ray height exceeds the clear aperture, vignetting occurs. The amount of vignetting is indicated by what fraction of the axial ray height will create a height which does not exceed the clear aperture when combined with the principal ray height.

9.5   Glare Stops, Cold Stops, and Baffles

A *glare stop* is essentially an auxiliary diaphragm located at an image of the aperture stop for the purpose of blocking out stray radiation. Depending on the system application, a glare stop may be called a *Lyot stop*, or in an infrared system, a *cold stop*. Figure 9.5 shows an erecting telescope in which the primary aperture stop is at the objective lens. Energy from sources outside the desired field of view, passing through the objective and reflecting from an internal wall, shield, or supporting member, can create a glare which reduces the contrast of the image formed by the system.

In a long wavelength infrared system, the housing itself may be a source of unwanted thermal radiation. This radiation can be blocked

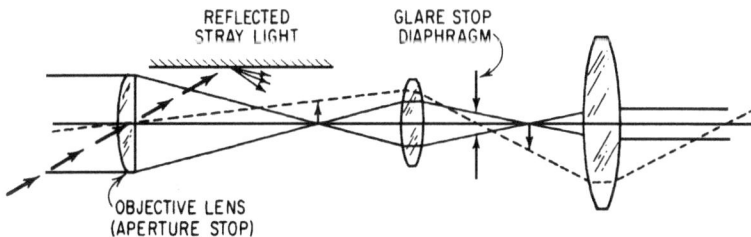

Figure 9.5   Stray light reflected from an inside wall of the telescope, is intercepted by the glare stop, which is located at the internal image of the objective lens.

out by an internal diaphragm which is an accurate image of the objective aperture. This "cold stop" is usually cooled and is located inside the evacuated detector Dewar.

Since the stray radiation will appear to be coming from the wall, and thus from outside the objective aperture, it will be imaged on the opaque portion of a diaphragm which is located at an accurate image of the objective aperture. Another glare stop could conceivably be located at the exit pupil of this particular system, since it is real and accessible; however, it would make visual use of the instrument quite inconvenient.

In most systems the aperture stop is located at or very near the objective lens. This location gives the smallest possible diameter for the objective, and since the objective is usually the most expensive component (per inch of diameter), minimizing its diameter makes good economic sense. In addition, there are often aberration considerations which make this a desirable location. However, there are some systems, such as scanners, where the need to minimize the size and weight of the scanner mirror makes it necessary to put the stop or a pupil at the scanner mirror rather than at the objective. This causes the objective to be larger, more costly, and more difficult to design.

In an analogous manner, field stops for this system could be placed at both internal images to further reduce stray radiation. The principle here is straightforward. Once the primary field and aperture stops of a system are determined, auxiliary stops may be located at images of the primary stops to cut out glare. If the glare stops are accurately located and are the same size as the images of the primary stops (or slightly larger), they do not reduce the field or illumination, nor do they introduce vignetting.

Baffles are often used to reduce the amount of radiation that is reflected from walls, etc., in a system. Figure 9.6 shows a simple radiometer consisting of a collector lens and a detector in a housing. Assume that radiation from a powerful source (such as the sun) outside the field of view reflects from the inner walls of the mount on to the detector and obscures the measurement of radiation from the desired target, as shown in the upper half of the sketch. Under these conditions, there is no possibility of using an internal glare stop (since there is no internal image of the entrance pupil) and the internal walls of the mount must be baffled as shown in the lower half of the sketch (although an external hood or sunshade could also be used if circumstances permit).

The key to the efficient use of baffles is to arrange them so that no part of the detector can "see" a surface which is *directly* illuminated. The method of laying out a set of baffles is illustrated in Fig. 9.7. The dotted lines from the rim of the lens to the edge of the detector indicate the necessary clearance space, into which the baffles cannot intrude without

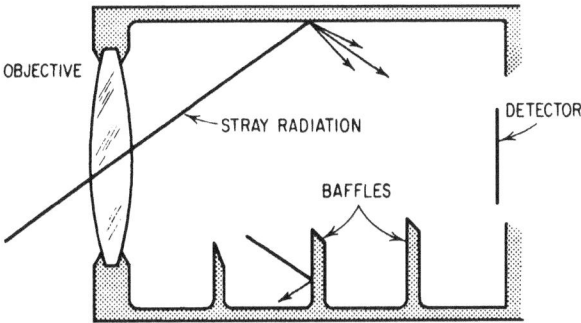

Figure 9.6 Stray (undesired) radiation from outside the useful field of this simple radiometer can be reflected from the inner walls of the housing and degrade the function of the system. Sharp-edged baffles, shown in the lower portion, trap this radiation and prevent the detector from "seeing" a *directly* illuminated surface.

obstructing part of the radiation from the desired field of view. The dashed line AA' is a "line of sight" from the detector to the point on the wall where the extraneous radiation begins. The first baffle is erected to the intersection of AA' with the dotted clearance line. Solid line BB' indicates the path of stray light from the top of the lens to the wall. The area from baffle #1 to B' is thus shadowed and "safe" for the detector to "see." The dashed line from B' to A is thus the safe line of sight, and baffle #2 located at the intersection of AB' and the clearance line will prevent the detector from "seeing" the illuminated wall beyond B'. This procedure is repeated until the entire side wall is protected. Note that the inside edges of the baffles should be sharp and their surfaces rough and blackened.

The cast and machined baffles shown in Fig. 9.6 are obviously expensive to fabricate. Less expensive alternatives include washers

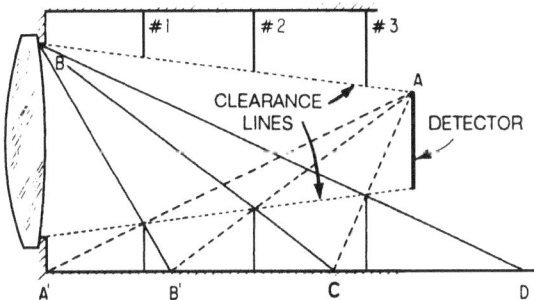

Figure 9.7 Construction for the systematic layout of baffles. Note that baffle #3 shields the wall back to point D; thus, all three baffles could be shifted forward somewhat, so that their coverages overlap.

constrained between spacers, or stamped, cup-shaped washers which can be cemented or press-fitted into place. This type of baffling is not necessary in all cases. Frequently, internal scattering can be sufficiently reduced by scoring or threading the offending internal surfaces of the mount. In this way, the reflections are broken up and scattered, reducing the amount of reflection and destroying any glare images. The use of a flat black paint is also highly advisable, although care must be taken to be sure that the paint remains both matte and black at near-grazing angles of incidence and at the application wavelength. Some black paints are light grey in the IR. Sandblasting to roughen the surface and blackening (for aluminum, black anodizing works well) is a simple and usually effective treatment. Another treatment is the application of black "flocked" paper. This can be procured in rolls, cut to size, and cemented to the offending surfaces; this is especially useful for large internal surfaces and for laboratory equipment. The source of glare light can be indentified by placing the eye at a location in the image which should be dark and looking back into the optics. For a projection system place the eye just outside the field.

Specialized flat black paints are available for specific applications and wavelengths. In the absence of special paints, Floquil brand flat black model locomotive paint usually can be found at the local hobby shop and makes a pretty good general-purpose flat black. A specialized anodizing process, Martin Optical Black (or Martin Infrablack for the infrared) is extremely effective (<0.2 percent reflective) but is very fragile.

9.6 The Telecentric Stop

A telecentric system is one in which the entrance pupil and/or the exit pupil is located at infinity. A telecentric stop is an aperture stop which is located at a focal point in an optical system. It is widely utilized in optical systems designed for metrology (e.g., comparators and contour projectors and in microlithography) because it tends to reduce the measurement or position error caused by a slight defocusing of the system. Figure 9.8a shows a schematic telecentric system. Note that the dashed principal ray is parallel to the axis to the left of the lens. If this system is used to project an image of a scale (or some other object), it can be seen that a small defocusing displacement of the scale does not change the height on the scale at which the principal ray strikes, although it will, of course, blur the image. Contrast this with Fig. 9.8b where the stop is at the lens, and the defocusing causes a proportional error in the ray height. The telecentric stop is also used where it is desired to project the image of an object with depth (along the axis), since it yields less confusing images of the edges of such an object.

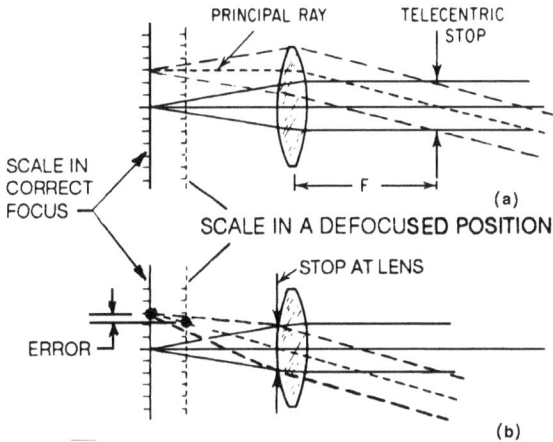

Figure 9.8 The telecentric stop is located at the focal point of the projection system shown, so that the principal ray is parallel to the axis at the object. When the object is slightly out of focus (dotted) there is no error in the size of the projected image as there is in the system with the stop at the lens, shown in the lower sketch.

9.7   Apertures and Image Illumination— ƒ-Number and Cosine-Fourth

ƒ-Number

When a lens forms the image of an extended object, the amount of energy collected from a small area of the object is directly proportional to the area of the clear aperture, or entrance pupil, of the lens. At the image, the illumination (power per unit area) is inversely proportional to the image area over which this object is spread. Now the aperture area is proportional to the square of the pupil diameter, and the image area is proportional to the square of the image distances, or focal length (f). Thus, the square of the ratio of these two dimensions is a measure of the relative illumination produced in the image.

The ratio of the focal length to the clear aperture of a lens system is called the relative aperture, ƒ-number, or "speed" of the system, and (other factors being equal), the illumination in an image is inversely proportional to the square of this ratio. The relative aperture is given by:

$$f/\# = f\text{-number} = \text{efl/clear aperture} \qquad (9.1)$$

As an example, an 8-in focal length lens with a 1-in clear aperture has an ƒ-number of 8; this is customarily written ƒ/8 or ƒ:8.

Another way of expressing this relationship is by the *numerical aperture* (usually abbreviated as N.A. or NA), which is the index of refraction (of the medium in which the image lies) times the sine of the half angle of the cone of illumination.

$$\text{Numerical aperture} = NA = n' \sin U' \qquad (9.2)$$

Numerical aperture and *f*-number are obviously two methods of defining the same characteristic of a system. Numerical aperture is more conveniently used for systems that work at finite conjugates (such as microscope objectives), and the *f*-number is appropriately applied to systems for use with distant objects (such as camera lenses and telescope objectives). For aplanatic systems (i.e., systems corrected for coma and spherical aberration) with infinite object distances, the two quantities are related by:

$$f\text{-number} = \frac{1}{2NA} \qquad (9.3)$$

The terms "fast" and "slow" are often applied to the *f*-number of an optical system to describe its "speed." A lens with a large aperture (and thus a small *f*-number) is said to be fast, or to have a high speed. A smaller aperture lens is described as slow. This terminology derives from photographic usage, where a larger aperture allows a shorter (or faster) exposure time to get the same quantity of energy to the film and may allow a rapidly moving object to be photographed without blurring.

It should be apparent that a system working at finite conjugates will have an object-side numerical aperture as well as an image-side numerical aperture and that the ratio NA/NA′ = (object-side NA)/ (image-side NA) must equal the absolute value of the magnification. The term "working *f*-number" is sometimes used to describe the numerical aperture in *f*-number terms. If we use the terms "infinity *f*-number" for the *f*-number defined in Eq. 9.1, then the image-side working *f*-number is equal to the infinity *f*-number times $(1 - m)$, where m is the magnification.

Another term that is occasionally encountered is the *T-stop,* or *T*-number. This is analogous to the *f*-number, except that it takes into account the transmission of the lens. Since an uncoated, many element lens made of exotic glass may transmit only a fraction of the light that a low-reflection coated lens of simpler construction will transmit, such a speed rating is of considerable value to the photographer. The relationship between *f*-number, *T*-number, and transmission is

$$T\text{-number} = \frac{f\text{-number}}{\sqrt{\text{transmission}}} \qquad (9.4)$$

Cosine-to-the-fourth

For off-axis image points, even when there is no vignetting, the illumination is usually lower than for the image point on the axis. Figure 9.9 is a schematic drawing showing the relationship between exit pupil and image plane for point A on axis and point H off axis. The illumination at an image point is proportional to the solid angle which the exit pupil subtends from the point.

The solid angle subtended by the pupil from point A is the area of the exit pupil divided by the square of the distance OA. From point H, the solid angle is the projected area of the pupil divided by the square of the distance OH. Since OH is greater than OA by a factor equal to $1/\cos \theta$, this increased distance reduces the illumination by a factor of $\cos^2 \theta$. The exit pupil is viewed obliquely from point H, and its projected area is reduced by a factor which is *approximately* $\cos \theta$. (This is a fair approximation if OH is large compared to the size of the pupil; for high-speed lenses used at large obliquities, it may be subject to significant errors. See Example A in Chap. 12 for an exact expression.)

Thus the illumination at point H is reduced by a factor of $\cos^3 \theta$. This is, however, true for illumination on a plane normal to the line OH (indicated by the dashed line in Fig. 9.9). We want the illumination in the plane AH. An illumination of x lumens per square foot on the dashed plane will be reduced on plane AH because the same number of lumens is spread over a greater area in plane AH. The reduction factor is $\cos \theta$, and combining all the factors we find that

$$\text{Illumination at } H = \cos^4 \theta \text{ (illumination at } A) \qquad (9.5)$$

The importance of this effect on wide-angle lenses can be judged from the fact that $\cos^4 30° = 0.56$, $\cos^4 45° = 0.25$, and $\cos^4 60° = 0.06$. It can be seen that the illumination on the film in a wide-angle camera may fall off quite rapidly.

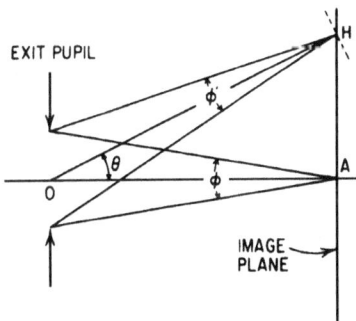

EXIT PUPIL

0

IMAGE
PLANE

Figure 9.9 Relationship between exit pupil and image points, used to demonstrate that the illumination at H is $\cos^4 \theta$ times that at A.

Note that the preceding has been based on the assumption that the pupil diameter is constant (with respect to θ) and that θ is the angle formed in image space (although it is often applied to the field angle in object space). The "cosine fourth law" can be modified if the construction of the lens is such that the apparent size of the pupil increases for off-axis points, or if a sufficiently large amount of barrel distortion is introduced to hold θ to smaller values than one would expect from the corresponding field angle in object space. Certain extreme wide-angle camera lenses make use of these principles to increase off-axis illumination. The $\cos^4$ effect is in addition to any illumination reduction caused by vignetting. It should be remembered that the cosine-fourth effect is *not* a "law" but a collection of four cosine factors which may or may not be present in a given situation.

9.8   Depth of Focus

The concept of depth of focus rests on the assumption that for a given optical system, there exists a blur (due to defocusing) of small enough size such that it will not adversely affect the performance of the system. The *depth of focus* is the amount by which the image may be shifted longitudinally with respect to some reference plane (e.g., film, reticle) and which will introduce no more than the acceptable blur. The *depth of field* is the amount by which the object may be shifted before the acceptable blur is produced. The size of the acceptable blur may be specified as the linear diameter of the blur spot (as is common in photographic applications) (Fig. 9.10) or as an angular blur, i.e., the angular subtense of the blur spot from the lens. Thus, the linear and angular blurs (B and β, respectively and the distance S *or* S' are related by

$$\beta = \frac{B}{S} = \frac{B'}{S'} \tag{9.6}$$

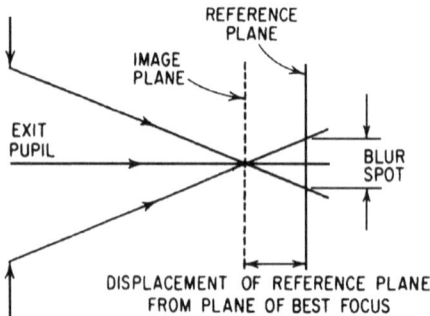

Figure 9.10   When an optical system is defocused, the image of a point becomes a blurred spot. The size of the blur is determined by the relative aperture of the system and the focus shift.

for a system in air, where the primed symbols refer to the image-side quantities.

Since depth of focus (δ) and depth of field (δ') are both longitudinal distances (in image and object spaces respectively) they are related by the longitudinal magnification, and

$$\delta' = \overline{m}\, \delta \approx m^2 \delta \tag{9.7}$$

The *hyperfocal distance* of a system is the distance at which the system must be focused so that the depth of field extends to infinity.

The idea of depth of focus was originally photographic, based on the concept that a defocus blur which is smaller than a silver grain in the film emulsion will not be noticeable. This concept also can be applied to pixel size in, for example, a charge-coupled device (CCD). If the acceptable blur diameter is B, then the depth of focus (at the image) is simply (see Fig. 9.10).

$$\delta' = \pm\, B(f\text{-number})$$

$$\delta' = \pm\, \frac{B}{2\mathrm{NA}} \tag{9.8}$$

The corresponding depth of field (at the object) is from D_{near} to D_{far}, where

$$D_{\text{near}} = \frac{fD\,(A + B)}{(fA - DB)} \tag{9.9}$$

$$D_{\text{far}} = \frac{fD\,(A - B)}{(fA + DB)} \tag{9.10}$$

and the hyperfocal distance is simply

$$D_{\text{hyp}} = \frac{-fA}{B} \tag{9.11}$$

where D = the nominal distance at which the system is focused (note
     that, by our sign convention, D is normally negative)
   A = the diameter of the entrance pupil of the lens
   f = the focal length of the lens
   B = the acceptable blur diameter

Note that there are several false assumptions here. We assume that the image is a perfect point, with no diffraction effects. We also assume that the lens has no aberrations and that the blurring on both sides of the focus is the same. None of these assumptions is correct, but the equations above *do* give a usable model for the depth of focus. In practice, the acceptable blur diameter B is usually determined empirically by examining a series of defocused images to decide the level of acceptability; the equations above are then fitted to the results.

9.9 Diffraction Effects of Apertures

Even if we assume that an infinitely small point source of light is possible, no lens system can form a true point image, even though the lens be perfectly made and absolutely free of aberrations. This results from the fact that light does not really travel in straight-line rays, but behaves as a wave motion, bending around corners and obstructions to a small but finite degree.

According to Huygen's principle of light-wave propagation, each point on a wave front may be considered as a source of spherical wavelets; these wavelets reinforce or interfere with each other to form the new wave front. When the original wave front is infinite in extent, the new wave front is simply the envelope of the wavelets in the direction of propagation.

As shown in Fig. 9.11, when a wavefront passes through an aperture the obscured portions of the wave no longer interact with the part of the wave which does go through the aperture. The result is that the wavefront changes shape by a small amount. According to a geometrical calculation, near the focus of a perfect lens the wavefront is a perfect sphere, and the rays (which are normal to the wavefront) all pass through the center of curvature of the sphere. But the diffraction effect of the aperture causes the wavefront to curve backward and the rays no longer all go through the point at the center of the sphere. In sum, for a circular aperture the illumination is distributed as shown in Fig. 9.15.

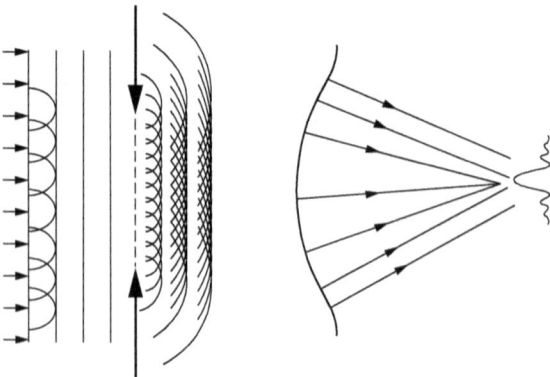

Figure 9.11 Diffraction of a wavefront by an aperture.

At an extreme, when the wave front is limited by an aperture to a very small size (say, to the order of a half wavelength), the new wave front becomes spherical about the aperture. Figure 9.12 shows a plane wavefront incident on a slit *AC,* which is in front of a perfect lens. The lens is focused on a screen, *EF.* We wish to determine the nature of the illumination on the screen. Since the lens of Fig. 9.12 is assumed perfect, the optical path lengths *AE, BE,* and *CE* are all equal and the waves will arrive in phase at *E,* reinforcing each other to produce a bright area. For Huygen's wavelets starting from the plane wave front in a direction indicated by angle α, the paths are different; path *AF* differs from path *CF* by the distance *CD.* If *CD* is an integral number of wavelengths, the wavelets from *A* and *C* will reinforce at point *F.* If *CD* is an odd number of half wavelengths, a cancellation will occur. The illumination at *F* will be the summation of the contributions from each incremental segment of the slit, taking the phase relationships into account. It can be readily demonstrated that when *CD* is an integral number of wavelengths, the illumination at *F* is zero, as follows: if *CD* is one wavelength, then *BG* is one-half wavelength and the wavelets from *A* and *B* cancel. Similarly, the wavelets from the points just below *A* and *B* cancel and so on down the width of the slit. If *CD* is *N* wavelengths, we divide the slit into 2*N* parts (instead of two parts) and apply the same reasoning. Thus, there is a dark zone at *F* when

$$\sin \alpha = \frac{\pm N\lambda}{w}$$

where *N* = any integer
  λ = the wavelength of the light
  w = the width of the slit

Thus, the illumination in the plane *EF* is a series of light and dark bands. The central bright band is the most intense, and the bands on either side are successively less intense. One can realize that the intensity should diminish by considering the situation when *CD* is

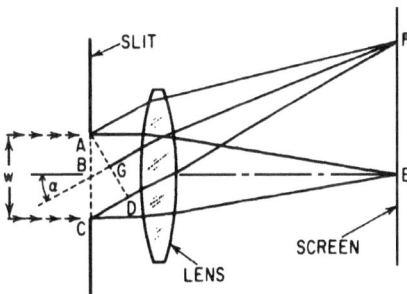

Figure 9.12 Plane wave front incident on a slit in front of a perfect lens.

1.5λ, 2.5λ, etc. When CD is 1.5λ, the wavelets from two-thirds of the slit can be shown (as in the preceding paragraph) to interfere and cancel out, leaving the wavelets from one-third of the aperture; when CD is 2.5λ, only one-fifth of the slit is uncanceled. Since the "uncanceled" wavelets are neither exactly in nor exactly out of phase, the illumination at the corresponding points on the screen will be less than one-third or one-fifth of that in the central band.

For a more rigorous mathematical development of the subject, the reader is referred to the references following this chapter. The mathematical approach is one of integration over the aperture, combined with a suitable technique for the addition of the wavelets which are neither exactly in nor exactly out of phase. This approach can be applied to rectangular and circular apertures as well as to slits.

For a rectangular aperture, the illumination on the screen is given by

$$I = I_0 \frac{\sin^2 m_1}{m_1^2} \cdot \frac{\sin^2 m_2}{m_2^2} \tag{9.12}$$

$$m_i = \frac{\pi w_i \sin \alpha_i}{\lambda} \qquad i = 1,2 \tag{9.13}$$

In these expressions λ is the wavelength, w the width of the exit aperture, α the angle subtended by the point on the screen, m_1 and m_2 correspond to the two principal dimensions, w_1 and w_2, of the rectangular aperture and I_0 is the illumination at the center of the pattern.

When the aperture is circular, the illumination is given by

$$I = I_0 \left[1 - \frac{1}{2}\left(\frac{m}{2}\right)^2 + \frac{1}{3}\left(\frac{m^2}{2^2 2!}\right)^2 - \frac{1}{4}\left(\frac{m^3}{2^3 3!}\right)^2 + \frac{1}{5}\left(\frac{m^4}{2^4 4!}\right)^2 - \cdots \right]^2$$

$$= I_0 \left[\frac{2J_1(m)}{m}\right]^2 \tag{9.14}$$

where m is given by Eq. 9.13 with the obvious substitution of the diameter of the circular exit aperture for the width, w, and $J_1(m)$ is the first order Bessel function. The illumination pattern consists of a bright central spot of light surrounded by concentric rings of rapidly decreasing intensity. The bright central spot of this pattern is called the *Airy disk*.

We can convert from angle α to Z, the radial distance from the center of the pattern, by reference to Fig. 9.13. If the optical system is reasonably aberration-free, then

$$l' = \frac{-w}{2 \sin U'}$$

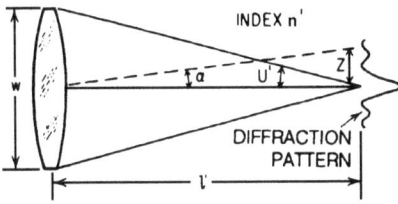

Figure 9.13 Illustrating the relations between α, U', Z, l' and w.

and to a close approximation, when α is small

$$Z = \frac{l'\alpha}{n'} = \frac{-\alpha w}{2n' \sin U'} \qquad (9.15)$$

The table of Fig. 9.14 lists the characteristics of the diffraction patterns for circular and slit apertures. The table is derived from Eqs. 9.12 and 9.14, but the data is given in terms of Z and $\sin U'$ rather than α and w. Note that $n' \sin U'$ is the numerical aperture (NA) of the optical system.

Notice that 84 percent of the energy in the pattern is contained in the central spot, and that the illumination in the central spot is almost 60 times that in the first bright ring. Ordinarily the central spot and the first two bright rings dominate the appearance of the pattern, the other rings being too faint to notice. The illumination distribution in a diffraction pattern is plotted in Fig. 9.15. One should bear in mind the fact that these energy distributions apply to perfect, aberration-free systems with circular or slit apertures which are uniformly transmitting and which are illuminated by wave fronts of uniform amplitude.

Ring (or band)	Circular Aperture			Slit Aperture	
	Z	Peak Illumination	Energy in Ring	Z	Peak Illumination
Central maximum	0	1.0	83.9%	0	1.0
1st dark ring	0.61 $\lambda/n'\sin U'$	0.0		0.5 $\lambda/n'\sin U'$	0.0
1st bright ring	0.82 $\lambda/n'\sin U'$	0.017	7.1%	0.72 $\lambda/n'\sin U'$	0.047
2d dark ring	1.12 $\lambda/n'\sin U'$	0.0		1.0 $\lambda/n'\sin U'$	0.0
2d bright ring	1.33 $\lambda/n'\sin U'$	0.0041	2.8%	1.23 $\lambda/n'\sin U'$	0.017
3rd dark ring	1.62 $\lambda/n'\sin U'$	0.0		1.5 $\lambda/n'\sin U'$	0.0
3rd bright ring	1.85 $\lambda/n'\sin U'$	0.0016	1.5%	1.74 $\lambda/n'\sin U'$	0.0083
4th dark ring	2.12 $\lambda/n'\sin U'$	0.0		2.0 $\lambda/n'\sin U'$	0.0
4th bright ring	2.36 $\lambda/n'\sin U'$	0.00078	1.0%	2.24 $\lambda/n'\sin U'$	0.0050
5th dark ring	2.62 $\lambda/n'\sin U'$			2.5 $\lambda/n'\sin U'$	0.0

Figure 9.14 Tabulation of the size of and distribution of energy in the diffraction pattern at the focus of a perfect lens.

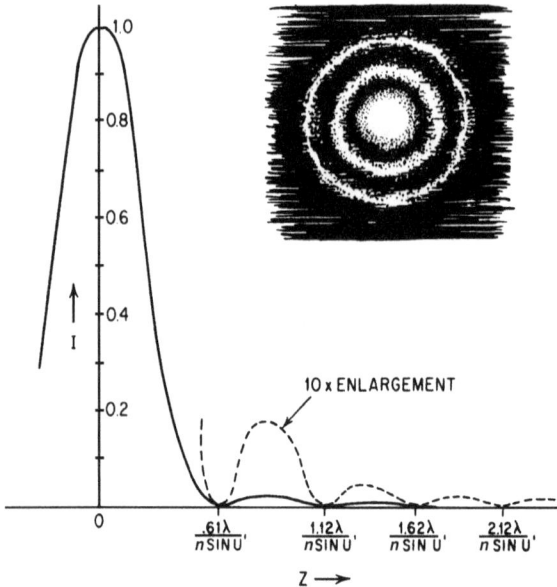

Figure 9.15 The distribution of illumination in the Airy disk. The appearance of the Airy disk is shown in the upper right.

The presence of aberrations will, of course, modify the distribution as will any nonuniformity of transmission or wave-front amplitude (see, for example, Sec. 9.11).

9.10   Resolution of Optical Systems

The diffraction pattern resulting from the finite aperture of an optical system establishes a limit to the performance which we can expect from even the best optical device. Consider an optical system which images two equally bright point sources of light. Each point is imaged as an Airy disk with the encircling rings, and if the points are close, the diffraction patterns will overlap. When the separation is such that it is just possible to determine that there are two points and not one, the points are said to be resolved. Figure 9.16 indicates the summation of the two diffraction patterns for various separations. When the image points are closer than $0.5\lambda/\text{NA}$ (NA is the numerical aperture of the system and equals $n' \sin U'$), the central maxima of both patterns blend into one and the combined patterns may appear to be due to a single source. At a separation of $0.5\lambda/\text{NA}$ the duplicity of the image points is detectable, although there is no minimum between the maxima from the two patterns. This is *Sparrow's criterion* for resolution. When

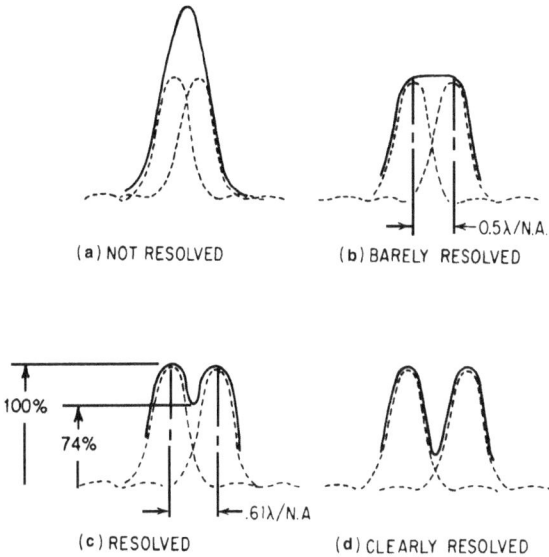

Figure 9.16 The dashed lines represent the diffraction patterns of two point images at various separations. The solid line indicates the combined diffraction pattern. Case (b) is the Sparrow criterion for resolution. Case (c) is the Rayleigh criterion.

the image separation reaches $0.61\lambda/NA$, the maximum of one pattern coincides with the first dark ring of the other and there is a clear indication of two separate maxima in the combined pattern. This is *Lord Rayleigh's criterion* for resolution and is the most widely used value for the limiting resolution of an optical system.*

From the tabulation of Fig. 9.14, we find that the distance from the center of the Airy disk to the first dark ring is given by

$$Z = \frac{0.61\lambda}{n' \sin U'} = \frac{0.61\lambda}{NA} = 1.22\lambda \ (f/\#) \qquad (9.16)$$

This is the separation of two image points corresponding to the Rayleigh criterion for resolution. This expression is widely used in determining the limiting resolution for microscopes and the like.

---

*The diffraction pattern of two point images will always differ somewhat from the diffraction pattern of a single point. It is thus possible to detect the presence of two points (as opposed to one) even in cases where the two points cannot be visually resolved or separated. This is the source of the occasional claims that a system "exceeds the theoretical limit of resolution." In Chap. 15 it is shown that there is a true limit on the resolution of a sinusoidal *line* target; the limit on the spatial frequency is $v_0 = 2NA/\lambda = 1/\lambda(f/\#)$.

For resolution at the image, the NA of the image cone is used; for res-olution at the object, the NA of the object cone is used.

To evaluate the performance limits of telescopes and other systems working at long object distances, an expression for the angular sepa-ration of the object points is more useful. Rearranging Eq. 9.15 and substituting the limiting value of Z from Eq. 9.16, we get, in radian measure,

$$\alpha = \frac{1.22\lambda}{w} \text{ radians} \tag{9.17}$$

For ordinary visual instruments, λ may be taken as 0.55 μm, and using $4.85 \cdot 10^{-6}$ radians for 1 second of arc, we find that

$$\alpha = \frac{5.5}{w} \text{ seconds of arc} \tag{9.18}$$

when w is the aperture diameter expressed in inches. By a series of careful observations, the astronomer Dawes found that two stars of equal brightness could be visually resolved when their separation was $4.6/w$ seconds. Notice that if the Sparrow criterion is used instead of the Rayleigh criterion in Eq. 9.18, the limiting resolution angle is $4.5/w$ seconds, which is in close agreement with Dawes' findings.

It is worth emphasizing here that the *angular* resolution limit is a direct function of wavelength and an inverse function of the aperture of the system. Thus, the limiting resolution is improved by reducing the wavelength or by increasing the aperture. Note that focal length or working distance do not directly affect the *angular* resolution. The *lin-ear* resolution is governed by the wavelength and the numerical aper-ture (NA or f-number), and *not* by the aperture diameter.

In an instrument such as a spectroscope, where it is desired to sep-arate one wavelength from another, the measure of resolution is the smallest wavelength difference, $d\lambda$, which can be resolved. This is usu-ally expressed as $\lambda/d\lambda$; thus, a resolution of 10,000 would indicate that the smallest detectable difference in wavelength was 1/10,000 of the wavelength upon which the instrument was set.

For a prism spectroscope, the prism is frequently the limiting aper-ture, and it can be shown that when the prism is used at minimum deviation, the resolution is given by

$$\frac{\lambda}{d\lambda} = B \frac{dn}{d\lambda} \tag{9.19}$$

where B is the length of the base of the prism and $dn/d\lambda$ is the dis-persion of the prism material.

A diffraction grating consists of a series of precisely ruled lines on a clear (or reflecting) base. Light can pass directly through a grating, but

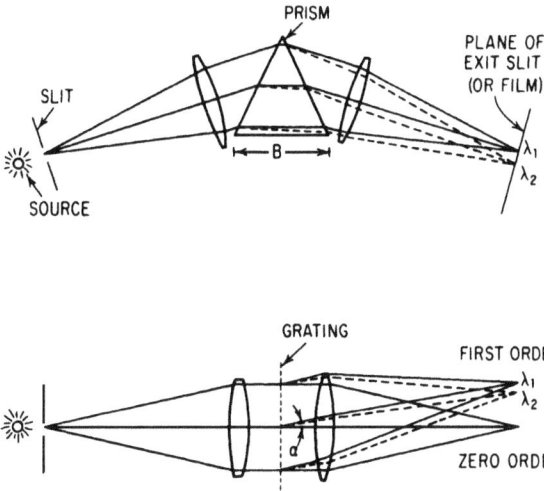

Figure 9.17 (Upper) Prism spectrometer. (Lower) Grating spectrometer.

it is also diffracted. As with the slit aperture discussed above, at certain angles the diffracted wavelets reinforce, and maxima are produced when

$$\sin \alpha = \frac{m\lambda}{S} \pm \sin I \qquad (9.20)$$

where λ is the wavelength, I is the angle of incidence, S is the spacing of the grating lines, m is an integer, called the *order* of the maxima, and the positive sign is used for a transmission grating, the negative for a reflecting. (Note that a sinusoidal grating has only a first order.) Since α depends on the wavelength λ, such a device can be used to separate the diffracted light into its component wavelengths. When used as indicated in Fig. 9.17, the resolution of a grating is given by

$$\frac{\lambda}{d\lambda} = mN \qquad (9.21)$$

where m is the order and N is the total number of lines in the grating (assuming the size of the grating to be the limiting aperture of the system).

9.11   Diffraction of a Gaussian (Laser) Beam

The illumination distribution in the image of a point as described in Secs. 9.9 and 9.10 was based on the assumptions that the optical system was perfect and that both the transmission and the wave-front amplitude were uniform over the aperture. Any change in the intensity distribution in the beam will change the diffraction pattern from that described above. Obviously, a similar change in the transmission of the aperture will produce the same effects.

A "gaussian beam" is one the intensity cross section of which follows the equation of a gaussian, $y = e^{-x^2}$. Laser output beams closely approximate gaussian beams. From mathematics we know that exponential functions, such as the gaussian are extremely resistant to transformations (consider, for example, the integral or differential of e^{-x}). Similarly, a gaussian beam tends to remain a gaussian beam (as long as it is "handled" by reasonably aberration-free optics) and the diffraction image of a point source also has a gaussian distribution of illumination.

The distribution of intensity in a gaussian beam is illustrated in Fig. 9.18 and can be described by Eq. 9.22.

$$I(r) = I_0 e^{-2r^2/w^2} \tag{9.22}$$

where $I(r)$ = the beam intensity at a distance r from the beam axis
$\quad\quad I_0$ = the intensity on axis
$\quad\quad r$ = the radial distance
$\quad\quad e$ = 2.718....
$\quad\quad w$ = the radial distance at which the intensity falls to I_0/e^2, i.e., to 13.5 percent of its central value. This is usually referred to as the beam width, although it is a semi-diameter. It encompasses 86.5 percent of the beam power.

Beam power

By integration of Eq. 9.22 we find the total power in the beam to be given by

$$P_{tot} = \tfrac{1}{2}\, \pi I_0 w^2 \tag{9.23}$$

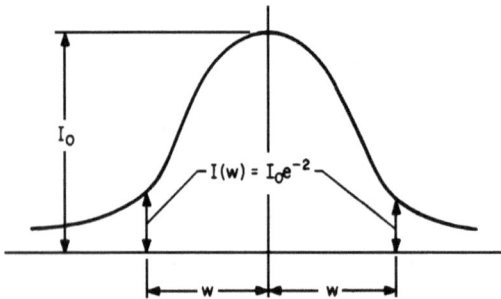

Figure 9.18 Gaussian beam intensity profile.

The power passed through a centered circular aperture of radius a is given by

$$P(a) = P_{tot}(1 - e^{-2a^2/w^2}) \tag{9.24}$$

The power passed by a centered slit of width $2s$ is given by

$$P(s) = P_{tot} \cdot \text{erf}\left(\frac{s\sqrt{2}}{w}\right) \tag{9.25}$$

where erf $(u) = \int_0^u e^{-t^2}\,dt =$ the error function, which is tabulated in mathematical handbooks.

Diffraction spreading of a gaussian beam

A gaussian beam has a narrowest width at some point, which is called the "waist." This point may be near where the beam is focused or near where it emerges from the laser. As the beam progresses away from the waist, it spreads out according to the following equation:

$$w_z^2 = w_0^2\left[1 + \left(\frac{\lambda z}{\pi w_0^2}\right)^2\right] \tag{9.26}$$

where $w_z =$ the semidiameter of the beam (i.e., to the $1/e^2$ points) at a longitudinal distance z from the beam waist.

$w_0 =$ the semidiameter of the beam (to the $1/e^2$ points) at the beam waist.

$\lambda =$ the wavelength

$z =$ the distance along the beam axis from the waist to the plane of w_z

At large distances it is convenient to know the angular beam spread. Dividing both sides of Eq. 9.26 by z^2, then, as z approaches infinity, we get

$$\frac{\alpha}{2} = \frac{w_z}{z}\Big|_{z\to\infty} = \frac{2\lambda}{\pi(2w_0)} \quad \text{or} \quad \alpha = \frac{4\lambda}{\pi(2w_0)} = \frac{1.27\lambda}{\text{diameter}} \tag{9.27}$$

where α is the angular beam spread in radians between the $1/e^2$ points. For many applications, the gaussian diffraction blur at the image plane can be found by simply multiplying α from Eq. 9.27 by the image conjugate distance (s' from Chap. 2).

The **Rayleigh Range** equals $\pm 4\lambda/\pi\theta^2$ where θ is the convergence/deviance angle of the beam. There the beam is 41 percent larger than at the waist.

Beam truncation

The effect of beam truncation, i.e., stopping down or cutting off the outer regions of the beam, is discussed by Campbell and DeShazer. They show that if the diameter of the beam is not reduced below $2(2w)$, where w is the beam semidiameter at the $1/e^2$ points, then the beam intensity distribution remains within a few percent of a true gaussian distribution. If the clear aperture is reduced below this value, it will introduce structure (i.e., rings) into the irradiance patterns, and the pattern gradually approaches Eq. 9.14 as the aperture is reduced.

A lens aperture large enough to pass a beam with a diameter of $4w$ is obviously very inefficient from a radiation transfer standpoint. For this reason, most systems truncate the beam, very often to the $1/e^2$ diameter, and the diffraction pattern is altered accordingly. If the beam is truncated down to 61 percent of the $1/e^2$ diameter, it is difficult to see the difference from a uniform beam, with the ring disk and rings.

Size and location of a new waist formed by a perfect optical system

When a gaussian beam passes through an optical system, a new waist is formed. Its size and location are determined by diffraction (and not by the paraxial equations of Chap. 2). The "waist" and "focus" are at different locations; in a weakly convergent beam, the separation may be large. The following equations allow calculation of the new waist size and location:

$$x' = \frac{-xf^2}{x^2 + \left(\dfrac{\pi w_1^2}{\lambda}\right)^2} \tag{9.28}$$

$$w_2^2 = \frac{f^2 w_1^2}{x^2 + \left(\dfrac{\pi w_1^2}{\lambda}\right)^2} = w_1^2\left(\frac{x'}{-x}\right) \tag{9.29}$$

where w_1 = the radius (to the $1/e^2$ points) of the original waist
$\quad\quad w_2$ = the radius of the new waist formed by the optical system
$\quad\quad f$ = the focal length of the lens
$\quad\quad x$ = the distance from the first focal point of the lens to the plane of w_1
$\quad\quad x'$ = the distance from the second focal point of the lens to the plane of w_2

Note that x and x' are usually negative and positive, respectively. Note also the similarity to the newtonian paraxial equation (Eq. 2.3).

Two points regarding the above are well worth emphasizing. First, laser researchers speak in terms of a "beam waist." Note that in the equations above and in common usage it is described as a radial dimension, not a diameter; the *diameter* of the waist is $2w$. Second, the waist and the focus are not the same thing, as a comparison of Eqs. 9.28 and 2.3 will indicate. In most circumstances the difference is trivial and gaussian beams may be handled by the usual paraxial equations. But when the beam convergence is small (i.e., with an *f*-number of a hundred or so), it is possible to distinguish both a focus and a separate beam waist. For example, if we project a 1-in laser beam (through a focusable beam expander) on a screen about 50 ft away, we can focus the beam to get the smallest possible spot on the screen. The *focus* is now at the screen. However, there is a location a few feet short of the screen at which a smaller beam diameter exists. This is the beam *waist*; it can be demonstrated by moving the screen (or a sheet of paper) toward the laser and observing the reduction of the spot size. Note that with the screen now at this beam waist position, the beam expander can be refocused to get a still smaller spot on the screen. Then there will be a new waist still closer to the laser, etc., etc., etc.

Note well that the *focus* is the smallest spot which can be produced on a surface at a given, fixed distance. The *waist* is the smallest diameter in the beam (see Gaskill, p. 435).

Note also that all the phenomena described in this section result from the gaussian distribution of beam intensity and *not* from the fact that the source may be a laser. The same effects could be produced by a radially graded filter placed over the aperture of the system. (The temporal and spatial coherence of a laser beam are, of course, what make it practical to demonstrate these effects.)

9.12   The Fourier Transform Lens and Spatial Filtering

In Fig. 9.19 we have a transparent object located at the first focal point of lens A. As indicated by the dashed rays in the figure, lens A images the object at infinity so that the rays originating at the axial point of the object are collimated. These rays are brought to a focus at the second focal plane of lens B, where the image of the object is located.

Now let us realize that the Fourier theory allows us to consider the object as comprised of a collection of sinusoidal gratings of different frequencies, amplitudes, phases, and orientations. If our object is a simple linear grating with but a single spatial frequency, it will deviate the light through an angle α according to Eq. 9.20, except that a

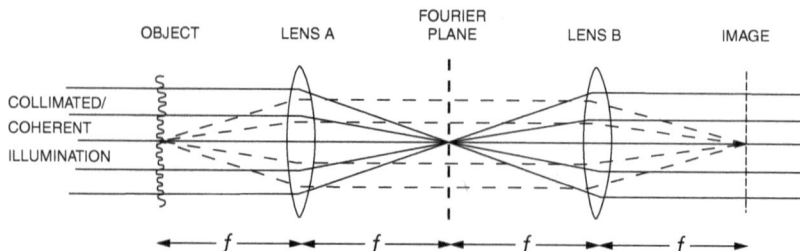

Figure 9.19   An illustration with a transparent object located at the first focal point of a lens.

sinusoidal grating has but a single diffraction order, the first. Now, if the object is illuminated by collimated/coherent light, that diffracted light will be focused as two points in the second focal plane of lens A (which is indicated as the Fourier plane, midway between the lenses in Fig. 9.19). The points will be laterally displaced by $\delta = f \tan \alpha$ from the nominal focus. Thus, if an annular zone in the Fourier plane is obstructed, all the spatial information of the frequency corresponding to the radius of the obstruction will be removed (filtered) from the final image. Thus it can be seen that the Fourier plane constitutes a sort of map of the spatial frequency content of the object and that this content can be analyzed or modified in this plane.

Bibliography

Campbell, J., and L. DeShazer, *J. Opt. Soc. Am.,* Vol. 59, 1969, pp. 1427–1429.

Gaskill, J., *Linear Systems, Fourier Transforms, and Optics,* New York, Wiley, 1978.

Goodman, J., *Introduction to Fourier Optics,* New York, McGraw-Hill, 1968.

Hardy, A., and F. Perrin, *The Principles of Optics,* New York, McGraw-Hill, 1932.

Jacobs, D., *Fundamentals of Optical Engineering,* New York, McGraw-Hill, 1943.

Jenkins, F., and H. White, *Fundamentals of Optics,* New York, McGraw-Hill, 1976.

Kogelnick, H., in Shannon and Wyant (eds.), "Laser Beam Propagation" *Applied Optics and Optical Engineering,* Vol. 7, New York, Academic, 1979.

Kogelnick, H., and T. Li, *Applied Optics,* 1966, pp. 1550–1567.

Pompea, S. M., and R. P. Breault, "Black Surfaces for Optical Systems," in *Handbook of Optics,* Vol. 2, New York, McGraw-Hill, 1995, Chap. 37.

Silfvast, W. T., "Lasers," in *Handbook of Optics,* Vol. 1, New York, McGraw-Hill, 1995, Chap. 11.

Smith, W., in W. Driscoll (ed.), *Handbook of Optics,* New York, McGraw-Hill, 1978.

Smith, W., in Wolfe and Zissis (eds.), *The Infrared Handbook,* Office of Naval Research, 1985.

Stoltzman, D., in Shannon and Wyant (eds.), *Applied Optics and Optical Design,* Vol. 9, New York, Academic, 1983.

Strong, J., *Concepts of Classical Optics,* New York, Freeman, 1958.

Walther, A., in Kingslake (ed.), "Diffraction" *Applied Optics and Optical Engineering,* Vol. 1, New York, Academic, 1965.

Exercises

1   Find the position and diameter of the entrance and exit pupil of a 100-mm focal length lens with a diaphragm located 20 mm to the right of the lens. The lens diameter is 15 mm and the diaphragm diameter is 10 mm.

ANSWER: Obviously the exit pupil is actually the diaphragm, so it is 10 mm in diameter and is located 20 mm to the right of the lens.

The entrance pupil is the image of the diaphragm formed by the lens. Using Eq. 2.5, and turning the system around, we have $s = -20$, and

$$s' = sf/(s + f) = -20 \times 100/(-20 + 100) = -2000/80 = -25$$

and if we turn the system back to its original orientation, the pupil is 25 mm to the right of the lens. The magnification is $m = s'/s = -25/-20 = +1.25$, so the entrance pupil diameter is $1.25 \times 10 = 12.5$.

2   What is the relative aperture of the lens of exercise #1 with light incident (a) from the left, and (b) from the right?

ANSWER: (a) f-number $= 100/12.5 = f/8$

(b) f-number $= 100/10 = f/10$

3   A telescope is composed of an objective lens, $f = 100$ in, diameter $= 1$ in and an eyelens $f = 1$ in, diameter $= 0.5$ in, which are 11 in apart. (a) Locate the entrance and exit pupils and find their diameters. (b) Determine the object and image fields of view in radians. Assume both object and image are at infinity.

ANSWER: (a) The entrance pupil is at the objective (which is the aperture stop), and has the same diameter as the objective (since a ray from the object passing through the rim of the objective will pass through the eyelens at $y = 0.1$ in, well within the eyelens diameter.) The exit pupil is the image of the aperture stop (objective) formed by the eyelens, and using Eq. 2.5 we find the image at

$$s' = sf/(s + f) = -11 \times 1/(-11 + 1) = +1.1 \text{ in (to the right of the eyelens—}$$
this is the eye relief)

The magnification $m = s'/s = 1.1/(-11) = -0.1$ and the exit pupil diameter is $h' = mh = -0.1 \times 1 = -0.1$ in, where the minus sign indicates an inverted image.

(b) For zero vignetting the ray from the bottom of the objective to the top of the eyelens determines the field. The slope of this ray is from -0.5 in at the objective to $+0.25$ at the eyelens in a distance of 11 in. Thus $u = 0.75/11 = 0.0681818 \ldots$ and in 10 in the ray rises 0.681818 in to a height of $0.681818 - 0.5 = 0.181818$ in above the axis. With a 10 in objective the "real" field is thus ±0.0181818 radian, and with a 1 in eyepiece, the apparent field is ±0.181818 radians.

For total vignetting, the ray from the top of the objective to the top of the eyelens will determine the field. Its slope is from 0.5 in at the objective to 0.25 in at the eyepiece in a distance of 11 in. Thus $u = -0.25/11 = -0.022727 \ldots$ and in 10 in it drops to a height of $0.5 - 10 \times 0.0227272 = 0.272727$ in. The "real" field is thus ± 0.0272727 and the "apparent" field is ± 0.272727.

For (approximately) 50 percent vignetting, one would use the principal ray through the center of the aperture stop.

4   A 4-in focal length $f/4$ lens is used to project an image at a magnification of four times $(m = -4)$. What is the numerical aperture (NA) in object space, and in image space?

ANSWER: The clear aperture of the lens is 4 in/4 = 1 in. Using Eq. 2.6, and solving for x and x' we get

$$x = f/m \quad \text{and} \quad x' = -fm \text{ which gives us}$$

$$x = 4/(-4) = -1 \text{ in} \quad \text{and} \quad x' = -4(-4) = +16 \text{ in}$$

$$\text{and} \quad s = x - f = -5 \text{ in} \quad \text{and} \quad s' = x' + f = 16 + 4 = +20 \text{ in}$$

If we ignore trigonometry and use NA = $\frac{1}{2}$CA/s, for object space the NA is $0.5/5 = 0.1$ and in image space NA = $0.5/20 = 0.025$.

5   An optical system composed of two thin elements forms an image of an object at infinity. The front lens has a 16-in focal length, the rear lens has an 8-in focal length and the lenses are separated by 8 in. If the exit pupil is located at the rear lens and there is no vignetting, what is the illumination at an image point 3 in from the axis, relative to the illumination on the axis?

ANSWER: To locate the image plane we can use (among many other possibilities) Eq. 4.6:

$$B = f_b(f_a - d)/(f_a + f_b - d) = 8(16 - 8)/(16 + 8 - 8) = 64/16 = 4 \text{ in.}$$

This is the stop to image distance. For a point 3 in from the axis the ray slope is arctan $(3/4) = 36.863898°$ and its cosine is 0.8. The cos^4 is 0.4096 and the relative illumination is 41 percent.

6   A 6-in diameter $f/5$ paraboloid is part of an infrared tracker which can tolerate a blur (due to defocusing) of 0.1 milliradians. (a) What tolerance must be maintained on the position of the reticle with respect to the focal point? (b) What is the tolerance if the system has a speed of $f/2$?

ANSWER: (a) The focal length of the paraboloid is $6 \times 5 = 30$ in and a 0.1 milliradian blur has a diameter of $30 \times 0.0001 = 0.003$ in. At a speed of $f/5$ a blur of 0.003 in will be generated by a defocus of $0.003 \times 5 = 0.015$ in.

(b) At a speed of $f/2$ the focal length is 12 in and the blur is $12 \times 0.0001 = 0.0012$ in. At $f/2$ the defocus is $2 \times 0.0012 = 0.0024$ in.

7   The hyperfocal distance of a 10-in focal length, $f/10$ lens is 100-in. (a) What is the diameter of the "acceptable" blur spot? (b) what is the closest distance at which an object is "acceptably" in focus? (c) Show that the answer to (b) is always one half the hyperfocal distance.

ANSWER: (a) The lens diameter is 10 in/10 = 1 in and per Eq. 2.3 the image of an object at 100 in from the lens is $x' = -f^2/x = -100/(-90) = +1.111 \ldots$ in from the second focal point. An object at infinity is imaged at the focal point, and at f/10 the blur is 1.111/10 = 0.1111 in.

(b) The closest "acceptable" distance object will be imaged δ beyond the image for the hyperfocal distance. By similar triangles: $\delta/0.1111 = (10 + 1.111 + \delta)/0.5$ in and $\delta = 1.3888$, or at 2.5 in beyond the focal point. The corresponding object distance is $x = -100/(-2.5) = 40$ in and the object distance from the lens is $40 + 10 = 50$ in, or one-half the hyperfocal distance.

(c) This exercise is left to the reader.

8   Compare the image illumination produced by an f/8 lens at a point 45° from the axis with that from an f/16 lens 30° off axis.

ANSWER: Illumination varies inversely with the square of the f-number and directly as the fourth power of the cosine of the angle of obliquity, or $\cos^4/(f/\#)^2$. For the f/8 lens this is $\cos^4 45°/64 = 0.25/64 = 0.003906$ and for the f/6 lens it is $\cos^4 30°/256 = 0.5625/256 = 0.002197$. Thus the f/8 lens illumination is $0.003906/0.002197 = 1.777 \ldots$, or 78 percent higher than the f/16 lens.

9   An optical system is required to image a distant point source as a spot 0.01 mm in diameter. Assuming that all the useful energy in the diffraction pattern will be within the first dark ring, for a wavelength of 550 nm what relative aperture and numerical aperture must the system have?

ANSWER: Per Eq. 9.20 the radius of the first dark ring of the diffraction pattern is equal to 1.22λ $(f/\#) = 0.61\lambda/\text{NA}$. So the NA must equal $2 \times 0.61 \times .00055/0.01 = 0.0671$ and the $f/\#$ is f/7.45.

10   A pinhole camera has no lens but uses a very small hole some distance from the film to form its image. If we assume that light travels in straight lines, then the image of a distant point source will be a blur whose diameter is the same as the pinhole. However, diffraction will spread the light into an Airy disk. Thus the larger the hole the bigger the geometrical blur, but the smaller the diffraction pattern. Assume that the sharpest picture will be produced when the geometrical blur is the same size as the central bright spot of the diffraction pattern. What size hole should be used when the film is 100 mm from hole? (*Hint:* Equate the hole diameter to the diameter of the first dark ring of the Airy pattern, as given by Eq. 9.20.)

ANSWER: Diameter = $2 \times 1.22\lambda$ $(f/\#)$ where $f/\# = 100$ mm/Diameter.

Thus $D = 2 \times 1.22 \times 0.00055 \times 100/D$

and $D = \sqrt{0.1342} = 0.366$ mm

11   What is the resolution limit (at the object) for a microscope objective whose acceptance cone has a numerical aperture of (a) 0.25, (b) 0.80, (c) 1.2 at a wavelength of 550 nm?

ANSWER: From Eq. 9.20   z = 0.61 × 0.00055/NA = 0.0003355/NA

(a) NA = 0.25, z = 0.001342 mm

(b) NA = 0.80, z = 0.000419 mm

(c) NA = 1.20, z = 0.000280 mm

12 (a) What diameter must a telescope objective have if the telescope is to resolve 11 seconds of arc? (b) If the eye can resolve 1 minute of arc, what power is needed to utilize this resolution?

ANSWER: (a) From Eq. 9.22   11 = 5.5/D and D = 5.5/11 = 0.5 in.

(b) For 11 second to be seen as 1 minute, the power must be 1 minute/11 seconds = 60 seconds/11 seconds = 5.5×.

10

Optical Materials

10.1 Reflection, Absorption, Dispersion

To be useful as an optical material, a substance must meet certain basic requirements. It should be able to accept a smooth polish, be mechanically and chemically stable, have a homogeneous index of refraction, be free of undesirable artifacts, and of course transmit (or reflect) radiant energy in the wavelength region in which it is to be used.

The two characteristics of an optical material which are of primary interest to the optical engineer are its transmission and its index of refraction, both of which vary with wavelength. The transmission of an optical *element* must be considered as two separate effects. At the boundary surface between two optical media, a fraction of the incident light is reflected. For light normally incident on the boundary the fraction is given by

$$R = \frac{(n' - n)^2}{(n' + n)^2} \tag{10.1}$$

where n and n' are the indices of the two media (a more complete expression for Fresnel surface reflection is given in Chap. 11).

Within the optical element, some of the radiation may be absorbed by the material. Assume that a 1-mm thickness of a filter material transmits 25 percent of the incident radiation at a given wavelength (excluding surface reflections). Then 2 mm will transmit 25 percent of 25 percent and 3 mm will transmit $0.25 \times 0.25 \times 0.25 = 1.56$ percent. Therefore, if t is the transmission of a unit thickness of material, the transmission through a thickness of x units will be given by

$$T = t^x \tag{10.2}$$

This relationship is often stated in the following form, where a is called the absorption coefficient and is equal to $-\log_e t$.

$$T = e^{-ax} \tag{10.3}$$

Thus, it can be seen that the total transmission through an optical element is approximately the product of its surface transmissions and its internal transmission. For a plane parallel plate in air, the transmission of the first surface is given (from Eq. 10.1) as

$$T = 1 - R = 1 - \frac{(n - 1)^2}{(n + 1)^2} = \frac{4n}{(n + 1)^2} \tag{10.4}$$

Now the light transmitted through the first surface is partially transmitted by the medium and goes on to the second surface, where it is partly reflected and partly transmitted. The reflected portion passes (back) through the medium and is partly reflected and partly transmitted by the first surface, and so on. The resulting transmission can be expressed as the infinite series

$$T_{1,2} = T_1 T_2 (K + K^3 R_1 R_2 + K^5 (R_1 R_2)^2 + K^7 (R_1 R_2)^3 + \cdots) \tag{10.5}$$

$$= \frac{T_1 T_2 K}{1 - K^2 R_1 R_2}$$

where T_1 and T_2 are the transmissions of the two surfaces, R_2 and R_1 are the reflectances of the surfaces, and K is the transmittance of the block of material between them. (This equation can also be used to determine the transmission of two or more elements, e.g., flat plates, by finding first $T_{1,2}$ and $R_{1,2}$, then using $T_{1,2}$ and T_3 together, and so on.)

If we set $T_1 = T_2 = 4n/(n + 1)^2$ from Eq. 10.4 into Eq. 10.5, and assume that $K = 1$, we find that the transmission, including all internal reflections, of a completely nonabsorbing plate is given by

$$T = \frac{2n}{(n^2 + 1)} \tag{10.6}$$

This is obviously the maximum possible transmission of an uncoated plate of index n.

Similarly, the reflection is given by

$$R = 1 - T = \frac{(n - 1)^2}{(n^2 + 1)} \tag{10.7}$$

It should be emphasized that the transmission of a material, being wavelength-dependent, may not be treated as a simple number over any appreciable wavelength interval. For example, suppose that a filter is found to transmit 45 percent of the incident energy between 1 and 2 μm. It cannot be assumed that the transmission of two such filters in series will be 0.45 × 0.45 = 20 percent unless they have a uniform spectral transmission (neutral density). To take an extreme example, if the filter transmits nothing from 1 to 1.5 μm and 90 percent from 1.5 to 2 μm, its "average" transmission will be 45 percent within the 1- to 2-μm band. However, two such filters, when combined, will transmit zero from 1 to 1.5 μm, and about 81 percent from 1.5 to 2 μm, for an "average" transmission of about 40 percent, rather than the 20 percent which two neutral density filters would transmit.

The *photographic density* of a filter is the log of its opacity (the reciprocal of transmittance), thus

$$D = \log \frac{1}{T} = -\log T$$

where D is the density and T is the transmittance of the material. Note that transmittance does not account for surface reflection losses; thus, density is directly proportional to thickness. To a fair approximation, the density of a "stack" of neutral density absorption filters is the sum of the individual densities.

Equation 10.3 can be written to the base 10 if desired. This is done when the term "density" is used to describe the transmission of, for example, a photographic filter. The equation becomes

$$T = 10^{-\text{density}}$$

so that a density of 1.0 means a transmission of 10 percent, a density of 2.0 means a transmission of 1 percent, etc. Note that densities can be added. A neutral absorbing filter with a density of 1.0 combined with a filter of density 2.0 will yield a density of 3.0 and a transmission of $0.1 \times 0.01 = 0.001 = 10^{-3}$.

Index dispersion

The index of refraction of an optical material varies with wavelength as indicated in Fig. 10.1 where a *very* long spectral range is shown. The dashed portions of the curve represent absorption bands. Notice that the index rises markedly after each absorption band, and then begins to drop with increasing wavelength. As the wavelength continues to increase, the slope of the curve levels out until the next absorption band is approached,

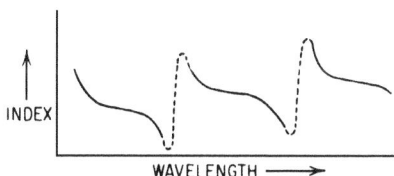

Figure 10.1 Dispersion curve of an optical material. The dashed lines indicate absorption bands. (Anomolous dispersion.)

where the downward slope increases again. For optical materials we usually need concern ourselves with only one section of the curve, since most optical materials have an absorption band in the ultraviolet and another in the infrared and their useful spectral region lies between the two.

Many investigators have attacked the problem of devising an equation to describe "the irrational variation of index" with wavelength. Such expressions are of value in interpolating between, and smoothing the data of, measured points on the dispersion curve, and also in the study of the secondary spectrum characteristics of optical systems. Several of these dispersion equations are listed below.

Cauchy $\qquad n(\lambda) = a + \dfrac{b}{\lambda^2} + \dfrac{c}{\lambda^4} + \cdots$ (10.8)

Hartmann* $\qquad n(\lambda) = a + \dfrac{b}{(c - \lambda)} + \dfrac{d}{(e - \lambda)}$ (10.9)

Conrady $\qquad n(\lambda) = a + \dfrac{b}{\lambda} + \dfrac{c}{\lambda^{3.5}}$ (10.10)

Kettler-Drude $\quad n^2(\lambda) = a + \dfrac{b}{c - \lambda^2} + \dfrac{d}{e - \lambda^2} + \cdots$ (10.11)

Sellmeier $\qquad n^2(\lambda) = a + \dfrac{b\lambda^2}{c - \lambda^2} + \dfrac{d\lambda^2}{e - \lambda^2} + \dfrac{f\lambda^2}{g - \lambda^2} + \cdots$ (10.12)

Herzberger $\qquad n(\lambda) = a + b\lambda^2 + \dfrac{e}{(\lambda^2 - 0.035)} + \dfrac{d}{(\lambda^2 - 0.035)^2}$ (10.13)

Old Schott $\qquad n^2(\lambda) = a + b\lambda^2 + \dfrac{c}{\lambda^2} + \dfrac{d}{\lambda^4} + \dfrac{e}{\lambda^6} + \dfrac{f}{\lambda^8}$ (10.14)

The new Schott catalog uses the Sellmeier equation (Eq. 10.12).

The constants (a, b, c, etc.) are, of course, derived for each individual material by substituting known index and wavelength values and solving the resulting simultaneous equations for the constants. The Cauchy equation obviously allows for only one absorption band at zero wavelength. The Hartmann formula is an empirical one but does allow absorption bands to be located at wavelengths c and e. The Herzberger expression is an approximation of the Kettler-Drude equation and is

---

*After an investigation, Arthur Cox concluded that the three term Hartman equation.

$$n(\lambda) = a + b/(c - \lambda)^{1.2}$$

"is as good as anything over 408 to 656 nm and 546 to 1014 nm."

reliable through the visible to about 1 μm in the near infrared. In his later work, Herzberger used 0.028 as the denominator constant. The Conrady equation is empirical and designed for optical glass in the visible region. All these equations suffer from the drawback that the index approaches infinity as an absorption wavelength is approached. Since little use is made of any material close to an absorption band, this is usually of small consequence.

Until recently Eq. 10.14 was used by Schott and other optical glass manufacturers as the dispersion equation for optical glass. It is accurate to about 3×10^{-6} between 0.4 and 0.7 μm, and to about 5×10^{-6} between 0.36 and 1.0 μm. The accuracy of Eq. 10.14 can be improved in the ultraviolet by adding a term in λ^4, and in the infrared by adding a term in λ^{-10}. More recently, glass manufacturers have switched to Eq. 10.12, the Sellmeier equation, in order to improve the accuracy.

The dispersion of a material is the rate of change of index with respect to wavelength, that is, $dn/d\lambda$. From Figs. 10.1 and 10.2, it can be seen that the dispersion is large at short wavelengths and becomes less at longer wavelengths. At still longer wavelengths, the dispersion increases again as the long-wavelength absorption band is approached. Notice in Fig. 10.2 that the glasses have almost identical slopes for wavelengths beyond 1 μm.

For materials which are used in the visible spectrum, the refractive characteristics are conventionally specified by giving two numbers, the

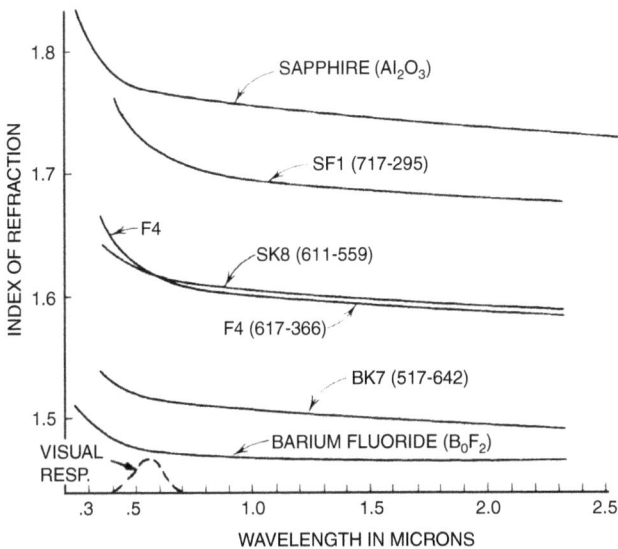

Figure 10.2 The dispersion curves for four optical glasses and two crystals.

index of refraction for the helium d line (0.5876 μm) and the Abbe V-number, or reciprocal relative dispersion. The V-number, or V-value, is defined as

$$V = \frac{n_d - 1}{n_F - n_C}$$ (10.15)

where n_d, n_F, and n_C are the indices of refraction for the helium d line, the hydrogen F line (0.4861 μm), and the hydrogen C line (0.6563 μm), respectively[*]. Note that $\Delta n = n_F - n_C$ is a measure of the dispersion, and its ratio with $n_d - 1$ (which effectively indicates the basic refracting power of the material) gives the dispersion relative to the amount of bending that a light ray undergoes.

For optical glass, these two numbers describe the glass type and are conventionally written $(n_d - 1):V$ as a six-digit code. For example, a glass with an n_d of 1.517 and a V of 64.5 would be identified as 517:645 or 517645.

For many purposes, the index and V-value are sufficient information about a material. For secondary spectrum work, however, it is necessary to know more, and the *relative partial dispersion*

$$P_C = \frac{n_d - n_C}{n_F - n_C}$$ (10.16)

is frequently used for this purpose. P_C is a measure of the rate of change of the slope of the index versus wavelength curve (i.e., the curvature or second derivative). Note that a relative partial dispersion can be defined for any portion of the spectrum and that most glass catalogs list about a dozen partials.

The index of refraction values conventionally given in catalogs, handbooks, etc., are those arrived at by measuring a sample piece in air, and are thus the index relative to the index of air at the wavelength, temperature, humidity, and pressure encountered in the measurement. Since the index is used in optical calculations as a relative number, this causes no difficulty if the index of air is assumed to be 1.0 exactly, and for all wavelengths (unless the optical system is to be used in a vacuum, in which case the catalog index must be adjusted for the index of air; see Sec. 1.2).

10.2 Optical Glass

Optical glass is almost the ideal material for use in the visual and near-infrared spectral regions. It is stable, readily fabricated, homogeneous, clear, and economically available in a fairly wide range of characteristics.

---

[*]The Fraunhofer lines of the solar spectrum are listed in Fig. 10.9.

Figure 10.3 gives some indication of the variety of the available optical glasses. Each point in the figure represents a glass whose n_d is plotted against its V-value; note that the V-values are conventionally plotted in reverse, i.e., descending, order. Glasses are somewhat arbitrarily divided into two groups, the *crown* glasses and the *flint* glasses, crowns having a V-value of 55 or more if the index is below 1.60, and 50 or more for an index above 1.60; the flint glasses are characterized by V-values less than these limits. The "glass line" in Fig. 10.3 is the locus of the ordinary optical glasses made by adding lead oxide to crown glass. These glasses are relatively cheap, quite stable, and readily available.

The addition of lead oxide to crown glass causes its index to rise, and its V-value to decrease, along the glass line. Immediately above the glass line are the barium crowns and flints; these are produced by the addition of barium oxide to the glass mix. In Fig. 10.3 these are identified by the symbol Ba for barium. This has the effect of raising the index without markedly lowering the V-value. The rare earth glasses are a completely different family of glasses based on the rare earths instead of silicon dioxide (which is the major constituent of the other glasses). These are identified by the symbol La in Fig. 10.3, signifying the presence of lanthanum.

The table of Fig. 10.4 lists the characteristics of the most common optical glass types. Each glass type in the table is available from the major glass manufacturers, so that all types listed are readily obtainable. The index data given are taken from the Schott catalog; the equivalent glasses from other suppliers may have slightly different nominal characteristics.

Recently the glass manufacturers have reformulated many optical glasses in order to eliminate toxic ingredients such as lead, cadmium, and arsenic. For the most part, the new formulations have refractive properties which are well within manufacturing tolerances of the glasses they replace, and their physical properties are often an improvement over the original glasses. Not all glasses have been reformulated, and many of the glasses for which lead is essential are still produced. There are a few glasses which have "disappeared". Since the field is still, at least in part, in a state of flux, we have elected not to attempt to update Figs. 10.3 and 10.4. In any case, one should *always* consult the vendor's current catalog or information for material data, since catalogs become obsolete almost as fast as textbooks.

Formerly, optical glass was made by heating the ingredients in a large clay pot, or crucible, stirring the molten mass for uniformity, and carefully cooling the melt. The hardened glass was broken into chunks which were then sorted to select pieces of good quality. Currently the molten glass is more likely to be poured into a large slab mold; this gives better control over the size of the pieces of glass available. Many barium

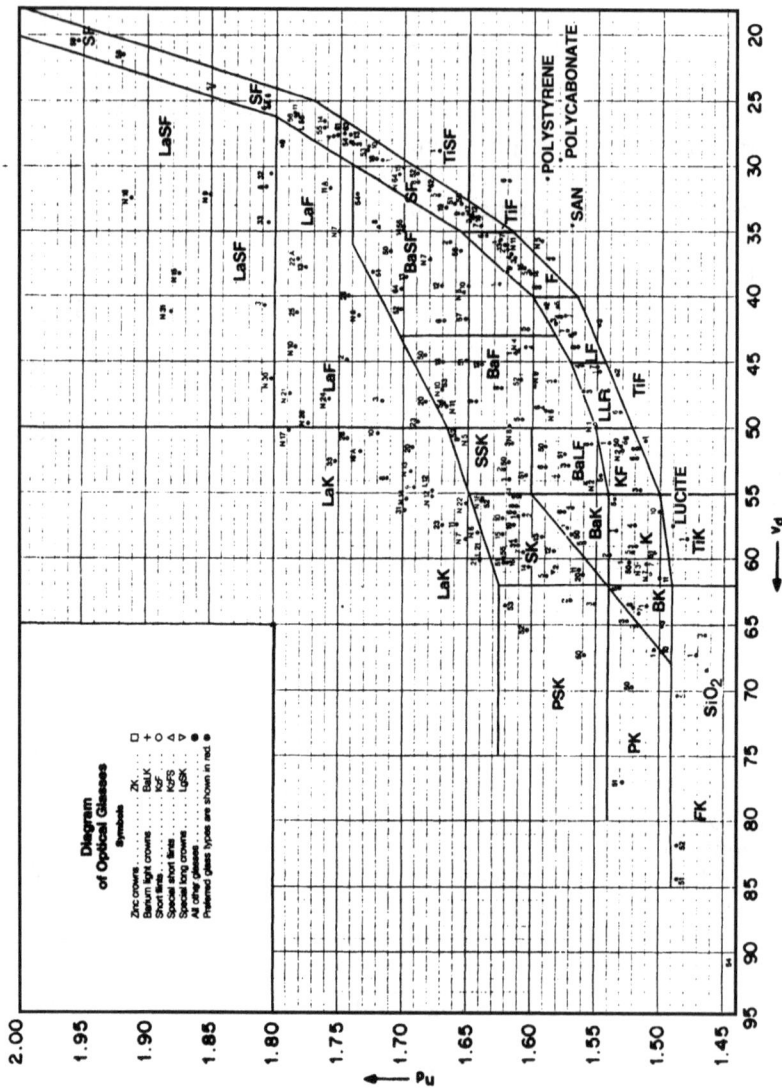

Figure 10.3 The "glass veil." Index (n_d) plotted against the reciprocal relative dispersion (Abbe V-value). The glass types are indicated by the letters in each area. The "glass line" is made up of the glasses of types K, KF, LLF, LF, and SF which are strung along the bottom of the veil. (Note that K stands for *kron*, German for "crown," S stands for *schuer*, or "heavy or dense.") (*Courtesy of Schott Glass Technologies, Inc., Duryea, Pa.*)

Type	n_d	V_d	$n_F - n_C$	n_r	n_C	n_F	n_g	n_h	CR	FR	SR	AR	$\alpha_{-30^\circ+70^\circ C}$ 10^{-6}/K	Tg, °C	D g/cm³	HK	τ_i
BK7 517642	1.51680	64.17	0.008054	1.51289	1.51432	1.52283	1.52669	1.53024	2	0	1	2.0	7.1	559	2.51	520	0.991
K 5 522595	1.52249	59.48	0.008784	1.51829	1.51982	1.52910	1.53338	1.53735	1	0	1	1.0	8.2	543	2.59	450	0.984
BaK 1 573575	1.57250	57.55	0.009948	1.56778	1.56949	1.58000	1.58488	1.58940	2	1	4	1.2	7.6	602	3.19	460	0.976
BaK 2 540597	1.53996	59.71	0.009043	1.53564	1.53721	1.54677	1.55117	1.55525	2	0	1	1.0	8.0	562	2.86	450	0.974
SK 4 613586	1.61272	58.63	0.010451	1.60774	1.60954	1.62059	1.62569	1.63042	3	2	51	2.0	6.4	643	3.57	500	0.973
SK 16 620603	1.62041	60.33	0.010284	1.61548	1.61727	1.62814	1.63312	1.63774	4(2.0)	4	52	3.0	6.3	638	3.58	490	0.970
SK N 18 639554	1.63854	55.42	0.011521	1.63308	1.63505	1.64724	1.65290	1.65819	3	4-5	52	2.2	6.4	643	3.64	470	0.93
KF 6 517522	1.51742	52.20	0.009913	1.51274	1.51443	1.52492	1.52984	1.53446	1	0	1	2.0	6.9	446	2.67	420	0.985
SSK 4 618551	1.61765	55.14	0.011201	1.61235	1.61427	1.62611	1.63163	1.63677	2	1	51	1.0	6.1	639	3.63	460	0.972
SSK N5 658509	1.65844	50.88	0.012940	1.65237	1.65456	1.66825	1.67471	1.68080	2	3	52	2.2	6.8	641	3.71	470	0.91
LaK N 7 652585	1.65160	58.52	0.011134	1.64628	1.64821	1.65998	1.66540	1.67042	4(2.3)	2	53(30)	4.2	7.1	618	3.84	460	0.960
-LaK 8 713538	1.71300	53.83	0.013245	1.70668	1.70898	1.72298	1.72944	1.73545	3	2	52	1.0	5.6	640	3.78	590	0.950
LaK 9 691547	1.69100	54.71	0.012631	1.68498	1.68716	1.70051	1.70667	1.71240	3	3	52	1.2	6.3	650	3.51	580	0.950

Figure 10.4 Characteristics of a selection of optical glasses. CR, FR, SR, and AR are codes indicating the resistance of the glass to staining or hazing due to environmental attack; the higher the number, the lower the resistance; α is the thermal expansion coefficient, and Tg is the transformation temperature. D is the density, HK is the Knoop hardness, and τ_i is the internal transmittance at 0.4 μm for a thickness of 25 mm.

glasses and all the rare earth glasses are processed in platinum cru-
cibles, since the highly corrosive molten glass tends to attack the walls
of a clay pot and the dissolved pot materials affect the glass characteristics. In extremely large volume production, a continuous process is used,
with the raw materials going in one end of the furnace and emerging as
extruded strip or rod glass at the other end. Raw glass is frequently
pressed into blanks, which are roughly the size and shape of the finished
element. The final stage before the glass is ready for use is annealing.
This is a slow cooling process, which may take several days or weeks,
and which relieves strains in the glass, assures homogeneity of index,
and brings the index up to the catalog value.

The characteristics of optical glass vary somewhat from melt to melt
(because of variations in composition and processing) and also due to
variations in annealing procedures. Ordinarily the lower index glasses
(to $n = 1.83$), are supplied to a tolerance of ±0.0005 on the catalog value
of n_d; the higher index glasses may vary ±0.0016 from the nominal
index. Similarly the V-value may vary from the catalog value by about
0.3 percent. Most glass manufacturers will select glass to closer tolerances at an increased price.

Optical glass may be obtained in hundreds of different types; complete
information is best obtained from the manufacturer's catalog.

Figure 10.5 gives an indication of the spectral transmission of optical
glasses. In general, most optical glasses transmit well from 0.4 to 2.0 μm.
The heavy flints tend to absorb more at the short wavelengths and
transmit more at the long wavelengths. The rare earth glasses also
absorb in the blue region. Since the transmission of a glass is affected
greatly by minute impurities, the exact characteristics of any given
glass may vary significantly from batch to batch, even when made by
the same manufacturer. In general the transmission values tend to
improve as the purity of the raw materials is improved over time.

Most optical glasses turn brown (or black) when exposed to nuclear
radiation because of increased absorption of the short (blue) wavelengths.
To provide glasses which can be used in a radiation environment,
the glass manufacturers have developed "protected" or "nonbrowning"
glasses containing cerium. These glasses will tolerate radiation doses
to the order of a million roentgens. Fused quartz glass, which is discussed in the next section, is almost pure SiO_2 and is extremely resistant to radiation browning.

Although not strictly "optical glass," ordinary window glass and
plate glass are frequently used when cost is an important factor. The
index of window glass ranges from about 1.514 to about 1.52, depending on the manufacturer. Ordinary window glass is slightly greenish,
due primarily to modest amounts of absorption in the red and blue
wavelengths; the red absorption continues to about 1.5 μm. Window

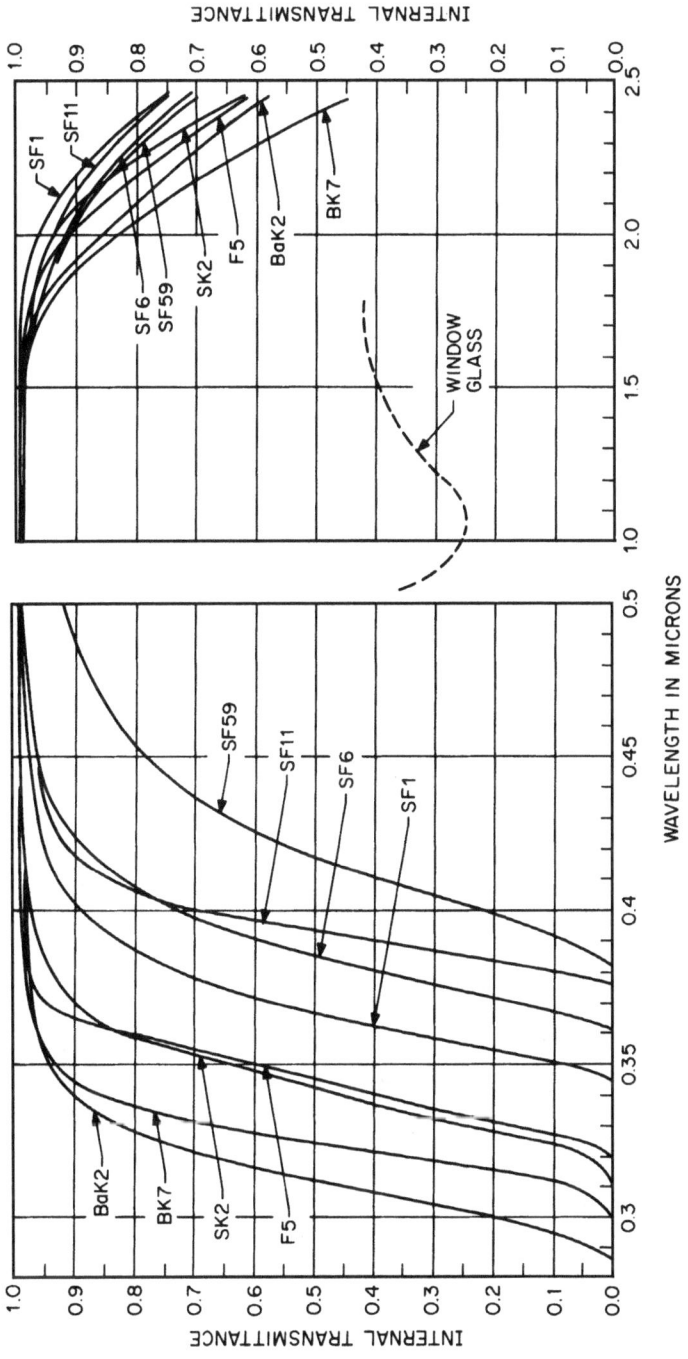

Figure 10.5 Internal transmittance of several representative optical glasses plus window glass, all for a thickness of 25 mm.

glass is also available in "water white" quality, without the greenish tint. For elements with one or two plane surfaces and with modest precision requirements, window glass can often be used without further processing; the accuracy of the plane surfaces is surprisingly good. By special selection, plane parallels can be obtained which meet fairly rigorous requirements. The secret here is to avoid pieces cut from the edge of the large sheets in which this type of glass is made; the center sections are usually far more uniform in surface and thickness. Note that the surface of "float glass" is significantly less smooth by a factor of 3 or 4, although recent process improvements have brought the surfaces up to that of window and plate glass.

10.3 Special Glasses

Several glasses are available which differ sufficiently from the standard optical glasses to deserve special mention.

Low-expansion glasses.   In applications where the elements of an optical system are subject to strong thermal shocks (as in projection condensers) or where extreme stability in the presence of temperature variations is necessary (such as astronomical telescope reflectors or laboratory instruments), it is desirable to use a material with a low thermal coefficient of expansion.

A number of borosilicate glasses are made with expansion coefficients which are less than half that of ordinary glass. Corning's Pyrex #7740 and #7760 have expansion coefficients between 30 and 40 $\times$ $10^{-7}/°C$. The index of refraction of these glasses is about 1.474 with a V-value of about 60, and their density is about 2.2. Unfortunately they are often afflicted with veins and striations so that they are suitable only for applications such as condensing systems when used as refracting elements. They are widely used for test plates and for mirrors. Some of these materials are yellowish or brownish, but others are available in a clear white grade.

Another low-expansion glass is fused quartz, which is also called fused silica glass. This material is essentially pure (more or less, depending on the grade and manufacturer) silicon dioxide (SiO_2) and has an extremely low expansion coefficient of $5.5 \times 10^{-7}/°C$. It was originally made by fusing powdered crystalline quartz. Fused quartz can be obtained in grades with homogeneity equal to that of optical glass. Fused quartz is a completely different material than crystalline quartz. Its index is 1.46 versus 1.55; it is amorphous (glassy) without crystalline structure; and it is not birefringent, as is quartz.

Fused quartz has excellent spectral transmission characteristics, extending further into both the ultraviolet and infrared than ordinary optical glass. For this reason it is frequently used in spectrophotometers, infrared equipment, and ultraviolet devices. The excellent thermal stability of fused quartz is responsible for its use where extremely precise reflecting surfaces are required. Large mirrors and test plates are frequently made from fused quartz for this reason. As previously mentioned, pure fused quartz is highly resistant to radiation browning. The index of refraction and transmission of fused quartz are given in Fig. 10.6. Note that the absorption bands indicated are not of the type indicated in Fig. 10.1, but are due to impurities and are thus subject to elimination, as indicated by the range of transmissions given.

A new class of materials, which are partially crystallized glasses, is available for use as extremely thermally stable mirror substrates, since

Wavelength, μm	Index at 24°C	Transmission 10 mm Thick (Incl. Refl. Losses)
0.17		0.0–0.56 (depending on purity)
0.1855	1.5746*	0.0–0.78
0.2026	1.54725*	0.3–0.84
0.2573	1.50384*	0.58–0.90
0.2749	1.49624*	0.88–0.92
0.35	1.47701	0.93
0.40	1.47021	0.93
0.45	1.46564	0.93
0.4861 (F)	1.46320	0.93
0.5	1.46239	0.93
0.55	1.45997	0.93
0.5893 (D)	1.45846	0.93
0.60	1.45810	0.93
0.6563 (C)	1.45642	0.93
0.70	1.45535	0.93
0.80	1.45337	0.93
1.0	1.45047	0.93
1.35	Absorption band	0.76–0.93
1.5	1.44469	0.93
2.0	1.43817	0.93
2.2	Absorption band	0.50–0.93
2.5	1.42991	0.93
2.7	Absorption band	0–0.8
3.0	1.41937	0.45–0.85
3.5	1.40601	0.6–0.7
4.0		0.1–0.15

*n at "room temperature" $V = 67.6$ $Pc = 0.301$

$\Delta n = 10^{-5} \Delta t$ (°C) visible, to $0.4 \times 10^{-5} \Delta t$ at 3.5 μm

Dispersion equation $n^2 = 2.978645 + \dfrac{0.008777808}{\lambda^2 - 0.010609} + \dfrac{84.06224}{\lambda^2 - 96.0}$

yields values about 0.00042 less than table.

Figure 10.6 Optical characteristics of fused quartz.

they can be fabricated with a zero thermal expansion coefficient. Owens-Illinois CER-VIT was the original material; Corning ULE and Schott ZERODUR have similar properties. These materials can be tailored to have a zero thermal expansion coefficient (plus or minus about 1×10^{-7}) at a given temperature. The zero thermal expansion coefficient results from the mixture of crystals (with a negative coefficient) and amorphous glass with a positive coefficient. These materials tend to be brittle, yellow or brown, and to scatter light, so they are not suitable for refracting optics.

Infrared transmitting glasses. A number of special "infrared" glasses are available. Some of these are much like extremely dense flint glasses, with index values of 1.8 to 1.9 and transmitting to 4 or 5 μm. The arsenic glasses transmit even further into the infrared. Arsenic-modified selenium glass transmits from 0.8 to 18 μm, but will soften and flow at 70°C. It has the following index values: 2.578 at 1.014 μm; 2.481 at 5 μm; 2.476 at 10 μm; 2.474 at 19 μm. Arsenic trisulfide glass transmits from 0.6 to 13 μm and is somewhat brittle and soft. Index values: 2.6365 at 0.6 μm; 2.4262 at 2 μm; 2.4073 at 5 μm; 2.3645 at 12 μm.

Gradient index glass. As indicated in Chap. 1, if the index of refraction is not uniform, light rays travel in curved paths rather than in straight lines. In visualizing this, it often helps to remember that the light rays curve toward the region of higher index. If the index varies in a controlled way, this property may be advantageously utilized. Glass can be doped by infusion with other materials, typically by the immersion of the glass into a bath of molten salts to effect an ion exchange which produces a changed index. A gradient also can be produced by fusing together layers of glasses with differing indexes. Several types of index gradient are useful in optical systems. A *radial gradient* has an index which varies with the radial distance from the optical axis. An *axial gradient* varies the index with the distance along the axis. A *spherical gradient* varies the index as a function of the radial distance from an axial point. An axial gradient at a spherical surface has an effect on the aberrations which is quite analogous to that of an aspheric surface. A radial gradient can produce lens power in a plano-plano element. For example, a plano element the index of which varies as a function of the radial distance r according to

$$n\,(r) = n_0(1 - Kr^2)$$

and has a length L will have a focal length given by

$$f = \frac{1}{n_0 \sqrt{2K} \sin (L \sqrt{2K})}$$

and a back focal length of

$$bfl = \frac{1}{n_0 \sqrt{2K} \tan (L \sqrt{2K})}$$

This effect is the basis of the GRIN rod lens and the SELFOC lens. K is called the *Gradient constant*, and is a function of wavelength and material.

10.4 Crystalline Materials

The valuable optical properties of certain natural crystals have been recognized for years, but in the past the usefulness of these materials was severely limited by the scarcity of pieces of the size and quality required for optical applications. However, many crystals are now available in synthetic form. They are grown under carefully controlled conditions to a size and clarity otherwise unavailable.

The table of Fig. 10.7 lists the salient characteristics of a number of useful crystals. The transmission range is indicated in micrometers for a 2-mm-thick sample; the wavelengths given are the 10 percent transmission points. Indices are given for several wavelengths in the transmission band.

Crystal quartz and calcite are infrequently used because of their birefringence, which limits their usefulness almost entirely to polarizing prisms and the like. Sapphire is extremely hard and must be processed with diamond powder. It is used for windows, interference filter substrates, and occasionally for lens elements. It is slightly birefringent, which limits the angular field over which it can be used. The halogen salts have good transmission and refraction characteristics, but their physical properties often leave much to be desired, since they tend to be soft, fragile, and occasionally hygroscopic.

Germanium and especially silicon are widely used for refracting elements in infrared devices. They are much like glass in their physical characteristics, and can be processed with ordinary glass-working techniques. Both are metallic in appearance, being completely opaque in the visible. Their extremely high index of refraction is a joy to the lens designer since the weak curvatures which result from the high index tend to produce designs of a quality which cannot be duplicated in comparable glass systems. Special low-reflection coatings are necessary since the surface reflection (per Eq. 10.1 et seq.) is very high, for example, 36 percent per uncoated germanium surface. Zinc sulfide, zinc selenide, and AMTIR are also widely used in infrared systems.

Worthy of special mention is calcium fluoride, or fluorite. This material has excellent transmission characteristics in both ultraviolet

Material	Transmission Range, μm	Index	Remarks
Crystal quartz (SiO₂)	0.12–4.5	n_o = 1.544, n_e = 1.553	Birefringent
Calcite (CaCO₃)	0.2–5.5	n_o = 1.658, n_e = 1.486	Birefringent
Rutile (TiO₂)	0.43–6.2	n_o = 2.62, n_e = 2.92	Birefringent
Sapphire (Al₂O₃)	0.14–6.5	1.834 @ 0.265, 1.755 @ 1.01, 1.586 @ 5.58	Hard, slightly birefringent
Strontium titanate (SrTiO₃)	0.4–6.8	2.490 @ 0.486, 2.292 @ 1.36, 2.100 @ 5.3	IR immersion lenses
Magnesium fluoride (MgF₂)	0.11–7.5	n_o = 1.378, n_e = 1.390	IR optics, low reflection coatings
Lithium fluoride (LiF)	0.12–9	1.439 @ 0.203, 1.38 @ 1.5, 1.109 @ 9.8	Prisms, windows, apochromatic lenses
Calcium fluoride (CaF₂)	0.13–12	See Fig. 7.10	Same as LiF
Barium fluoride (BaF₂)	0.25–15	1.512 @ 0.254, 1.468 @ 1.01, 1.414 @ 11.0	Windows
Sodium chloride (NaCl)	0.2–26	1.791 @ 0.2, 1.528 @ 1.6, 1.175 @ 27.3	Prisms, windows, hygroscopic
Silver chloride (AgCl)	0.4–28	2.096 @ 0.5, 2.002 @ 3., 1.907 @ 20.	Ductile, corrosive, darkens
Potassium bromide (KBr)	0.25–40	1.590 @ 0.404, 1.536 @ 3.4, 1.463 @ 25.1	Prisms, windows, soft, hygroscopic
Potassium iodide (KI)	0.25–45	1.922 @ 0.27, 1.630 @ 2.36, 1.557 @ 29	Soft, hygroscopic
Cesium bromide (CsBr)	0.3–55	1.709 @ 0.5, 1.667 @ 5, 1.562 @ 39	Hygroscopic, prisms and windows
Cesium iodide (CsI)	0.25–80	1.806 @ 0.5, 1.742 @ 5, 1.637 @ 50	Prisms and windows
Silicon (Si)	1.2–15	3.498 @ 1.36, 3.432 @ 3, 3.418 @ 10	IR optics
Germanium (Ge)	1.8–23	4.102 @ 2.06, 4.033 @ 3.42, 4.002 @ 13	IR optics, absorbs at higher temp., subject to thermal runaway @ 40°C
Zinc Selenide (ZnSe)	0.5–22	2.489 @ 1, 2.430 @ 5, 2.406 @ 10, 2.366 @ 15	
Zinc Sulfide (ZnS)	0.5–14	2.292 @ 1, 2.246 @ 5, 2.200 @ 10, 2.106 @ 15	
AMTIR (Ge/As/Se)	0.7–14	2.606 @ 1, 2.511 @ 5, 2.497 @ 10, 2.482 @ 14	
Gallium Arsenide (GaAs)	1–15	3.317 @ 3, 3.301 @ 5, 3.278 @ 10, 3.251 @ 14	
Cadmium Telluride (CdTe)	0.2–30	2.307 @ 3, 2.692 @ 5, 2.680 @ 10, 2.675 @ 12	
Magnesium Oxide (MgO)	0.25–9	1.722 @ 1, 1.636 @ 5, 1.482 @ 8	

Transmission range wavelengths are the 10 percent transmission points for a 2-mm thickness. n_o and n_e are indices for the ordinary and extraordinary rays.

Figure 10.7 Characteristics of optical crystals.

Wavelength, μm	Index	Absorption Coefficient, cm^{-1}
0.2	1.49531	—
0.3	1.45400	—
0.4	1.44186	—
0.4861 (F)	1.43704	—
0.5893 (D)	1.43384	—
0.6563 (C)	1.43249	—
1.014	1.42884	—
2.058	1.42360	—
3.050	1.41750	—
4.0	1.40963	—
5.0	1.39908	—
7	—	0.02
8	—	0.16
8.84	1.33075	—
9	—	0.64
10	—	1.8

$V = 95.3 \quad P_c = 0.297$
$\Delta n = -10^{-5} \Delta T \ (°C)$

Figure 10.8 Index and transmission of calcium fluoride (CaF$_2$) for various wavelengths.

and infrared, which make it valuable for instrumentation purposes. In addition, its partial dispersion characteristics are such that it can be combined with optical glass to form a lens system which is free of secondary spectrum. Its physical properties are not outstanding since it is soft, fragile, resists weathering poorly, and has a crystal structure which sometimes makes polishing difficult. In exposed applications, the fluorite element can sometimes be sandwiched between glass elements to protect its surfaces. The table of Fig. 10.8 lists selected index and transmission values for fluorite. Natural fluorite has been used in microscope objectives for many, many years. The FK glasses, especially FK51, FK52, and FK54, share many of fluorite's characteristics and are very useful in correcting the secondary spectrum.

10.5 Plastic Optical Materials

Plastics are rarely used for high precision optical elements. A great deal of effort was made to develop plastics for optical systems during the Second World War, and a few systems incorporating plastics were produced. Since then, the technology of fabrication of plastic optics has advanced significantly, and today, in addition to novelty items such as toys and magnifying glasses, plastic lenses can be found in a multitude of optical applications, including inexpensive, disposable camera lenses, many zoom lenses, projection TV lenses, and some high quality camera lenses. The low cost of mass-produced plastic optics is one important factor in this popularity; another is the ease of production of aspheric surfaces. Once the aspheric mold has been fabricated, an aspheric surface is as easy to make as is a spherical surface (in marked contrast to glass optics). The rule of thumb that the introduction of an aspheric surface allows the elimination of an element from the system attests to the value of optical plastic materials. This aspheric capability largely offsets the unfortunate fact that the number of suitable optical plastics is very small and that there are only relatively low index materials in that number.

In considering a venture into the plastic optics arena, one is well advised to seek out a specialist in making plastic optics. Not only is the typical injection molder incapable of making good optics, but he or she usually has no conception of what is required to do so. The successful fabricators have developed good, reliable sources of consistently high-quality raw materials and material handling techniques, and they have molding machines which have been adapted to the special requirements of optical work. Temperature control is extremely critical, and a longer cycle time is necessary to achieve an optical level of precision. I encountered an extreme case a few years ago. I had designed a visual system for a client who insisted (against my advice) not only on patronizing an inexperienced (in optics) injection molder but also insisted on using an unusual material. The result was a system which you literally could not see through.

In addition to the general, smooth aspheric capability, plastics are widely used to make Fresnel lenses, where fine steps are necessary. The condenser system in overhead projectors and the field lenses in the viewfinders of single-lens reflex cameras are examples of plastic Fresnel lenses. Another currently popular application is in diffractive optics (discussed at greater length in Chaps. 12 and 16), where the diffractive surface is basically a Fresnel surface the step height of which is on the order of a half wavelength.

Another advantage in mass production is the ability to mold both the lens element and its mounting cell in one shot. The cells of an assembly can in fact be designed so that the lens assembly simply snaps together, and a drop of a suitable solvent or adhesive can make the assembly permanent.

The obvious advantages of plastic—that it is light and relatively shatterproof—are offset by a number of disadvantages. It is soft and scratches easily. Except by molding, it is difficult to fabricate. Styrene plastic is frequently hazy, scatters light, and is occasionally yellowish. Plastics tend to soften at 60 to 80°C. In some plastics the index is unstable and will change as much as 0.0005 over a period of time. Most plastics will absorb water and change dimensionally; almost all are subject to cold flow under pressure. The thermal expansion coefficient is almost 10 times that of glass, being 7 or 8 $\times$ 10^{-5}/°C.

The change of index with temperature for plastics is very large (about twenty times that of glass) and *negative*. Thus, maintaining focus over a range of temperature is a significant problem for plastic optics. Often they must be athermalized as well as achromatized. The density of plastics is low, usually to the order of 1.0 to 1.2. The characteristics of some of the most widely used optical plastics are summarized in Fig. 10.9.

Another optical application for plastics is in *replication*. In this process a precisely made master mold is vacuum-coated with a release, or parting layer, plus any required high- or low-reflection coatings. (The nature of the release layer is usually considered proprietary, but very thin layers of silver, salt, silicon, or plastic have been publicly mentioned.) Next, a few drops of low-shrinkage epoxy are pressed out into a thin (ideally about 0.001- or 0.002-in-thick) layer between the master and a closely matching substrate. The substrate may be Pyrex, ceramic, or *very* stable aluminum (for reflector optics), or glass (for refracting optics). When the epoxy has cured, the master is removed

Wavelength, μm	Acrylic (Lucite)	Polystyrene	Polycarbonate	Copolymer Styrene-Acrylonitrile (SAN)
	492:574	590:309	585:299	567:348
1.01398 *t*	1.483115	1.572553	1.567248	1.551870
0.85211 *s*	1.484965	1.576196	1.570981	1.555108
0.70652 *r*	1.487552	1.581954	1.576831	1.560119
0.65627 *C*	1.489201	1.584949	1.579864	1.562700
0.64385 *C'*	1.489603	1.585808	1.580734	1.563438
0.58929 *D*	1.491681	1.590315	1.585302	1.567298
0.58756 *d*	1.491757	1.590481	1.585470	1.567440
0.54607 *e*	1.493795	1.595010	1.590081	1.571300
0.48613 *F*	1.497760	1.604079	1.599439	1.579000
0.47999 *F'*	1.498258	1.605241	1.600654	1.579985
0.43584 *g*	1.502557	1.615446	1.611519	1.588640
0.40466 *h*	1.506607	1.625341	1.622447	1.597075
0.36501 *i*	1.513613	1.643126	1.643231	1.612490
Thermal expansion coefficient °C^{-1}	68 $\times$ 10^{-6}	70 $\times$ 10^{-6}	66 $\times$ 10^{-6}	65 $\times$ 10^{-6}
d*n*/d*t*, °C^{-1}	-105×10^{-6}	-140×10^{-6}	-107×10^{-6}	-110×10^{-6}
Service temperature °C	83°	75	120	90
Density	1.19	1.06	1.20	1.09'

Figure 10.9 Properties of several optical plastics. (*From Lytle and Altman.*) Note that index values may vary significantly from one manufacturer to another.

and a reasonably precise (negative) replica is left on the substrate. This process has several advantages. For example, any surface (including aspherics) for which a master can be made can be replicated relatively inexpensively, since the master can be used over and over. Other advantages are that a mirror can be made an integral part of its mount, the bottom of a blind hole can have an optical polish and figure, and extremely thin and lightweight parts can be produced. In many cases these things are effectively impossible with standard optical fabrication techniques. The limitations to replicated parts are the inherent softness of the epoxy and the change in the surface figure from that of the mold.

10.6 Absorption Filters

Absorption filters are composed of materials which transmit light selectively; that is, they transmit certain wavelengths more than others. A small percentage of the incident light is reflected, but the major portion of the energy which is not transmitted through the filter is absorbed by the filter material. Obviously, every material discussed in the preceding sections of this chapter is, in the broadest sense, an absorption filter, and occasionally these materials are introduced into optical systems as filters. However, most optical glass filters are made by the addition of metallic salts to clear glass or by dyeing a thin gelatin film to produce a more selective absorption than is available in "natural" materials.

The prime source of dyed gelatin filters is the Eastman Kodak Company, the line of Wratten filters of which is widely used for applications where the versatility of dyed gelatin is required and the environmental requirements are not too severe. Gelatin filters are usually mounted between glass to protect the soft gelatin from damage.

The number of coloring materials which are suitable for use in optical filter glass is limited, and the types of filter glass available are thus not as extensive as one might desire. In the visible region, there are several main types. The red, orange, and yellow glasses all transmit the red and near-infrared and have a fairly sharp cutoff, as indicated in Fig. 10.10. The position of this cutoff determines the apparent color of the filter. Green filters tend to absorb both the red and blue portions of the spectrum. Their transmission curves often resemble the spectral sensitivity curve of the eye. Blue optical glass filters can be a disappointment, since they occasionally transmit not only blue light, but some green, yellow, orange, and frequently a sizable amount of red light as well. The purple filters transmit both the red and blue ends of the spectrum, with fair suppression of the yellow and green spectral regions. Filter glass is manufactured by most optical glass companies

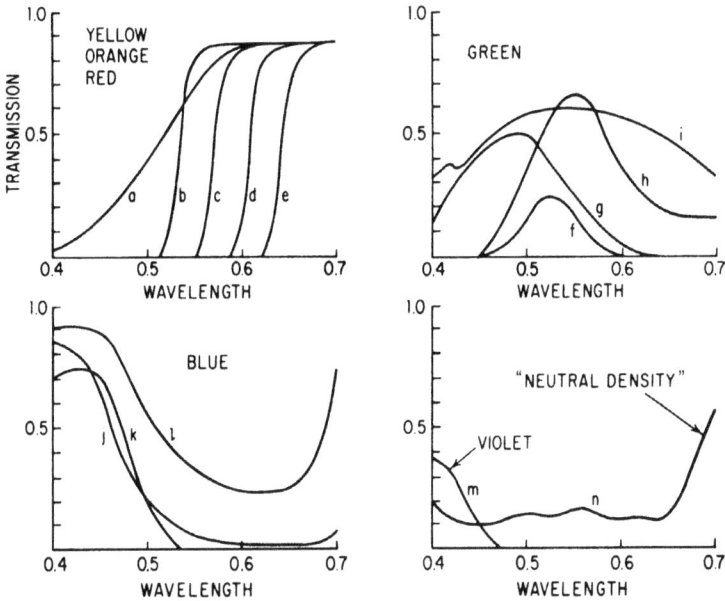

Figure 10.10 Spectral transmission curves for several optical glass filters.

as well as a number of establishments which make commercial colored glass (as opposed to "optical" glass, which is more carefully controlled).

The transmission characteristics of glass filters vary from melt to melt for any given type. If a filter application requires that the transmission be accurately controlled, it is frequently necessary to adjust the finished thickness of the filter to compensate for these variations. The red filters are probably the most variable; since they are sensitive to heat, some red glasses cannot be re-pressed into blanks. Spectral transmission data for filters is usually given for a specific thickness and includes the losses due to Fresnel surface reflections. To determine the transmission for thicknesses other than the nominal value, the transmittance, that is, the "internal" transmission of the piece without the reflection losses, must be determined. In most cases, it is sufficient to divide the transmission by Eq. 10.4 to get the transmittance. Then Eq. 10.2 or 10.3 can be used to determine the transmittance of the new thickness. This transmittance times the T of Eq. 10.4 will then give the total transmission for the filter to a reasonable accuracy.

This process is greatly simplified by the use of a log-log plot of the transmittance. The Schott catalog of filter glass makes use of this type of scale. A transparent overlay makes it possible to evaluate instantly the effect of a thickness change. A study of Fig. 10.11 will indicate the utility of this type of a transmittance plot; the same filter is shown in

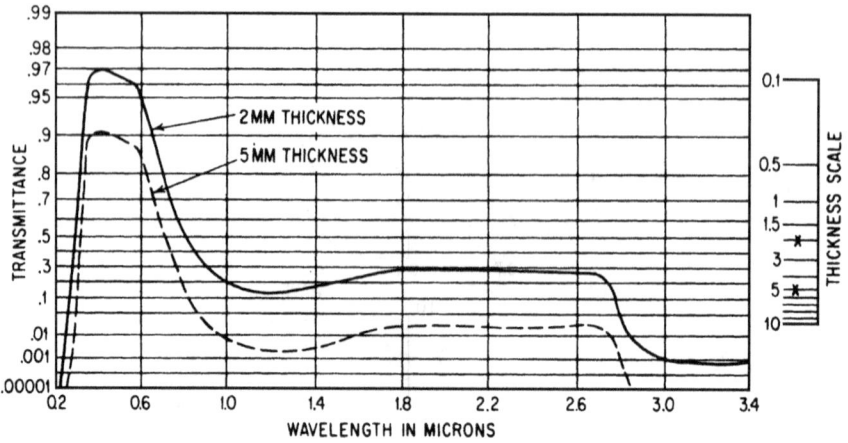

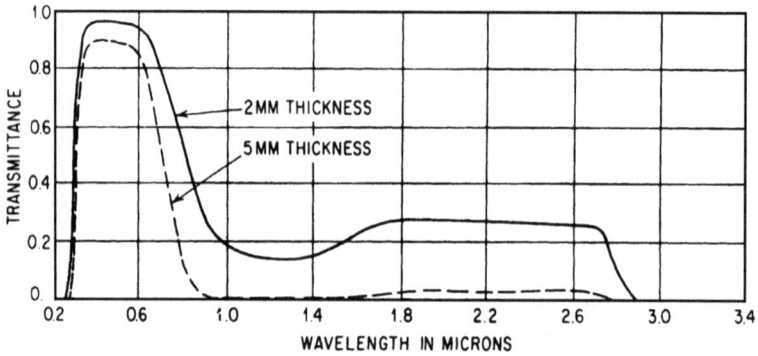

Figure 10.11 Spectral transmittance of Schott KG2 heat-absorbing filter glass. The upper graph is plotted on a log-log scale. Note that the vertical spacing between the two plots is equal to the distance from 2 to 5 on the thickness scale at the right. The same data is plotted on a conventional linear scale in the lower figure for comparison.

two thicknesses on a log-log scale in the upper figure and on a linear scale in the lower. Against the log-log scale, the thickness change is effected by a simple vertical displacement of the plot. The amount of the displacement is given by the thickness scale at the right. Notice how much more information this type of plot can give (and how much more is required to prepare one!). The data plotted in this form is transmittance; to determine the total transmission of the filter, the surface reflection losses must be taken into account, either by Eq. 10.4 or 10.5.

Glass filters are also available to transmit either the ultraviolet or infrared regions of the spectrum without transmitting the visible. Typical transmission plots for these filters are shown in Fig. 10.12. Heat-absorbing glasses are designed to transmit visible light and

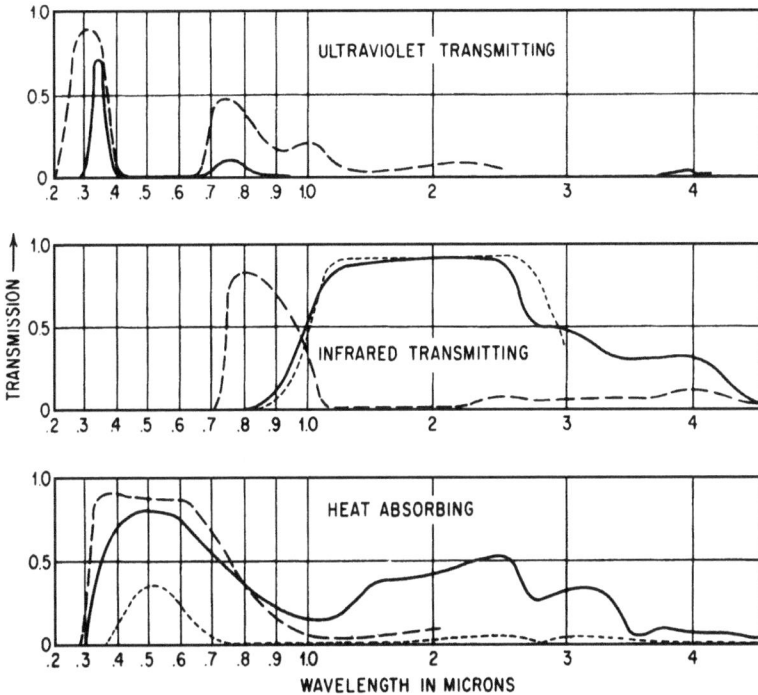

Figure 10.12 Transmission characteristics of special-purpose glass filters. UV transmitting: solid line, Corning 7-60; dashed, Corning 7-39. IR transmitting solid, Corning 7-56 (#2540); dashed, Corning 7-69; dotted, Schott UG-8. Heat absorbing: solid, Corning I-59 extra light Aklo; dashed, Pittsburgh Plate Glass #2043 Phosphate—2 mm; dotted, Corning I-56 dark shade Aklo.

absorb infrared energy. These are frequently used in projectors to protect the film or LCD from the heat of the projection lamp. Since they absorb large quantities of radiant energy, they become hot themselves and must be carefully mounted and cooled to avoid breakage from thermal expansion. From the spectral transmission characteristics given in Fig. 10.12, it is apparent that the phosphate heat-absorbing glass is more efficient than the Aklo; the phosphate glass is subject to large bubbles and inclusions which do not, however, prevent its use in most applications. See also the discussion of "hot" and "cold" mirrors in Chap. 11.

10.7 Diffusing Materials and Projection Screens

A piece of white blotting paper is an example of a (reflecting) diffusing material. Light which strikes its surface is scattered in all directions; as a result, the paper appears to have almost the same brightness regardless of the angle at which it is illuminated or the angle from which it is viewed. A perfect, or lambertian, diffuser is one which has the same apparent brightness from any angle; thus the radiation emitted per unit area in the surface is given by $I_0 \cos \theta$, where θ is the angle to the surface normal and I_0 is the intensity of an element of area in a direction perpendicular to the surface.

There are a number of quite good reflecting diffusers with relatively high efficiencies. Matte white paper is a very convenient one and reflects 70 to 80 percent of the incident visible light. Magnesium oxide and magnesium carbonate are frequently used in photometric work since their efficiencies are high, to the order of 97 or 98 percent.

The brightness (luminance) of a perfectly diffuse reflector is proportional to the illumination falling on it and to its reflectivity. If the illumination is measured in footcandles, multiplication by the reflectivity yields the brightness in foot-lamberts. The brightness in lamberts is given by the illumination in lumens per cm^2 times the reflectivity, and if this product is divided by π, the result is the brightness in candles per cm^2, or in lumens per steradian per cm^2. (See Chap. 12 for more material on photometric considerations.)

As indicated above, a perfectly diffuse surface appears to have the same brightness regardless of the angle at which it is viewed. A projection screen which is not perfectly diffuse can have a brightness ranging from zero to that of the projector light source. For example, consider a perfect mirror screen in the shape of an ellipsoid, with the viewer's eye placed at one focus and the projector at the other. All of the light will be reflected to the eye; none will be scattered. From this eye position the screen will have the same brightness as if one looked directly into the projection lens; when viewed from any other location, the screen will appear completely dark. The *gain* of a projection screen is the ratio of its brightness to that of a perfectly diffuse (or lambertian) screen, which by definition has a gain of 1.0. A diffuse screen can be viewed from any direction, and its brightness, while low, is independent of the viewing angle. The higher the gain of a screen, the smaller the angle over which it has its rated gain. Beaded screens and facetted, lenticular screens are used to concentrate and distribute the light in a controlled manner. Aluminum paints are used to coat screens which must maintain polarization, and with a smooth curved surface can achieve gains as high as 4.0 in commercial products. Beaded screens can achieve a gain as high as 10, but only over an extremely restricted angle. Many projection screens are rated at a gain of about 2.0.

Transmitting diffusers are used for such applications as rear projection screens and to produce even illumination. The most commonly used are opal glass and ground glass (Fig. 10.13). Opal glass contains a suspension of minute colloidal particles and diffuses by multiple

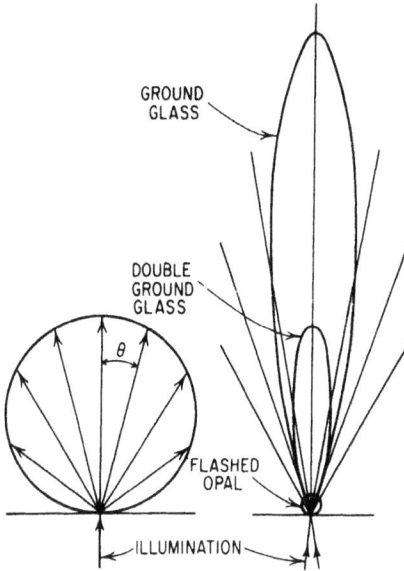

Figure 10.13 Polar intensity plots of diffusing materials. (*Left*) For a "perfect diffuser", the intensity of a unit area of the surface varies with cos θ. (*Right*) The relative intensities of single- and double-ground glass and flashed opal glass.

scattering from these particles. The transmitted light is slightly yellowish since the shorter wavelengths are scattered more than the longer. Opal glass is ordinarily used as *flashed opal,* which is a thin layer of opal glass fused to a supporting sheet of clear glass. The diffusion of flashed opal is quite good. When illuminated normally, the brightness at 45° from the normal is about 90 percent of what one would expect from a perfect diffuser. Its total transmission is quite low, about 35 or 40 percent. It should be noted that, since good diffusion means that the incident light is scattered into 2π steradians, the axial brightness of a rear-illuminated screen of good diffusion is very low when compared with a poor diffuser.

Ground glass is produced by fine grinding (or etching) the surface of a glass plate to produce a large number of very small facets which refract the incident light more or less randomly. The total transmission of ground glass is about 75 percent. This transmission is quite strongly directional, and ground glass is far from a perfect diffuser. Its characteristics vary somewhat, depending on the coarseness of the surface. Typically, for a normally illuminated surface, the brightness at 10° from the normal is about 50 percent of the normal brightness; at 30°, the brightness is about 2.5 percent of the brightness at the normal. This characteristic is of course quite useful when partial diffusion is desired. By combining two sheets of ground glass (with the ground faces in contact), the transmission is lowered about 10 percent but the diffusion is improved; at 20° to the normal, the brightness is about 20 percent; at 30°, about 7 percent. With two sheets, the diffusion can be increased by spacing them apart, although this will destroy their utility as a projection screen.

A sheet of tracing paper has diffusion characteristics quite similar to ground glass, and there are several commercially available plastic screen materials which are somewhat better diffusers than ground glass. The plastic surface also can be shaped to control the beam spread.

A rear projection screen, when used in a lighted room, is illuminated from both sides. The room light reduces the contrast of the projected image. This situation is sometimes alleviated by introducing a sheet of gray glass (that is, a neutral filter) between the diffusing screen and the observer. When this is done, the light from the projector is reduced by a factor of T, the transmission of the gray glass, but the room light is reduced by T^2, since the room light must pass through the gray glass twice to go from the room to the diffuser and back to the observer's eye.

10.8 Polarizing Materials

Light behaves as a transverse wave in which the waves vibrate perpendicular to the direction of propagation. If the wave motion is considered as a vector sum of two such vibrations in perpendicular planes, then plane polarized light results when one of the two components is removed from a light beam. Plane polarized light can be produced by passing the radiation from an ordinary source through a polarizing prism, several types of which are available. These prisms depend on the birefringent characteristic of calcite ($CaCO_3$), which has a different index of refraction for the two planes of polarization. Since light of one polarization is refracted more strongly than the other, it is possible to separate them either by total internal reflection (as in the Nicol and Glan-Thompson prisms) or by deviation in different directions (as in the Rochon and Wollaston prisms).

Such prisms are large, heavy, and expensive. Sheet polarizers, which are made by aligning microscopic crystals in a suitable base, are thin, light, relatively inexpensive, useful over a wide field of view, and simple to fabricate into an almost unlimited range of sizes and shapes. Thus, despite the fact that they are not quite as efficient as a good prism polarizer and are not effective over as large a wavelength range, they have largely supplanted prisms for the great majority of applications where polarization is required. Several companies, produce a number of types of sheet polarizers. For work in the visible region, several types are available, depending on whether optimum transmission or optimum extinction (through crossed polarizers) is desired. Special types are available for use at high temperatures and also for use in the near-infrared (0.7 to 2.2 µm). Polaroid also produces circular (as opposed to plane) polarizers in sheet form.

Since a plane polarizer will eliminate half the energy, it is obvious that the maximum transmission of a "perfect" polarizer in a beam of unpolarized light will be 50 percent. Practical values range from 25 to 40 percent for sheet polarizers, depending on the type. If two polarizers are "crossed," that is, oriented with their polarizing axes at 90°, the transmission will be zero if the polarization is complete. This can be achieved with Nicol prisms, but the sheet polarizers have a residual transmission ranging from 10^{-6} to 5×10^{-4}, again dependent on the type. The transmission characteristics of sheet polarizers are wavelength-dependent as well.

When two polarizers are placed in a beam of unpolarized light, the transmission of the pair depends on the relative orientation of their polarization axes. If θ is the angle between the axes, then the transmission of the pair is given by:

$$T = K_0 \cos^2 \theta + K_{90} \sin^2 \theta \qquad (10.17)$$

where K_0 is the maximum transmission and K_{90} is the minimum. Typical value pairs for K_0 and K_{90} are 42 percent and 1 or 2 percent; 32 percent and 0.005 percent; 22 percent and 0.0005 percent.

Reflection from the surface of a glass plate may also be used to produce plane polarized light. When light is incident on a plane surface at *Brewster's angle,* one plane of polarization is completely transmitted (if the glass is perfectly clean) and about 15 percent of the other is reflected. This occurs when the reflected and refracted rays are at 90° to each other; thus, *Brewster's angle* is given by

$$I = \arctan \frac{n'}{n} \qquad (10.18)$$

The reflected beam is thus completely polarized and the transmitted beam partially so. The percentage of polarized light in the transmitted beam can be increased by using a stack of thin plates all tilted to Brewster's angle. For an index of 1.52, Brewster's angle is 56.7°. Note that Brewster's angle is the angle at which the tangent term in Eq. 11.1 goes to zero.

The subject of polarized light is treated at greater length in texts devoted to physical optics, to which the reader is referred. Two additional points are worth noting: one, interference filters (Chap. 11) are usually obliquely polarizing and are occasionally used as polarizers; and two, opal glass and other diffusers are excellent depolarizers, as are integrating spheres.

10.9 Cements and Liquids

Optical cements are used to fasten optical elements together. Two main purposes are served by cementing: the elements are held in accurate alignment with each other independent of their mechanical mount, and the reflections from the surfaces (especially those from TIR) are largely eliminated by cementing. Ordinarily the layer of cement used is extremely thin and its effect on the optical characteristics of the system can be totally neglected; some of the newer plastic cements, designed to withstand extremes of temperature, are used in thicknesses of a few thousandths of an inch (which could affect the performance of an optical system under critical conditions where the light rays have large slopes).

Canada balsam is made from the sap of the balsam fir. It is available in a liquid form (dissolved in xylol) and in stick or solid form. Elements to be cemented are cleaned and placed together on a hot plate. When the elements are warm enough to melt the balsam, the stick is rubbed on the lower element. The upper element is replaced and the excess cement and any entrapped air bubbles are worked out by oscillating or rocking the upper element. The elements are then placed in an alignment fixture to cool. Balsam cement has an index of refraction of about 1.54 and a V-value of about 42. These are conveniently midway between the refractive characteristics of crown and flint glasses. Unfortunately, Canada balsam will not withstand high or low temperatures. It softens when heated and splits at low temperatures and is thus unsuited for rigorous thermal environments. Balsam is very rarely used today.

A great number of plastic cements have been developed to withstand extremes of both temperature and shock. For the most part, these are thermosetting (heat-curing) or ultraviolet light-curing plastics, although a few thermoplastic (heat-softening) materials are used. Cements are available which will withstand temperatures from 82°C down to −65°C without failure when properly used. In general the thermosetting cements are supplied in two containers (sometimes refrigerated), one of which contains a catalyst which is mixed into the cement prior to use. A drop of cement is placed between the elements to be cemented, the excess cement and air bubbles are worked out, and the elements are placed in a fixture or jig for a heating cycle which cures the cement. Once the cement has set, it is exceedingly difficult to separate the components; the customary technique is to shock them apart by immersion in hot (150 to 200°C) castor oil. The index of refraction of plastic cements ranges from 1.47 to 1.61, depending on the type, with most cements falling between 1.53 and 1.58 with a V-value between 35 and 45. Epoxies and methacrylates are widely used. Because of the variety of types and characteristics which are available, one should consult the manufacturer's literature for specific details regarding any given cement.

A rarely used method of fastening optical elements together is by what is called *optical contact*. Both pieces must be scrupulously cleaned (often the final cleaning is with a cloth slightly stained with polishing rouge) and laid together. If the surface shapes match well enough, as the air is pressed out from between the pieces a molecular attraction will cause them to adhere in a surprisingly strong bond, which will withstand a force of about 95 lb/in^2. Usually the only way properly contacted surfaces can be separated is by heating one of them and allowing thermal expansion to break the contact (it often breaks the glass as well). Occasionally, soaking in water will separate the pieces.

Optical liquids are used primarily for microscope immersion fluids and for use in index measurement (in critical-angle refractometers). For microscopy, water ($n_d = 1.33$), cedar oil ($n_d = 1.515$), and glycerin (ultraviolet $n = 1.45$) are frequently utilized. For refractometers alpha-bromonaphthalene ($n = 1.66$) is the most commonly used liquid. Methylene iodide ($n = 1.74$) is used for high index measurement (since the liquid index must be larger than that of the sample to avoid total internal reflection back into the sample).

Bibliography

American Institute of Physics Handbook, 3d ed., New York, McGraw-Hill, 1972.

Ballard, S., K. McCarthy, and W. Wolfe, *Optical Materials for Infrared Instrumentation,* Univ. of Michigan, 1959 (Supplement, 1961).

Bennett, J. M., "Polarization," in *Handbook of Optics,* Vol. 1, New York, McGraw-Hill, 1995, Chap. 5.

Bennett, J. M., "Polarizers," in *Handbook of Optics,* Vol. 2, New York, McGraw-Hill, 1995, Chap. 3.

Conrady, A., *Applied Optics and Optical Design,* Oxford, 1929. (This and Vol. 2 were also published by Dover, New York.)

Driscoll, W. (ed.), *Handbook of Optics,* New York, McGraw-Hill, 1978.

Hackforth, H., *Infrared Radiation,* New York, McGraw-Hill, 1960.

Handbook of Chemistry and Physics, Chemical Rubber Publishing Co., published annually.

Hardy, A., and F. Perrin, *The Principles of Optics,* New York, McGraw-Hill, 1932.

Herzberger, M., *Modern Geometrical Optics,* New York, Interscience, 1958.

Jacobs, D., *Fundamentals of Optical Engineering,* New York, McGraw-Hill, 1943.

Jacobs, S., in Shannon and Wyant (eds.), *Applied Optics and Optical Engineering,* Vol. 10, San Diego, Academic, 1987 (dimensional stability).

Jacobson, R., in Kingslake (ed.), *Applied Optics and Optical Engineering,* Vol. 1, New York, Academic, 1965 (projection screens).

Jamieson, J., et al., *Infrared Physics and Engineering,* New York, McGraw-Hill, 1963.

Jenkins, F., and H. White, *Fundamentals of Optics,* 4th ed., New York, McGraw-Hill, 1976.

Kreidl, N., and J. Rood, in Kingslake (ed.), *Applied Optics and Optical Engineering,* Vol. 1, New York, Academic, 1965 (materials).

Lytle, J. D., "Polymetric Optics," in *Handbook of Optics,* Vol. 2, New York, McGraw-Hill, 1995, Chap. 34.

Meltzer, R., in Kingslake (ed.), *Applied Optics and Optical Engineering,* Vol. 1, New York, Academic, 1965 (polarization).

Moore, D. T., "Gradient Index Optics," in *Handbook of Optics,* Vol. 2, New York, McGraw-Hill, 1995, Chap. 9.

Palmer, J. M., "The Measurement of Transmission, Absorption, Emission and Reflection," in *Handbook of Optics,* Vol. 2, New York, McGraw-Hill, 1995, Chap. 25.

Paquin, R. A., "Properties of Metals," in *Handbook of Optics,* Vol. 2, New York, McGraw-Hill, 1995, Chap. 35.

Parker, C., in Shannon and Wyant (eds.), *Applied Optics and Optical Engineering,* Vol. 7, New York, Academic, 1979 (refractive materials).

Photonics Buyers Guide, Optical Industry Directory, Laurin Publishers, Pittsfield, MA (published annually).

Pompea, S. M., and R. P. Breault, "Black Surfaces for Optical Systems," in *Handbook of Optics,* Vol. 2, New York, McGraw-Hill, 1995, Chap. 37.

Scharf, P., in Kingslake (ed.), *Applied Optics and Optical Engineering,* Vol. 1, New York, Academic, 1965 (filters).

Strong, J., *Concepts of Classical Optics,* New York, Freeman, 1958.

Tropf, W. J., M. Thomas, and T. J. Harris, "Properties of Crystals and Glasses," in *Handbook of Optics,* Vol. 2, New York, McGraw-Hill, 1995, Chap. 33.

Welham, B., in Shannon and Wyant (eds.), *Applied Optics and Optical Engineering,* Vol. 7, New York, Academic, 1979 (plastics).

Wolfe, W., in W. Driscoll (ed.), *Handbook of Optics,* New York, McGraw-Hill, 1978 (materials).

Wolfe, W., in Wolfe and Zissis (eds.), *The Infrared Handbook,* Washington, D.C., Office of Naval Research, 1985 (materials).

Exercises

1 (a) What is the transmission of a stack of thin plane parallel plates of glass ($n = 1.5$) at normal incidence?

(b) What percentage of the incident light is transmitted directly, i.e., without any intervening reflections?

ANSWER:

(a) Using Eq. 10.6 $T_1 = 2n/(n^2 + 1) = 3/3.25 = 0.92307692$

  Using Eq. 10.7 $R_1 = (n - 1)^2/(n^2 + 1) = 0.25/3.25 = 0.07692308$

  Using Eq. 10.5 $T_{1.2} = T_1 T_2 K/(1 - K^2 R_1 R_2)$ and:
  $T_1 = T_2; R_1 = R_2; K = 1.0$
  $T_{1.2} = 0.8571428$
  $R_{1.2} = 1 - T_{1.2} = 0.1428572$

  Using Eq. 10.5 $T_{1.2.3} = T_{1.2} T_3 K/(1 - K^2 R_{1.2} R_3)$ and: $T_3 = T_1; R_3 = R_1$
  $T_{1.2.3} = 0.80$

(b) Neglecting multiple reflections:

  Using Eq. 10.7 $R = (n - 1)^2/(n^2 + 1) = 0.04$
  $T = 1 - R = 0.96$
  Through 6 surfaces: $T = 0.96^6 = 0.782758$

2 If a 1-cm thickness of a material transmits 85 percent and a 2-cm thickness transmits 80 percent, (a) what percentage will a 3-cm-thick piece transmit? (b) What is the absorption coefficient of the material? (Neglect all *multiple* reflections.)

ANSWER: (a)   Each additional cm of material reduces the transmission by a factor of 0.80/0.85 = 0.941176.

Thus for 3 cm we get $T = 0.8 \times 0.941176 = 0.752941$

(b)   Neglecting all multiple reflections $T = T_{\text{SURFACE}} \times T_{\text{ABSORBTION}}$ and by Eq. 10.3, $T_A = e^{-ax}$, so:

$$T = T_s e^{-at} \quad t = 1.0 \quad 0.85 = T_s e^{-a}$$
$$t = 2.0 \quad 0.80 = T_s e^{-2a}$$

$$T_s = 0.85 e^{+a} = 0.80 e^{+2a}$$
$$e^{-a} = 0.80/0.85 = 0.941176$$
$$a = -\log_e 0.941176 = 0.06062462 \text{ cm}^{-1}$$

3   Determine the coefficients for the dispersion equation, Eq. 10.8 given in Sec. 10.1 for BK7 glass as given in Fig. 10.4, using the d(.5876), C(.6563), and F(.4861) lines. Calculate the index for the r (.7065), g (.4358), and h (.4047) lines and compare with the entries in Fig. 10.4.

ANSWER: To get the coefficients, solve three simultaneous equation versions of Eq. 10.8 using the wavelengths given above and the indices from Fig. 10.4. Put the coefficients into Eq. 10.8 and determine the indices using the wavelengths for r, g, and h as given above.

4   Plot the spectral transmission curve which will result if the filters c and f shown in Fig. 10.10 are combined.

ANSWER: From the figure, determine the transmission of each filter for several wavelengths between 450 nm and 600 nm. Multiply the transmissions at each wave length to get the transmission of the combination at that wavelength.

Optical Coatings

11.1 Dielectric Reflection and Interference Filters

The portion of the light reflected (*Fresnel reflection*) from the surface of an ordinary dielectric material (such as glass) is given by

$$R = \frac{1}{2} \left[\frac{\sin^2 (I - I')}{\sin^2 (I + I')} + \frac{\tan^2 (I - I')}{\tan^2 (I + I')} \right] \tag{11.1}$$

where I and I' are the angles of incidence and refraction, respectively. The first term of Eq. 11.1 gives the reflection of the light which is polarized normal to the plane of incidence (s-polarized), and the second term the reflection for the other plane of polarization (p-polarized). As indicated in Sec. 10.1, at normal incidence Eq. 11.1 reduces to

$$R = \frac{(n' - n)^2}{(n' + n)^2} \tag{11.2}$$

The variation of reflection from an air-glass interface as a function of the angle of incidence (I) is shown in Fig. 11.1, where the solid line is R, the dashed line is the sine term, and the dotted line is the tangent term. Notice that the dotted line drops to zero reflectivity at Brewster's angle (Eq. 10.18).

The reflection from more than one surface can be treated as indicated by Eq. 10.5 when the separation between the surfaces is large compared to the wavelength of light. However, when the surface-to-surface separation is small, then interference between the light reflected from

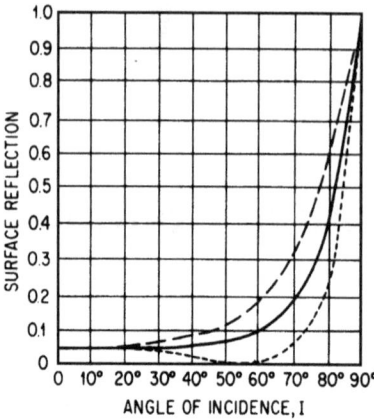

Figure 11.1 The reflection from a single air-glass interface (for an index of 1.523). Solid line is the reflection of unpolarized light. The fine dashed line is the reflection of p-polarized light, with the electric field vector parallel to the plane of incidence. The heavier dashed line is for the s-polarization. (Note that the "plane of polarization" was originally defined to be at right angles to what we now call the plane of polarization/vibration.)

the various surfaces will occur and the reflectivity of the stack of surfaces will differ markedly from that given by Eq. 10.5. (At this point the reader may wish to refer to the discussion of interference effects contained in the first chapter.)

Optical coatings are thin (usually a fractional wavelength thick) films of various substances which are vacuum deposited in layers on an optical surface for the purpose of controlling or modifying the reflection and transmission characteristics of the surface. The Table 11.1 lists a number of materials which have been used in optical coatings.

TABLE **11.1**   **Table of Optical Coating Materials**

Material	Formula	Index	Material	Formula	Index
Aluminum oxide	Al_2O_3	1.62	Zirconium dioxide	ZrO_2	2.2
Cadmium telluride	CdTe	2.69	Zinc selenide	ZnSe	2.44
Cerium dioxide	CeO_2	2.2	Zinc sulfide and	ZnS	2.3
Cerium fluoride	CeF_3	1.60	in the infrared		
Cryolite	Na_3AlF_6	1.35	Germanium	Ge	4.0
Hafnium oxide	HfO_2	2.05	Lead telluride	PbTe	5.1
Lanthanum trifluoride	LaF_3	1.57	Silicon and as	Si	3.5
Magnesium fluoride	MgF_2	1.38	metallic reflectors		
Neodimium fluoride	NdF_3		Aluminum		
Silicon dioxide	SiO_2	1.46	Silver		
Silicon monoxide	SiO	1.86	Gold		
Tantalum pentoxide	Ta_2O_5	2.15	Copper		
Thorium fluoride	ThF_4	1.52	Chrome		
Titanium dioxide	TiO_2	2.3	Rhodium		
Yttrium oxide	Y_2O_3	1.85			

With the exception of the reflecting films, such films have an optical thickness (the *optical thickness* is the physical thickness times the index) which is measured in wavelengths, typically one-quarter or one-half wavelength. The deposition of thin films is carried out in a vacuum and is done by heating the material to be deposited to its evaporation temperature and allowing it to condense on the surface to be coated. The thickness of the film is determined by the rate of evaporation (or more accuratelly condensation) and the length of time the process is allowed to continue. Since interference effects produce colors in the light reflected from thin films, just as in oil films on wet pavements, it is possible to judge the thickness of a film by the apparent color of light reflected from it. Simple coatings can be controlled visually by utilizing this effect, but coatings consisting of several layers are often monitored photoelectrically, using monochromatic light, so that the sinusoidal rise and fall of the reflectivity can be accurately assessed and the thickness of each layer controlled. By using two different wavelengths (often from lasers), this technique can achieve high precision. Another popular monitoring technique utilizes a quartz crystal of the type used to control radio broadcast frequencies. The oscillation frequency of such a crystal varies with its mass or thickness. By depositing the coating directly on the crystal and measuring its oscillation frequency, the coating thickness can be accurately monitored.

Let us first consider a single-layer film the optical thickness of which (nt) is exactly one-quarter of a wavelength. For light entering the film at normal incidence, the wave reflected from the second surface of the film will be exactly one-half wavelength out of phase with the light reflected from the first surface when they recombine at the first surface, resulting in destructive interference (assuming that there is no phase change by reflection). If the amount of light reflected from each surface is the same, a complete cancellation will occur and no light will be reflected. Thus, if the materials involved are nonabsorbing, all the energy incident on the surface will be transmitted. This is the basis of the "quarter-wave" low-reflection coating which is almost universally used to increase the transmission of optical systems. Since low-reflection coatings reduce reflections, they tend to eliminate ghost images as well as the stray reflected light which reduces contrast in the final image. Before the invention of low-reflection coatings, optical systems which consisted of many separate elements were impractical because of the transmission losses incurred in surface reflections and the frequent ghost images. Even complex lenses were usually limited to only four air-glass surfaces. A magnesium fluoride coating has an additional benefit in that it is actually (when properly applied) a protective coating; the chemical stability of many glasses is enhanced by coating.

The reflectivity of a surface coated with one thin film is given by the equation

$$R = \frac{r_1^2 + r_2^2 + 2r_1r_2 \cos X}{1 + r_1^2 r_2^2 + 2r_1r_2 \cos X} \qquad (11.3)$$

where

$$X = \frac{4\pi n_1 t_1 \cos I_1}{\lambda} \qquad (11.4)$$

$$r_1 = \frac{-\sin (I_0 - I_1)}{\sin (I_0 + I_1)} \quad \text{or} \quad \frac{\tan (I_0 - I_1)}{\tan (I_0 + I_1)} \qquad (11.5)$$

$$r_2 = \frac{-\sin (I_1 - I_2)}{\sin (I_1 + I_2)} \quad \text{or} \quad \frac{\tan (I_1 - I_2)}{\tan (I_1 + I_2)} \qquad (11.6)$$

and λ is the wavelength of light; t is the thickness of the film; n_0, n_1, and n_2 are the refractive indices of the media; and I_0, I_1, and I_2 are the angles of incidence and refraction. Figure 11.2 shows a sketch of the film and indicates the physical meanings of the symbols. The sine or tangent expressions for r_1 and r_2 are chosen depending on the polarization of the incident light as in Eq. 11.1; for unpolarized light, which is composed equally of both polarizations, R is computed for each polarization and the two values are averaged. If we assume nonabsorbing materials, the transmission T equals $(1 - R)$. At normal incidence $I_0 = I_1 = I_2 = 0$, and r_1 and r_2 reduce to

$$r_1 = \frac{n_0 - n_1}{n_0 + n_1} \qquad (11.7)$$

$$r_2 = \frac{n_1 - n_2}{n_1 + n_2} \qquad (11.8)$$

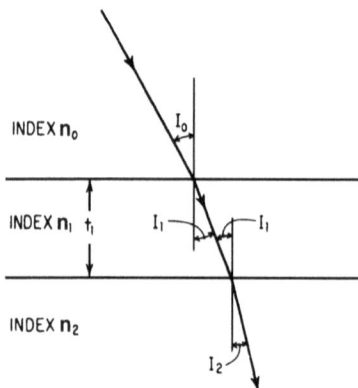

INDEX n_0

INDEX n_1 t_1

INDEX n_2

Figure 11.2 Passage of light ray through a thin film, indicating the terms used in Eq. 11.3.

Using Eqs. 11.7 and 11.8 for r_1 and r_2, Eq. 11.3 can be solved for the thickness which yields a minimum reflectance. As the preceding discussion would lead one to expect, this occurs when the optical thickness of the film is one-quarter wavelength, that is,

$$n_1 t_1 = \frac{\lambda}{4} \qquad (11.9)$$

At normal incidence the reflectivity of a quarter-wave film is thus equal to

$$\left[\frac{(n_0 n_2 - n_1^2)}{(n_0 n_2 + n_1^2)} \right]^2 \qquad (11.10a)$$

and the film index which will produce a zero reflectance is

$$n_1 = \sqrt{n_0 n_2} \qquad (11.10b)$$

Thus, to produce a coating which will completely eliminate reflections at an air-glass surface, a quarter-wave coating of a material whose index is the square root of the index of the glass is required. Magnesium fluoride (MgF_2) with an index of 1.38 is used for this purpose; its ability to form a hard durable film which will withstand weathering and frequent cleaning is the prime reason for its use, despite the fact that its index is higher than the optimum value for almost all optical glasses. Equation 11.10b indicates that the magnesium fluoride, with its index of 1.38, would be an ideal low-reflection coating material for a substrate with an index of $1.38^2 = 1.904$. Thus it is a much more efficient low-reflection coating for high-index glass than for ordinary glass of a lower index. The measured white light reflection of a low-reflection coating on various index materials is shown in Fig. 11.3.

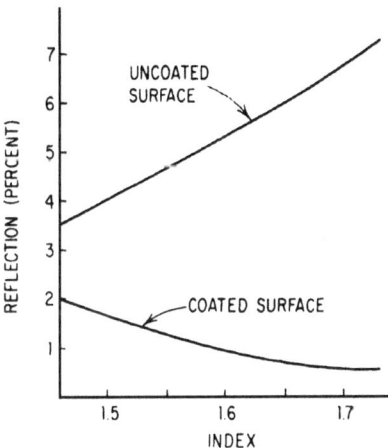

Figure 11.3 The measured reflection of white light from an uncoated surface and from a surface coated with a quarter-wave MgF_2 low-reflection coating, as a function of the index of the base material.

From Eq. 11.3 it is apparent that the reflectivity of a coated surface will vary with wavelength. Obviously a quarter-wave coating for one wavelength will be either more or less than a quarter-wave thick for other wavelengths, and the interference effects will be modified accordingly. Thus a low-reflection coating designed for use in the visible region of the spectrum will have a minimum reflectance for yellow light, and the reflectance for red and blue light will be appreciably higher. This is the cause of the characteristic purple color of single-layer low-reflection coatings. Figure 11.4a indicates this variation.

With more than one layer, more effective antireflection coatings can be constructed. Theoretically, two layers allow the reduction of the reflection to zero, provided that materials of suitable index are available; frequently, three layers are used for this purpose. Such a coating achieves a zero reflectivity at a single wavelength at the expense of a much higher reflectivity on either side. Because of the shape of the reflectivity curve, this is called a V-coating. It is widely used for monochromatic systems, such as those utilizing lasers as light sources.

With three or more layers, a broad-band, higher-efficiency, low-reflection coating may be achieved as shown in Fig. 11.4b. Such a coating may have two, or three minima, depending on the complexity of the coating design. A typical reflection over the visual spectrum is to the order of 0.25 percent, sometimes with another 0.25 percent lost to scattering and absorption.

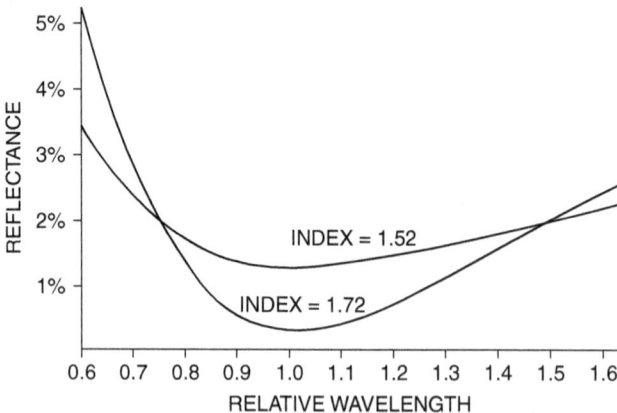

Figure 11.4a The reflectivity of a quarter-wave MgF_2 ($n = 1.38$) coating on high ($n = 1.72$) and low ($n = 1.52$) index substrates. The wavelength scale is normalized to the wavelength at which the coating has an optical thickness of one quarter-wave. Note that as the wavelength approaches infinity, the reflectivity will approach the value of the uncoated substrate (4.26 percent for the 1.52 index, and 7.01 percent for the 1.72 index), as it will also at the normalized wavelength of 0.5.

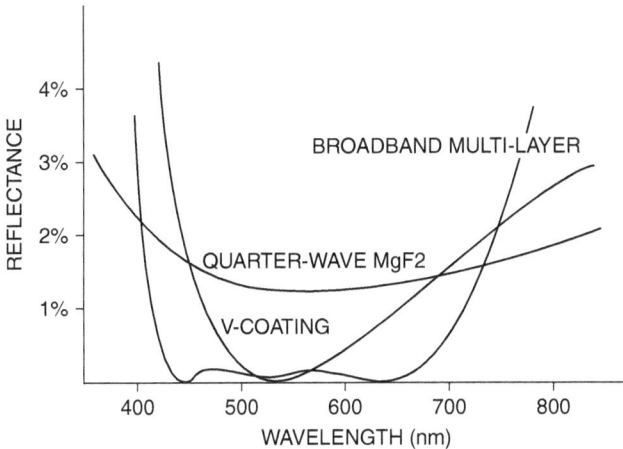

Figure 11.4b The reflectivity of three typical coatings: a single quarter-wave layer of magnesium fluoride; a two (or more) layer "V-coating"; and   a three (or more) layer broadband ultra-low reflectivity coating.

Thin-film computations

The following equations can be used to calculate the reflection and transmission of an interference coating of any number of layers. The equations can be used at oblique angles and will accommodate absorbing materials. They do require a knowledge of complex arithmetic; if not already familiar with the subject, the interested reader may wish to consult a basic text on complex arithmetic. These equations are the basis of most of the computer programs used in the design and evaluation of thin films. The formulas given here are taken from Peter Berning, in G. Hass (ed.), *Physics of Thin Films,* vol. 1, Academic, 1963.

The reflection and transmission characteristics of a "stack" of several thin films can be expressed in explicit equations; however, their complexity increases rapidly with the number of films, and the following recursion expressions are usually preferable. The physical thickness of each film is represented by t_j and the index by $n_j = N_j - iK_j$ (n is the complex index, N is the ordinary index of refraction, and K is the absorption coefficient, which is zero for nonabsorbing materials). The angle of incidence within the jth film is ϕ_j; and the "effective" refractive index is $u_j = n_j \cos \phi_j$ or $u_j = n_j/\cos \phi_j$ (for light polarized with the electric vector perpendicular to [s], or parallel to [p], the plane of incidence, respectively). Thus, for oblique incidence the calculations are carried out for both polarizations and the results are averaged (assuming the incident light to be unpolarized and to consist of equal parts of each polarization).

Since most calculations are carried out at normal incidence ($\phi_j = 0$) and for nonabsorbing materials ($K_j = 0$), one may ordinarily use $u_j = n_j = N_j$.

The subscript notation is $j = 0$ for the substrate, $j = 1$ for the first film, $j = 2$ for the second, etc., $j = p - 1$ for the last film and $j = p$ for the final medium, which is usually air. For each film g_j, the effective optical thickness, in radians, is computed from

$$g_j = \frac{2\pi n_j t_j \cos \phi_j}{\lambda} \tag{11.11}$$

where λ is the wavelength of light for which the calculation is made.

Starting with $E_1 = E_0{}^+ = 1.0$ and $H_1 = u_0 E_0{}^+ = u_0$, the following equations are applied iteratively at each surface, with the subscript j advancing from $j = 1$ to $j = p - 1$.

$$E_{j+1} = E_j \cos g_j + \frac{iH_j}{u_j} \sin g_j \tag{11.12}$$

$$H_{j+1} = iu_j E_j \sin g_j + H_j \cos g_j \tag{11.13}$$

where $i = \sqrt{-1}$ and the other terms have been defined above. Readers familiar with matrix notation may prefer to manipulate the equivalent matrix form

$$\begin{pmatrix} E_{j+1} \\ H_{j+1} \end{pmatrix} = \begin{pmatrix} \cos g_j & \frac{i}{u_j} \sin g_j \\ iu_j \sin g_j & \cos g_j \end{pmatrix} \begin{pmatrix} E_j \\ H_j \end{pmatrix} \tag{11.14}$$

When Eqs. 11.12 and 11.13 (or 11.14) have been applied to the entire stack, we have the values of E_p and H_p, which will generally be complex numbers of the form $z = x + iy$. These are substituted into

$$E_p{}^+ = \frac{1}{2}\left(E_p + \frac{H_p}{u_p}\right) = x_2 + iy_2 \tag{11.15}$$

$$E_p{}^- = \frac{1}{2}\left(E_p - \frac{H_p}{u_p}\right) = x_1 + iy_1 \tag{11.16}$$

and the reflectance of the thin-film system is found from

$$R = \left|\frac{E_p{}^-}{E_p{}^+}\right|^2 \tag{11.17}$$

where the symbol $|z|$ indicates the modulus of a complex number z, so that

$$|z| = |x + iy| = \sqrt{x^2 + y^2}$$

and

$$R = |z|^2 = x^2 + y^2 = \left| \frac{x_1 + iy_1}{x_2 + iy_2} \right|^2 = \frac{x_1^2 + y_1^2}{x_2^2 + y_2^2}$$

If the computation has been for normal incidence through nonabsorbing materials, the transmission is given by

$$T = 1 - R \tag{11.18}$$

Otherwise, the transmission is given by

$$T = \frac{n_0 \cos \phi_0}{n_p \cos \phi_p} \left| \frac{E_0^+}{E_p^+} \right|^2 \tag{11.19a}$$

or

$$T = \frac{n_0 \cos \phi_p}{n_p \cos \phi_0} \left| \frac{E_0^+}{E_p^+} \right|^2 \tag{11.19b}$$

where Eq. 11.19a is used for light polarized with the electric vector perpendicular to [s] and Eq. 11.19b for the electric vector parallel to [p] the plane of incidence.

A discussion of the design of multilayer coatings is beyond the scope of this volume; the interested reader should pursue the subject in the references listed at the end of this chapter. By suitable combinations of thin films of different indices and thicknesses a tremendous number of transmission and reflection effects can be created. Among the types of interference coatings which are readily available are long- or short-pass transmission filters, bandpass filters, narrow bandpass (spike filters), achromatic extra-low-reflection coatings as well as the reflection coatings described in the next section. An extremely valuable property of thin-film coatings is their spectral versatility. Once a combination of films has been designed to produce a desired characteristic, the wavelength region can be shifted at will by simply increasing or decreasing all the film thicknesses in proportion. For example, a spike filter designed to transmit a very narrow spectral band at 1 μm can be shifted to 2 μm by doubling the thickness of each film in the coating. This, of course, is limited by the absorption characteristics of the substrate and the film materials.

The characteristics of a number of typical interference coatings are shown in Fig. 11.5. Note that the wavelength scale is plotted in arbitrary units, with a central wavelength of 1, since (within quite broad limits)

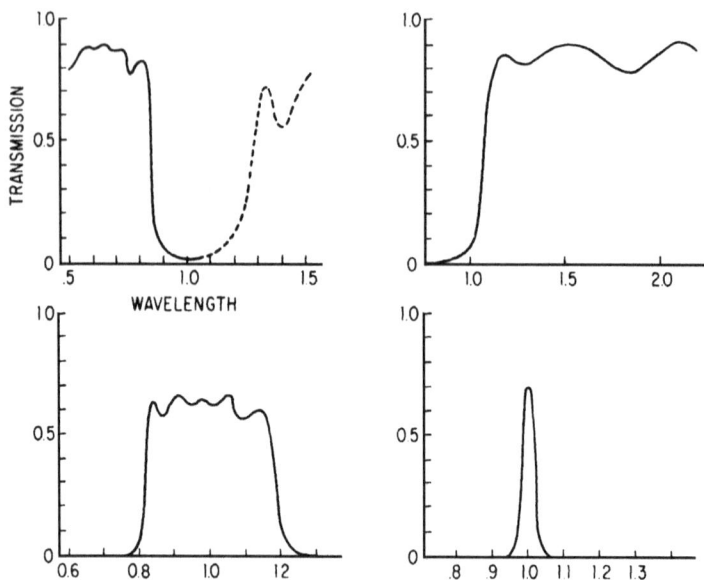

Figure 11.5 Transmission of typical evaporated interference filters plotted against wavelength in arbitrary units. (*Upper left*) Short-pass filter (note that dashed portion of curve must be blocked by another filter if low long wavelength transmission is necessary). (*Upper right*) Long-pass filter. (*Lower left*) Bandpass filter. (*Lower right*) Narrow bandpass (spike) filter.

the characteristics can be shifted up or down the spectrum as described in the preceding paragraph. Most interference filters are very nearly 100 percent efficient, so that the reflection for a film is equal to one minus the transmission (except in regions where the materials used become absorbing). Since the characteristics of an interference filter depend on the thickness of the films, the characteristics will change when the angle of incidence is changed. This is in great measure due to the fact that the optical path through a film is increased when the light passes through obliquely. For *moderate obliquity* angles the effect is usually to shift the spectral characteristics to a slightly *shorter* wavelength. The wavelength shift with obliquity is approximated by

$$\lambda_\theta = \frac{\lambda_0}{n} \sqrt{n^2 - \sin^2\theta}$$

where λ_θ is the shifted wavelength at an angle of incidence of θ, λ_0 is the wavelength for normal incidence, and n is the "effective index" for the coating stack (n is typically in the range of 1.5 to 1.9 for most coatings). Coatings also shift wavelength effects with temperature; this shift is to the order of one- or two-tenths of an angstrom per degree Celsius.

Coatings consisting of a few layers are for the most part reasonably durable and can withstand careful cleaning. However, coatings consisting of a great number of layers (and coatings consisting of 50 or more layers are occasionally used) tend toward delicacy, and must be handled with due respect.

Some multilayer coatings are quite effective polarizers when used obliquely (and as such, are occasionally responsible for "mysterious" happenings). This is notably true in systems using linearly polarized laser beams. One must also exercise care in photometric or radiometric applications (e.g., spectrophotometers), since polarization effects can introduce significant errors.

11.2 Reflectors

Although polished bulk metals are occasionally used for mirror surfaces, most optical reflectors are fabricated by evaporating one or more thin films on a polished surface, which is usually glass. Obviously the interference filters described in the preceding section can be used as special-purpose reflectors in instances where their spectral characteristics are suitable. However, the workhorse reflector material for the great majority of applications is an aluminum film deposited on a substrate by evaporation in vacuum. Aluminum has a broad spectral band of quite high reflectivity and is reasonably durable when properly applied. Almost all aluminum mirrors are "overcoated" with a thin protective layer of either silicon monoxide or magnesium fluoride. This combination produces a first-surface mirror which is rugged enough to withstand ordinary handling and cleaning without undue scratching or other signs of wear.

The spectral reflectance characteristics of several evaporated metal films are shown in Fig. 11.6. With the exception of the curve for rhodium, the reflectivities given here can seldom be attained for practical purposes; the silver coating will tarnish and the aluminum film will oxidize, so that the reflectances tend to decrease with age, especially at shorter wavelengths. The high reflectivity of silver is only useful when the coating is properly protected.

Figure 11.7 indicates the variety of characteristics which are available in commercial aluminum mirrors. A run-of-the-mill protected aluminum mirror can be expected to have an average visual reflectance of about 88 percent. Two, four, or more interference films may be added to improve the reflectance where the additional cost can be accepted. This enhanced reflectivity within the bandpass of the mirror is obtained at the expense of a lowered reflectivity on either side, as can be seen from the dashed curve in Fig. 11.7.

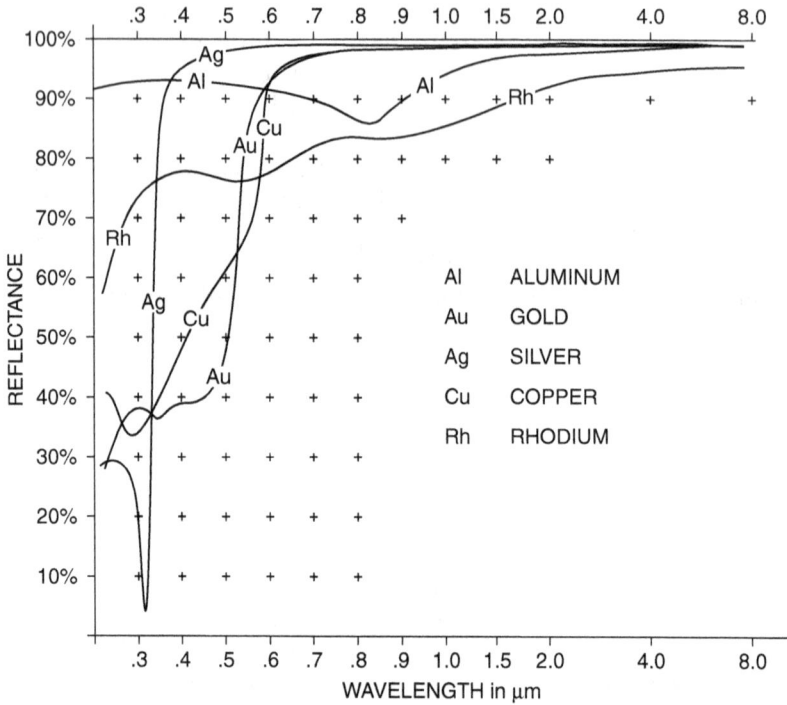

Figure 11.6 Spectral reflectance for evaporated metal films on glass. Data represent new coatings, under ideal conditions.

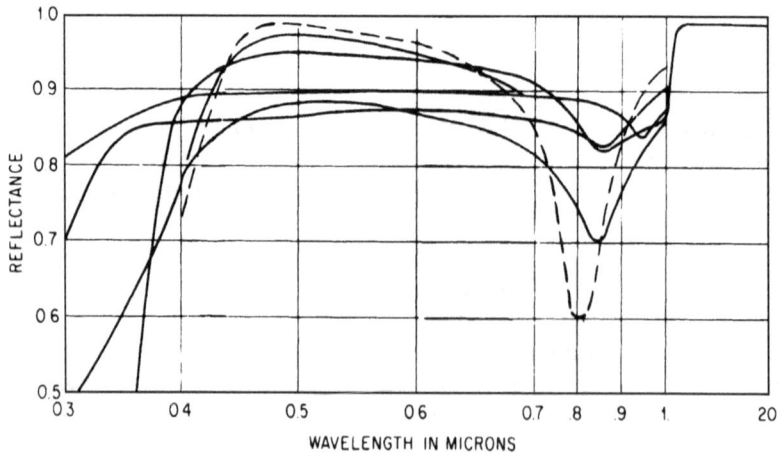

Figure 11.7 Spectral reflectance of aluminum mirrors. The solid curves are for aluminum films with various types of thin film overcoatings—either for protection or for increased reflectivity. The dashed line is an extra-high-reflectance multilayer coating. All coatings shown are commercially available.

Dichroics and semireflecting mirrors constitute another class of reflector. Both are used to split a beam of light into two parts. A dichroic reflector splits the light beam spectrally, in that it transmits certain wavelengths and reflects others. A dichroic reflector is often used for heat control in projectors and other illuminating devices. A *hot mirror* is a dichroic which transmits the visible region of the spectrum and reflects the near infrared. A *cold mirror* does just the reverse, in that it transmits the infrared and reflects the visible. For example, a cold mirror introduced into the optical path will allow undesired heat in the form of infrared radiation to be removed from the beam by transmitting it to a heat dump. These mirrors have the advantage over heat-absorbing filter glass in that they do not themselves get hot and thus do not require a fan for cooling. A semireflecting mirror is, nominally at least, spectrally neutral; its function is to divide a beam into two portions, each with similar spectral characteristics. Figure 11.8 shows the characteristics of a variety of these partial reflectors.

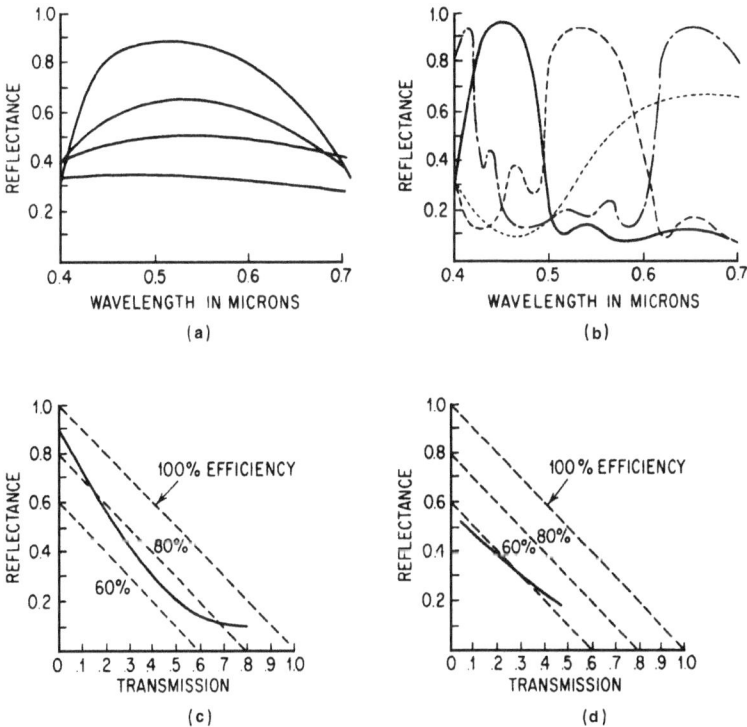

Figure 11.8 Characteristics of partial reflectors. (a) Multilayer "neutral" semireflectors (efficiency better than 99 percent). (b) Dichroic multilayer reflectors—blue, green, red, and yellow reflection. (c) Visual efficiency of aluminum semireflectors. (d) Visual efficiency of chrome semireflectors.

11.3 Reticles

A reticle is a pattern used at or near the focus of an optical system, such as the cross hairs in a telescope. For a simple cross-hair pattern, fine wire or spider (web) hair is occasionally used, stretched across an open frame. However, a pattern which is supported on a glass (or other material) substrate offers considerably more versatility, and most reticles, scales, divided circles, and patterns are of this type.

The simplest type of reticle is produced by scribing, or scoring, the glass surface with a diamond tool. A line produced this way, while not opaque, modifies the glass sufficiently so that under the proper type of illumination the line will appear dark. Where clear lines in an opaque background are desired, the glass can be coated with an opaque coating, such as evaporated aluminum, and the lines scribed through the coating with a diamond or hardened steel tool, depending on the type of line desired. Scribing produces very fine lines.

Another old technique is to etch the substrate material. A waxy resist is coated on the substrate and the desired pattern cut through the resist. The exposed portion of the substrate is then etched (with hydrofluoric acid in the case of glass) to produce a groove in the material. The groove can be filled with titanium dioxide (white), or lamp black in a waterglass medium, or evaporated metal. Etched reticles are durable and have the advantage that they can be edge-lighted if illumination is necessary. Any substrate that is readily etched can be used. This process is used for many military reticles and also for accurate metrology scales on steel.

The most versatile processes for production of reticles are based on the use of a photoresist, or photosensitive material. Photoresists are exposed like a photographic emulsion, either by contact printing through a master or by photography. However, when the photoresist is "developed," the exposed areas are left covered with the resist and the unexposed areas are completely clear. Thus, an evaporated coating of any of a number of metals (aluminum, chrome, inconel, nichrome, copper, germanium, etc.) can be deposited over the resist. In the clear areas the coating adheres to the substrate; when the resist is removed, it carries away the coating deposited upon it, leaving a durable pattern which is an exact duplicate of the master. The precision, versatility, ruggedness, and suitability for mass production of this technique have earned it a prominent place in the field of reticle manufacture.

The photoresist technique may also be combined with etching, where the material to be etched is either a metal substrate or an evaporated metal film.

Where the reticle pattern must be nonreflecting, the glue silver process or the black-print process is used. The technique is similar to that used in producing the photoresist pattern, except that the

TABLE 11.2   Resolution and Accuracy of Reticle Producing Techniques

Method	Finest line width, in	Dimensional repeatability, in	Minimum figure height, in
Scribing	0.00001	±0.00001	
Etch (and fill)	0.0002–0.0004	±0.0001	0.004
Photo-resist (evaporated metal)	0.0001–0.0002	±0.00005	0.002
Glue silver	0.00003–0.0002	±0.00005–0.0005	0.002
Black print	0.001	±0.0001	0.005
Emulsion	0.00005–0.0001	±0.00005	0.001

photosensitive material is opaque. The clear areas are free of emulsion. Glue silver reticles are fragile but capable of very high resolution of detail. The black-print process is more durable. Occasionally an extremely high resolution photographic emulsion is used for a reticle pattern; however, the presence of emulsion in the clear areas of the pattern is ordinarily a drawback.

Table 11.2 indicates the resolution and accuracy possible with these techniques. These figures represent the highest level of quality that reticle manufacturers are capable of at the present time; if cost is a factor, one is well advised to lower one's requirements an order of magnitude or so below the levels indicated here.

Bibliography

American Institute of Physics Handbook, 3d ed., New York, McGraw-Hill, 1972.
Barnes, W., in Shannon and Wyant (eds.), *Applied Optics and Optical Engineering,* Vol. 7, New York, Academic, 1979 (reflective materials).
Baumeister, P., in Kingslake (ed.), *Applied Optics and Optical Engineering,* Vol. 1, New York, Academic, 1965 (coatings).
Berning, P., in Hass (ed.), *Physics of Thin Films,* Vol. 1, New York, Academic, 1963 (calculations).
Dobrowolski, J., "Optical Properties of Films and Coatings," in *Handbook of Optics,* Vol. 1, New York, McGraw-Hill, 1995, Chap. 42.
Dobrowolski, J., in W. Driscoll (ed.), *Handbook of Optics,* New York, McGraw-Hill, 1978 (coatings).
Hass, G., in Kingslake (ed.), *Applied Optics and Optical Engineering,* Vol. 3, New York, Academic, 1975 (mirror coatings).
Heavens, O., *Optical Properties of Thin Films,* London, Butterworth's, 1955.
Holland, L., *Vacuum Deposition of Thin Films,* New York, Wiley, 1956.
Macleod, H., in Shannon and Wyant (eds.), *Applied Optics and Optical Engineering,* Vol. 10, San Diego, Academic, 1987 (coatings).
Macleod, H., *Thin Film Optical Filters,* 2d ed., New York, McGraw-Hill, 1988.
Palmer, J. M., "The Measurement of Transmission, Absorption, Emission and Reflection," in *Handbook of Optics,* Vol. 2, New York, McGraw-Hill, 1995, Chap. 25.
Paquin, R. A., "Properties of Metals," in *Handbook of Optics,* Vol. 2, New York, McGraw-Hill, 1995, Chap. 35.
Photonics Buyers Guide, Optical Industry Directory, Laurin Publishers, Pittsfield, MA (published annually).
Pompea, S. M., and R. P. Breault, "Black Surfaces for Optical Systems," in *Handbook of Optics,* Vol. 2, New York, McGraw-Hill, 1995, Chap. 37.

Rancourt, J., *Optical Thin Films*, New York, McGraw-Hill, 1987.
Scharf, P., in Kingslake (ed.), *Applied Optics and Optical Engineering*, Vol. 1, New York, Academic, 1965 (filters).
Thelen, A., *Design of Optical Interference Coatings*, New York, McGraw-Hill, 1988.
Vasicek, A., *Optics of Thin Films*, Amsterdam, North Holland, 1960.

Exercises

1 Plot, in the manner of Fig. 11.1, the curve of reflection against angle of incidence for a single surface of glass ($n = 1.52$), coated with a quarter wavelength of magnesium fluoride ($n = 1.38$).

ANSWER: Use Eq. 11.3 to 11.6 with the following:

$$n_0 = 1.0$$
$$n_1 = 1.38$$
$$n_2 = 1.58$$

Eq. 11.4 $X = (4\pi n_1 t_1 \cos I_1)/\lambda = [4\pi(\lambda/4) \cos I_1]/\lambda = \pi \cos I_1$

$\sin I_1 = (n_0/n_1) \sin I_0 = (1.0/1.38) \sin I_0$

$\sin I_2 = (n_0/n_2) \sin I_0 = (1.0/1.52) \sin I_0$

Using Eq. 11.5 to determine the two values for r_1 and Eq. 11.6 for the two values of r_2 and Eq. 11.3 to determine the reflectance for both planes of polarization, we get:

angle I_0	$R(\perp)$	$R(\|)$	R average
0°	1.26%	1.26%	1.26%
20°	1.56%	1.01%	1.28%
40°	3.11%	0.32%	1.71%
50°	5.31%	0.03%	2.67%
51°	5.64%	0.02%	2.83%
52°	5.99%	0.03%	3.01%
53°	6.37%	0.04%	3.21%
55°	7.23%	0.11%	3.67%
60°	10.08%	0.60%	5.34%
80°	44.94%	24.32%	34.63%
90°	100.0%	100.0%	100.0%

12

Principles of Radiometry and Photometry

12.1 Introduction

In concept, both radiometry and photometry are quite straightforward; however, both have been cursed with a jungle of changing and often bewildering terminology. Radiometry deals with radiant energy (i.e., electromagnetic radiation) of any wavelength. Photometry is restricted to radiation in the visible region of the spectrum. The basic unit of power (i.e., rate of transfer of energy) in radiometry is the watt; in photometry, the corresponding unit is the lumen, which is simply radiant power as modified by the relative spectral sensitivity of the eye (Fig. 8.8 and 8.9 in Chap. 8) per Eq. 12.18. Note that watts and lumens have the same dimensions, namely energy per time.

All radiometry must take into account the variation of characteristics with wavelength. Examples are the spectral variation of emission, the variation of transmission of the atmosphere and optics with wavelength, and the differences in detector and film response with wavelength. A convenient way to deal with this is to multiply, wavelength by wavelength, all such factors together so as to arrive at one unified spectral weighting function. Thus, all radiometry is spectrally weighted and it should be apparent that photometry is simply one particular spectral weighting. See Sec. 12.9.

The principles of radiometry and photometry are readily understood when one thinks in terms of the basic units involved, rather than the special terminology which is conventionally used. The next five sections will discuss radiation in terms of watts; the reader should remember that the discussion is equally valid for photometry, if lumens are read for watts.

12.2   The Inverse Square Law; Intensity

Consider a hypothetical point (or "sufficiently" small) source of radiant energy, which is radiating uniformly in all directions. If the rate at which energy is radiated is P watts, then the source has a radiant *intensity* J of $P/4\pi$ watts per steradian,* since the solid angle into which the energy is radiated is a sphere of 4π steradians. Of course there are no truly "point" sources and no practical sources which radiate uniformly in all directions, but if a source is quite small relative to its distance, it can be treated as a point, and its radiation, in the directions in which it does radiate, can be expressed in watts per steradian.

If we now consider a surface which is S cm from the source, then 1 cm² of this surface will subtend $1/S^2$ steradians from the source (at the point where the normal from the source to the surface intersects the surface, if S is large). The *irradiance H* on this surface is the incident radiant power per unit area and is obtained by multiplying the intensity of the source in watts per steradian by the solid angle subtended by the unit area. Thus, the irradiance is given by

$$H = J\frac{1}{S^2} \tag{12.1}$$

The units of irradiance are watts per square centimeter (W/cm²). Equation 12.1 is, of course, the "inverse square" law, which is conventionally stated: the illumination (irradiance) on a surface is inversely proportional to the square of the distance from the (point) source.

Thus, if our uniformly radiating point source emits energy at a rate of 10 W, it will have an intensity $J = 10/4\pi = 0.8$ W ster⁻¹, and the radiation falling on a surface 100 cm away would be 0.8×10^{-4} W/cm², or 80 µW/cm². If the surface is flat, the irradiance will, of course, be less than this at points where the radiation is incident at an angle, since the solid angle subtended by a unit of area in the surface will be reduced. From Fig. 12.1 it can be seen that the source-to-surface distance

---

*A steradian is the solid angle subtended (from its center) by $1/4\pi$ of the surface area of a sphere. Thus, a sphere subtends 4π (12.566) steradians from its center; a hemisphere subtends 2π steradians. The size of a solid angle in steradians is found by determining the area of that portion of the surface of a sphere which is included within the solid angle and dividing this area by the square of the radius of the sphere. For a small solid angle, the area of the included flat surface normal to the "central axis" of the angle can be divided by the square of the distance from the surface to the apex of the angle to determine its size in steradians. One can visualize a steradian as a cone with an apex angle of about 65.5°, or 3283 square degrees.

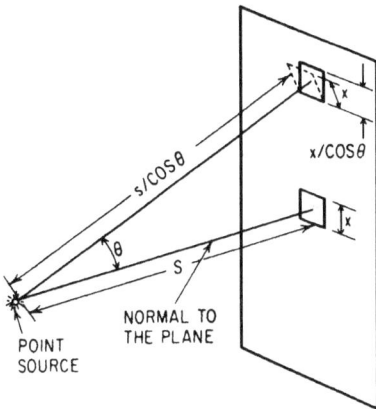

Figure 12.1 Geometry of a point source irradiating a plane, showing that irradiance (or illumination) varies with $\cos^3 \theta$.

is increased to $S/\cos \theta$ and that the effective area (normal to the direction of the radiation) is reduced by a $\cos \theta$ factor. Thus, the solid angle subtended, and the irradiance, are reduced by a $\cos^3 \theta$ factor.

12.3   Radiance and Lambert's Law

An extended source, that is, one, the dimensions of which are significant, must be treated differently than a point source. A small area of the source will radiate a certain amount of power per unit of solid angle. Thus, the radiation characteristics of an extended source are expressed in terms of power per unit solid angle per unit area. This is called *radiance*; the usual units for radiance are watts per steradian per square centimeter (W ster^{-1} cm^{-2}) and the symbol is N. Note that the area is measured normal to the direction of radiation, not in the radiating surface.

Most extended sources of radiation follow, at least approximately, what is known as Lambert's law of intensity,

$$J_\theta = J_0 \cos \theta \tag{12.2}$$

where J_θ is the intensity of a small incremental area of the source in a direction at an angle θ from the normal to the surface, and J_0 is the intensity of the incremental area in the direction of the normal. For example, a heated metal disk with a total area of 1 cm^2 and a radiance of 1 W ster^{-1} cm^{-2} will radiate 1 W/ster in a direction normal to its surface. In a direction 45° to the normal, it will radiate only 0.707 W/ster ($\cos 45° = 0.707$).

Notice that although radiance is given in terms of watts per steradian per square centimeter, this should not be taken to mean that the radiation is uniform over a full steradian or over a full square centimeter. Consider a source consisting of a 0.1-cm square incandescent filament in a 20-cm-diameter envelope. Assume that the bulb is painted so that only a 1-cm square transmits energy, and that the source radiates one-fiftieth of a watt through this square. (We assume, for convenience, that the radiation intercepted by the painted envelope is thereby totally removed from consideration.) Now the filament has an area of 0.01 cm² and is radiating 0.02 W into a solid angle of (approximately) 0.01 steradian. Therefore, it has a radiance of 200 W ster⁻¹ cm⁻², but only within the solid angle subtended by the window! Outside this angle the radiance is zero. This concept of radiance over a limited angle becomes important in dealing with the radiance of images and must be thoroughly understood.

There are several interesting consequences of Lambert's law that are worthy of consideration, not only for their own sake but because they illustrate the basic techniques of radiometric calculations. *The radiance of a surface is conventionally taken with respect to the area of a surface normal to the direction of radiation.* It can be seen that, although the emitted radiation per steradian falls off with cos θ according to Lambert's law, the "projected" surface area falls off at exactly the same rate. The result is that *the radiance of a Lambertian surface is constant with respect to θ.* In visual work the quantity corresponding to radiance is brightness (or luminance), and the above is readily demonstrated by observing that the brightness of a diffuse source is the same regardless of the angle from which it is viewed.

12.4 Radiation into a Hemisphere

Let us determine the total power radiated from a flat diffuse source into a hemisphere. If the source has a radiance of N W ster⁻¹ cm⁻², one might expect that the power radiated into a hemisphere of 2π steradians would be $2\pi N$ W/cm². That this is twice too large is readily shown. With reference to Fig. 12.2, let A represent the area of a small source

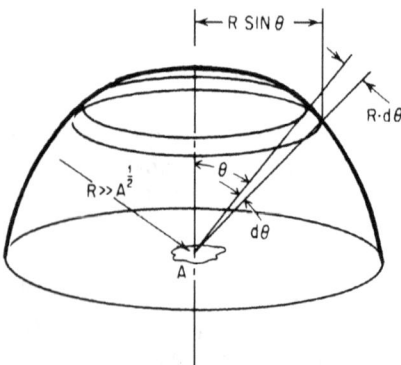

Figure 12.2 Geometry of a lambertian source radiating into a hemisphere.

with a radiance of N W ster^{-1} cm^{-2} and an intensity of $J_\theta = J_0 \cos \theta = NA \cos \theta$ W/ster. The incremental ring area on a hemisphere of radius R has an area of $2\pi R \sin \theta \cdot R \, d\theta$ and thus subtends (from A) a solid angle of $2\pi R^2 \sin \theta \, d\theta/R^2 = 2\pi \sin \theta \, d\theta$ steradians. The radiation intercepted by this ring is the product of the intensity of the source and the solid angle, or

$$dP = J_\theta \, 2\pi \sin \theta \, d\theta = 2\pi NA \sin \theta \cos \theta \, d\theta \qquad (12.3)$$

Integrating to find the total power radiated into the hemisphere from A, we get

$$P = \int_0^{\pi/2} 2\pi NA \sin \theta \cos \theta \, d\theta = 2\pi NA \left[\frac{\sin^2 \theta}{2} \right]_0^{\pi/2} = \pi NA \text{ watts} \qquad (12.4)$$

Dividing by A to get watts emitted per square centimeter of source, we find the radiation into the 2π steradian of the hemisphere to be πN W/cm^2, not $2\pi N$. *This is the basic relationship between radiance and the power emitted from the surface.*

12.5 Irradiance Produced by a Diffuse Source

It is frequently of interest to determine the irradiance produced at a point by a lambertian source of finite size. Referring to Fig. 12.3, assume that the source is a circular disk of radius R and that we wish to determine the irradiance at some point X which is a distance S from the source and is on the normal through the center of the source. (Note that we will determine the irradiance on a plane parallel to the plane of the source.) The radiant intensity of a small element of area dA in the direction of point X is given by Eq. 12.2 as

$$J_\theta = J_0 \cos \theta = N \, dA \cos \theta$$

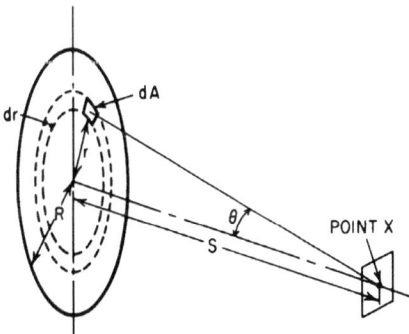

Figure 12.3 Geometry of a circular source irradiating point X.

where N is the radiance of the source. Since the distance from dA to X is $S/\cos\theta$, and the radiation arrives at an angle θ, the incremental irradiance at X produced by dA is

$$dH = J_0 \cos\theta \left[\frac{\cos^3\theta}{S^2} \right] = \frac{N\, dA\, \cos^4\theta}{S^2} \tag{12.5}$$

The same irradiance is produced by each incremental area making up a ring of radius r and a width dr, so that we can substitute the area of the ring, $2\pi r\, dr$, for dA in Eq. 12.5 to get the incremental irradiance from the ring.

$$dH = \frac{2\pi r\, dr\, N\, \cos^4\theta}{S^2} \tag{12.6}$$

To simplify the integration, we substitute

$$r = S\tan\theta$$

$$dr = S\sec^2\theta\, d\theta$$

into Eq. 12.6 to get

$$dH = \frac{2\pi S\tan\theta S\sec^2\theta\, d\theta N\cos^4\theta}{S^2}$$

$$= 2\pi N\tan\theta \cos^2\theta\, d\theta = 2\pi N\sin\theta \cos\theta\, d\theta$$

Integrating to determine the irradiance from the entire source, we get

$$H = \int_0^{\theta_m} 2\pi N\sin\theta \cos\theta\, d\theta = 2\pi N \left[\frac{\sin^2\theta}{2} \right]_0^\theta$$

$$H = \pi N\sin^2\theta_m \text{ watt/cm}^2 \tag{12.7}$$

where H is the irradiance produced at a point by a circular source of radiance N W ster^{-1} cm^{-2} which subtends an angle of $2\theta_m$ from the point (when the point is on the "axis" of the source). Note well that θ_m is the angle defined by the source diameter.

Unfortunately noncircular sources do not readily yield to analysis. However, *small* noncircular sources may be approximated with a fair degree of accuracy by noting that the solid angle subtended by the source from X is

$$\Omega = 2\pi (1 - \cos\theta) = 2\pi \frac{\sin^2\theta}{(1 + \cos\theta)}$$

and for small values of θ, $\cos\theta$ approaches unity and

$$\omega = \pi \sin^2\theta$$

Thus, if the angle subtended by the source is moderate, we can substitute into Eq. 12.7 and write

$$H = N\omega \tag{12.8}$$

If the point X does not lie on the "axis" (the normal through the center of the circular source), then the irradiance would be subject to the same factors outlined in the discussion of the "cosine-fourth" rule in Sec. 9.7. Thus, if the line from the point X_ϕ to the center of the circle makes an angle ϕ to the normal, the irradiance at X_ϕ is given by

$$H_\phi = H_0 \cos^4 \phi \tag{12.9}$$

where H_0 is the irradiance along the normal given by Eq. 12.7 or 12.8 and H_ϕ is the irradiance at X_ϕ (measured in a plane parallel to the source). (See the note in Example 12.1 regarding the inaccuracy of the cosine-fourth rule when the angles θ and ϕ are large.)

It is apparent that Eqs. 12.8 and 12.9 may be used in combination to calculate the irradiance produced by any conceivable source configuration, to whatever degree of accuracy that time (or patience) allows.

12.6   The Radiometry of Images

The conservation of radiance

When a source is imaged by an optical system, the image has a radiance, and it may be treated as a secondary source of radiation. However, one must always keep in mind that the radiance of an image differs from the radiance of an ordinary source in that the radiance of an image exists *only* within the solid angle subtended from the image by the clear aperture (or exit pupil) of the optical system. Outside of this angle, the radiance of the image is zero.

At first consideration the *conservation of radiance* (or *brightness*) seems quite counterintuitive. Ordinarily, the solid angle of radiation accepted by an optical system from a source is quite small, as is the fraction of the total power which passes through the lens and forms the image. It is difficult to accept that the image formed by this small fraction of the source power will have the same radiance as does the source. We can easily demonstrate this, using only the first-order optics from Chap. 2.

Let us assume a small source of radiance N with an area A. The source thus has an intensity of AN. The source is imaged by an optical system with an area P which is located a distance S from the source. The solid angle subtended by the lens from the source is thus P/S^2, and the power intercepted by the lens and formed into the image is ANP/S^2.

The lens will form an image with a magnification M, and the area of the image will thus be AM^2. The image distance will be MS, and the solid angle subtended by the lens from the image will be P/M^2S^2 ster. Thus the power in the image (ANP/S^2) is spread over the image area (AM^2) and exists only over the solid angle (P/M^2S^2). The image radiance is power per unit area per solid angle; combining the expressions above, we get (neglecting any transmission losses)

$$\text{Image radiance} = \text{power/area} \cdot \text{solid angle}$$

$$= \frac{(ANP/S^2)}{(AM^2)\,(P/M^2S^2)}$$

We can cancel A, P, S, and M, leaving us with

$$\text{Image radiance} = N \text{ (the object radiance)}$$

which is a statement of the *conservation of radiance* (or *brightness*).

The *conservance of radiance* (or brightness/luminance) states that the radiance of an image formed by an optical system is equal to the radiance of the object, mutiplied by the transmission of the system. More precisely, the radiance divided by the square of the index is the invariant quantity. Thus (with the object and image both in air) we have

$$N' = tN \tag{12.10a}$$

and more generally,

$$N' = tN\,(n'/n)^2 \tag{12.10b}$$

where N and N' are the radiance and n and n' are the indices of object and image space respectively, and t is the system transmission. Another way of expressing Eq. 12.10a is: *In air, the radiance of an image cannot exceed the radiance of the object.* Note that the index factor $(n'/n)^2$ can also be applied to Eq. 12.11 for the irradiance H.

The irradiance of an image

By the application of exactly the same integration technique used in Sec. 12.5, it can be shown that the *irradiance* produced in the plane of an image is given by

$$H = T\pi N \sin^2 \theta' \text{ watt/cm}^{-2} = TN\omega \text{ (for small angles)} \tag{12.11}$$

where T is the system transmission, N (W ster^{-1} cm^{-2}) is the object radiance, and θ' is the half angle subtended by the exit pupil of the optical system from the image. Small or noncircular exit pupils and cylindrical lens systems can be handled by substituting the solid angle ω for $\pi \sin^2 \theta'$ (just as in Eq. 12.8); image points off the optical axis are subject to the cosine-fourth law in addition to any losses due to vignetting (Eq. 12.9 and Sec. 9.7).

The similarity between the equations for the irradiance produced by a diffuse source and by an optical system makes it apparent that, when it is viewed from the image point, the aperture of the optical system takes on the radiance of the object it is imaging. This is an extremely useful concept; for radiometric purposes, a complex optical system can often be treated as if it consisted solely of a transmission loss and an exit pupil with the same radiance as the object. Similarly, when an optical system produces an image of a source, the image can be treated as a new source of the same radiance (less transmission losses). Of course, the direction that radiation is emitted from the image is limited by the aperture of the system.

When an object is so small that its image is a diffraction pattern (Airy disk), then the preceding techniques, which apply to extended sources, cannot be used. Instead, the power intercepted by the optical system, reduced by transmission losses, is spread into the diffraction pattern. To determine the irradiance (or the radiance) of the image, we note that 84 percent of the power intercepted and transmitted by the lens is concentrated into the central bright spot (the Airy disk). A precise determination of irradiance requires that one integrate the relative irradiance-times-area product over the central disk and equate this to 84 percent of the image power. If P is the total power in the Airy pattern, H_0 the irradiance at the center of the pattern, and z the radius of the first dark ring, a numerical integration of Eq. 9.14 over the central disk yields

$$0.84P = 0.72H_0 z^2$$

Rearranging and substituting the value of z given by Eq. 9.16, we get

$$H_0 = 1.17 \frac{P}{z^2} = \pi P \left(\frac{\text{NA}}{\lambda} \right)^2$$

where λ is the wavelength and NA is $n' \sin U'$, the numerical aperture. The irradiance for points not at the center of the pattern is then found by Eq. 9.14. Note that the preceding assumes a circular aperture; for rectangular apertures, the process would be based on Eq. 9.12.

Example 12.1

In Fig. 12.4, A is a circular source with a radiance of 10 W ster^{-1} cm^{-2} radiating toward plane BC. The diameter of A subtends 60° from point B. The distance AB is 100 cm and the distance BC is 100 cm. An optical system at D forms an image of the region about point C at E. Plane BC

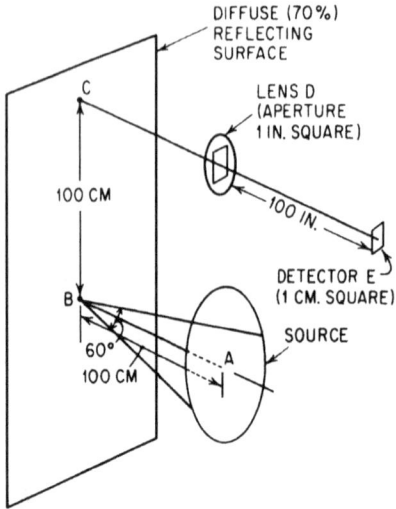

Figure 12.4 Example 12.1.

is a diffuse (lambertian) reflector with a reflectivity of 70 percent. The optical system (*D*) has a 1-in-square aperture and the distance from *D* to *E* is 100 in. The transmission of the optical system is 80 percent. We wish to determine the power incident on a 1-cm square photodetector at *E*.

We begin by determining the irradiance at *B*, using Eq. 12.7; the source radiance is 10 W ster^{-1} cm^{-2} and the half angle θ is 30°, giving

$$H_B = \pi N \sin^2 \theta = \pi \cdot 10 \cdot \left(\frac{1}{2}\right)^2 = 7.85 \text{ W/cm}^2$$

Since angle *BAC* is 45°, we can find the irradiance at *C* from Eq. 12.9, noting that cos 45° is 0.707

$$H_C = H_B \cos^4 45° = 7.85 \times (0.707)^4 = 1.96 \text{ W/cm}^2$$

(Note that the cosine-fourth effect derived in Sec. 9.7 included one cosine term which was approximate; its accuracy depended on the distance from the pupil to the image surface being *much* larger than the pupil diameter. In Example 12.1 this approximation is quite poor. P. Foote, in the *Bulletin of the Bureau of Standards* 12, 583 (1915), gave the following expression for the irradiance, which is accurate even when the source is large compared with the distance.

$$H = \frac{\pi N}{2}\left[1 - \frac{(1 + \tan^2 \phi - \tan^2 \theta)}{[\tan^4 \phi + 2 \tan^2 \phi(1 - \tan^2 \theta) + 1/\cos^4\theta]^{1/2}}\right]$$

If we compare the irradiance from this equation with that from Eqs. 12.7 and 12.8 for the angles ϕ and θ from Example 12.1, we find that this irradiance is 42 percent greater than the cosine-fourth result. This is, of course, a rather extreme case.)

It is now necessary to determine the radiance of the surface at C. The diffuse surface at C reradiates 70 percent of the incident 1.96 W/cm² into a full hemisphere; the total power reradiated is thus 1.37 W/cm². In Sec. 12.4 it was shown that a source of radiance N radiated πN W/cm² into a hemisphere. Thus the radiance at point C is given by

$$N_C = \frac{RH}{\pi} = \frac{0.7 \times 1.96}{\pi} = \frac{1.37}{\pi} = 0.44 \text{ W ster}^{-1} \text{ cm}^{-2}$$

The irradiance at E can now be determined from Eq. 12.11, noting that the solid angle subtended by the aperture of the lens system is $1/(100)^2$, or 10^{-4} ster, and substituting this for $\pi \sin^2 \theta$ in Eq. 12.11,

$$H_E = T_D \pi N_C \sin^2 \theta = T_D N_C \omega$$
$$= 0.8 \times 0.44 \times 10^{-4} = 0.35 \times 10^{-4} \text{ W/cm}^2$$

Since the photodetector at E has an area of 1 cm², the radiant power falling on it is just 0.35×10^{-4} W, or 35 μW.

12.7 Spectral Radiometry

In the preceding discussion, no mention has been made of the spectral characteristics of the radiation. It is apparent that every radiant source has some sort of spectral distribution of its radiation, in that it will emit more radiation at certain wavelengths than at others.

For many purposes, it is necessary to treat intensity (J), irradiance (H), radiance (N), etc. (in fact, all the quantities listed in Fig. 12.5) as

Name	Symbol	Description	Units
Radiant power (flux)	P (ϕ)	Rate of transfer of energy	W (Joule/sec)
Radiant intensity	J (I)	Power per unit solid angle from a source	W/ster¹
Radiance	N (L)	Power per unit solid angle per unit area from a source	W ster⁻¹ cm⁻²
Irradiance	H (E)	Power per unit area incident on a surface	W/cm²
Radiant energy	U		Joule
Radiant emittance	W (M)	Power per unit area emitted from a surface	W/cm²

Figure 12.5 Radiometric terminology. The names, symbols, descriptions, and preferred units for quantities in radiometric work.

functions of wavelength. To do this we refer to the above quantities per unit interval of wavelength. Thus, if a source emits 5 W of radiant power in the spectral band between 2 and 2.1 μm, it emits 50 W per micrometer (W/μm) in this region of the spectrum. The standard symbol for this type of quantity is the symbol given in Fig. 12.5 subscripted with a λ, and the name is preceded by "spectral." For example, the symbol for spectral radiance is N_λ and its units are watts per steradian per square centimeter per micrometer (W ster^{-1} cm^{-2} μm^{-1}).

In many applications it is absolutely necessary to take the spectral characteristics of sources, detectors, optical systems, filters, and the like into account. This is accomplished by integrating the particular radiation product function over an appropriate wavelength interval. Since most spectral characteristics are not ordinary functions, the process of integration is usually numerical, and thus laborious. As a brief example, suppose that the irradiance in an image is desired. The spectral radiance of the object can be described by some function $N(\lambda)$ and the transmission of the atmosphere, the optical system, and any filters can be combined in a spectral transmission function $T(\lambda)$. Equation 12.11 will give the irradiance of the image (for any given wavelength); for use over an extended wavelength interval, we must write

$$H = \int_{\lambda_1}^{\lambda_2} T(\lambda)\, \pi N(\lambda) \sin^2 \theta\, d\lambda = \pi \sin^2 \theta \int_{\lambda_1}^{\lambda_2} T(\lambda)\, N(\lambda)\, d\lambda \text{ W/cm}^2 \quad (12.12)$$

where λ_1 and λ_2, the limits of the integration, may be zero and infinity, but are usually taken as real wavelengths which encompass the region of interest. In practice, it is usually necessary to perform the integration numerically; this process is represented (for this particular example) by the summation:

$$H = \pi \sin^2 \theta \sum_{\lambda = \lambda_1}^{\lambda_2} T(\lambda)\, N(\lambda)\, \Delta\lambda \text{ W/cm}^2 \quad (12.13)$$

The spectral response of a detector is included in a calculation in the same manner. For example, the effective power falling on a detector with an area of A and a relative spectral response $R(\lambda)$, when the detector is located in the image plane of the system above, would be (provided that the image completely covered the detector)

$$P = A\pi \sin^2 \theta \int_{\lambda_1}^{\lambda_2} R(\lambda)\, T(\lambda)\, N(\lambda)\, d\lambda \text{ W}$$

12.8 Blackbody Radiation

A perfect blackbody is one which totally absorbs all radiation incident upon it. The radiation characteristics of a heated blackbody are subject to known laws, and since it is possible to build a close approximation to an ideal blackbody, a device of this type is a very useful standard source for the calibration and testing of radiometric instruments. Further, most sources of thermal radiation, i.e., sources which radiate because they are heated, radiate energy in a manner which can be readily described in terms of a blackbody emitting through a filter, making it possible to use the blackbody radiation laws as a starting point for many radiometric calculations.

Planck's law describes the spectral radiant emittance of a perfect blackbody as a function of its temperature and the wavelength of the emitted radiation.

$$W_\lambda = \frac{C_1}{\lambda^5 \, (e^{C_2/\lambda T} - 1)} \qquad (12.14)$$

where W_λ = the radiation emitted into a hemisphere by the blackbody in power per unit area per wavelength interval (W cm^{-2} μm^{-1})

λ = the wavelength (μm)

e = the base of natural logarithms (2.718...)

T = the temperature of the blackbody in Kelvin (K = °C + 273)

C_1 = a constant = 3.742×10^4 when area is in square centimeters and wavelength in micrometers

C_2 = a constant = 1.4388×10^4 when square centimeters and micrometers are used

Figure 12.6 indicates the shape of the curve of W_λ plotted against wavelength. Note that the spectral radiance (N_λ) is given by W_λ/π.

If we integrate Eq. 12.14, we can obtain the total radiation at all wavelengths. The resulting equation is known as the *Stefan-Boltzmann law,*

$$W_{\text{TOT}} = 5.67 \times 10^{-12} T^4 \text{ W/cm}^2 \qquad (12.15)$$

and indicates that the total power radiated from a blackbody varies as the fourth power of the absolute temperature.

If we differentiate Planck's equation (12.14) and set the result equal to zero, we can determine the wavelength at which the spectral emittance (W_λ) is a maximum and also the amount of W_λ at this wavelength. *Wien's displacement law* gives the wavelength for maximum W_λ as

$$\lambda_{\text{max}} = 2897.8 T^{-1} \, \mu\text{m} \qquad (12.16)$$

and W_λ at λ_{max} as

$$W_{\lambda, \, \text{max}} = 1.286 \times 10^{-15} T^5 \text{ W/cm}^2 \cdot \mu\text{m}^{-1} \qquad (12.17)$$

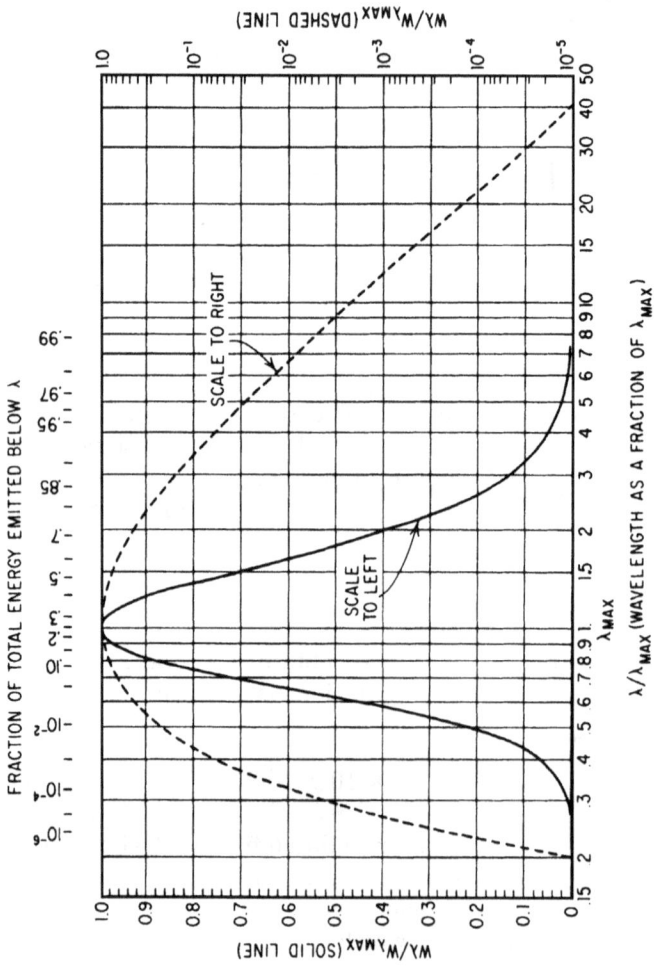

Figure 12.6 Spectral distribution of blackbody radiation (normalized).

Notice that the higher the temperature, the shorter the wavelength at which the peak occurs and that W_λ at the peak varies as the fifth power of the absolute temperature.

Before the advent of the electronic calculator, Planck's equation was very awkward to use and for this reason a number of tables, charts, and slide rules are available which allow the user to simply look up the values of W_λ for the appropriate temperature and wavelength. Figure 12.6 may be used for this purpose when the precision required is relatively modest.

The use of Fig. 12.6 is quite simple: First the total energy (W_{TOT}), the peak wavelength (λ_{max}), and the maximum spectral radiant emittance ($W_{\lambda,\,max}$) are calculated for the desired temperature by Eqs. 12.15, 12.16, and 12.17, respectively. The graph in Fig. 12.6 is of $W_\lambda/W_{\lambda,\,max}$ plotted against relative wavelength. Thus, if W_λ for a particular wavelength (λ) is desired, the value of $W_\lambda/W_{\lambda,\,max}$ corresponding to the appropriate value of λ/λ_{max} is selected and multiplied by the value of $W_{\lambda,\,max}$ from Eq. 12.17.

Across the top of Fig. 12.6 is a scale which indicates the fraction of the total energy emitted at all wavelengths below that corresponding to the point on the scale. Note that exactly 25 percent of the energy from a blackbody is emitted at wavelengths shorter than λ_{max}. If it is necessary to determine the amount of power emitted in a spectral band between two wavelengths (λ_1 and λ_2), the wavelengths are converted to relative wavelengths (λ_1/λ_{max} and λ_2/λ_{max}) and the fractions corresponding to them are selected from the scale at the top of the figure. The total power (W_{TOT}) from Eq. 12.15 times the difference between the two fractions will give the amount of power emitted in the wavelength interval.

Example 12.2

For a blackbody at a temperature of 27°C (80.6°F), T is $273 + 27 = 300$ K, and the total emitted radiation is given by Eq. 12.15

$$W_{TOT} = 5.67 \times 10^{-12}(300)^4 = 4.59 \times 10^{-2} \text{ W/cm}^2$$

The wavelength at which W_λ is a maximum is given by Eq. 12.16

$$\lambda_{max} = 2897.9\,(300)^{-1} = 9.66 \ \mu\text{m}$$

and the radiant emittance at this wavelength is obtained from Eq. 12.17

$$W_{\lambda,\,max} = 1.288 \times 10^{-15}\,(300)^5 = 3.13 \times 10^{-3} \text{ W cm}^{-2}\ \mu\text{m}^{-1}$$

As an aside, note that this (300 K) is a reasonable value for the ambient temperature and that our result indicates that the earth and most things on it are strongly emitting at a wavelength of 10 μm. This is the basis of the "see in the dark" FLIR systems which are sensitive to this spectral region; most such systems use germanium optics, which transmit well in the 8- to 14-μm region (which also happens to be a good transmission window of the atmosphere). Thus there is no such thing as darkness if you can detect 10-μm radiation.

Suppose we wish to know the characteristics of this blackbody in the wavelength region between 4 and 5 μm. We express these wavelengths in terms of λ_{max} as 4/9.66 = 0.414 and 5/9.66 = 0.518. From Fig. 12.6, the corresponding values of $W_\lambda/W_{\lambda, max}$ are 0.07 and 0.25; these values, multiplied by $W_{\lambda, max} = 3.13 \times 10^{-3}$ W cm^{-2} μm^{-1} give us the spectral radiant emittances for these wavelengths

At 4 μm:

$$W_\lambda = 0.22 \times 10^{-3} \text{ W cm}^{-2} \text{ μm}^{-1}$$

At 5 μm:

$$W_\lambda = 0.78 \times 10^{-3} \text{ W cm}^{-2} \text{ μm}^{-1}$$

Using the fraction scale across the top of the chart, we find that about 0.011 of the radiation is emitted below 5 μm (rel. λ = 0.518) and about 0.0015 below 4 μm. Thus, approximately 1 percent of the total radiation (W_{TOT}), amounting to about 4×10^{-4} W/cm^2, is emitted in this spectral band. The radiance of the surface will be $4 \times 10^{-4}/\pi$ W ster^{-1} cm^{-2} in this spectral band. If the blackbody is a foot square, with an area of about 1000 cm^2, it will radiate about 0.4 W between 4 and 5 μm into a hemisphere of 2π ster.

Most thermal radiators are not perfect blackbodies. Many are what are called gray-bodies. A gray-body is one which emits radiation in exactly the same spectral distribution as a blackbody at the same temperature, but with reduced intensity. The *total emissivity* (ϵ) of a body is the ratio of its total radiant emittance to that of a perfect blackbody at the same temperature. Emissivity is thus a measure of the radiation and absorption efficiency of a body. For a perfect blackbody ϵ = 1.0, and most laboratory standard blackbodies are within a percent or two of this value. The table of Fig. 12.7 lists the total emissivity for a number of common materials. Note that emissivity varies with both wavelength and with temperature.

Radiation incident on a substance can be transmitted, reflected (or scattered), or absorbed. The transmitted, reflected, and absorbed fractions obviously must add up to 1.0. The absorbed fraction is the emissivity. Thus a material with either a high transmission or a high reflection must have a low emissivity.

Material	Total Emissivity	
Tungsten	500 K	0.05
	1000 K	0.11
	2000 K	0.26
	3000 K	0.33
	3500 K	0.35
Polished silver	650 K	0.03
Polished aluminum	300 K	0.03
Polished aluminum	1000 K	0.07
Polished copper		0.02–0.15
Polished iron		0.2
Polished brass	4–600 K	0.03
Oxidized iron		0.8
Black oxidized copper	500 K	0.78
Aluminum oxide	80–500 K	0.75
Water	320 K	0.94
Ice	273 K	0.96–0.985
Paper		0.92
Glass	293 K	0.94
Lampblack	273–373 K	0.95
Laboratory blackbody cavity		0.98–0.99

Figure 12.7   The *total* emissivity of a number of materials.

When dealing with gray-bodies, it is necessary to insert the emissivity factor ϵ into the blackbody equations. Planck's law (Eq. 12.14), the Stefan-Boltzmann law (Eq. 12.15), and the Wien displacement law (Eq. 12.17) should be modified by multiplying the right-hand term by the appropriate value of ϵ. For many materials the emissivity is a function of wavelength. This is apparent from the fact that many substances (glass, for example) have a negligible absorption, and consequent low emissivity, at certain wavelengths, while they are almost totally absorbent at other wavelengths. In regions of the spectrum where this occurs, emissivity becomes spectral emissivity (ϵ_λ) and is treated just as any other spectral function. For many materials, emissivity will decrease as wavelength increases. It should also be noted that most materials show a variation of emissivity with temperature as well as wavelength, and precise work must take this into account. Emissivity usually increases with temperature.

Note that not all sources are continuous emitters. Gas discharge lamps at low pressure emit discrete spectral lines; the plot of spectral radiant emittance for such a source is a series of sharp spikes, although there is usually a low-level background continuum. In high-pressure arcs, the spectral lines broaden and merge into a continuous background with less pronounced spikes.

Color temperature

Before leaving the subject of blackbody radiation, the concept of color temperature should be mentioned. The color temperature of a source of light is a colorimetric concept related to the apparent visual color of a source, not its temperature. For a blackbody, the color temperature is equal to the actual temperature in Kelvin. For other sources, the color temperature is the temperature of the blackbody which has the same apparent color as the source. Thus, exceedingly bright or dim sources may have the same color temperature, but radically different radiances or intensities. Color temperature usually runs about 150 K higher than filament temperature. Color temperature is extremely important in colorimetry and in color photography where fidelity of color rendition is important, but is little used in radiometry.

12.9   Photometry

Photometry deals with *luminous radiation,* that is, radiation which the human eye can detect. The basic photometric unit of radiant power is the lumen, which is defined as a luminous flux emitted into a solid angle of one steradian by a point source the intensity of which is $\frac{1}{60}$ of that of 1 cm^2 of a blackbody at the solidification temperature of platinum (2042 K). From the preceding section, we know that a blackbody radiates energy throughout the entire electromagnetic spectrum. Chapter 8 indicated that the eye was sensitive to only a small interval of this spectrum and that its response to different wavelengths within this interval varied widely. Thus, if a source of radiation has a spectral power function $P(\lambda)$ (W μm^{-1}), the visual effect of this radiation is obtained by multiplying it by $V(\lambda)$,* the visual response function which is tabulated in Fig. 8.9. The effective visual power of a source is, therefore,

---

*Note that $V(\lambda)$ is customarily the photopic (normal level of illumination and brightness) visual response curve. Under conditions of complete dark adaptation, the visual response for scotopic vision would be used. The conversion constant in Eq. 12.18 becomes about 1746 instead of 680.

the integral (or summation) of $P(\lambda)\, V(\lambda)\, d\lambda$ over the appropriate wavelength interval. From the definition of the lumen, it can be determined that one watt of radiant energy at the wavelength of maximum visual sensitivity (0.555 μm) is equal to 680 lumens. Therefore, the luminous flux emitted by a source with a spectral power of $P(\lambda)$ W μm^{-1} is given by*

$$F = 680 \int V(\lambda)\, P(\lambda)\, d\lambda \text{ lumens} \qquad (12.18)$$

The unit of luminous intensity is called the candle (or "candela") and is so named because the original standard of intensity was an actual candle. A point source of one candlepower is one which emits one lumen into a solid angle of one steradian. A source of one candle intensity which radiates uniformly in all directions emits 4π lumens. From the definition of the lumen, it is apparent that a 1-cm^2 blackbody at 2042 K has an intensity of 60 candles.

Illumination, or *illuminance,* is the luminous flux per unit area incident on a surface. The most widely used unit of illumination is the foot-candle. One footcandle is one lumen incident per square foot. The misleading name footcandle resulted from the fact that it is the illumination produced on a surface one foot away from a source of one-candle intensity. The photometric term illuminance corresponds to irradiance in radiometry.

The term *brightness,* or *luminance,* corresponds to the term *radiance.* Brightness is the luminous flux emitted from a surface per unit solid angle per unit of area (projected on a plane normal to the line of sight). There are several commonly used units of brightness. The candle per square centimeter is equal to one lumen emitted per steradian per square centimeter. The lambert is equal to $1/\pi$ candles per square centimeter. The foot-lambert is equal to $1/\pi$ candles per square foot. The foot-lambert is a convenient unit for illuminating engineering work, since it is the brightness which results from one footcandle of illumination falling on a "perfect" diffusing surface. (Since one lumen is incident on the 1-ft^2 area under an illumination of one footcandle, the total flux radiated into a hemisphere of 2π ster. from a perfectly diffuse (lambertian) surface is just one lumen. As pointed out in Sec. 12.4 and Example 12.1, the resulting brightness is $1/\pi$ lumen ster^{-1} ft^{-2},

---

*Since the constant 680 in Eq. 12.18 is derived by numerical integration of a table of measured values, this number is not an exact constant; a value of 683 is also used.

not $1/2\pi$ lumen ster^{-1} ft^{-2}). The brightness of a number of sources is tabulated in Fig. 12.8 and natural illumination and reflectance levels are tabulated in Fig. 12.9.

The terminology of photometry has grown through engineering usage, and is thus far from orderly. Special terms have derived from

Source	Brightness, candles cm^{-2}
Sun (zenith) through atmosphere	1.6×10^5 cd/cm^2
Sun (zenith) above atmosphere	2.75×10^5
Sun (horizon)	6×10^2
Blue sky	0.8
Dark cloudy sky	4×10^{-3}
Night sky	5×10^{-9}
Moon	0.25
Exteriors—daylight (typical)	1
Exteriors—night (typical)	10^{-6}
Interiors—daylight (typical)	10^{-2}
Mercury arc—laboratory	10
Mercury arc—high pressure	5×10^5
Xenon arc	1.5×10^4 to 1.5×10^5
Carbon arc	10^4 to 10^5
Tungsten—3655 K (melting point)	5.7×10^3
3500 K	4.2×10^3
3000 K	1.3×10^3
Tungsten filament – ordinary lamp	5×10^2
– projection lamp	3×10^3
Blackbody—2042 K	60.0 (by definition)
—4000 K	2.5×10^4
—6500 K	3×10^5
Fluorescent lamp	0.6
Sodium lamp	6
Flame—candle, kerosene	1
Least perceptible brightness	5×10^{-11}
Least perceptible point source	2×10^{-8} cd @ 3 m distance
Star Sirius	1.5×10^6
Atom bomb	10^8
Lightning	8×10^6
Ruby laser	10^{14}
Metal halide lamp	4×10^4

Figure 12.8 Typical values for the brightness (luminance) of a number of sources.

Source	Illumination, footcandles
Direct sunlight	10,000
Open shade	1,000
Overcast/dark day	10–100
Twilight	0.1–1.0
Full moon	0.01
Starlight	0.0001
Dark night	0.00001

(a)

Material	Reflectance
Asphalt	0.05
Trees, grass	0.20
Red brick	0.35
Concrete	0.40
Snow	0.85
Aluminum building	0.65
Glass window wall	0.70
Parking lot with cars	0.40

(b)

Figure 12.9 (a) Illumination levels produced by sources in nature. (b) Reflectance of a number of exteriors.

special usages, and many such terms have survived. A tabulation of photometric units is given in Fig. 12.10.

Photometric calculations may be carried out exactly as are radiometric calculations, using the relationships presented in Secs. 12.2 through 12.6. If lumens are substituted for watts in all the expressions, the computations are straightforward. When the starting and final data must be expressed in the special terminology of photometry (as opposed to what one might term the rational units of lumens, steradians, and square centimeters), then conversion factors may be necessary for each relationship. A very simple way of avoiding this difficulty is to convert the starting data to lumens, steradians, and square centimeters, complete the calculation, and then convert the results into the desired units.

For convenience, the basic relationships are repeated here in both radiometric (left column) and photometric (right column) form:

Flux (Symbol F)
  lumen   defined in text, Eq. 12.18

Intensity (Symbol I) (also *luminous pointance*)
  candela (colloq: "candle")   one lumen per steradian emitted from a point source.
                                    1/60 of the intensity of one cm^2 of a blackbody at 2042 K.
  "old candle"   1.02 candela
  carcel   9.6 candles
  hefner   0.9 candles

Illumination (Symbol E) (also *illuminance, luminous incidance, lum. areance*)
  footcandle   one lumen per ft^2 incident on a surface = 10.76 lux =
                                    1/929 phot
  phot (centimeter-candle)   one lumen per cm^2 = 10^4 lux = 929 ft cd
  lux (meter-candle)   one lumen per m^2 = .0929 ft cd
  nox   0.001 lux

Brightness (Symbol B) (also *luminance, luminous sterance*)
  candle per sq. cm.   one lumen emitted per steradian per cm^2 projected area
                                    normal to direction
  stilb   one cd/cm^2 = 929 cd/ft^2 = π lamberts = 929 ft lamberts
  lambert   (1/π) candles per cm^2
  foot-lambert   (1/π) candles per m^2 = (1/929) lambert = (1/π 929) stilb
  meter-lambert   = apostilb = 1000 skot = 10^{-4} lamberts
  nit   one lux per ster = .0001 stilb = candel/m^2
  skot   3.18 $\times$ 10^{-8} stilb = 9.29 $\times$ 10^{-5} ft lamberts = 2.957 $\times$
                                    10^{-5} cd/ft^2

The *Troland* is an annoying unit of retinal illumination produced by viewing a surface of
one cd/m^2 brightnass when the pupil area is one mm^2. It equals 0.0035 lumens/m^2; It equals
3.5 nano-lumens per cm^2; It equals the number of *nits* times the pupil area in mm^2

Figure 12.10 Photometric quantities.

Radiant Intensity: $J = P/\Omega$

J is radiant intensity

P is the radiant power emitted
into solid angle Ω

Irradiance: $H = J/S^2 = J\Omega$

H is the irradiance incident on a
surface a distance S from a point
source of intensity J. Ω is the solid

Luminous Intensity: $I = F/\Omega$

I is luminous intensity

F is the luminous flux emitted into
solid angle Ω

Illumination (illuminance):
$E = I/S^2 = \Omega$

E is the illumination incident on a
surface a distance S from a point
source of intensity I. Ω is the solid

angle subtended by a unit area of the surface from the source.

$$H = \pi N \sin^2 \theta$$

H is the irradiance produced by a diffuse circular source of radiance N at a point from which the source diameter subtends 2θ.

$$H = N\omega$$

H is the irradiance produced by a diffuse source of radiance N at a point from which the area of the source subtends the solid angle ω.

$$H = T\pi N \sin^2 \theta$$

$$(H = TN\omega)$$

H is the irradiance at an image formed by an optical system of transmission T the exit pupil diameter (area) of which subtends an angle 2θ (solid angle ω) from the image point when object radiance is N.

Radiance: $N = P/(\pi A)$

N is the radiance of a diffuse source of area A which emits radiant power P into a hemisphere of 2π steradians.

---

angle subtended by a unit area of the surface from the source.

$$E = \pi B \sin^2 \theta$$

E is the illumination produced by a diffuse circular source of brightness (luminance) B at a point from which the source diameter subtends 2θ.

$$E = B\omega$$

E is the illumination produced by a diffuse source of brightness B at a point from which the area of the source subtends the solid angle ω.

$$E = T\pi B \sin^2 \theta = T\pi B/4(f/\#)^2 (m + 1)^2$$

$$(E = TB\omega) \qquad \left[m = \left(\frac{s'}{f} - 1 \right) \right]$$

E is the illumination at an image formed by an optical system of transmission T the exit pupil diameter (area) of which subtends an angle 2θ (solid angle ω) from the image point when the object brightness is B.

Brightness (luminance):
$B = F/(\pi A)$

B is the brightness of a diffuse source of area A which emits luminous flux F into a hemisphere of 2π steradians.

Example 12.3

It may be instructive to repeat Example 12.1 in photometric terms and to indicate at each step in the calculation the conversions to the various photometric units. We will use Fig. 12.4 again; the only change in the starting data will be that the source A will be assumed to have a brightness of 10 lumens ster^{-1} cm^{-2}.

From Fig. 12.10, we note that the source brightness may also be expressed as 10 candles cm^{-2}, as 10 stilb, as 10π lamberts, or as 9290π foot-lamberts.

The illumination produced at point B is calculated from Eq. 12.7 (after rewriting it in photometric symbols)

$$H = \pi N \sin^2 \theta$$

$$E = \pi B \sin^2 \theta$$

$$= \pi (10L \text{ ster}^{-1} \text{ cm}^{-2}) \left(\frac{1}{2}\right)^2$$

$$= 7.85 \text{ lumen cm}^{-2}$$

Applying the cosine-fourth law, we find the illumination at C

$$E_C = E_B \cos^4 45°$$

$$= 7.85 \times (0.707)^4$$

$$= 1.96 \text{ lumen cm}^{-2}$$

Since there are 929 cm^2 per square foot

$$E_C = 929 \times 1.96 = 1821 \text{ lumens ft}^{-2}$$

$$= 1821 \text{ footcandles}$$

Since the surface BC has a diffuse reflectivity of 70 percent, we can multiply the illumination in footcandles by 0.7 to obtain the brightness in foot-lamberts

$$B = 0.7 \times 1821 = 1275 \text{ foot-lamberts}$$

Similarly 0.7 times the illumination in lumens cm^{-2} will yield the brightness in lamberts

$$B = 0.7 \times 1.96 = 1.37 \text{ lamberts}$$

Or we can retain the lumen units, and determine that, with 1.96 lumen cm^{-2} falling on a surface 70 percent reflectivity, 1.37 lumen cm^{-2} will be emitted into a hemisphere, and, following our previous reasoning, compute the brightness as

$$B = \frac{1.37}{\pi}$$

$$= 0.44 \text{ lumen ster}^{-1} \text{ cm}^{-2}$$

$$= 0.44 \text{ candle cm}^{-2}$$

The illumination at E is determined from Eq. 12.11 as before

$$H = TN\pi \sin^2 \theta$$

$$= TN\omega$$

$$E = TB\omega$$

$$= 0.8 \times 0.44 \times 10^{-4}$$

$$= 0.35 \times 10^{-4} \text{ lumen cm}^{-2}$$

$$= 929 \times 0.35 \times 10^{-4} = 0.032 \text{ footcandles}$$

12.10 Illumination Devices

Searchlight

A *searchlight* is one of the simpler, and at the same time one of the least understood, illuminating devices. It consists of a source of light (usually small) placed at the focal point of a lens or reflector. The image of the source is thus located at infinity. A common misconception is that the beam of light produced is a "collimated parallel bundle" which extends out to infinity with a constant diameter and a constant power density. A little consideration of the matter will reveal the fallacy: the rays from any *point* on the source do indeed form a collimated parallel bundle, etc. However, a geometrical point on any source of finite brightness must emit zero energy, since a point has zero area, and therefore the "collimated bundle" of rays has zero energy.

With reference to Fig. 12.11, which shows a source S at the focal point of lens L, the image (S') will be located at infinity. Since source S subtends an angle α from lens L, the image S' will also subtend α. Now the illumination at a point on the axis will be determined by the

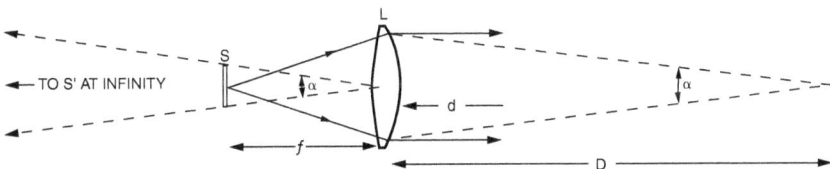

Figure 12.11 The optics of a searchlight.

brightness of the image and the solid angle subtended by the image. Thus, for points *near the lens,* the illumination is given by

$$E = TB\omega \tag{12.19}$$

which the reader will recognize as Eq. 12.8 rewritten in photometric symbols with a transmission constant (T) added. B is the brightness of source S (since the brightness of an image equals the brightness of the object) and ω is the solid angle subtended by the image. (We have tacitly assumed ω to be small.) Now for a point at the lens, it is obvious that the solid angle ω subtended by the image S' is exactly equal to the solid angle subtended by the source S from the lens. Since S' is at infinity, this angle will not change as we shift our reference point a short distance along the axis away from the lens, and the illumination will remain constant in this region. However, at a distance $D = $ (lens diameter)/α, the source image will subtend the same angle as the diameter of the lens, and for points more distant than D, the size of the solid angle subtended by the source of illumination will be limited by the lens diameter. This solid angle will obviously be equal to (area of lens)/d^2 and the illumination beyond distance D will fall off with the square of the distance (d) to the lens. Thus, the equations governing the illumination produced by a searchlight are

$$D = \frac{\text{lens diameter}}{\alpha} \tag{12.20}$$

$$\text{for } d \leq D\text{: } E = TB\omega = \text{(a constant)} \tag{12.21}$$

$$\text{for } d \geq D\text{: } E = \frac{TB \text{ (lens area)}}{d^2} \tag{12.22}$$

The general technique used here is applicable to almost any illumination problem, and we can restate it in general terms as follows:

To determine the illumination at a point, the size and position of the source image, as seen from the point, are calculated. The pupils and windows of the system (again, as seen from the point) are determined. Then the illumination at the point is the product of the system transmission, the source brightness and the solid angle subtended by that area of the source which can be seen from the point through the pupils and windows of the system, multiplied by the cosine of the angle of incidence.

Note that for points (which lie within the beam) beyond the critical distance D, the searchlight acts as if it were a source of a diameter equal to that of the searchlight lens and a brightness TB. As mentioned in Sec. 12.6, this concept is quite useful in evaluating the illumination at an image point; here we find that it occasionally can be applied to points which are not image points.

The *beam candle power* of a searchlight is simply the intensity of the (point) source which would produce the same illumination at a great distance. A point source with an intensity of I candles will emit I lumens per steradian. A one-square-foot area placed d feet from the point source will subtend $1/d^2$ steradians from the source, and will thus be illuminated by I/d^2 lumens per square foot (footcandles). We can determine the necessary candle power for I by equating this illumination to that produced by the searchlight according to Eq. 12.22.

$$E = \frac{I}{d^2} = \frac{TB \text{ (lens area)}}{d^2} \qquad (12.23)$$

and beam candlepower:

$$I = TB \text{ (lens area)}$$

where I is the beam candle power in lumens per steradian (or candles). Note that the lens area should be specified in the same units as the source brightness.

Projection condenser

The second illumination device we shall consider is the *projection condenser,* which is schematically diagrammed in Fig. 12.12. The purpose of the projector is to produce a bright and evenly illuminated image of the film on the screen. This could be achieved by placing a sheet of diffusing material behind the film and illuminating this diffuser. The resultant image would be dim, because the maximum brightness which the image

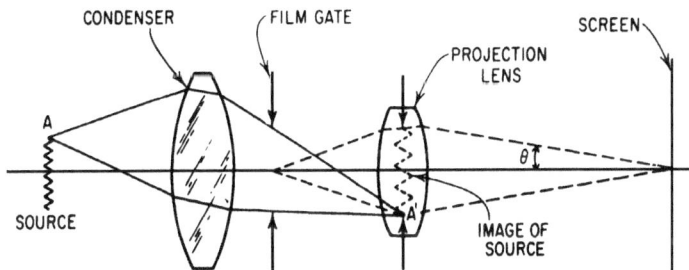

Figure 12.12 Schematic of a projection condenser system. The condenser forms an image of the source (lamp filament) in the aperture of the projection lens.

could achieve would be that of the diffuser, which would be considerably less than that of the lamp. The function of the condenser is to image the source in the pupil of the projection lens so that the lens aperture has the same brightness as the source. When this is done, the screen is illuminated according to Eq. 12.11, where the solid angle is that subtended by the source image (in the projection lens) from the screen. It is apparent that the maximum value for the screen illumination is limited by the size of the projection lens aperture. Therefore, the maximum screen illumination is achieved when the image of the source completely fills the aperture of the lens. This is required for all points within the field of view, and the condenser diameter must be sufficiently large so that it does not vignette, if maximum illumination at the edge of the picture is required. In this regard, note that the ray from the corner of the film to the opposite edge of the lens aperture is the most demanding. The cosine-fourth rule will, of course, reduce the illumination at points off the axis.

From the above, one might conclude that with a condenser of sufficient magnification, the image of a very small source could be magnified enough to fill the pupil of the projection lens. The necessary illuminating cone angle is determined by the film gate and its distance to the lens pupil (i.e., to the image of the source). In Chap. 2 we found that the magnification was given by $m = h'/h = u/u'$. The Abbe sine condition uses $m = \sin u/\sin u'$ for systems of reasonable image quality. Since u' in this case is fixed by the film gate, it is apparent that a large magnification will require a large value of u. The largest value that u can have is 90° with a sine of one; this establishes the limit on the magnification that can be attained. This limit can be expressed as

$$\left| \frac{P\alpha}{nS} \right| \le 1.0 \tag{12.24}$$

where P is the aperture of the projection lens, α is the half-field angle of projection, n is the index in which the source is immersed (usually $n = 1.0$ for air), and S is the size of the source. It is impossible for Eq. 12.24 to exceed a value of 1.0; a value of 0.5 is typical of many systems. Note that a value of 0.5 corresponds to a working speed of $f/1.0$ and that a value of 1.0 would require a working speed of $f/0.5$. (Eq. 12.24 is analogous to Eq. 13.22 for detector systems.)

When the source is irregular in shape, as in "V" filament lamps for example, the solid angle for Eq. 12.11 is determined just as one might expect, by dividing the area of the actual image of the filament by the square of the distance to the screen. Condenser design is discussed in Sec. 17.4.

Telescope brightness

The apparent brightness of an image as seen by the eye is a function of the diameter of the pupil of the eye, since it determines the illumination of the retina, in accordance with Eq. 12.11a. When the eye is used with an optical instrument, such as a telescope, the exit pupil of the instrument enters the picture. If the exit pupil is larger than that of the eye, then the apparent brightness of the object seen through the instrument is equal to the brightness of the object (modified by transmission losses and index effects), since the solid angle subtended by the pupil from the retina is unchanged. When the instrument exit pupil is smaller than that of the eye, then the apparent brightness of the object is reduced in proportion to the relative areas of the pupils. The exception to these brightness relationships of object and image occurs when the object is smaller than the diffraction limit of the optical system (e.g., a star). Since this is not an extended source, all the energy in the retinal image is concentrated on a few retinal receptors, and when the magnification and aperture of a telescope are increased so that its exit pupil diameter stays the same, its effective collection area is increased (at the objective) so that more energy is concentrated on the same retinal cells (because the size of the retinal image is the same, being governed by the diffraction limit), resulting in an increase in the apparent brightness of the source. For example, if a high enough power telescope of large aperture is used, stars may be seen in daylight, since their apparent brightness is increased while that of the sky (as an extended object) is not.

Integrating sphere

An *integrating sphere* is often used in the measurement of light and light sources, and also as a uniform lambertian (diffuse) source of light. It is a hollow sphere, coated on the inside with a highly reflective white diffuse paint. If spot A on the inside of the sphere is illuminated, the light reflected from this spot produces an illumination at some other point B on the inside of the sphere. This illumination varies with the cosines of angles ϕ and θ made by the line connecting A and B with the normals to the sphere surface at A and B. Thus the illumination at B varies as

$$\frac{\cos \theta \cos \phi}{D^2} \tag{12.25}$$

where D is the distance from A to B. This expression, for the inside of a sphere, is a constant. Thus the entire inner surface of the sphere is uniformly illuminated by the light reflected from the illuminated spot.

If we cut two small holes in the sphere, one to admit light and the other (in a location not directly illuminated by the first hole) for a light sensor, we have a device which can read the amount of radiation admitted into the sphere without any variation of sensitivity resulting from the direction of the light, the size of the beam, or the position of the beam in the admitting hole. The total radiation emitted by a lamp or other source which is placed inside the sphere can readily be measured. Conversely, if the light sensor is replaced by a source of light, then the other hole becomes an almost perfect, uniform, unpolarized, lambertian source of radiation. The area of the holes should not exceed 2 percent of the area of the sphere, the integrating sphere is an excellent device to measure the transmission of a lens.

Bibliography

American Institute of Physics Handbook, New York, McGraw-Hill, 1963.

Carlson, F., and C. Clark, in Kingslake (ed.), *Applied Optics and Optical Engineering,* Vol. 1, New York, Academic, 1965 (light sources).

Eby, J., and R. Levin, in Shannon and Wyant (eds.), *Applied Optics and Optical Engineering,* Vol. 7, New York, Academic, 1979 (light sources).

Hackforth, H., *Infrared Radiation,* New York, McGraw-Hill, 1960.

Hardy, A., and F. Perrin, *The Principles of Optics,* New York, McGraw-Hill, 1932.

Jamieson, J., et al., *Infrared Physics and Engineering,* New York, McGraw-Hill, 1963.

Kingslake, R., *Applied Optics and Optical Design,* Vol. 2, New York, Academic, 1965 (illumination).

Kingslake, R., *Optical System Design,* San Diego, Academic, 1983.

LaRocca, A., "Artificial Sources," in *Handbook of Optics,* Vol. 1, New York, McGraw-Hill, 1995, Chap. 10.

LaRocca, A., in Wolfe and Zissis (eds.), *The Infrared Handbook,* Washington, D.C., Office of Naval Research, 1985 (sources).

Nicodemus, F., "Radiometry," in Kingslake (ed.), *Applied Optics and Optical Engineering,* Vol. 4, New York, Academic, 1967.

Norton, P., "Photodetectors," in *Handbook of Optics,* Vol. 1, New York, McGraw-Hill, 1995, Chap. 15.

Snell, J., in W. Driscoll (ed.), *Handbook of Optics,* New York, McGraw-Hill, 1978 (radiometry).

Suits, G., in Wolfe and Zissis (eds.), *The Infrared Handbook,* Washington, Office of Naval Research, 1985 (sources).

Teele, R., in Kingslake (ed.), *Applied Optics and Optical Design,* Vol. 1, New York, 1965 (photometry).

Walsh, J., *Photometry,* New York, Dover, 1958.

Wolfe, W., in Shannon and Wyant (eds.), *Applied Optics and Optical Engineering,* Vol. 8, New York, Academic, 1980 (radiometry).

Wolfe, W. L., and P. W. Kruse, "Thermal Detectors," in *Handbook of Optics,* Vol. 1, New York, McGraw-Hill, 1995, Chap. 19.

Wolfe, W., *Optical Engineer's Desk Reference,* Bellingham and Washington, SPIE and OSA, 2003.

Zalewski, E. F., "Radiometry and Photometry," in *Handbook of Optics,* Vol. 2, New York, McGraw-Hill, 1995, Chap. 24.

Zissis, G., and A. LaRocca, in W. Driscoll (ed.), *Handbook of Optics,* New York, McGraw-Hill, 1978 (sources).

Zissis, G., in Wolfe and Zissis (eds.), *The Infrared Handbook,* Washington, Office of Naval Research, 1985.

Exercises

1 A point source emits 10 W/steradian toward a 4 in diameter optical system. How much power is collected by the optical system when its distance from the source is (a) 10 ft, (b) one mile?

ANSWER: (a) At 10 feet the solid angle subtended by the 4 in aperture is area/distance^2 = $\pi 2^2/120^2$ = 0.00087266 ster. The power equals the angle times the intensity = 0.00087266 × 10 = 0.0087266 W.

(b) At a mile, the distance is 5280 × 12 = 63,360 in and the solid angle = $\pi 2^2/63,360^2$ = 3.130 × 10^{-9} ster. Thus the power collected is 3.130 × 10^{-8}.

2 A 10 candlepower point source illuminates a perfectly diffusing surface which is tilted at 45° to the line of sight to the source. What is the brightness of the surface if it is 10 ft from the source?

ANSWER: A 10 candlepower source emits 10 lumens per steradian.
    At a distance of 10 ft, one square foot subtends $1/10^2$ = 0.01 steradian.
    The illumination on a surface normal to the propogation equals the intesity times the solid angle = 10 lumens × 0.01 ster = 0.1 lumens per square foot. Because the surface is tilted at 45°, the area within 0.01 steradian is increased by 1/cos 45°, and the illumination is 0.0707 lumens per square feet, which is 0.0707 footcandles. On a perfect diffuser this produces a brightness of 0.0707 foot lamberts.
    Using an alternate approach, 0.0707 lumens per square feet will be emitted/reflected in a lambertian fashion into a hemisphere of 2π ster to produce a brightness of $0.0707/\pi$ = 0.022508 lumens ster^{-1}ft^{-2}. Since there are 929 cm^2 in a square feet this is 0.24228 × 10^{-4} lumen ster^{-1}cm^{-2} or 0.24 × 10^{-4} stilb.

3 A fluorescent lamp 10 in long and 1 in wide illuminates a slit, parallel to the lamp and 10 in from the lamp. If the lamp has a brightness of 0.5 candles/cm^2, (a) what is the illumination at the center of the slit, and (b) at the ends of the slit? (Hint: divide the lamp into 10 one-inch-square sources.)

ANSWER: At normal incidence and emission a one inch square of 0.5 cd/cm^2 brightness which is 10 in away will produce an illumination

$$E_0 = B \times \omega = 0.5 \text{ cd/cm}^2 \times (1^2/10^2) = 0.005 \text{ lumens/cm}^2$$

At an angle θ from the normal this becomes

$$E_\theta = \cos^4\theta \times 0.005 \text{ lm/cm}^2$$

At the center of the lamp the two adjacent squares have their centers displaced 0.5 in from the center. The next squares are 1.5 in off center, the next 2.5 in, the 3.5 in, and 4.5 in. These squares make tilt angles from the centerline equal to

$$\theta = \arctan (\text{displacement}/10 \text{ in})$$

and we can tabulate the $\cos^4\theta$ for displacements of 0.5 in to 4.5 in [and to 9.5 in for use in part (b)] as follows:

displacement	$\cos^4\theta$
0.5 in	0.955019
1.5 in	0.956474
2.5 in	0.885813
3.5 in	0.800620
4.5 in	0.651560
5.5 in	0.589447
6.5 in	0.494192
7.5 in	0.409600
8.5 in	0.337040
9.5 in	0.276281

At the center $E = 2 \times 0.005 \times \Sigma \cos^4\theta$ where the summation is from displacement 0.5 in to 4.5 in

$$= 2 \times 0.005 \times 4.329476 = 0.04329 \text{ lumens/cm}^2$$

which is equal to $929 \times 0.04329 = 40.22$ footcandles

(b) At the end of the slit we remove the factor of 2 and sum from 0.5 in to 9.5 in, which gives us 0.03218 lumens/cm², which equals 29.90 footcandles.

4 A 16-mm movie projector uses a 2 in f/1.6 projection lens and a lamp with a filament brightness of 3000 candles/cm2. If the condenser fills the pupil of the lens with the filament image, what is the illumination produced on a screen 20 ft from the lens? (Assume transmission of 95 percent for the lens and 85 percent for the condenser.)

ANSWER: Using Eq. 12.11 in photometric symbols

$$E = t \, \pi \, B \sin^2\theta$$

The diameter of the aperture for a 2 in f/1.6 lens is $2/1.6 = 1.25$ in. At a distance of 20 ft $= 240$ in the sine of the half-angle subtended by the lens aperture is $\frac{1}{2} \times 1.25/240 = 0.0026404$ and $\sin^2\theta = 0.000006782$

$$E = 0.95 \times 0.85 \times \pi \times 3000 \times 0.000006782$$

$$= 0.051612 \text{ lumens/cm}^2$$

$$= 929 \times 0.051612 = 47.9 \text{ lumens/ft}^2 = 47.9 \text{ footcandles}$$

5 (a) What is the spectral radiant emittance of a 1000-K blackbody in the region of 2000 nm wavelength? What is the radiance? (b) If an idealized bandpass filter, transmitting 100 percent between 1950 and 2020 nm, is used, what is the total power falling on a 1 cm² detector placed one meter from a 1 cm² 1000K blackbody? (Use Eq. 12.7)

ANSWER: (a) Per Eq. 12.16 the wavelength for peak emittance is $2897.8/T$ μm $= 2.8978$ μm and per Eq. 12.17 the peak emittance is $1.286 \times T^5/10^{15} = 1.286$ W/cm²μm. Our wavelength as a fraction of the peak is $2.0/2.8978 = 0.69$, and from Fig. 12.6, at this fractional wavelength the emittance equals 0.7 of the peak, or 0.9 W/cm² μm.

(b) The total emittance in a 100 nm band is thus $0.1 \times 0.9 = 0.09$ W/cm² emitted into 2π steradians. The radiance is $0.09/\pi = 0.0286$ W/cm² ster. The irradiance equals $N\omega = 0.0286 \times (1^2/100^2) = 2.86 \times 10^{-6}$ W/cm². The flux equals area × irradiance $= 1 \times 2.86 \times 10^{-6}$ W.

Alternately we can consider the blackbody as a point source emitting 1.0 cm² × 0.0286 W/ster cm² $= 0.0286$ W/ster. With the detector subtending $1^2/100^2 = 0.0001$ ster, it intercepts $0.0001 \times 0.0286 = 2.86 \times 10^{-6}$ W.

6 Show that, for long projection distance, the maximum lumen output of a projector is given by

$$F = \pi ABT/4(f/\#)^2 \text{ lumens}$$

where A is the area of the film gate, B is the source brightness, T is the transmission of the system, and $(f/\#)$ is the relative aperture of the projection lens.

ANSWER: Illumination $E = T\pi B \sin^2\theta$

$\sin\theta = (d/2)/D = d/2D$ [D is the projection distance and d is the lens clear aperture]

$$E = T\pi Bd^2/4D^2$$

$d = efl/(f/\#)$ where efl is the focal length of the projection lens

$$E = T\pi B \text{ efl}^2/(f/\#)^2 4D^2$$

For long throws the magnification of the film gate to the screen is approximately D/efl, and thus the area of the screen is $A \times (D/\text{efl})^2$

The total lumens on the screen is the product of the illumination E and the area, or

Flux $= T\pi B \text{ efl}^2/(f/\#)^2 4D^2$ times $A \times (D/\text{efl})^2$

$\qquad = \pi ABT/4(f/\#)^2$

$\qquad$ QED

13

Optical System Layout

This chapter will be devoted to the first-order optics of several typical optical systems. The number of systems covered here is, of necessity, limited, and the emphasis is placed on those fundamental principles which are applicable to a broad range of optical systems. The rather straightforward algebraic manipulations and the considerations of image size and position which follow are quite typical of those encountered in the preliminary stages of optical system design. Constructional details of the optical components have been deliberately omitted and are discussed at considerable length in later chapters. Note that the system diagrams in this chapter show the components as simple lenses. These could equally well be mirrors instead of lenses, and typically are fairly complex assemblies of lens elements.

13.1 Telescopes, Afocal Systems

The primary function of a telescope is to enlarge the apparent size of a distant object. This is accomplished by presenting to the eye an image which subtends a larger angle (from the eye) than does the object. The magnification, or power, of a telescope is simply the ratio of the angle subtended by the image to the angle subtended by the object.* Nominally, a telescope works with both its object and image located at infinity; it is referred to as an afocal instrument, since it has no focal length. In the following material, a number of basic relationships for telescopes and afocals will be presented, all based on systems with both object and image located at infinity. In practice, small

---

*For large angles, the magnification is the ratio of the tangents of the half-angles.

departures from these infinite conjugates are the rule, but for the most part they may be neglected. However, the reader should be aware that the fact that the object and/or the image are not at infinity will occasionally have a noticeable effect and must then be taken into account. This is usually important only with low-power devices. See also the comments on instrument myopia in Sec. 8.4.

There are three major types of telescopes: astronomical (or inverting), terrestrial (or erecting), and Galilean. An astronomical or Keplerian telescope is composed of two positive (i.e., converging) components spaced so that the second focal point of the first component coincides with the first focal point of the second, as shown in Fig. 13.1a. The objective lens (the component nearer the object) forms an inverted image at its focal point; the eyelens then reimages the object at infinity where it may be comfortably viewed by a relaxed eye. Since the internal image is inverted, and the eyelens does not reinvert the image, the view presented to the eye is inverted top to bottom and reversed left to right.

In a Galilean, or "Dutch," telescope, 13.1b, the positive eyelens is replaced by a negative (diverging) eyelens; the spacing is the same, in that the focal points of objective and eyelens coincide. In the Galilean scope, however, the internal image is never actually formed; the object

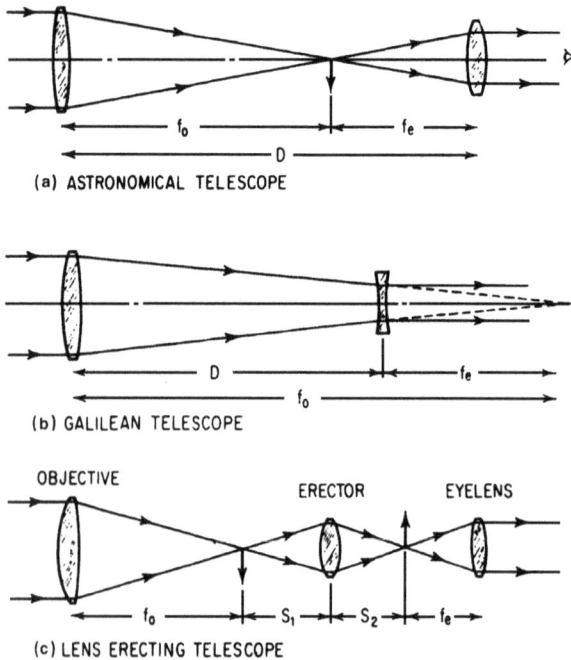

(a) ASTRONOMICAL TELESCOPE

(b) GALILEAN TELESCOPE

(c) LENS ERECTING TELESCOPE

Figure 13.1   The three basic types of telescope.

for the eyelens is a "virtual" object, no inversion occurs, and the final image presented to the eye is erect and unreversed. Since there is no real image formed in a Galilean telescope, there is no location where cross hairs or a reticle may be inserted.

Assuming the components of the telescope to be thin lenses, we can derive several important relationships which apply to *all* telescopes and afocal systems and which are of great utility. First, it is readily apparent that the length (*D*) of a simple telescope is equal to the sum of the focal lengths of the objective and eyelens.

$$D = f_o + f_e \tag{13.1}$$

Note that in the Galilean telescope, the spacing is the difference between the absolute values of the focal lengths since f_e is negative.

The magnification, or magnifying power, of the telescope is the ratio between u_e, the angle subtended by the image, and u_o, the angle subtended by the object. The size (*h*) of the internal image formed by the objective will be

$$h = u_o f_o \tag{13.2}$$

and the angle subtended by this image from the first principal point of the eyelens will be

$$u_e = \frac{-h}{f_e} \tag{13.3}$$

Combining Eqs. 13.2 and 13.3, we get the magnification

$$\text{MP} = \frac{u_e}{u_o} = \frac{-f_o}{f_e} \tag{13.4}$$

and

$$f_e = D/(1 - \text{MP})$$
$$f_o = \text{MP}D/(1 - \text{MP})$$

The sign convention here is that a positive magnification indicates an erect image. Thus, if objective and eyelens both have positive focal lengths, MP is negative and the telescope is inverting. The Galilean scope with objective and eyelens of opposite sign produces a positive MP and an erect image.

Note that u_o can represent the *real* angular field of view of the telescope and u_e the *apparent* angular field of view, and that Eq. 13.4 defines the relationship between the real and apparent fields for small angles. For large angles, the tangents of the half-field angles should be substituted in this expression.

From Chap. 9 we recall that the exit pupil of a system is the image (formed by the system) of the entrance pupil. In most telescopes the objective clear aperture is the entrance pupil and the exit pupil is the image of the objective as formed by the eyelens. Using the newtonian expression relating object and image sizes ($h'=hf/x$), and substituting CA_e (the exit pupil diameter) and CA_o (the entrance pupil diameter) for h' and h, f_e for f, and $-f_o$ for x, we get

$$\frac{CA_o}{CA_e} = \frac{-f_o}{f_e} = MP \tag{13.5}$$

While the above derivation has assumed the entrance pupil to be at the objective, Eq. 13.5 is valid regardless of the pupil location, as is obvious from the rays sketched in Fig. 13.1.

We also can get a simple expression for the eye relief of the Kepler telescope as follows:

$$R = (MP - 1)\, f_e/MP$$

The amount of motion of the eyepiece needed to focus the telescope for someone who is nearsighted or farsighted is given by

$$\delta = Df_e^2/1000$$

where δ is in millimeters and D is in diopters.

Equations 13.4 and 13.5 can be combined to relate the external characteristics (magnifications, fields of view, and pupils) of *any* afocal system, regardless of its internal construction

$$MP = \frac{u_e}{u_o} = \frac{CA_o}{CA_e} \tag{13.6}$$

The erecting telescope, Fig. 13.1c, consists of positive objective and eyelenses with an erecting lens between the two. The erector reimages the image formed by the objective into the focal plane of the eyelens. Since it inverts the image in the process, the final image presented to the eye is erect. This is the form of telescope ordinarily used for observing terrestrial objects, where considerable confusion can result from an inverted image. (An erect image may also be obtained by the use of an erecting prism as discussed in Chap. 7.) The magnification of a terrestrial telescope is simply the magnification that the telescope would have without the erector, multiplied by the linear magnification of the erector system

$$MP = -\frac{f_o}{f_e} \cdot \frac{s_2}{s_1} \tag{13.7}$$

where s_2 and s_1 are the erector conjugates as indicated in Fig. 13.1c. For a scope as shown, f_o, f_e, and s_2 are positive signed quantities and s_1 is negative. The resulting MP is thus positive, indicating an erect image.

An afocal system is the basis of the *laser beam expander*. The beam diameter of a laser is enlarged by a factor equal to the MP when the laser beam is sent into the eyepiece end of the telescope. Expansion of the beam reduces the beam divergence. The Galilean form (Fig. 13.1b) is usually preferred because there is no focus (which can cause a breakdown of the air if the laser is powerful) and the optical design characteristics are more favorable. However, the Keplerian form (Fig. 13.1a) is used when a spatial filter (a pinhole at the focus) is necessary.

An afocal system can also be used to change the power, focal length, and/or the field of view of another system by inserting it in a space in the system where the light is collimated (i.e., where the object or image is at infinity.) (See Sec. 17.4 and Fig. 17.34.)

Note that an afocal system can be used to image objects which are not at an infinite distance. For example, the exit pupil of a telescope is the image of the aperture stop, which is usually at the objective lens. Again, a consideration of the rays diagramed in Fig. 13.1 will indicate that the linear magnification m is the same, regardless of where the object and image are located. The magnification $m = h'/h$ is equal to the reciprocal of the angular magnification, MP. Thus, $m = h'/h = 1/\text{MP}$. Note that if the aperture stop is placed at the internal focus, then an afocal system becomes telecentric in both object and image space.

13.2 Field Lenses and Relay Systems

In a simple two-element telescope as shown in Fig 13.2a, the field of view is limited by the diameter of the eyelens (as was discussed at greater length in Chap. 9). In the sketch, the solid rays indicate the largest field angle that a bundle may have and still pass through the telescope without vignetting; for the bundle represented by the dashed rays, only the ray through the upper rim of the objective gets through, and vignetting is effectively complete.

The function of a *field lens* is indicated in Fig. 13.2b. If the field lens is placed exactly at the internal image, it has no effect on the power of the telescope, but it bends the ray bundles (which would otherwise miss the eyelens) back toward the axis so that they pass through the eyelens. In this way the field of view may be increased without increasing the diameter of the eyelens. Note that the exit pupil is shifted to the left, closer to the eyelens, by the introduction of a positive field lens. The distance from the vertex of the eyelens to the exit pupil is called the "eye relief" (since the eye must be placed at the pupil to see the full field of view). The necessity for a positive eye relief obviously limits the

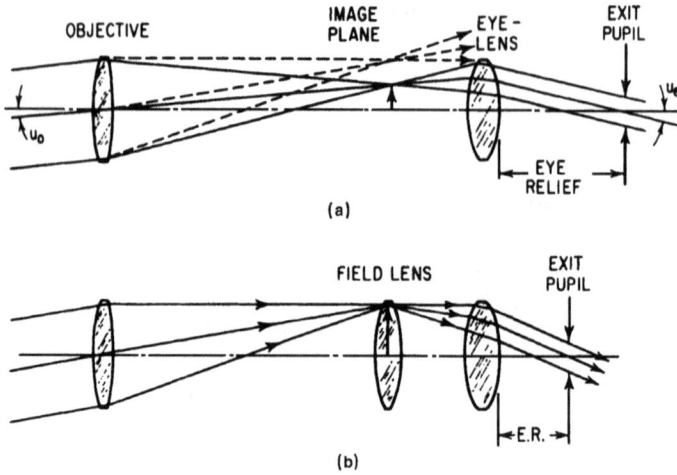

Figure 13.2 The action of the field lens in increasing the field of view.

strength of the field lens that can be used. In practice, field lenses are rarely located exactly at the image plane, but either ahead of or behind the image, so that imperfections in the field lens are out of focus and are not visible.

Periscopes and endoscopes

When it is desired to carry an image through a relatively long distance and the available space limits the diameter of the lenses which can be used, a system of *relay lenses* can be effective. In Fig. 13.3, the objective lens forms its image in field lens A. The image is then relayed to field lens C by lens B which functions like an erector lens. The image is then relayed again by lens D. The power of field lens A is chosen so that it forms an image of the objective at lens B; similarly, field lens C forms an image of lens B in lens D. In this way, the entrance pupil (which, in this example, is at the objective) is imaged at each of the relay lenses in turn and the image of the object is passed through the

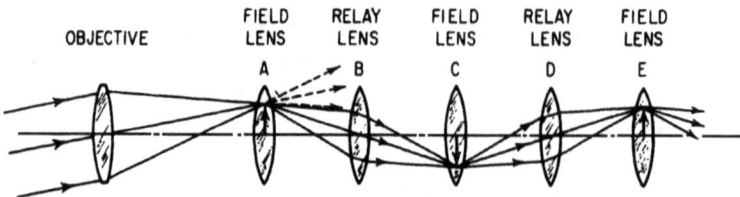

Figure 13.3 A system of relay lenses.

system without vignetting. The dashed rays emerging from lens A will indicate the large diameters which would otherwise be necessary to cover the same field of view. This type of system is used in periscopes and endoscopes.

An optimum arrangement for most optical systems is often the layout with the least total amount of lens power. In a periscope system the minimum power system is simple to design. Given the maximum lens diameter (which is determined by the available space) the image at the field lenses is arranged to fill this diameter, and the clear aperture of the relay lens is filled with the beam. Thus, with reference to Fig. 13.3, the focal length of the objective is set equal to the field lens CA divided by the total field of view, and the distance from A to B is the product of the relay lens CA times the f-number of the objective lens. Lenses B, C, D, etc., all have the same focal length, which is half the distance from A to B, and lenses B, C, D, etc., are all working at unit magnification ($m = -1$). This arrangement yields the minimum lens power for the system; this is the best layout for a periscope system.

An *endoscope* is a miniature periscope used to examine the inside of a cavity through a small orifice; they are widely used in medical applications. The size of the optics in a medical endoscope is on the order of 2 or 3 mm in diameter. The *equivalent air path* is the actual physical path divided by the index of refraction. In an endoscope or periscope, the number of relay stages is determined by the length of the instrument. If the airspaces are filled with glass, the equivalent air path is shortened by a factor equal to the index of the glass, and the number of relay stages is thereby reduced. Rather than simply fill the spaces with rods of glass, the relay lenses are typically made as cemented doublets, with the flint (negative) element made thick enough to fill the space. The outer surface of the flint is made convex so that it functions as the field lens. This is often referred to as a *rod-lens endoscope*. The reduction in the number of relay components both reduces the cost of the endoscope and improves the image quality (especially by reducing the secondary spectrum and the Petzval field curvature).

13.3 Exit Pupils, the Eye, and Resolution

Since almost all telescopes are visual instruments, they must be designed to be compatible with the characteristics of the human eye. In Chap. 8, we saw that the pupil of the eye varied in diameter from 2 mm to about 8 mm, depending on the age of the viewer and the brightness of the scene being viewed. Since the pupil of the eye is, in effect, a stop of a telescopic system, its effect must be considered. For ordinary use, an exit pupil of 3-mm diameter will fill the pupil of the eye and no increase in retinal illumination will be obtained by providing a larger exit pupil. From Eq. 13.5, it is apparent that the maximum *effective* clear aperture for an ordinary telescope objective is thus limited to a diameter of about 3-mm times the magnification. In practice, this is, however, a fairly flexible situation. In surveying instruments exit pupils of 1.0 to 1.5 mm are common, since size and weight are at a premium and resolution is the most desired characteristic. In ordinary binoculars, a 5-mm pupil is usually provided; the added pupil diameter makes it much easier to align the binocular with the eyes. For the same reason, rifle scopes usually have exit pupils ranging in size from 5 to 10 mm. Telescopes and binoculars designed for use at low light levels (such as night glasses) usually have 7- or 8-mm exit pupils in order to obtain the maximum retinal illumination possible when the pupil of the eye is large.

In Chap. 8, it was indicated that the resolution of the eye was at best about one minute of arc; Chap. 9 indicated that the angular resolution of a perfect optical system was $(5.5/D)$ seconds of arc when the clear aperture of the system (D) was expressed in inches. One or both of these limitations will govern the effective performance of any telescope, and for the most efficient design of a telescope, both should be taken into account. If two objects which are to be resolved are separated by an angle α, after magnification by a telescope their images will be separated by $(MP)\alpha$. If $(MP)\alpha$ exceeds one minute of arc, the eye will be able to separate the two images; if $(MP)\alpha$ is less than one minute, the two objects will not be seen as separate and distinct. Thus, the magnification of a telescope should be chosen so that

$$MP > \frac{1}{\alpha} \qquad (\alpha \text{ in minutes})$$

$$> \frac{0.0003}{\alpha} \qquad (\alpha \text{ in radians}) \qquad (13.8)$$

where α is the angle to be resolved. For critical work, a magnification value considerably larger than indicated in Eq. 13.8 is often selected in order to minimize the visual fatigue of the viewer.

From the opposite point of view, since the resolution of a telescope (in object space) is limited to $(5.5/D)$ seconds, it is apparent that the smallest resolved detail in the image presented to the eye will subtend an angle of $(MP)(5.5/D)$ seconds, and if this angle equals or exceeds one minute, the eye can discern all of the resolved details. Equating this angle to one minute (60 seconds), we find that the maximum "useful" power for a telescope is

$$MP = 11D \qquad (13.9)$$

(when D is in inches). Magnification in excess of this power is termed *empty magnification,* since it produces no increase in resolution. *However, it is not unusual to utilize magnifications two or three times this amount to minimize visual effort.* The upper limit on effective magnification usually occurs at the point when the diffraction blurring of the image becomes a distraction sufficient to offset the gain in visual facility.

Example 13.1

As numerical examples to illustrate the preceding sections, we will determine the necessary powers and spacings to produce a telescope with the following characteristics: a magnification of 4× and a length of 10 in. We will do this in turn for an inverting telescope, a Galilean telescope, and an erecting telescope, and will discuss the effects of arbitrarily limiting the element diameters to 1 in.

For a telescope with only two components, it is apparent that Eqs. 13.1 and 13.4 together determine the powers of the objective and eyelens. Thus, we have

$$D = f_o + f_e = 10 \text{ in}$$

and

$$\text{MP} = \frac{-f_o}{f_e} = \pm 4\times$$

where the sign of the magnification will determine whether the final image is erect (+) or inverted (−). Combining the two expressions and solving for the focal lengths, we get

$$f_o = \frac{(\text{MP})\, D}{(\text{MP}) - 1}$$

$$f_e = \frac{D}{1 - (\text{MP})}$$

For the inverting telescope, we simply substitute MP = −4 and D = 10 in, to find that the required focal length for the objective is 8 in; for the eyelens, it is 2 in. Since the lens diameters are to be 1 in, the exit pupil diameter is 0.25 in (from Eq. 13.5). The position of the exit pupil can be determined by tracing a ray from the center of the objective through the edge of the eyelens or by use of the thin-lens equation (Eq. 2.4), as follows:

$$\frac{1}{s'} = \frac{1}{f} + \frac{1}{s} = \frac{1}{f_e} + \frac{1}{(-D)} = \frac{1}{2} - \frac{1}{10} = 0.4$$

$$s' = 2.5 \text{ in}$$

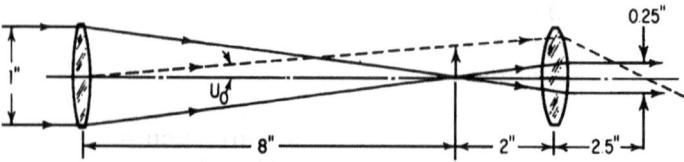

Figure 13.4   The inverting telescope of Example 13.1.

Thus, the eye relief of our simple telescope is $2\frac{1}{2}$ in.

The field of view of this telescope is not clearly defined, since it is determined by vignetting at the eyelens, as consideration of Fig. 13.4 will indicate. The aperture will be 50 percent vignetted at a field angle such that the principal (or chief) ray passes through the rim of the eyelens. Under these conditions

$$u_o = \frac{\text{dia. eyelens}}{2D} = \frac{1}{2 \times 10} = \pm 0.05 \text{ radians}$$

and the real* field of view totals 0.1 radians, or about 5.7°.

This is a poor representation of what the eye will see, however, since the vignetted exit pupil at this angle closely approximates a semicircle 0.25 in in diameter and can thus completely fill a 3-mm eye pupil. The field angle at which no rays get through the telescope is a somewhat more representative value for the field of view. If we visualize the size of u_o in Fig. 13.4 as being slowly increased, it is apparent that the ray from the bottom of the objective will be the first to miss the eyelens and the ray from the top of the objective will be the last to be vignetted out. For the example we have chosen, with both lenses 1 in in diameter, it is apparent that the limiting diameter of the internal image will also be 1 in. (For differing lens diameters, it is a simple exercise in proportion to determine the height at which this ray strikes the internal focal plane.) The half field of view for 100 percent vignetting is then the quotient of the semidiameter of the image divided by the objective focal length, or ± 0.0625 radians; the total real field is 0.125 radians, or about 7.1°.

Thus, for an exit pupil of 0.25 in, the field of view is totally vignetted at 0.125 rad, 50 percent vignetted at 0.1 rad, and unvignetted at 0.075 rad. These three conditions are illustrated in Fig. 13.5, and it is

---

*The *real* field of a telescope is the (angular) field in the object space. The *apparent* field is the (angular) field in the image (i.e., eye) space.

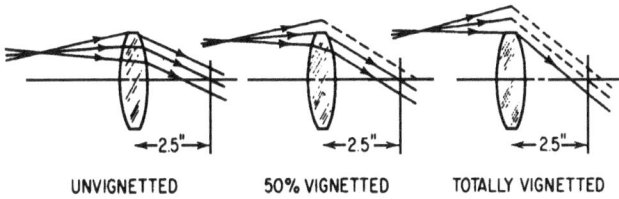

Figure 13.5 The vignetting action of the eyelens determines the field of view in an astronomical telescope.

apparent that the "effective" position of the exit pupil shifts inward as the amount of vignetting increases.

Let us now determine the minimum power for a field lens which will completely eliminate the vignetting at a field angle of ±0.0625 rad. From Fig. 13.6, it can be seen that the field lens must bend the rays from the objective so that ray B strikes no higher than the upper rim of the eyelens. The slope of ray B is equal to 1 in (the difference in the heights at which it strikes the objective and the field lens) divided by 8 in (the distance from field lens to objective), or +0.125. After passing through the field lens, we desire the slope to be zero (in this case) as indicated by the dashed ray B'. Using Eq. 4.1, we can solve for the power of the field lens as follows:

$$u' = u - y\phi_f$$

$$0.0 = +0.125 - (0.5)\,\phi_f$$

$$\phi_f = +0.25$$

$$f_f = \frac{1}{\phi} = 4 \text{ in}$$

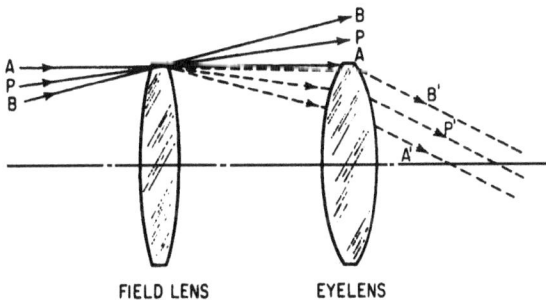

Figure 13.6 Ray diagram used to determine field lens power in Example 13.1.

We can now determine the new eye relief by tracing a principal ray from the center of the objective through the field and eye lenses.

$$u'_o = \frac{y_f}{f_o} = +0.0625 = u_f$$

$$u'_f = u_f - y_f \phi_f = +0.0625 - 0.5\,(0.25) = -0.0625$$

$$y_e = y_f + u'_f f_e = 0.5 - 0.0625\,(2) = 0.375$$

$$u'_e = u'_f - y_e \phi_e = -0.0625 - 0.375\,(0.5) = -0.25$$

$$l'_e = \text{eye relief} = \frac{-y_e}{u'_e} = \frac{-0.375}{-0.25} = 1.5 \text{ in}$$

Note that u'_e and u_o are still related by the magnification, as in Eq. 13.4, where

$$\text{MP} = \frac{u'_e}{u_o} = \frac{-0.25}{+0.0625} = -4\times$$

since the power of the system has not been changed by the introduction of the field lens located exactly at the focal plane. If we desire to locate the field lens slightly out of the focal plane, the general approach would be the same; the distances, ray heights, etc., in the computations would, of course, be modified accordingly. The power of the telescope would be increased if the field lens were placed to the right of the focus, and vice versa. In either case the scope is slightly shortened.

For the Galilean version of our telescope, we solve for the component focal lengths by substituting $+4\times$ for the magnification in the equations in the second paragraph of Example 13.1 and get

$$f_o = \frac{(\text{MP})\,D}{(\text{MP}) - 1} = \frac{(+4)\,10}{+4 - 1} = +13.33 \text{ in}$$

$$f_e = \frac{D}{1 - (\text{MP})} = \frac{10}{1 - (+4)} = -3.33 \text{ in}$$

If we assume the aperture stop to be at the objective lens of a Galilean telescope, the exit pupil will be found to be inside the telescope, and we obviously cannot put the viewer's eye there. Thus in a Galilean scope the aperture stop is not the objective lens but is the pupil of the user's eye, and the exit pupil is wherever the eye is located. This is usually about 5 mm behind the eyelens. To determine the field of view, we must trace a principal ray through the center of the pupil and passing through the edge of the objective, as indicated in Fig. 13.7. This can be done by assuming some arbitrary value for u_e and tracing

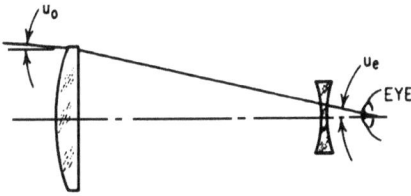

Figure 13.7 In a Galilean telescope, the field of view is determined by the diameter of the objective lens and the location of the exit pupil, which is usually the pupil of the observer's eye.

the ray through, then scaling the ray data by an appropriate constant (as indicated in Chap. 9) to make the ray height at the objective equal to one-half its clear aperture. To simplify matters, we will assume here that the pupil is coincident with the eyelens; thus, u_e is equal to half the objective diameter divided by the spacing between the lenses, or 0.05 radians in this instance. Since $MP=u_e/u_o$ per Eq. 13.4, we can solve for $u_o=0.05/4=0.0125$ radians. The total real field is 0.025 radians (about 1.5°), considerably less than that of the inverting telescope discussed above. Note that the same type of field vignetting considerations as discussed related to the eyelens of the astronomical telescope may be applied to the objective of the Galilean telescope. One must also bear in mind that the *direction* of the Galilean field of view can be changed by a lateral shift of the viewer's eye; this is not true for a telescope with a real internal image when the field stop is located at the image.

For the erecting telescope example, we will lay out a telescopic rifle sight, with a magnification of +4×, a length of 10 in, and a maximum lens diameter of 1 in, as before. For small-caliber (.22) rifles, a 2-in eye relief is acceptable; for heavier guns, eye reliefs of 3 to 5 in are common. Let us assume that we desire an eye relief of 4 in and design the telescope accordingly. The entrance pupil (at the objective) has a diameter of 1 in; by Eq. 13.7, the exit pupil diameter is thus 0.25 in. Again by Eq. 13.6, the apparent field at the eyepiece (u_e) is equal to $4u_o$, where u_o is the real field. With reference to Fig. 13.8, it is apparent that u_e is limited by the diameter of the eyelens and that for an *unvignetted*

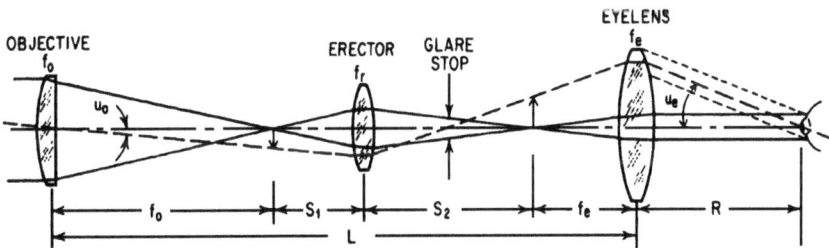

Figure 13.8 Optics of a simple erecting telescope.

pupil and a 1-in-diameter eyelens, the 4-in eye relief R limits us to an apparent field as follows:

$$u_e = 4u_o = \pm \frac{1}{2R} \text{ (eyelens dia. } - \text{ pupil dia.)}$$

$$= \pm \frac{1}{2 \times 4} (1 - 0.25) = \pm 0.09375$$

$$u_o = \pm 0.0234 = (\pm 1.3°)$$

To determine the spacing and powers of the components, we note that the length will be

$$L = f_o - s_1 + s_2 + f_e$$

and the magnification will be

$$M = \frac{-f_o s_2}{f_e s_1}$$

We can combine these expressions and derive equations for s_1, s_2, and f_r in terms of M, L, f_o, and f_e as follows:

$$s_1 = \frac{-f_o (L - f_o - f_e)}{(Mf_e + f_o)}$$

$$s_2 = \frac{-s_1 Mf_e}{f_o} = \frac{Mf_e (L - f_o - f_e)}{(Mf_e + f_o)}$$

$$f_r = \frac{s_1 s_2}{s_1 - s_2} = \frac{Mf_e f_o (L - f_o - f_e)}{(Mf_e + f_o)^2}$$

At this point, we are faced with a situation which is very common in the layout stages of optical design. We can elect to proceed algebraically to find an expression for f_o and f_e which will yield a scope with the desired eye relief R, or we can proceed numerically. In general, for a one-time solution, the numerical approach is usually the better choice, especially if the system under consideration is well understood. If one is likely to design a number of systems of the same type with various parameters, or if one is "exploring" and wishes to locate *all* possible solutions, the often tedious labor of an algebraic solution may be well repaid.

The preceding equations indicate that we have two choices (or degrees of freedom) which we can make, namely f_o and f_e, and arrive at a 4× scope of 10-in length; we have not, however, included the eye relief in these equations. To resolve this situation numerically, we would now assume some reasonable value for f_o, then proceed to test various values of f_e, selecting the value of f_e which yields the desired value for the eye relief R. Since R is not a critical dimension, a graphic solution (after a few values of f_e have been tried), plotting R versus f_e would be quite adequate for our purpose. Repeating the process for several additional values of f_o would then indicate the range of solutions available.

To arrive at a solution analytically, we would proceed as follows: a principal ray, starting at the center of the objective lens with some arbitrary slope angle would be ray-traced by thin-lens equations (4.1, 4.2, and 4.3), using the symbolic values for the spacings and lens powers derived from the three equations immediately preceding. The symbolic values for the powers and spacings involved would thus be:

$$\text{First airspace} = f_o - s_1 = f_o + \frac{f_o\,(L - f_o - f_e)}{(Mf_e + f_o)}$$

$$\text{Erector power } \phi_r = \frac{1}{f_r} = \frac{(Mf_e + f_o)^2}{Mf_e f_o\,(L - f_o - f_e)}$$

$$\text{Second airspace} = s_2 + f_e = f_e + \frac{Mf_e\,(L - f_o - f_e)}{(Mf_e + f_o)}$$

$$\text{Eyelens power } \phi_e = \frac{1}{f_e}$$

The expression for the final intercept length of this ray, $l'_e = -y_e/u'_e$ is then equated to the eye relief R, and a solution for f_e expressed in terms of f_o, M, L, and R is extracted. As can be imagined, the procedure is lengthy and the probability of making an error in the derivation is approximately unity for the first few attempts. Careful work and frequent checking are not only advisable, they are mandatory. When the smoke has cleared away, one finds that

$$f_e = \frac{M^2 RL - f_o\,(M^2 R + L)}{M^2\,(R + L) - f_o\,(M - 1)^2}$$

and that for any chosen value for f_o (which is less than L and more than zero), a set of powers and spacings can be obtained which will satisfy our original conditions for power M, length L, and eye relief R.

We are now faced, regardless of whether we have arrived via numbers or symbols, with the problem of determining what is a suitable value for f_o upon which to base our solution. There are a number of criteria by which to judge the value of a given solution. In general, one desires to minimize the power of the components in any given system; in subsequent chapters, it will become apparent that it is often advisable to minimize one or all of the following: $\Sigma|\phi|$, $\Sigma|y\phi|$, $\Sigma|y^2\phi|$ (where the symbol $|x|$ indicates the absolute value of x), ϕ is the component power, and y represents the height of either the axial or principal ray on the component, or the element semiclear aperture.

Avoiding, for a few chapters at least, the rationale behind these desiderata, we shall proceed to indicate the technique. For a number of arbitrarily chosen values of f_o, we determine the required values for f_r and f_e (as well as s_1 and s_2). Then the values of the component powers ϕ_o, ϕ_r, and ϕ_e (where $\phi = 1/f$) as well as $\Sigma|\phi| = |\phi_o| + |\phi_r| + |\phi_e|$ are plotted against f_o, resulting in a graph as shown in Fig. 13.9. Note that the minimum $\Sigma|\phi|$ occurs in the region of $f_o = 3.5$; for want of a better criterion, this is a reasonable choice.

To carry the matter a bit further, we can trace an axial ray and a principal ray through each solution. The axial ray has starting data (at the objective) of $y = 0.5$ and $u = 0$; the principal ray starting data is $y_p = 0$ and $u_p = 0.0234375$, chosen on the basis of eye relief and eyelens diameter considerations as discussed several paragraphs above. From these ray traces, we can determine the axial ray height y at each lens, y^2, and the necessary minimum clear diameter at each lens $D = 2(|y| + |y_p|)$ to pass the full bundle of rays at the edge of the field. It turns out that *under the conditions we have established,* the diameter for the objective and eyelens must be 1 in, and the diameter of the erector lens is 0.3125 in for all values of f_o. From this information, a graph as shown in Fig. 13.10 can be plotted. The choice of which of the four minima to select must be made on the basis of

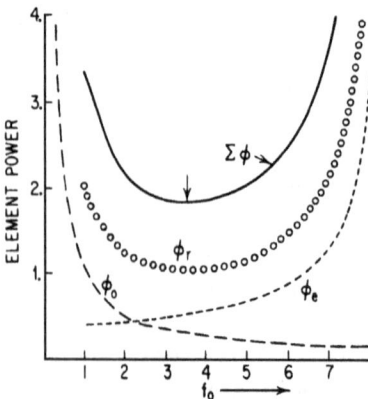

Figure 13.9 Plot of the element powers for a 10-in-long erecting telescope with 4-in eye relief versus the arbitrarily chosen objective focal length. ϕ_0, ϕ_r, and ϕ_e are the powers of the objective, erector, and eyelens, respectively.

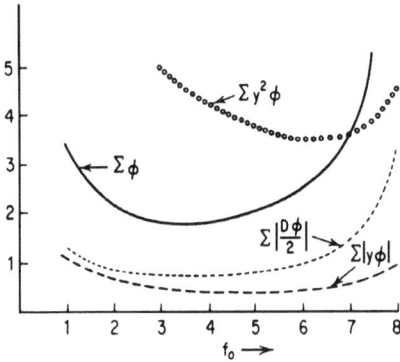

Figure 13.10 The sum of the element power as a function of the choice of the objective focal length.

material which is contained in subsequent chapters. In general, however, a minimum $\Sigma|\phi|$ in this example would reduce the Petzval curvature of field, a minimum $\Sigma|D\phi|$ would reduce the cost of making the optics, and minimum $\Sigma|D\phi|$, $\Sigma|y\phi|$, or $\Sigma|y^2\phi|$ would tend to reduce other aberrations, the choice being dependent upon which aberration one most desired to reduce.

Assuming that we have chosen $f_o = +4$, the values of the lens powers and spacings would be determined as follows:

$$f_o = +4$$

$$f_e = \frac{4 \times 4 \times 4 \times 10 - 4\,(4 \times 4 \times 4 + 10)}{4 \times 4\,(4 + 10) - 4\,(4 - 1)\,(4 - 1)} = +1.8298$$

$$s_1 = \frac{-4\,(10 - 4 - 1.8298)}{(4 \times 1.8298 + 4)} = -1.4737$$

$$s_2 = \frac{-(-1.4737) \times 4 \times 1.8298}{4} = +2.6965$$

$$f_r = \frac{-(-1.4737) \times 4 \times 1.8298}{(4 \times 1.8298 + 4)} = +0.9529$$

13.4 The Simple Microscope or Magnifier

A microscope is an optical system which presents to the eye an enlarged image of a near object. The image is enlarged in the sense that it subtends (from the eye) a greater angle than the object does when viewed at normal viewing distance. The "normal viewing distance" is conventionally considered to be about 10 in; this represents an average value for the distance at which most people see detail most clearly.

(Obviously, very young people can see detail in objects a few inches from the eye and mature persons whose visual accommodation is failing may have difficulty focusing on objects several feet away.) The magnification or magnifying power of a microscope is defined as the ratio of the visual angle subtended by the image to the angle subtended by the object at a distance of 10 in from the eye.

The simple microscope or magnifying glass consists of a lens with the object located at or within its first focal point. In Fig. 13.11, the object h, a distance s from the magnifier, is imaged at a distance s' with a height h'. As shown, the image is virtual and both s and s' are negative quantities according to our sign convention. We can readily determine the magnification by using the first-order equations (2.4 and 2.7) as follows. The object and image distance equation

$$\frac{1}{s'} = \frac{1}{f} + \frac{1}{s}$$

is solved for s

$$s = \frac{fs'}{f - s'}$$

and substituted into the equation for the image height

$$h' = \frac{hs'}{s} = \frac{h(f - s')}{f}$$

Now if the eye is located at the lens, the angle subtended by the image is given by

$$\alpha' = \frac{h'}{s'} = \frac{h(f - s')}{fs'}$$

If the unaided eye were to view the object at a distance of -10 in, the angle subtended would be

$$\alpha = \frac{-h}{10 \text{ in}}$$

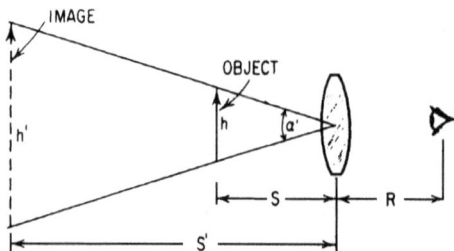

Figure 13.11 The simple microscope, or magnifier, forms an erect, virtual image of the object.

The magnifying power is the ratio between these two angles

$$\mathrm{MP} = \frac{\alpha'}{\alpha} = \frac{h\,(f - s')}{fs'} \times \frac{(-10 \text{ in})}{h}$$

$$= \frac{10 \text{ in}}{f} - \frac{10 \text{ in}}{s'} \tag{13.10}$$

Thus we find that the magnification produced by a simple microscope depends not only on its focal length but on the focus position chosen. If one adjusts the object distance so that the image is at infinity (i.e., $s = -f$ and $s' = \infty$) and can be viewed with a relaxed eye, then the magnification becomes simply

$$\mathrm{MP} = \frac{10 \text{ in}}{f} \tag{13.10a}$$

If the focus is set so that the image appears to be 10 in away (i.e., $s' = -10$ in) then

$$\mathrm{MP} = \frac{10 \text{ in}}{f} + 1 \tag{13.10b}$$

The value of MP given by Eq. 13.10a is conventionally used to express the power of magnifiers, eyepieces, and even compound microscopes.

The preceding assumed that the eye was located at the lens. If the image is not located at infinity, the magnifying power will be reduced as the eye is moved away from the lens. If R is the lens-to-eye distance, the magnification becomes

$$\mathrm{MP} = \frac{10\,(f - s')}{f\,(R - s')} \tag{13.10c}$$

Note that if the dimensions are in millimeters, the constant 10 becomes 254, or 250.

13.5 The Compound Microscope

As illustrated in Fig. 13.12, a compound microscope consists of an objective lens and an eyelens. The objective lens produces a real

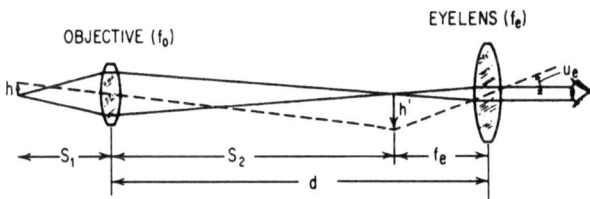

Figure 13.12 The compound microscope.

inverted image (usually enlarged) of the object. The eyelens reimages the object at a comfortable viewing distance and magnifies the image still further. The magnifying power of the system can be determined by substituting the value of the combined focal length of the two components (as given by Eq. 4.5) into Eq. 13.10a

$$f_{eo} = \frac{f_e f_o}{f_e + f_o - d} \tag{13.11}$$

$$\text{MP} = \frac{10 \text{ in}}{f_{eo}} = \frac{(f_e + f_o - d) \, 10 \text{ in}}{f_e f_o}$$

The more conventional way to determine the magnification is to view it as the product of the objective magnification times the eyepiece magnification. With reference to Fig. 13.12, this approach gives

$$\text{MP} = M_o \times M_e = \frac{s_2}{s_1} \cdot \frac{10 \text{ in}}{f_e} \tag{13.12}$$

Equations 13.11 and 13.12 yield exactly the same value of magnification, as can be shown by substituting $(d - f_e)$ for s_2; determining s_1 in terms of d, f_e, and f_o (from Eq. 2.4); and substituting in Eq. 13.12 to get Eq. 13.11.

An ordinary laboratory microscope has a tube length of 160 mm. The tube length is the distance from the second (i.e., internal) focal point of the objective to the first focal point of the eyepiece. Thus, by Eq. 2.6, the objective magnification is $160/f_o$, and rewriting Eq. 13.12 for millimeter measure, we get

$$\text{MP} = \frac{-160}{f_o} \cdot \frac{254}{f_e} \tag{13.13}$$

Standard microscope optics are usually referred to by their power. Thus, a 16-mm focal length objective has a power of $10\times$ and an 0.5-in focal length eyepiece has a power of $20\times$. The combination of the two would have a magnifying power of $200\times$, or 200 diameters.

The resolution of a microscope is limited by both diffraction and the resolution of the eye in the same manner as in a telescope. In the case of the microscope, however, we are interested in the linear resolution rather than angular resolution. By Rayleigh's criterion, the smallest separation between two object points that will allow them to be resolved is given by Eq. 9.16

$$Z = \frac{0.61\lambda}{\text{NA}}$$

where λ is wavelength and $\text{NA} = n \sin U$, the numerical aperture of the system. Note that the index n and the slope of the marginal ray U are those at the object. Because of the importance of the numerical aperture in this regard, microscope objectives are usually specified by power and numerical aperture; for example, a 16-mm objective is usually listed as a 10×NA 0.25.

At a distance of 10 in, the visual resolution of one minute of arc (0.0003 radians) corresponds to a linear resolution of about 0.003 in, or 0.076 mm. When the object is magnified by an optical system, the *visual* resolution at the object is thus

$$R = \frac{0.003 \text{ in}}{\text{MP}} = \frac{0.076 \text{ mm}}{\text{MP}} \tag{13.14}$$

If we now equate the visual resolution R with the diffraction limit Z and solve for the magnification, we find that

$$\text{MP} = \frac{0.12 \text{ NA}}{\lambda} \tag{13.15}$$

with λ in millimeters, is the magnifcation at which the diffraction limit and visual limit match. At this power the eye can resolve all the detail present in the image, and setting $\lambda=0.55$ μm, any magnification beyond 225 NA is "empty magnification." However, as with telescopes, magnifications several times this amount are regularly used, as discussed in Sec. 13.3.

13.6 Rangefinders

Figure 13.13 is a schematic diagram of a simplified triangulation rangefinder. The eye views the object by two paths; directly through semitransparent mirror M_1 and by an offset path via M_1 and fully

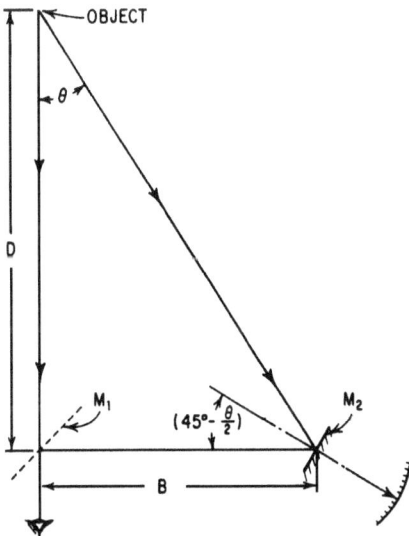

Figure 13.13 Basic rangefinder optical system. The eye views the object directly through semi-reflector M_1 and also through movable mirror M_2. The angular setting of M_2 which brings both views into coincidence determines the range.

reflecting mirror M_2. The angular position of one of the mirrors is adjusted until both images coincide. In the rudimentary instrument shown here, a pointer attached to mirror M_2 can be used to read the value of $\theta/2$; the distance to the object is found from

$$D = \frac{B}{\tan \theta} \qquad (13.16)$$

where B is the base length of the instrument. In actual rangefinders, a telescope is often combined with the mirror system to increase the accuracy of the reading, and any one of a number of devices may be used to determine θ; the distance is usually read directly from a suitable range scale so that no calculation is necessary.

The accuracy of the value of D depends on how accurately θ can be measured. For large ratios of D/B, we can write

$$D = \frac{B}{\theta} \qquad (13.17)$$

and differentiating with respect to θ, we get

$$dD = -B\theta^{-2}\,d\theta \qquad (13.18a)$$

Substituting $\theta=B/D$ into Eq. 13.18a, we find that the error in D due to a setting error of $d\theta$ is

$$dD = \frac{-D^2}{B}\,d\theta \qquad (13.18b)$$

Now $d\theta$ is primarily limited by how well the eye can determine when the two images are in coincidence. This is essentially the vernier acuity of the eye and is about 10 seconds of arc (0.00005 radians). If the magnification of the rangefinder optical system is M, then $d\theta$ is $0.00005/M$ radians, and the ranging error is

$$dD = \pm\frac{5 \times 10^{-5}D^2}{MB} \qquad (13.18c)$$

Thus, the greater the base B and the greater the magnification M, the more accurate the value of the range D.

A few of the devices encountered in rangefinders are illustrated in Fig. 13.14. In Fig. 13.14a the end mirrors are replaced by penta-prisms (or "penta"-reflectors), which are constant-deviation devices, bending the line of sight 90° regardless of their orientation. The reason for their use is to remove a source of error, since no change in the relative angular position of the two images is produced by misalignment of the penta-prisms as would be the case with simple 45° mirrors. A double

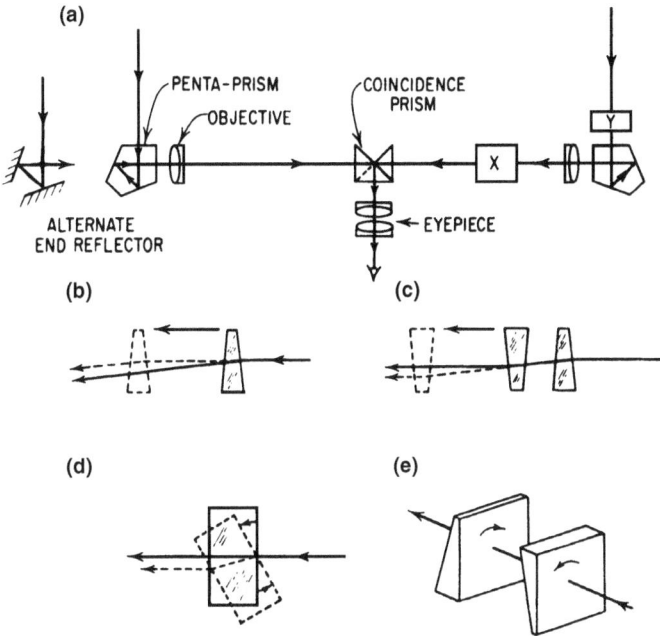

Figure 13.14 Typical rangefinder optical devices. (a) A telescopic rangefinder with coincidence prism and penta-prism end reflectors. (b) Sliding prism used at X to establish coincidence. (c) Pair of sliding prisms used at X. (d) Rotating parallel plate used at X. Optical micrometer. (e) Counter-rotating prisms used at Y to establish coincidence. Risley prisms.

telescope is built into the system to provide magnification; the power of each branch of the telescope must be carefully matched to avoid errors. The coincidence prism is provided to split the field of view into two halves, with a sharply focused dividing line between. In the system as shown, the final image is inverted; an erecting system, either prism or lens, is frequently included. Actual coincidence prisms are usually much more complex than that shown here.

A great variety of devices may be utilized to bring the two images into coincidence. Those shown in Fig. 13.14b to d are located between the objective and eyelens, usually in the region marked X in Fig. 13.14a. The sliding prism of Fig. 13.14b produces a displacement at the image plane which increases with its distance from the image; it is usually an achromatic prism. Figure 13.14c shows two identical prisms with variable spacing, which displace but do not deviate the rays. The rotating block in Fig. 13.14d operates on the same principle. All of the above tend to introduce astigmatism (that is, a difference of focal position in vertically and horizontally aligned images) since they are tilted surfaces in a convergent beam. The counterrotating wedges of Fig. 13.14e can be located in parallel light (region Y in Fig. 13.14a) and thus avoid this difficulty. Note that as one wedge turns clockwise, the other must rotate counterclockwise through exactly the same angle; in this way the vertical deviation is maintained at zero while the horizontal deviation can be varied plus or minus twice the deviation of an individual wedge. These are sometimes called Risley prisms.

Another device to produce a variable angle of deviation consists of a fixed plano concave lens and a movable plano convex lens of the same radius with their curved surfaces nested together. When the convex lens is located so that its plane surface is parallel to that of the concave lens, the pair produces no angular deviation. However, if the convex lens is rotated (about its center of curvature), the pair effectively becomes a prism and will produce an angular deviation. This device can be executed with spherical surfaces or with cylindrical surfaces.

Single-lens reflex (SLR) cameras often incorporate a split-image rangefinder which is based on an entirely different principle than the coincidence rangefinder described above. The viewfinder of an SLR camera consists of the camera objective lens, a field lens, and an eyelens. The field lens is divided into three zones as indicated in Fig. 13.15b. The outer zone functions as a straightforward field lens, redirecting the light at the edge of the field so that it passes through the eyelens. It is made in the form of a plastic *Fresnel lens,* in which the curved surface of a lens is collapsed in annular zones to a thin plate, as shown in Fig. 13.15a. This has the refracting effect of the lens without its thickness or weight. Such Fresnel lenses are also used as condensers in overhead projectors, as well as in spotlights and signal lamps. The center zone of the SLR field lens is split into two halves. Each half is a wedge prism; the two prisms are oriented in opposite directions. If the image formed by the objective lens is in focus, it is located in the plane of the wedges and the two halves of the image line up with each other. If the image is out of focus, the image through one-half of the split wedge is deviated in one direction; through the other half the deviation is in the other direction and the image is split. The intermediate zone of the field lens has a surface comprised of tiny pyramidal prisms which deviate and break up an out-of-focus image so as to exaggerate the out-of-focus blurring.

For many applications the optical rangefinder has been superceded by the laser rangefinder. This is essentially optical radar, where the distance to the target is obtained by measuring the travel time for a pulse of light to reflect from the target and return. In military applications a high-power laser is used; in surveying applications a cooperative target such as a retrodirector (corner-cube prism) is used and a much lower power source is adequate.

(a)

(b)

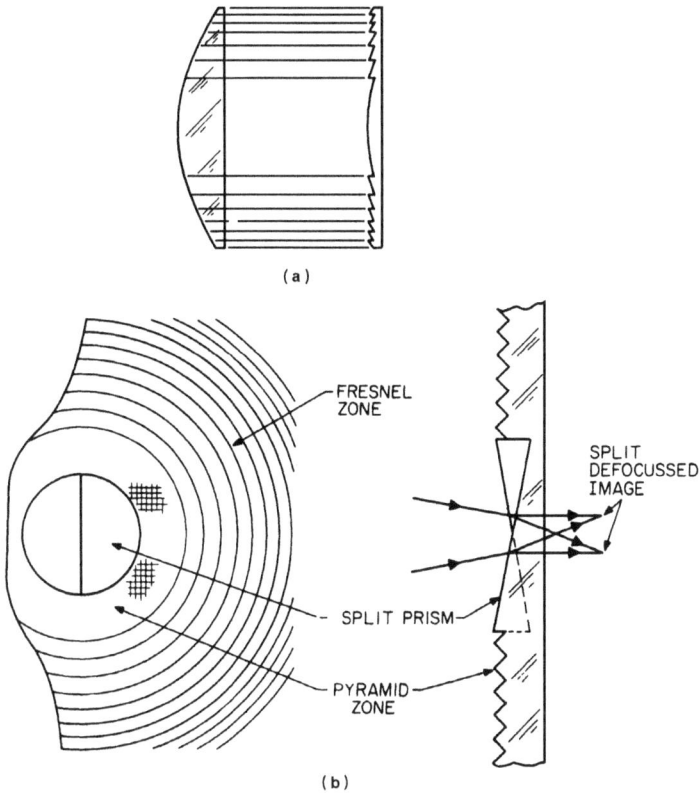

Figure 13.15 (a) A Fresnel lens is shown with the equivalent lens from which it is derived. Each annular zone of the Fresnel lens has the same surface slope as the corresponding zone of the lens. (b) The split-prism rangefinder of a 35-mm SLR camera splits an out-of-focus image in two by means of oppositely oriented wedge prisms in its central zone. If the image is focused on the wedge surface, it is not deviated or split. The area surrounding the split prism is comprised of tiny pyramidal prisms which break up an out-of-focus image and exaggerate its blur. The outer zone is a Fresnel lens acting as a field lens for the camera viewfinder.

13.7 Radiometer and Detector Optics

A radiometer is a device for measuring the radiation from a source. In a simple form, it may consist of an objective lens (or mirror) which collects the radiation from the source and images it on the sensitive surface of a detector capable of converting the incident radiation into an electrical signal. A "chopper," which may be as simple as a miniature fan blade, is usually interposed in front of the detector to provide an alternating signal for the benefit of the electronic circuitry which must amplify and process the detector output.

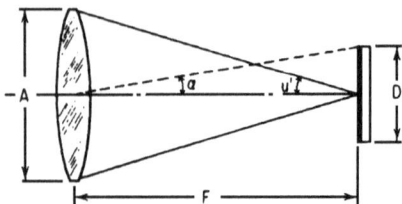

Figure 13.16 A simple radiometer with an objective lens which forms an image of the radiation source directly on the detector cell.

The radiometer is widely used for the purpose its name would seem to imply, to measure radiation. However, it is also the basis of many other applications. The receiver in a communications system by which one talks over a beam of light is a sort of radiometer the output of which is converted into audible form. The seeker head of an infrared homing air-to-air missile (e.g., the Sidewinder) is basically a radiometer the output of which is arranged to indicate whether the hot exhaust of an enemy jet is on or off the line of sight.

A simple radiometer is sketched in Fig. 13.16. The detector, with a diameter D, is located at the focus of an objective with a focal distance F and a diameter A. The half-field of view of the system is α, and since the detector is at the focus of the system, it is apparent that the half-field of view is given by

$$\alpha = \frac{D}{2F} \tag{13.19}$$

In the various applications of radiometers, the following characteristics are frequently desirable in the optical system

1. In order to collect a large quantity of power from the source, the diameter A of the system should be as large as possible.

2. In order to increase the signal-to-noise ratio, the size D of the detector should be as small as possible.

3. In order to cover a practical field of view, the field angle α should be of reasonable size (and often, should be as large as possible).

The relationship between A and F is, as we have previously noted, a limited one. If the optical system is to be aplanatic* (that is, free of spherical aberration and coma), the second principal surface (or principal "plane") must be spherical; for this reason, the effective diameter A cannot exceed twice the focal distance F, and the slope of the

---

*The frequent assumption of aplanatic systems in the analysis of radiometric systems is based (1) on the usual need for good image quality and (2) on the fact that the image illumination (irradiance) produced by an aplanatic system cannot be exceeded, so that the assumption provides a limiting case.

marginal ray at the image cannot exceed 90°. This limits the numerical aperture of the system to NA = n' sin 90° = n'; for systems in air with distant sources the limiting relative aperture becomes $f/0.5$. There are other limits imposed on the speed of the objective lens; the design of the system may be incapable of whatever resolution is required at large aperture ratios, or physical limitations (or predetermined relationships) may limit the acceptable speed of the objective.

We can introduce the effective $f/\#$ of the objective by multiplying both sides of Eq. 13.19 by A; setting $(f/\#) = F/A$ and rearranging, to get, for systems in air,

$$(f/\#) = \frac{D}{2A\alpha} \tag{13.20}$$

or for systems with the final image in a medium of index n'

$$NA = n' \sin u' = \frac{A\alpha}{D} \tag{13.21}$$

Equation 13.21 can also be demonstrated by setting the optical invariant (Eq. 4.14) at the objective ($I = A\alpha/2$) equal to the invariant at the image ($I = \frac{1}{2}Dn'u'$) and substituting sin u' for u' (in accordance with our requirement for aplanatism).

Since the $(f/\#)$ cannot be less than 0.5 and sin u' cannot exceed 1.0, it is apparent that the objective aperture A, half-field angle α, and detector size D, are related by

$$\left| \frac{A\alpha}{n'D} \right| \leq 1.0 \tag{13.22}$$

It should be noted that Eq. 13.22, since it can be derived by way of the optical invariant with no assumptions as to the system between object and detector, is valid for all types of optical systems, including reflecting and refracting objectives with or without field lenses, immersion lenses, light pipes, etc. *It is thus quite futile to attempt a design with the left member of Eq. 13.22 larger than unity; in fact, it is sometimes difficult to exceed (efficiently) a value of 0.5 when good imagery is required.* This limit is applicable to *any* optical system, no matter how simple or complex. Equation 13.22 is exactly analogous to Eq. 12.24 for projection or illumination systems.

As an example of the application of Eq. 13.22, let us determine the largest field of view possible for a radiometer with a 5-in aperture and a 1-mm (0.04-in) detector. If the detector is in air ($n'=1.0$) we then have, from Eq. 13.22,

$$\frac{5\alpha}{0.04} \leq 1.0 \text{ or } \alpha \leq 0.008 \text{ radians}$$

and the absolute maximum total field (0.016 radians) is a little less than one degree (0.01745 radians). An immersion lens at the detector (described below) with an index n' would increase the maximum field angle to $0.016n'$.

An *immersion lens* is a means of increasing the numerical aperture of an optical system by a factor of the index n of the immersion lens, usually without modifying the characteristics of the system. Another way of considering the immersion lens is to think of it as a magnifier which enlarges the apparent size of the detector. The most frequently utilized form of immersion lens is a hemispherical element in optical contact with the detector. In Fig. 13.17, a concentric immersion lens of index n' has reduced the size of the image to h'/n'. Since the first surface of the immersion lens is concentric with the axial image point, rays directed toward this point are normal to this surface and are not refracted. For this reason, neither spherical aberration nor axial coma nor axial chromatic is introduced. The optical invariant at the image is $h'n'u'$, and since u' is not changed by the immersion lens, it is apparent that as n' increases, h' must decrease.

In the use of immersion lenses, one must beware of reflection (especially total internal reflection) at the plane surface. Ideally, the detector layer should be deposited directly on the immersion lens. Since immersion lenses are usually resorted to in cases where the angles of incidence are large, total internal reflection can occur if the immersion lens index is high and a low-index layer (air or cement, for example) separates it from the detector.

In the application of radiometer-type systems, it is not unusual that one wishes to use an objective of relatively low speed with a small detector and still cover a large field of view. This is readily accomplished by means of a field lens. The field lens is located at (or more frequently, near) the image plane of the objective system and redirects the rays at the edge of the field toward the detector, as indicated in Fig. 13.18. As can be seen from a brief consideration of the figure, the field lens actually images the clear aperture of the objective on the surface of the detector. The optimum arrangement

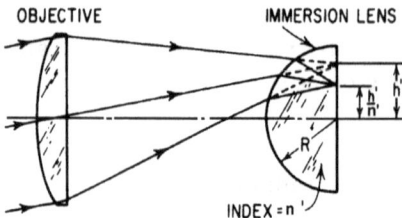

Figure 13.17 A hemispherical immersion lens concentric with the focus of an optical system reduces the linear size of the image by a factor of its index.

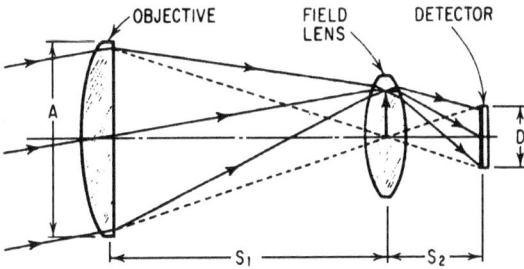

Figure 13.18 Radiometer with field lens to increase the field of view with a small detector.

is when the image of the objective aperture is the same size as the detector and

$$\frac{s_1}{s_2} = (-)\frac{A}{D}$$

This arrangement not only makes a larger field angle possible, but has the advantage of providing an even illumination over a large portion of the detector surface. Most detectors vary in sensitivity from point to point over their surface; with a field lens of focal length given by

$$f = \frac{s_1 s_2}{s_1 - s_2}$$

the same area of the detector is illuminated regardless of where the source is imaged in the field of view. Field lenses and immersion lenses are frequently combined. Note that the insertion of a field lens in a radiometer does not change the limitations of Eqs. 13.21 and 13.22; it simply permits the use of an objective system with a low numerical aperture by raising the numerical aperture at the detector.

Another device to enlarge the field of view of a radiometer with a small detector is the light pipe, or cone channel condenser. In Fig. 13.19, a principal ray from the objective is shown being reflected from the walls of a tapered light pipe. Note that without the light pipe, the ray would completely miss the detector.

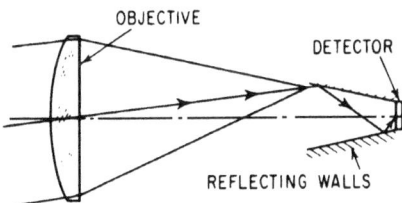

Figure 13.19 The action of a reflecting light pipe in increasing the field of view of a radiometer.

It is instructive to consider the "unfolded" path of a ray through such a system, as indicated in Fig. 13.20. The actual reflective walls of the light pipe are shown as solid lines; the dashed lines are the images of the walls formed by reflection from each other. This layout is analogous to the prism unfolding technique explained in Chap. 7 as a "tunnel diagram" and allows us to draw the path of a ray through the system as a straight line. Note that ray A in the figure undergoes three reflections before it reaches the detector end of the pipe. Ray B, entering at a greater angle, never does reach the detector, but is turned around and comes back out the large end of the pipe. This is a limit on the effectiveness of the pipe and is analogous to the $f/\#$ or numerical aperture limit on ordinary optical systems discussed above in the derivation of Eqs. 13.20 et seq.

A light pipe may be constructed as a hollow cone or pyramid with reflective walls in the manner indicated in Figs. 13.19 and 13.20. It is also common to construct them out of a solid piece of transparent optical material. The walls may then be reflective coated or one may rely on total internal reflection if the angles are properly chosen. Note that with a solid light pipe, total internal reflection may occur at the exit face; this can be avoided by "immersing" the detector at the exit end of the pipe. The use of a solid pipe effectively increases its acceptance angle by a factor of the index n of the pipe material; the effect on the system is exactly analogous to the use of an immersion lens, and the total radiometer system is still governed by Eq. 13.22 as before. Light pipes may be used with field lenses; the most common arrangement is to put a convex spherical surface on the entrance face of a solid pipe.

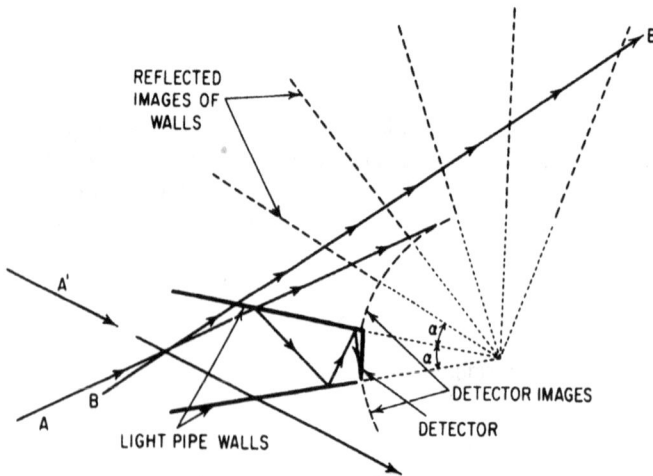

Figure 13.20 Ray tracing through a light pipe by means of an "unfolded" diagram.

If one were to look into the large end of a pyramidal light pipe, one would see a sort of checkerboard multiple image of the exit face (or detector), as indicated in Fig. 13.20 for a two-dimensional case. The checkerboard is wrapped around a sphere centered on the apex of the pyramidal pipe. This image is, of course, the effective size of the ("magnified") detector, and the cone of light from the objective, as indicated by rays A and A' is spread out over this array. This effect is occasionally useful in decorrelating the point-for-point relationship between the detector surface and the objective aperture which is established when a field lens is used. The effect is even more pronounced in a conical pipe.

The discussion in this section has been devoted to condensing radiation on to a small detector. The tables can be turned. If we replace the detector with a small source of radiation, devices such as field lenses and light pipes can be used to increase the apparent size of the source and to reduce the angle through which it radiates (or vice versa).

A common application of the light pipe is in *illumination systems,* especially where extremely uniform illumination is required and the source is very nonuniform, such as a high-pressure mercury or xenon or metal halide arc lamp. If the light pipe is made with parallel sides (either as a cylinder or with a square or rectangular cross section) as shown in Fig. 13.21, the image of the light source can be focused on one end of the pipe; the other end is then quite uniformly illuminated. As can be seen from the figure, the multiple reflections of the source form a checkerboard array of images which is effectively a new light source, and the illumination across the exit end of the pipe is quite uniform. Of course there is no reason that a tapered pipe cannot be used in this way, and this is occasionally done. Note that the proportions of the light pipe (length, diameter) and the convergence of the imaging beam

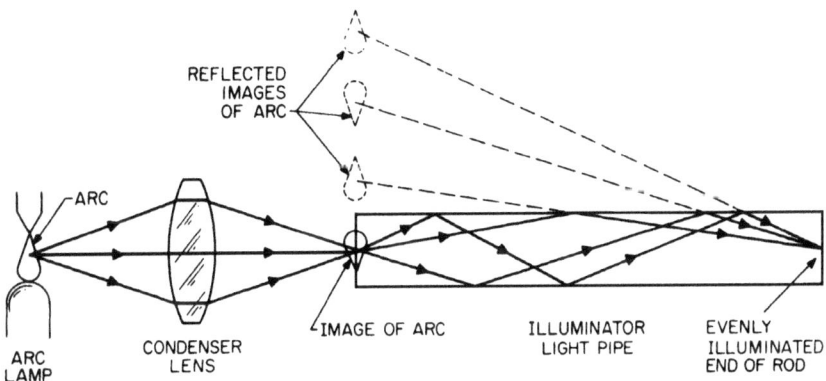

Figure 13.21 A light pipe can be used to produce very uniform illumination at its exit face when a light source is focused on the other end. The multiple images produced by reflections from the pipe walls become the illuminating source for the exit face.

will determine the number of reflections and the number of reflected source images.

13.8 Fiber Optics

A long, polished cylinder of glass can transmit light from one end to the other without leakage, provided that the light strikes the walls of the cylinder with an angle of incidence greater than the critical angle for total internal reflection. The path of a meridional ray through such a cylinder is shown in Fig. 13.22. The geometric optics of meridional rays through such a device are relatively simple. For a cylinder of length L, the path traveled by the meridional ray has a length given by

$$\text{Path length} = \frac{L}{\cos U'} \tag{13.23}$$

and the number of reflections undergone by the ray is

$$\text{No. reflections} = \frac{\text{path length}}{(d/\sin U')} = \frac{L}{d}\tan U' \pm 1 \tag{13.24}$$

where U' is slope of the ray inside the cylinder, d is the cylinder diameter, and L its length. For the light to be transmitted without reflection loss, it is necessary that the angle I exceed the critical angle

$$\sin I_c = \frac{n_2}{n_1}$$

where n_1 is the index of the cylinder and n_2 the index of the medium surrounding the cylinder. From this one can determine that the maximum external slope of a meridional ray which is to be totally reflected is

$$\sin U = \frac{1}{n_0}\sqrt{n_1^2 - n_2^2} \tag{13.25}$$

This "acceptance cone" of a cylinder is often specified as a numerical aperture; by rearranging Eq. 13.25, we get

$$NA = n_0 \sin U = \sqrt{n_1^2 - n_2^2} \tag{13.26}$$

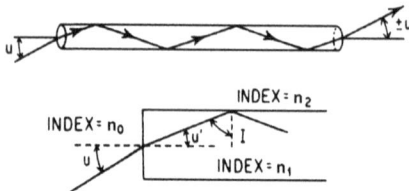

Figure 13.22 Light is transmitted through a long polished cylinder by means of total internal reflection.

This is the minimum value for the numerical aperture; as indicated below and in Fig. 13.23, skew rays have a larger NA than do meridional rays.

Again, with reference to Fig. 13.22, it is apparent that if the meridional ray had entered the cylinder well above or well below the axis, it would have emerged with a slope angle of $-U$. The path of a pair of skew rays is indicated (in an end-on view) in Fig. 13.23. Note that a skew ray is rotated with each reflection and that the amount of rotation depends on the distance of the ray from the meridional plane. Thus, a bundle of parallel rays incident on one end of a cylinder will emerge from the other end as a hollow cone of rays with an apex angle of $2U$. If the diameter of the cylinder is small, diffraction effects may diffuse the hollow cone to a great extent. It is also worth noting that since the skew rays strike the surface of the cylinder at a greater angle of incidence than the meridional rays, the numerical aperture for skew rays is larger than that for meridional rays.

If the light-transmitting cylinder is bent into a moderate curve, a certain amount of light will leak out the sides of the cylinder. However, the major portion of the light is still trapped inside the cylinder, and a simple curved rod is occasionally a convenient device to pipe light from one location to another.

Optical fibers are extremely thin filaments of glass or plastic. Typical diameters for the fibers range from 1 to 2 μm to 25 μm or more. At these small diameters, glass is quite flexible, and a bundle of optical fibers constitutes a flexible light pipe. Figure 13.24 shows a few of the applications of fiber optics. Figure 13.24a indicates the basic property of an oriented, or "coherent," bundle of fibers in transmitting an image from one end of the fiber to the other. If the bundle is constrained at both ends so that each fiber occupies the same relative position at each end, then the fiber rope may literally be tied in knots without affecting its image-transmitting properties. Fiber bundles with lengths of many feet are obtainable with surprisingly high transmissions.

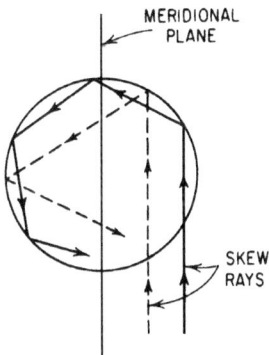

Figure 13.23 The path of skew (nonmeridional) rays through a reflecting cylinder is a sort of helix. The amount of rotation a ray undergoes in traversing a given length depends on its entrance position.

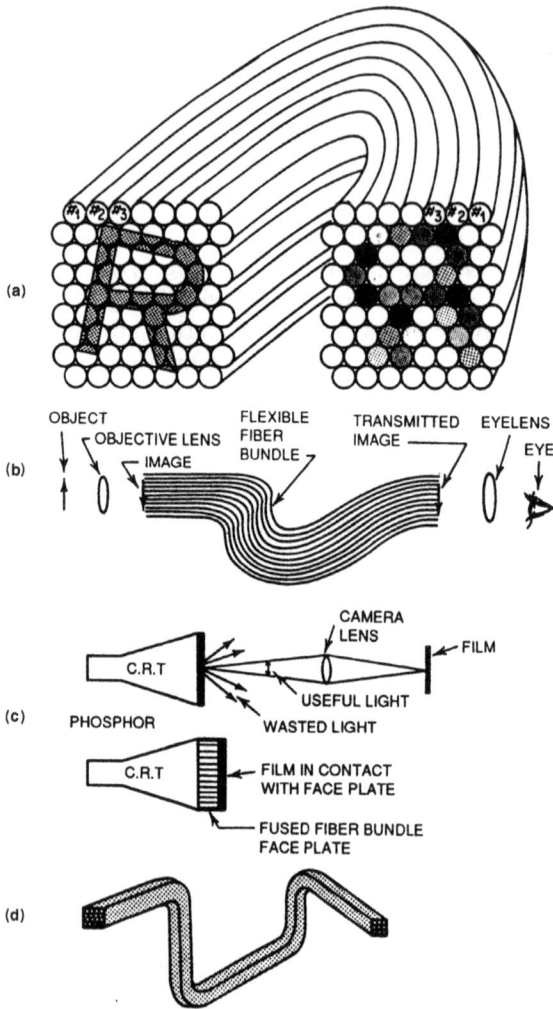

Figure 13.24 Fiber optics.

The limiting resolution (in line pairs per unit length) of a coherent fiber bundle is approximately equal to half the reciprocal of the fiber diameter; by synchronously oscillating or scanning both ends of the fiber, this resolution can be doubled. When the fibers are tightly packed, their surfaces contact each other and leakage of light from one fiber to the next will occur. Moisture, oil, or dirt on the fiber surface can also interfere with total internal reflection. This is prevented by coating or "cladding" each fiber with a thin layer of lower-index glass or plastic. For example, the core glass may have $n_1 = 1.72$ and the cladding $n_2 = 1.52$, yielding a numerical aperture according to Eq. 13.26 of the order of 0.8. Since the total internal reflection (TIR) occurs at the core-cladding interface, moisture or contact between the outer surfaces does not frustrate the TIR if the cladding is thick enough.

Figure 13.24b shows a flexible gastroscope or sigmoidoscope. An objective lens forms an image of the object on one end of a coherent fiber bundle; at the other end the transmitted image is viewed with the aid of an eyepiece or video camera.

Ordinary photography of a cathode ray tube face is an inefficient process. The phosphor radiates in all directions and a camera lens intercepts only a small portion of the radiated light. A tube face composed of a hermetically fused fiber array (Fig. 13.24c) can transmit all the energy radiated into a cone defined by its NA to a contacted photographic film with negligible loss. Fused fibers are always clad with low-index glass to separate the fibers; frequently an absorbing layer or absorbing fibers are added to prevent contrast reduction by stray light which is emitted at angles larger than the numerical aperture of the fibers. Fiber optics are also available as optical conduit, that is, rigid fused bundles, for efficient transmission of light through labyrinthian paths, as shown in Fig. 13.24d.

Flexible plastic fibers with diameters on the order of 0.5 in are used as single fibers in illumination systems.

A tapered, coherent, fused-fiber bundle can be used as either a magnifier or minifier (depending on whether the original object is placed at the small or large end of the taper). By twisting a coherent bundle of fibers, either fused or not, an image erector can be made which will carry out the function of the erector prisms described in Chap. 7. These are often found in image-intensifier systems such as those used in night vision goggles.

Hollow glass fibers in diameters from 0.5 to 1.0 mm, internally coated, are moderately flexible and have been used to transmit radiation in the 10-μm wavelength region. These fibers do a reasonable job of maintaining the gaussian distribution of the laser light.

Gradient index fibers

The preceding descriptions have dealt with fibers the principal function of which was to transmit power from one end to the other, with little or no concern for any coherence; energy incident on one end of the fiber is effectively homogenized or scrambled and transmitted to the other end. But if the index of the fiber is made high in the center, gradually changing to low at the outside, then the ray paths through the fiber will be curved rather than straight lines. If the index gradient is properly chosen (i.e., approximately a function of the reciprocal of the square of the radial distance from the center of the fiber), the ray paths are sinusoidal as shown in Fig. 13.25. This has two significant effects. Rays originating from a point are brought to a focus periodically along the fiber; thus the fiber is capable of forming an image just as a lens is. This is the basis of the GRIN or SELFOC rod. For example, if the index is given as a function of the radial distance r as

$$n\ (r) = n_0\ (1 - kr^2/2)$$

then the focal length of a rod with an axial length of t is

$$efl = \frac{1}{n_0\ \sqrt{k}\ \sin\ (t\ \sqrt{k})}$$

and the back focus is

$$bfl = \frac{1}{n_0\ \sqrt{k}\ \tan\ (t\ \sqrt{k}\ )}$$

The "pitch" of the sinusoidal ray path is $2\pi/\sqrt{k}$.

Since the focusing effect is continuous along the length of the rod, such a device is the equivalent of the periscope system of relay and field lenses described in Sec. 13.2. A length of rod corresponding to two relay lenses and one intermediate field lens as shown in Fig. 13.25 will thus produce an erect image of an area approximately equal to the rod diameter. A row, or a double row, of such rods is the basis of compact table top (scanning) copy machines. Obviously, a long GRIN rod can function as an endoscope and a short rod (less than a quarter of the length shown in Fig. 13.25) will function like an ordinary lens. This latter is called a *Wood lens*.

The other significant aspect of such an index gradient is that because the light rays travel in sinusoidal paths, they never reach the walls of the fiber and do not depend on reflection at a low-index cladding

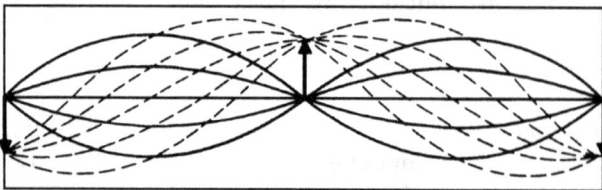

Figure 13.25 In a gradient index rod or fiber (GRIN or SELFOC rod), light rays travel in sinusoidal paths because the index is high at the center of the rod and lower at the edge. Such a rod can form an image just as a lens does. The rod length shown is the equivalent of two relay lenses and an intermediate-field lens. A short length of rod will act like a single lens element, and a longer length can act like a periscope.

layer to confine them to the fiber. Also, the optical path (index times distance) is the same for all paths; obviously the axial path is the shortest, but it is at the highest index. This constancy of optical path means that the travel time is the same for all paths over the full numerical aperture; contrast this with the path length given by Eq. 13.23, which varies with the cosine of the ray slope angle.

Fibers for communications

Another application for optical fibers is in communication. Using light as an extremely high frequency carrier wave, the data transmission rate can be very, very high. Fibers can be made with extremely low absorption (less than 0.1-dB loss per kilometer) so that transmission of information over distances of several miles becomes practical. However, if the lengths of the possible ray paths differ from each other, the elapsed time for light to travel from one end of the fiber to the other will vary from ray to ray. At high data rates, only a small amount of travel time difference is enough to introduce a phase shift sufficient to reduce the signal modulation to a useless level. Again, Eq. 13.23 indicates the path length variation involved. The fibers used for telephone and data transmission are typically single-mode fibers (with core diameters on the order of 10 μm and an index difference of about 1 percent at the boundary) which will not support propagation of a light wave except directly down the length of the fiber. In addition to the variation of path length, another source of trouble results from the fact that in most materials the index varies with wavelength, and thus, even with a constant path length, the optical path would vary with wavelength. Communication fiber materials, in addition to low absorption, are characterized by a very low dispersion in the (narrow) region of the spectrum in which they are used. Silica (SiO_2) fibers are made with near zero dispersion at 1.3 μm wavelength and very low absorption at 1.55 μm. Multilayer cladding can shift the zero dispersion to 1.55 μm and flatten it, to make 1.3 to 1.6 μm useful.

13.9   Anamorphic Systems

An anamorphic optical system is one which has a different power or magnification in one principal meridian than in the other. Such devices usually make use of either cylinder lenses or prisms. With reference to Fig. 13.26a, consider the fan of rays shown in the figure. The left-hand cylindrically surfaced lens is the equivalent of a plane parallel plate for these rays. However, the right-hand lens refracts these rays just as a spherical lens would, because its cylinder axes are at 90° to the left lens. The magnification of this fan of rays is about −0.5× as

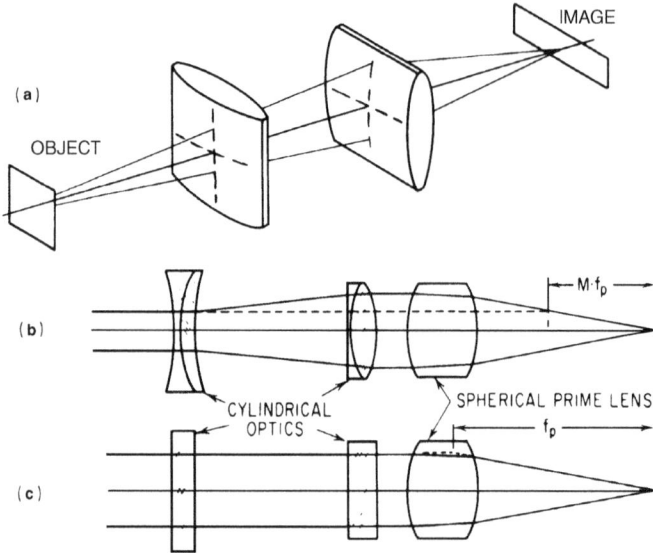

Figure 13.26 Cylindrical anamorphic systems.

drawn. If we consider a fan of rays in the other prime meridian, however, the situation is reversed; the lens effect occurs at the left lens and
the magnification is about $-2.0\times$. Thus the square object figure is
imaged as a rectangle four times as wide as it is high. Since the focusing
effect of a cylinder varies as the square of the cosine of the angle that
a ray fan makes to its power meridian, if both prime meridians are in
focus, then all meridians are in focus.

An interesting aspect of cylindrical anamorphics is the if both prime
meridians are in focus, the all meridians are in focus. Of course this is
necessary for a good image. This is easily understood if we recall the
power of a cylinder at an angle θ from its power meridian is given by
$\phi_\theta = \theta_0 \cos^2\theta$. When we have two cylinders at 90° to each other, the
angle θ for one is $(\theta - 90°)$ for the other, sin for one is cos for the other,
and $\sin^2 + \cos^2 = 1.0$.

Note also that a pair of crossed slits with variable spacing make a
zoom anamorphic pinhole lens.

Another typical anamorphic system consists of an ordinary spherical
objective lens combined with a Galilean telescope composed of cylinder
lenses, as indicated in Fig. 13.26b and c. In the upper sketch (b), it is
apparent that the cylindrical afocal combination serves to shorten the
focal length of the prime lens and thus widen its field of view (for a
given film size). In the other meridian (Fig. 13.26c), the cylinder lenses
are equivalent to plane parallel plates of glass and do not affect the
focal length or coverage of the prime lens. Thus, the system has a focal

length equal to that of the prime lens f_p in one direction and a focal length equal to the magnification of the attachment times the prime lens focal length Mf_p in the other. In Fig. 13.26 the system is shown as a reversed Galilean telescope with a magnification of less than unity, and Mf_p is less than f_p. This is the type of system used in many wide-screen motion picture processes. The wide angular field is used to compress a large horizontal field of view into a normal film format. The distorted picture which results is expanded to normal proportions by projecting the film through a projection lens equipped with a similar attachment. Note that these attachments are used with ordinary camera and projector equipment.

Note that because an anamorphic system has a different equivalent focal length in each meridian, if it is to be focused at a finite distance, it will require a different shift of the lens to focus in each meridian. Thus the prime (spherical) lens must be focused separately from the cylindrical attachment (which is then focused by changing the space between the two components). This type of focusing has the unfortunate effect of changing the anamorphic ratio in a way which makes the face in a closeup appear fatter than it actually is. This is not a popular effect among the acting profession. There are two alternatives to this. One is to put a focusing component in front of the system. This is usually a pair of weak spherical elements, one positive and one negative, so that when closely spaced their power is zero; as the spacing between them is increased, their power becomes positive and the system is focused on a close distance. This is, in effect, a collimator for the object. The other alternative is called a Stokes lens, which consists of a pair of weak cylinders of equal but opposite powers, placed between the two components of the afocal cylindrical attachment, with their axes tilted at 45° to the axes of the attachment. When the two Stokes cylinders are counterrotated, both meridians of the system are focused at the same time.

A *Bravais system* is the finite conjugate analog of an afocal power changer. Figure 13.27 shows the principle of a Bravais system inserted into the image space of an optical system for the purpose of increasing

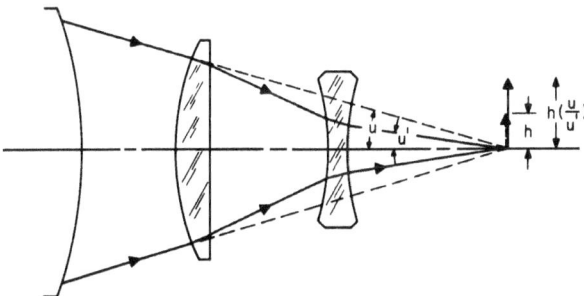

Figure 13.27 Bravais system.

the size of the image without changing the image location. The component powers of this type of system can be determined from Eqs. 4.9 and 4.10 by setting the object to image distance T (the "track length") equal to zero. (Note that the arrangement shown here is usually much more satisfactory than that with the component powers reversed, which reduces the image size.)

The equations yeild the following:

$$\phi_a = (m - 1)(1 - k)/kms = (m - 1)(1 - k)/md$$

$$\phi_b = (1 - m)/k(1 - k)s = (1 - m)/(1 - k)d$$

where m is the magnification, d is the space between the components, s is the distance from the a component to the image, and $k = d/s$.

If a Bravais system is made with cylindrical optics, the image can be enlarged in one meridian and not in the other. This is of course an anamorphic system and has been successfully used for motion picture work. The value of such a "rear" anamorphic attachment is that its size is much less than that of the equivalent afocal attachment placed in front of the lens; this feature is especially important for use with long-focus zoom lenses, where the necessary size for a "front" anamorph can be overwhelming. In addition, there is no focus problem and no "fat" problem.

Cylinder lenses are also used to produce line images where a narrow slit of light is required. The image of a small light source formed by a cylinder lens is a line of light parallel to the axes of the cylindrical surfaces of the lens. The width of the line is equal to the image height given by the first-order optical equations; the length of the line is limited by the length of the lens, or as shown in Fig. 13.26c, it may be controlled by another cylindrical lens oriented at 90° to the first.

A prism may also be used to produce an anamorphic effect. In Sec. 13.1 (Eqs. 13.5 and 13.6), we saw that the magnification of an afocal optical system was given by the ratio of the diameters of its entrance and exit pupils. A refracting prism, used at other than minimum deviation, has different-sized exit and entrance beams and thus produces a magnification in the meridian in which it produces a deviation. Thus a single prism may be used as an anamorphic system. To eliminate the angular deviation, two prisms, arranged so that their deviations cancel and their magnifications combine, are usually used. Figure 13.28 illustrates the action of a single prism and also shows a compound anamorphic attachment made up of two prisms. Since the anamorphic "magnification" of a prism is a function of the angle at which the beam enters the prism, a variable-power anamorphic can be made by simultaneously rotating both prisms in such a way that their deviations always cancel. Prism anamorphic systems are "in focus" and free of

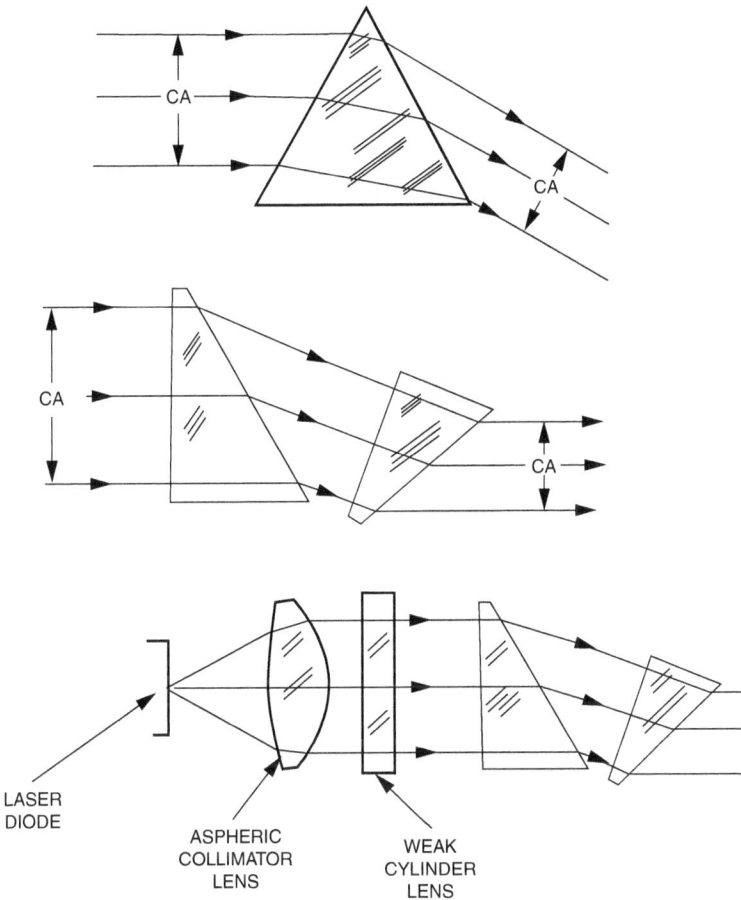

Figure 13.28   The anamorphic action of refracting prisms.

axial astigmatism only when used in parallel (collimated) light. Unlike cylindrical systems, they cannot be focused by changing the space between elements. For this reason, prism anamorphics are frequently preceded by a focusable pair of spherical elements which collimate the light from the object.

For use in systems which are not monochromatic, the prisms must be achromatized (as discussed in Sec. 7.5). Prism anamorphs have been used to project wide-screen (anamorphosed) movies; in this application, each achromatic prism component typically consisted of two or three prism elements. The useful field of such a device is rather small; being completely unsymmetric, it has all (both odd and even) orders of aberrations, including some unusual kinds of lateral color and distortion.

A laser diode is a useful light source, but it has two properties which ordinarily are a handicap: The output beam is not circular in cross section, but elliptical, and the source itself has a small but significant amount of astigmatism so that instead of appearing as a simple point, it appears as a point in different longitudinal locations for each meridian. The lower sketch in Fig. 13.28 shows a laser diode collimator, consisting of an aspheric surfaced collimator singlet, a weak cylindrical lens to cancel out the source astigmatism, and an anamorphic prism pair to convert the elliptical beam to one with a circular cross section. Note that the nearly monochromatic character of the output radiation makes achromatism unnecessary.

Refractive systems are also used to de-anamaphose devices such as laser diodes. One consists of a pair of plano-convex cylindrical elements cemented together at their plano surfaces with their cylinder axes at 90°. If the one nearer the diode has a hyperbolic cylinder and the other a surface that is nearly an elliptical cylinder the pair can be anamorphic and free of spherical aberration. They can be airspaced if the spacing from their thicknesses is not sufficient. Another device is an element with parallel cylindrical surfaces, one convex, the other concave.

13.10   Variable-Power (Zoom) Systems

The simplest variable-power system is a lens working at unit power. If the lens is shifted toward the object, the image will become larger and will move further from the object. If the lens is moved away from the object, the image will become smaller and will again move away from the object. Thus one may find any number of conjugate pairs for which the object-to-image distance is the same but which have magnifications which are reciprocals of each other.

Figure 13.29 indicates the relationships involved in this arrangement. The algebraic expressions shown can be derived readily by manipulation of the thin-lens equation (Eq. 2.4).

The applicability of this particular zoom system is limited, since the commercial demand for variable-power systems at unit magnification is quite modest. However, by combining the moving element with one or two additional elements (usually of opposite sign), the zoom system can be made to operate at any desired set of conjugates. Several such arrangements are shown in Fig. 13.30. Note that in each system the moving lens passes through a point at which it works at unit magnification. By adding either a positive or negative eyelens or by simply adjusting the power of the last lens of the system, as indicated in the lower sketch, a telescope or afocal attachment may be made.

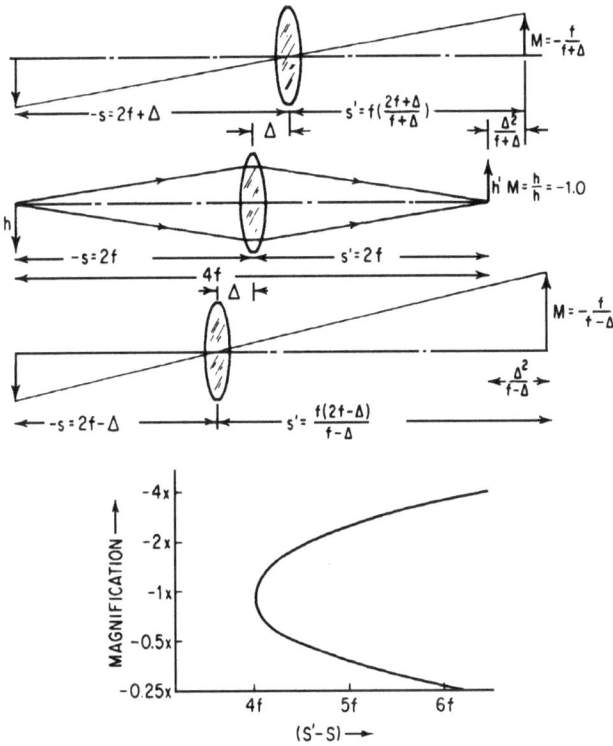

Figure 13.29 The basic unit power zoom lens. The graph indicates the shift of the image as the lens is moved to change the magnification.

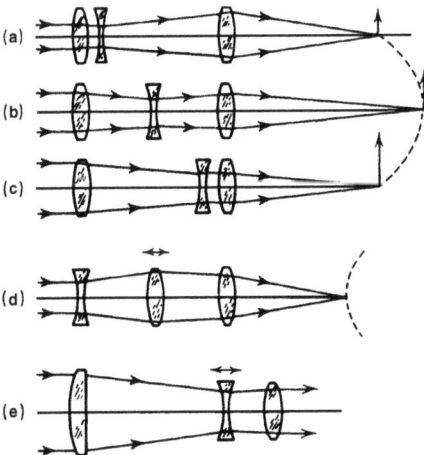

Figure 13.30 Zoom systems based on the unit power principle.

A system which is in focus only at two different magnifications is called a *bang-bang zoom*. It can be quite useful if what is wanted is a system with just two magnifications (and a continuous "zooming" action is not necessary). Since a bang-bang system is much easier and cheaper to design and build than a continuously in-focus zoom, it is often well worth considering whether a true zoom is really needed in a given application, or whether a simple choice of two magnifications, focal lengths, or powers would be sufficient.

All variable-power systems with a single moving component have the same characteristic relationship between image shift and magnification (or focal length). Thus for an uncompensated "single-lens" zoom system, there can be at most two magnifications at which the image is in exact focus. At all other powers, the image will be defocused. This situation can be alleviated in two ways. A "mechanically compensated" zoom system is one in which the defocusing is eliminated by introducing a compensating shift of one of the other elements of the system, as exemplified by Fig. 13.31. Since the motion of the compensating element is nonlinear, it is usually effected by a cam arrangement, hence the name "mechanically compensated."

In a zoom system, the motion of the elements will, of course, cause the ray heights, angles, etc. to change. It is apparent that the chromatic contributions of a single element (which are proportional to $y^2\phi/V$ and $yy_p\phi/V$ for axial and lateral chromatic, respectively) will vary accordingly. Thus, in order to achieve a *fully* achromatic system through the zoom, each component must be individually achromatized. However, since a small amount of chromatic often can be tolerated, singlet components are not uncommon.

The formulas for a thin-lens layout of this type of system are shown in Fig. 13.31 and can be derived by manipulation of the first-order expressions of Chap. 2. To use the formulas, one may arbitrarily select

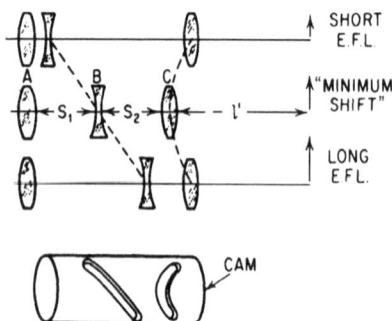

Figure 13.31 Mechanically compensated zoom system. Motion of lens C is to hold the distance from lens A to the focal point at a constant value as lens B is moved.

a value for ϕ_A, the power of the first element, then determine ϕ_B, ϕ_C, and the spacings for the "minimum shift" setting. To find the spacings for other positions of the moving lens, choose a value for one space and solve for the position of the compensating element to maintain the final focus at the same distance from the fixed element.

It should be apparent that despite the use of three components in the preceding discussion, only two components are necessary to make a mechanically compensated zoom lens. Given any two components, if we change the space between them, Eqs. 4.4 and 4.5 indicate that the effective focal length will be changed. Of course, the back focal length will also change (according to Eq. 4.6), and the entire system will have to be shifted to maintain the focus. It usually turns out to be advantageous if one component is positive and the other negative. There are thus two possible arrangements, depending on which power comes first, and one's choice can be based on size and focal-length considerations. Many of the newer 35-mm camera zoom lenses are of this type.

Many of the newer zoom lens designs have more than two moving components. The extra motion may be used to improve the image quality through the zoom or to stabilize the image quality when the lens is focused at a near distance.

The other technique for reducing the focus shift in a variable-power system is called optical compensation. If two (or more) *alternate* lenses are linked and moved together with respect to the lenses between them, the powers and spaces can be so chosen that there are more than two magnifications at which the image is in exact focus. Two systems of this type are shown in Fig. 13.32. In the upper sketch, the first and third elements are linked and move to produce the varifocal effect. The second element, the other elements, and the film

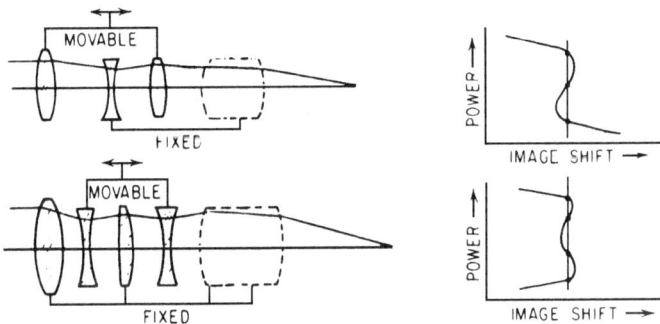

Figure 13.32 Optically compensated zoom systems. The upper system has three "active" components and three points of compensation as indicated in the upper graph. The lower system has four "active" components and four compensation points.

plane are all held in a fixed relationship with each other. The image motion produced by this type of system is a cubic curve, as shown in the upper graph. It is thus possible to arrange the powers and spaces so that the image is in exact focus for three positions in the zoom. The defocusing between these points is greatly reduced in comparison with the simpler systems described above, and if the range of powers is modest and the focal length of the system is short, a nonlinear compensating motion of one of the elements is not necessary. In the second system of Fig. 13.32, the motion of the image is described by a still-higher-order curve, and four points of exact compensation are possible; the residual image shift is about one-twentieth of the shift of the upper system. It turns out that the maximum number of points of exact compensation is equal to the number of variable airspaces. (Note that in Fig. 13.30 this number is 2, and the image motion is parabolic with two possible points of compensation.)

Originally it was thought that the fabrication of a mechanically compensated zoom lens would be almost impossibly difficult, requiring an unattainable level of precision, which could not be maintained as the cams, etc., wore with use. This turned out to be an incorrect assumption, and mechanically compensated zoom systems are widely used for almost all applications. Optical compensation is rare for several reasons. The requirements of the power and space layout to achieve optical compensation are extremely stringent and restrict the lens designer's ability to maintain the correction of the lens system throughout the zoom range. In addition, the size of the optically compensated system is significantly larger than the equivalent mechanically compensated system. Despite the fact that the optically compensated lens with its simple and undemanding mechanics is less expensive to fabricate, provided size is not a problem, the optically compensated zoom is effectively obsolete.

In zoom systems the focal lengths of a stationary first element and of the elements following the last moving lens may be changed at will, *provided the relationship between the focal points of the elements is maintained.* Such changes modify the focal length (or power) of the overall system and, in the case of the following elements, the amount of image shift as well. However, since a change in object position will shift the focus point of the first element with respect to the other elements, a zoom system is sensitive to object position. In order to maintain precise compensation, most zoom lenses are focused by moving an element of the first component with respect to the rest to offset this effect. As with the anamorphic systems discussed in Sec. 13.9, the leading component serves to collimate the light from the object.

13.11   The Diffractive Surface

The diffractive surface (or "kinoform" or "binary surface") as used in imaging optics is discussed at some length in connection with the design of telescope objectives in Chap. 16. In this section we are concerned not with the kinoform's Fresnel surface modulo 2π but with those surfaces which operate on the basis of diffraction in order to introduce a controlled diffusion or to produce a message or a pattern from a simple laser beam. Often these surfaces are simple two-, four-, or eight-level patterns with randomized surface elevations. These devices are made feasible by the recent advances in fabrication technology which make it possible to produce the microscopic wavelength-sized surface details required to produce these effects.

To those who think in terms of the phasefront or wavefront, the form of such a device is derived by describing the phasefront which will produce the desired effect and then determining the surface contour which will impose this phasefront on the input beam. However, for those who think geometrically, this is a less than satisfying explanation of how such a surface functions. The following are not elegant depictions, but they do serve the purpose of taking some of the mystery out of such devices.

The diffusing surface can be visualized as a surface randomly covered with microscopic lenses of a scale on the order of several wavelengths, either concave or convex, the ratio of diameter to focal length of which equals the diffusion angle. Such diffusers are commercially available in diffusions of $\frac{1}{2}°$, $1°$, etc. They can be useful in a number of applications, such as where one desires to destroy the spatial coherence in a laser system in order to eliminate interference patterns. The surface lens concept is not necessary; the same result can be produced by a stepped surface which locally alters the phase of the wavefront.

The pattern-generating surface is a little more difficult. Visualize a surface covered with weak prisms, each of which directs its portion of the incoming laser beam in a direction which will form a specific part of the desired pattern. When such a surface is produced on a microscopic wavelength scale, there are many, many tiny prisms in the area covered by the beam, and when the beam is translated across the surface, there are always enough prisms within the beam to produce the pattern. The bigger the beam diameter, the more prisms will be involved, and the better the definition of the pattern will be. There is an inherent "speckle" produced in this process which shows up as a random pattern of dots in the final image. Again, the effect can be produced by stepped surfaces which alter the wavefront diffractively to produce the desired patterns.

Bibliography

Allard, F. (ed.), *Fiber Optics Handbook,* New York, McGraw-Hill, 1990.

Benford, J., and H. Rosenberger, "Microscopes" in Kingslake (ed.), *Applied Optics and Optical Engineering,* Vol. 4, New York, Academic, 1967.

Bergstein, L., and L. Motz, *J. Opt. Soc. Am.,* Vol. 52, April 1962, pp. 363–388 (zoom lenses).

Brown, T. G., "Optical Fibers and Fiber-Optic Communication," in *Handbook of Optics,* Vol. 2, New York, McGraw-Hill, 1995, Chap. 10.

Habell, K., and A. Cox, *Engineering Optics,* Pitman, 1948.

Inoue, S., and R. Oldenboug, "Microscopes," in *Handbook of Optics,* Vol. 2, New York, McGraw-Hill, 1995, Chap. 17.

Jacobs, D., *Fundamentals of Optical Engineering,* New York, McGraw-Hill, 1943.

Johnson, R. B., "Lenses," in *Handbook of Optics,* Vol. 2, New York, McGraw-Hill, 1995, Chap. 1.

Keck, D., and R. Love, "Fiber Optics for Communications" in Kingslake, R., and B. Thompson (eds.), *Applied Optics and Optical Engineering,* Vol. 6, New York, Academic, 1980.

Kingslake, R., *Optics in Photography,* SPIE Press, 1992.

Kingslake, R., *Optical System Design,* San Diego, Academic, 1983.

Kingslake, R., "The Development of the Zoom Lens," *J. Soc. Motion Picture and Television Engrs.,* Vol. 69, August 1960, pp. 534–544.

Legault, R., in Wolfe and Zissis, *The Infrared Handbook,* Washington, Office of Naval Research, 1985 (reticles).

Melzer, J., and K. Moffitt, *Head Mounted Displays,* New York, McGraw-Hill, 1997.

Moore, D. T., "Gradient Index Optics," in *Handbook of Optics,* Vol. 2, New York, McGraw-Hill, 1995, Chap. 9.

Patrick, F., "Military Optical Instruments" in Kingslake (ed.), *Applied Optics and Optical Engineering,* Vol. 5, New York, Academic, 1969.

Siegmund, W., "Fiber Optics" in Kingslake (ed.), *Applied Optics and Optical Engineering,* Vol. 4, New York, Academic, 1967.

Siegmund, W., in W. Driscoll (ed.), *Handbook of Optics,* New York, McGraw-Hill, 1978 (fiber optics).

Smith, W. J., *Practical Optical System Layout,* New York, McGraw-Hill, 1997.

Smith, W. J., "Techniques of First-Order Layout," in *Handbook of Optics,* Vol. 1, New York, McGraw-Hill, 1995, Chap. 32.

Smith, W., in W. Driscoll (ed.), *Handbook of Optics,* New York, McGraw-Hill, 1978.

Smith, W., in Wolfe and Zissis (eds.), *The Infrared Handbook,* Washington, Office of Naval Research, 1985.

Strong, J., *Concepts of Classical Optics,* New York, Freeman, 1958.

Wetherell, W. B., "Afocal Systems," in *Handbook of Optics,* Vol. 2, New York, McGraw-Hill, 1995, Chap. 2.

Wetherell, W. B., in Shannon and Wyant (eds.), *Applied Optics and Optical Engineering,* Vol. 10, San Diego, Academic, 1987 (afocal systems).

Wolfe, W., in Wolfe and Zissis (eds.), *The Infrared Handbook,* Washington, Office of Naval Research, 1985 (scanners).

Exercises

1 (a) What focal lengths are required for the eyelens and objective of a 20 × astronomical (Keplerian) telescope which is 10 in long?

(b) What is the eye relief?

(c) What is the minimum objective diameter if the diffraction limit is to match the resolution of the eye?

(d) What is the maximum "real" field of this telescope if the eyepiece diameter is 0.5 in?

ANSWER: (a) The length of the scope is $f_o + f_e = 10$ in and for $20 \times f_o = 20f_e$ so 10 in $= 21 f_e$ and $f_e = 10/21 = 0.4762$ in. For a 20 power scope $f_o = 20 \times 0.4762 = 9.5238$ in.

(b) The exit pupil is the image of the aperture stop (which is the objective) formed by the eyelens. The distance to the objective (s) is -10 in so by Eq. 2.5 we get

$$ER = s' = sf/(s + f) = -10 \times 0.4762/(-10 + 0.4762) = 0.5 \text{ in.}$$

(c) Assuming the eye resolution is one minute of arc (0.0003), a telescope resolution of one twentieth of this (0.000015 or 3 seconds) when magnified by the $20\times$ power of the scope will match that of the eye. The Rayleigh limit according to Eq. 9.18 requires a diameter of $5.5/3 = 1.833$ in for a resolution of three seconds.

(d) With an objective diameter fo 1.833 in and an eyepiece diameter of 0.5 in, a ray from the top of the objective (at $y = 0.5 \times 1.833 = 0.9167$ in) to the top of the eyelens (at $y = 0.25$ in) will have a slope of $(0.9167 - 0.25)/10 = 0.06667$. At the focus of the objective the ray height will be $0.9167 - 9.5238 \times 0.06667 = 0.2817$ in. The half field angle is arctan $0.2817/9.5238 = $ arctan $0.0296 = 1.69°$. The full field is $3.39°$.

2 It is desired to add an afocal attachment in front of a 10 in $f/10$ camera lens to convert it to a focal length of 5 in.

(a) What element powers are necessary for a 3-in long reversed Galilean to accomplish this?

(b) What diameter must the front (negative) element have if the vignetting is not to exceed 50 percent for an object field of $\pm60°$?

(c) Sketch the system. Is this a reasonable diameter?

ANSWER: (a) MP $= 5$ in/10 in $= 0.5\times = -f_o/f_e$, thus $f_e = -2f_o$ Since the length $= 3$ in $= f_o + f_e$ we can solve to get $f_o = -3$ in and $f_e = +6$ in.

(b) The combined FOV is $\pm60°$. Since the MP is $0.5\times$, the image field will be $0.5 \times 60 = \pm30°$. (Rigorously, this should be done with tangents.) If we assume that the aperture stop is at the eye/rear lens, then the principal ray will have a slope of $30°$, and the principal ray height at the objective/front lens will be 3 in $\times$ tan $30° = 1.732$ in. If the principal ray gets through, the vignetting will be 50 percent. The diameter will be 3.464 in.

(c) If you draw the front lens, its edge thickness will be about $1\frac{1}{4}$ in for an equi-concave shape. With an equi-concave shape, it would appear that the ray may be close to total internal reflection at the outer surface. Either splitting the lens into two plano-convex elements or bending it to a meniscus form might be wise.

3 A microscope is required to work at a distance of 3 in from the object to the objective lens. If the objective and the eyepiece both have 2 in focal lengths, what is the length of the microscope and what is its power?

ANSWER: A 2 in focal length objective placed 3 in from the object will form an image at $s' = sf/(s + f)$ (per Eq. 2.5) $= -3 \times 2/(-3 + 2) = 6$ in. For a final image distance of infinity, the internal image should fall at the first focal point of the eyelens, so the spacing (length) of the microscope will be 8 in. The objective magnification per Eq. 2.7a is $m = s'/s = -6/3 = -2\times$ and the eyepiece power (per Eq. 13.10a) *is MP* $= 10$ in/$f = 10$ in/2 in $= 5\times$. Thus the power of the microscope is MP $= -2 \times 5 = -10\times$.

4 What is the magnification produced by a telescope made up of a 5-in focal length objective and a 5-in focal length eyepiece (and thus *nominally* of unit power) when it is set at minus one diopter (i.e., the image of an infinitely distant object is at -40 in from the eyepiece)?

ANSWER: The objective-to-eyelens spacing must be such that the eyepiece forms its image at $s' = -40$ in. Using an equation from Sec. 2.4, $s = s'f/(f - s')$ we get $s = -40 \times 5/(5 + 40) = -200/45 = -4.444$ in and the lens spacing must be 9.444 in.

If the object subtends an angle θ, then the size of the image formed by the objective is $f/\theta = 5\theta$ in. If this image is viewed by the eyepiece located 4.444 in away, it subtends an angle of $5\theta/4.444 = 1.125\theta$. So, if the eye is located at the eyelens, the magnification is MP $= 1.125\theta/\theta = 1.125\times$.

The eyerelief is found by Eq. 2.5 $s' = sf/(s + f) = -9.444 \times 5/(-9.444 + 5)$ $= 10.625$ in.

The final image formed by the eyelens is given as -40 in (one diopter) from the eyelens, and since its object (the internal image) is 4.444 in away the eyepiece magnification is $-40/-4.444 = +9.0$ and size of the image is $9 \times 5\theta$ in. From a distance of $40 + 10.625$ it subtends an angle of $9 \times 5\theta/50.625 = 0.8889\theta$. So, with the eye at the exit pupil, the magnification is MP $= 0.8889\theta/\theta = 0.8889\times$.

Alternatively, per Eq. 13.10c, the magnification of the eyepiece is MP $= 10(f - s')/f(R - s') = 10(5 - (40))/5(10.625 - (40)) = 450/253.125 = 1.7778\times$. If the object subtends an angle of θ, then the image formed by the objective is 5θ in. When magnified by the eyepiece it appears to be $5 \times 1.7778 = 8.8889\theta$ in as if viewed from 10 in. If viewed from 10 in, it subtends $8.889\theta/10 = 0.8889\theta$, and the magnification is MP $= 0.8889\theta/\theta = 0.8889\times$.

5 What base length must a rangefinder have to measure a range of 2000 m to an accuracy of ± 0.5 percent, if it incorporates a 20$\times$ telescope?

ANSWER: Using Eq. 13.18c and solving for B, we get $B = 5 \times 10^{-5} D^2/MdD$

$$D = 2000 \text{ m} \qquad dD = 0.5\% \times 2000 = 10 \text{ m} \qquad M = 20\times$$
$$B = 5 \times 10^{-5} \times 2000^2/20 \times 10 = 200/200 = 1 \text{ m}.$$

6 Determine the focal length, diameter, and position relative to the detector for a radiometer field lens. The objective is a 5 in diameter $f/4$ paraboloid and the detector is 0.2 in^2. The field to be covered is ± 0.02 radians.

ANSWER: To fill the detector with the image of the objective, the field lens magnification will be $0.2/5 = 0.04\times$. The objective focal length is 4×5 in $= 20$ in, so to get a magnification of $0.04\times$ the image distance must be 0.04×20 in $= 0.8$ in and by Eq. 2.5 $f = ss'/(s - s') = -20 \times 0.8/(-20 - 0.8) = -16/-20.8 = +0.7692$ in.

The minimum diameter for the field lens is 2 × 0.02 × 20 = 0.8 in. Summarizing: f = 0.7692 in, the field lens diameter = 0.8 in, and the field lens is 0.8 in from the detector.

7 The entrance opening of a tapered hollow light pipe is twice the exit opening. What is the largest angle that a ray through the center of the entrance opening can make with the axis of the pipe and still emerge from the exit end of the pipe?

ANSWER: Referring to Fig. 13.20, if the pipe apertures are 2× and ×, then the center of the dashed circle must be a distance equal to the pipe length L from the pipe exit, and the radius of the circle must equal the pipe length L. Thus if we construct a right triangle with the hypotenuse equal to the length plus the distance to the circle center (= $2L$) and a normal to ray A in the figure as the triangle side, the angle of ray A = arcsin ($L/2L$) = arcsin 0.5 = 30°.

8 A plano-convex hemi-cylindrical rod has a cylindrical radius of 2.5 mm, is 20-mm long, and is located 50 mm from a 1-mm-square source of light. At the "focus," what is the size of the illuminated area? (Assume n = 1.5 for the rod.)

ANSWER: The focal length of the rod per Eq. 3.25 is $2R$ = 5.0 mm. Per Eq. 2.5 the focus in the power meridian is s' = $sf'/(s + f)$ = -50 × 5/(-50 + 5) = $+5.555$ in and the magnification (per Eq. 2.7a) m = s'/s = 5.555/-50 = -0.1111. The height of the illuminated area (per Eq. 2.7a) is h' = mh = 0.1111 × 1 mm = 0.1111 mm. The width (in the zero power meridian) is determined by the beam spread. It is defined by the rod length (20 mm) at a distance of 50 mm, yielding a spread angle of 20/50 = 0.4. At a distance of 50 + 5.555, the beam spreads out to 0.4 × 55.5555 = 22.2222 mm.

14

Case Studies in System Layout

14.1 Introduction

This chapter consists of about a dozen tasks that are quite typical of those faced by an optical engineer who must produce an optical system layout to solve some problem. The solutions to the problems presented here are fully worked out, with complete references to the applicable sections and to the equations which are necessary to produce the layout. A system layout is simply a specification of the component powers, diameters and their spacings which will produce the desired image (i.e., one with the desired size, location and orientation).

The diameters which we determine in this chapter are actually clear apertures. Of course the real diameters will need to be a bit larger to accomodate the necessary mounting shoulders.

The next step in designing an optical system, after the component powers and diameters have been worked out (as we do in this chapter), is to make a crude, reasonably representative, sketch of the system. We can draw singlet elements as equi-convex (or equi-concave), or as plano-convex (or plano-concave). If we assume an index of 1.50 and a zero thickness, then the radii for the equi-convex form will equal the focal length. If we assume an index of 1.75, then the radii will equal 1.5 times the focal length. For the plano-convex form, the radius will be half of these values. If an achromatic component will be necessary, we can assume BK7 (517642) and SF1 (717295) glasses, and the radii of the crown element will be ±0.56 times the focal length, and the third radius (on the flint) will be convex and 1.7 times the focal length. An optical software program is very handy in making such drawings, and in selecting realistic thicknesses. Note that the thin lens spacings must be adjusted to take the element thickness into account. And of course, one use of the lens drawings is that if the sketch looks too fat or too thick, it indicates that a split into two elements may be necessary.

14.2 Telephoto Lens

Project: For an object at Infinity, we need a system which is 220-mm long (front lens to image), which has a 100-mm working distance (rear lens to image), has a focal length (efl) of 300 mm, a speed of $f/4.0$, covers a field of 11.4° (±0.1 radians) with zero vignetting, and has its aperture stop at the front lens.

Find: The component locations are effectively given, so we only need to determine the component powers and their diameters.

Solution:

The space between components is simply the front-lens-to-image distance (220 mm) minus the working distance (100 mm), or 120 mm. With reference to Eqs. 4.7 and 4.8 the component locations define the back focus $B = 100$ mm and the spacing $d = 220 - 100 = 120$ mm. The focal length of the system f_{ab} is required to be 300 mm. So, using Eq. 4.7 for the focal length of the front component, and Eq. 4.8 for the rear component, we have

$$f_a = df_{ab}/(f_{ab} - B) = 120 \times 300/(300 - 100) = 36,000/200 = 180 \text{ mm}$$

$$f_b = -dB/(f_{ab} - B - d) = -120 \times 100/(300 - 100 - 120)$$

$$= 12,000/80 = -150 \text{ mm}$$

With the aperture stop at the front component its diameter must equal the focal length (300 mm) divided by the $f/\#$ (4), so we have a diameter of $300/4 = 75$ mm for component a. At the rear component the diameter of an axial bundle of rays is determined by the back focus

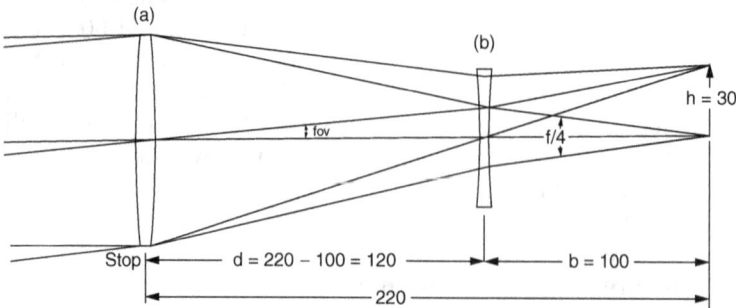

Figure 14.1 System for Sec. 14.2.

(100 mm) and the system speed ($f/4$), so the diameter of an axial bundle of rays originating at an object point will be $100/4 = 25$ mm. For an object point which is 0.1 radians off axis, the principal ray will pass through the center of the front component undeviated and will strike the rear component at a height equal to its slope times the spacing = 0.1×120 mm = 12 mm. A 25-mm-diameter bundle centered at the height of the principal ray (12 mm) will require a semi-aperture of 12 + 25/2 = 24.5 mm, or a diameter of 49 mm.

14.3 Retrofocus Lens

Project: We need a system which is the same as the system in Sec. 14.2, *except* that the focal length must be 50 mm (instead of 300 mm), and the aperture stop must be at the rear component (instead of at the front).

Find: The powers (or focal lengths) and diameters of the components.

Solution:

Again, using Eqs. 4.7 and 4.8, we get the focal lengths of the components:

$$f_a = df_{ab}/(f_{ab} - B) = 120 \times 50/(50 - 100) = 6000/(-50) = -120 \text{ mm}$$

$$f_b = -dB/(f_{ab} - B - d) = -120 \times 100/(50 - 100 - 120)$$

$$= -12,000/(-170) = +70.5888 \ldots \text{ mm}$$

The diameter of the on-axis beam at the front component will be the system focal length (50) divided by the $f/\#$ (4), or $50/4 = 12.5$ mm. For the rear element, the axial beam diameter is the back focus (100) divided by the $f/\#$, or $100/4 = 25$ mm. Since the stop is at the rear component, this (25 mm) is its diameter. With a field of ± 0.1 radians, the image height will be equal to the field (± 0.1) times the focal length (50), or $\pm 0.1 \times 50 = \pm 5.0$ mm. Thus the principal ray through the center of

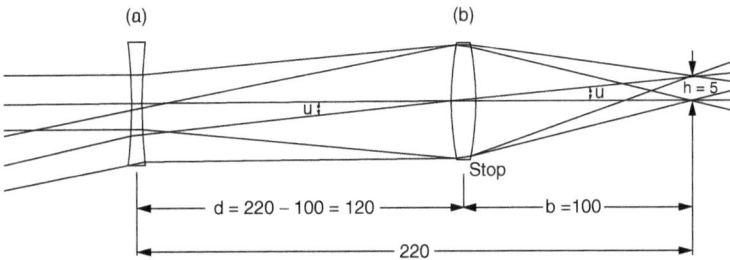

Figure 14.2   System for Sec. 14.3.

the stop (at the rear lens) will have a slope of the image height divided by the back focus, or $\pm 5/100 = \pm 0.05$. Since this ray will not be deviated by the rear lens, the slope between the elements is ± 0.05 and the ray will strike the front element at a height of $\pm 0.05 \times 120 = \pm 6.0$ mm. A beam diameter of 12.5 mm centered at 6.0 mm off the axis will require a semi-diameter of $6.0 + 12.5/2 = 12.25$ mm, so the front component must have a diameter of 24.5 mm.

14.4 Relay System

Project: (a) With the image located 200 mm from the object, we need a magnification of $+0.5$, an object-to-first-lens distance of 50 mm, and an object-to-second-lens distance of 150 mm.

(b) We want the same system as in (a), except that we need the image to be inverted instead of upright.

Find: The component powers.

Solution:

(a) We use Eqs. 4.9 and 4.10 because the object and image are both at finite distances. With respect to these equations, the object distance is s and it equals -50 mm. The image distance s' equals the object to image distance (200 mm) minus the object-to-second-lens distance (150 mm), or $s' = 200 - 150 = +50$. The spacing d equals the object-to-second-lens distance (150 mm) minus the object-to-first-lens distance (50 mm), or $d = 150 - 50 = +100$. The magnification m is $+0.5$, and the image is erect. Using Eq. 4.9 for ϕ_a and Eq. 4.10 for ϕ_b, we get:

$$\phi_a = (ms - md - s')/msd = (0.5 \times (-50) -0.5 \times 100 - 50)/0.5$$
$$\times (-50) \times 100 = -125/(-2{,}500) = +0.05 \qquad (f_a = +20.0)$$

$$\phi_b = (d - ms + s')/ds' = [100 - 0.5 \times (-50) + 50]/100 \times 50$$
$$= +0.035 \qquad (f_b = +28.57142957\ldots)$$

$$\text{Total Power} = \phi_a + \phi_b = +0.085 \text{ mm}^{-1}$$

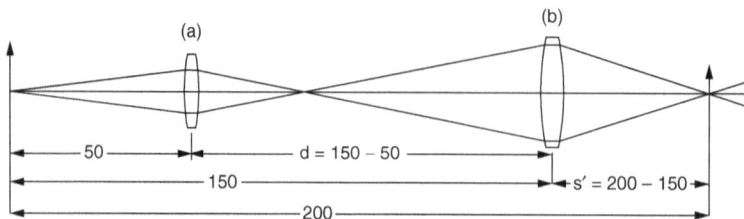

Figure 14.3 System for Sec. 14.4a.

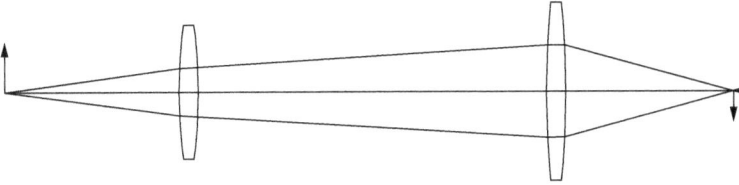

Figure 14.4 System for Sec. 14.4b.

(b) The data is the same except the magnification $m = -0.5$ instead of $+0.5$.

$$\phi_a = [-0.5 \times (-50) - (-0.5) \times 100 - 50]/(-0.5) \times (-50) \times 100$$
$$= +25/+2500 = +0.01 \qquad (f_a = +100)$$
$$\phi_b = [100 - (-0.5) \times (-50) + 50]/100 \times 50 = +125/+5000$$
$$= +0.025 \qquad (f_b = +40 \text{ mm})$$
$$\text{Total Power} = \phi_a + \phi_b = +0.035 \text{ mm}^{-1}$$

Note that, whether the image is inverted or erect makes a big difference in both the total power (0.085 vs. 0.035) and the individual component (0.05 vs. 0.01 and 0.035 vs. 0.025) powers needed to do the job. Obviously the choice of erect image versus inverted image is not an insignificant one.

14.5 Aperture Stop for Relay System of Sec. 14.4

Project: We want the system of Sec. 14.4b to work at a numerical aperture (NA = $n \sin U$) of 0.25 and to cover an image size of ± 7.5 mm without vignetting. The aperture stop must be midway between the components.

Find: Determine the necessary component diameters.

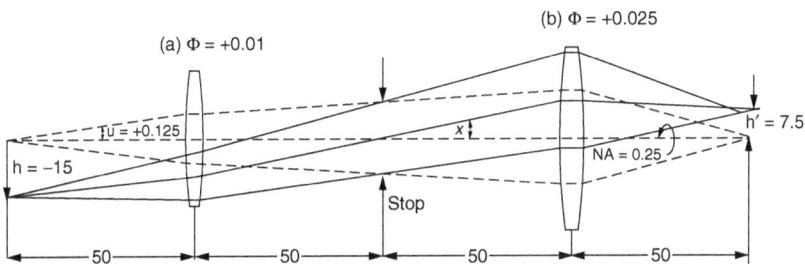

Figure 14.5 System for Sec. 14.5.

Solution:

For an on-axis image point the axial beam diameter at the lens (b) is determined by the NA and the image distance. The axial beam diameter will be equal to twice the ray slope (NA = 0.25) times the image distance (s') or $2 \times 0.25 \times 50 = 25$ mm.

Since the magnification is -0.5, the slope of the axial ray in object space (per Eq. 4.16) must be $u = -(u' = -0.25) \times (m = -0.5) = +0.125$. The axial beam diameter at lens (a) will be $2 \times$ (object distance = 50) $\times 0.125 = 12.5$ mm.

Now we must determine the slope of the principal ray which passes through the center of the aperture stop (located midway between the components) and intersects the image plane at a height of ± 7.5 mm. To do this we can trace a ray with a slope of "X" through the aperture stop. The principal ray height on lens (b) will be the ray slope times the distance, $y = 50x$. Using Eq. 4.1, we find the slope of this ray after passing through lens (b)

$$u' = u - y\phi = x - 50x(\phi = 0.025) = -0.25x$$

and, using Eq. 4.2, the ray will strike the image plane at a height of:

$$h' = y_2 = y_1 + du'_1 = 50x + (d = 50)(-0.25x) = 37.5x$$

The image height must be $\pm 7.5 = 37.5x$, so the necessary slope for the principal ray is $x = \pm 7.5/37.5 = \pm 0.20$. With a slope of ± 0.2 the principal ray height on the components will be $\pm 0.2 \times 50 = \pm 10$ mm.

To find the necessary diameters for zero vignetting we add twice the principal ray heights to the axial beam diameters, and find:

For lens (a), diameter = $20 + 12.5 = 32.5$ mm.

For lens (b), diameter = $20 + 25 = 45.0$ mm.

14.6   Short Range Telescope

Project: A "telescope" 250-mm long is to view an object which is 1000 mm from the objective. We want the image to appear to be 10 times as large as if it were viewed from a distance of 1000 mm.

Find: The component powers for the "telescope."

Solution:

We assume an object height of "h." The object will subtend an angle $h/1000$ from the distance of 1000 mm. If the image is to appear ten times as large, the image must subtend an angle of $h/100$. There are several ways of finding a solution:

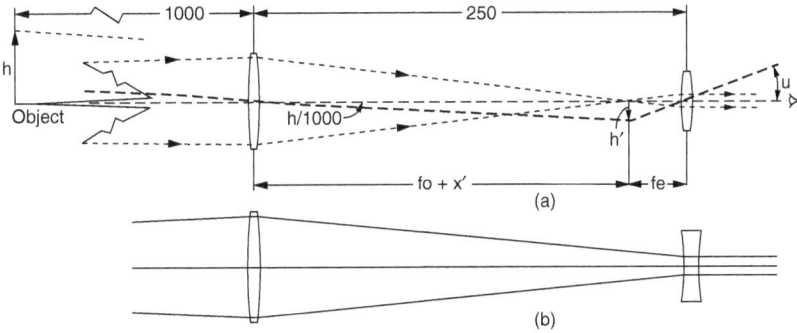

Figure 14.6 Systems for Sec. 14.6 (a) Inverted image (b) erect image.

Alternate Approach #1:

Using Newton's equation (Eq. 2.3), the internal image formed by the objective lens will be located at $s' = f + x' = f - f^2/x = f - f^2/(f - 1000)$, where f is the objective lens focal length.

The height of the image will be equal to the slope of the principal ray ($u = h/1000$) times the image distance s', or $h' = (h/1000)[f - f^2/(f - 1000)]$.

From the eyepiece, the internal image will subtend an angle of h'/f_e which must be equal to $h/100$ to achieve the required 10x magnification. So we have:

$$\text{Scope length} = f_e + s' = 250$$

Solving for f_e we get $f_e = 250 - s = 250 - f + f^2/(f - 1000)$

The 10x magnification requires that the angle subtended at the eyepiece equal $h/100 = h'/f_e = hf/f_e(f - 1000)$, and substituting for f_e we get:

$$\pm 0.01h = hf/(f - 1000)[250 - f + f^2/(f - 1000)].$$

Solving this for f is **not** a trivial exercise.

Alternate Approach #2:

The internal image is at s', and $s = -1000$. Thus the magnification is $m = s'/s = -s'/1000$, and the internal image height is $h' = hm = -hs'/1000$.

To get a length of 250 mm, $f_e = 250 - s'$

The magnification requires that $\pm 0.01h = h'/f_e = hf/1000f$, and substituting for f_e we get: $\pm 0.01h = -h's'/1000(250 - s')$.

We solve this for s': $\pm 10(250 - s') = -s'$

$$2500 - 10s' = \pm s'$$
$$2500 = 10s' \pm s' = 11s' \text{ or } 9s'$$
$$s' = 227.27 \text{ or } -27.77$$

Then find f_e:

$$f_e = 250 - s' = 22.727 \text{ or } -27.77$$

And calculate f:

$$f = ss'/(s - s')$$
$$= 1000s'/(-1000 - s')$$
$$= 185.185 \text{ or } 217.391$$

Figure 14.6 shows the two solutions.

Alternate Approach #3

Trial #1)   We make a guess for the eyepiece focal length, say $f_e = 50$. Then $s' = 250 - f_e = 200$. The objective magnification $= s'/s = 200/(-1000) = -0.2$. Then $h' = mh = -0.2h$ and the angle subtended at the eyepiece, $u = -h'/50 = 0.2h/50 = 0.004h$. This is smaller than the required $0.01h$ by a factor of 2.5.

Trial #2)   We guess again, changing the starting value of f_e by the factor of 2.5 to $f_e = 50/2.5 = 20$. Then $s' = 230$; $m = -0.230$; $h = -0.230h$, and the value of u is $0.230h/20 = 0.0115h$. This is 15 percent too large.

Trial #3)   Using the 15 percent factor, we guess again: $f_e = 20 \times 1.15 = 23$. Then $s' = 227$; $m = -0.227$; $h' = -227h$, and the value of u is $0.227h/23 = 0.00987h$, only $0.00013h$ too small.

For preliminary system layout work this result is entirely adequate, but this process can be continued until the value of u is as near to the desired value as necessary.

Note: This approach gave us the positive eyepiece solution because we started the numerical approximation process with a value closer to that solution than to the negative eyepiece solution. If we had chosen a negative value for f_e as our starting point, we would have found the negative eyepiece solution.

This sort of numerical approach is often the best choice if you have a good understanding of the system and only need to get a solution once. It will only give you the solution that you have visualized. However, the algebraic approach will indicate whether there are multiple solutions. It's usually more work, but it has the edge if you want to make several repeated solutions, as for example in a parametric study.

14.7   Field Lens for Sec. 14.6

Project: Using the positive eyelens solution of the previous project in Sec. 14.6:

(a) Determine the location and size of the exit pupil. Assume the stop is at the objective and is 50 mm in diameter.
(b) Determine the focal length and diameter for a field lens located at the internal image plane, which will allow the system to image an object which is 88 mm in diameter without vignetting by an eyepiece which is 15 mm in diameter.

Find:

(a) The eye relief and exit pupil diameter.

(b) The diameter and power of the field lens.

Solution:

(a) The eyelens will image the aperture stop (the objective) per Eq. 2.5 at:

$$s' = fs/(f + s) = 22.727\,(-250)/(22.727 - 250)$$
$$s' = +25.0 \text{ mm (this is the eye relief)}$$

The diameter of the exit pupil per Eq. 2.7a will be:

$$h' = mh = s'h/s = 25 \times 50/(-250)$$
$$h' = -50 \text{ mm} \qquad \text{(The negative sign simply means that the image is inverted.)}$$

(b) The diameter of the internal image is equal to the objective magnification $[m = s'/s = 227.27/(-1000) = -0.22727]$ times the object height (88 mm). Thus the image height (and the necessary clear diameter of the field lens) is $h' = mh = -0.22727 \times 88 = 20$ mm.

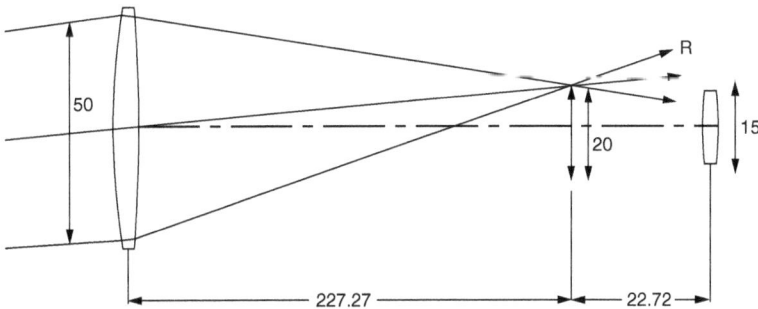

Figure 14.7   System for Sec. 14.7.

The refractive power of the field lens is the power which will bend the ray R (which starts at the bottom of the objective and ends at the top of the field lens) so that this ray will pass through the top of the 15-mm-diameter eyelens. The slope of this ray is equal to the height it rises (25 + 10 = 35 mm) in traveling 227.27 from the objective to the field lens. The slope is thus 35/227.27 = +0.154.

For this ray to strike the eyelens at its edge (7.5 mm above the axis), it must be refracted to have a downward slope of 10 − 7.5 = 2.5 mm in a distance of 22.727, or $u' = -2.5/22.727 = -0.11$.

We now have a ray with a slope of $u = +0.154$, striking the field lens at a height of $y = +10$, which must be deviated to a slope of $u' = -0.11$. The deviation of a ray is given by Eq. 4.2 as: $u' - u = -y\, \phi$. Solving for the power ϕ, we get:

$$\phi = (u - u')/y = (0.154 + 0.11)/10 = +0.0264$$

and the field lens focal length is

$$f = 1/\phi = 1/0.0264 = +37.878 \text{ mm}$$

14.8   Raytrace of Sec. 14.7

Project: Locate the exit pupil for the final system of the project in Sec. 14.7, and determine its diameter, by tracing an axial ray and a principal ray, using Eqs. 4.1 and 4.2.

Find: The eye relief and the exit pupil diameter by ray tracing.

Solution:

We organize the system data so as to systematically carry out the raytrace as follows:

System data						
efl f	object		185.18	37.88		22.73
Power $\phi = 1/f$		+0.0054		+0.0264		+0.044
space d	1000		227.273		22.727	
Axial raytrace						
ray height y	0.0		25	0.0		−2.5
ray slope u		+.025		−0.11	−0.11	0.0
Principal raytrace						
ray height y	−44		0.0	10		5.
ray slope u		+.044		+.044	−.22	−.44

Note that this arrangement places the ray heights in columns under the lens powers, and places the ray slopes under the airspaces.

From the principal ray data we can locate the exit pupil where the ray intersects the axis at $I_p' = -y_p/u_p' = -5.0/(-0.44) = +11.36$ mm.

This is the eye relief distance from the eyelens. The exit pupil diameter can be determined from the principal ray data as well. Its diameter is the diameter of the image of the objective lens. The magnification is $u/u' = .044/(-.44) = -0.1$ and the image size is the exit pupil diameter which is 0.1×50 mm = 5.0 mm.

If we use the axial ray data, the pupil diameter is simply $2 \times 2.5 = 5.0$ mm.

Comparing these results with the results calculated in Sec. 14.7, we note that the insertion of the field lens has shifted the exit pupil toward the eyelens by $(25. - 11.36) = 13.63$ mm, but the pupil diameter is unchanged. This is because we located the field lens at an internal focal plane.

14.9   125 Power Microscope

Project: Lay out a 125× microscope which has a length of 200 mm.

Find: The component powers and locations.

Solution:

A 125× microscope will have an effective focal length of 10 in/MP, or 250 mm/MP = 250/125 = 2 mm. Since we want an ordinary microscope, the image will be inverted and the focal length will be negative, so its efl = -2 mm.

For a two component system with a focal length of -2 and a length of 200, the component focal lengths can be determined by using Eqs. 4.7 and 4.8:

$$f_a = df_{ab}/(f_{ab} - B) = 200 \times (-2)/(-2 - B) = 400/(2 + B)$$

$$f_b = -db/(f_{ab} - B - d) = -200B/(-2 - B - 200) = 200 \, B/(202 + B)$$

Thus we have a free variable, which is the working distance B. So we will do a parametric study, choosing a few reasonable values for B and calculating the component powers which result from that choice. The results are tabulated below:

	f_a	ϕ_a	f_b	ϕ_b	$(\phi_a + \phi_b)$
$B = 5$	57.1	.0175	4.83	.207	.2245
$B = 10$	33.33	.030	9.43	.106	.136
$B = 15$	23.52	.0425	13.82	.0723	.1148
$B = 20$	18.18	.055	18.02	.0555	.105
$B = 25$	14.81	.0675	22.03	.0454	.1129
$B = 30$	12.5	.08	23.86	.0387	.1186
$B = 35$	10.81	.0925	29.54	.03385	.1264

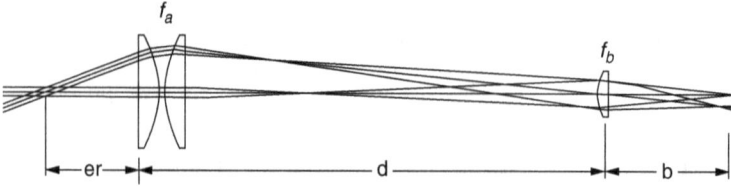

Figure 14.8 System for Sec. 14.9.

A reasonable way to determine the best choice for the value of B might be to use the value which minimizes the total power of the system ($\phi_a + \phi_b$). This choice often produces the system with the least amount of aberrations. In our case this would be about $B = 20$ mm, which yields an objective focal length of 18.18 mm and an eyepiece focal length of 18.02 mm, nearly the same.

Depending on what the application was, it might be desirable to select a long working distance, or perhaps a long eye relief. In Figure 14.8 the eyepiece (component a) is shown as two elements because the equiconvex form had an extreme amount of spherical aberration of the pupil.)

14.10   Brueke 125× Magnifier

Project: Repeat the project of Sec. 14.9, except with an erect image.

Find: The component powers and locations.

Solution:

We simply use a focal length of +2 instead of −2 as in Sec. 14.9. For a two component system with a length of 200 and a focal length of +2, the component focal lengths are given by Eqs. 4.7 and 4.8:

$$f_a = df_{ab}/(f_{ab} - B) = 200 \times 2/(2 - B) = 400/(2 - B)$$

$$f_b = -dB/(f_{ab} - B - d) = -200B/(2 - B - 200) = 200B/(198 + B)$$

Again, we do a parametric study by choosing a few reasonable values for B and calculating the resulting component powers, as tabulated below:

	f_a	ϕ_a	f_b	ϕ_b	$(-\phi_a + \phi_b) = \Sigma \ |\phi|$
$B = 5$	−133.	−.0075	4.93	.203	.2105
$B = 10$	−50.	−.02	9.62	.104	.124
$B = 15$	−30.8	−.0325	14.0	.071	.1035
$B = 20$	−22.2	−.0425	18.35	.0545	.0995
$B = 25$	−17.4	−.0575	22.42	.0446	.1041
$B = 30$	−14.3	−.07	26.32	.0380	.108
$B = 35$	−12.1	−.0825	30.04	.0333	.1158

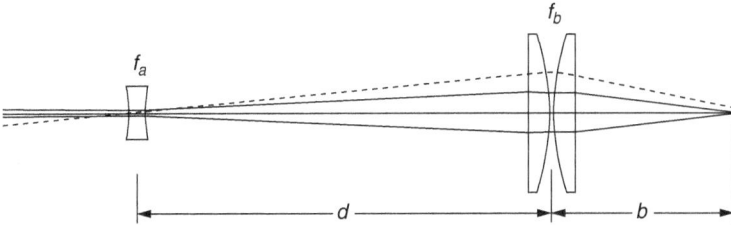

Figure 14.9   System for Sec. 14.10.

Again, a choice for the lowest sum of the *absolute* powers leads us to about $B = 20$ mm, yielding an objective focal length of $+18.35$ mm and a negative eyepiece focal length of -22.2 mm. The power sum here is 0.0995; in Sec. 14.9 the sum was 0.105, almost the same. However in the present case one component is positive and one is negative, whereas both were positive in Sec. 14.9. The result is that this system can be corrected for aberrations more easily, since the aberrations of a negative component will tend to cancel out the aberrations of the positive component. In Sec. 14.9 the aberrations would tend to add. Note that this system has no internal focus, so that a reticle or crosshair is not a possibility. In order to get a reasonable field the objective must have a large diameter; we split it into two elements to control its spherical aberration. The field of view tends to be small in this sort of system, just as in the Galilean telescope. The aperture stop (and pupil) must be at the negative eyelens. The aperture stop in a visual system of this type is often the pupil of the user's eye. The eye relief tends to be quite small (actually negative), since the exit pupil is inside the device. This configuration (called a Brueke magnifier) can yield a powerful magnifier with a conveniently long working distance.

14.11   A 4× Mechanically Compensated Zoom Lens

Project: Make a thin lens layout for a three component mechanically compensated zoom lens with a zoom range $R = 4\times$ and with the focal length ranging from 1 to 4 in. Use the $(+ - +)$ arrangement shown in Figure 14.10, with the middle lens moving to vary the focal length and the rear lens moving to compensate for the focus shift.

There are many ways to approach this project. One can write equations describing the desired focal lengths and the required image location in terms of the element powers and spacings. The equations are then solved for the necessary powers and spaces. If an optical software program is available, its damped least squares routine can be used to

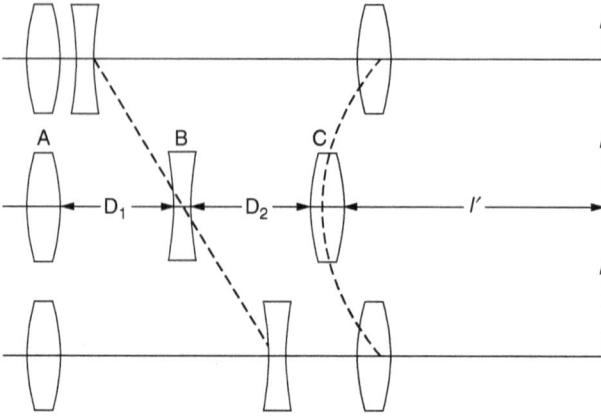

Figure 14.10 Schemat of three components, mechanically compensate zoom for Sec. 14.11.

determine the powers and spacings. Here we elect to take a simple, straightforward **numerical** approach as follows:

We begin by assuming that the center lens has a focal length of minus one, i.e. $\phi_B = -1.0$ (*This arbitrary choice is unlikely to give us the focal lengths which we desire, so we will need to scale or adjust our results after we achieve a 4× zoom with focus compensation.*) Our first task is to determine the conjugates for the middle lens which will yield magnifications of $m = -\sqrt{R} = -2.0$ and $m = -\sqrt{1/R} = -0.5$. This will give us the desired zoom range, R = 4. We simply solve the Gauss equation (Eq. 2.4) for the object and image distances which give us the desired magnifications. We multiply Eq. 2.4 $(1/s') = (1/s) + (1/f)$ by fs' to clear out the fractions and get:

$$f = s' + f(s'/s)$$

and substituting for the magnification, $m = s'/s$, and solving for s', we get

$$s' = f(1 - m)$$

Dividing this by s and setting $(s'/s) = m$ yields this equation for s:

$$s = f(1 - m)/m$$

Since we assumed $f = -1$, we get as the desired conjugates:

for $m = -0.5$, $s = 3.0$ and $s' = -1.5$
for $m = -2.0$, $s = 1.5$ and $s' = -3.0$

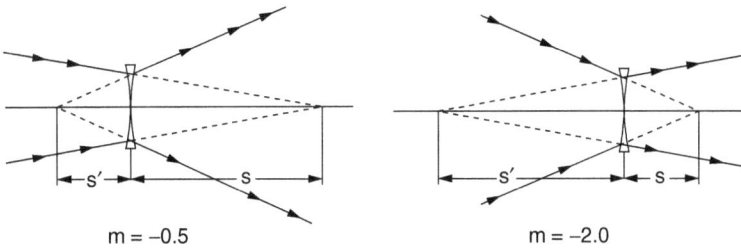

Figure 14.11 Conjugates distances (s and s') for middle zoom element at the extrems of its travel.

as shown in Fig. 14.11. Note that a positive value for s indicates that the object is to the right of the lens, where it is projected by the front lens. The negative value for s' indicates a virtual image.

In order to have a value of $+3$ for s, the focal length of the front (A) lens must be $f_A = +3.0$ or longer. Since we are working with "thin" (i.e., zero thickness) lenses, we should allow enough extra space so that we can later give these lenses real thicknesses without interference or bumping. So we choose a power for the first lens of $\phi_A = 0.3$, and the focal length of 3.333 in will leave a space of 0.333 in between the A and B elements at the left end of the B lens travel.

The motion of the B lens will be the difference between the values for s for the extreme values for m; thus $\delta s = 3.0 - 1.5 = 1.5$. Using $f_A = 3.333$ in, we make a few trials and select $f_C = +1.25$ in for the third or compensator element.

Note that the minimum distance between the object and image of a thin lens is four times its focal length, and if we select too long a focal length for the third (C) lens, the result may be imaginary or physically impossible.

We now raytrace using the two extreme values of the front space $d_1 = 0.333$ in and 1.833 in, with the corresponding values of $d_2 = 2.333$ in and 0.833 in. These values are the spacings which will give the proper values of s and s' for lens B (as determined above) to yield the long and short focal lengths for the system when combined with the A and C lenses which we have elected to try. We also raytrace $d_1 = 1.333$ in and $d_2 = 1.333$, in which will give us the data of a mid-range system (before we shift the C lens to compensate the focus). This particular spacing puts the B lens at unit (-1) magnification and it will require the maximum shift of the C lens to focus the system.

Note that the choice of d_2 is (within limits) arbitrary.

Here we show the raytrace for these spacings, using Eqs. 4.1 and 4.2:

φ	(A) + 0.3	(B) −1.0	(C) + 0.8		
d		0.333	2.333		f=0.80645
y	1.0	0.9	2.3		l' = 1.85484
u	0.0	−0.3	+0.6	−1.24	$\Sigma d + l'$ = 4.5215
d		1.333	1.333		f = 2.0
y	1.0	0.6	1.0		l' = 2.0
u	0.0	−0.3	+0.3	−0.50	Σ = 4.666
d		1.833	0.833		f = 3.22581
y	1.0	0.45	0.575		l' = 1.85484
u	0.0	−0.3	+0.15	−0.31	Σ = 4.5215

We note that, as expected, the first and third raytraces yield equal image distances of $l' = 1.85484$ in, and that their focal lengths, at 0.80645 and 3.22581 give a zoom range of 4.0, as desired.

Now we must determine the motion of the compensating lens which will maintain focus through the zoom. We use the third lens (C) as the compensator. (*Note that the first lens could serve as the compensator if we preferred.*) For both the first and third raytraces, the distance from the front lens to the focus Σ = 4.5215 in. In the second raytrace (d_1 = 1.33) the image distance after the second lens (B) is $-y_B/u'_B = -0.6/0.3$ = −2.0. Thus the distance from this image to the **desired** focus position is equal to $[\Sigma - d_1 - (-2)] = 4.52150 - 1.3333 + 2.0 = 5.188176$. This is the required object to image distance for lens (C). Using x for d_2 the conjugates for lens (C) are $s = -2 - x$ and $s' = 3.188172 - x$. Substituting into the Gauss equation (Eq. 2.4), we get

$$1/(3.188 - x) = 1/(-2 - x) + 0.8$$

and clearing of fractions and dividing by 0.8, we get:

$$0 = x^2 - 1.188172x + 0.1088708;\text{ solving this for x gives;}$$
$$x = d_2 = 1.0881177 \text{ or } 0.100542.$$

Using the more practical value of 1.0881177 for d_2, our raytrace is now:

φ	(A) + 0.3	(B) −1.0	(C) +0.8		
d		1.333	1.0881177		f = 2.2668119
y	1.0	0.6	0.9263252		l' = 2.1000545
u	0.0	−0.3	+0.3	−0.4411482	Σ = 4.5215055

Repeating this process for two more intermediate values for d_1, we get the following table:

d_1	d_2	f	l'	$d_1 + d_2 + l'$
0.333...	2.333...	0.806452	1.854839	4.521505
0.8	1.743190	1.266634	1.978315	4.521505
1.333...	1.088118	2.668119	2.100055	4.521505
1.6	0.892737	2.831886	2.028768	4.521505
1.833...	0.833...	3.225806	1.854839	4.521505

These calculated motions of the lenses as they zoom are plotted in Fig. 14.12.

Our zoom range is now $R = 3.225806/0.806452 = 4.0$, but the system focal lengths are not 1.0 in and 4.0 in as desired. There are several ways of changing the system to get the right focal lengths.

1. We can simply scale the whole system by a factor of $4/3.225806 = 1.24\times$.

2. We can change f_A to $3.333 \times 1.24 = 4.1333 \ldots$ and change d_1 to $4.1333 - 3.0 = 1.13333$ (to hold s for lens B equal to 3.0 for the short focal length position).

3. Or, within limits, we can change the focal length and/or the location of lens C.

Note that this project requires a very large amount of calculation. For most of us the probability of error is significantly more than unity. Pay very careful attention to the sign of a quantity, because it may change sign as you progress from one step in the calculation to the next. It often helps to make a rough sketch of the lenses and rays in the course of the work. The process used above can be helpful as a template for other projects of this type.

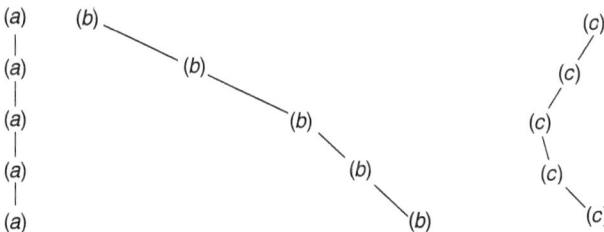

Figure 14.12 Component motions during zoom.

14.12 Doing System Layout by Computer

The task of Sec. 14.11 (or almost any layout effort) can also be accomplished using an optical software program. It is, however a wise idea to have a good grasp of the manual approach(es) described above, and a thorough understanding of the type of system at hand; this may keep you out of trouble with the computer program.

Thin lenses can be simulated by zero thickness elements. Using an index of 2.0 is a covenience, because a plano-convex or -concave element will have a radius equal to the focal length. (Or if you prefer, the radii of an equi-convex or equi-concave element are equal to twice the local length.) One simply sets up a system which resembles the desired one. If it forms an image of the right orientation, within an order of magnitude of the right size, somewhere in the desired neighborhood, and not near a singularity, you can probably use the computer DLS (damped least squares) routine to find a solution. The variables are the surface curvatures of the elements, and also any spacings which are not fixed.

The merit function one uses is very important. It should include all of the desired system characteristics including the focal length, image location and orientation, size, spatial, and configuration limitations. If the object is at infinity, y_1/u'_k is the focal length, and u'_k will control the focal length. For a finite conjugate system u_1/u'_k will control the magnification and image orientation. Overall system lengths and element positions are easily specified and controlled. Angle-solves and height-solves are sometimes useful. If the entire project specification is based on paraxial optics, the system will be very robust and "blow-up" proof, because paraxial raytraces do not fail by TIR or by missing the surface, as "real" rays do.

The DLS system is especially useful in the preliminary layout of zoom systems. For example, in project 14.11 one would set up three configurations; each would have the same optical elements, but different spacings. The curvatures and spacings would be variables, and the merit function would include the image location w.r.t. the first (or the fixed) lens, the system focal lengths, and if desired, the system length (which we did not consider in 14.11). In a zoom system the merit function must include "<" and ">" operands to prevent the elements from getting too close to each other.

Note that this computer approach, and most "numerical" approaches (including the one which we used above) share the limitation that the system which is found will almost always be the type of system which the designer envisioned at the start. A new and unexpected solution is a very unlikely result. This illustrates one advantage of an algebraic approach, in that if there is more than one solution, algebra will find it, and usually flag it with a quadratic or higher order expression.

A convenience of a computer layout is that once you find a solution it is a simple matter to insert reasonable thickness, achromatize, and/or split the elements as necessary to get good performance, as outlined in Chaps. 16 to 19; your system is already set up.

14.13   An Athermalized Mid-IR System with an External Cold Stop

In this section we discuss the layout and athermalization of a mid-IR system which requires an external cold stop in a cooled Dewar vacuum flask. The specifications for the system are as follows:

Optical system specifications

1) efl:	150 mm
2) Aperture:	31 mm
3) Length:	260 mm
4) Wavelength	4.5 μm to 5.1 μm
5) Field:	3.0°(±1.5°)
6) bfl:	>23 mm clearance
7) Cold stop:	17.0 mm from detector array
8) Image quality:	
central 1.5°FOV:	50% energy within 50 μm diameter
within 3.0°FOV:	50% energy within 75 μm diameter
9) Distortion:	<20%
10) Vignetting:	none
11) Thermal compensation:	Passive
12) Packaging:	Requires a mirror fold

The requirement for an external pupil (cold stop) means that the system will have (at least) two widely separated components, with the second acting as a relay lens, and that the focal length will be negative.

Finding the initial layout

There are five specifications (efl, length, bfl, stop position, and "fold-ability") which will be determined by the component powers and the spacings. In a two component system there are only three variables (two component powers and their spacing) with which to achieve these five requirements. If we are fortunate, two of the requirements may fall into place when we have controlled the other three. Otherwise a more complex construction for the individual components (such as telephoto or retrofocus component types), or an additional component, would be required in order to meet the specifications.

The requirements for specifications 1, 3, 6, 7, and 12 can be attacked, either by way of a algebraic solution, or by a numerical trial and error approach.

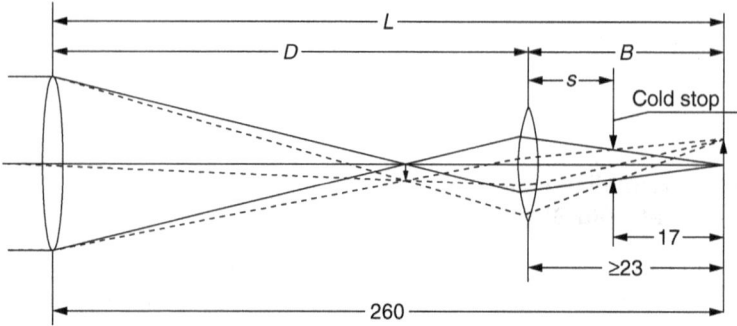

Figure 14.13 System layout for Sec. 14.13.

If the same problem is to be solved several times, the algebraic approach is usually well worth the added effort that it requires. The algebraic method may also serve to indicate unusual or unexpected solutions. A numerical method is usually quicker at finding a familiar, well-understood solution. However, in using it one may easily overlook an unforeseen or unfamiliar solution.

While we will not do so in this particular project, in general it is very wise to look for the first-order layout with the minimum amount of component power. The less the component power, the fewer elements are required for the component, the smaller the residual aberrations, and the less sensitive the system will be to misalignments and fabrication errors.

The algebraic method

The algebraic method can be based on paraxial raytracing as follows: Rays which define the specific quantities can be traced, using symbols (rather than numbers) for the component powers and spacings. Equations for the required focal length, pupil position, etc. can be derived from the ray trace results. These equations can then be solved for the component powers and positions which will satisfy the specifications.

For the case at hand, we could derive:

Eq. 4.4	for efl	$1/\text{efl} = \phi = \phi_a + \phi_b - D\phi_a\phi_b$
Eq. 4.6a	for bfl	$B = (1 - D\phi_a)(\text{efl}) = L - D$
Eq. 2.5	for stop location	$S = D/(D\phi_b - 1) = B - 17$

A simultaneous solution of these three equations would yield the spacing D and the component powers ϕ_a and ϕ_b which would satisfy the requirements for the system length ($L = 260$), the focal length (efl $= -150$) and the pupil position ($B - S = 17$).

A semi-numerical method

Another way (which is much more popular among those who dread algebra) is to use the simultaneous solution of Eqs. 4.4 and 4.6a above for the values of ϕ_a and ϕ_b which will provide a given B, D, L, and efl.

Eq. 4.7 $\phi_a = (F - B)/DF$
Eq. 4.8 $\phi_b = (L - F)/DB$

where $L = B + D$ and $F = $ efl. Once ϕ_b is known, we can use Eq. 2.5 to calculate the pupil position which results for this set of powers.

Using $L = 260 = B + D$, and $F = -150$, we choose D as the free variable in our parametric study. Selecting a few reasonable values for D, we evaluate Eq. 4.8 and Eq. 2.5 and tabulate the results:

D = 230   B = 260 − D = 30 ϕ_b = .0594203   S = 18.16   B − S = 11.84
  = 220   = 40   = .0465909   = 23.78   = 16.22
  = 210   = 50   = .0390476   = 29.17   = 20.83

We interpolate between d = 220 and 210 and get:

D = 218.3   B = 41.7 ϕ_b = 0450396   S = 24.72   B − S = 16.98,

which gives us a value of 16.98 for the cold stop location, (B − S). This is more than close enough to the specified 17 mm for a preliminary this lens layout. Finally, from Eq. 4.7 we find that $\phi_a = +.0058543$.

Characteristics of the initial layout

We now trace (paraxial) axial and chief rays using Eqs. 4.1 and 4.2 through the system and determine that the minimum clear apertures for the components are 31 mm for "a" and 20 mm for "b". These sizes seem reasonable.

Assuming that the elements are to be made of silicon ($n = 3.427$, $V = 1511.$), and using the raytrace data, we can estimate the axial chromatic aberration blur (using Eq. 6.3r)

$$\Sigma \, TAchA = \Sigma y^2 \phi/Vu'_k = .0144 \text{ mm}$$

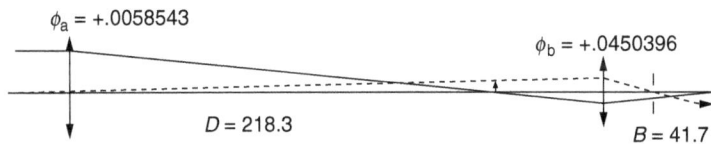

Figure 14.14   Thin lens axial and principal raytrace of the thin lens system.

This seems compatible with the specified 50 μm blur diameter for 50 percent of the energy in a point image. We suspect that it may not be necessary to achromatize the system.

Thermal defocusing

The thermal effect on the power of an element is given in Chap. 16 by:

$$\delta\phi/\delta t = \phi[(\delta n/\delta t)/(n - 1) - \alpha] = \phi \cdot T$$

where $\delta n/\delta t$ is the change of index with temperature, n is the index of refraction, α is the thermal expansion coefficient, ϕ is the element power, and T is the quantity in the square brackets [ ]. For silicon, with $n = 3.427$, $\alpha = 2.62e - 06$, and $\delta n/\delta t = 159e - 06$, we calculate T as $6.289e - 05$, and for a 100°C temperature change, the element power becomes

$$\phi_{100} = \phi(1 + 100T) = 1.00629\phi$$

Assuming that the mechanical structure of the system will be fabricated of aluminum with a CTE of $\alpha = .000024$, we have for the nominal system and for the $\delta t = 100°C$ system, in columns (A) and (B) respectively,

(A) Nominal	(B) $\delta t = 100°C$
$\phi_a = +.0058543$	x1.00629 = +.0058911
$D = 218.3$	x1.00240 = 218.82392
$\phi_b = +.0450396$	x1.00629 = +.0453229
$B = 41.7$	x1.00240 = 41.8006
	and, by Eqn. 4.6a $B_{100} = 40.0858$

where the factors 1.00629 and 1.00240 are $(1 + 100T)$ for silicon and $(1 + 100\alpha)$ for aluminum. Calculating the back focus B from the $\delta t = 100°C$ power and space data above, we get $B_{100} = 40.0858$, indicating that the focus has shifted by $(40.0858 - 41.8006) = -1.7148$ away from the location of the detector. A defocus of this distance will produce a blur spot diameter of $1.71/f/no = 1.71(31/150) = 0.35$ mm, which is clearly far greater than the specified maximum 50% blur size of 50 μm.

Athermalizing the system

Athermalizing, or eliminating the thermal focus shift, is analogous to correcting the chromatic aberration of the system; it requires at least two materials with different properties. To correct chromatic aberration we utilize differing Abbe V–values; for athermalization we need different T–values. If we can find an appropriate combination of both V– and T–values, we can simultaneously correct both the thermal and chromatic conditions.

The properties of several materials suitable for use in the 4.5 μm to 5.1 μm wavelength region are tabulated below.

Silicon	$T = 6.29e{-}05$	$V = 1511$	$1/V = 6.62e{-}04$
Germanium	13.22e–05	673	14.86e–04
Amtir	2.93e–05	642	15.04e–04
Zinc Selenide	3.03e–05	342	28.42e–04
Zinc Sulfide	3.57e–05	915	11.15e-04

If we plot T vs. $(1/V)$, a line drawn thru the points for two materials and extended to the T–axis will indicate the equivalent T–value of an achromatic doublet made from the two materials. From the figure it is apparent that silicon-germanium makes an interesting combination. To *fully* correct the system we would have to athermalize and achromatize *both* components of our system. But since the undercorrected chromatic aberration calculated for the initial layout appeared acceptable, we consider the possibility of athermalizing by adding a single negative germanium element to component "a". This should have an overcorrecting effect on the chromatic aberration and also provide a

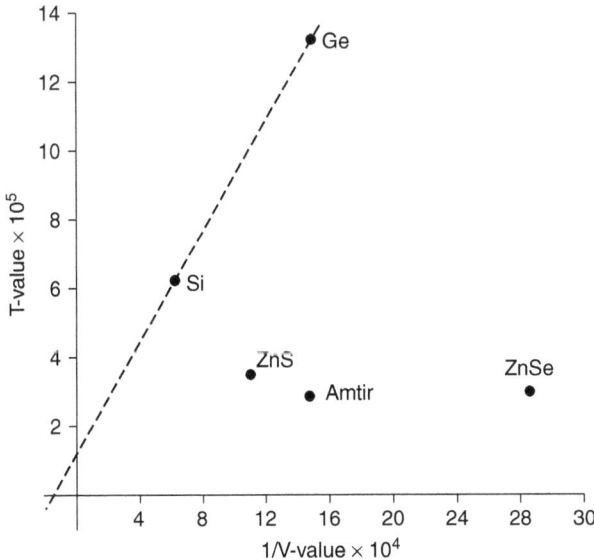

Figure 14.15 "T value" vs. reciprocal "V value" for candidate material.

mechanism to correct the spherical aberration in the final lens design process.

We could solve for the power of the added germanium element needed to athermalize the system by using an algebraic approach. But instead we shall once again elect the numerical, trial and error technique. Assuming that the "a" component will consist of a silicon-germanium doublet with the same power (ϕ_a = +.0058543) as in the layout above, and that the individual element powers for the "a" component are ϕ_{a1} and ϕ_{a2}, we choose the power of the germanium element (ϕ_{a2}) as the free variable. We then determine the silicon element power from

$$\phi_{a1} = \phi_a - \phi_{a2} = .0058453 - \phi_{a2}$$

Then we can determine the power of the doublet component at $\delta t = 100°C$ from

$$\phi_{aT} = 1.00629\ \phi_{a1} + 1.01322\ \phi_{a2}$$

where the constants are $(1 + 100T)$, ($\equiv \theta$), for silicon and germanium. Following the procedure used above to determine the thermal focus shift for the initial layout, we get the following tabulation for several selected values of ϕ_{a2}:

Trial	#1	#2	#3	#4	Final #5
ϕ_{a1}	+.005854	+.010854	+.015854	+.020854	+.017453
ϕ_{a2}	0	−.005	−.010	−.015	−.011599
$\theta \cdot \phi_{a1}$	+.005891	+.010922	+.015954	+.020986	+.017563
$\theta \cdot \phi_{a2}$	0	−.005066	−.010132	−.015198	−.011752
ϕ_{aT}	+.005891	+.005856	+.005822	+.005787	+.005811
$1.0024 \cdot D$			218.8239		
$\theta \cdot \phi_b$			+.045323		
B_{100}	40.0854	40.7834	41.5424	42.3783	41.8013
Shift	−1.7153	−1.0173	−0.2583	+0.5776	+0.0006
Blur	.3545	.2102	.0534	.1194	.0001
TachC	.0144	.0059	−.0048	−.0144	−.0079

The value of ϕ_{a2} for the final (#5) trial was found by interpolation between trials #3 and #4. The thermal shift is effectively zero for trial #5; in practice one would not need to be so close to zero for a preliminary thin lens layout; the introduction of the actual element thicknesses will cause changes, and also the final computer lens design process will be set up to control the thermal defocusing along with the other characteristics and aberrations. Note that the chromatic aberration has changed sign and is now overcorrected, and that the chromatic blur has been reduced to about half of that for the system with simple silicon elements. This is an added benefit of an informed choice of materials.

We have, as hoped, more or less fortuitously found a system which satisfies all five first-order requirements by using just three variables. We controlled the focal length, the length and the stop position; the clearance (back focus) and the foldability requirements were satisfied without the need for any additional effort. The thermal stability was achieved by adding a germanium element which also served to reduce the chromatic aberration. This completes the system layout.

Some comments on the lens design process

The next step is the lens design process. Here, the considerations of image blur size, distortion and pupil aberrations are addressed, while maintaining the first-order requirements for focal length, stop position, length, clearance, and thermal compensation. Note that these are "maintained". Computer lens design software (i.e., so-called "automatic lens design") usually requires that the starting design which is given to the computer be close to the form which satisfies the specification, in order for it to find a solution. This is why we need to find a reasonable first-order solution as a starting point before undertaking the lens design process.

In doing the lens design of this system, the control of distortion and the pupil aberrations (among others) required that the relay component "b" be made of three thin elements rather than one. (It is also possible that two thick, strongly meniscus elements might be adequate. See the Able IR telescope in Chap. 19.)

The final design has 50% of the energy within 23 μm and 27μm (vs. the 50μm and 75 μm of the specification), indicating that the fabrication tolerances should be reasonable. The external pupil made the distortion correction difficult; however the final distortion is only 1.5%, and the pupil distortion is an acceptable 3.5%.

As a final note, we must acknowledge that we *could* have found the first-order solution using a computer automatic lens design program:

STEP #1 The components would be entered as zero thickness convex-plano singlets. The curvature of the convex surfaces and the spacing would be allowed to vary. The "merit function" of the program would include targets for EFL, L, and S (based, just as we did

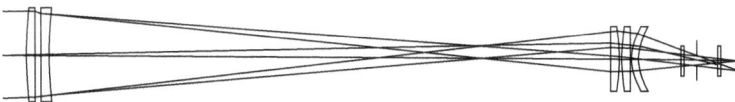

Figure 14.16   The final design of the lens system.

here, on paraxial ray traces). Given a reasonable starting system, a program will quickly find our solution.

STEP #2 A similar process could then be used to athermalize the system.

STEP #3 Having established this much, one could then proceed to the final lens design stage.

Two caveats are in order. First, note well that the computer process, like any numerical convergence scheme, requires a starting point that is within some limited distance from the desired solution. If this is *not* the case, the computer may come up with a totally impossible solution, or perhaps no solution at all. Thus, at least part of the process we have discussed here is required in order to find the starting point. Second, we should also note that a stepped process is an almost necessary one. The reason is that if the first-order properties of a system are not well established *before* the computer aberration correction is begun, the program may find a local optimum in the merit function wherein the aberrations are well corrected but the first order properties are way off the mark.

Wave-Front Aberrations and MTF

15.1 Introduction

In previous chapters and in appendix A we discuss the means by which ray paths are traced through an optical system and how the numerical values of the image aberrations may be determined. In this chapter, we will consider the interpretation of the results of such computations. The basic question to which we address ourselves is: "What effect does a given amount of aberration have on the performance of the optical system?"

We have seen that raytracing yields an incomplete picture of the image-forming characteristics of a system, since the image formed by a "perfect" lens or mirror is not the geometric point that raytracing might lead us to expect, but a finite-sized diffraction pattern—the Airy disk and the surrounding rings. For modest departures from perfection (i.e., aberrations which cause a deformation of the wave front amounting to less than one or two wavelengths) it is thus appropriate to consider the manner in which an aberration affects the distribution of energy in the diffraction pattern. For larger amounts of aberration, however, the illumination distribution as described by raytracing can yield a quite adequate representation of the performance of the system. Thus, it is convenient to divide our considerations into (1) the effects of small amounts of aberration, which we treat in terms of the wave nature of light, and (2) the effects of large amounts of aberration, which may be treated geometrically.

15.2 Optical Path Difference: Focus Shift

We will begin our discussion of small amounts of aberration by determining the optical path difference (OPD) or wave-front deformation introduced by a longitudinal shift of the reference point. Figure 15.1 shows a spherical wave front (solid line) emerging from the pupil of a "perfect" optical system with a focus at point F. We wish to determine the OPD with respect to a reference point at R, which is some arbitrary distance δ from F. If we construct a reference sphere (dashed), centered on R, which coincides with the wave front at the axis, then the OPD for a given zone (of radius Y) is the distance* from the reference sphere to the wave front measured along the radius of the reference sphere, as indicated in Fig. 15.1.

From the figure we can see that, for modest amounts of OPD, the path difference is equal to the radius of the reference sphere $(l + \delta)$ minus the radius of the wave front (l) all less $\delta \cos U$.

$$\frac{\text{OPD}}{n} = (l + \delta - \delta \cos U - l)$$

$$= \delta (1 - \cos U)$$

To an approximation sufficient for our purposes, we can make the substitution

$$\cos U \approx 1 - \tfrac{1}{2} \sin^2 U$$

and the optical path difference resulting from a shift of the reference point by an amount δ is given by

$$\text{OPD} = \tfrac{1}{2} n\delta \sin^2 U \tag{15.1}$$

A longitudinal shift of the reference point is equivalent to defocusing the system; by use of Rayleigh's quarter-wave criterion we can establish a rough allowance for the tolerable depth of focus. Setting the OPD equal to a quarter wavelength of light and solving for the permissible focus shift,

$$\text{Depth of focus } \delta = \pm \frac{\lambda}{2n \sin^2 U_m} = 2\lambda \, (f/\#)^2 \tag{15.2a}$$

where λ is the wavelength of light, n is the index of the final medium, and U_m is the final slope of the marginal ray through the system. Note that U_m is used because the maximum amount of OPD occurs at the edge of the wave front. We can convert this to transverse measure by

---

*Times the index n of the final medium, if the final medium is not air.

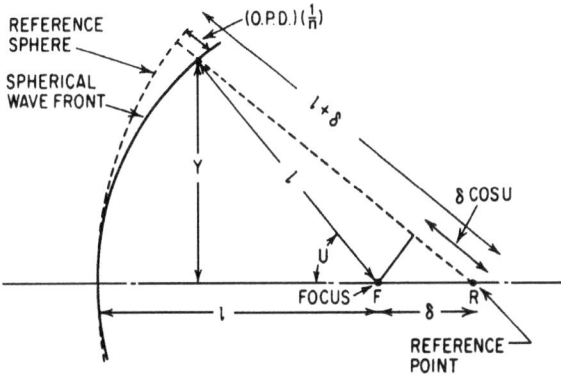

Figure 15.1 The optical path difference (OPD) introduced by a small longitudinal displacement (δ) of the reference point is equal to the index (n) times [the radius of the reference sphere ($l + \delta$) minus the radius of the wave front (l) minus $\delta \cos U$].

multiplying by the ray slope; using $\sin U_m$ as a close enough approximation for the slope, we get
Transverse $\lambda/4$ defocus,

$$H' = \frac{0.5\lambda}{n \sin U_m} = \frac{0.5\lambda}{\text{NA}} = \lambda \; (f/\#) \qquad (15.2b)$$

where $\text{NA} = n \sin U_m$ and $(f/\#) = f$-number.

15.3 Optical Path Difference: Spherical Aberration

We begin by determining the OPD with respect to a reference sphere centered at the paraxial focus. In Fig. 15.2, the deformed wave front is shown as a solid line; the ray (normal to the wave front) from zone Y intersects the axis at point M. The reference sphere, centered at P, is shown dashed, and the OPD is, as before, the radial distance between the two surfaces, times the index. Since the wave front is shown lagging behind the reference sphere, the sign of the OPD is shown negative, to be consistent.

The ray is normal to the wave front and the radius is normal to the reference sphere; thus the angle α between the surface normals is also the angle between the surfaces, and, as indicated in the lower sketch, the change in OPD corresponding to a small change in height dY is given by the relation

$$\alpha = \frac{(-d\text{OPD})}{n \; dY}$$

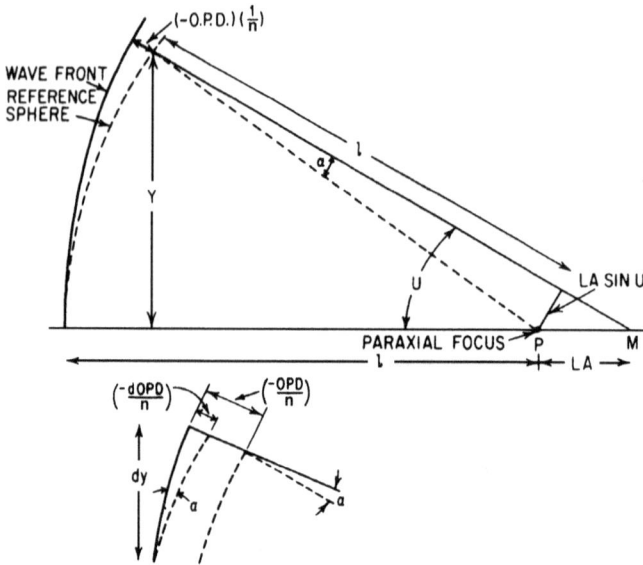

Figure 15.2 The OPD (with reference to the paraxial focal point) produced by spherical aberration. The small diagram indicates the relationship $\alpha = (-1/n)\, d\text{OPD}/dy$. In the upper sketch, it is apparent that $\alpha = \text{LA} \sin U/l$.

But the angular aberration α is also related to the spherical aberration by

$$\alpha = \frac{(\text{LA}) \sin U}{l} = \frac{(\text{LA})\, Y}{l^2}$$

Combining and solving for $d\text{OPD}$ we get

$$d\text{OPD} = \frac{-Y n\, (\text{LA})\, dY}{l^2}$$

Now the longitudinal spherical aberration is a function of Y and can be represented by the series

$$\text{LA} = aY^2 + bY^4 + cY^6 + \cdots \tag{15.3}$$

Making this substitution and integrating

$$\text{OPD} = -\int_0^Y \frac{nY}{l^2}\,(aY^2 + bY^4 + cY^6 + \cdots)\, dY$$

$$= -\frac{n}{l^2}\left(\frac{aY^4}{4} + \frac{bY^6}{6} + \frac{cY^8}{8} + \cdots\right)\Bigg|_0^Y$$

$$= \frac{-nY^2}{2l^2} \left(\frac{aY^2}{2} + \frac{bY^4}{3} + \frac{cY^6}{4} + \cdots \right)$$

$$\text{OPD} = -\tfrac{1}{2}\, n\, \sin^2 U \left(\frac{aY^2}{2} + \frac{bY^4}{3} + \frac{cY^6}{4} + \cdots \right) \qquad (15.4)$$

Now Eq. 15.4 is the OPD with respect to the paraxial focus of the system. It is reasonable to expect that a more desirable reference point than the paraxial focus exists. Thus, by combining Eqs. 15.1 and 15.4, we get

$$\text{OPD} = \tfrac{1}{2}\, n\, \sin^2 U \left[\delta - \left(\frac{aY^2}{2} + \frac{bY^4}{3} + \frac{cY^6}{4} + \cdots \right) \right] \qquad (15.5)$$

which is the OPD with respect to an axial point a distance δ from the paraxial focus.

Third-order spherical aberration.   In many optical systems, the spherical aberration is almost entirely third-order; this is true for almost all systems composed of simple positive elements, and very nearly true for many other systems. Under such circumstances, Eq. 15.3 reduces to

$$\text{LA} = aY^2 \qquad (15.6)$$

and Eq. 15.5 reduces to

$$\text{OPD} = \tfrac{1}{2}\, n\, \sin^2 U \left[\delta - \tfrac{1}{2}\, aY^2 \right] \qquad (15.7)$$

Now at the edge of the aperture $Y = Y_m$ and $\text{LA} = \text{LA}_m$; substituting these values into Eq. 15.6, we find that (for third-order spherical)

$$a = \frac{\text{LA}_m}{Y_m{}^2}$$

and that

$$\text{OPD} = \frac{1}{2} n\, \sin^2 U \left[\delta - \frac{1}{2}\, \text{LA}_m \left(\frac{Y}{Y_m} \right)^2 \right] \qquad (15.8)$$

To determine the value of δ which will result in the smallest amount of OPD, we can try several values of δ in Eq. 15.8 and plot the OPD for each as a function of Y. This has been done for shifts of $\delta = 0$, $\tfrac{1}{2}\text{LA}_m$, and LA_m; the results are plotted in Fig. 15.3. It is apparent that the smallest departure from the spherical reference surface occurs when the OPD is zero at the margin. The corresponding shift of the reference point is $\text{LA}_m/2$. Therefore, from the standpoint of wave-front aberration, the best focus is midway between the marginal and paraxial focal points.

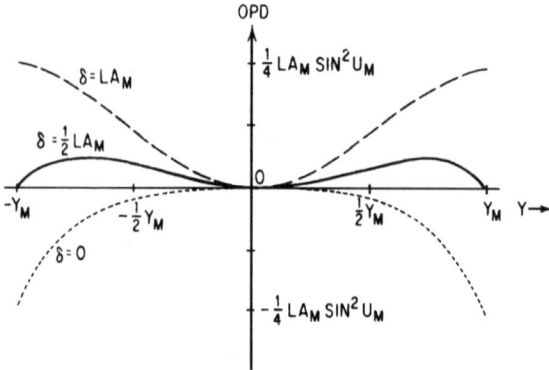

Figure 15.3 The OPD of a system with third-order spherical aberration, plotted as a function of Y for three positions of the reference point.

Two things should be noted at this point: One is that the best RMS focus shift for third order spherical is the same as the one cited here as producing the smallest departure from the reference sphere. Another is that while defocus produces a tilt of the ray intercept plot (or H-tan U curve) as discussed in Sec. 5.8, in the corresponding OPD plot, defocus changes the shape of the curve quadratically, e.g., so that a simple defocus produces a parabolic shaped plot.

If we now substitute $\delta = LA_m/2$ into Eq. 15.8, we find (by differentiating with respect to Y and setting the result equal to zero) that the maximum OPD occurs at $Y = Y_m \sqrt{0.5} = 0.707\ Y_m$ and is given by

$$OPD = \frac{LA_m}{16}\ n \sin^2 U_m$$

This is one-quarter of the OPD at the paraxial focus.

Applying Rayleigh's criterion by setting the OPD equal to one-quarter wavelength, we find the amount of marginal spherical aberration corresponding to this OPD is

$$LA_m = \frac{4\lambda}{n \sin^2 U_m} = 16\lambda\ (f/\#)^2 \tag{15.9a}$$

Again, making an approximate conversion to transverse aberration by multiplying by $\sin U_m$, we get

$$TA_m = \frac{4\lambda}{n \sin U_m} = \frac{4\lambda}{NA} = 8\lambda\ (f/\#) \tag{15.9b}$$

Fifth-order spherical aberration. When the spherical aberration consists of third and fifth order (and this includes the vast majority of all optical systems), we can write

$$LA = aY^2 + bY^4$$

Substituting $LA = LA_m$ at $Y = Y_m$ and $LA = LA_z$ at $Y = 0.707\ Y_m$, we find that the constants a and b are related to the marginal and zonal spherical by the following expressions:

$$LA_m = aY_m^2 + bY_m^4$$

$$LA_z = \frac{aY_m^2}{2} + \frac{bY_m^4}{4}$$

$$a = \frac{4LA_z - LA_m}{Y_m^2}$$

$$b = \frac{2LA_m - 4LA_z}{Y_m^4}$$

The OPD is represented by truncating Eq. 15.5

$$OPD = \frac{1}{2}\ n\ \sin^2 U\left(\delta - \frac{aY^2}{2} - \frac{bY^4}{3}\right)$$

and the graph of OPD versus Y is a curve of the type shown in the upper plot of Fig. 15.4. The exact shape of the curve is, of course, dependent on the values of a, b, and δ.

The best focus occurs when

$$\delta = \frac{-3a^2}{16b} = \frac{-3\ (4LA_z - LA_m)^2}{32\ (LA_m - 2LA_z)} = \frac{3}{4}\ LA_{max} \qquad (15.10)$$

since at this point the OPD is zero for three values of Y as shown in the middle plot of Fig. 15.4. At this focus, the OPD at the margin is

$$OPD_m = \frac{1}{2}\ n\ \sin^2 U_m\left(\frac{-3a^2}{16b} - \frac{aY_m^2}{2} - \frac{bY_m^4}{3}\right) \qquad (15.11)$$

and at the maximum (point x), which occurs at

$$Y = Y_m\ \sqrt{-\frac{a}{4b}}$$

the OPD is given by

$$OPD_x = \frac{na^3\ \sin^2 U_m}{96b^2 Y_m^2} \qquad (15.12)$$

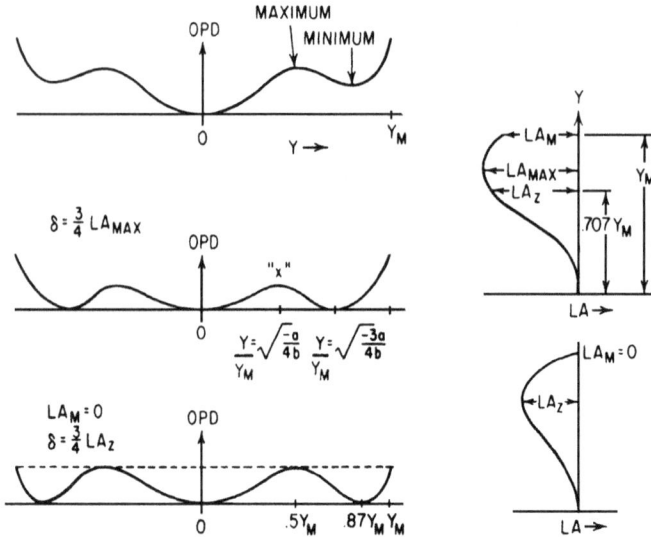

Figure 15.4 OPD vs. Y in the presence of third- and fifth-order aberration. Upper: OPD is a sixth-order function of Y, its shape depending on the aberration coefficients, a and b, and the position of the reference point (δ). Middle: OPD vs. Y when $\delta = (^3/_4)\text{LA}_{\text{max}}$ Lower: OPD is minimized when $\text{LA}_m = 0$ and $\delta = (^3/_4)\text{LA}_z$.

If the marginal spherical aberration of the system is corrected (so that $\text{LA}_m = 0$) then the values of OPD at the margin and at point X are equal, as indicated in the lower plot of Fig. 15.4. This is the condition for minimum OPD in the presence of fifth-order spherical. Then the shift of the reference point is given by

$$\delta = {}^3/_4 \text{LA}_z$$

indicating that the best focus is three-fourths of the way from the paraxial focus to the zonal focus.

The equation for this plot is OPD $= K[y^2 - 24y^4 + 16y^6]$ where K is a scaling factor (equal to one in Fig. 15.4). Note that, depending on whether the OPD is plotted versus y, or $\tan U$, or $\sin U$, or another parameter which indicates the ray position in the aperture, the optimum refocus may differ slightly. A value of 0.8LA_z is often cited as the optimum defocus on an RMS basis.

The residual OPD is given by

$$\text{OPD}_m = \text{OPD}_x = \frac{n\text{LA}_z \sin^2 U_m}{24} \qquad (15.13)$$

This is one-eighth of the OPD at the paraxial focus. Equating this to one-quarter wavelength, we find that the Rayleigh criterion allows a residual zonal spherical of

$$\mathrm{LA}_z = \frac{6\lambda}{n \sin^2 U_m} \tag{15.14a}$$

To make an approximate conversion to transverse aberration, we multiply by sin U_z, which is approximately equal to 0.7 sin U_m, and we get

$$\mathrm{TA}_z = \frac{4.2\lambda}{n \sin U_m} = \frac{4.2\lambda}{\mathrm{NA}} \tag{15.14b}$$

The wave aberration polynomial

Equations 5.1 and 5.2 presented a power series expansion which expressed the transverse ray aberration as a function of h, s, and θ (see Fig. 5.1 for the meaning of these terms). A similar expression can be derived for the wave-front aberration, or OPD as follows.

$$
\begin{aligned}
\mathrm{OPD} = {}& A'_1 s^2 + A'_2 sh \cos\theta \\
& + B'_1 s^4 + B'_2 s^3 h \cos\theta + B'_3 s^2 h^2 \cos^2\theta + B'_4\, s^2 h^2 + B'_5\, sh^3 \cos\theta \\
& + C'_1 s^6 + C'_2 s^5 h \cos\theta + C'_4\, s^4 h^2 + C'_5\, s^4 h^2 \cos^2\theta + C'_7\, s^3 h^3 \cos\theta \\
& + C'_8\, s^3 h^3 \cos^3\theta + C'_{10}\, s^2 h^4 + C'_{11}\, s^2 h^4 \cos^2\theta + C'_{12}\, sh^5 \cos\theta \\
& + D'_1\, s^8 + \cdots
\end{aligned}
$$

Note that although the constants here correspond to those in Eqs. 5.1 and 5.2, they are not numerically the same. However, the expressions are related by

$$y' = \mathrm{TA}_y = \frac{-l}{n}\frac{\partial \mathrm{OPD}}{\partial y} \qquad \text{and} \qquad x' = \mathrm{TA}_x = \frac{-l}{n}\frac{\partial \mathrm{OPD}}{\partial x}$$

where l is the pupil-to-image distance and n is the image space index. Note that the exponent of the semiaperture term s is larger by one in the wave-front expression than in the ray-intercept equations. This equation allows us to determine the shape of the wave front for any combination of aberrations.

15.4   Aberration Tolerances

The preceding sections form a basis for the establishment of what are usually referred to as *aberration tolerances*. We should note, however, that the use of the word "tolerance" in this connection does not carry the same go, no-go connotation that it does in matters mechanical, where parts may suddenly cease to fit or function when tolerances are exceeded. *Any* amount of aberration degrades the image; a larger amount simply degrades it more. Thus, it might be more accurate to call this section "Aberration Allowances."

The Rayleigh criterion, or limit, allows not more than one-quarter wavelength of OPD over the wave front with respect to a reference sphere about a selected image point, in order that the image may be "sensibly" perfect. For convenience, we will use the term one Rayleigh limit to mean an OPD of one-quarter wavelength. We have previously (Chap. 9) noted that the image formed by a perfect lens is a diffraction pattern which contains 84 percent of its energy in a central disk, the remaining 16 percent being distributed in the rings of the pattern. When the OPD is less than several Rayleigh limits, the size of the central disk is basically unchanged, but a noticeable shift of energy from the central disk to the rings takes place.

RMS OPD

The preceding discussions have measured the OPD in terms of its maximum departure from the reference sphere. This is often referred to as peak-to-peak or peak-to-valley (P-V) OPD. It correlates well to image quality when the shape of the wave front is relatively smooth. However, it is inadequate if the wave front is abruptly irregular. In such circumstances the RMS OPD is a better measure of the effect of the wave-front deformation. RMS stands for "root mean square," and is the square root of the average (or mean) of the squares of all the OPD values sampled over the full aperture of the system. Consider, for example, an otherwise perfect optical system with a bump on one surface. If the bump covers only a very small area, its effect on the image will be correspondingly small, even if the P-V OPD of the bump in the wave front is quite large. In this sort of case the RMS OPD would be very small and would represent the effect of the bump on the image much more accurately than the P-V OPD would. The relationship between RMS OPD and P-V OPD for the case of the very smooth wave-front deformation caused by defocusing is

$$\text{RMS OPD} = \frac{\text{P-V OPD}}{3.5}$$

For a less smooth wave-front deformation the denominator in this expression will be larger; this is especially true for deformations caused by high-order aberrations or by random fabrication errors. Most workers assume a denominator of 4 or 5 in the above expression when dealing with random errors. Thus the Rayleigh quarter-wave criterion corresponds to an RMS OPD of a fourteenth- or a twentieth-wave. The fact that a twentieth-wave sounds much more impressive than a quarter-wave may have contributed to the popularity of RMS OPD among suppliers of optical systems.

Strehl ratio

The *Strehl ratio* is the illumination at the center of the Airy disk for an aberrated system, expressed as a fraction of the corresponding illumination for a perfect system, as shown in Fig. 15.5. It is a good measure of image quality when the optical system is well corrected. A Strehl ratio of 80 percent corresponds to a quarter-wave P-V OPD (exactly for defocus, approximately for most aberrations). For modest amounts of OPD, the relationship between the Strehl ratio and the RMS OPD is well approximated by

$$\text{Strehl ratio} = e^{-(2\pi\omega)^2}$$

where ω is the RMS OPD in waves, or by Strehl ratio $= 1 - 3.2(\text{P-V})^2$.

For various amounts of OPD, the several measures of image quality are related as indicated in the following table. It assumes that the OPD is due to defocusing. The P-V OPD is given in both Rayleigh limits (RL) and wavelengths. The *Marechal criterion* for image quality is a Strehl ratio of 0.80, which corresponds to the Rayleigh limit for defocusing but is otherwise more general than the quarter-wave limit.

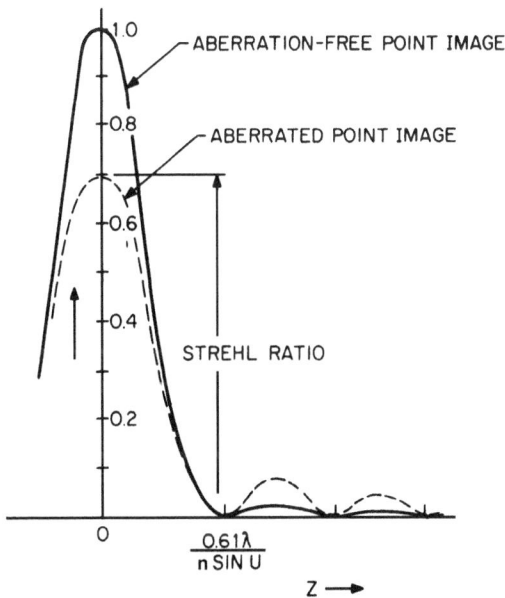

Figure 15.5 The Strehl ratio is the ratio of the central illumination in an aberrated image to the central illumination in an aberration image.

Relation of Image Quality Measures to OPD

P-V OPD	RMS OPD	Strehl ratio	% energy in Airy disk	Rings
0.0	0.0	1.00	84	16
$0.25\text{RL} = \lambda/16$	0.018λ	0.99	83	17
$0.5\text{RL} = \lambda/8$	0.036λ	0.95	80	20
$1.0\text{RL} = \lambda/4$	0.07λ	0.80	68	32
$2.0\text{RL} = \lambda/2$	0.14λ	0.4*	40	60
$3.0\text{RL} = 0.75\lambda$	0.21λ	0.1*	20	80
$4.0\text{RL}=\lambda$	0.29λ	0.0*	10	90

*The smaller values of the Strehl ratio do not correlate well with image quality.

Thus it is apparent that an amount of aberration corresponding to one Rayleigh limit does cause a small but appreciable change in the characteristics of the image. For most systems, however, one may assume that, if the aberrations are reduced to the Rayleigh limit, the performance will be first class and that it will take a determined investigator a considerable amount of effort to detect the resultant difference in a performance. An occasional system does require correction to a fraction of the Rayleigh limit. Microscopes and telescopes are usually corrected to meet or better the Rayleigh criterion, on the axis at least; photographic lenses approach this level of correction only infrequently.

The following tabulation indicates the amount of aberration corresponding to one Rayleigh limit (OPD $= \lambda/4$) when the reference point is chosen to minimize the P-V OPD.

Out of focus

Longitudinal:

$$\Delta l' = \frac{\lambda}{2n \sin^2 U_m} \tag{15.15a}$$

If $\lambda = 0.5$ μm, then $\Delta l' = \pm (f/\#)^2$ in μm.

Transverse:

$$H' = \frac{0.5\lambda}{\text{NA}} \tag{15.15b}$$

Angular aberration:

$AA = \pm\lambda/nD$ radians, where D is the exit pupil diameter. (15.15c)

Third-order marginal spherical

Longitudinal:

$$\text{LA}_m = \frac{4\lambda}{n \sin^2 U_m} \tag{15.16a}$$

Transverse:

$$\text{TA}_m = \frac{4\lambda}{\text{NA}} \tag{15.16b}$$

Angular aberration:

$$\text{AA} = \pm 8\lambda/nD \text{ radians} \tag{15.16c}$$

Zonal residual spherical ($\text{LA}_m = 0$)

Longitudinal:

$$\text{LA}_z = \frac{6\lambda}{n \sin^2 U_m} \tag{15.17a}$$

Transverse:

$$\text{TA}_z = \frac{4.2\lambda}{\text{NA}} \tag{15.17b}$$

Angular aberration:

$$\text{AA} = \pm 8.4\lambda/nD \text{ radians} \tag{15.17c}$$

Tangential coma

$$\text{Coma}_T = \frac{1.5\lambda}{\text{NA}} \tag{15.18}$$

Angular aberration:

$$\text{AA} = \pm 3\lambda/nD \text{ radians} \tag{15.18b}$$
$$\text{for Strehl} = 0.8 \times 1.28$$

Chromatic aberration

Axial color:

$$\text{LAch} = L'_F - L'_C = \frac{\lambda}{n \sin^2 U_m} \tag{15.19a}$$

$$\text{TAch} = \frac{\lambda}{\text{NA}} \tag{15.19b}$$

Lateral color:

$$\text{TchA} = H'_F - H'_C = \frac{0.5\lambda}{\text{NA}} \tag{15.20}$$

The symbols are λ, the wavelength of light; n, the index of the medium in which the image is formed; U_m, the slope angle of the marginal axial ray at the axial image; H, the image height; $\text{NA} = n \sin U_m$, the numerical aperture; D is the exit pupil diameter for an afocal system.

The allowance for longitudinal color is derived from the out-of-focus allowance; if the reference point is midway between the long- and short-wavelength focal points, it is apparent that they may be separated by twice the out-of-focus allowance before the Rayleigh limit is exceeded.

For the chromatic aberrations these amounts are less significant in terms of their effect on the image quality (e.g., MTF) than are the quarter-wave amounts of the monochromatic aberrations. This is because only the extreme wavelengths (e.g., C and F) are a quarter-wave off the nominal wave front; all the other wavelengths are at less than a quarter-wave. Since for most systems the spectral response is at least somewhat peaked up for the central wavelengths, this means that for chromatic aberrations in amounts corresponding to Eqs. 15.19 and 15.20, more than half of the effective illumination has less than an eighth-wave OPD. Thus for ordinary chromatic one can assume that 1.8 to 2.5 (depending on whether the system spectral response is flat or peaked) times the amounts indicated above will produce about the same effect on the image as the quarter-wave amounts for the monochromatic aberrations. If the chromatic is in the form of secondary spectrum, factors of 2.5 to 4.5 are appropriate. Note that the human visual response is quite peaked and factors approaching the larger ones above are suitable for visual systems.

The allowance for coma is frequently exceeded, since it is extremely difficult to correct a system to this level of quality over an appreciable field. Conrady suggests that an OSC of ± 0.0025 or less is an appropriate coma allowance for telescope objectives, and a smaller amount (± 0.0010) for camera lenses.

The out-of-focus allowance is, of course, applicable to curvature of field, and values of z_s and z_t (x_s and x_t) should (ideally, at least) be less than twice this amount. However, it is a rare system that can be corrected to this level, and most optical systems which cover an extended field exceed this allowance many times over.

Example 15.1

For a visual optical system with a relative aperture of $f/5$, $\sin U_m = 0.10$ and $\lambda = 0.55$ µm $= 0.00055$ mm. The aberration allowances corresponding to one-quarter wave OPD are thus given by:

$$\text{Out of focus} = \pm \frac{0.00055}{2\,(0.1)^2} = \pm 0.0275 \text{ mm}$$

$$\text{Marginal spherical} = \pm \frac{4\,(0.00055)}{(0.1)^2} = \pm 0.22 \text{ mm}$$

$$\text{Zonal spherical} = \pm \frac{6\,(0.00055)}{(0.1)^2} = \pm 0.33 \text{ mm} \ (\text{LA}_m = 0)$$

$$\text{Tangential coma} = \pm \frac{1.5\,(0.00055)}{0.1} = \pm 0.00825 \text{ mm}$$

$$\text{Axial chromatic} = \pm \frac{0.00055}{(0.1)^2} = \pm 0.055 \text{ mm} \ (= \pm 0.13 \text{ mm realistically})$$

15.5   Image Energy Distribution (Geometric)

When the aberrations exceed the Rayleigh limit by many times, diffraction effects become relatively insignificant, and the results of geometric raytracing can be used to predict the appearance of a point image with a fair degree of accuracy. This can be done by dividing the entrance pupil of the optical system into a large number of equal areas and tracing a ray from the object point through the center of each of the small areas. The intersection of each ray with the selected image plane is plotted, and since each ray represents the same fraction of the total energy in the image, the density of the points in the plot is a measure of the power density (irradiance, illuminance) in the image. Obviously the more rays that are traced, the more accurate the representation of the geometrical image becomes. A ray intercept plot of this type is called a *spot diagram*. Figure 15.6 indicates several methods of placing

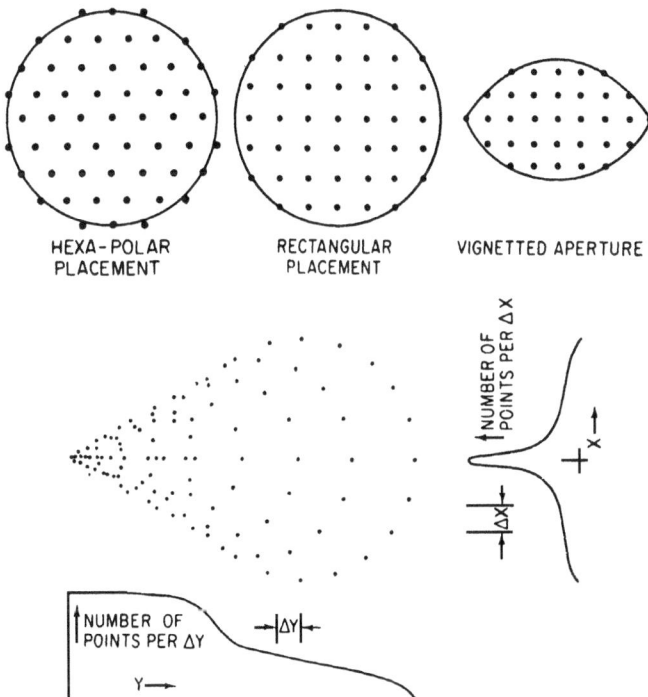

Figure 15.6   The upper sketches show the placement of rays in the entrance pupil so that each ray "represents" an equal area. Shown below are a spot diagram (for a system with pure coma) and the line spread functions (below and to the right) obtained by counting the number of points between parallel lines separated by a small distance, ΔY or ΔX. The "rectangular" ray placement, although less elegant than the hexa-polar, is widely used because the data is more easily utilized in wave-front and MTF calculation.

the rays in the entrance pupil and shows an example of a spot diagram. The rectangular ray placement is the most used, being the easiest to do and also having utility in OPD and MTF calculations.

The preparation of a spot diagram obviously entails a great amount of raytracing. As pointed out in Sec. A.3, the rays on each side of the meridional plane are mirror images of each other; this reduces the necessary raytracing by 50 percent. The number of rays to be traced can be reduced markedly by an interpolation process. To produce a spot diagram which faithfully reproduces the image, several hundred ray intersections are required. However, if 20 or 30 rays are traced, it is possible to fit an interpolation equation to their intercept coordinates so that the required (larger) number of points can be computed from the equation. Equations such as Eqs. 5.1 and 5.2 are suitable for this purpose. However, the high computation speed now available in most desktop computers makes this unnecessary, and most spot diagrams are made by simply tracing several hundred rays through the system.

For an accurate analysis, the effects of color on the energy distribution must also be included. This is accomplished by tracing additional rays at different wavelengths; the variation of system sensitivity with wavelength may be taken into account by tracing fewer rays in the less-sensitive wavelengths or by an appropriate weighting scheme. For devices with appreciable fields of view, spot diagrams must also be prepared for several obliquities.

Focusing must also be taken into account. Since it is difficult to predict in advance the exact position of the plane of best focus, spot diagrams are often prepared for several positions of the image plane and the best is selected. One way of accomplishing this efficiently is to hold the final ray data (intercepts and directions) in the computer memory and to calculate a new set of intercepts for each focus shift.

15.6   Spread Functions—Point and Line

The image of a point (whether the data are derived from a spot diagram or from an exact diffraction calculation) can be considered from a three-dimensional point of view to be a sort of illumination mountain, as sketched in Fig. 15.7. The *point spread function* can be described two dimensionally by a series of cross sections through the three-dimensional solid. The solid corresponding to a line image is also shown in Fig. 15.7. The cross section of the line solid is called the *line spread function* and can be obtained by integrating the point solid along sections parallel to the direction of the line, since the line image is simply the summation of an infinite number of point images along its length. The lower part of Fig. 15.6 shows a spot diagram for a system with pure third-order coma and the line spread functions derived from it.

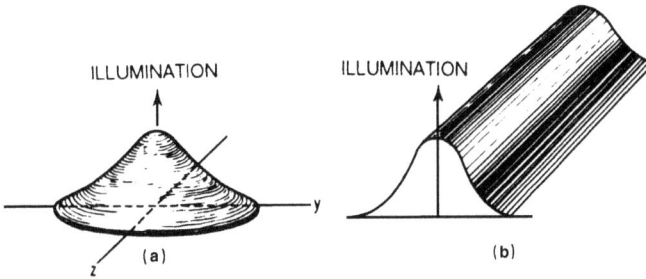

Figure 15.7 The energy distribution in the image of a point (a) and a line (b). The line image (b) is generated by summing an infinite number of point images (a) along its length. The line spread function is the cross section of (b).

A *knife-edge trace* is a plot of the energy which passes a knife edge versus the position of the knife edge as the knife is scanned laterally through the image of a point. The slope, or derivative, of the knife-edge scan is equal to the value of the line spread function. This relationship is often used to measure the line spread function in order to measure the MTF.

The RMS *spot size* is a convenient metric for the quality of the image of a point. Many rays are traced from a point, and the "center of gravity" of all the image plane intersections is determined. Then the RMS spot size is given by

$$\text{RMS} = \sqrt{\Sigma R_i^2 / n}$$

where R_i is the radial distance of spot i from the "center of gravity".

15.7 Geometric Spot Size Due to Spherical Aberration

Third-order spherical aberration

The meridional spread of an image can, of course, be read directly from a ray intercept curve (see Fig. 5.24, for example). For points on the axis, the image blur is symmetrical and it is possible to obtain simple expressions for the size of the blur spot.

Figure 15.8 shows the ray paths near the image plane of a system afflicted with underconnected third-order spherical aberration. It is apparent that the minimum diameter blur spot for this system occurs at a point between the marginal focus and the paraxial focus. This point is three-quarters of the way from the paraxial focus to the marginal focus, and the diameter of the spot at this point is given by:

$$B = \frac{1}{2} \text{LA}_m \tan U_m$$
$$= \frac{1}{2} \text{TA}_m$$

$$(15.21)$$

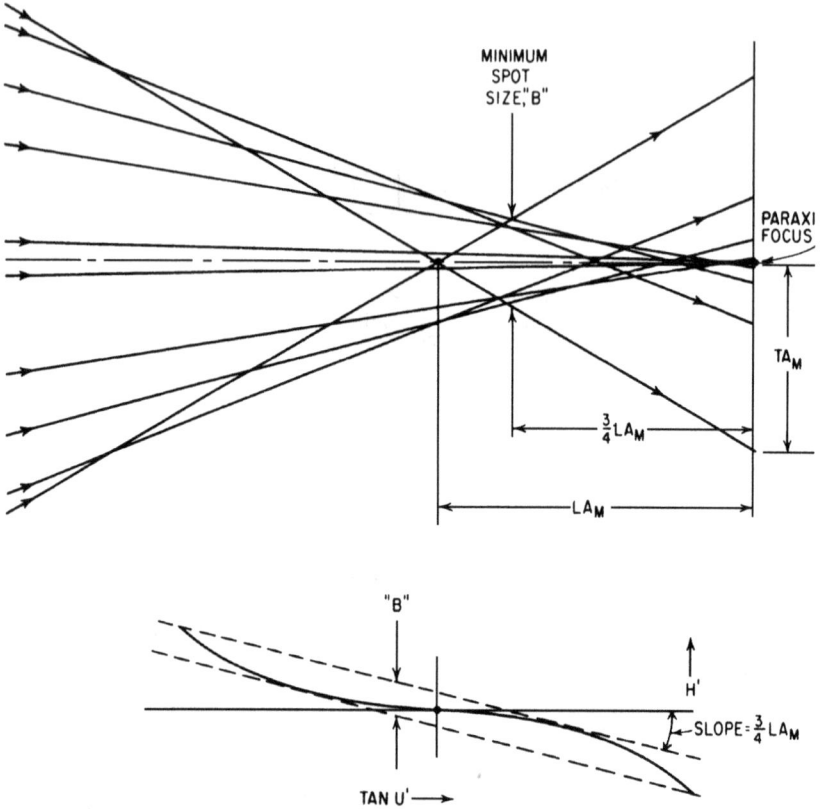

Figure 15.8 The upper figure shows the ray paths near the focus of a system with third-order spherical aberration. The smallest blur spot occurs at 0.75LA_m from the paraxial focus. The lower figure is a ray intercept curve (H' vs. $\tan U'$) for the same case; the slope of the dashed lines ($dH'/d \tan U'$) equals 0.75LA_m and their separation indicates the diameter of the blur spot.

Fifth-order spherical aberration

When the spherical aberration consists of both third and fifth orders, the situation is more complex. From a geometric standpoint, the minimum spot size can be shown to occur when the marginal spherical is equal to two-thirds of the (0.707) zonal spherical, or

$$\text{LA}_z = 1.5\text{LA}_m$$

and LA = zero at $y = 1.12Y_m$. For most systems, this means that both LA_m and LA_z are undercorrected when the minimum geometric spot size is desired.

Then the "best" focus occurs at

$$\delta = 1.25\text{LA}_m = 0.83\text{LA}_z$$

and the size of the blur spot is

$$B = \tfrac{1}{2}\text{LA}_m \tan U_m$$
$$= \tfrac{1}{3}\text{LA}_z \tan U_m \tag{15.22}$$

The best RMS spot size occurs when $\text{LA}_z = 1.75\ \text{LA}_m$ (vs. 1.5) and the defocus is

$$\delta = 1.5\ \text{LA}_m = 0.857\ \text{LA}_z$$

However, if the marginal spherical is corrected to zero, then the "best" geometric focus is at

$$\delta = 0.42\text{LA}_z$$

and for small values of U, the minimum blur spot size is

$$B = 0.84\text{LA}_z \tan U_m \tag{15.23}$$

The "best" focus positions described above are not necessarily those one would select visually, and the reader may have noticed that they differ from those selected on the basis of OPD in Sec. 15.3. Figure 15.9 shows a ray intercept curve for fifth-order spherical with the marginal spherical corrected to zero. The slope of the two solid lines indicates the amount of focus shift required to minimize the blur spot. (Remember that the slope $\Delta H/\Delta \tan U$ is equivalent to a focus shift, and that the vertical separation of the lines indicates the size of the blur.) However, the *dashed* pair of lines (which enclose the ray intercepts from about 80 percent of the aperture) indicate a focus position at which there is a much higher concentration of light within a much smaller spot, and this is usually the preferred focus, even though the *total* spread of the image is greater by a factor of almost 2.

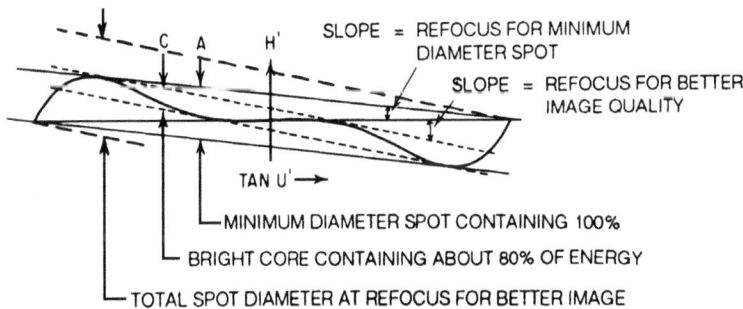

Figure 15.9 The image blur spot size for third- and fifth-order spherical aberration, balanced for $\text{LA}_m = 0$, illustrating the effects of various focus settings.

The concept of minimum blur size is little used in optical systems for visual or photographic work, since the minimum geometric blur position is seldom, if ever, chosen as the focus. However, in systems which use photodetectors, one frequently wishes to determine the smallest detector that will collect all the energy in the image. Under such circumstances, the blur spot sizes given by Eqs. 15.21, 15.22, and 15.23 are extremely useful. The geometric spot minimum is often a consideration when a system's performance is well below that of a "diffraction-limited" system.

Example 15.2

A visual system, working at $f/5$ (sin $U_m = 0.1$), which has an undercorrected third-order longitudinal spherical aberration of 0.22 mm, will have its minimum diameter blur spot $0.75 \times 0.22 = 0.165$ mm ahead of the paraxial focus, and by Eq. 15.21 the size of this blur spot will be equal to

$$B = \frac{1}{2} \times 0.22 \times 0.1005 = 0.011 \text{ mm}$$

It is interesting to note that on the basis of the OPD analysis, the best focus should occur $0.5 \times 0.22 = 0.11$ mm ahead of the paraxial focus and that the diameter of the central disk of the Airy pattern is equal to

$$\frac{1.22\lambda}{n \sin U} = \frac{1.22 \ (0.00055)}{0.1} = 0.0066 \text{ mm}$$

This central disk should contain about 68 percent of the energy in the image, since a marginal spherical of 0.22 mm is equal to just one Rayleigh limit (as shown in Example 15.1).

If an $f/5$ system has third- and fifth-order spherical with a corrected marginal and a zonal residual of 0.33 mm (again in longitudinal measure), the smallest geometric spot size would be found at about $0.42 \times 0.33 = 0.14$ mm from the paraxial focus and the spot size would be

$$B = 0.84 \times 0.33 \times 0.1005 = 0.028 \text{ mm}$$

Here the comparison with the OPD analysis is less fortuitous. The zonal spherical of 0.33 mm is again equivalent to one Rayleigh limit; we would expect the central disk of the diffraction pattern to be 0.0066 mm as above, and the best focus to be about $0.75 \times 0.33 = 0.25$ mm from the paraxial focus. The agreement with geometry is somewhat better if we use the focus indicated by the dashed lines of Fig. 15.9; the position of "best focus" is almost exactly the same as the OPD best focus and the diameter of the intense center spot of the geometric pattern is to the order of 0.01 mm.

15.8 The Modulation Transfer Function

A type of target commonly used to test the performance of an optical system consists of a series of alternating light and dark bars of equal width, as indicated in Fig. 15.10a. Several sets of patterns of different spacings are usually imaged by the system under test and the finest set in which the line structure can be discerned is considered to be the

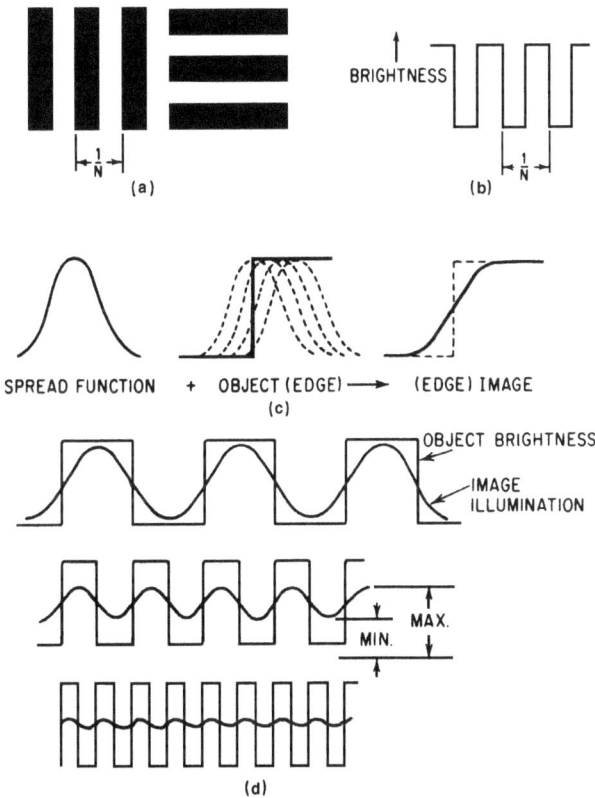

Figure 15.10 The imagery of a bar target. (a) A typical bar target used in testing optical systems consists of alternating light and dark bars. If the pattern has a frequency of N lines per millimeter, then it has a period of $1/N$ millimeters, as indicated. (b) A plot of the brightness of (a) is a square wave. (c) When an image is formed, each line is imaged as a blur, with an illumination distribution described by line the spread function. The image then consists of the summation of all the spread functions. (d) As the test pattern is made finer, the contrast between the light and dark areas of the image is reduced.

limit of resolution of the system, which is expressed as a certain number of lines per millimeter.* When a pattern of this sort is imaged by an optical system, each geometric line (i.e., of infinitesimal width) in the object is imaged as a blurred line, the cross section of which is the line spread function. Figure 15.10b indicates a cross section of the brightness of the bar object, and Fig. 15.10c shows how the image spread function "rounds off" the "corners" of the image. In Fig. 15.10d, the effect of the image blur on progressively finer patterns is indicated. It is apparent that when the illumination contrast in the image is less than the smallest amount that the system (e.g., the eye, film, or photodetector) can detect, the pattern can no longer be "resolved."

If we express the contrast in the image as a "modulation," given by the equation

$$\text{Modulation} = \frac{\text{max.} - \text{min.}}{\text{max.} + \text{min.}}$$

(where max. and min. are the image illumination levels as indicated in Fig. 15.10d), we can plot the modulation as a function of the number of lines per millimeter in the image, as indicated in Fig. 15.11a. The intersection of the modulation function line with a line representing the smallest amount of modulation which the system sensor can detect will give the limiting resolution of the system. The curve indicating the smallest amount of modulation detectable by a system or sensor (i.e., the threshold) is often called an AIM curve, where the initials stand for the Aerial Image Modulation required to produce a response in the system or sensor. The response characteristics of the eye, films, image tubes, CCDs, etc., are appropriately described by an AIM curve. Note that the modulation threshold usually rises with spatial frequency, although there are exceptions. Figure 8.4 is effectively an AIM curve for the eye; note that at very low angular frequencies the contrast threshold of the eye rises (for physiologic reasons).

It should be apparent that the limiting resolution does not fully describe the performance of the system. Figure 15.11b shows two modulation plots with the same limiting resolution, but with quite different performances. The plot with the greater modulation at the lower frequencies is obviously superior, since it will produce crisper, more contrasty images. Unfortunately, the type of choice one is usually faced with

---

*Note that in optical work the convention is to consider a "line" to consist of one light bar and one dark bar, i.e., one cycle. In television parlance, both light and dark lines are counted. Thus, 10 "optical" lines indicate 10 light and 10 dark lines, whereas 10 "television" lines indicate 5 light and 5 dark lines. To avoid confusion, "optical" lines are frequently referred to as line pairs, e.g., 10 line pairs per millimeter.

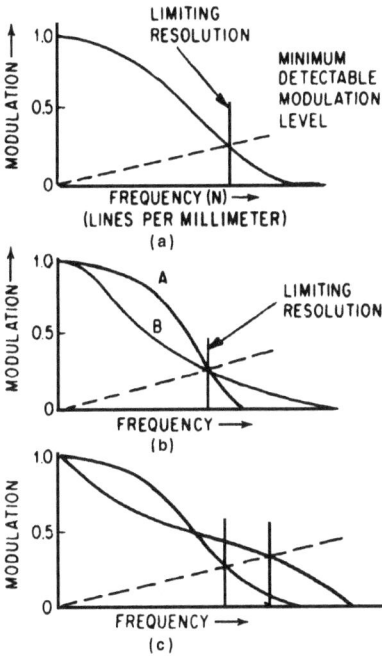

Figure 15.11 (a) The image modulation can be be plotted as a function of the frequency of the test pattern. When the modulation drops below the minimum that can be detected, the target is not resolved. (b) The system represented by (a) will produce a superior image, although both (a) and (b) have the same limiting resolution.

in deciding between two systems is less obvious. Consider Fig. 15.11c, where one system shows high limiting resolution and the other shows high contrast at low target frequencies. In cases of this type, the decision must be based on the relative importance of contrast versus resolution in the function of the system.*

The preceding discussion has been based on patterns the brightness distribution of which is a "square wave" (Fig. 15.10b) and the image illumination distribution of which is distorted or "rounded off" by characteristics of the optical system, as indicated in Fig. 15.10d. However, if the object pattern brightness distribution is in the form of a sine wave, the distribution in the image is also described by a sine wave, regardless of the shape of the spread function. This fact has led to the widespread use of the modulation transfer function to describe the performance of a lens system. The modulation transfer function is the

---

The Strehl definition is the ratio of the light intensity at the peak of the diffraction pattern of an aberrated image to that at the peak of an aberration-free image, and is one of the many criteria that have been proposed for image evaluation. It can be computed by calculating the volume under the (three-dimensional) modulation transfer function and dividing by the volume under the curve for an aberration-free lens (Sec. 15.10). A similar criterion for quick general evaluation of image quality is the normalized area under the modulation transfer curve.

ratio of the modulation in the image to that in the object as a function of the frequency (cycles per unit of length) of the sine-wave pattern.

$$\text{MTF } (v) = \frac{M_i}{M_o}$$

A plot of MTF against frequency v is thus an almost universally applicable measure of the performance of an image-forming system and has been applied not only to lenses but to films, phosphors, image tubes, the eye, and even to complete systems such as camera-carrying aircraft.

One particular advantage of the MTF is that it can be cascaded by simply multiplying the MTFs of two or more components to obtain the MTF of the combination. For example, if a camera lens with an MTF of 0.5 at 20 cycles per millimeter is used with a film with an MTF of 0.7 at this frequency, the combination will have an MTF of $0.5 \times 0.7 = 0.35$. If the object to be photographed with this camera has a contrast (modulation) of 0.1, then the image modulation is $0.1 \times 0.35 = 0.035$, close to the limit of visual detection.

One should note, however, that MTFs do not cascade between optical components which are directly coherently "connected," i.e., lenses which are not separated by a diffuser of some sort. This is because the aberrations of one component may compensate for the aberrations in another, and thus produce an image quality for the combination which is superior to that of either component. Any "corrected" optical system illustrates this point.

In the past, the MTF has been referred to as *frequency response, sine wave response,* or *contrast transfer function.*

If we assume an object consisting of alternating light and dark bands, the brightness (luminance, radiance) of which varies according to a cosine (or sine) function, as indicated by the upper part of Fig. 15.12, the distribution of brightness can be expressed mathematically as

$$G (x) = b_0 + b_1 \cos (2\pi vx) \qquad (15.24)$$

where v is the frequency of the brightness variation in cycles per unit length, $(b_0 + b_1)$ is the maximum brightness, $(b_0 - b_1)$ is the minimum brightness, and x is the spatial coordinate perpendicular to the bands. The modulation of this pattern is then

$$M_0 = \frac{(b_0 + b_1) - (b_0 - b_1)}{(b_0 + b_1) + (b_0 - b_1)} = \frac{b_1}{b_0} \qquad (15.25)$$

When this line pattern is imaged by an optical system, each point in the object will be imaged as a blur. The energy distribution within this blur will depend on the relative aperture of the system and the aberrations present. Since we are dealing with a linear object, the image of each line element can be described by the line spread function

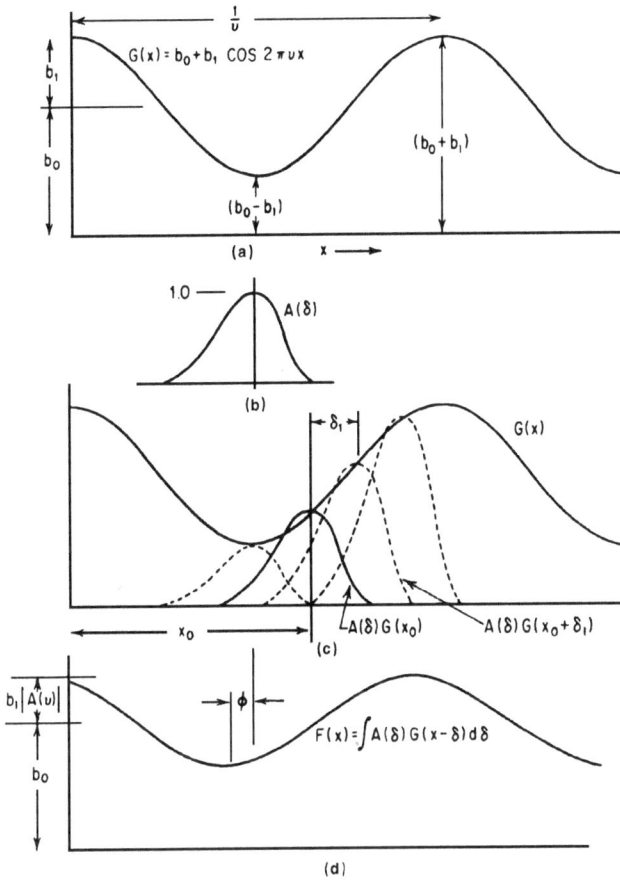

Figure 15.12 Convolution of the object brightness distribution function $G(x)$ with the line spread function $A(\delta)$. (a) The object function, $G(x) = b_0 + b_1 \cos (2\pi v x)$, plotted against x. (b) The line spread function $A(\delta)$. Note the asymmetry. (c) Illustrating the manner in which $G(x)$ is modified by $A(\delta)$. A point (or more accurately, a line element) at x_0 is imaged by the system as $G(x_0)$ times $A(\delta)$. Similarly at $x_0 + \delta_1$, the image of the line element is described by $A(\delta)G (x_0 + \delta_1)$. Thus the image function at a given x has a value equal to the summation of the contributions from all the points the spreadout images of which reach x. (d) The image function $F(x) = \int A(\delta)G(x - \delta) \, d\delta$ has been shifted by ϕ and has a modulation $M_i = M_0 |A(v)|$.

(Sec. 15.5, Fig. 15.7) indicated in Fig. 15.12 as $A(\delta)$. We now assume (for convenience) that the dimensions x and $(1/v)$ in Eq. 15.24 are the corresponding dimensions in the image. It is apparent that the image energy distribution at a position x is the summation of the product of $G(x)$ and $A(\delta)$ and can be expressed as

$$F(x) = \int A(\delta) \, G \, (x-\delta) \, d\delta \qquad (15.26)$$

Combining Eqs. 15.24 and 15.26, we get

$$F(x) = b_0 \int A(\delta) \, d\delta + b_1 \int A(\delta) \cos [2\pi v \, (x-\delta)] \, d\delta \qquad (15.27)$$

After normalizing by dividing by $\int A(\delta) \, d\delta$, Eq. 15.27 can be transformed to

$$F(x) = b_0 + b_1 \, |A(v)| \, \cos (2\pi vx - \phi)$$
$$= b_0 + b_1 A_c(v) \cos (2\pi vx) + b_1 A_s(v) \sin (2\pi vx) \qquad (15.28)$$

where

$$|A(v)| = [A_c^2(v) + A_s^2(v)]^{1/2} \qquad (15.29)$$

and

$$A_c(v) = \frac{\int A(\delta) \cos (2\pi v \delta) \, d\delta}{\int A(\delta) \, d\delta} \qquad (15.30)$$

$$A_s(v) = \frac{\int A(\delta) \sin (2\pi v \delta) \, d\delta}{\int A(\delta) \, d\delta} \qquad (15.31)$$

$$\cos \phi = \frac{A_c(v)}{|A(v)|} \qquad (15.32)$$

$$\tan \phi = \frac{A_s(v)}{A_c(v)} \qquad (15.33)$$

Note that the resulting image energy distribution $F(x)$ is still modulated by a cosine function of the same frequency v, demonstrating that a cosine distribution object is always imaged as a cosine distribution image. If the line spread function $A(\delta)$ is asymmetrical, a phase shift ϕ is introduced. This is a lateral shift of the location of the image (at this frequency).

The modulation in the image is given by

$$M_i = \frac{b_1}{b_0} |A(v)| = M_0 \, |A(v)| \qquad (15.34)$$

and $|A(v)|$ is the modulation transfer function.

$$\text{MTF}(v) = |A(v)| = \frac{M_i}{M_0}$$

The *optical transfer function* (OTF) is the complex function which describes this process. It is a function of the spatial frequency v of the sine-wave pattern. The real part of the OTF is the *modulation transfer function* (MTF) and the imaginary part is the *phase transfer function* (PTF). If the PTF is linear with frequency, it is, of course, just a simple lateral displacement of the image (as, for example, distortion), but if it is nonlinear, it can have an effect on the image quality. A phase shift of 180° is a reversal of contrast, in that the image pattern is light where it should be dark, and vice versa. See Fig. 20.24 for example.

15.9   Square-Wave vs. Sine-Wave Targets

Once the MTF has been determined and plotted for a range of frequencies, it is possible to determine an analogous function for the modulation transfer of a square wave pattern, i.e., a bar target of the type shown in Fig. 15.10. This is done by resolving the square wave into its Fourier components and taking the sine wave response to each component. Thus, for a given frequency v, the square wave modulation transfer $S(v)$ is given by the following equation [in which MTF(v) is written $M(v)$ for clarity].

$$S(v) = \frac{4}{\pi} \left[M(v) - \frac{M(3v)}{3} + \frac{M(5v)}{5} - \frac{M(7v)}{7} + \cdots \right] \quad (15.35a)$$

The inverse of this function is

$$M(v) = \frac{\pi}{4} \left[S(v) + \frac{S(3v)}{3} - \frac{S(5v)}{5} + \frac{S(7v)}{7} - \cdots \right] \quad (15.35b)$$

Practical resolution considerations

A rough indication of the practical meaning of resolution can be gained from the following, which lists the resolution required to photograph printed or typewritten copy.

Excellent reproduction (reproduces serifs, etc.) requires 8 resolution line pairs per the height of a lowercase letter e.

Legible (easily) reproduction requires 5 line pairs per letter height.

Decipherable (e, c, o partly closed) requires 3 line pairs per letter height.

Point sizes of type (where P is the point size) are
Height of an upper case letter = 0.22P mm = 0.0085P in
Height of a lower case letter = 0.15P mm = 0.006P in

The correlation between resolution in cycles per minimum dimension (height, length of military targets) and certain functions (often referred to as *Johnson's law*) is

Detect	1.0 line pairs per dimension
Orient	1.4 line pairs per dimension
Aim	2.5 line pairs per dimension
Recognize	4.0 line pairs per dimension
Identify	6–8 line pairs per dimension
Recognize with 50% accuracy	7.5 line pairs per height
Recognize with 90% accuracy	12 line pairs per height

The smallest distance in the image which is resolved (SIDR) is given by

$$\text{SIDR} = \frac{0.5}{R} \text{ mm}$$

where R is the resolution in lpm.

The MTF values regarded as *"good"* for 35 mm cameras may be of interest. On-axis MTF values of 30 percent at 50 line pairs per millimeter (lpm) and 50 percent at 30 lpm have been suggested as acceptable levels of performance. For an $f/1.8$ double Gauss, 65 percent at 40 lpm and 40 percent at 80 lpm have been proposed as suitable performance on the axis. It should be apparent that off-axis performance will be degraded by additional off-axis aberrations. Another consideration is that the relative area of the corners of a picture is very small, and that the importance of the imagery in the corners is (usually) quite trivial. Another suggestion is that the MTF should exceed 20 percent at 30 lpm over 90 percent of the field with the aperture wide open.

15.10 Special Modulation Transfer Functions: Diffraction-Limited Systems

Sections 15.5 to 15.7 discussed performance in geometric terms; the spot-diagram techniques set forth there are applicable only when the aberrations are large. When they are small, the interactions between the diffraction effects of the system aperture and the aberrations become very complex. If there are no aberrations present, the MTF of a system is related to the size of the diffraction pattern (which is a function of the numerical aperture of the system and the wavelength of the light used). For a "perfect" optical system, the MTF is

$$\text{MTF}(v) = \frac{2}{\pi} (\phi - \cos \phi \sin \phi) \qquad (15.36)$$

where

$$\phi = \cos^{-1}\left(\frac{\lambda v}{2\text{NA}}\right)$$

and v is the frequency in cycles per millimeter, λ is the wavelength in millimeters, NA is the numerical aperture ($n' \sin U'$), and $\cos^{-1}(x)$ means the angle whose cosine is x.*

---

*Equation 15.36 applies to uniformly illuminated and transmitting circular apertures. For apertures of *any* other shape, the diffraction MTF is equal to the (normalized) area common to the aperture and the aperture displaced. Equation 15.36 is thus the (normalized) area common to two circles of radius R, as their centers are separated by an amount equal to $2vR/v_0$. For a rectangular aperture the plot of MTF would thus be a straight line. The cutoff frequency v_0 is computed from Eq. 15.37 in each case using the aperture size (i.e., the $f/\#$ or NA) in the direction of the resolution.

It is apparent that MTF(v) is equal to zero when ϕ is zero; thus, the "limiting resolution" for an aberration-free system, often called the *cutoff frequency,* is

$$v_0 = \frac{2NA}{\lambda} = \frac{1}{\lambda \, (f/\#)} \tag{15.37}$$

where λ is in millimeters, $f/\#$ is the relative aperture of the system, and v_0 is in cycles per millimeter. Notice that an optical system is a low-pass filter which cannot transmit information at a higher spatial frequency than the cutoff frequency v_0.

For an afocal system (or one with the image at infinity), the cutoff frequency is given by

$$v_0 = D/\lambda \text{ cycles/radian}$$

where D is the pupil diameter.

A plot of Eq. 15.36 is shown in Fig. 15.13; the frequency scale is in terms of v_0, the limiting frequency given by Eq. 15.37. It should be noted that for *ordinary* systems, this level of performance cannot be exceeded. A geometric MTF curve derived from the *raytrace* data (and neglecting diffraction) of a well-corrected lens will sometimes exceed the values of Fig. 15.13; such results are, of course, incorrect and derive from the fact that the light ray concept only partially describes the behavior of electromagnetic radiation. Note also that aberrations always reduce the MTF.

The effects of small amounts of defocusing on the diffraction limited MTF are shown in Fig. 15.14. Note that curve B corresponds to the depth of focus allowed by one Rayleigh limit as discussed in

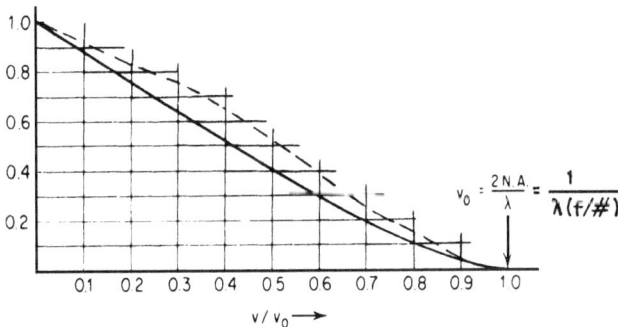

Figure 15.13 The modulation transfer function of an aberration-free system (solid line). Note that frequency is expressed as a fraction of the cutoff frequency. The dashed line is the modulation factor for a square wave (bar) target. Both curves are based on diffraction effects and assume a system with a uniformly transmitting circular aperture.

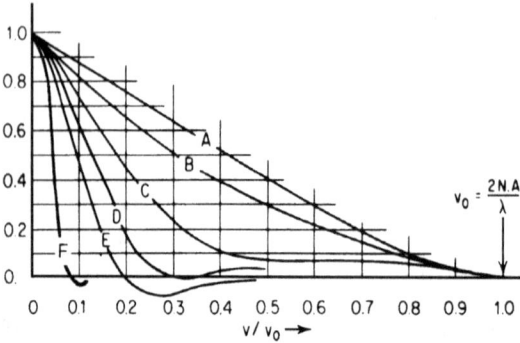

Figure 15.14 The effect of defocusing on the modulation transfer function of an aberration-free system.

(a) In focus		OPD = 0.0
(b) Defocus	$= \lambda/(2n \sin^2 U)$	OPD = $\lambda/4$
(c) Defocus	$= \lambda/(n \sin^2 U)$	OPD = $\lambda/2$
(d) Defocus	$= 3\lambda/(2n \sin^2 U)$	OPD = $3\lambda/4$
(e) Defocus	$= 2\lambda/(n \sin^2 U)$	OPD = λ
(f) Defocus	$= 4\lambda/(n \sin^2 U)$	OPD = 2λ

(Curves are based on diffraction effects—not on a geometric calculation.)

Secs. 15.2 and 15.4. The small effect produced by an OPD of one-quarter wavelength indicates the astuteness of Rayleigh's selection of this amount as one which would not "sensibly" affect the image quality.

The effect of defocusing on the MTF is well approximated by:

$$\text{Defocused MTF} = (\text{MTF per Eq. 15.36}) \, [2J_1(x)/x].$$

where $x = 2\pi\delta \, \text{NA} \, v(v_0 - v)/v_0$. The maximum error is about 0.017 at $v = \frac{1}{2} v_0$.

By way of comparison, Fig. 15.15 shows the MTF plots which would be obtained by geometrical calculations of a perfect system defocused by the same amounts. The agreement between Fig. 15.14, the curves of which are derived from wave-front analysis, and Fig. 15.15 is poor for small amounts of OPD. However, when the defocusing is sufficient to introduce an OPD of one wavelength or more, the agreement becomes much better. Note that all the curves of Fig. 15.15 are of the same family and that one can be derived from another by a simple ratioing of the frequency scale. These curves are representations of

$$\text{MTF}(v) = \frac{2J_1(\pi B v)}{\pi B v} \approx \frac{J_1(2\pi\delta \text{NA} v)}{\pi\delta \text{NA} v} \qquad (15.38)$$

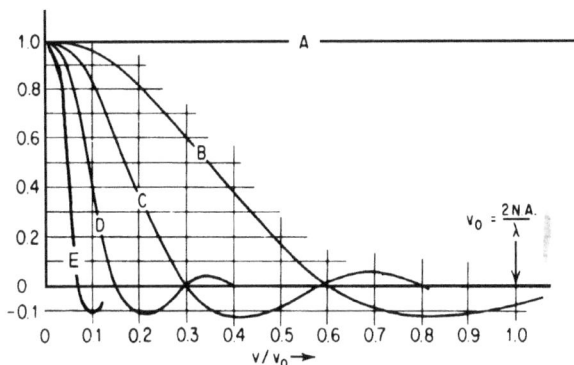

Figure 15.15 The effect of defocusing on the geometrically calculated modulation transfer function of an aberration-free system.

(a) In focus	$OPD = 0.0$	
(b) Defocus $= \lambda/(2n \sin^2 U)$	$OPD = \lambda/4$	Blur $= \lambda/NA = 2\lambda(f/\#)$
(c) Defocus $= \lambda/(n \sin^2 U)$	$OPD = \lambda/2$	Blur $= 2\lambda/NA = 4\lambda/NA(f/\#)$
(d) Defocus $= 2\lambda/(n \sin^2 U)$	$OPD = \lambda$	Blur $= 4\lambda/NA = 8\lambda/NA((f/\#)$
(e) Defocus $= 4\lambda/(n \sin^2 U)$	$OPD = 2\lambda$	

These geometrically derived plots are in poor agreement with the exact diffraction plots of Fig. 15.14 when the defocusing is small. The agreement at $OPD = \lambda$ (curve D above, curve E in Fig. 15.14 is fair; the match at $OPD = 2\lambda$ is quite good). The effective "cut-off" frequency is $v_0 = 1.2/\text{Blur}$.

where $J_1(\pi B v)$ indicates the first-order Bessel function,* B is the diameter of the blur spot produced by defocusing, δ is the longitudinal defocusing, NA is the numerical aperture, and v is the frequency in cycles per unit length.

Note that in Figs. 15.14 and 15.15, some of the curves show a negative value for the MTF. This indicates that the phase shift in the image (ϕ in Eq. 15.33) is 180° and that the image is light where it should be dark and vice versa. This is known as spurious resolution (since a line pattern can be seen, but it is not a true image of the object) and is a phenomenon which is frequently observed in defocused, well-corrected lenses or in lenses the defocused image of a point of which is a nearly uniformly illuminated circular blur. See Fig. 20.24 for example.

In Fig. 15.16, the effects of third-order spherical aberration on MTF are shown. Note once again that the effect of an amount of aberration corresponding to the Rayleigh limit ($OPD = \lambda/4$) is quite modest. The situation here is quite similar to the defocusing case, in that MTF

$$*J_n(x) = \sum_{k=0}^{x} \frac{(-1)^k x^{n+2k}}{2^{(n+2k)} k!(n+k)!} \qquad J_1(x) = \frac{x}{2} - \frac{(x/2)^3}{1^2 2} + \frac{(x/2)^5}{1^2 2^2 3} - \cdots$$

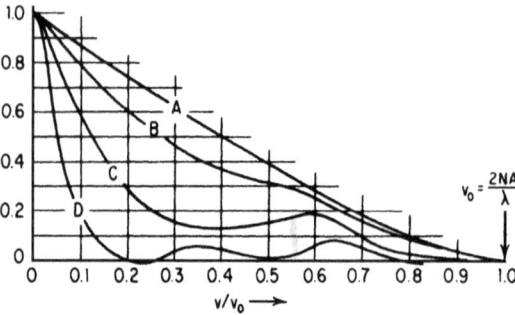

Figure 15.16 The effect of third-order spherical aberration on the modulation transfer function.

(a) $LA_m = 0.0$ $OPD = 0$

(b) $LA_m = 4\lambda/(n \sin^2 U)$ $OPD = \lambda/4$

(c) $LA_m = 8\lambda/(n \sin^2 U)$ $OPD = \lambda/2$

(d) $LA_m = 16\lambda/(n \sin^2 U)$ $OPD = \lambda$

These curves are based on diffraction wave-front computations for an image plane midway between the marginal and paraxial foci.

curves based on geometrical calculations are in poor agreement with Fig. 15.16 where the aberration is small, but in quite reasonable agreement where the aberration is to the order of one or two wavelengths of OPD.

Figure 15.17 shows the effect of a central obstruction in the aperture of a diffraction-limited system. Note that the introduction of a disk into the aperture* drops the response at low frequencies but raises it slightly at high frequencies (although it cannot change v_0, the cutoff frequency). Thus, a system of this type tends to show greatly reduced contrast on coarse targets and a somewhat higher limit of resolution (when used with a system which requires a modulation of more than zero to detect resolution). This is the result of shifting light from the Airy disk to the rings of the diffraction pattern.

Note that, for an image which is off the axis, the effective numerical aperture is reduced. The oblique distance from the stop to the image is greater than the axial distance; this reduces the NA by a factor equal to the cos θ. In addition, the tangential lines are spread out on the film or image plane because of the oblique incidence of the beam. This reduces the lpm on the surface by cos θ. The projected area of the pupil

---

Apodization is the use of a variable transmission filter or coating at the aperture to modify the diffraction pattern. Coatings which reduce the transmission at the center of the aperture tend to "favor" the response at high frequencies; coatings which reduce transmission at the edge of the aperture tend to favor the lower frequencies.

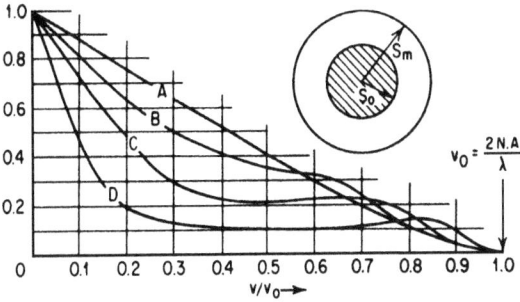

Figure 15.17 The effect of a central obscuration on the modulation transfer function of an aberration-free system.

(a) $s_0/s_m = 0.0$

(b) $s_0/s_m = 0.25$

(c) $s_0/s_m = 0.5$

(d) $s_0/s_m = 0.75$

when viewed obliquely is reduced by a factor of approximately cos θ. So the NA for sagittal lines is reduced by cos θ, and the NA for tangential lines is reduced by cos³ θ.

The calculation of MTF can be done by autocorrelating the aberrated wave front against itself, shifted laterally by a fraction of the aperture, the shift corresponding to the spatial frequency. If the lens is perfect, with no wave-front deformation, then the process amounts to simply shifting the aperture contour (e.g., a circle) against itself. The normalized area common to both circles, for example, is then the MTF. This is shown Fig. 15.18a for a circular aperture; Fig. 15.18b for a rectangular aperture; and Fig. 15.18c for an aperture with a central obstruction. For a rectangular aperture the common area will plot against displacement as a straight line. A central obscuration in the

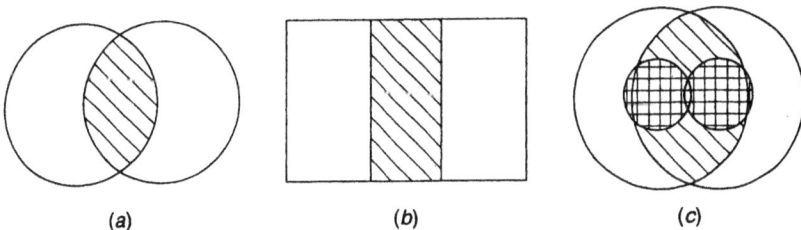

(a)   (b)   (c)

Figure 15.18 For a perfect lens, the MTF is equal to the (normalized) area common to the aperture and the aperture displaced laterally by a distance corresponding to the spatial frequency. The cutoff frequency v_0 corresponds to the shift where the common area is zero. (a) A circular aperture. (b) A rectangular aperture. (c) A circular aperture with a central obstruction. The MTF plots for these apertures are shown in Fig. 15.19.

aperture will reduce the MTF at low frequencies and raise it slightly at high frequencies. If the aperture transmission varies, or if the intensity of the wave front is not uniform (as in a laser beam), a similar technique will work with the aperture areas weighted appropriately. The MTF plots corresponding to these four cases are shown in Fig. 15.19. Obviously the displacement at which the common area is zero corresponds to the cutoff frequency.

MTF with coherent and semi-coherent illumination

All the preceding discussions (except that dealing with a central obstruction of the aperture) have assumed a uniformly illuminated and uniformly transmitting aperture. When the illumination system is arranged so that only the central part of the aperture is illuminated (and this can be done with Koehler illumination if a projection condenser images the source at a size smaller than the pupil of the projection lens), then the MTF plot is modified in a way which is nearly the reverse of that shown in Fig. 15.17.

Fourier theory tells us that we can consider the brightness distribution of an object to be the sum of many sinusoidal brightness distributions of differing frequencies, intensities, and orientations. To simplify matters, let us assume that we are projecting the image of a simple sinusoidal grating. Remembering that a sinusoidal grating has only the first diffraction order, consider the system shown in Fig. 15.20.

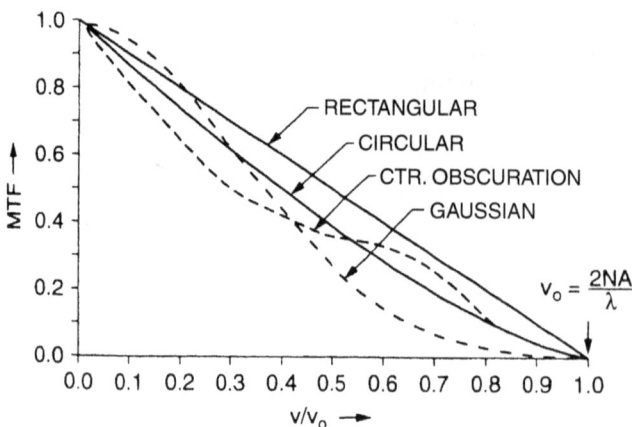

Figure 15.19 The MTF plots for a perfect lens with: (a) a circular aperture; (b) a rectangular aperture; (c) a circular aperture with a central obstruction; and (d) an apodized aperture, or one illuminated by a gaussian laer beam.

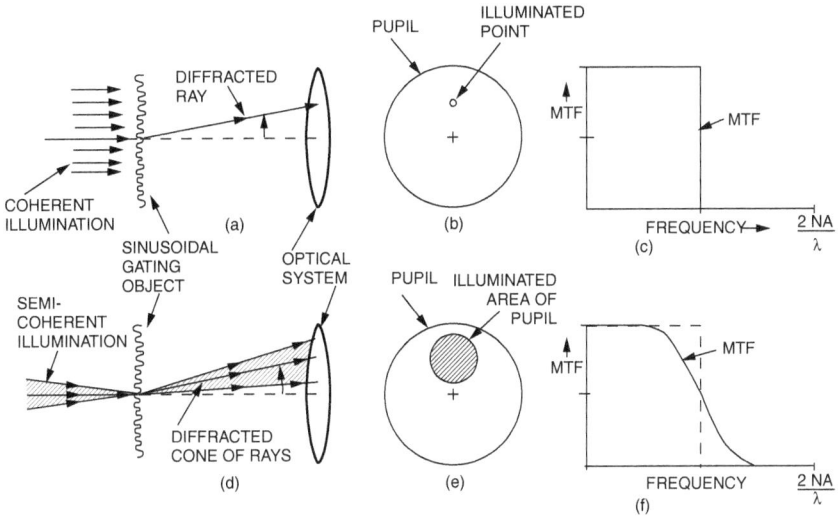

Figure 15.20 (a–c) The MTF with coherent illumination. (d–f) The MTF with semi-coherent illumination (which partially fills the pupil).

If the illumination is *coherent* (i.e., collimated), the light from a point in the grating will be diffracted into the first order, as illustrated in Fig. 15.20a. If the angle of diffraction is less than that of the numerical aperture (NA) of the projection lens, the full power will be projected into the image. But if the grating frequency is high enough (so that $v \geq NA/\lambda$), the diffracted ray will pass outside the lens aperture, and no light corresponding to this frequency will make it into the image. The result of this situation is an MTF plot as shown in Fig. 15.20c, with 100 percent MTF for spatial frequencies of NA/λ or less and zero MTF for all higher frequencies. Note that NA/λ is just half the cutoff frequency ($v_0 = 2NA/\lambda$), as given in Eq. 15.37 for the incoherent illumination case.

If the illumination is *semicoherent,* the projection lens pupil will be partially filled, as indicated in Fig. 15.20d, e. As we consider an increasing grating frequency, the location of the illuminated area in the pupil will move toward the edge. However, at the edge of the pupil, the cutoff is gradual rather than abrupt, as in the coherent case described above, and we get an MTF plot of the sort shown in Fig. 15.20f.

Figure 15.21 shows the effect on the MTF for several values of the illumination system NA, expressed as a fraction of the lens NA. These partial coherence effects are useful in both microlithography and microscopy. Note that decentering or tilting the illuminating beam can be used to get directional effects and that ring illumination can emphasize a particular frequency.

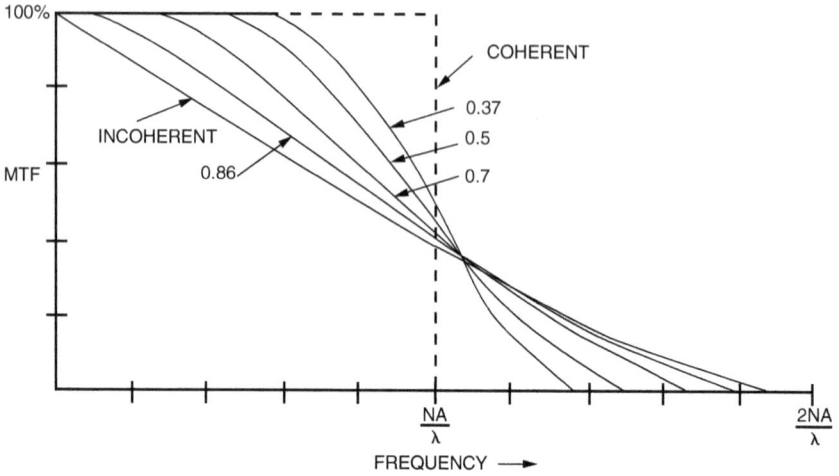

Figure 15.21 MTF vs. frequency for a partially filled pupil (semicoherent illumination). Numbers are the ratio of illuminating system NA to optical system NA.

As mentioned previously, MTFs have been applied to image-receiving systems which are not imaging systems. Figure 15.22 shows the MTF curves for a number of photographic emulsions. Since the MTF of a film is computed on the basis of equivalent relative exposures derived from density measurements on films exposed to sinusoidal test patterns, it is possible to have a film MTF greater than unity. This results from the chemical effects of development of the film on adjacent areas and will be noticed at the low-frequency end of the curves in Fig. 15.22. An AIM curve, as described in Section 15.8, can also be used to represent the response characteristics of nonimaging devices and sensors such as films.

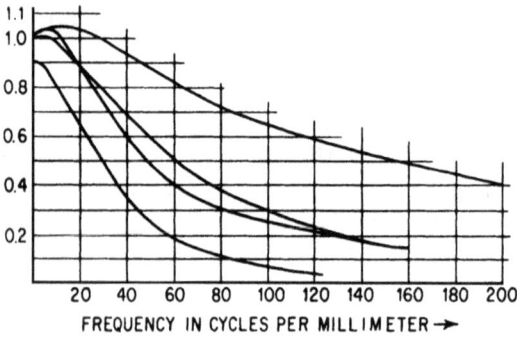

Figure 15.22 Modulation transfer functions of several photographic emulsions.

15.11   Radial Energy Distribution

The data of a point spread function or a spot diagram can be presented in the form of a *radial energy distribution plot.* If the blur spot is symmetrical, it is apparent that a small circular aperture centered in the image would pass a portion of the total energy and block the rest. A larger aperture would pass a greater portion of the energy and so on. A graph of the encircled fraction of the energy plotted against the radius (semidiameter) of the aperture is called the radial energy distribution curve.

A radial energy distribution curve, such as the example shown in Fig. 15.23, can be used to compute the MTF for an optical system by means of the summation equation

$$\text{MTF}(v) = \sum_{i=1}^{i=m} \Delta E_i J_0(2\pi v \overline{R}_i)$$

where v is the frequency in cycles per unit length, ΔE_i is the difference $(E_i - E_{i-1})$ between two values of E, the fractional energy, $\overline{R}_i$ is the average $\frac{1}{2}(R_i + R_{i-1})$ of the corresponding values of the radius and $J_0(\ )$ indicates the zero-order Bessel function.*

Although this radial energy distribution relationship is (strictly speaking) valid only for point images which have rotational symmetry, i.e., for images on the optical axis, it can be used to predict approximate averaged resolution for off-axis points. This procedure, while it cannot yield separate radial and tangential values for resolution, does serve to give the designer a rough idea of the state of correction of the system.

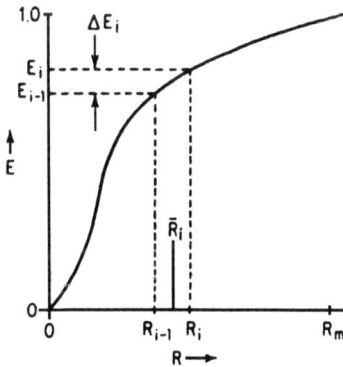

Figure 15.23   Radial energy distribution plot. The curve indicates the fraction E of the total energy in an image pattern which falls within a circle of radius R. Thus all the energy is encompassed by a circle of radius R_m; E_i of the energy by a circle of radius R_i.

$$*J_0(x) = 1 - \left(\frac{x}{2}\right)^2 + \frac{\left(\frac{x}{2}\right)^4}{1^2 2^2} - \frac{\left(\frac{x}{2}\right)^6}{1^2 2^2 3^2} + \cdots$$

15.12   Point Spread Functions for the Primary Aberrations

The figures of this section illustrate the effects of the primary aberrations on the point spread function (PSF) of an optical system. Figures 15.24 through 15.29 each show four point spread functions, the first for a peak-to-valley OPD (wave-front deformation) of an eighth-wave, the second for a quarter-wave (which is the Rayleigh criterion), the third for a half-wave, and the fourth for a full wavelength of OPD. The caption for each figure also gives the RMS (root mean square) OPD and the Strehl ratio for each PSF (see Sec. 15.4 and Fig. 15.5).

Figure 15.24 shows the effect of simple defocusing on the PSF. Note that for defocusing, the Rayleigh criterion (which is the OPD equal to a quarter-wave) is identical to the Marechal criterion (Strehl ratio equal to 0.80). In Fig. 15.25, which shows the effect of simple third-order

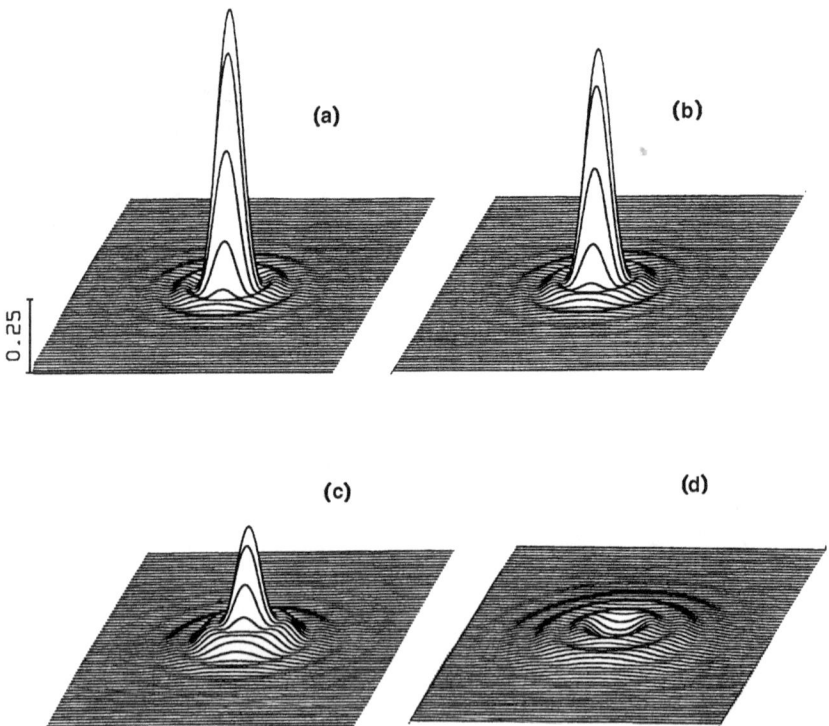

Figure 15.24   Point spread functions for different amounts of defocus. (a) 0.125 wave (P-V); 0.037 wave RMS; 0.95 Strehl. (b) 0.25 wave (P-V); 0.074 wave RMS; 0.80 Strehl. (c) 0.50 wave (P-V); 0.148 wave RMS; 0.39 Strehl. (d) 1.00 wave (P-V); 0.297 wave RMS; 0.00 Strehl.

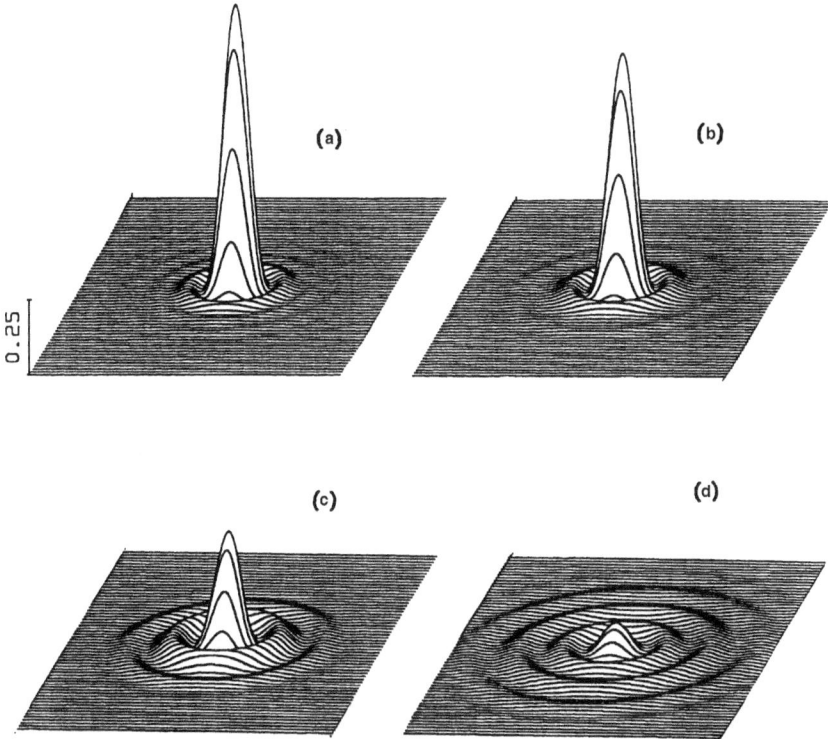

Figure 15.25 Point spread functions for different amounts of third-order spherical aberration. (a) 0.125 wave (P-V); 0.040 wave RMS; 0.94 Strehl. (b) 0.25 wave (P-V); 0.080 wave RMS; 0.78 Strehl. (c) 0.50 wave (P-V); 0.159 wave RMS; 0.37 Strehl. (d) 1.00 wave (P-V); 0.318 wave RMS; 0.08 Strehl. *Note:* Reference sphere centered at $0.5LA_m$ (midway between marginal and paraxial foci).

spherical aberration, the PSF for an eighth-wave is almost identical to that for defocusing, and the quarter-wave PSF is very similar. But when we compare the half- and full-wave plots, the diffe-rences are quite apparent, despite the fact that the effects on the MTF and resolution are still comparable.

The coma PSF in Fig. 15.26, however, is noticeably different even at an OPD of an eighth-wave, where the unsymmetrical rings in the diffraction pattern are already apparent. At one wave of OPD the PSF is clearly showing the "comma-" or "comet-shaped" figure that one gets from a geometric optics spot diagram (see Fig. 15.6 for example).

Figure 15.27 may be a bit surprising to some readers. Most discussions of astigmatism (including that in Sec. 5.2 of this text) which are based on geometric optics indicate that the blur spot between the sagittal and tangential focal lines is an ellipse or a circle, depending

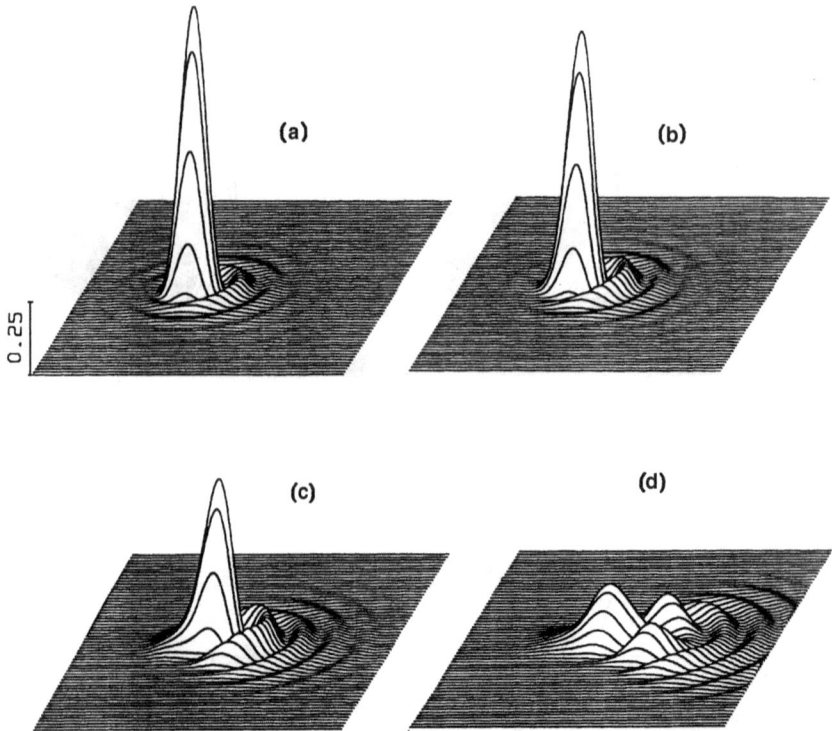

Figure 15.26 Point spread functions for different amounts of third-order coma. (a) 0.125 wave (P-V); 0.031 wave RMS; 0.96 Strehl. (b) 0.25 wave (P-V); 0.061 wave RMS; 0.86 Strehl. (c) 0.50 wave (P-V); 0.123 wave RMS; 0.65 Strehl. (d) 1.00 wave (P-V); 0.25 wave RMS; 0.18 Strehl. *Note:* P-V OPD reference sphere centered at 0.25Coma_T from chief ray intersection point. RMS OPD reference sphere centered at 0.226Coma_T from chief ray intersection point.

on where the image is examined. However, in the PSF for either half- or full-wave OPD we can easily see that the blur spot is not circular but has a definite four-sided aspect. Anyone who has microscopically examined the image of a point source formed by a lens with astigmatism as its major aberration will have observed this (and probably has wondered where the square-shaped image blur came from). It may help to understand this phenomenon to realize that the two focal lines are effectively acting as apertures, and the diffraction effect of this is to introduce the cross-shaped illumination distribution.

The most customary balance between third-order and fifth-order spherical aberration is with the aberration of the marginal ray corrected to zero. This state of correction produces the least peak-to-valley OPD (as demonstrated in Sec. 15.3). In Fig. 15.28, the eighth-wave

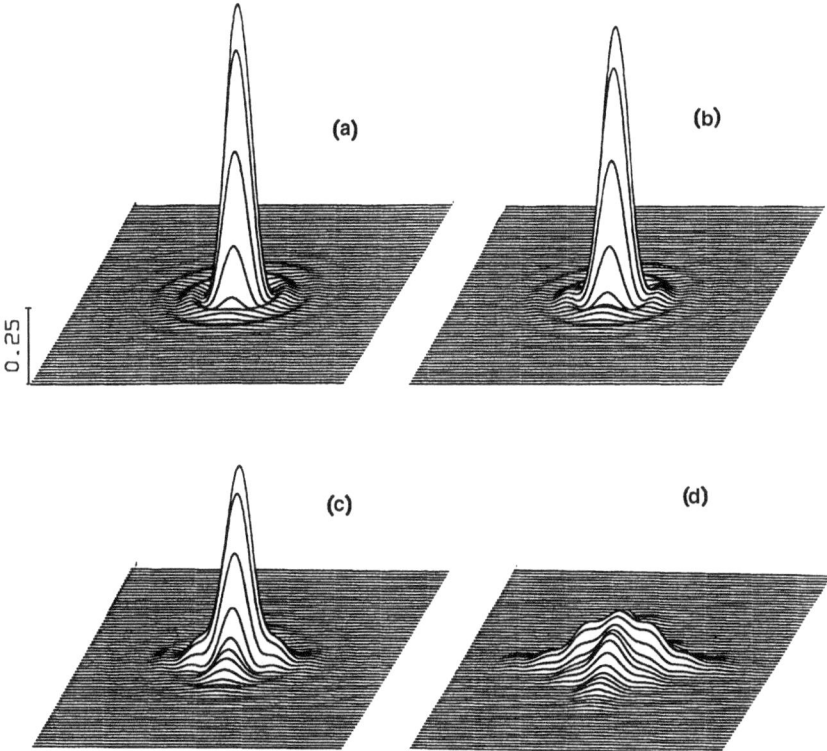

Figure 15.27 Point spread functions for different amounts of astigmatism. (a) 0.125 wave (P-V); 0.026 wave RMS; 0.97 Strehl. (b) 0.25 wave (P-V); 0.052 wave RMS; 0.90 Strehl. (c) 0.50 wave (P-V); 0.104 wave RMS; 0.65 Strehl. (d) 1.00 wave (P-V); 0.207 wave RMS; 0.18 Strehl. *Note:* Reference sphere centered midway between sagittal and tangential foci.

PSF is not very different from that for pure third-order spherical, or even defocus, but at a quarter-wave one begins to notice that the rings are more pronounced than they are in Figs. 15.24 and 15.25. This effect is quite noticeable in a star test, where heavy rings in the diffraction pattern are an indication of a zonal spherical residual.

The final figure in this series, Fig. 15.29, compares the PSF for the various aberrations, each of which is set at a value which equals the Marechal criterion (a Strehl ratio of 80 percent). There are differences apparent if one looks very closely at the defocus and spherical plots, and there are obvious differences for astigmatism and coma. However, the net effect on the image quality is surprisingly similar. This is, of course, the reason that the Rayleigh criterion of a quarter-wave (peak-to-valley) OPD and the Marechal criterion of 0.80 Strehl are so widely accepted by lens designers as a standard of image quality.

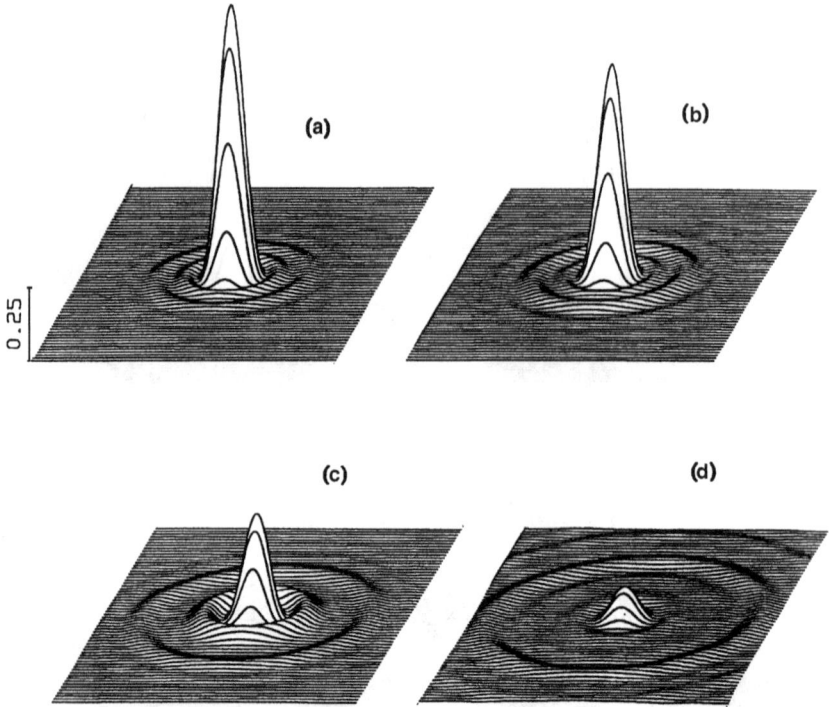

Figure 15.28 Point spread functions for different amounts of zonal spherical aberration (third- and fifth-order spherical balanced so that marginal spherical equals zero). (a) 0.125 wave (P-V); 0.042 wave RMS; 0.93 Strehl. (b) 0.25 wave (P-V); 0.085 wave RMS; 0.75 Strehl. (c) 0.50 wave (P-V); 0.208 wave RMS; 0.35 Strehl. (d) 1.00 wave (P-V); 0.403 wave RMS; 0.09 Strehl. *Note:* Reference sphere centered at $0.75LA_z$ for P-V and at $0.8LA_z$ for RMS.

Note: These figures were prepared by applying an optical software program to systems which were set up to show only the particular aberration under consideration. A paraboloidal reflector was the obvious choice for the defocusing PSF because its axial image is totally aberration-free. The spherical aberration plots were created by deforming the paraboloid with a fourth-order deformation term for the third-order spherical plot and fourth- and sixth-order deformations for the third- and fifth-order spherical plot. The coma PSF was calculated using a paraboloidal reflector with its aperture stop at the focal plane (which eliminates astigmatism), as in Fig. 18.2a. The image was put on a curved surface which approximated a sphere of radius equal to the focal length of the reflector and which was centered at the center of curvature of the paraboloid. The astigmatism PSF was produced by introducing an additional cylindrical parabolic reflector.

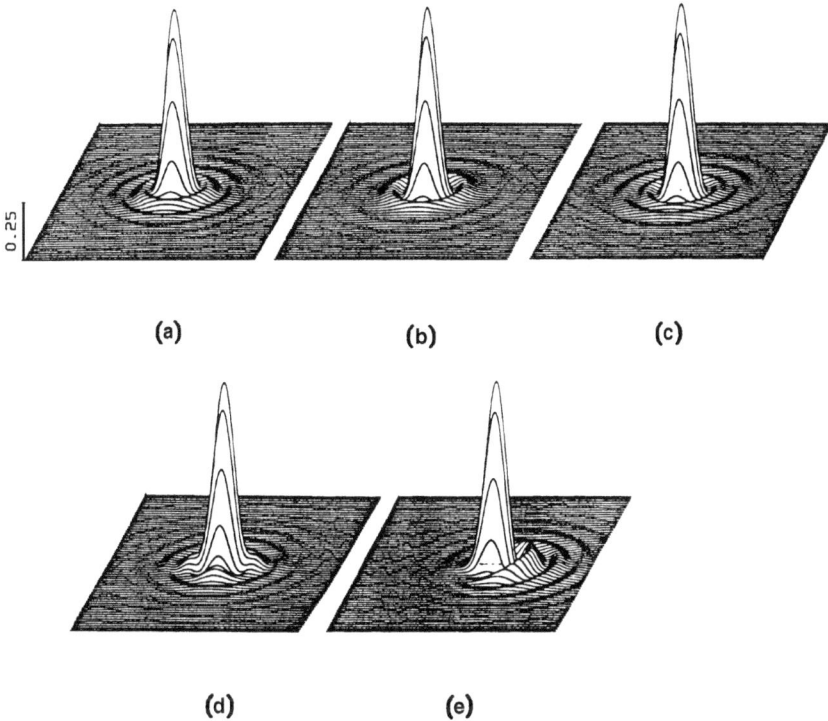

0.25

(a)   (b)   (c)

(d)   (e)

Figure 15.29 Point spread functions for five different aberrations, each with a Strehl ratio of 0.80 (the Marechal criterion). In each case the center of the reference sphere is located to minimize the RMS OPD, which is 0.075 wave for all five aberrations. (a) Defocus: 0.25 wave (P-V). (b) Third-order spherical: 0.235 wave (P-V). (c) Balanced third- and fifth-order spherical: 0.221 wave (P-V). (d) Astigmatism: 0.359 wave (P-V). (e) Coma: 0.305 wave (P-V).

Bibliography

Altman, J. H., "Photographic Films," in *Handbook of Optics,* Vol. 1, New York, McGraw-Hill, 1995, Chap. 20.

Boreman, G. D., "Transfer Function Techniques," in *Handbook of Optics,* Vol. 2, New York, McGraw-Hill, 1995, Chap. 32.

Born, M., and E. Wolf, *Principles of Optics,* New York, Pergamon Press, 1999.

Conrady, A., *Applied Optics and Optical Design,* Oxford, 1929. (This and Vol. 2 were also published by Dover, New York.)

Gaskill, J., *Linear Systems, Fourier Transforms, and Optics,* New York, Wiley, 1978.

Goodman, J., *Introduction to Fourier Optics,* New York, McGraw-Hill, 1968.

Herzberger, M., *Modern Geometrical Optics,* New York, Interscience, 1958.

Hopkins, H., *Wave Theory of Optics,* Oxford, 1950.

Levi, L., and R. Austing, *Applied Optics,* Vol. 7, Washington, Optical Society of America, 1968, pp. 967–974 (defocused MTF).

Linfoot, E., *Fourier Methods in Optical Design,* New York, Focal, 1964.

Marathay, A. S., "Diffraction," in *Handbook of Optics,* Vol. 1, New York, McGraw-Hill, 1995, Chap. 3.

O'Neill, E., *Introduction to Statistical Optics,* Reading, Mass., Addison-Wesley, 1963.

Perrin, F., *J. Soc. Motion Picture and Television Engrs.*, Vol. 69, March–April 1960 (MTF, with extensive bibliography).

Selwyn, E., in Kingslake (ed.), *Applied Optics and Optical Engineering*, Vol. 2, New York, Academic, 1965 (lens-film combination).

Smith, W., in W. Driscoll (ed.), *Handbook of Optics*, New York, McGraw-Hill, 1978.

Smith, W., in Wolfe and Zissis (eds.), *The Infrared Handbook*, Washington, Office of Naval Research, 1985.

Suits, G., in Wolfe and Zissis (eds.), *The Infrared Handbook*, Washington, Office of Naval Research, 1985 (film).

Wetherell, W., in Shannon and Wyant (eds.), *Applied Optics and Optical Engineering*, Vol. 8, New York, Academic, 1980 (calculation of image quality).

Exercises

1   To a third order approximation, the longitudinal spherical aberration of a spherical reflector is equal to $y^2/8f$. What is the maximum diameter that a 36-in focal length spherical reflector may have without exceeding an OPD of one-quarter wavelength for visual light? (Use $\lambda = 20 \times 10^{-6}$ in, and Eq. 15.16)

ANSWER: Eq. 15.16a: $LA_m = 4\lambda/n \sin^2 U_m$ is the amount of spherical for a quarter-wave of OPD. Substituting y/f for sin U and clearing, we get $LA_m = 4\lambda f^2/ny^2$ and set it equal to $y^2/8f$, and solve for y, getting

$$y = 32\ \lambda f^3/n = 32 \times 2 \times 10^{-5} \times 36^3/1.0 = 29.85984$$

$$y = 2.337609 \text{ in}$$

and the diameter = 4.675218 in for a speed of $f/7.7$.

2   The third-order sagittal coma of a paraboloid reflector is given by $-y^2\theta/4f$, where θ is the half-field angle in radians. What field will a 5-in diameter $f/8$ reflector cover without exceeding the Rayleigh limit? (Use $\lambda = 0.000020$ in)

ANSWER: Eq. 15.18a $Coma_T = 1.5\lambda/NA$ is the quarter-wave OPD amount of coma. Substituting y/f for NA and equating it to three times $-y^2\theta/4f$ (to get the tangential coma = $3\times$ sagittal coma), we get

$$1.5\lambda f/y = 3(-y^2\theta/4f), \text{ and solving for } \theta$$
$$\theta = 2\lambda f^2/y^3$$

The focal length is $5 \times 8 = 40$ in, the semi = aperture $y = 2.5$ in and the wavelength is 0.00002. So we have $\theta = 2 \times 0.00002 \times 40^2/2.5^3 = 0.004096$ radians, and the full field is 0.008192 radians, or 0.4694°.

3   An $f/5$ system is defocused by 0.05 mm. What is the modulation transfer factor for a "sine wave" target with a spatial frequency (at the image) of 120 cycles per millimeter? Use Fig. 15.14, Eq. 15.2, and assume $\lambda = 0.5\ \mu$m.

ANSWER: Eq. 15.2a: $\lambda/4$ Depth of Focus $\delta = 2\lambda(f/\#)^2 = 2 \times 0.0005 \times 5^2 = 0.025$ mm.

Thus a defocus of 0.5 mm corresponds to an OPD of $\lambda/2$, and curve C of Fig. 15.14 will apply.

The MTF cutoff frequency per Eq. 15.37 $v_0 = 1/\lambda(f/\#) = 1/0.0005 \times 5 = 400$ cycles per millimeter and 120 cycles per millimeter is $120/400 = 0.3\ v_0$. At this frequency, curve C of Fig. 15.14 indicates an MTF of about 0.23.

16

The Basics of Lens Design

16.1 Introduction

In the immediately preceding chapters, we have been concerned with the *analysis* of optical systems, in the sense that the constructional parameters of the system were given and our object was the determination of the resultant performance characteristics. In this chapter we take up the *synthesis* of optical systems; here the desired performance is given and the constructional parameters are to be determined. A large part of the synthesis process is, of course, concerned with analysis, since optical design is in great measure a systematic application of the cut-and-try process.

There is absolutely no "direct" method of optical design for original systems; that is, there is no sure procedure that will lead (without foreknowledge) from a set of performance specifications to a suitable design. However, when it is known that a certain type of design or configuration is capable of meeting a given performance level, it is a fairly straightforward process for a competent designer to produce a design of the required type. Further, modest improvements to existing designs can almost always be effected by well-established techniques. Thus, it is apparent that a good portion of the ammunition in a lens designer's arsenal consists of an intimate and detailed knowledge of a wide range of designs, their characteristics, limitations, idiosyncrasies, and potentials. Here is one part of the art in optical design; basically it consists of the choice of the point at which the designer begins.

The electronic computer, in the course of little more than a decade, radically modified the techniques used by optical designers. Previously a designer resorted to all manner of ingenious techniques to avoid tracing rays because of the great expenditure of time and effort involved. The personal computer (PC) has reduced raytracing time by about ten orders of magnitude, and it is now easier to trace rays through a system than it is to speculate, infer, or interpolate from incomplete data. A computer can even be made to carry through the entire design process from start to finish, more or less without human intervention. The results produced by such a process are nonetheless intricately dependent on the starting point elected (as well as the manner in which the computer has been programmed), so that a great deal of art (if perhaps somewhat less personal satisfaction) is still present in even the most automatic technique.

The ordinary design process can be considered to have four stages, as follows: first, the selection of the type of design to be executed, i.e., the number and types of elements and their general configuration. Second, the determination of the powers, materials, thicknesses, and spacings of the elements. These are usually selected to control the chromatic aberrations and the Petzval curvature of the system, as well as the focal length (or magnifying power), working distances, field of view, and aperture. (Choices made at this stage may affect the performance of the final system tremendously, and can mean the difference between success and failure in many cases.) In the third stage, the shapes of the elements or components are adjusted to correct the basic aberrations to the desired values. The fourth stage is the reduction of the residual aberrations to an acceptable level. If the choices exercised in the first three stages have been fortuitous, the fourth stage may be totally unnecessary. At the other extreme, the end result of the first three stages may be so hopeless that a fresh start from stage 1 is the only alternative.

In fully automatic computer design procedures, a portion of stage 1 and all of stages 2, 3, and 4 may be accomplished more or less simultaneously (using an approach that might take a human computer a lifetime or two to slog through). Computer design techniques are discussed in Sec. 16.8.

The basic principles of optical design will be illustrated by three detailed examples in the following sections. A simple meniscus (box) camera lens will be used to show the effects of bending and stop shift techniques, as well as the handling of a simplified exercise in satisfying more requirements than there are available degrees of freedom. An achromatic telescope objective will introduce material choice, achromatism, and multiple bending techniques. An air-spaced (Cooke) triplet anastigmat will illustrate the problem of controlling all the first- and third-order aberrations in a system with just a sufficient number of degrees of freedom to accomplish this and will further illustrate the technique of material selection. The design characteristics of several additional types of optical systems are discussed in Chaps. 17 and 18.

At this point it should be emphasized that the design procedures implied by the discussions in Secs. 16.2, 16.4 to 16.6, and to some extent 16.7, while perfectly valid, are presented here primarily as a way of explaining the principles, relationships, limitations, etc., involved in the design. These procedures are rarely used today; the computer, especially the desktop personal computer, or PC, has enough computing power so that every designer can have access to some sort of automatic lens design program. Nonetheless, a knowledge of these procedures and principles is of great utility to a designer, even when using an automatic design program. For example, such knowledge helps in selecting a good starting design for the computer and, among other things, often helps in figuring out what went wrong when the designer has asked the computer to do the optically impossible.

16.2   The Simple Meniscus Camera Lens

There are just two elements to work with in the design of a meniscus camera lens, the lens itself and the aperture stop. If, for the moment, we restrict ourselves to a thin, spherical-surfaced element, the parameters which we may choose or adjust are the material of the lens, its focal length, its shape (or bending), the position of the stop, and the diameter of the stop. With these degrees of freedom we must design a lens which will produce an acceptable image on a given size of film. This implies that all the aberrations of the system must be "sufficiently" small. It is immediately apparent that the spherical aberration will be undercorrected and that the Petzval curvature will be inward-curving (and equal to $-h^2\phi/2n$); these are the immutable characteristics of a simple lens. Thus, the element power, the size of the aperture, and the field of view must be chosen small enough so that the effects of these aberrations are tolerable. The lens material usually chosen is common crown glass or acrylic plastic, on the basis of cost, since a box camera lens must be inexpensive. A high-index crown does not produce enough improvement in the Petzval curvature to warrant its increased cost; a flint glass would introduce increased chromatic aberrations.

We find ourselves with just two uncommitted degrees of freedom, namely the bending of the lens and the position of the stop. Now in a simple undercorrected system it is axiomatic that for a given (i.e., fixed) shape of the lens (or lenses), the position of the stop (the "natural" stop position—see Sec. 5.4) for which the coma is zero is also the position for which the astigmatism is the most overcorrected (i.e., most backward-curving). Since the Petzval surface will be inward (toward the lens) curving, some overcorrected astigmatism is desirable.

Thus the design technique is straightforward: we choose (arbitrarily) a shape for the lens, determine the stop position at which coma is zero, and evaluate the aberrations. By repeating this process for several bendings and graphing the aberrations as a function of the shape, we can then choose the best design.

There are several ways in which this can be accomplished. Since this is a simple lens of moderate aperture and field, the third-order aberrations are quite representative of the system and one would be quite safe in relying on them. The design could also be handled by trigonometric raytracing. For this example we will work out the design using the thin-lens (G-sum) third-order aberration equations of Sec. 6.4 and then check the results by raytracing.

Assuming that the glass has an index of 1.50 and a V-value of 62.5, we will set up the G-sum equations for a focal length of 10, an aperture diameter of 1.0, and an image height of 3 (all in arbitrary units and all subject to scaling and adjustment later). Thus, the element power $\phi = \frac{1}{10} = 0.1$, and the total curvature $c = c_1 - c_2 = \phi/(n - 1) = 0.2$. With the object at infinity, $u_1 = 0$. Using the G-values of Eq. 6.3u, we find that the spherical and coma (stop at the lens) given by Eqs. 6.3m and 6.3n are

$$\text{TSC} = -0.145833C_1^2 + 0.05C_1 - 0.005625$$

$$\text{CC} = -0.0625C_1 + 0.01125$$

Now the position of the stop can be determined by solving Eq. 6.3g for Q when CC* is zero.

$$\text{CC*} = 0 = \text{CC} + Q \cdot \text{TSC}$$

$$Q = \frac{-\text{CC}}{\text{TSC}}$$

Equations 6.3o, p, and r give us

$$\text{TAC} = -0.0225$$

$$\text{TPC} = -0.015$$

$$\text{TAchC} = -0.008$$

and by substituting the above into Eqs. 6.3h, j, and l, we get the following expressions for the third-order astigmatism, distortion, and lateral color with the stop as defined by Q above.

$$\text{TAC*} = -0.0225 + 2Q \cdot \text{CC} + Q^2 \cdot \text{TSC}$$

$$\text{DC*} = -0.0825Q + 3Q^2\text{CC} + Q^3\text{TSC}$$

$$\text{TchC*} = -0.008Q$$

Having established the above relationships, we now select several values for C_1 and evaluate the third-order aberrations for each. The results are indicated in the tabulation of Fig. 16.1 and the graph of Fig. 16.2. Note that $X_s = PC* + AC*$ and $X_t = PC* + 3AC*$. [Here we revert to the older symbol (X) for field curvature rather than the currently popular Z.]

A study of Fig. 16.2 can be quite rewarding. First, we note that there are two regions which appear most promising, namely the meniscus shapes at either side of the graph. On the left, the lens is concave to the incident light and (since Q is positive) the stop is in front of the lens. To the right the lens is convex to the incident light and the stop is behind the lens. Both forms have more undercorrected spherical aberration than the less strongly bent shapes, but both have their field curvature "artificially" flattened by overcorrected astigmatism. Note that the form with the least spherical aberration (where CC = 0 and the stop is in contact with the lens) has the most strongly inward curving field. *This inward-curving field is characteristic of any thin optical system with the stop in contact,* since by Eqs. 6.3p and 6.3h

$$\text{Stop in contact } X_T = PC* + 3AC* = \frac{-h^2\phi\,(3n + 1)}{2n}$$

Selecting the bending $C_1 = -0.2$ for further investigation, we note that $Q = +1.11$ (from Fig. 16.1). Since $Q = y_p/y$ and $y = 0.5$, we find $y_p = 0.555$. The slope of the principal ray in object space which will

C_1	-0.4	-0.2	0.0	$+0.2$	$+0.4$	$+0.6$	$+0.8$
ΣSC	-0.98	-0.43	-0.11	-0.03	-0.18	-0.56	-1.18
ΣCC	$+0.036$	$+0.024$	$+0.011$	-0.001	-0.014	-0.026	-0.039
Q	$+0.74$	$+1.11$	$+2.00$	-0.86	-1.53	-0.93	-0.66
l_p	-1.23	-1.84	-3.33	$+1.43$	$+2.55$	$+1.56$	$+1.26$
$\Sigma AC*$	$+0.087$	$+0.077$	0.00	-0.429	-0.028	$+0.040$	$+0.059$
X_s	-0.21	-0.22	-0.30	-0.73	-0.33	-0.26	-0.24
X_t	-0.04	-0.07	-0.30	-1.59	-0.38	-0.18	-0.12
ΣDC	-0.02	-0.03	-0.08	$+0.07$	$+0.06$	$+0.03$	$+0.02$
% Dist.	-0.7%	-1.1%	-2.5%	$+2.3\%$	$+2.1\%$	$+1.0\%$	$+0.7\%$
$\Sigma TchC$	-0.006	-0.009	-0.016	$+0.007$	$+0.012$	$+0.007$	$+0.005$

Figure 16.1   Tabulation of the third-order aberrations of a thin lens with the stop at the coma-free position, for various values of C_1.

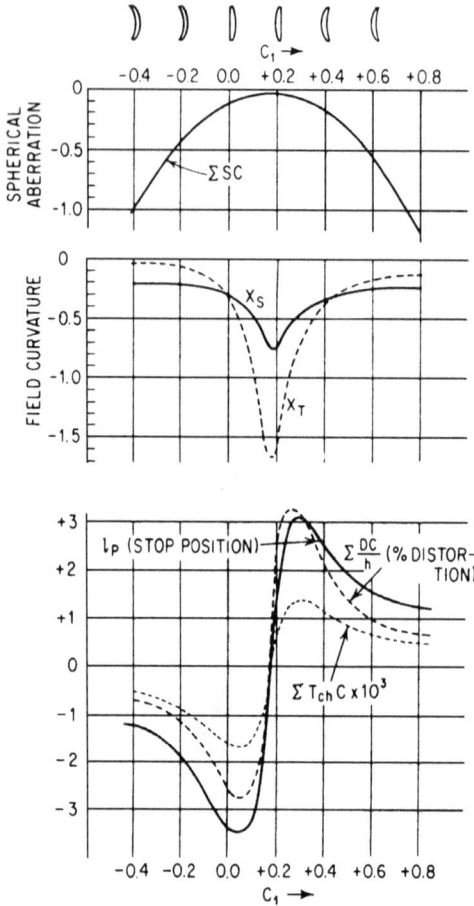

Figure 16.2 The third-order aberrations of a thin lens ($f = 10$, $y = 0.5$, $h = 3$, $n = 1.5$) with the stop at the coma-free position, plotted as a function of the curvature of the first surface (C_1).

yield an image height $h = +3$ with a focal length of $+10$ is $u_p = +0.3$. The stop position is thus

$$l_p = \frac{-y_p}{u_p} = \frac{-0.555}{+0.3} = -1.85$$

or 1.85 units to the left of the lens.

We must of course convert our thin lens to a real lens. A ray with a slope of $+0.3$ through the upper edge of the stop (diameter $= 1.0$) will strike the lens at a height of 1.05, and we shall assume a diameter of twice this for the lens. We determine the curvature of the second surface from $C_2 = C_1 - C = -0.2 - 0.2 = -0.4$, and compute the sagittal heights of the surfaces for the diameter of 2.10. Thus for our lens to have an edge thickness of 0.1, it must have a center thickness of $CT = ET + SH_1 - SH_2 = 0.1 - 0.11 + 0.23 = 0.22$. We now trace an oblique fan of four equally spaced meridional rays through the system and calculate two values of coma (by Eq. A.5d), one from the upper three rays and one from the lower three. By linear interpolation between the two overlapping three ray bundles, we find that a bundle with a chief ray axial intercept of $L_{pr} = -1.664$ will have zero coma. This is the stop position for the *thick* lens (vs. $l_{pr} = -1.85$ for the *thin* lens.)

The results of a raytrace analysis are shown in Fig. 16.3. The field curvature and spherical aberration forecast by the thin-lens third-order

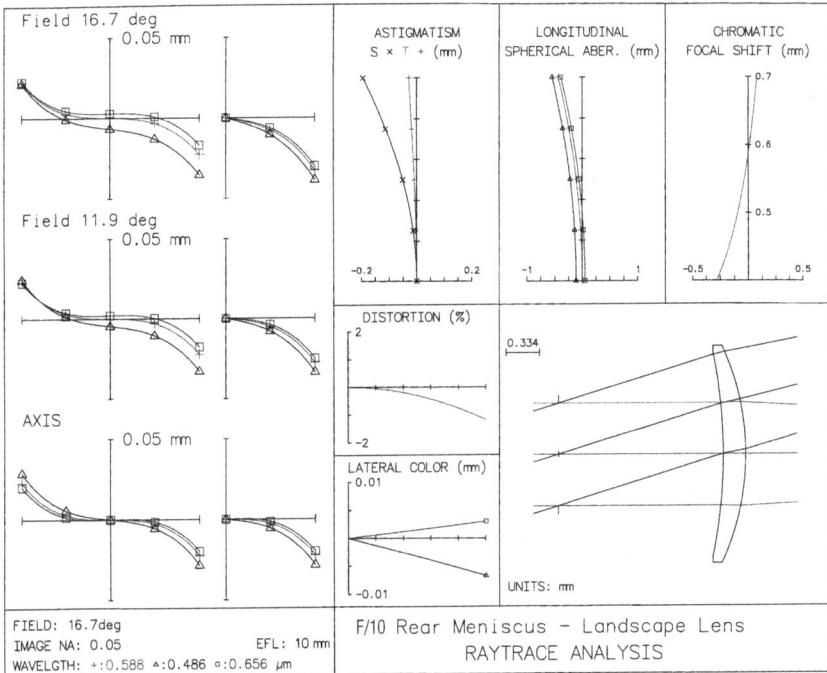

Figure 16.3 The final rear meniscus *Landscape Lens* design. Construction data and aberration plots for a focal length of 10.0 and an aperture of 1.0. Note that the exact trigonometricly calculated aberrations shown here are very close to the values given by the thin-lens third-order (G-sum) calculations in the text.

computations can be compared here, and the agreement with the actual raytrace is quite good. Note that complete TOA plots could be derived from our knowledge of the manner in which the TOA vary with aperture and image height (see the tabulation of Fig. 5.16). For example, knowing that (longitudinal) third-order spherical varies as Y^2 and that SC = -0.429 for $Y = 0.5$, we could determine that SC = -0.107 for $Y = 0.25$ and plot it accordingly.

To complete the design we would next scale the entire system to the actual focal length desired. (Note that all the linear dimensions of any system, including the aberrations, may be multiplied by the same constant to effect a change in scale. No additional computation is necessary.) Next an appropriate size for the aperture would be selected, i.e., one which would reduce the aberration blurs to sizes commensurate with the intended application.

The lens form that we have elected to design in this example has the aperture stop in front, i.e., to the left of the lens. This is often referred to as the *rear-meniscus* form. From Fig. 16.2 it is apparent that there is a similar *front-meniscus* form with the stop behind (to the right of) the lens. *Question:* Which is the better design? On the basis of aberration correction, the rear meniscus is slightly better. However, there are several points on which the front meniscus is superior. In a camera, the length of the camera will be approximately equal to the lens focal length for the front meniscus, whereas for the rear meniscus we must add the distance to the stop, resulting in a significantly longer camera. Further, in an inexpensive camera, the shutter is usually a simple spring-driven blade located at the aperture stop. Thus, for the rear meniscus, the shutter mechanism is exposed to the environment; in the front meniscus, the lens acts as a protective window. Finally, and perhaps most important, in the front meniscus, the lens is out in front and quite visible to the customer, whereas in the rear meniscus, all the customer ever sees is the less appealing stop and shutter mechanism. These latter "commercial" reasons are why the front-meniscus form has been universally used for inexpensive cameras since the 1940s. Apparently there is more to optical engineering than aberration correction.

At the start of this section we assumed that the lens would be thin and its surfaces spherical. If we increase the thickness of a meniscus lens and maintain its focal length at a constant value by adjusting one of the radii, it is apparent from the thick-lens focal-length equation (Eq. 3.21) that we must either reduce the power of the convex surface or increase the power of the concave surface to maintain the focal length as the thickness is increased. Either change will have the effect of reducing the inward Petzval curvature of field. This principle (i.e., separation of positive and negative surfaces, elements, or components in order to reduce the Petzval sum) is a powerful one and is the basis of all anastigmat designs.

The value of aspheric surfaces is limited in a design as simple as the box camera lens. However, if the lens is molded from plastic, an aspheric surface is as easy to produce as a spherical one; many simple cameras now have aspheric plastic objectives. The aspheric surface affords the designer additional freedom to modify the system to advantage. A diffractive surface could be used to achromatize the lens (and affect the other aberrations as well).

16.3   The Symmetrical Principle

In an optical system which is *completely* symmetrical, coma, distortion, and lateral color are identically zero. To have complete symmetry a system must operate at unit magnification and the elements behind the stop must be mirror images of those ahead of the stop. This is a principle of great utility, not only for systems working at unit power, but even for systems working at infinite conjugates. This is due to the fact that, although coma, distortion, and lateral color are not completely eliminated under these conditions, they tend to be drastically reduced when the elements of any system are made symmetrical, or even approximately so. For this reason many lenses which cover an appreciable field with low distortion and low coma tend to be generally symmetrical in construction.

If we were to apply this principle to the meniscus camera lens, we would simply use two identical menisci equidistant on either side of the stop. The resulting lens would be practically free of coma, distortion, and lateral color. Symmetry, plus the thick meniscus principle (to flatten the field) achieves a very remarkable astigmatic field coverage of ±67° for the Hypergon lens, which is also shown in Fig. 16.4. This is accomplished at the expense of a heavily undercorrected spherical aberration which limits its useful speed to about $f/30$ or $f/20$. Note that, at a speed of $f/30$, a quarter-wave of longitudinal spherical is ±7.9 mm, and the quarter-wave depth of focus is ±1.0 mm.

16.4   Achromatic Telescope Objectives
(Thin-Lens Theory)

An achromatic doublet is composed of two elements, a positive crown glass element and a negative flint glass element. (Stated more generally, an achromatic doublet consists of a low-relative-dispersion

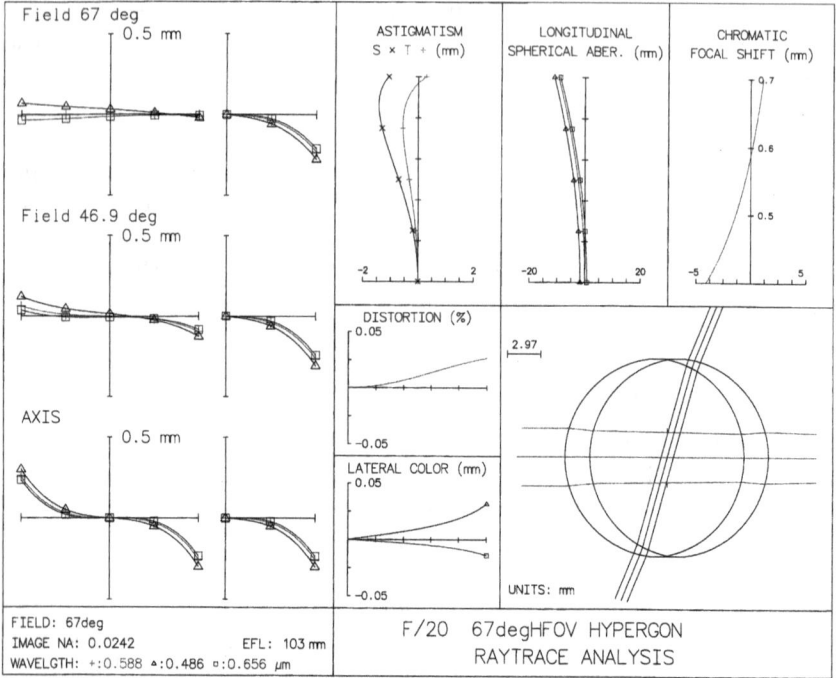

| Field 67 deg | | | |
| 0.5 mm | ASTIGMATISM S × T + (mm) | LONGITUDINAL SPHERICAL ABER. (mm) | CHROMATIC FOCAL SHIFT (mm) |

FIELD: 67deg
IMAGE NA: 0.0242   EFL: 103 mm
WAVELGTH: +:0.588 ▲:0.486 □:0.656 µm

F/20   67degHFOV HYPERGON
RAYTRACE ANALYSIS

An extreme form of symmetrical meniscus construction is the *Hypergon* shown here; it covers a field of about 135° at a speed usually slower than *f*/20. The inner and outer radii of this Hypergon differ by only 0.7 percent; as a result the Petzval field curvature is near zero. The steep, thin, shells make this a difficult lens to fabricate. Note that the obliquity of the beam at the stop severely reduces the beam width and the image illumination at the edge of the field.

F/20 Hypergon Lens, USP#706,650–1902

radius	space	mat'l		s-a
8.57	2.20	BK1	510635	8.52
8.63	6.90			8.52
STOP	6.90			2.18
−8.63	2.20	BK1	510635	8.52
−8.57	92.92			8.52

EFL   = 103.15
BFL   = 92.92
NA   = 0.0242 (*f*/20.6)
GIH   = 243.4 (HFOV = 67°)
VL   = 18.2
PTZ/F = −17.69

Figure 16.4 Symmetrical (simple) meniscus lenses. The *Periscopic* lens is a symmetrical pair of meniscus elements each similar to the lens shown in Fig. 16.3. The Periscopic is occasionally used in inexpensive cameras. Its symmetrical construction eliminates the distortion and lateral color which degrade the performance of the single meniscus *Landscape Lens.*

element of the same sign power as the doublet and a high-relative-dispersion element of opposite sign.) As degrees of freedom we have the choice of glass types for the elements, the powers of the two elements, and the shapes of the two elements.

We assume here that we are designing a telescope objective, that the stop or pupil will be located at the lens, and that the lens will be thin. The astigmatism of a thin lens in contact with the stop is fixed, regardless of the number of elements, their index, or their shapes. Equation 6.3o indicates TAC = $(h^2\phi u'_k)/2$ for a single element. Since the power of a doublet is simply the sum of the powers of the elements, this equation applies to a doublet as well as a singlet. Thus we cannot affect the astigmatism (and can do very little about the Petzval curvature). The field will be strongly inward-curving.

With reference to Fig. 16.5, it is apparent that we have only four variable parameters with which to correct the aberrations. Actually, one parameter must always be assigned to control the focal length in any lens design. Thus we have three variables left; we will use them to correct spherical aberration, coma, and axial chromatic aberration.

Since the lens is to be free of chromatic aberration, we must assign the element powers to the determination of focal length and the control of chromatic aberration. Again we begin by using the thin-lens third-order aberration equations; assigning the subscripts a and b to the two elements, Eq. 6.3r gives us

$$\sum\text{TAchC} = \text{TAchC}_a + \text{TAchC}_b = \frac{Y_a^2\phi_a}{V_a u'_k} + \frac{Y_b^2\phi_b}{V_b u'_k}$$

Since the elements are to be cemented together or very nearly in contact, we can substitute $y_a = y_b = y$ and $u'_k = -y/f$ to get

$$\sum\text{TAchC} = -fy\left[\frac{\phi_a}{V_a} + \frac{\phi_b}{V_b}\right] \tag{16.1}$$

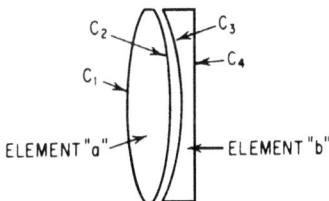

Figure 16.5 Achromatic doublet.

We now set ΣTAchC $= 0$ (or some other value, if desired) and make a simultaneous solution of Eq. 16.1 with

$$\frac{1}{f} = \phi_a + \phi_b \tag{16.2}$$

to get the necessary powers for the elements. For zero chromatic, we get

$$\phi_a = \frac{V_a}{f(V_a - V_b)} \tag{16.3}$$

$$\phi_b = \frac{V_b}{f(V_b - V_a)} = \frac{-\phi_a V_b}{V_a} \tag{16.4}$$

Having determined ϕ_a and ϕ_b, we can now write thin-lens equations for the third-order spherical and coma in terms of the shapes of the elements [after tracing a marginal (thin-lens) paraxial ray to determine the values for u'_k of the combination and v (or v') for each element]. Since the aperture stop will be at the lens, $Q = 0.0$ and the coma will be given by Eq. 6.3n. After the appropriate substitutions for h, y, $C_a = \phi_a/(n_a - 1)$, $C_b = \phi_b/(n_b - 1)$, and the G-factors, we arrive at an equation of the following general form for coma:

$$\Sigma CC = CC_a + CC_b = K_1 C_1 + K_2 + K_3 C_3 + K_4$$
$$= K_1 C_1 + K_3 C_3 + (K_2 + K_4) \tag{16.5}$$

where C_1 and C_3 are the curvatures of the first surfaces of the elements (Fig. 16.5), and K_1 through K_4 are constants. (Note that by using the alternate form of Eq. 6.3n for element a, the equation could be written in C_2 and C_3, the curvatures of the adjacent inner surfaces). Now for any desired value of ΣCC, we find that

$$C_3 = \frac{\Sigma CC - K_1 C_1 - K_2 - K_4}{K_3}$$

or, combining constants

$$C_3 = K_5 C_1 + K_6 \tag{16.6}$$

Thus for any shape of element a, Eq. 16.6 indicates the unique shape for element b which will give the desired amount of coma.

In similar fashion we can write an expression for the thin-lens third-order spherical (using Eq. 6.3m) in the following form:

$$\Sigma TSC = TSC_a + TSC_b$$
$$= K_7 C_1^2 + K_8 C_1 + K_9 + K_{10} C_3^2 + K_{11} C_3 + K_{12} \tag{16.7}$$

By substituting the value for C_3 from Eq. 16.6 into 16.7, and combining constants, we get a simple quadratic equation in C_1 of the form

$$0 = C_1^2 + K_{13}C_1 + K_{14} \qquad (16.8)$$

which can be solved for the value of C_1. When used with the value of C_3 given by Eq. 16.6, this will yield a doublet with spherical and coma of the desired amounts. (Note that because Eq. 16.8 is a quadratic, there may be one, two, or no solutions.)

For a first try, one would use the above procedure with $\sum$TAchC, $\sum$TSC, and $\sum$CC equal to zero (or whatever values are desired). Next, appropriate thicknesses are inserted, and the system tested by raytracing to determine the actual values of spherical, coma (or OSC), and axial color. If these are not within tolerable limits, the thin-lens solution can be repeated using (for the desired $\sum$TAchC, $\sum$TSC, and $\sum$CC) the negatives of the corresponding values determined by raytracing. This process converges to a solution very rapidly.

While the above procedure is useful in understanding the nature of the doublet telescope objective, a designer with an optical software computer program could handle this project very easily and quickly. The four surface curvatures would be declared as variables, and the merit function would consist of targets for the actual ray-traced values of marginal spherical aberration, coma, and chromatic aberration plus the effective focal length. Given a reasonable starting lens form, the task is trivial, and the nearest solution to the starting form is found immediately.

16.5 Achromatic Telescope Objectives (Design Forms)

Depending on the choice of glass, the relative aperture, the desired values of the aberrations, and also on which solution to the quadratic was selected, the procedure outlined in Sec. 16.4 will result in an objective with one of the forms sketched in Fig. 16.6. In general the edge contact form and, for lenses of modest (up to 3- or 4-in) diameter, the cemented form is preferred, primarily because the relationship between the elements (as regards mutual concentricity about the axis and freedom from tilt) can be more accurately maintained in fabrication. The crown-in-front forms are more commonly used because the front element is more frequently exposed to the rigors of weather; crown glasses are in general more resistant to weathering than flint glasses.

The Fraunhofer and Steinheil forms represent one root of the quadratic of Eq. 16.8, and the Gauss form is the other root. Whether one

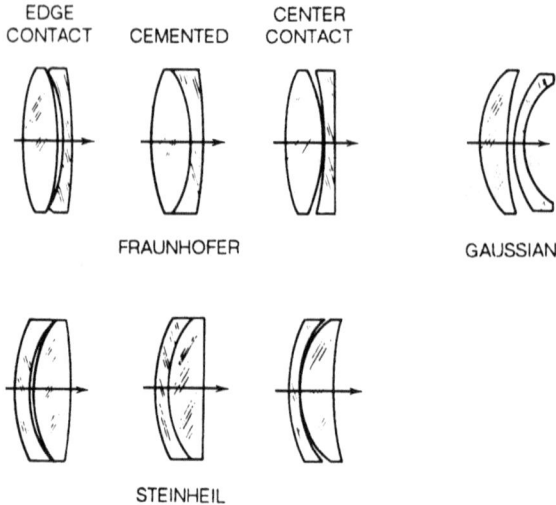

Figure 16.6 Various forms of achromatic doublets. The upper row are crown-in-front doublets and the lower row are flint-in-front. The curvatures are exaggerated for clarity. The center contact form is usually avoided because it is more difficult to manufacture. The shapes indicated are for lenses corrected for a distant object to the left.

gets the Fraunhofer or the Steinheil form simply depends on whether the left-hand element is of crown or flint glass. From an image-quality standpoint, there is little difference between them although the Steinheil radii are significantly stronger than the Fraunhofer form. However the Gauss objective is very different. The Gauss lens has about an order-of-magnitude more zonal spherical aberration residual and slightly (about 20 percent) more secondary spectrum than the Fraunhofer. However, it has only about half the spherochromatism. Another difference is that there is no solution for the Gauss form if the lens elements are too thick; thus the speed is limited to about $f/5$ or $f/7$ to avoid thick elements. The Fraunhofer and Steinheil forms can be corrected at speeds faster than $f/3$ (although the residual aberrations are of course quite large at high speeds).

If one followed the procedure of Sec. 16.4, a design resulting in a cemented doublet (i.e., $C_2 = C_3$) would be a lucky accident. When a cemented interface is necessary, an alternate procedure is followed. The spherical and coma contribution equations are written in C_2 and C_3 (instead of C_1 and C_3) and C_2 is set equal to C_3, resulting in equations in C_2 (or C_3) which may then be solved for either the desired coma or spherical. If these values are plotted as a function of the shape of the doublet (i.e., versus C_1 or C_2 or C_4) the resulting graph will look like one of those in Fig. 16.7, in which $\sum TSC$ is a parabola and $\sum CC$ is

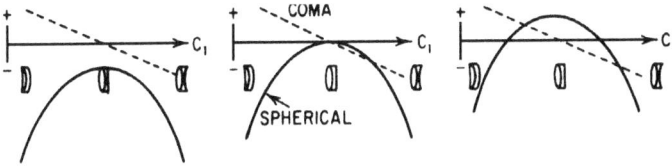

Figure 16.7 The variation of spherical aberration (solid line) and coma (dashed) as a function of the shape of a cemented achromatic doublet. Depending on the materials used there may be two forms with zero spherical (right), one form (center), or no form (left). The center graph is the preferred type since spherical and coma are both corrected. The left hand plot usually results from a big difference in V-value, and the right hand plot results when the V-value difference is small.

a straight line. In the left plot there is no solution for spherical, in the center plot the solutions for spherical and coma occur at the same bending, and on the right there are two possible solutions for spherical with equal and opposite-signed amounts of coma, and often with pronounced meniscus shapes. (These latter solutions are valuable if one desires to utilize the doublets in a symmetrical combination about a central stop, e.g., as an erector or a rapid rectilinear photo lens; the coma can then be used to reduce or overcorrect the astigmatism per Eq. 6.3h.) The exact form obtained is dependent primarily on the types of glass chosen. In general, the spherical aberration parabola can be raised by selecting a new flint glass with a lower index and higher V-value, or by selecting a new crown with a higher index and lower V. Thus the strongly meniscus solutions of the right-hand plot in Fig. 16.7 result from a glass pair with a small difference in V-value. Results approximating those in the middle graph of Fig. 16.7 can be obtained with BK7 (517:642) and SF2 (648:339). The best glass choice depends on the aperture ($f/\#$) of the lens.

Figure 16.8 shows the spherical aberration and the spherochromatism of a typical cemented doublet. As previously noted, the field curvature of a thin system with stop in contact is strongly inward and cannot be modified unless the stop is shifted. Thus, systems of this type are limited to applications which require good imagery over relatively small fields (a few degrees from the axis).

It is occasionally desirable to produce a doublet objective with both the zonal and marginal spherical simultaneously corrected. This can be accomplished by using the airspace of a broken contact doublet as an added degree of freedom. The design is begun exactly as in Sec. 16.4, except that two (or more) thick-lens solutions are derived, one with a minimum airspace and the other(s) with an increased space. The calculated zonal spherical is then plotted against the size of the airspace, and the airspace with LA_z equal to zero is selected; this form will usually have little or no zonal OSC. Speeds of $f/6$ or $f/7$ can be attained

RADIUS   THICKNESS   GLASS

+69.281

7.0   BaK2   540:597

-40.903

2.5   SF2   648:339

-130.529

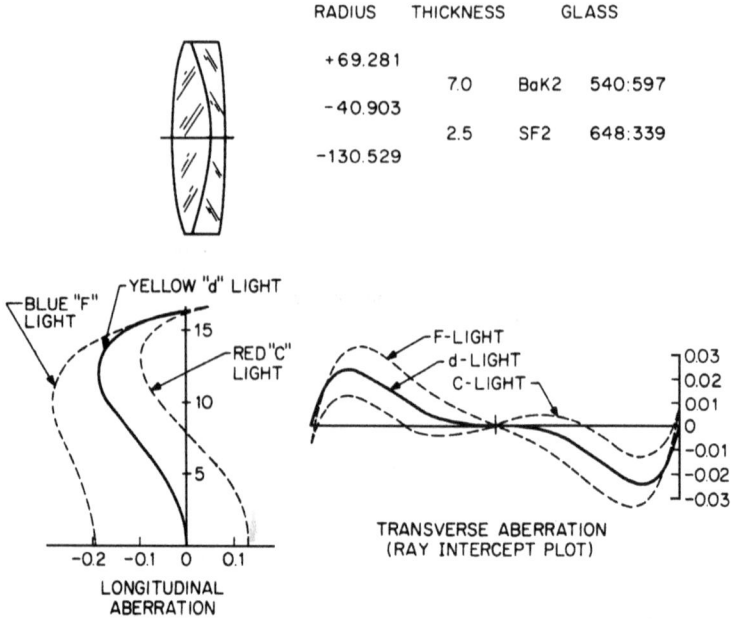

Figure 16.8 The spherical aberration and spherochromatism of a cemented achromatic doublet, efl = 100, f/3.0. Note that the chromatic is corrected at the margin. This is good practice if the spherochromatism is large; otherwise the image shows a blue flare. For small amounts, correction at the 0.7 zone is often a better choice.

with practically no spherical or axial coma over the entire aperture. Good glass choices are a light barium crown combined with either a dense flint or an extradense flint; either crown-in-front or flint-in-front forms are possible. In this type of lens the residual axial aberration consists almost solely of secondary spectrum.

Spherochromatism, which is the variation of spherical aberration as a function of wavelength, can be corrected by a change in the spacing between elements (or components) which differ in the sign of their contributions to spherical and chromatic aberration. This general principle may be applied to the doublet achromat in a manner paralleling the use of the airspace to correct zonal spherical; indeed, the basic principle is the same for both aberrations.

The source of spherochromatism can be understood by realizing that (in a cemented doublet) the two outer surfaces contribute under corrected spherical aberration, while the cemented interface contributes overcorrected spherical. The amount of the contribution varies directly with the size of the index change, or "break," across the surface. The contributions are in balance for the nominal wavelength. At a shorter wavelength all the indices are higher; because of its greater dispersion, the index of the negative (flint) element increases about twice as rapidly as that of the positive (crown) element. The index break at all three surfaces is larger at the shorter wavelength. However, the index break at the outer surfaces is $(n - 1)$, whereas at the cemented surface it is $(n' - n)$; as the wavelength and the indices change, $(n' - n)$ changes proportionately more than does $(n - 1)$. Thus as we go to a shorter wavelength, the overcorrecting contribution of the cemented surface is increased more than the undercorrection from the outer surfaces. The result is that the short-wavelength light is overcorrected compared to the central or longer wavelength. This is spherochromatism.

Now, if the airspace between elements is increased, as indicated in Fig. 16.9b, the blue marginal ray, having been refracted more strongly than the red ray by the crown element, will strike the flint element at a lower height than will the red ray. Thus the refraction of the blue ray at the flint will be lessened relative to the red, and its overcorrection reduced accordingly.

A very similar argument can be applied to the reduction of an undercorrected *zonal spherical* (which is caused by an overcorrected fifth-order spherical) by use of an increased airspace. The increased airspace affects the zonal spherical because the undercorrected spherical of the positive element bends the marginal ray toward the axis disproportionately more than the zonal ray. Thus, when the airspace is increased, the ray height at the overcorrecting negative element is reduced proportionately more for the marginal ray than for the zonal ray. The result is that the overcorrection is reduced more at the margin than at the zone, and, when the element shapes are readjusted to correct the marginal aberration, the zonal spherical is reduced. An airspaced doublet with reduced spherochromatism and reduced zonal spherical is shown in Fig. 16.10. *Both principles are applicable to more complex lenses as well.*

One method of effecting a simultaneous elimination of both spherochromatism and zonal spherical is indicated in Fig. 16.9(c). The

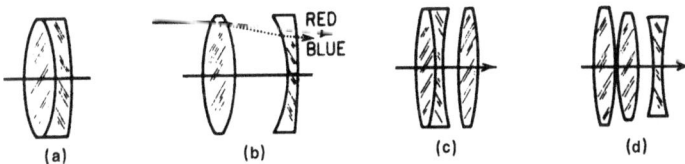

Figure 16.9 The ordinary spherochromatism of a doublet can be corrected by increasing the airspace [shown highly exaggerated in (b)]. This reduces the height at which the blue ray strikes the flint by a greater amount than for the red ray, thus reducing the overcorrection of the marginal blue ray. Sketches (c) and (d) show triplet forms which can be used to correct spherochromatism and spherical zonal residuals simultaneously. The configuration shown in (c) is 50% less sensitive to decentration than (d).

RADIUS	THICKNESS	GLASS
+47.819		
	7.0	BK7 517:642
−56.354		
	3.853	
−46.852		
	2.5	SF1 717:295
−160.303		

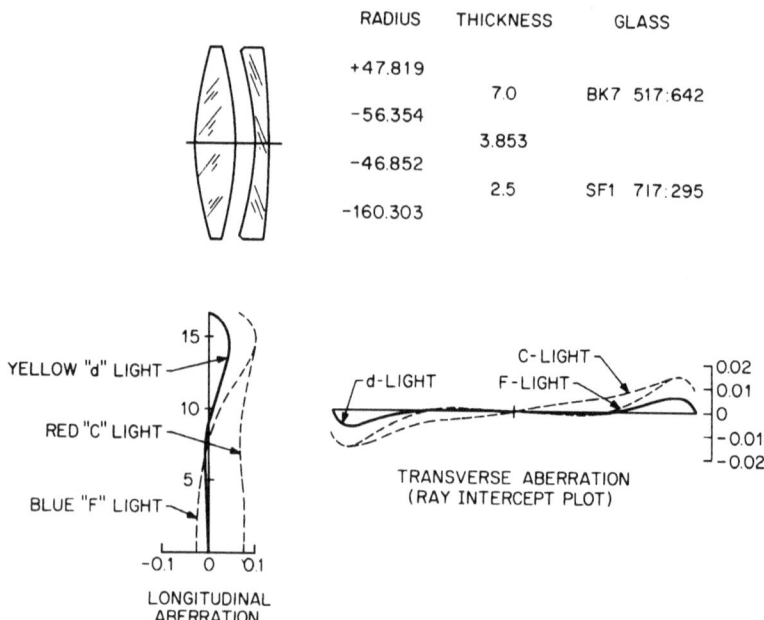

YELLOW "d" LIGHT

RED "C" LIGHT

BLUE "F" LIGHT

LONGITUDINAL ABERRATION

C-LIGHT
d-LIGHT   F-LIGHT

TRANSVERSE ABERRATION
(RAY INTERCEPT PLOT)

Figure 16.10 The spherical aberration and spherochromatism of an airspaced achromatic doublet, efl = 100, f/3.0. The size of the airspace used here is a compromise between the value which would minimize the zonal spherical aberration and that which would minimize the spherochromatism. Compare the residual aberrations with those of the cemented doublet in Fig. 16.8.

doublet plus singlet configuration (in any of several arrangements of the elements) introduces still another degree of freedom, namely the balance of positive (crown) power between the two components, which can be used with the airspace to bring about the correction. The airspaced triplet shown in Fig. 16.9(d) is also capable of very good correction, but is more difficult to manufacture. Figure 17.39 illustrates the reduction of spherical by element splitting. Figure 19.1 shows an airspaced triplet telescope objective.

The *secondary spectrum* (SS) contribution of a thin lens is given by Eq. 6.3t; combining this with the requirements for achromatism (Eqs. 16.3 and 16.4), we find that the secondary spectrum of a thin achromatic doublet is given by

$$SS = \frac{f\,(P_b - P_a)}{(V_a - V_b)} = \frac{-f\,\Delta P}{\Delta V} \tag{16.9}$$

For any of the ordinary glass combinations used in doublets, the ratio $\Delta P/\Delta V$ is essentially constant, and the visual secondary spectrum is about 0.0004 to 0.0005 of the focal length. Similarly, the secondary spectrum of any achromatized combination of two separated components can be shown to be

$$SS = \frac{\Delta P}{D\,\Delta V}\,[f^2 + B\,(L-2f)] \qquad (16.10)$$

where D is the airspace, B the back focus, and $L = B + D$ is the length from front component to the focal point. Again it is apparent that the ratio $\Delta P/\Delta V$ is the governing factor. Note that in this case the secondary color of two spaced positive lenses is less than that of a thin doublet of the same focal length; conversely, the secondary color of a telephoto lens (positive front component, negative rear component) or reversed telephoto is greater than the corresponding thin doublet, as is any anastigmat. Approximate values for the secondary spectrum of some "typical" designs using ordinary glass are tabulated below:

Achromatic doublet	SS = efl/2200
Petzval projection lens	SS = efl/3000
Cooke triplet	SS = efl/1300
Telephoto	SS = efl/1700

There are a few glasses which will reduce the secondary spectrum, for example, FK51, 52, or 54 used with a KzFS glass or an LaK glass as the flint element will reduce the visual secondary spectrum to a small fraction of the ordinary value. Note, however, that for most of these pairs $V_a - V_b$ is small, and the powers of the individual elements required for achromatism are higher than with an ordinary pair of glasses. This increase in element power causes a corresponding increase in the other residual aberrations. These glasses, with unusual partial dispersions, generally work poorly in the shop, lack chemical stability, and cannot withstand severe thermal shock. They are also difficult to melt (and stir) to high optical quality.

As mentioned in Chap. 10, calcium fluoride (CaF_2, fluorite) may be combined with an ordinary glass (selected so that $P_a = P_b$) to make an achromat that is essentially free of secondary spectrum. It is also worth noting that there are no ordinary glass pairs which will form a useful achromat in the 1.0- to 1.5-μm spectral band; fluorite can be combined with a suitable glass to make an achromat for this region. Silicon and germanium are useful for achromats at longer wavelengths, as are BaF_2, CaF_2, ZnS, ZnSe, and AMTIR.

A triplet achromat can be used to reduce the secondary spectrum without the necessity of exactly matching the partial dispersions as in the doublet. If one plots the partial dispersion P against the V-value,

most glasses fall along a straight line. What is needed to correct secondary spectrum is a pair of glasses with the same partial P, but with a significant difference in V-value. It turns out that in this sort of plot one can synthesize a glass anywhere along a line connecting two glass points by making a doublet of the two glasses. Thus one can arrange a triplet so that two of the elements synthesize a glass with exactly the same partial as the third glass. Some useful Schott glass combinations are (PK51, LaF21, SF15), (FK6, KzFS1, SF15), (PK51, LaSFN18, SF57); the power arrangement for these combinations is usually plus, minus, and weak plus, respectively. Other glass manufacturers have equivalent glass combinations. The thin lens element powers for a triplet apochromat can be found from the following equations, which are for a unit power ($f = 1.0$) system.
Define:

$$X = V_a (P_b - P_c) + V_b (P_c - P_a) + V_c (P_a - P_b)$$

Note that X is equal to twice the area of the triangle formed by the three glasses in the P vs. V plot, and since X appears in the denominator of the following equations for the element powers, the bigger the value of X, the smaller the powers (and the better the design).
  Then:

$$\phi_a = V_a (P_b - P_c)/X$$

$$\phi_b = V_b (P_c - P_a)/X$$

$$\phi_c = V_c (P_a - P_b)/X = 1.0 - \phi_a - \phi_b$$

See Fig. 19.3 for an example of an apochromatic triplet telescope objective.

A lens in which three wavelengths are brought to a common focus is called an *apochromat*. Often this term also implies that the spherical aberration is corrected for two wavelengths as well. By properly balancing the glass combinations given above one can achromatize the triplet for four wavelengths; such lenses are called *superachromats*.*

Airspaced achromat (dialyte)

A widely airspaced doublet can be made achromatic, but the chromatic correction will vary with the object distance; it will be achromatic only

---

*See M. Herzberger and N. McClure, "The Design of Superachromatic Lenses," *Applied Optics*, vol. 2, June 1963, pp. 553–560.

for the design distance. The following equations will yield a separated achromatic doublet which is corrected for an object at infinity.

$$\phi_A = \frac{V_A B}{f (V_A B - V_B f)}$$

$$\phi_B = \frac{-V_B f}{B (V_A B - V_B f)}$$

$$D = \frac{(1 - B/f)}{\phi_A}$$

where f is the focal length, D is the airspace, and B is the back focal length.

Note that, although the dialyte (per these equations) is corrected for axial chromatic, unless both components are individually corrected for chromatic, the system will have lateral color.

Athermalization

When the temperature of a lens element is changed, two factors affect its focus or focal length. As the temperature rises, all the dimensions of the element are increased; this, by itself, would lengthen the effective and back focal lengths. The index of refraction of the lens also changes with temperature. For many glasses the index rises with temperature; this effect tends to shorten the focal lengths.

The change in the power of a thin element with temperature is given by

$$\frac{d\phi}{dt} = \phi \left[\frac{1}{(n-1)} \frac{dn}{dt} - \alpha \right]$$

where dn/dt is the differential of index with temperature, and α is the thermal expansion coefficient for the lens material. Thus for a thin doublet

$$\frac{d\Phi}{dt} = \phi_A T_A + \phi_B T_B$$

where

$$T = \left[\frac{1}{(n-1)} \frac{dn}{dt} - \alpha \right]^*$$

---

*For a diffractive surface, as described in Sec. 16.6, the term "T" is simply equal to 2α.

and Φ is the doublet power. For an athermalized doublet (or for one with some desired $d\Phi/dt$), we can solve for the element powers

$$\phi_A = \frac{(d\Phi/dt) - \Phi T_B}{T_A - T_B}$$

$$\phi_B = \Phi - \phi_A$$

To get an athermalized *achromatic* doublet, we can plot T against $1/V$ for all the glasses under consideration. Then a line drawn between two glass points is extended to intersect the T axis. The value of the $d\Phi/dt$ for the achromatic doublet is equal to the doublet power times the value of T at which the extended line intersects the T axis. Thus one desires a pair of glasses with a large V-value difference and a small or zero T-axis intersection.

An athermal achromatic triplet can be made with three glasses as follows:

$$\phi_A = \frac{\Phi V_A \, (T_B V_B - T_C V_C)}{D}$$

$$\phi_B = \frac{\Phi V_B \, (T_C V_C - T_A V_A)}{D}$$

$$\phi_C = \frac{\Phi V_C \, (T_A V_A - T_B V_B)}{D}$$

where $D = V_A \, (T_B V_B - T_C V_C) + V_B \, (T_C V_C - T_A V_A) + V_C \, (T_A V_A - T_B V_B)$, V_n is the V-value of element n, and T is defined above.

16.6   The Diffractive Surface in Lens Design

A *diffractive surface* as used in lens design is a *fresnel* surface (as shown in Fig. 13.15) "modulo 2π." In other words, it is a fresnel surface where the height of each step is such that the wave front is retarded or stepped by exactly one wavelength. Thus the step height is $\lambda/(n - 1)$, assuming that the surface is bounded by air. For a glass or plastic surface ($n \approx 1.5$), this is a step height of about two wavelengths, as opposed to a step height on the order of several tenths of a millimeter for an ordinary plastic fresnel. The slope and shape of the fresnel facets can be as defined by a sphere or an aspheric. Note that similar results can be obtained with a local variation of the index of refraction.

Diffraction efficiency

The term *kinoform* indicates a fresnel surface with smooth facets. A curved-surface kinoform theoretically can have 100 percent efficiency.

A "linear" (cone-shaped) kinoform can be 99 percent efficient. A "binary" surface approximates the smooth fresnel facets with a stair-step contour produced by a high-resolution photolithographic process. The surface relief is created by exposure through a series of masks. The number of levels produced equals 2^n, where n is the number of masks used, hence the name *binary*. The efficiency (i.e., the percentage of light which goes in the desired direction) of a binary surface is limited by the number of levels which are used to approximate the ideal smooth contour of the fresnel facet. A one-mask, 2-level surface is 40.5 percent efficient; a two-mask, 4-level surface is 81.1 percent efficient; a three-mask, 8-level surface is 95.0 percent efficient; a four-mask, 16-level surface is 98.7 percent efficient; and an M-level surface is $[\sin(\pi/M)/(\pi/M)]^2$ efficient. The theoretical efficiency of any diffraction surface, whether kinoform or binary, will be reduced by any fabrication departures from the ideal shape, such as rounding of sharp corners, etc.

Since the wave front is stepped or retarded at each diffractive fresnel step by exactly one wavelength for the nominal wavelength, it is apparent that the coherent behavior of the system is preserved only for the nominal wavelength. At this wavelength, the phase from the top of one zone exactly matches that from the bottom of the preceding zone. The surface is less efficient for other wavelengths, and thus the spectral bandwidth over which a diffractive surface is useful is limited. This limitation may show up as inefficiency or as unwanted diffractive orders, ghosts, stray light, low contrast, etc. The efficiency at other than the nominal wavelength (λ_0) is

$$E = [\sin \pi (1 - \lambda_0/\lambda) / \pi (1 - \lambda_0/\lambda)]^2$$

Over a bandwidth of ($\Delta\lambda$), the average efficiency is

$$\text{ave } E \approx 1 - [\pi (\Delta\lambda)/6 \lambda_0]^2$$

Manufacturability

The following expressions allow an estimate of the practicality or manufacturability of a diffractive lens. As indicated above, the step height is $\lambda/(n - 1)$. The radial spacing distance from one fresnel step to the next is approximately

$$\text{Spacing} \approx R\lambda/Y (n - 1) = F\lambda/Y$$

where R is the diffractive surface radius of curvature, F is its focal length, and Y is the radial distance from the axis. The minimum spacing (at the edge of the diffractive lens) is

$$\text{Min spacing} \approx 2\lambda (f/\#) = \lambda/\text{NA}$$

where $f/\# = F/2Y_{max}$ = the relative aperture, and NA $= n \sin u =$ the numerical aperture. The total number of fresnel steps or zones is

$$\text{Number of steps} \approx D^2/8\lambda F$$

where D is the lens diameter. It is apparent that the longer the wavelength and the weaker the power of the diffractive surface, the wider and deeper are the steps, and the easier is the fabrication task. Techniques used for fabrication include single-point diamond turning (especially good for long-wave IR), ion-beam machining, electron-beam writing, laser-beam writing, and photolithography (which is extremely difficult on curved surfaces but effective on plano surfaces). For large commercial quantities, injection-molded plastic elements are an economical choice. Another useful process is epoxy replication. Applications of diffractive optics include hybrid (combined refractive and diffractive) lenses, microlens (size about 50 μm) arrays, anamorphic arrays, prisms, beamsplitters, beam multiplexers, filters, etc.

The Sweatt model

From a lens design standpoint, an easy way to handle and understand the use of a diffractive surface is through the *Sweatt model.* W. C. Sweatt* showed that a raytrace model consisting of a very high index, zero-thickness lens could be used to predict the effect of a diffractive surface; the higher the index, the closer the results of the raytrace come to matching the exact diffraction results. An index of about 10,000 is a reasonable value to use. Since the diffractive effect is a direct function of wavelength, the index of the model should vary as the wavelength, and

$$n (\lambda) = 1 + (n_0 - 1) (\lambda/\lambda_0)$$

where λ_0 and n_0 are the nominal wavelength and index, respectively.
   Thus, for the visual region, using d, F, and C light, we have for

d-light at 0.5875618 μm,

$$n_d = 10,001.00$$

F-light at 0.4861327 μm,

$$n_F = 8,274.73$$

---

*J. Opt. Soc. Am., vol. 67, 1977, p. 803; vol. 69, 1979, p. 486; Appl. Opt., vol. 17, 1978, p. 1220.

C-light at 0.6562725 μm,

$$n_C = 11{,}170.42$$

and the Abbe V-value,

$$V = (n_d - 1)/(n_F - n_C) = -3.45$$

The negative V-value results from the fact that the index rises with wavelength instead of dropping as in ordinary refractive materials. The partial dispersion is $P = (n_F - n_d)/(n_F - n_C) = 0.5962$. These extremely unusual values make the diffractive surface a most singular optical material. This low-V-value (i.e., high dispersion) characteristic of a diffractive device indicates that there will be very large amounts of chromatic aberration when a diffractive surface is used over a significant spectral bandwidth.

The achromatic diffractive singlet

If we assume a single element of BK7 ($n_d = 1.5168$, $V = 64.2$, $P = 0.6923$), we can apply Eqs. 16.3 and 16.4 to determine the powers of the singlet and the diffractive element which will produce an achromat. The result is a power of $\phi_a = V_a\Phi/(V_a - V_b) = +0.949\Phi$ for the BK7 element and $\phi_b = +0.051\Phi$ for the diffractive element (where Φ is the desired power of the achromat). The negative V-value of the diffractive surface produces an achromat where both elements have positive power. If we allow the diffractive surface to be aspheric (in the actual surface this is done by making the slope and shape of the fresnel facets correspond to those of an aspheric surface), we can produce a singlet of the desired power which is corrected for spherical aberration, chromatic aberration, and coma. The necessary four degrees of freedom are the power and bending of the singlet and the power and fourth-order asphericity (or conic constant) of the diffractive surface.

The resulting design is shown in Fig. 16.11. The residual aberrations (zonal spherical, spherochromatism, and secondary spectrum) can be compared with those of the ordinary achromatic doublet shown in Fig. 16.8. Note that the sign of the secondary spectrum is reversed from that of an ordinary doublet (because of the unusual P and V of the diffractive surface) and that the spherochromatism is large, more than twice that of the doublet of Fig. 16.8 (and is also of reversed sign). The spherochromatism can be corrected by aspherizing the first surface with a fourth-order deformation term in a manner analogous to adjusting the airspace of the doublet in Fig. 16.10 (i.e., we change the relative heights at which the red and blue rays strike the diffractive surface). The zonal spherical can be removed with a sixth-order deformation

RADIUS	THICKNESS	MATERIAL		
59.307733	4.50	BK7	1.5168	64.17
–972.261092	0.00	DIFF	10,001.	–3.45
–972.186542*	97.415	AIR		

*Conic constant kappa = –0.186743

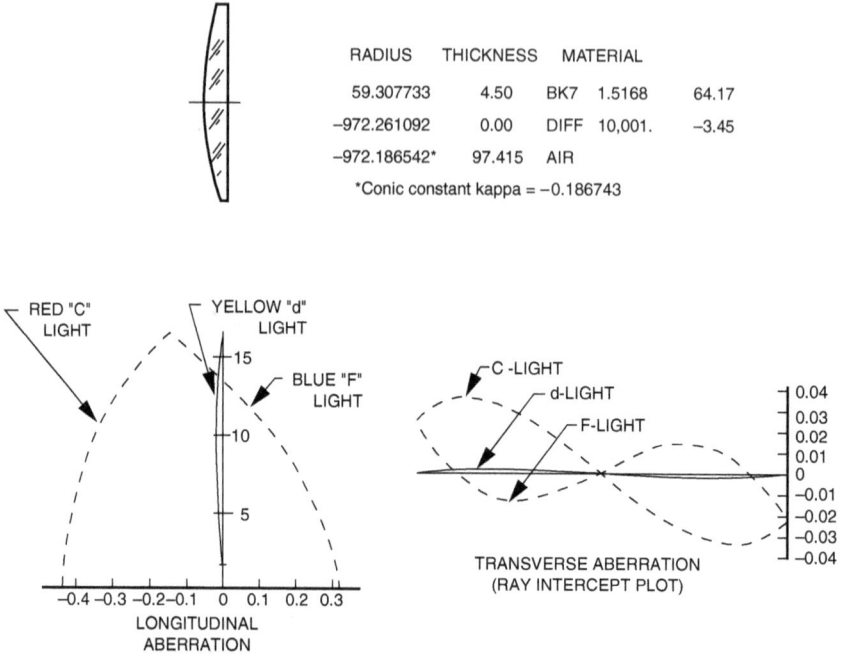

Figure 16.11 The spherical aberration and spherochromatism of a hybrid refractive-diffractive singlet, efl = 100, f/3.0. Compare with the doublet of Fig. 16.8 (but note that the scales for LA are different). Both the spherochromatism and secondary spectrum are larger and of the opposite sign from Fig. 16.8. As indicated in the text, the spherochromatism and the zonal spherical can be eliminated easily by aspherizing the first surface (which would be quite a feasible option if the lens were injection-molded from acrylic).

term on the first surface. The use of an aspheric first surface is an economically practical move, assuming that the lens is to be injection-molded from plastic. The result is a lens the only axial aberration of which is about 0.17 mm of secondary spectrum.

Alternately, because photolithographic fabrication is most conveniently done on a flat surface, one might want to limit the lens shape to a plano-convex form and use as degrees of freedom the lens index, its radius, the power of the diffractive surface, and its asphericity. The optimal index is about 1.55 for this lens. If the lens material is acrylic ($n = 1.492$), and if we elect to control focal length, spherical, and chromatic (neglecting coma), the tangential coma at one degree off axis is -0.0156; if the material is polystyrene ($n = 1.590$), it is $+0.0101$.

Achromatic diffractive singlets have been very satisfactorily used in eyepieces, magnifiers, zoom camera lenses, and many other applications where the object field is of relatively uniform brightness. Their compactness and light weight as compared with a glass achromat make them very desirable for many applications such as head-mounted displays.

Diffractive surfaces sometimes have proven less satisfactory for systems where there is a high brightness source in (or near) the field of view or a wide spectral bandwidth.

The apochromatic diffractive doublet

Since the unusual V-value and partial dispersion of the diffractive surface are so far from the line of normal glasses in a P versus V plot, we can easily produce an apochromatic lens using two ordinary glasses plus a diffractive surface to eliminate the secondary spectrum.

The element powers for a three-element apochromat can be found using the following equations:

$$X = V_a \, (P_b - P_c) + V_b \, (P_c - P_a) + V_c \, (P_a - P_b)$$

$$\phi_a = \Phi V_a \, (P_b - P_c) \, / X$$

$$\phi_b = \Phi V_b \, (P_c - P_a) \, / X$$

$$\phi_c = \Phi V_c \, (P_a - P_b) \, / X$$

where Φ is the power of the apochromatic triplet, V_i is the V-value, and P_i is the partial dispersion of the ith element.

If we use acrylic ($n = 1.4918$, $V = 57.45$, $P = 0.7014$) and polystyrene ($n = 1.5905$, $V = 30.87$, $P = 0.7108$) for elements a and b, and the diffractive surface ($n = 10,001$, $V = -3.45$, $P = 0.5962$) for element c, we get the following starting powers for the elements:

$$\phi_a = +1.9544\Phi \qquad \text{(acrylic)}$$

$$\phi_b = -0.9640\Phi \qquad \text{(polystyrene)}$$

$$\phi_c = +0.0096\Phi \qquad \text{(diffractive)}$$

The lens can be corrected for marginal and zonal spherical aberration, coma, chromatic, spherochromatic, and secondary spectrum using the techniques outlined above. A drawback for this particular lens is that the secondary spectrum varies with aperture and is fully corrected only at one zone.

16.7   The Cooke Triplet Anastigmat: Third-Order Theory

The process of designing a Cooke triplet anastigmat which is described in this section is not one that is used today. In fact, it is unlikely that you can find a lens designer under the age of 85 who has ever attempted to use this sort of process. The process, although very long and very laborious, does work, although many trials, iterations, and variations may be needed to produce a good design, one where the higher order aberrations are reasonable.

Nonetheless, a careful reading of this section is well worthwhile, since it sheds light on the many, complex interrelationships in what is actually a quite simple lens. It tells, for example, why it is necessary to use the center element glass type as a variable if you want the best lens, and why the best flint glass is one from along the "glass line."

Of course most, if not all, lens design today is done with the ubiquitous PC, using one of the optical software programs, e.g., OSLO, ZEMAX, CodeV. Sec. 16.8 will cover this aspect of Cooke triplet design.

The Cooke triplet is composed of two outer positive crown elements and an inner negative flint element, with relatively large airspaces separating the elements. This type of lens is especially interesting because there are just enough available degrees of freedom to allow the designer to correct all of the *primary* aberrations. The basic principle used to flatten the field curvature (i.e., the Petzval sum) is quite simple: the contribution that an element makes to the power of a system is proportional to $y\phi$, and the contribution to the chromatic varies with $y^2\phi$. However, the contribution to the Petzval curvature is a function of ϕ alone and is independent of y. Now in a thin (compact) system, all the elements have essentially the same value of y and the powers of the elements are determined by the requirements of focal length and chromatic correction; consequently, the Petzval radius (ρ) of a thin doublet is often about $-1.4f$, and its radius rarely exceeds 1.5 or 2 times the focal length. However, when the negative elements of a system are spaced away from the positive elements (so that the ray height y at the negative elements is reduced), the power of the negative elements must be increased to maintain the focal length and chromatic correction of the system. As a result, the overcorrecting contribution of the negative element to the Petzval curvature is increased. Thus by the proper choice of spacing, the Petzval radius can be lengthened to several times the system focal length and the field proportionately "flattened."

From Fig. 16.12, which shows a schematic triplet, we can determine the available degrees of freedom. They are

1. Three powers (ϕ_a, ϕ_b, ϕ_c)
2. Two spaces (S_1, S_2)
3. Three shapes (C_1, C_3, C_5)
4. Glass choice
5. Thicknesses

Of these, items 1, 2, and 3 will be of immediate interest; they total eight variables. Item 4, glass choice, is an extremely important tool, but we

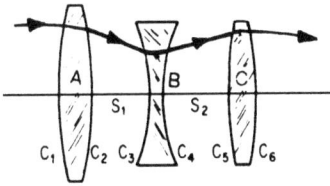

Figure 16.12 The Cooke triplet anastigmat.

will reserve its discussion until later. Item 5, element thickness, is only marginally effective; in regard to the primary corrections, its effect duplicates that of the spacings.

With these eight degrees of freedom, the designer wishes to correct (or control) the following primary characteristics and aberrations.

1. Focal length

2. Axial (longitudinal) chromatic aberration

3. Lateral chromatic aberration

4. Petzval curvature

5. Spherical aberration

6. Coma

7. Astigmatism

8. Distortion

Thus, there are just the necessary eight degrees of freedom to control the eight primary corrections.

Note that the fact that there are eight variable parameters does not guarantee that there is a solution. The relationships involved are, in several instances, nonlinear, as the thin-lens equations (Eqs. 6.3) indicate. It is entirely possible to choose a set of desired aberration values and/or glass types for which there is no solution. On the other hand, it is also possible that there are as many as eight solutions, as will be seen in the following paragraphs.

Power and spacing solution. The first four items listed immediately above can be seen (by reference to the thin-lens third-order aberration equations) to be functions of element power and ray height (which is a function of spacing); they are independent of element shape. Thus, it is necessary that the powers and spaces be chosen to satisfy these four conditions, which may be expressed as follows:

Power:

$$\text{Desired } \Phi = \frac{1}{f} = \frac{1}{y_a} \sum y\phi \qquad (16.11)$$

Axial color:

$$\text{Desired } \sum\text{TAchC} = \frac{1}{u'_k} \sum \frac{y^2 \phi}{V} \tag{16.12}$$

Lateral color:

$$\text{Desired } \sum\text{TchC*} = \frac{1}{u'_k} \sum \frac{yy_p \phi}{V} \tag{16.13}$$

Petzval sum:

$$\text{Desired } \sum\text{PC} = \frac{h^2}{2} \sum \frac{\phi}{n} \tag{16.14}$$

where the summation is over the three elements. These expressions are essentially the same as those of Sec. 6.4, and the meanings of the symbols are given there.

The four conditions above must be satisfied by the choice of five variables (three powers plus two spacings). There is one more variable than necessary; this "extra" is utilized in a later step to control one of the remaining aberrations (usually distortion). There are almost as many ways of solving this set of equations as there are designers. Stephens* has worked out the algebraic solution for the triplet, and his paper gives explicit equations for the values of the powers and spaces. An iterative approximation technique (which may be easily modified to apply to systems with more than three components) along the following lines is an alternate method. Its description will help to understand the limits and interrelationships involved in this design.

1. Assume a value for the ratio of the powers of elements c and a. This will be the "extra" degree of freedom mentioned above. ($K = \phi_c/\phi_a = 1.2$ is a typical value.)

2. Choose a value (arbitrary) for ϕ_a. (In the absence of prior experience, $\phi_a = 1.5\Phi$ is suitable.) This determines ϕ_c, since from step 1, $\phi_c = K\phi_a$ and also determines ϕ_b, since Eq. 16.14 can be solved for ϕ_b when ϕ_a, ϕ_c, h, n, and $\sum\text{PC}$ are known or assumed.

3. Choose a value for S_1 (one-fifth to one-tenth of the focal length is suitable).

4. Solve for the value of S_2 which will satisfy Eq. 16.12 (assume that u'_k is equal to Φy_a). This is done by tracing a ray through elements a and b to determine y_a, y_b, and u'_b. Then find S_2 to yield the value of y_c, which satisfies Eq. 16.12. (Note that S_2 may have a negative value on the first try.)

---

*R. E. Stphens, *J. Opt. Soc. Am.*, vol. 38, 1948, p. 1032.

5. Trace a principal ray (thin-lens paraxial) through the desired stop position, which may be conveniently placed at element b to minimize the labor. Again assume u'_k as in 4 and determine $\sum \text{TchC}^*$.

6. Repeat from step 3 with a new choice for S_1 until $\sum \text{TchC}^*$ is as desired. (As a second guess for S_1, try the average of S_1 and S_2 from the first try.)

7. Determine the system power Φ. If not as desired, scale the value of ϕ_a used in step 2 and repeat from step 2 until a solution is obtained.

Graphs of the relationships between S_1 and $\sum \text{TchC}^*$ and between ϕ_a and Φ are useful in steps 6 and 7.

Element shape solution.   When the element powers and spacings have been determined, there are three uncommitted degrees of freedom, namely the shapes of the three elements (plus the "extra," K, mentioned in step 1 above). These variables must be adjusted so that the spherical, coma, astigmatism, and distortion are corrected to their desired values. Referring to the thin-lens contribution equations of Sec. 6.4, the aberrations can be seen to be quadratic functions of the element shapes; thus, a simultaneous algebraic solution cannot be used and some sort of successive approximation procedure is necessary.

Thin-lens paraxial marginal and principal rays are traced through the three elements. The principal ray is traced so that the aperture stop is at lens b; both y_p and Q for lens b will be zero.

1. Assume an (arbitrary) value for C_1 and calculate TAC^*_a (the astigmatism contribution) for element a by Eq. 6.3h (a value of $C_1 = 2.5\Phi$ is a reasonable first choice).

2. Since the stop is located at element b, TAC_b will not change with bending (Eq. 6.3o). Now solve Eq. 6.3h for the shape of element c, that is, the value of C_5, which will give TAC^*_c which will yield the desired $\sum \text{TAC}$ when combined with AC^*_a and AC_b. Normally there are two solutions for C_5 and the more reasonable one is used.

3. Now CC^*_a and CC^*_c (the coma contributions) are calculated from Eq. 6.3g. Since the equation for CC_b is linear in C_3 (Eq. 6.3n, since $Q_b = 0$), it can be solved for the unique value of C_3 which will yield the desired $\sum \text{CC}^*$.

4. The value of $\sum \text{TSC}$ (spherical aberration) is now determined from Eq. 6.3m.

5. The procedure is repeated from step 1 with a new value of C_1, and a graph of $\sum \text{TSC}$ is plotted against C_1. The shape C_1 for which $\sum \text{TSC}$ is equal to the desired value is chosen and the corresponding values

of C_3 and C_5 are determined so that $\sum TSC$, $\sum TAC^*$, and $\sum CC^*$ are simultaneously as desired.

6. If $\sum DC^*$ (distortion) is within acceptable limits, well and good; if not, a new power and space solution must be made with a different value of $K = \phi_c/\phi_a$. The value of $\sum DC^*$ can be plotted for several values of K as an aid in effecting a solution.

Note that in step 5, there may be two, one, or no solutions for the desired $\sum TSC$. The best triplets seem to result from cases where the parabolic plot of $\sum TSC$ just barely reaches the desired level. Step 6 also may have no, one, or two solutions. Thus, with two possible solutions in each of steps 2, 5, and 6, there are, theoretically at least, eight possible solutions for the thin-lens Cooke triplet. As indicated above, it is also possible that for a given set of conditions, there may be no solution. Usually, however, there is only one "reasonable" solution; occasionally there are two.

The next step is the addition of thickness to the design. Center thicknesses for the crown elements are chosen to give workable edge thicknesses; the second surface curvatures (C_2, C_4, and C_6) are adjusted to hold the thick-element powers exactly to the thin-lens powers. Airspaces are chosen so that the principal points of the elements are spaced apart by the thin-lens spacings. In this way, the thick-lens triplet will have exactly the same focal length as the thin-lens version.

The thick lens is now submitted to a trigonometric raytrace analysis and the values of the seven primary aberrations are determined. If (as is likely) the aberrations are not as desired, a new round of design is initiated, with the new "desired" thin-lens aberration values adjusted to offset the difference between the raytracing results and the desired final values. For example, if the original "desired" $\sum TSC$ was -0.2 and the raytracing yielded a marginal spherical, $TA_m = +0.2$, the new "desired" $\sum TSC$ would be set at -0.4, assuming that the desired end result was $TA_m = 0.0$.

Initial choice of desired aberration values. In general, the initial choice for the "desired" third-order aberration sums should be small, undercorrected amounts, since the higher-order aberrations are usually overcorrecting. Spherical, Petzval, and axial chromatic follow this rule. Since the Cooke triplet is relatively symmetrical, the residuals of distortion, coma, and lateral color are small, and initial "desired" values of zero are appropriate. The desired Petzval sum should be definitely negative. For high-speed lenses, the Petzval radius is frequently as short as two or three times the focal length; moderate-aperture systems ($f/3.5$) usually have $\rho = -3f$ to $-4f$; slow systems may have $\rho = -5f$ or longer. One reason for this relationship is that the flatter (less undercorrected) the Petzval surface is made, the higher the element powers; hence the higher the residual aberrations, especially zonal spherical. The value chosen for the desired $\sum PC$ is also an important factor in determining whether or not there is a solution for step 5 in the curvature determination process. The "desired" astigmatism sum is best set slightly positive, between zero and about one-third the absolute value of the Petzval sum, so that the inward curvature of the Petzval surface is offset by the overcorrected astigmatism.

Glass choice.   The choice of the glass to be used in the triplet is one of the most important design factors. From field (Petzval) curvature considerations, it is desirable that the positive elements have a high index of refraction and the negative element a low one to reduce $\sum \phi/N$. As usual, the V-value of the positive elements should be high and that of the negative element low in order to effect chromatic correction. For the positive elements, one of the dense barium crowns is the usual choice, although the light barium crowns on one hand and the rare earth (lanthanum) glasses on the other are also used. Although triplet designs are possible with ordinary crown glass or even plastics, their performance is relatively poor.

It turns out that the interrelated requirements of Eqs. 16.11 through 16.14 lead to long systems (i.e., S_1 and S_2 are large) when the difference between the V-values of the positive and negative elements is large. A lens with a large vertex length will, at any given diameter, vignette at a smaller angle than will a short lens. Further, it turns out that the longer the lens: (1) the smaller the spherical zonal and (2) the smaller the field coverage (i.e., the higher-order astigmatism and coma are greater and limit the angle over which a good image can be obtained when the lens is long). Thus, long systems are appropriate for high-speed, small-angle systems; short systems for small-aperture, wide-angle applications. As a very rough rule of thumb, the vertex length of a triplet is frequently equal to the diameter of the entrance pupil.

The length of the triplet can be controlled by the choice of the glasses used. For example, if a shorter system is desired, the substitution of a flint with a higher V-value (or a crown with a lower V-value) will produce the necessary change. To get a longer system, use a higher V-value crown and/or a lower V-value flint. (However, note that a system which is *too* long will have no solution for the element shapes. The ray height on the negative element may be so low that its overcorrecting contribution to the spherical aberration is insufficient to offset the undercorrection of the positive elements simultaneously with the requirements for coma and astigmatism correction as well as chromatic and Petzval.)

Some other length considerations. If the Petzval curvature is made more inward curving (negative), the system becomes longer. If the ratio $y_A\phi_A/y_C\phi_C$ equals 1.0, the system is shorter than if the ratio is 2.0 or 0.5. The index of the glasses have little effect on the length, although high index glass for elements A and C produce lower element powers. It is a trueism that lenses "want" to be as large as you allow them to be.

Interestingly enough, *this relationship between vertex length and zonal spherical and field coverage is a general one and applies to most ordinary anastigmats.** Thus, if an anastigmat design has too much zonal spherical and more than enough angular coverage, one can simply choose new glasses to lengthen the system and strike the desired balance between field and aperture, or vice versa. There are, of course, limits to the effectiveness of this technique.

In general, the higher the index of the crown (positive) elements and the lower the index of the flint, the better the design will be. In other words, with all else equal, a triplet with a more positive index difference (n crown $- n$ flint) will have a smaller zonal spherical and/or a wider field coverage. See also Figs. 17.38 and 17.39 for the effect of index on aberrations.

Figure 6.2 showed a triplet of modest aperture and field coverage. Figures 16.13, 16.14 and 16.15 illustrate triplets of reduced vertex length and increasingly smaller aperture and wider fields of view. Needless to say, the Cooke triplet is best designed using an automatic computer lens design program of the type described in Sec. 16.8. However, the automatic design program can be better utilized and better results will be obtained if the designer has mastered the information in this section. Figures 19.9 and 19.10 also show Cooke triplet designs. Figure 19.39 shows a triplet with an aspheric field corrector, suitable for use in a point-and-shoot camera, and Fig. 19.41 shows an infrared (8–14 μm) triplet. Figure 19.42 is another IR triplet-based lens of very high speed ($f/0.55$).

If one were to use the technique described above, it is not likely that the resulting lens would have the trigonometric correction that was desired. One possibility would be to adjust the desired level of the third-order aberrations by amounts equal (and opposite in sign) to the amount by which one desires to adjust the trigonometricly calculated aberrations. This works well because the high order aberrations are quite stable, and the change in the total aberrations is usually very close to the changes made in the third-order sums.

A differential solution can be effected if the partials of the aberrations with respect to the construction parameters are determined and

---

*See W. Smith, *J. Opt. Soc. Am.*, vol. 48, 1958, pp. 98–105.

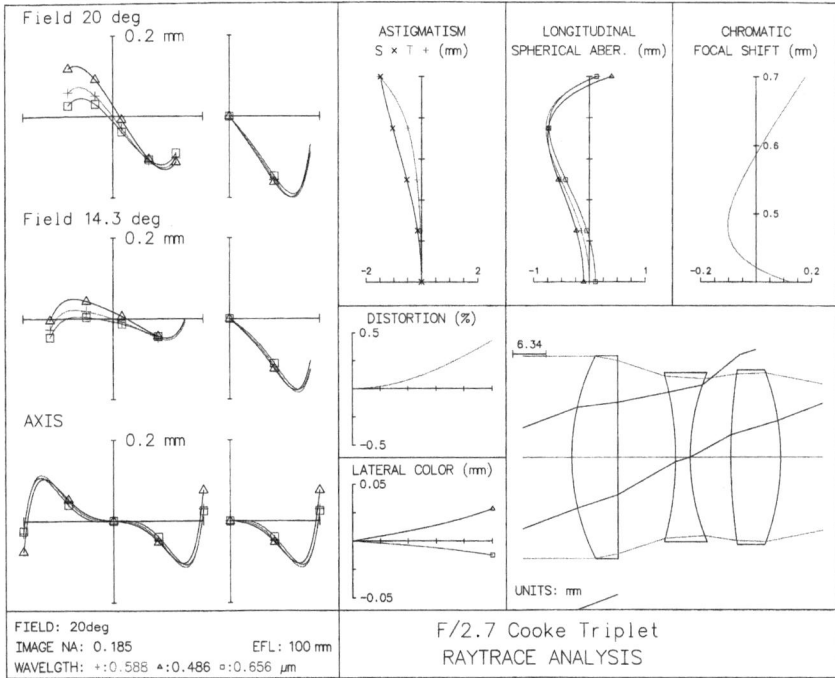

Field 20 deg
0.2 mm

ASTIGMATISM
S × T + (mm)

LONGITUDINAL
SPHERICAL ABER. (mm)

CHROMATIC
FOCAL SHIFT (mm)

Field 14.3 deg
0.2 mm

AXIS
0.2 mm

DISTORTION (%)

LATERAL COLOR (mm)

UNITS: mm

FIELD: 20deg
IMAGE NA: 0.185   EFL: 100 mm
WAVELGTH: +:0.588 ▲:0.486 □:0.656 μm

F/2.7 Cooke Triplet
RAYTRACE ANALYSIS

F/2.7 Cooke Triplet USP#2, 453, 260–1948

radius	space	mat'l		s-a
40.72	8.74	SK6	614564	18.6
plano	11.05			18.6
−55.65	2.78	SF2	648338	15.5
39.75	7.63			15.5
107.00	9.54	SK6	614564	16.0
−43.23	79.25			16.0

$$\text{EFL} = 100.0$$
$$\text{BFL} = 79.25$$
$$\text{NA} = 0.185 \ (f/2.7)$$
$$\text{GIH} = 36.4 \ (\text{HFOV} = 20°)$$
$$\text{VL} = 39.74$$
$$\text{PTZ/F} = -2.11$$

Figure 16.13 A Cooke triplet of high speed and modest angular field. Compare aperture, field, **flint glass, and length** with Figs. 16.14 and 16.15. USP#2, 453, 260–1948, Pestrocov. This design was for a 1-in focal length 16-mm camera lens. Construction data and aberration plots for a focal length of 100.

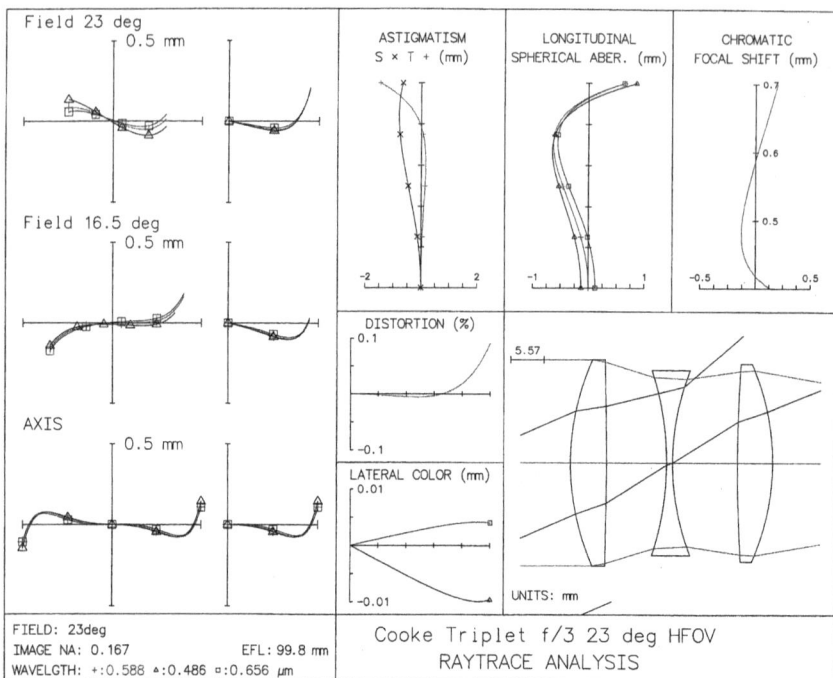

Field 23 deg
0.5 mm

ASTIGMATISM
S × T + (mm)

LONGITUDINAL
SPHERICAL ABER. (mm)

CHROMATIC
FOCAL SHIFT (mm)

Field 16.5 deg
0.5 mm

AXIS
0.5 mm

DISTORTION (%)

LATERAL COLOR (mm)

UNITS: mm

FIELD: 23deg
IMAGE NA: 0.167   EFL: 99.8 mm
WAVELGTH: +:0.588 ▲:0.486 □:0.656 μm

Cooke Triplet f/3 23 deg HFOV
RAYTRACE ANALYSIS

F/3 Cooke Triplet EP#155, 640–1919

radius	space	mat'l		s-a
40.10	6.00	SK4	613586	16.7
−537.00	10.0			16.7
−47.00	1.0	FN11	621362	15.0
40.00	10.8			15.0
234.50	6.0	SK4	613586	16.0
−37.90	85.26			16.0

EFL   = 99.76
BFL   = 85.26
NA   = 0.167 (f/3.0)
GIH   = 42.3 (HFOV = 23°)
VL   = 33.8
PTZ/F = −2.45

Figure 16.14 A Cooke triplet of moderate speed and coverage. Compare aperture, field, **flint glass, and length** with Figs. 16.13 and 16.15. English Patent #155, 640–1919. This design is of the type made for use in 35 mm slide projectors. Construction data and aberration plots for a focal length of 100.

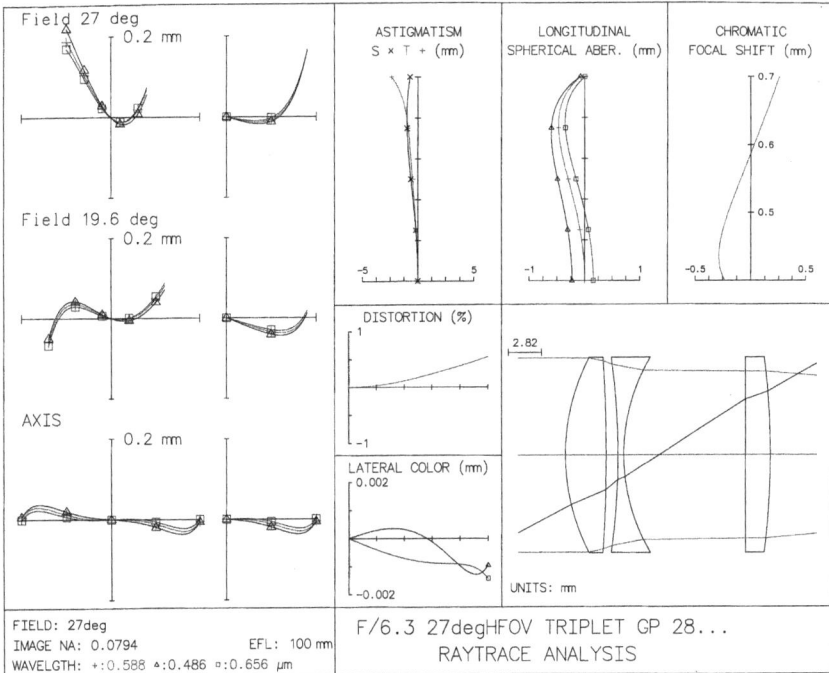

FIELD: 27deg
IMAGE NA: 0.0794   EFL: 100 mm
WAVELGTH: +:0.588 ▵:0.486 □:0.656 μm

F/6.3 27degHFOV TRIPLET GP 28...
RAYTRACE ANALYSIS

F/6.3 Cooke Triplet DRP#287, 089–1913

radius	space	mat'l		s-a
16.80	3.50	SK3	609589	8.0
−116.90	1.00			8.0
−56.30	0.50	LLF1	548457	8.0
15.40	3.00			8.0
STOP	7.30			6.89
plano	2.10	SK3	609589	8.0
−37.90	85.41			

EFL   = 100.24
BFL   = 85.41
NA   = 0.079 (f/6.3)
GIH   = 51.1 (HFOV = 27°)
VL   = 17.4
PTZ/F = −3.76

Figure 16.15 A Cooke triplet of small aperture and wide field coverage. Compare aperture, field, **flint glass, and length** with Figs. 16.13 and 16.14. German Patent #287, 089–19913. Construction data and aberration plots for a focal length of 100.

a set of simultaneous equations of the form of Eq. 16.15 are set up and solved. In Eq. 16.15, ΔA_n is the desired change in aberration A_n, ΔC_i is the change in the parameter C_i required to produce the change, and $(\delta A_n/\delta C)_i$ is the partial of the aberration A_n with respect to C_i. If available graphs of the aberrations versus the parameter changes can help to avoid problem areas, especially when quadratic (or higher) relationships are involved.

$$\Delta A_n = \sum_{i=1}^{i=K} \left(\frac{\delta A_n}{\delta C} \right)_i \Delta C_i \qquad (16.15)$$

16.8 Automatic Design by Electronic Computer

A simple linear solution

The fantastically high computation speed of the electronic computer makes it possible to perform a major portion of the optical design task on an "automatic" basis. One possible program is essentially a duplication of the process that a designer goes through in correcting the primary aberrations of a system. The computer is presented with an initial prescription and a set of desired values for a limited set of aberrations. The program then computes the partial differentials of the aberrations with respect to each parameter (curvature, spacing, etc.) which is to be adjusted, and establishes a set of simultaneous equations (Eq. 16.15), which it then solves for the necessary changes in the parameters. Since this linear solution is an approximate one, the computer then applies these changes to the prescription (assuming that the solution is an improvement) and continues to repeat the process until the aberrations are at the desired values. When there are more variable parameters than system characteristics to be controlled, there is no unique solution to the simultaneous equations; in this case, the computer will add another requirement, namely that the sum of the squares of the (suitably weighted) parameter changes be a minimum. This allows a solution to be found and has the added advantage that it holds the system close to the original prescription. Since the solution of simultaneous equations may call for excessively large changes to be applied, the computer is usually instructed to scale down the changes if they exceed a certain predetermined value.

This "simultaneous" technique is a useful one. Even modest-sized computers are capable of handling this problem without difficulty and several inexpensive computer programs of this type are available, often based on third-order aberration contributions. Since the designer is in rather close control of the situation, this technique is, in effect, simply an automation of conventional methods as described in the preceding section. Thus, the designer should have a fairly good knowledge of the system, and the system must have a solution reasonably close to the initial prescription. This type of approach is very efficient for making modest changes in designs or for touching-up a design. It also makes easy work of systems with exceedingly complex interrelationships of the variables, such as the older meniscus anastigmats of the Dagor or Protar type.

Fully automatic lens design optimization

Of course it is not really *automatic*, but that is what it is called. To illustrate and explain the functioning of a modern optical design program we will describe the steps that a lens designer might follow in order to design a simple Cooke triplet anastigmat. We assume that the design form has been chosen (i.e., the triplet) and that the requirements such as field coverage, relative aperture, resolution, etc. are reasonable and appropriate for the design type.

The first step is to produce a suitable **starting design**. Very often the starting form is simply one taken from the literature, patent data, or the lens library of the design program software. It can also be prepared using the interactive lens drawing capability of the program. The designer inputs a very rough design, guessing at the radii, thicknesses, glasses and spacings. When the lens is drawn on the monitor screen, the dimensions are adjusted so that the element thicknesses, edge and center, are appropriate, and the vignetting and the ray paths are reasonable. The system data is adjusted or scaled so that the focal length is very close to the desired focal length and any other spatial requirements are close to being met. It is extremely important that the starting form be close to the desired solution, or the program may find an "optimum" form which does not meet the basic dimensional requirements. In the starting form for a triplet, a reasonable vertex length may equal the clear aperture. Setting the program "pick-ups" so that the last surface curvature equals minus the first, and so that the flint is equi-concave can be a convenience. What is important at this stage is that the lens have the right focal length and that it look like a triplet.

The glass types must be selected. The outer crown glasses are chosen from the upper-left area of the glass map (dense barium crowns or lanthanum types) based on factors such as cost, durability, high index and high V-value. From a design standpoint there is no reason not to use the same glass type for both crowns. The center flint glass must be a variable in order that the vertex length can be at the best value; to start, one may choose any reasonable flint from along the glass line. If the field of view is relatively small and the NA is to be large, a low V-value flint is a good start. For a wide field, slow speed lens a higher V-value is a good choice.

A **merit function** is required. The merit function is typically the sum of the squares of calculated characteristics which are deemed undesirable. These may be aberrations, departures from desired dimensions and the like. They are often referred to as *operands*, or targets. Obviously a more appropriate name for the merit function would be **error function**, or **defect function**. The aim is to define and weight the merit function so that, as a single number, its value accurately represents the "worth" of the system. Representing the performance of a lens system with a single number is a tricky business, and care must be exercised in designing the merit function. The smaller the value of the merit function, the better the system. Most software programs have two or three **default** merit functions available to the user. One type uses calculated aberrations as its operands. This type might have as operands the marginal and zonal spherical, axial and transverse chromatic, field curvature (x_S and x_T), spherical and coma at several field angles, plus the slopes and the curvatures of the ray intercept plots. This type is more efficient, but less robust than a second type of merit function which uses RMS spot sizes or RMS OPD. The rays used are chosen and weighted based of a form of Gaussian quadrature, so that a very exact value for the RMS operand can be obtained from a limited number of rays.

For a start it is sometimes worthwhile to optimize the first-order characteristics and the third-order aberrations before starting on the OPD or blur spot merit functions. The reason this added step is worthwhile is that "real" rays can fail if a ray misses a surface or if the ray is totally reflected at a surface. A ray failure *may* cause the optimization process to fail. However paraxial rays do not fail, no matter how ridiculous the lens you start with, and the third-order aberrations are based on paraxial rays. One simply defines a merit function with the focal lengths and the third-order aberrations as the operands; the DLS (see below) program will take care of it nicely.

The default merit function usually requires some modification to suit the application. Such things as focal length, object to image distance, back focus, aperture sizes, and the like may need to be added to the default merit function. [There is a useful alternative to putting the focal length in the merit function. An "angle solve" on the last surface will solve for the surface curvature which will yield the angle (u'_K) of the axial marginal paraxial ray which produces the desired focal length and NA. This works well if the last surface is not close to the focus.] Further along in the design process the operand weightings may be adjusted, or more operands may be added to the merit function to adjust the aberration balance or to control other characteristics. For our triplet design a focal length control is mandatory.

Next, the **variables** are defined. These usually include the surface curvatures, the airspaces and some glasses (which must be constrained to a realistic area of the glass map). Thicknesses may be varied, subject to constraints on edge thickness and the like. Most programs have "pick-ups" which will set the curvature, thickness or spacing to ± that of a preceding surface. For our triplet we would likely vary all the curvatures, both airspaces, and the glass of the center element (constraining it to lie along the "glass line"). For a "standard" triplet the thickness of the elements tends to be minimal, and (unless they are extreme) they are ineffective variables. Variables can be "bounded" to limit them to reasonable values.

Next (after saving the lens and merit function in a separate file as insurance against accidents) the lens is subjected to one or two iterations of the **damped least squares (DLS)** optimization routine. Using a set of equations like Eq. 16.15, the merit function operands are defined as the amount by which each ΔA departs from a desired value (which is usually the amount which will reduce A to zero). The least squares program solves for the set of the variables which will produce the smallest merit function. (The "standard" least squares equations can be found in most mathematics handbooks.) The solution process assumes that the relationships between the variables and the operands are linear. A quick look at the material in Chaps. 5 and 6, is sufficient to indicate that even in the low-order aberration region, non-linearity is the order of the day. (Rudolf Kingslake once wisely observed that "A lens is about as nonlinear as anything in physics.") In order to prevent the nonlinear relationships from producing ridiculous and physically impossible solutions, the least squares solution is modified by adding to the merit function the suitably weighted squares of the changes in the variables. This serves to damp down the size of the changes to the variables, especially the big changes. The nonlinear relationships mean that the solution will be an approximate one; the process is repeated until it converges to a minimum value for the merit function.

If the changes to the initial design produced by the DLS routine are not suitable, it is likely that an adjustment to the starting data and/or to the merit function will be necessary. Listing or printing out the prescription, the merit function, the variables (and their bounds) usually locates the problem(s). A drawing of the lens plus a few ray traces on the monitor can sometimes identify the trouble spot. If one or two variables are headed for extreme values, a suitable boundary may fix the problem.

If the initial design changes are reasonable, the DLS process is continued until a (local) optimum is reached. The optimum is a design where any small variable changes (or any possible combination of small changes) will cause the merit function value to become larger. This may or may not be a good solution to the design problem. There may be many local optima even for a very simple lens. For example, consider the meniscus "landscape lens" which has two obvious optima, one with the aperture stop in front and the other with the aperture stop in the rear. Or consider the Gauss and Fraunhofer forms of telescope objectives which represent different solutions to a quadratic equation. The simple triplet anastigmat has many, many local optima, most of them very bad ones. One of the reasons for trying to start a design task with a good looking prescription is that it gets you into a good solution neighborhood and keeps you away from the bad ones.

One way to sharpen up the search for a better, perhaps different, optimum is to switch merit functions. For example: change from an RMS type to an aberration type; or change from a spot diagram merit function to an OPD merit function. (Incidentally an RMS OPD merit function tends to produce better MTF results, and the spot size RMS merit function will produce more compact blur spots.) Another technique is to make large, arbitrary changes to one or two variables and then reoptimize, freezing them for a few cycles and then releasing them to vary again. Sometimes a series of these shocks to the system will produce a very significant improvement in the performance after re-optimizing. It is just a way of finding a different local optimum.

One can visualize the merit function as a location in an n dimensional space, where the value of the merit function is a function of the n variables. For our triplet we have six curvatures, two airspaces and the flint glass type as the variables. Thus our merit function is in a nine dimension space, and our task is to create a nine dimensional solution vector. This is quite difficult to visualize. It is much easier if we visualize a two variable system, where the merit function is analogous to a landscape, and the variables are the coordinates, and the merit function value is the elevation of a point in the landscape. We are trying to locate the lowest point in the landscape (without any knowledge of the geography except in the immediate neighborhood of our starting point). If we simply travel downhill in the direction of the steepest slope we will get to a local optimum (where any direction is up). Unfortunately we have no idea of the nature of the landscape other than that in the immediate vicinity of our solution. There may be a different, splendid optimum nearby. The only way we can find out is to repeat our process from a large number of different starting points.

This latter is basically what the "global optimization" programs do. Most of these programs are modifications or improvements of a process called "generalized simulated annealing." Here, the program **randomly** selects all of the (variable) lens dimensions (within a limited range and according to a given probability distribution). The resulting prescription is evaluated, and if it is better than the previous form it is unconditionally accepted, and the process goes on. If the new version is worse, it may be accepted on the basis of random chance, weighted by a probability function which reduces the chance of acceptance in proportion to the amount that the new lens is worse than the old. Where the DLS process will very efficiently home in on the nearest optimum, this random choice approach allows the possibility of escape from a local optimum. This process works, but you have to kiss an awful lot of frogs before you find a prince. Typically these programs are allowed to run for several hours (or even days) and they find a large number of different solutions, which are then touched up with a DLS program.

In any case, it is apparent that since the DLS design program will seek out the nearest minimum, the selection of the starting point for the process is vitally important. In fact, once the merit function is defined and weighted, the starting design form uniquely defines a single minimum. Obviously the choice of the starting form is a critical factor. Fortunately, it seems that with most merit functions, most nonsimple design types have relatively broad, flat minima, and one can choose a starting point over a fairly large volume in solution space and expect a reasonably good result. An experienced lens designer uses knowledge of successful design types and features to direct the computer to good starting points. The novice designer should study the standard, classical design forms as an aid in selecting appropriate starting points.

The mathematics of the DLS process are written up in many places. Two which explain the basic operations are G. Spencer, "A Flexible Automatic Lens Correction Procedure," *Applied Optics,* vol. 2, 1963, pp. 1257–1264, and W. Smith, in W. Driscoll (ed.), *Handbook of Optics,* New York, McGraw-Hill, 1978. Simulated annealing is discussed in Jones A.E.W. and G.W. Forbes, J. Global Optim., 6, 1–37, 1995.

16.9   Practical Considerations

The following is a partial list of certain design characteristics which, although they may be quite beneficial to the performance of a design, tend to have an undesirable effect on the difficulty and cost of fabrication. Thus, unless you enjoy being unpopular with the opticians who must execute your designs, this list represents things which you should assiduously avoid if at all possible.

1. Materials which are soft and easily abraded.

2. Materials which are thermally fragile and which may split from a mild thermal shock, such as that encountered in blocking or washing under a hot or cold water tap.

3. Materials with low acid resistance or high stain characteristics.

4. Expensive materials. (Often you can find a similar, cheaper glass which is nearly as good.)

5. Thin elements, i.e., those with a large ratio of diameter to the average thickness. Such elements can deform under the stress of blocking or polishing, making an accurate surface geometry almost impossible to produce. Note that a negative element with a substantial edge thickness often can tolerate a center thickness which would be too thin for a weaker element.

6. Thin-edged elements chip easily and, if processed at a diameter larger than the finished one, may become sharp-edged during fabrication. Also a thin-edged element is difficult to mount satisfactorily.

7. A very thick element obviously requires more material and may require an awkward arrangement when blocked. Visualize Fig. 20.2 if the elements are as thick as the diameter. A thin lens with the same radius can have more lenses blocked on a tool because they can be placed closer together at the surface; with the thick lens, there are large gaps between the elements at the surface which make polishing difficult.

8. Very "strong" curves (i.e., with a large diameter-to-radius ratio) lead to fewer elements blocked per tool and the correspondingly increased processing costs, difficulty in polishing surfaces accurately, and difficulty in testing the surface accuracy with a test plate or interferometer.

9. Meniscus elements the surfaces of which are concentric or nearly concentric with each other. A monocentric element must be ground and polished so that the two surfaces are properly aligned during these operations; it cannot be "centered" after polishing as an ordinary element can.

10. Nearly equiconvex or equiconcave elements can cause trouble in assembly because it is difficult to tell one side from the other, and the element is liable to be mounted backward.

11. Weakly curved, nearly plane surfaces are more expensive to tool and fabricate than a plane surface. It is almost always possible to force such a design to a plane surface with little or no sacrifice in image quality.

12. Precision bevels. If possible, avoid mounting from a beveled surface. Use a loosely toleranced 0.5 mm by 45° chamfer to eliminate sharp edges; this kind of edge break is almost free.

13. Avoid odd-angle precision bevels. Many shops are tooled for 45°, 30°, or 60°; other angles may require new tooling.

14. Cemented triplets and quadruplets are unpopular in some shops.

15. Tight scratch and dig specifications on surfaces which are not visible to the ultimate customer are usually a waste of money. With a few exceptions (such as surfaces near an image plane or the optics of a high-powered laser system), scratch and dig considerations are purely cosmetic and have no functional effect (unless the lens aperture is so small that a dig can actually obstruct a significant fraction of the beam area).

16. Tight tolerances in general. See Chap. 20 for a discussion of efficient tolerance budgeting.

17. Avoid thin, narrow airspaces between surfaces, especially if the ray slopes are large in the space. Note that the space may be correcting a high-order aberration created elsewhere in the system. Such spaces often need extremely tight fabrication tolerances.

18. In general, just avoid tight tolerances if you can.

19. Another fabrication consideration arises in the course of the design process from the choice of the center thickness bounds. Obviously the thickness bound for a positive element must consider the edge thickness, and the bound for a negative element must not be so small that the element cannot be fabricated. But too large a value placed on the center (or edge) thickness bound may force the design toward elements of large diameter, volume, and cost. This occurs because the element thickness pushes the far surface further away from the stop; the ray beams spreading from the stop then require larger clear apertures for the elements, which then require increased center thickness, and so on in a vicious circle. Violating an arbitrary minimum thickness limit by a bit may improve the situation markedly. It is a sort of snowball or domino effect; a thinner element allows a smaller diameter, which allows a thinner lens and smaller diameters for subsequent elements, which in turn can be thinner, and so on.

Bibliography

Conrady, A., *Applied Optics and Optical Design,* Oxford, 1929. (This and Vol. 2 were also published by Dover, New York.)
Cox, A., *A System of Optical Design,* Focal, 1965 (lens construction data).
Dictionary of Applied Physics, Vol. 4, London, Macmillan, 1923.
Farn, M. W., and W. B. Veldkamp, "Binary Optics," in *Handbook of Optics,* Vol. 2, New York, McGraw-Hill, 1995, Chap. 8.

Fischer, R. (ed.), *Proc. International Lens Design Conf.,* S.P.I.E., Vol. 237, 1980.

Greenleaf, A., *Photographic Optics,* New York, Macmillan, 1950.

Herzberger, M., *Modern Geometrical Optics,* New York, Interscience, 1958.

Jacobs, D., *Fundamentals of Optical Engineering,* New York, McGraw-Hill, 1943.

Kidger, M., *Fundamental Optical Design,* SPIE, 2002.

Kidger, M., *Intermediate Optical Design,* SPIE, 2004.

Kingslake, R. (ed.), *Applied Optics and Optical Engineering,* Vol. 3, New York, Academic, 1965 (lens design).

Kingslake, R., *Lens Design Fundamentals,* New York, Academic, 1978.

Kingslake, R., *Lenses in Photography,* Garden City, 1952.

Linfoot, E., *Recent Advances in Optics,* London, Clarendon, 1955.

Martin, L., *Technical Optics,* New York, Pitman, 1950.

Merte, W., *Das Photographische Objektiv,* Parts 1 and 2, translation, CADO, Wright-Patterson AFB, Dayton, 1949.

Merte, Richter, and von Rohr. *Handbuch der Wissenschaftlichen und Angewandten Photographie,* Vol. 1, 1932; *Erganzungswerke,* 1943, Vienna, Springer. Reprinted by Edwards Brothers, 1944 and 1946 (lens construction data).

Merte, *The Zeiss Index of Photographic Lenses,* Vols. 1 and 2, CADO, Wright-Patterson AFB, Dayton, 1950 (lens construction data).

MIL-HDBK-141, *Handbook of Optical Design,* 1962.

Peck, W., in Shannon and Wyant (eds.), *Applied Optics and Optical Engineering,* Vol. 8, New York, Academic, 1980 (automatic lens design).

Rodgers, P., and M. Roberts, "Thermal Compensation Techniques," in *Handbook of Optics,* Vol. 1, New York, McGraw-Hill, 1995, Chap. 39.

Rosin, S., "A New Thin Lens Form," *J. Opt. Soc. Am.,* Vol. 42, 1952, pp. 451–455.

Sinclair, D. C., "Optical Design Software," in *Handbook of Optics,* Vol. 1, New York, McGraw-Hill, 1995, Chap. 34.

Smith, W. J. (ed.), *Lens Design,* S.P.I.E., Vol. CR41, 1992.

Smith, W. J., *Modern Lens Design,* New York, McGraw-Hill, 1992.

Smith, W., in W. Driscoll (ed.), *Handbook of Optics,* New York, McGraw-Hill, 1978.

Smith, W., in Wolfe and Zissis (eds.), *The Infrared Handbook,* Washington, Office of Naval Research, 1985.

Taylor, W., and D. Moore (eds.), *Proc. International Lens Design Conf.,* S.P.I.E., Vol. 554, 1985.

Exercises

The exercises for this chapter take the form of suggestions for individual design projects; as such, there can be no "right" answers, and none are given. The effort involved in each exercise is considerable, and it is likely that only those interested in obtaining first-hand experience in optical design will wish to undertake these exercises. The casual reader will, however, be amply rewarded by mentally reviewing the steps he or she would follow in attempting the exercises.

1 Design a symmetrical double-meniscus objective of the periscopic type. Select a bending (a ratio of 3:2 for the curvatures is appropriate), determine the proper spacing for a flattened field, and calculate the thin-lens third-order aberrations for the combination. Analyze the final design by raytracing and compare the results with the third-order calculations. The student may wish to repeat the process for several additional bendings, perhaps including the Hypergon (Fig. 16.4), and to compare the results of each, noting the variations of aperture and coverage.

2 Design an airspaced (edge contact) achromatic doublet objective using BK7 (517:642) and SF2 (648:339). Correct the spherical aberration for an aperture of f/3.5. Raytrace marginal and zonal rays in C, d, and F light to evaluate the axial image. Compare the coma obtained by raytracing an oblique fan with the OSC calculation. Do a Fraunhofer and a Steinheil lens, vary the airspace to correct zonal spherical and spherochromatism.

3 Design an f/3.5 telescope objective lens consisting of a BK7 singlet and a doublet of BK7 and SF2. Vary the distribution of powers and the spacing to optimize the correction of zonal spherical and spherochromatic. Note that there are four possible configurations, which is best?

4 Design a Cooke triplet anastigmat. For a minimal exercise, duplicate the design of Fig. 16.14 using the same glasses and the same power and space layout as a starting point. For a more ambitious project, design the same lens, but derive the power and space layout without recourse to the data of the figure. Do a design with N-LaSF31 (881410) for the crowns. (Note: use the same field and f-number as in Fig. 16.14.)

17

Lens Design for Eyepieces, Microscopes, Cameras, etc.

17.1 Telescope Systems and Eyepieces

The design of a telescopic system begins with a first-order layout of the powers and spacings of the objective, erectors, field lenses, prisms, and eyepiece, as required to produce the desired magnification, field of view, aperture (pupil), eye relief, and image orientation. Then the individual components are designed so that the telescope, as an entire system, is corrected. Often the eyepiece is designed first; the design is carried out as if the eyepiece were imaging an infinitely distant object through an aperture stop located at the system exit pupil. That is, the rays are traced in the reverse direction from the direction in which the light travels in the actual instrument. Usually a principal ray is traced from the objective (or the aperture stop) through the eyepiece to locate the exit pupil, then an oblique bundle can be traced in the reversed direction (from the eye) to evaluate the off-axis imagery. Almost all optical design is done in this manner, by tracing rays from long conjugate to short, largely for convenience, because the focus variations (due to aberrations and small power changes) are smaller and more readily managed at the short conjugate.

The erectors, if there are any, are usually designed next; their design is frequently included in the eyepiece design by considering the erector and eyepiece as a single unit. (Alternatively, the erector may be considered as a part of the objective; the choice is usually determined by the location of the reticle.) Usually the objective is designed last and its spherical and chromatic aberrations are adjusted to compensate for any undercorrection of the eyepiece. Note that prisms must be included in the design process if they are "inside" the system, since they contribute aberrations which must be offset by the objective and eyepiece. Prisms can be introduced into the calculation as plane parallel plates of appropriate thickness. (See Sec. 7.8)

An eyepiece is a rather unusual system, in that it must cover a fairly wide field of view through a relatively small aperture (the exit pupil) which is *outside* the system. The external aperture stop and wide field force the designer to use care with regard to coma, distortion, lateral color, astigmatism, and curvature of field; the first three mentioned can become unusually difficult, since even approximate symmetry about the stop (which is used in many lens systems to reduce these aberrations) is not possible. On the other hand, the small relative aperture of an eyepiece tends to hold spherical and axial chromatic aberrations to reasonable values. Typically an eyepiece is fairly well corrected for coma for one zone of the field (a fifth-order coma of the y^2h^3 type is common in wide-angle eyepieces) and the field is sometimes artificially flattened by overcorrected astigmatism which offsets the undercorrected Petzval curvature. Lateral color may or may not be well corrected; frequently some undercorrection exists to offset the effect of prisms. There is almost always some pincushion distortion apparent (note that when an eyepiece is traced from long to short conjugate, the sign of the distortion is reversed). An eyepiece can be considered "reasonably" corrected for distortion if it has 3 to 5 percent; 8 to 12 percent distortion is not uncommon in eyepieces covering total fields of 60° or 70°. One way to eliminate this distortion is by the use of aspheric surfaces, a not very attractive solution unless molded surfaces are used. One should remember that, in many applications, the function of the outer portion of the field of view is to orient the user and to locate objects which are then brought to the center of the field for more detailed examination. Thus, eyepiece correction off axis need not be as good as that of a camera lens, for example.

Spherical aberration of the pupil can be a problem. This is especially so in wide angle eyepieces. The exit pupil is the image of the aperture stop (usually at the objective) formed by the eyepiece. Undercorrected pupil spherical causes the pupil to be imaged closer to the eyepiece for images at the edge of the field, so that in order to see the outer field the eye must be moved in toward the eyelens. In severe cases, when the eye is located in a compromise position, the center and edge of the field are fully illuminated, but the zonal field is not. This shows up as a dark "kidney bean" shaped area in the zonal field.

This problem is exacerbated by the fact that, as the eyeball is rotated to view the outer field, the pupil of the eye swings off the axis, so that the eye must be translated to see the edge of the field. This pupil translation is a problem in wide field eyepieces whether there is pupil spherical or not.

Because the eyepiece is subject to a final evaluation by a visual process, it is sometimes difficult to predict, from raytracing results alone just what the visual impression will be. For this reason, it is frequently useful to begin an eyepiece design on the lens bench, by mocking up an eyepiece out of available stock elements. A series of mockups will yield a good grasp of the more promising orientations and arrangements of the elements. The designer can then use these as starting points for the design effort with reasonable assurance that the visual "feel" of the finished design will be acceptable.

Note that the conventional correction of distortion (where $h = f \tan \theta$) causes the apparent angular size of an image to change as it is scanned across the field. A distortion which yields the relationship $h = f\theta$ will give a constant angular size; this is a common type of distortion found in many eyepieces.

Field curvature (defocus) causes a "swimming" effect of the image as the eye is scanned across the system pupil. Usually a field curvature of about 2 diopters or less (at the eye) is considered good; 4 diopters is about the maximum acceptable although 6 diopters is not uncommon.

The Huygenian eyepiece.   The Huygenian eyepiece (Fig. 17.1a) consists of two plano-convex elements, an eyelens and a field lens, with the plane surface of each toward the eye. The focal plane is between the elements. For a given set of powers of the elements, the spacing can be adjusted

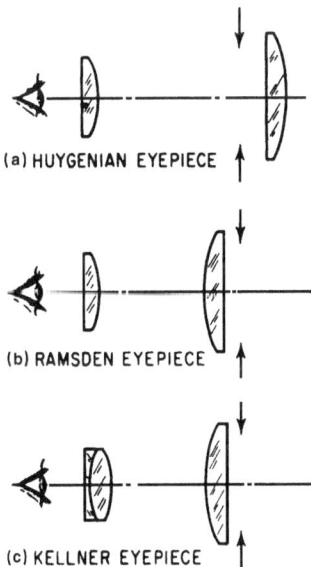

(a) HUYGENIAN EYEPIECE

(b) RAMSDEN EYEPIECE

(c) KELLNER EYEPIECE

Figure 17.1   Three basic eyepiece forms.

to eliminate lateral color. The required spacing is approximately equal to the average of the focal lengths of the elements. The only remaining degree of freedom is the ratio of powers between the elements. This is used to eliminate coma (and thus artificially flatten the field via the "natural" stop position, as discussed in Sec. 16.2). Since the image plane is between the lenses and is viewed by the eyelens alone, it is not well corrected and is unsuitable for use with a reticle. The eye relief of the Huygenian is often uncomfortably short.

The Ramsden eyepiece.   The Ramsden eyepiece (Fig. 17.1b) also consists of two plano-convex elements, but the plane surface of the field lens faces away from the eye. The spacing is made about 30 percent shorter than the Huygenian to allow an external focal plane, and for this reason lateral color is not fully corrected. Coma is corrected as in the Huygenian by varying the ratio of field lens power to eyelens power. The Ramsden eyepiece can be used with a reticle.

The Kellner eyepiece.   The Kellner (Fig. 17.1c) is simply a Ramsden eyepiece with an achromatized eyelens to reduce the lateral color. It is frequently used in low-cost binoculars.

The relative characteristics of the three simple eyepieces described above are summarized in the table of Fig. 17.2. They are almost invariably made in plano-convex form and little is gained by departing from this form. Since these eyepieces are chiefly noted for their low cost, the usual material for the single elements is common crown; indeed, they

Relative characteristics	Huygenian	Ramsden	Kellner
Spherical	1	0.2	0.2
Chromatic (axial)	1	0.5	0.2
Lateral color (CDM)	0.0	0.01	0.003
Distortion	1	0.5	0.2
Field curvature	1	~0.7	~0.7
Eye relief	1	1.5 to 3	1.5 to 3
Coma	0.0	0.0	0.0
MP tolerance*	1	5	5
efl ratio, high power†	2.3	1.4	0.8
efl Ratio, low power†	1.3	1.0	0.7

*The MP tolerance is the relative ability to retain the desired state of correction when used at magnifications other than that for which the eyepiece was originally designed. †Ratio of the focal length of the field lens to the focal length of the eyelens; high and low power refer to the power of the telescope with which the eyepiece is to be used.

Figure 17.2   The relative characteristics of three simple eyepieces.

are frequently made from selected window glass by grinding and polishing only the convex surface. In the Kellner eyelens, the index difference across the cemented face is critical; usually a light barium crown is used to keep the overcorrection of the astigmatism from becoming too large when a wide field of view is desired. Departure from the plano-convex form, in favor of a biconvex shape, is not uncommon in the Kellner eyepiece. The half-field covered by these eyepieces is to the order of $\pm 15°$, more or less, depending on the performance required.

The compensating eyepiece. In some optical systems, especially high powered microscope objectives with aplanatic front elements, lateral color cannot be corrected. The compensating microscope eyepiece is designed to have laterral color which matches and corrects for the lateral color of the objective. There are many design forms; some are modifications of the standard eyepieces. One form has a doublet eyelens and a meniscus shaped field lens made of flint glass. The field stop is between the lenses (as in the Huygenian).

The orthoscopic eyepiece. The orthoscopic eyepiece (Fig. 17.3a) consists of a single-element eyelens (usually plano-convex) and a cemented triplet (usually symmetrical). The eyelens is frequently of light barium crown or light flint glass and the triplet is composed of borosilicate

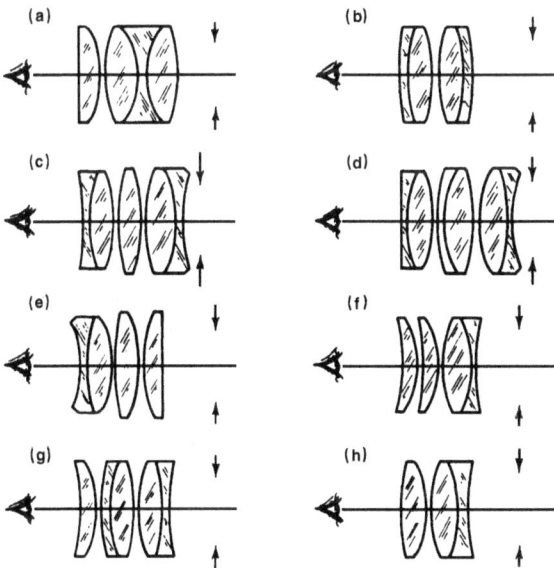

Figure 17.3 Eyepiece designs. (a) Orthoscopic; (b) symmetrical; (c) Erfle; (d) Erfle; (e) Berthele; (f) see also Fig. 19.8 for an example of this design; (g) an Erfle modification; (h) Koenig.

crown and dense flint glass. This is a better eyepiece than the preceding simple types and is used for half-fields of $\pm20°$ to $\pm25°$. The Petzval curvature is about 20 percent less than that of the Ramsden or Kellner, although higher-order astigmatism causes a strongly backward-curving tangential field at angles of more than 18° or 20° from the axis. (This high-order astigmatism is the characteristic which limits the field coverage of most eyepieces; some control can often be achieved by lowering the index difference across cemented surfaces.) The eye relief is long, to the order of 80 percent of focal length. Distortion correction is quite good.

The symmetrical, or Plössl, eyepiece. This excellent eyepiece is composed of two achromatic doublets (usually identical) with their crown elements facing each other (Fig. 17.3b). It is usually executed in borosilicate crown (517:642) and extra dense flint (649:338) glass, although it can be improved a bit by raising the index of both elements. It shares the long eye relief ($0.8F$) and field characteristics of the orthoscopic, but is in general a somewhat superior eyepiece, except that its distortion is typically 30 to 50 percent greater than the orthoscopic. It is widely used in military instruments and as a general-purpose eyepiece of moderate (to $\pm25°$) field. A similar eyepiece with both flints facing the eye is occasionally used. See Fig. 19.8 for an example of a symmetrical eyepiece.

The Erfle eyepiece. This eyepiece (Fig. 17.3c) is probably the most widely used wide-field ($\pm30°$) eyepiece. The eye relief is long ($0.8F$), but working distance is quite short. The Petzval sum is about 40 percent less than the orthoscopic or symmetrical types because of the field-flattening effect of the concave field lens surface, and distortion is about the same as the orthoscopic (for the same angular field). The type shown in Fig. 17.3c usually has undercorrected lateral color (for use with erecting prisms) which can be reduced by use of an achromatic center lens as in Fig. 17.3d. Glasses used are usually dense barium crown and extra dense flint. An example of an Erfle eyepiece is shown in Fig. 19.9.

Magnifiers. Magnifiers and viewer lenses are basically the same as eyepieces, with one notable exception: There is no fixed exit pupil. This means that the eye is free to take almost any position in space and therefore the aberrations of the magnifier must be insensitive to pupil shift. For this reason, magnifiers tend to be symmetrical in configuration. Two plano-convex lenses with convex surfaces facing or a symmetrical (Plössl) construction are common for better-grade magnifiers. Where cost is important and a single element must be used, the following arrangements are good. If the eye is always close to the magnifier, use a plano-convex form with the plano surface toward the eye. If the eye is always far from the magnifier, use a plano-convex form with the convex surface toward the eye. If the eye position is variable, as in a general-purpose magnifier, an equiconvex form is probably the best compromise. Figure 19.5 is an example of a doublet magnifier.

Triple aplanat (Steinheil or Hastings triplets). A very high quality magnifier or "loupe" is the triple aplanat, often attributed to Hastings or Steinheil. It is a symmetrical cemented triplet, consisting of a strong central equi-convex crown (522595 or 517642) element with identical outer meniscus flint (717295, 649338, or 620364) elements. Its diameter is usually about 60 percent of the focal length, and the vertex thickness is large, from 45 percent to 60 percent of the focal length. The working distance is about 80 percent to 90 percent of the focal length. The image quality is high, with spherical, coma and chromatic corrected, and minimal distortion. The magnifying power ranges from 5 to 20.

Note that the eyepieces of instruments which use an electronic image tube, such as the Sniperscope, fall into the category of magnifiers, since they are used to view a diffuse image on the phosphor surface of the image tube. As such they must be designed so that they perform well with the eye in a wide range of locations.

The optics of tabletop slide viewers, "head-up displays," or HUDs, and many simulators not only fall into this category but also share the requirement that both eyes view the image through a single optical system. Such systems are called *biocular* (as opposed to *binocular* systems, in which both eyes are used, but in which each eye views the image through a separate optical train). In a biocular system one must not only be concerned about the effects of eye motion but must also be concerned about any disparity between the images as seen by the two eyes. The convergence, divergence, and dipvergence (the vertical difference in direction) required of the eyes as they view the image must be carefully considered in designing the system. Thus a biocular device is designed for a pupil large enough to encompass both eyes plus any head motion, although the image sharpness and resolution are determined by the aberrations of a pupil the size of which is defined by that of the viewer's eye.

Diopter adjustment (focusing) of eyepieces. In binocular systems, one eyepiece is usually focusable to compensate for any difference in focus between the two eyes. The motion of the eyepiece is

$$\delta = 0.001 f^2 D \text{ m}$$

or

$$\delta = 0.0254 f^2 D \text{ in}$$

where f is the eyepiece focal length, and D is the shift of the image position in diopters (relative to the second focal point of the eyepiece—where the eye is presumed to be located). The usual adjustment range is ± 4 diopters.

Erectors.   Erector systems come in all sizes and shapes. Occasionally a single element may serve as an erector, or two simple elements in the general form of a Huygenian eyepiece may be used, as in the terrestrial eyepiece shown in Fig. 17.4a. This form of eyepiece is widely used in surveying instruments, occasionally with an achromatic eyelens. A popular erector for gun scopes is illustrated in Fig. 17.4b and consists of a single element plus a low-power, overcorrecting doublet, often meniscus in shape. Photographic type lenses are occasionally used as erectors, near symmetrical forms of the Cooke triplet, the Dogmar, or the double-Gauss being the most popular. Probably the most widely used erector consists of two achromats, often meniscus, crown elements facing, with a modest spacing between them.

As previously mentioned, erectors are usually designed in conjunction with either the eyepiece or objective of a telescopic system. Considerable care should be taken in the first-order layout of any telescope in order to be certain that the work load placed on the erector is not impossibly large. The introduction of suitable field lenses is often necessary to reduce the height of the principal ray at the erector,

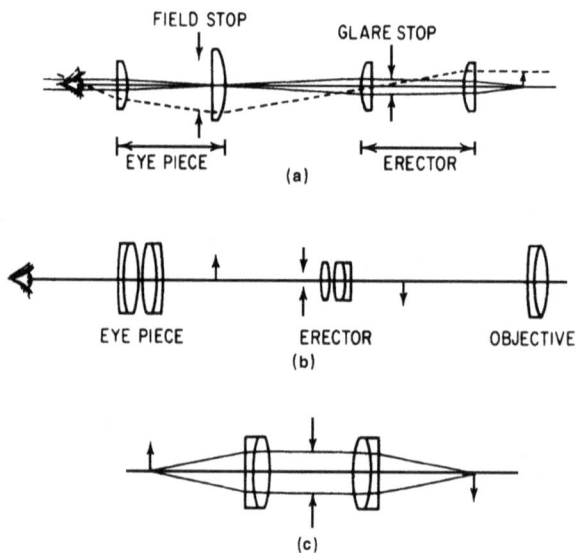

Figure 17.4   Erector systems. (a) The four-element terrestrial erecting eyepiece. (b) Typical gunsight optical system. (c) Symmetrical doublet erector.

although this does produce an undesirable increase in the Petzval curvature. Note that many erectors have external pupils, often in the form of a glare stop.

Objective systems. For most telescopic systems, the objective will be an ordinary achromatic doublet, or one of the variations described in Sec. 16.5. A photographic-type objective may be used where a wide field is desired, Cooke triplets and Tessars being the most commonly used. A Petzval objective is useful when high relative apertures are necessary; the construction of a Petzval objective (Sec. 17.3) is such that its rear lens acts as a sort of field lens, and this characteristic is occasionally useful. For high-power telescopes where it is desirable to keep the system as short as possible, a telephoto type of construction is valuable. The front component is an achromatic doublet and the rear is a negative lens, either simple or achromatic. The focal length is usually 20 to 50 percent longer than the overall length of the objective. Either the Petzval or telephoto type of objective can be used as an internal focusing objective (Fig. 17.5), where focusing is accomplished by shifting the rear (inside) component, making a more easily sealed instrument. Surveying instruments and theodolites conventionally use the telephoto form with the focusing lens located about two-thirds of the way from the front component to the focal plane so that the stadia "constant" will remain constant as the instrument is focused. Alignment telescopes use a positive focusing lens of high power placed near the focal plane at infinity focus; thus, a modest shift of the focusing lens toward the front component allows the system to be focused at extremely short distances, or even on the objective lens itself. Note

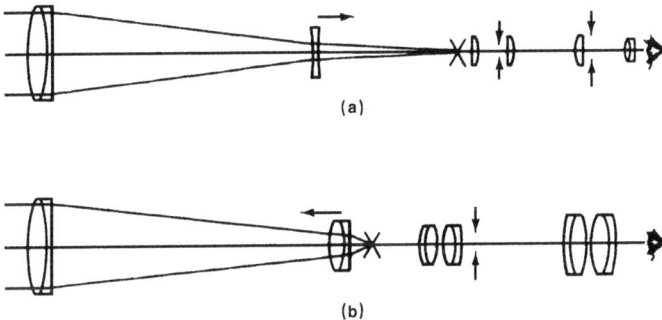

(a)

(b)

Figure 17.5 Telescopic systems. (a) Typical surveying telescope with negative focusing lens and terrestrial eyepiece. Note that the objective is telephoto, in that its effective focal length is longer than the objective. (b) Alignment telescope. The strong positive focusing lens, when shifted forward, allows the instrument to focus at extremely short distances.

that any system which works over a wide range of magnifications (as this type of focusing lens does) should be designed so that the change of aberration contribution is small as the magnification is varied.

17.2 Microscope Objectives

Microscope objectives (Fig. 17.6) may be divided into three major classes: those designed to work with the object under a cover glass, those designed to work with no cover glass, and immersion objectives, which are designed to contact a liquid in which the object is immersed. All types are designed by raytracing from the long conjugate to the short; the effects of the cover glass (when used) must be taken into account by including it in the raytrace analysis. Standard cover glass thickness is 0.18 mm (0.16 to 0.19 mm, $n = 1.523 \pm 0.005$, $v = 56 \pm 2$).

Microscope objectives are designed to work at specific conjugates, and their correction will suffer if they are used at other distances. For cover glass objectives and immersion objectives, the standard distance from object plane to image plane is 180 mm. For metallurgical types (no cover glass), the standard distance is 240 mm. The chief effect of changing the tube length or cover glass thickness from its nominal value is to overcorrect or undercorrect the spherical aberration; an

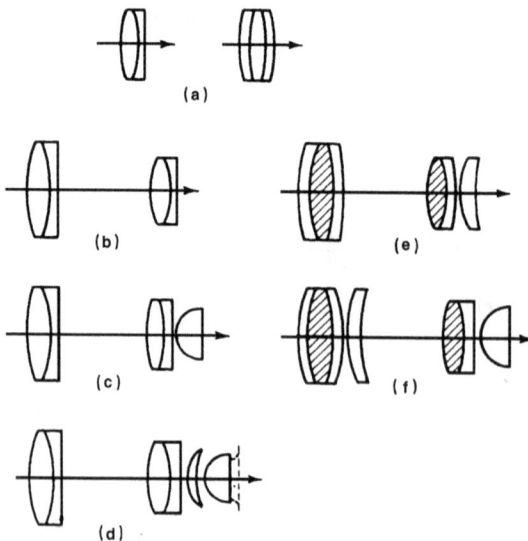

Figure 17.6 Microscope objectives. (a) Low-power achromatic doublet or triplet. (b) 10 × NA 0.25. (c) Amici objective 20 × NA 0.5 to 40 × NA 0.8. (d) Immersion objective. (e) Apochromatic 10 × NA 0.3. Shading indicates fluorite (CaF_2). (f) Apochromatic 50 × NA 0.95.

objective which has been improperly adjusted at the factory may be reclaimed by using a nonstandard tube length or cover glass if the defect is not too serious. Currently, each manufacturer seems to use a different distance.

Note that ordinary microscope objectives are designed to yield an essentially perfect image, and aberrations (on axis at least) should be reduced to well below the Rayleigh limit if at all possible. Micro-objectives for projection or photography may be corrected with more emphasis on the outer portions of the field, depending on the exact application for which they are intended.

Low-power objectives. These are usually ordinary achromatic doublets, or occasionally three-element systems, as shown in Fig. 17.6a. The 32-mm NA 0.10 or 0.12 is the most common and produces a magnification of about 4×. A 48-mm NA 0.08 is also occasionally encountered. These may be designed in exactly the same manner as the achromatic telescope objective discussed in Secs. 16.4 and 16.5, except that the "object" will be located at 150 mm (more or less) instead of at infinity.

Medium-power objectives. As shown in Fig. 17.6b, these are usually composed of two widely spaced achromatic doublets. The most common objective is the 10×, 16 mm, which is available in several forms. The ordinary achromatic 10× objective has an NA of 0.25 and is probably the most widely used of all objectives. The divisible or separable (Lister) version is designed so that it can be used as a 16-mm or, by removing the front doublet, as a 32-mm objective. This is accomplished at the sacrifice of astigmatism correction, since both components must be independently free from spherical and coma and thus no correction of astigmatism is possible. An apochromatic 16-mm objective is also available with an NA of 0.3; fluorite (CaF_2) is used in place of crown glass to reduce the secondary spectrum.

The power layout for this type of objective is usually arranged so that the product $y\phi$ is the same for each doublet; in this way the "work" (bending of the marginal ray) is evenly divided. Conventionally the second doublet is placed midway between the first doublet and the image formed by the first doublet. (Note that the preceding refers to raytracing sequence—in use the "second" doublet is near the object to be magnified and the "first" doublet is nearer the actual image.) This relatively large spacing allows the cemented surface of the second doublet to overcorrect the astigmatism and flatten the field (assuming the stop to be at the first doublet). This layout leads to a thin-lens arrangement with the space about equal to the focal length of the objective, the focal length of the first doublet approximately twice that of the objective, and that of the second doublet about equal to that of the objective. Note that this arrangement is similar to that of a high-speed Petzval-type projection lens (see Fig. 17.26).

Ordinarily three sets of shapes for the two components can be found for which spherical and coma are corrected. One form will be that of the divisible objective, with the spherical and coma zero for each doublet; this is usually the form with the poorest field curvature.

Aplanatic surfaces. If the surface contribution equation for the spherical aberration of a single surface is solved for zero spherical, three solutions are found. One case occurs when the object and image are at the surface and is of little interest. A second is of more value; when object and image both lie at the center of curvature, there is no spherical aberration introduced (and the axial rays are not deviated). The third case, usually called the aplanatic case, allows the convergence of a cone of rays to be increased (or decreased) by a factor equal to the index without the introduction of spherical aberration. It occurs when any of the following relationships are satisfied.

$$L = R\left(\frac{n' + n}{n}\right) \tag{17.1}$$

$$L' = R\left(\frac{n' + n}{n'}\right) = \frac{n}{n'}L \tag{17.2}$$

$$U = I' \tag{17.3}$$

$$U' = I \tag{17.4}$$

$$\frac{n'}{n} = \frac{\sin U'}{\sin U} \tag{17.5}$$

Note that if any of the above are satisfied, all are satisfied, and that, since no spherical is introduced, if $L = l$, then $L' = l'$. It is also worth noting that coma is zero for all three cases and that astigmatism is zero for the first and third cases and overcorrecting between. Figure 17.7 and 17.8 illustrate the three aplanatic cases.

High-power objectives. This principle is used in the "aplanatic front" of an oil-immersion microscope. The object is immersed in an oil the index of refraction of which matches that of the first lens. R_1 (as shown in Fig. 17.9) is chosen to satisfy Eq. 17.1; this results in a hyperhemispheric form for the first element. R_2 is chosen so that the image formed by R_1 is at its center of curvature; R_3 is chosen to satisfy Eq. 17.1. Note that $\sin U$ is reduced by a factor of n at each element, and that the "aplanatic front" reduces the numerical aperture of the

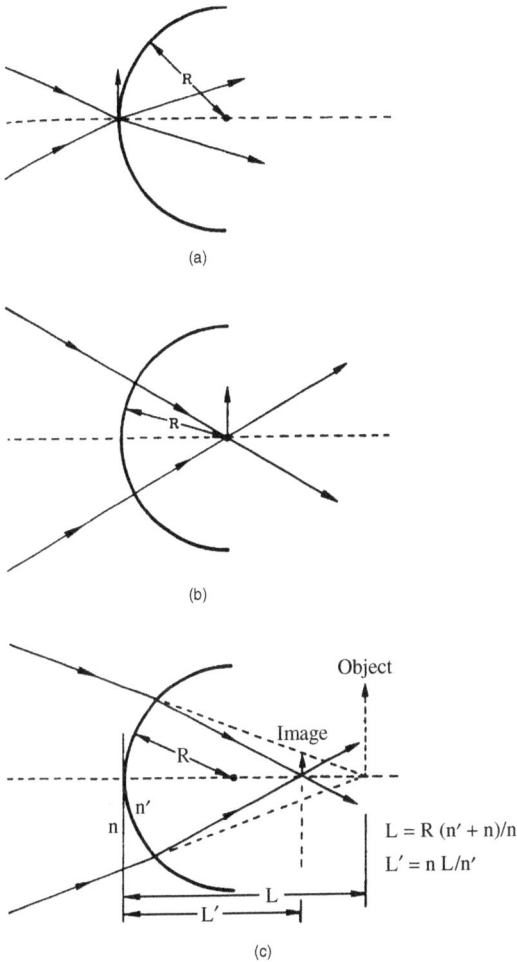

(a)

(b)

Image

Object

$$L = R (n' + n)/n$$
$$L' = n L/n'$$

(c)

Figure 17.7 Aplanatic surfaces. Surfaces for which both spherical aberration and coma are zero. (*a*) Object and image lie at the surface. (*b*) The object and image both lie at the center of curvature of the surface. (*c*) The object and image positions indicated produce a useful magnification, equal to the ratio of the indices (n/n')

cone of rays from a large value (as high as NA = n sin U = 1.4) to a value which a more conventional "back" system can handle.

The Amici objective (Fig. 17.6c) consists of a hyperhemispheric front element combined with a Fig. 17.6b (Petzval) type of back combination. Since the Amici is usually a dry objective, the radius of the hyperhemisphere is frequently chosen somewhat flatter than that called for by the aplanatic case to partially offset the spherical introduced by the

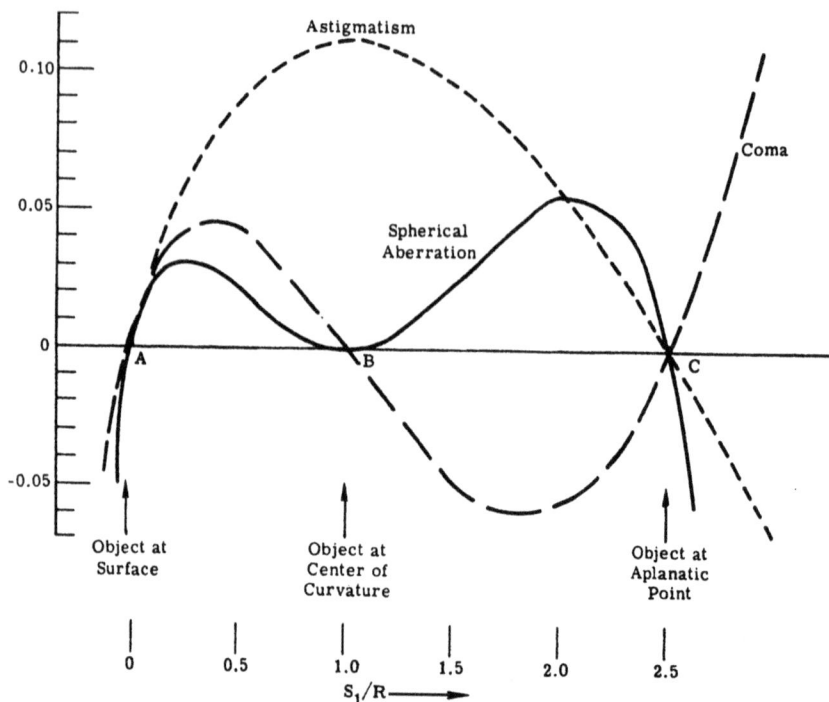

Figure 17.8 Aplanatic surface. The aberrations are plotted versus S_1/R, the ratio of object distance to surface radius.

dry plano surface. The space between the hyperhemisphere and the adjacent doublet is kept small to reduce the lateral color introduced by the front element. The standard 4-mm 40× NA 0.65 to 0.85 objectives are usually Amici objectives. The working distance (object to front surface) is quite small in the Amici, to the order of a half millimeter. Since there is a direct relationship between zonal spherical and working

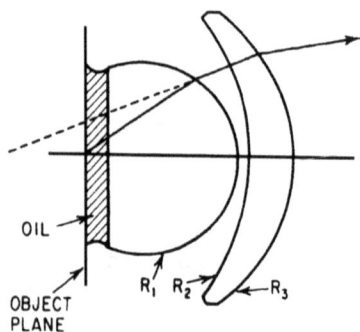

Figure 17.9 The aplanatic front. The object is immersed in a fluid the index of which matches that of the hyperhemispheric first element. R_1 is an aplanatic surface. The image formed by R_1 is at the center of curvature of R_2. R_3 is an aplanatic surface of the same type as R_1.

distance in this type of objective, the higher-NA versions tend to have very short working distances.

The oil-immersion objective utilizes the full "aplanatic front" and may be combined with a Fig. 17.6b type of back, as shown in Fig. 17.6d, or a more complex arrangement. Both the Amici and immersion types are frequently designed with fluorite (CaF_2) crowns to reduce or eliminate secondary spectrum. Some of the new FK glasses can serve the same purpose.

Note that although the aplanatic front is a classic textbook case, departures from the exact aplanatic form are common. For example, it is possible to find a meniscus lens of higher power than the aplanatic case which will introduce overcorrected spherical. This not only reduces the ray-bending work that the back elements must accomplish, but also reduces the correction load as regards spherical aberration (but not chromatic). Aplanatic-front objectives have a residual lateral color resulting from the separation of the chromatically undercorrected front and the overcorrecting back. Special *compensating eyepieces* with opposite amounts of lateral color are used to correct this situation. (See Sec. 17.1.)

Flat-field microscope objectives.   The objectives shown in Fig. 17.6 are all afflicted with a strongly inward-curving field. Such objectives can yield extremely sharp images in the center of the field, but the deep field curvature and/or astigmatism severely limit the resolution toward the edge of even the relatively small field of the microscope. Many flat-field types of objectives have their Petzval curvature reduced by a thick-meniscus negative component placed in the long conjugate. This may be an achromatized doublet as shown in Fig. 17.10, or simply a thick singlet. The field-flattening effect is greater if the negative-power element or surface is a large distance from the positive-power member. Often the balance of the objective is simply a stack of positive components. The improvement in image quality at the edge of the field is quite marked when compared to the standard type of objective. Another desirable feature of this form of objective is a long working

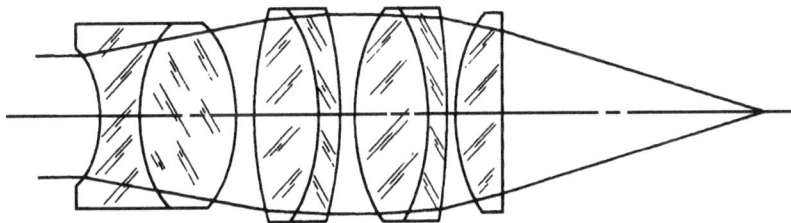

Figure 17.10   Achromatized negative doublet in a flat-field microscope objective.

distance from object to front lens. Note that this configuration is the analog of the retrofocus or reversed telephoto camera lens. Many flat-field objectives incorporate a construction similar to the thick-meniscus doublets of the double-Gauss or Biotar form (see Fig. 17.16) as a field-flattening device. Another technique is to convert the aplanatic hemispheric or hyperhemispheric front element to a meniscus element. The concave surface is close to the object plane and acts as a "field flattener." Its power contribution ($y\phi$) is small because the marginal ray height (y) is small when close to the object plane, but the concave surface introduces a significant positive, backward-curving Petzval contribution. The commercial brand names of microscope objectives of this type usually incorporate the letters "plan" in some form. Figures 19.31, 19.57, and 19.58 show high-power flat-field objectives.

Reflecting objectives. Objectives for use in the ultraviolet or infrared spectral regions are frequently made in reflecting form, because of the difficulty of finding suitable refracting materials for these spectral regions. The central obscuration required by such a construction will modify the diffraction pattern of the image, significantly reducing the contrast of coarse targets and improving the contrast slightly for fine details, as indicated in Chap. 15.

The basic construction of a reflecting objective is shown in Fig. 17.11a; it consists of two monocentric (or nearly monocentric) spherical mirrors in the Schwarzschild configuration (see Sec. 18.5). If both mirrors have a common center of curvature at the aperture stop, the system can be made free of third-order spherical, coma, and astigmatism; the focal surface is then a sphere centered on the aperture. The infinite conjugate case can be described by the following expressions (for a focal length f):

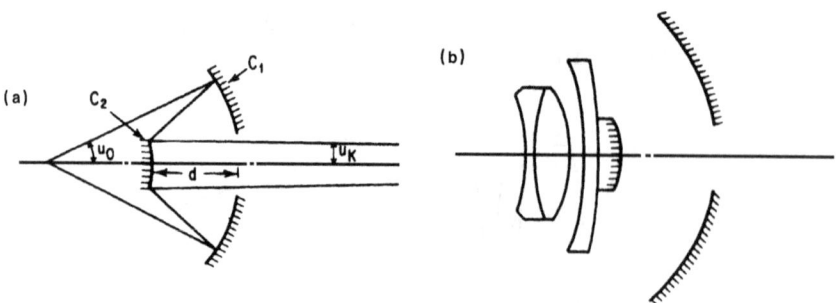

Figure 17.11 Reflecting microscope objectives. (a) Concentric 30 × NA 0.5. (b) Ultraviolet 50 × NA 0.7. Fused quartz and calcium fluoride are used for the refracting elements. (*Courtesy of D. Grey.*)

Space between mirrors

$$d = 2f \tag{17.6}$$

Convex radius

$$R_2 = (\sqrt{5} - 1)f \tag{17.7}$$

Concave radius

$$R_1 = (\sqrt{5} + 1)f \tag{17.8}$$

R_2-to-focus distance

$$= (\sqrt{5} + 2)f \tag{17.9}$$

R_1 clear aperture

$$y_1 = (\sqrt{5} + 2)y_2 \tag{17.10}$$

Fractional area obscuration

$$= 20\% \tag{17.11}$$

There are a number of variations on this basic form, some with less obscuration, some with reduced high-order spherical aberration.

The resulting system not only has zero third-order spherical, but even the higher orders tend to be exceedingly small; by proper choice of parameters, a delightfully simple but nonetheless useful objective can be obtained. The two-mirror system is limited to about 35× at NA = 0.5. For higher magnifications and numerical apertures, it is necessary to introduce additional refracting elements to maintain correction, as indicated in the sketch of the 50× NA 0.7 ultraviolet objective in Fig. 17.11b. Aspheric surfaces have also been utilized. The added elements can also serve to reduce the central obscuration or to flatten the field. A field flattener at the focal plane, with x = 1.5, $R_3 = -0.45f$, and $R_4 = +1.25\,f$ has been proposed.

17.3   Photographic Objectives

In this section, we will outline the basic design principles of the photographic objective, and for this purpose we will classify objectives according to their relationship to, or derivation from, a few major categories: (a) meniscus types, (b) Cooke triplet types, (c) Petzval types, and (d) telephoto types. These categories are quite arbitrary and are chosen for their value as illustrations of design features rather than for any historic or generic implications.

Meniscus anastigmats. In this category, we include those objectives which derive their field correction primarily from the use of a thick meniscus. As mentioned in Secs. 16.1 and 16.2, a thick-meniscus element has a greatly reduced inward Petzval curvature in comparison with a biconvex element of the same power; indeed, the Petzval sum can be overcorrected if the thickness is made great enough. The simplest example of this type of lens is the Goerz Hypergon (Fig. 16.4) which consists of two symmetrical menisci. Because the convex and concave radii are nearly equal, the Petzval sum is very small, and the fact that the surfaces are very nearly concentric about the stop enables the lens to cover an extremely wide (135°) field, although at a very low aperture ($f/30$).

To obtain an increased aperture, it is necessary to correct the spherical and chromatic aberrations. This can be accomplished by the addition of negative flint elements, as in the Topogon lens, Fig. 17.12. Note that the construction of this lens is also very nearly concentric about the stop; lenses of this type cover total fields of 75° to 90° at speeds of $f/6.3$ to $f/11$. The illumination fall off is worse than cosine-, fourth in this type of lens.

Attempts to design a system consisting of symmetrical cemented meniscus doublets in the latter half of the Nineteenth Century were only partially successful. If the spherical aberration was corrected by means of a diverging (i.e., with negative power) cemented surface, the higher-order overcorrected astigmatism necessary to artificially flatten the tangential field tended to become quite large at wide angles. If a high-index crown and low-index flint were used to reduce the Petzval field curvature, the resulting *collective* cemented surface was incapable of correcting the spherical. In 1890, Rudolph (Zeiss) designed the *Protar,* Fig. 17.13, which used a low-power "old" achromat (i.e., low-index crown, high-index flint) front component and a "new" achromat (high-index crown and low-index flint) rear component. The dispersive cemented surface of the front component was used to correct the spherical, while the collective cemented surface of the rear kept the astigmatism in control. Note that the components are thick menisci, which allows reduction of the Petzval sum, while the general symmetry helps

Figure 17.12 The Topogon lens (U.S. Patent 2,031,792-1936) covers 90° to 100° at a speed of $f/8$.

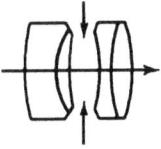

Figure 17.13 The Zeiss Protar (U.S. Patent 895,045-1908).

to control the coma and distortion. Lenses of the Protar type cover total fields of 60° to 90° at speeds of $f/8$ to $f/18$.

A few years later, Rudolph and von Hoegh (Goerz), working independently, combined the two components of the Protar into a single cemented triplet component, which contained both the required dispersing and collective cemented surfaces. The Goerz *Dagor* is shown in Fig. 17.14, and is composed of a symmetrical pair of cemented triplets. Each half of such a lens can be designed to be corrected independently so that photographers were able to remove the front component to get two different focal lengths. A great variety of designs based on this principle were produced around the turn of the century, using three, four, and even five cemented elements in each component, although very little was gained from the added elements. Protars and Dagors are still used for wide-angle photography because of the fine definition obtained over a wide field, especially when used at a reduced aperture. See Fig. 19.17 for an example of a Dagor design.

The additional degree of freedom gained by breaking the contact of the inside crowns and raising the crown index of the Dagor construction proved to be of more value than additional elements. Lenses of this type (Fig. 17.15) are probably the best of the wide-angle meniscus systems and cover fields up to 70° total at speeds of $f/5.6$ (or faster for smaller fields). The Meyer *Plasmat,* the Ross *W. A. Express,* and the Zeiss *Orthometar* are of this construction, and recently excellent 1:1 copy lenses (symmetrical) have been designed for photocopy machines. Note that the broken contact allows the inner crown to be made of a higher-index glass and still correct the spherical aberration.

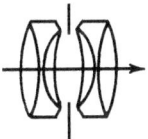

Figure 17.14 The Goerz Dagor (U.S. Patent 528,155,1894). The glasses used are 613:563, 568:560, and 515:547, from the left. The construction is symmetrical about the stop. See also Fig. 19.17.

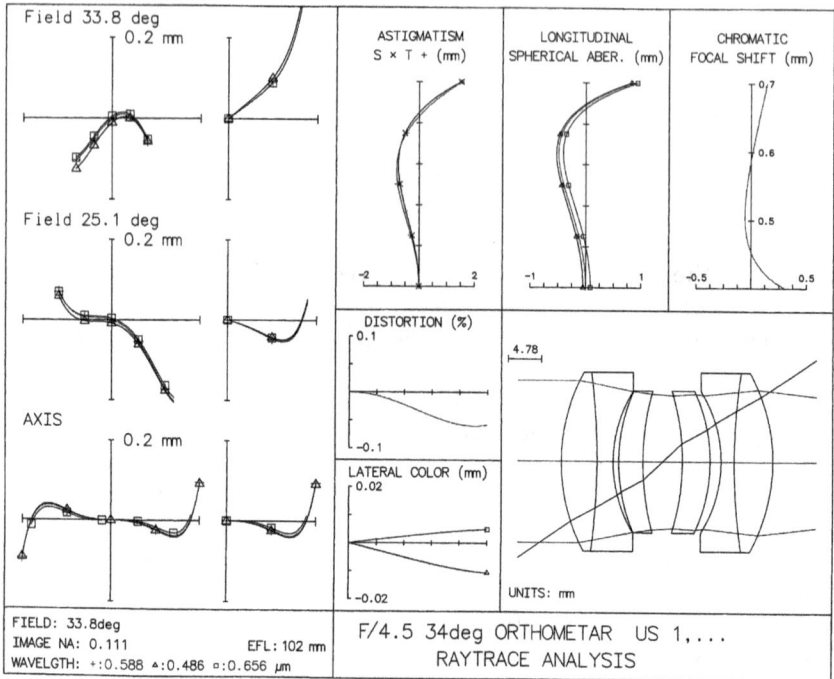

| Field 33.8 deg | ASTIGMATISM S × T + (mm) | LONGITUDINAL SPHERICAL ABER. (mm) | CHROMATIC FOCAL SHIFT (mm) |

FIELD: 33.8deg
IMAGE NA: 0.111
WAVELGTH: +:0.588 ▲:0.486 □:0.656 μm
EFL: 102 mm

F/4.5 34deg ORTHOMETAR US 1,...
RAYTRACE ANALYSIS

DISTORTION (%)
LATERAL COLOR (mm)
UNITS: mm

F/4.5 Orthometar USP# 1,792,917

radius	space	mat'l		s-a
25.90	5.10	SK8	611559	12.5
−96.10	2.30	LLF2	541472	12.5
18.40	0.80			9.9
23.80	3.50	SK20	560612	9.9
35.50	2.85			9.9
Stop	2.85			9.38
−33.10	4.00	SK20	560612	9.9
−22.70	1.60			9.9
−18.10	1.90	LLF2	541472	9.9
77.40	5.60	SK8	611559	12.5
−25.40	88.29			12.5

EFL   = 101.63
BFL   = 88.29
NA   = 0.111 (F/4.5)
GIH   = 68.09 (HFOV = 33.8°)
PTZ/F = −5.83
VL   = 30.5
PTZ/F = −5.74

Figure 17.15 The Plasmat, or Euryplan, or Orthometar, USP#1,792,917. Constructional data and aberration curves for a focal length of 101.6.

The design of the thick-meniscus anastigmats is a complex under-taking because of the close interrelationship of all the variables. In general the exterior shape and thickness are chosen to control the Petzval sum and power, and the distance from the stop can be used to adjust the astigmatism. However, the adjustment of element powers to correct chromatic inevitably upsets the balance, as does the bending of the entire meniscus to correct spherical. What is necessary is one simultaneous solution for the relative powers, thicknesses, bendings, and spacings; an approach of the type described in Chap. 16 for the simultaneous solution of the third-order aberrations is ideally suited to this problem, and the automatic computer design programs make easy work of it.

The efforts of designers in this direction over the past 75 years have been well spent, and it is exceedingly difficult to improve on the best representative designs in this category unless one utilizes the newer types of optical glass (e.g., the rare earth glasses).

The *double-Gauss* (Biotar) (Fig. 17.16) and the Sonnar types (Fig. 17.17) of objectives both make use of the thick-meniscus principle, although they differ from the preceding meniscus types in that they are used at larger apertures and smaller fields. The Biotar objective in its basic form consists of two thick negative-meniscus inner doublets and two single positive outer elements as shown in Fig. 17.16. This is an exceedingly powerful design form, and many high-performance lenses are modifications or elaborations of this type. If the vertex length is made short and the elements are strongly curved about the central stop, fairly wide fields may be covered. Conversely, a long system with flatter curves will cover a narrow field at high aperture.

Common elaborations of the Biotar format include compounding the outer elements into doublets or triplets, or converting the meniscus doublets into triplets. Frequently the outer elements are split (after shifting some power from the inner crowns) in order to increase the speed. Some recent designs have advantageously broken the contact at the cemented surface, especially in the front meniscus.

One may also double up on the inner meniscus doublets. In extreme cases all the elements of the Biotar can be duplicated, leading to a 12-element design with two front singlets, two front inner doublets, two rear inner doublets, and two rear singlets. Another interesting variation (the principle of which can be used in any design with a large enough airspace) is the insertion of a low- or zero-power doublet into the center airspace. The glasses of this doublet are chosen to have the same or nearly the same index and V-value, but significantly different partial dispersions. The low-power and matching index and V-value mean that the effect on most aberrations is negligible, but the partial dispersion difference can be arranged so that the secondary spectrum

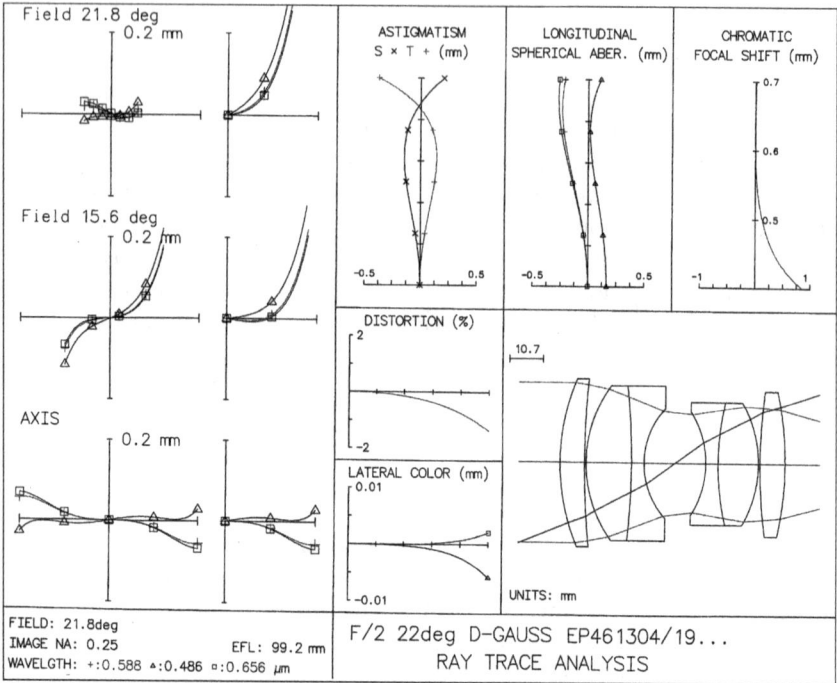

F/2.0 Double Gauss EP# 461, 304–1936

radius	space	mat'l		s-a
63.90	7.90	SK8	611559	26.0
240.30	0.50			26.0
39.50	14.50	SK10	623569	24.0
−220.50	4.00	F15	606378	24.0
24.50	9.95			17.0
STOP	9.95			15.86
−28.70	4.00	F15	606378	16.0
78.80	12.90	SK10	623569	19.0
−37.90	0.50			19.0
161.90	8.00	SK10	623569	22.5
−103.20	64.00			22.5

$$EFL = 99.22$$
$$BFL = 64.00$$
$$NA = 0.250 \ (F/2.0)$$
$$GIH = 39.69 \ (HFOV = 21.8°)$$
$$VL = 72.7$$
$$PTZ/F = -5.41$$

Figure 17.16 The Double Gauss (Biotar) objective. USP#2, 117, 252–1938; EP#461, 304–1936. Constructional data and aberration curves for a focal length of 99.

Figure 17.17   The Sonnar-type objective.

of the lens is reduced. There are several pairs of dense flint glasses which are suitable for this purpose.

As indicated above, the double-Gauss (Biotar) is an extremely powerful and versatile design form. It is the basis of most normal focal length 35-mm camera lenses and is found in many applications where extremely high performance is required of a lens. It can be made into a wide-angle lens or can be modified to work at speeds in excess of $f/1.0$ with equal facility. Additional examples of double-Gauss designs are presented in Chap. 19.

A "macro", close-focusing version of the Biotar has been proposed which either: (a) fixes the rear singlet and shifts the other elements to focus, or (b) shifts only the rear singlet to focus. Here, of course the idea is to reduce the change in aberrations which occurs when the entire lens is shifted to refocus.

Very high speed lenses (such as double Gauss objectives) frequently have a lot of zonal spherical aberration which causes the focus to shift when the lens is stopped down. These aberrations of these lenses are often balanced with the marginal spherical overcorrected. When the lens is stopped down there is very little focus shift and also very little zonal spherical, so the image quality is good under the normal illumination setting of the aperture. When the lens is wide open, the resolution remains high, but the image contrast is low due to the overcorrection. This background haze is not as serious as it might seem, because it effectively moves the exposure along the toe of the H & D curve, doing less damage than one might expect. (This is similar to the practice of pre-exposing film to increase its speed rating.)

Airspaced anastigmats.   These are systems which utilize a large separation between positive and negative components to correct the Petzval sum. Although it is historically incorrect in several instances, from a design standpoint it is useful to view these lenses as derivatives from the Cooke triplet, Fig. 17.18 (see also Secs. 16.7 and 16.8).

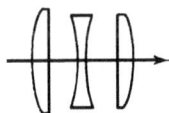

Figure 17.18   The Cooke triplet.

The *Tessar* (although actually derived from a meniscus-type lens) may be regarded as a triplet with the rear positive element compounded; the classic form of the Tessar is shown in Fig. 17.19. The additional freedom gained by compounding may be regarded as simply a means of artificially generating an unavailable glass type by combining two available glasses; alternatively, one may utilize the refractive characteristics of the cemented interface to control the course of the upper rim ray, which is affected strongly by this surface. The Tessar formulation, either as shown, or with the doublet reversed, or even with the front element compounded, is utilized when a performance a bit beyond that of the Cooke triplet is required. The reversed doublet form is usually better when high-index rare earth glasses are utilized. Figures 19.19 and 19.20 are additional examples of Tessar designs.

A further example of the compounding of the elements of the basic triplet is the Pentac (or Heliar) type, Fig. 17.20, which is simply a symmetrical extension of the Tessar principle. A Heliar design is shown in Fig. 19.21. In the Hektor (Fig. 17.21), all three elements are compounded and the speed can be raised to $f/1.9$ with fields to the order of $\pm 20°$. Many "compounded triplets" make use of what is sometimes called a "Merté" surface; the cemented surface of the negative component of the Hektor is an example of such a surface. This is a strongly curved (usually cemented) collective surface so arranged that the angle of incidence increases rapidly toward the margin of the lens. Such a surface contributes a modest amount of undercorrecting spherical to the rays near the axis, since the index break across the surface is not large. As the angle of incidence rises (and it may approach 45°), because of the nonlinearity of Snell's law, the spherical aberration contribution rises even more rapidly, and the undercorrecting effect dominates the marginal zone. The result is a spherical aberration curve which shows not only negative third- and positive fifth-order aberration, but a sizeable amount of negative seventh order as well. The spherical aberration shown in Fig. 17.21 is a rather extreme example of this technique. This is an approach which obviously must be used with discretion, since large amounts of high-order aberration are delicately balanced. Such a surface is best located near the stop to minimize the disparity of its effects on the upper versus lower rim rays; otherwise, the off-axis ray intercept curves may tend toward a very unpleasant asymmetry. A similar design is shown in Fig. 19.22.

Notice that in both Figs. 17.19 and 17.21, the doublets are composed of a positive crown with a higher index than the negative flint. The inward-curving Petzval contribution of such a doublet is much less than that of a single-lens element. And of course the undercorrected chromatic of a singlet is reduced or eliminated, since the doublet is at

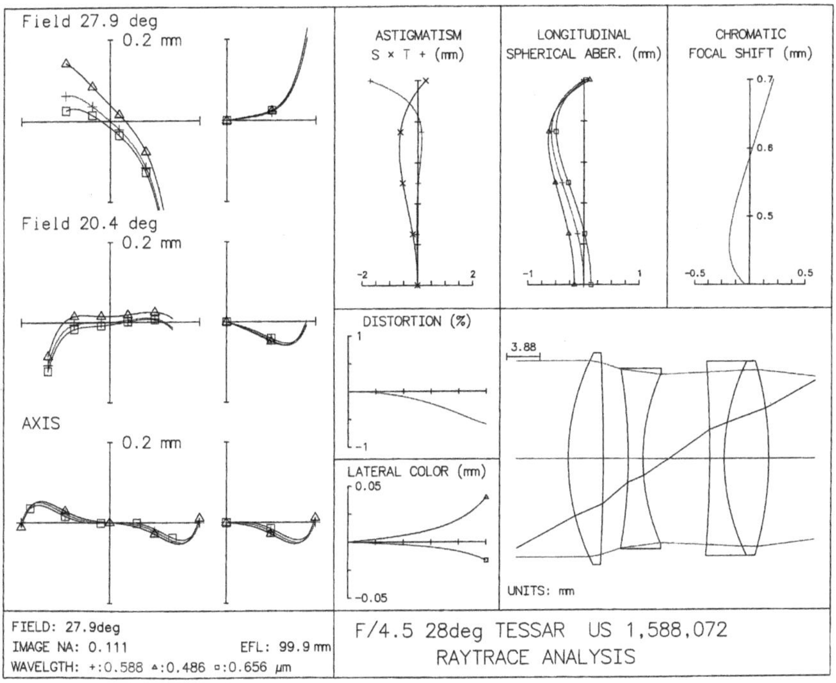

FIELD: 27.9deg
IMAGE NA: 0.111   EFL: 99.9 mm
WAVELGTH: +:0.588 ▲:0.486 □:0.656 μm

F/4.5 28deg TESSAR   US 1,588,072
RAYTRACE ANALYSIS

F/4.5 Tessar USP 1,588,072

radius	space	mat'l		s-a
25.90	4.10	SK2	607566	12.0
−253.00	2.90			12.0
−60.60	1.70	LLF4	561452	10.2
23.60	3.00			9.7
STOP	4.90			9.58
−144.00	1.70	LLF4	561452	11.0
24.90	5.10	SSK4	617551	11.0
−39.40	88.31			11.0

EFL = 99.88
BFL = 88.31
NA = 0.111 (F/4.5)
GIH = 52.94 (HFOV = 27.9°)
VL = 23.4
PTZ/F = 3.33

Figure 17.19 The Tessar objective USP#1,588,073–1922. Construction data and aberration curves for a focal length of 100.

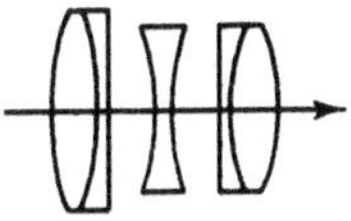

Figure 17.20 The Pentac-Heliar anastigmat. In an earlier version the outer doublets were reversed.

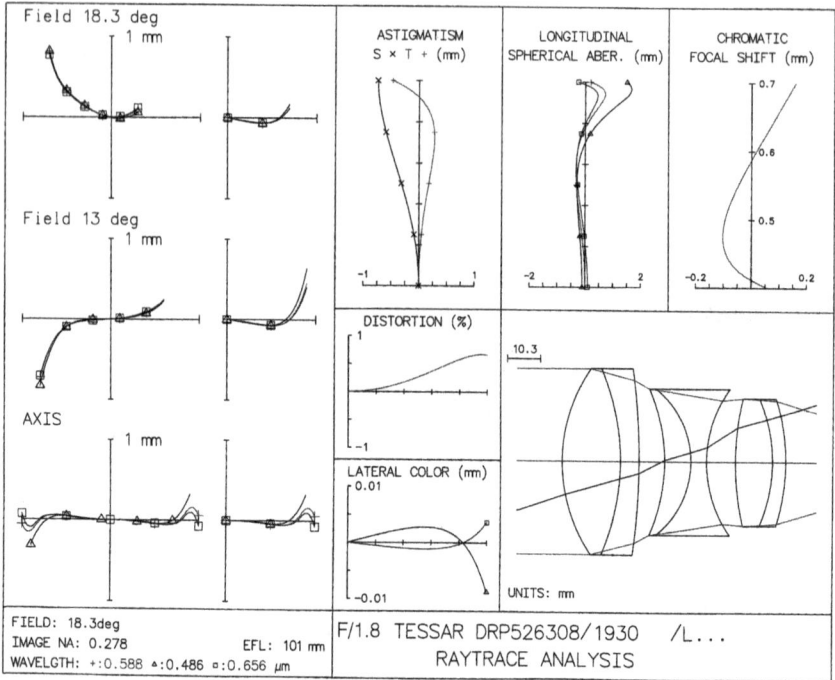

F/1.8 Hektor

radius	space	mat'l		s-a
48.80	18.00	SK15	623581	28.0
−69.30	5.70	F13	622360	28.0
−208.40	7.00			28.0
STOP	0.90			19.76
−53.70	8.20	SK15	623581	22.0
−27.60	4.90	LF6	567428	22.0
37.10	9.00			22.0
81.70	11.40	SK15	623581	19.2
−47.80	4.10	LLF2	541472	19.2
−63.20	60.59			19.2

EFL = 100.61
BFL = 60.59
NA   = 0.278 (F/1.8)
GIH = 33.20 (HFOV = 18.3°)
VL   = 69.7
PTZ/F = −2.19

Figure 17.21 The Hektor anastigmat, German Patent #526,308–1930. Note that the spherical aberration plot shows a large seventh-order component which originates at the strongly curved fifth surface (called a Merte' surface).

least partially achromatized. Remembering that the Petzval contribution is proportional to ϕ/n, it is apparent that the compounding of these elements produces a component which is equivalent to a singlet with both a high index and a high V-value. (This is true for the positive doublets; of course, the reverse is true for a negative doublet.)

Note that in almost all cases where a doublet is used in an anastigmat, it is a "new achromat," with the crown index higher than the flint index, yielding a converging cemented surface. This construction tends to have at least some of the above-mentioned "Merté" effect on the higher-order spherical, but the cemented surface does *not* correct the third-order spherical as the diverging cemented surface of the "old achromat" doublet does.

Another basic technique for the reduction of the residual aberrations involves splitting the individual elements into two (or more) elements. A single crown element has about five times as much undercorrected spherical as a two-element lens of equivalent power and aperture when both elements are shaped for minimal spherical (see Fig. 17.39). Thus, a split allows the contributions of the other elements of the system to be reduced, resulting in a corresponding decrease in higher-order aberrations. Ordinarily the crown elements of a triplet are split when a larger aperture is desired; Figs. 17.22 and 17.23 are examples of this technique. Since it requires a fairly long system and high speed to make this technique effective, the angular coverage of such systems is usually modest. However, by *compounding* the split elements, excellent combinations of aperture and field have been obtained from these forms. Splitting the front crown is usually more profitable than splitting the rear, since the astigmatism at the edge of the field is better controlled in the split-front types, and the meniscus shape is beneficial for the Petzval field curvature. Although less frequently encountered, element splitting can also be effective to a limited degree in extending field coverage. Additional variants on the split-crown triplets can be found in Chap. 19.

Split-flint triplets (Fig. 17.24) should really be regarded as thick-meniscus systems with an air lens separating the crown and flint of each half; indeed this was their historical derivation. This form is not especially notable for reduced spherical zonal as are the split-crown types, but some of the finest general-purpose photographic objectives (e.g., the $f/4.5$ Dogmar and Aviar lenses) have been of this construction. The general symmetry of this design lends itself to a wider angular coverage than do the split-crown types, although, as in most "triplet-derived" forms, the limit of coverage is often sharply defined and image quality tends to fall off rapidly beyond the stigmatic node. (This last comment is less true of systems where the crown-flint spacing is small, since these types are closer to the meniscus lenses than to triplets.) Figure 19.18 is another example of the Dogmar. Many excellent

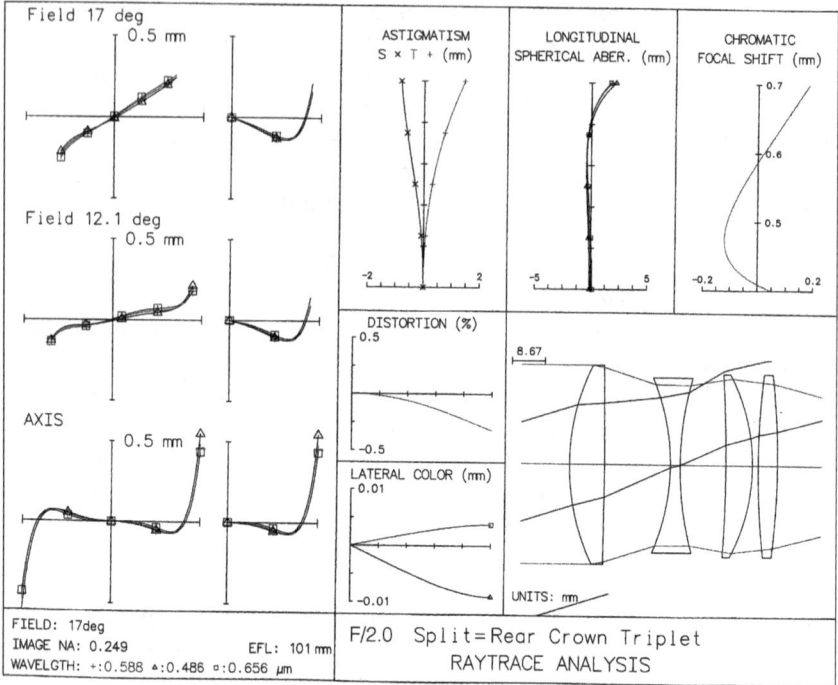

Within the figure:

Field 17 deg
0.5 mm

ASTIGMATISM
S × T + (mm)

LONGITUDINAL
SPHERICAL ABER. (mm)

CHROMATIC
FOCAL SHIFT (mm)

Field 12.1 deg
0.5 mm

DISTORTION (%)

AXIS
0.5 mm

LATERAL COLOR (mm)

UNITS: mm

FIELD: 17deg
IMAGE NA: 0.249   EFL: 101 mm
WAVELGTH: +:0.588 ▲:0.486 □:0.656 μm

F/2.0 Split=Rear Crown Triplet
RAYTRACE ANALYSIS

F/2.8 Split Rear Crown Triplet

radius	space	mat'l		s-a
56.00	9.00	SK3	609589	25.0
−1.5e+03	17.00			25.0
−55.00	2.50	SF5	673322	22.0
80.0	12.00			22.0
−373.00	7.00	SK3	609589	23.0
−48.00	1.50			23.0
194.00	5.00	SK3	609589	23.0
−194.00	79.23			23.0

EFL = 100.57
BFL = 79.23
NA   = 0.249 (F/2.0)
GIH = 30.75 (HFOV = 17.0°)
VL   = 54.0
PTZ/F = −1.83

Figure 17.22 The split-rear crown triplet. EP#237, 212–1925. Construction data and aberration curves for a focal length of 100.).

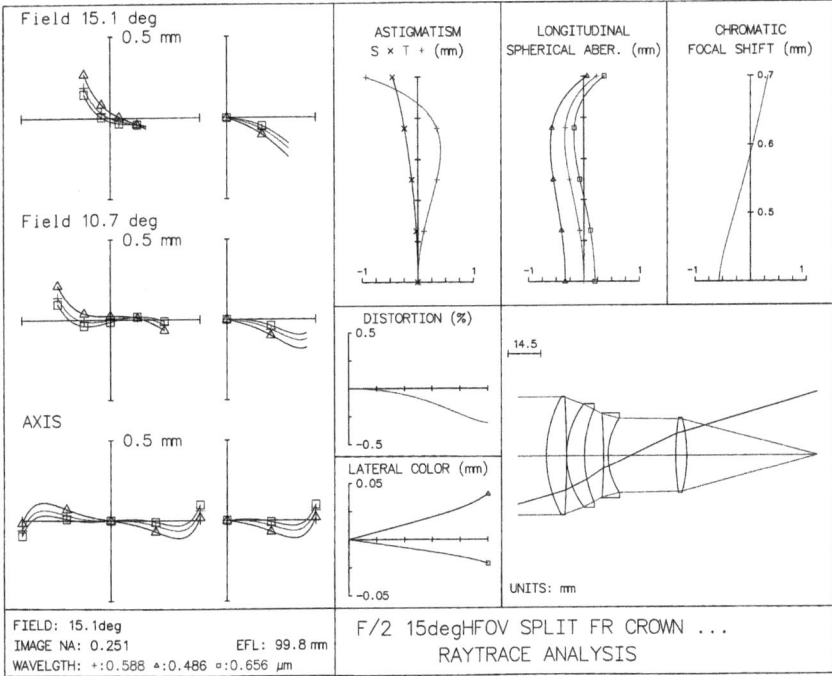

Field 15.1 deg	ASTIGMATISM S × T + (mm)	LONGITUDINAL SPHERICAL ABER. (mm)	CHROMATIC FOCAL SHIFT (mm)

FIELD: 15.1deg
IMAGE NA: 0.251   EFL: 99.8 mm
WAVELGTH: +:0.588 ▲:0.486 □:0.656 μm

F/2 15degHFOV SPLIT FR CROWN ...
RAYTRACE ANALYSIS

F/2.0, 150° HFOV Split Front Crown Triplet

radius	space	mat'l		s-a
51.00	8.80	SK11	564608	25.1
−441.00	0.03			25.1
35.30	7.80	SK11	564608	22.0
47.80	8.40			20.0
−254.80	2.00	SF2	648338	18.0
28.30	10.00			16.0
STOP	19.40			15.69
107.80	4.90	SK11	564608	16.0
−60.30	56.89			16.0

EFL = 99.79
BFL = 56.89
NA   = 0.251 (F/2.0)
GIH = 26.94 (HFOV = 15°)
VL   = 61.3
PTZ/F = −2.25

Figure 17.23 The split-front crown triplet. EP#237, 212–1925. Construction data and aberration curves for a focal length of 100.

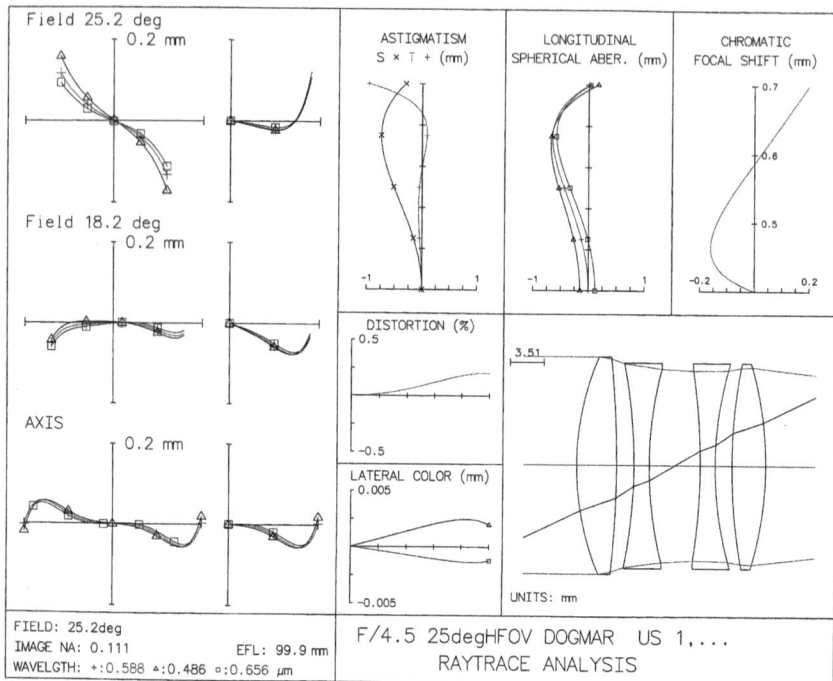

FIELD: 25.2deg
IMAGE NA: 0.111   EFL: 99.9 mm
WAVELGTH: +:0.588 ▲:0.486 □:0.656 μm

F/4.5 25degHFOV DOGMAR   US 1,...
RAYTRACE ANALYSIS

F/4.5, 25° HFOV Dogmar USP 1,108, 307–1914

radius	space	mat'l		s-a
27.70	4.20	SK6	614564	11.2
−103.10	1.80			11.2
−53.90	1.60	LF6	567428	10.5
37.70	2.70			9.9
STOP	2.70			9.81
−63.30	1.60	LLF1	548457	9.9
35.10	1.80			10.5
53.20	3.60	SK6	614564	10.5
−35.70	88.56			10.5

EFL   = 99.86
BFL   = 88.56
NA   = 0.111 (F/4.5)
GIH   = 46.93 (HFOV = 25°)
VL   = 20.2
PTZ/F = −3.11

Figure 17.24 The Dogmar anastigmat USP#1,108, 307–1914. Construction data and aberration curves for a focal length of 100.

process and enlarging lenses are based on this format. Process lenses of this type are made with glasses of unusual partial dispersions in order to correct or reduce secondary spectrum. Such lenses usually have the letters "apo" in their trade names to denote apochromatic or semiapochromatic correction.

Lenses for close conjugate work, such as enlarger lenses, are often airspaced anastigmats. They differ from camera objectives primarily in that they are designed for low magnification ratios, rather than for infinite object distances. Most camera objectives maintain their correction down to object distances to the order of 25 times their focal length, and a few do well at even shorter distances. Enlargers, however, are frequently used at magnifications approaching unity, and enlarging lenses are usually designed at conjugate ratios of 3, 4, or 5. A lens which is approximately symmetrical (such as the Dogmar) makes a good enlarger lens since it is a bit less sensitive to object-image distance changes. Compounded triplets of approximately symmetrical construction are also used, and the Tessar formula is widely used because of its wide field of coverage and relatively simple and inexpensive construction.

Petzval lenses. The original Petzval portrait lens (Fig. 17.25) was a relatively close-coupled system consisting of two back to back achromatic doublets, the rear doublet with broken contact, with a sizeable airspace between. It covered a modest field at a speed of about $f/3.7$. The modern version, often referred to as the Petzval projection lens because of its widespread use as a motion picture projection objective, utilizes a larger airspace (almost equal to its focal length) and covers half-field of $\pm5°$ to $\pm10°$ at speeds up to $f/1.6$. This type of system (Fig. 17.26) is noted for the excellence of its correction on axis, and also for its strongly inward-curving field. The field is artificially flattened by overcorrected astigmatism which is introduced at the cemented surface of the rear doublet. A typical formulation has a thin-lens spacing about equal to the focal length, a front doublet with twice the focal length of the system, and a rear doublet with a focal length equal to that of the system. Thus, the (thin-lens) back focus is about half the focal length, and the front vertex-to-focal plane distance is about 1.5 times the focal length. If the airspace is appreciably shortened, it may be necessary to break contact or increase the index break at the rear

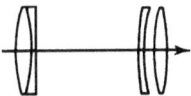

Figure 17.25   The Petzval portrait lens.

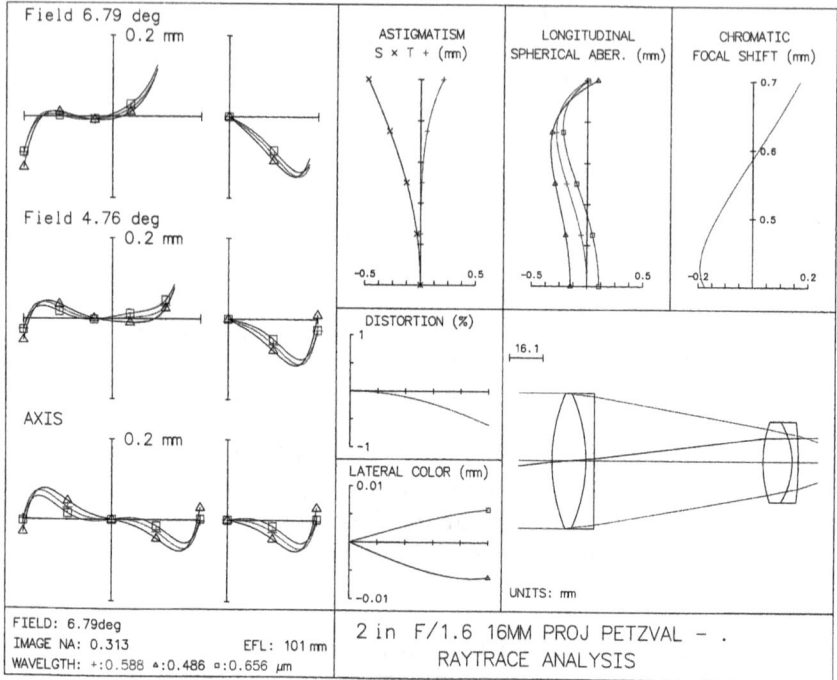

FIELD: 6.79deg
IMAGE NA: 0.313   EFL: 101 mm
WAVELGTH: +:0.588 ▲:0.486 □:0.656 μm

2 in F/1.6 16MM PROJ PETZVAL – .
RAYTRACE ANALYSIS

2 in F/1.6 16 mm Projection Lens

radius	space	mat'l		s-a
81.50	16.40	C1	523586	31.8
−76.72	4.00	F4	617366	31.8
plano	82.00			31.8
57.32	14.00	C1	523586	19.0
−36.00	3.00	F4	617366	19.0
−196.70	38.44			19.0

EFL = 1.1.34
BFL = 38.44
NA = 0.313 (F/1.6)
GIH = 12.06 (HFOV = 6.8°)
VL = 119.4
PTZ/F = −0.93

Figure 17.26 The Petzval projection lens. USP#1, 843, 519–1932. Construction data and aberration curves for a focal length of 101.3.

doublet to maintain the overcorrected astigmatism. Note that the Petzval projection lens as shown in Fig. 17.26 is basically the same design form as that of a 10×, NA 0.25 microscope objective. The Petzval projection lens construction inherently has low spherochromatism, low secondary spectrum, and a relatively small zonal spherical aberration.

The inward-curving Petzval surface can be corrected by the use of a negative "field flattener" element near the focal plane, Fig. 17.27. In this location the power contribution ($y\phi$) of the element is low, but the Petzval field is nicely flattened, and a lens of beautiful definition over a small field can be obtained. The drawback to this is the location of the element near the image plane, where dust and dirt can become quite noticeable. Note that the field flattener is made of flint glass, which helps the correction of the chromatic aberration.

The glasses used in the Petzval lens are usually an ordinary crown and common dense flint. Occasionally higher-index glass is used and one or both doublets are of the broken contact type.

An interesting variation on the field-flattener Petzval is shown in Fig. 17.28, in which the rear negative element does double duty, serving both as the rear flint and as the field flattener as well. The broken contact in the front doublet is necessary to correct the aberrations. This lens has a tendency toward increased zonal spherical as well as fifth-order coma of the y^4h type, which is introduced by the airspaced front doublet. This aberration is frequently encountered in other design types as well, when a strong negative "air lens" is used in this manner to correct spherical aberration. The glasses used in this lens are dense barium crowns (SK4) and dense flints (SF1).

The already small spherical zonal of the Petzval lens can be reduced still further by splitting the rear doublet into two doublets as indicated in Fig. 17.29 or by the introduction of a meniscus element into the central airspace, Fig. 17.30. One Petzval modification achieved a speed of $f/1.0$ (with an almost spherical image surface) by splitting off a sizeable part of the power of each crown element into separate plano-convex elements. Other modifications have made use of strongly meniscus front correctors to reduce the spherical zonal, or of thick rear concentric meniscus elements to improve the field. Two recent designs which are used as 2-in $f/1.4$ projection lenses for 16-mm motion pictures are shown in Fig. 17.31. Additional variations on the Petzval theme are shown in Chap. 19.

Telephoto lenses. Telephoto lenses are arbitrarily defined as lenses the length from front vertex to film plane of which is less than the focal length. The *telephoto ratio* is this vertex length divided by the focal length; a lens with a ratio of one or less is considered a telephoto lens.

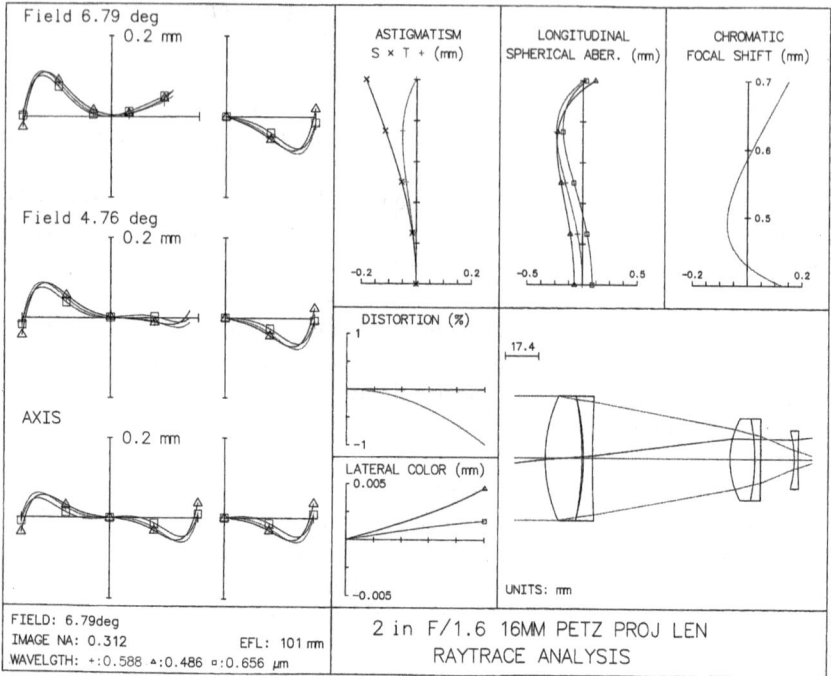

FIELD: 6.79deg
IMAGE NA: 0.312   EFL: 101 mm
WAVELGTH: +:0.588 ▲:0.486 □:0.656 μm

2 in F/1.6 16MM PETZ PROJ LEN
RAYTRACE ANALYSIS

2 in F/1.6 16 mm Projection Lens

radius	space	mat'l		s-a
75.66	19.40	DBC1	611588	31.8
−158.30	0.7534			31.8
−129.78	4.80	EDF3	720293	31.8
600.68	72.70			31.8
43.50	13.00	DBC1	611588	21.0
−103.82	3.00	EDF3	720293	21.0
plano	17.86			21.0
−54.78	2.00	F4	617366	15.0
243.12	13.20			15.0

EFL = 101.48
BFL = 13.20
NA = 0.312 (F/1.6)
GIH = 12.08(HFOV = 6.8°)
VL = 133.5134
PTZ/F = −2.99

Figure 17.27 Petzval projection lens with field flatttener. Construction data and aberration curves for a focal length of 101.5.

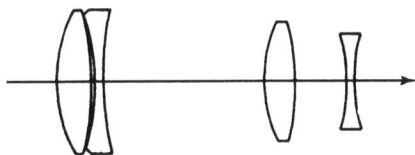

Figure 17.28 $f/1.6$ Petzval lens with field-flattening effect achieved by large airspace between rear crown and flint.

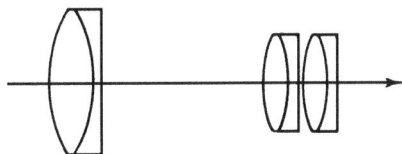

Figure 17.29 $f/1.3$ Petzval lens with two rear doublets to reduce spherical zonal. (U.S. Patent 2,158,202-1939).

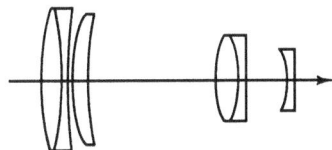

Figure 17.30 Field-flattener Petzval lens with front crown split into two elements to reduce spherical zonal. (U.S. Patent 2,541,484-1951).

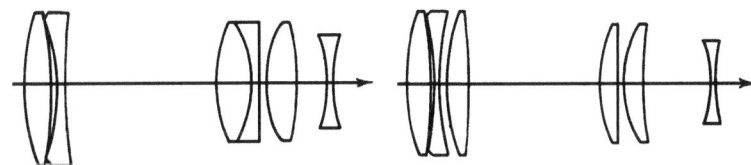

Figure 17.31 High-performance 2-in, $f/1.4$, 16-mm motion picture projection lenses. (Left is U.S. Patent 2,989,895-1961. Right is U.S. Patent 3,255,664-1966).

This is achieved by a positive front component separated from a negative rear component, as indicated in Fig. 17.32. Several forms of telephoto lenses are also shown; distortion correction is usually achieved by splitting the rear component. A common difficulty of the telephoto and reverse telephoto lenses is a strong inclination toward an overcorrected Petzval sum and a backward-curving field when extreme ratios are obtained. Figures 19.15 and 19.16 shows a typical telephoto lens designs.

Reverse telephoto (retrofocus) lenses. By reversing the basic power arrangement of the telephoto, a back focal length which is longer than the effective focal length may be achieved. This (Fig. 17.33) is a useful form when prisms or mirrors are necessary between the lens and the image plane; it also allows the use of a short-focallength projection lens with a condenser designed for longer lenses, since the pupil position is well away from the image plane. The construction was originally a strong negative achromat in front, combined with a modification of a standard objective. Biotars, triplets, and Petzvals have all been used for the rear member. It is usually necessary to split the negative achromat and bend it concave to the rear member to achieve good correction. In extreme forms ("sky-lenses" or "fish-eye"

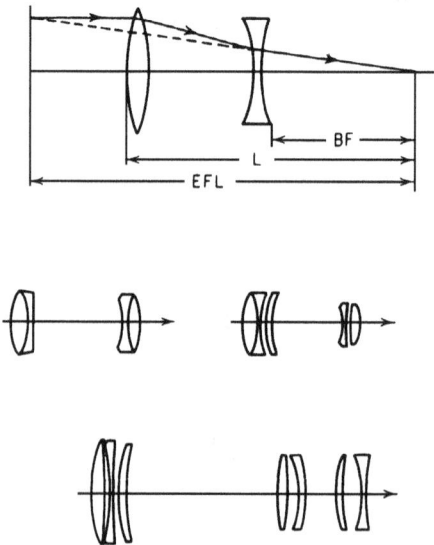

Figure 17.32 Telephoto lenses. A focal length which is greater than the physical length of the lens is achieved by a positive front member widely separated from a negative rear member.

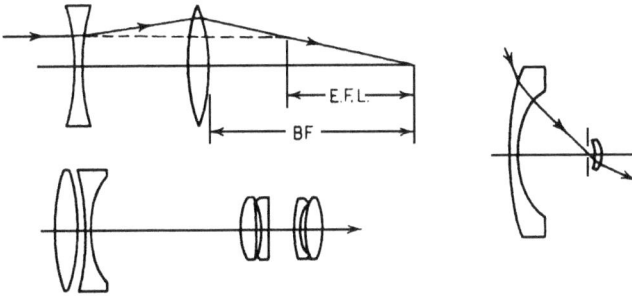

Figure 17.33 The reverse telephoto is characterized by a long back focus which is useful for short-focal-length lenses. In extreme forms (right-hand sketch) the coverage can be made to exceed 180°.

lenses) coverage can exceed ±90° with a very strongly meniscus negative front element. Obviously in order to image 180° or more on a finite-sized flat film, a large amount of barrel distortion is unavoidable.

The retrofocus lens has found wide use with the popularity of the single-lens reflex 35-mm camera, which requires a long back focus to clear the viewfinder mirror as it swings up out of the way when the exposure is made. All of the short-focus, wide-angle SLR lenses are of this type. The retrofocus has evolved into a very powerful design form in its own right and can no longer be regarded as a standard camera lens with a negative lens out in front. After all, since the front negative component more than corrects the Petzval curvature, it makes little sense to overdo the correction with an already field-flattened standard design type. Figures 19.13 and 19.14 show a retrofocus and a "fish-eye" lens, respectively.

If one examines the ray path in the right-hand sketch of Fig. 17.33, it is apparent that the negative element is reducing the angular coverage required of the positive element. This idea is the basis of many wide-angle camera lenses; this type consists of a collection of positive components surrounded by meniscus negative elements. The *Angulon* and several other designs are of this type. Figures 19.40 and 19.41 show examples of this type of wide-angle lens.

Afocal attachments. These usually take the form of Galilean or reversed-Galilean telescopes as indicated in Fig. 17.34. The focal length of the "prime" lens is multiplied by the magnification of the telescopic attachment. The field of view limits the power of the telephoto types to about 1.5×, but the wide-angle type of attachment is useful to about 0.5×. Such systems are, of course, designed to use an external stop (that of the prime lens) and frequently require quite a bit

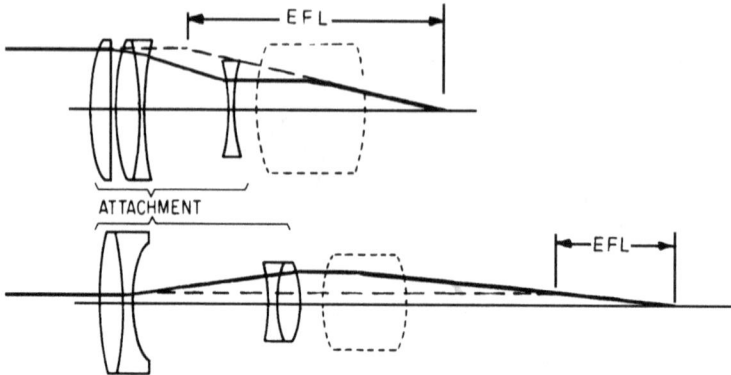

Figure 17.34 The focal length of a prime lens can be modified by the use of an afocal attachment, which is basically a Galilean telescope. The upper sketch shows a "telephoto" attachment which increases the focal length. The lower system is a "wide-angle" which reduces the focal length.

of "stopping-down" to achieve satisfactory imagery, especially in the simpler constructions.

An afocal attachment can be added to almost any optical system in order to change its focal length or field or magnification. The idea is obviously most applicable where the object or image is at a distance so that the afocal is working in collimated light. For noncollimated applications a Bravais system (see Sec. 13.9) can serve the same function.

17.4 Condenser Systems

The condenser in a projection system is quite analogous to the field lens in a telescope or radiometer. The function of the condenser is illustrated in Fig. 17.35. The upper sketch shows a projection system without a condenser. It is apparent that for the axial object point *A,* only about half the lens area can be used, for point *B* only an even smaller fraction of the lens is utilized, and that no light from the lamp passing through point *C* can pass through the projection lens. The result is that the illumination at the projected image is not as high as it might be and drops off rapidly away from the axis. This can be alleviated somewhat by moving the lamp closer to the film, and, in a very few cases, this solution is satisfactory, if inefficient. However, the filament is usually not uniform enough to allow it to be projected directly without producing objectionable nonuniformity of illumination at the image.

The "Koehler" projection condenser shown in the lower sketch of Fig. 17.35 images the lamp filament directly into the aperture of the projection lens. If the image size is equal to (or greater than) the lens

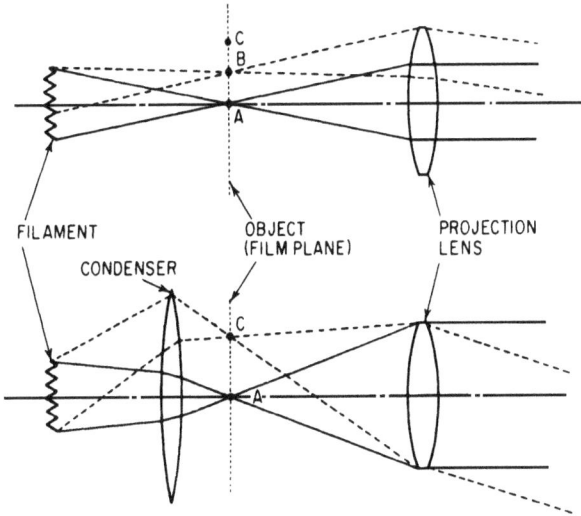

Figure 17.35 The projection condenser produces an image of the source in the pupil of the projection lens. Note that the minimum condenser diameter for optimum illumination at the image of point C is determined by a line through C and the opposite rim of the pupil.

aperture size, the illumination is optimized, and if the condenser has a sufficient diameter, the illumination over the full image field is as uniform as possible. The requirements for an ideal condenser may be expressed as follows: The image of the filament must completely fill the projection lens aperture through a small pinhole placed anywhere in the field (i.e., at the film plane). The photometric aspects of condensers are discussed in Sec. 12.10.

The chief aberrations of concern in condenser systems are usually spherical and chromatic aberrations; coma, field curvature, astigmatism, and distortion are of secondary importance in ordinary systems. Figure 17.36 is an exaggerated sketch of a condenser afflicted with

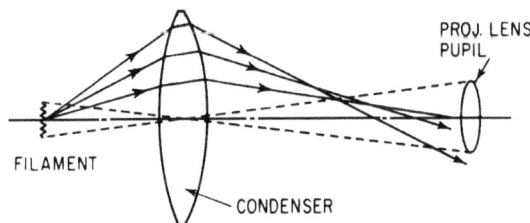

Figure 17.36 Spherical aberration in a condenser can cause the rays through the margin of the condenser to completely miss the aperture of the projection lens.

spherical aberration. Note that the filament image formed by the marginal zone of the condenser completely misses the projection lens aperture, resulting in a marked falloff in illumination at the edge of the field. This situation could be alleviated by reducing the condenser power so that the marginal ray focus was at the lens; however, in difficult cases this can result in a dark zonal ring in the field because at least some of the zonal rays will then miss the aperture. The effects of chromatic aberration are similar, except that one end of the spectrum (red or blue) may miss the aperture and cause an unevenly colored field of view, especially noticeable at the field boundary.

Except in unusual cases (e.g., some microscope condensers) chromatic effects can be held to a tolerable level without achromatizing. Spherical aberration is controlled by splitting the condenser into two or three elements of approximately equal power and bending each element toward the "minimum spherical" shape, as indicated in Fig. 17.37a and b. An aspheric surface can be molded on one of the elements to reduce the spherical aberration, as in Fig. 17.37c. The aspheric is often a simple paraboloid, and a molded surface can be sufficiently precise to meet the requirements of a condenser system.

When the light source is uniformly bright, it can be imaged directly on the film gate. In arc-lamp motion picture projectors, an ellipsoidal mirror is used for this purpose, as shown in Fig. 17.37d. Note that for

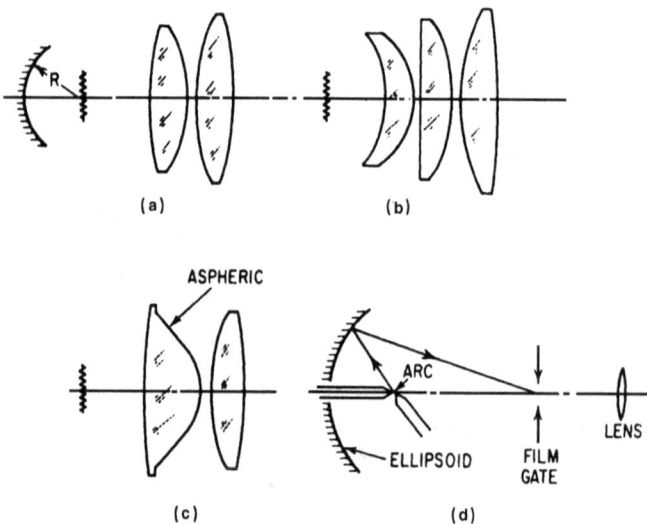

Figure 17.37 Condensing systems. (a) Two-element design with reflector concentric to source. (b) Three elements, shaped to minimize spherical. (c) Aspheric surfaces can be used to reduce spherical. (d) The crater of a carbon arc can be imaged directly at the film gate by an ellipsoidal mirror.

full illumination the mirror must be large enough to accept the ray from the bottom of the projection lens aperture through the top of the film gate. The ellipsoidal mirror is used since it has no spherical aberration when the arc is at one focus of the ellipse and the image (film gate) is at the other. Note that an ellipsoid does have a substantial coma, however, and thus off-axis imagery through the margin of the mirror may depart considerably from that predicted by first-order optics.

Some projection lamps incorporate a reflector inside the glass bulb which functions in the same manner as the ellipsoidal mirror of Fig. 17.37d. This allows the system to push the limits on the smallness of the source (as described in Sec. 12.10) and makes for efficient usage of a small, low-wattage lamp filament. The mirrors in this type of projection lamp are often faceted; this allows some control over the magnification produced by each zone of the reflector, and also allows the direction in which the light is reflected to be adjusted in order to provide the most desirable distribution in the film gate.

Another construction uses a small lamp integrally fabricated with a molded, faceted, and much larger reflector. The lamp filament is located close to the focal point of the reflector, and the condenser images the entire reflector in the projection lens aperture in what is effectively a Koehler configuration, treating the entire reflector as the source.

Most condensing systems can be significantly improved by the addition of a spherical reflector behind the light source, as indicated in Fig. 17.37a. If the source is at the center of curvature, the mirror images the source back on itself, effectively increasing its average brightness. With a lamp filament of relatively open construction, such as a V shape, or two parallel coils, the increase in illumination may approach the reflectivity of the reflector, i.e., 80 to 90 percent. The gain is much less in a tightly packed source, but even a biplane filament will gain 5 or 10 percent from a properly aligned reflector.

It should be noted that if the projection lens aperture is only partially filled by the filament image, the diffraction effects will differ from those associated with a fully illuminated aperture. For example, if only the center of the aperture is illuminated, this "semicoherent illumination" causes the MTF at low frequencies to be increased, and the MTF at high frequencies to be reduced. If a two-coil filament is imaged with the coil images at the extreme edges of the aperture and the center unilluminated, not only is the MTF balance between high and low frequencies changed, but the imagery in one meridian (i.e., line orientation) is quite different from that in the other, often giving the impression of an astigmatic image. See also Figs. 15.18, 15.19, and 15.20 in Sec. 15.10.

17.5 Aberration Characteristics of Simple Lenses

Figure 17.38 shows the angular spherical aberration blur of a single positive element with the object at infinity, as a function of the shape and the index of the lens. The angular blur size β is given by:

$$\beta = y^3 \, f^3 \, [n^2 - (2n + 1)K + (n + 2)K^2/n]/4(n - 1)^2$$

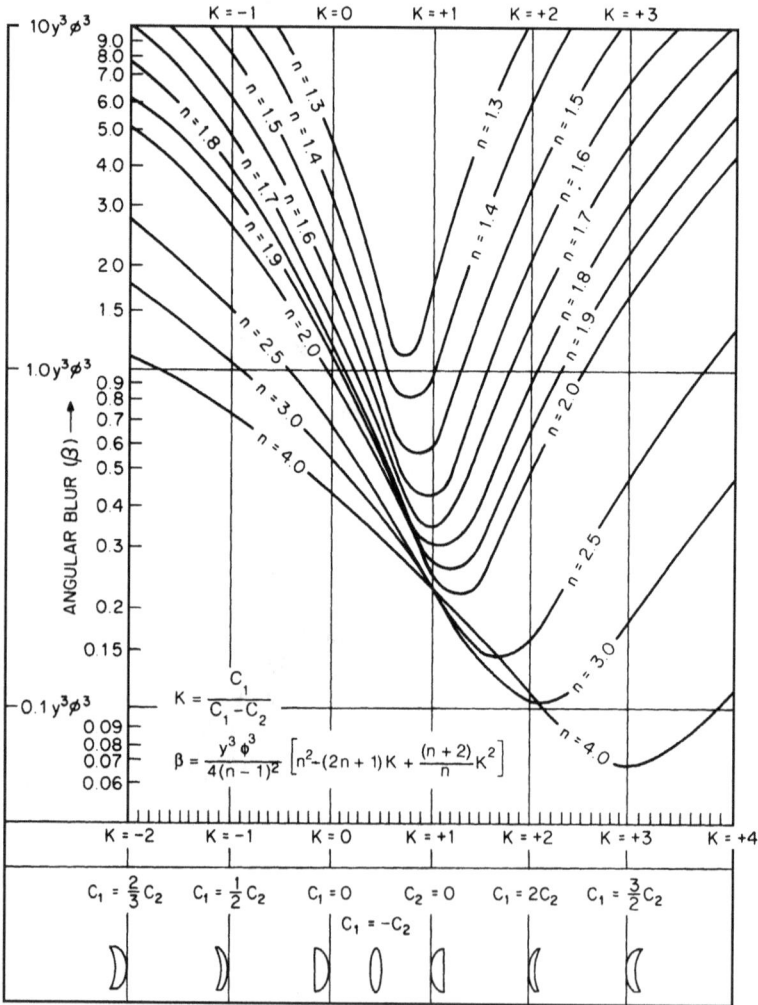

Figure 17.38 The angular spherical aberration blur β of a single-lens element as a function of shape for various values of the index of refraction. ϕ is the element power; y is the semiaperture. The angular blur can be converted to longitudinal spherical by $LA = -2\beta/y\phi^2$. (Object at infinity.)

where y is the semiaperture, ϕ is the element power, n is the index, and $K = C_1/(C_1 - C_2) = R_2/(R_2 - R_1)$

The angular blur can be converted to the blur spot diameter by multiplying by the focal length of the element ($1/\phi$), or by dividing by the power ϕ.

The longitudinal spherical aberration is equal to: $LA = -2\beta/y\phi^2$ where β is the angular blur.

The minimum blur is $\beta = n(4n + 1)/[128 \, (f/\#)^3 \, (n - 1)^2 \, (n + 2)]$.

The shape at which the spherical aberration is a minimum is given by:

$$R_2/R_1 = (2 \, n^2 + n)/(2 \, n^2 - n - 4)$$

or $$C_1/(C_1 - C_2) = K = n(2n + 1)/2(n + 2)$$

For a plano-convex lens the angular blur $\beta = y^3\phi^3n^2/4(n - 1)^2$ and $LA = y^3\phi \, n^2/2(n - 1)^2$.

Figure 17.38 plots the spherical aberration blur as a function of element shape and index of refraction. This plot makes the effect of a change of index quite apparent. As the index is increased, the spherical aberration is reduced, and the shape of the element which yields the minimum amount of spherical becomes more and more meniscus. This illustrates why lens designers use high-index glasses to improve a lens design. Note that the minimum spherical for an index of 4.0 is the same as that of a spherical mirror.

For an equi-convex lens working at $m = -1$ (i.e., at 1:1)

$$\text{Transverse spherical} = \text{TSC} = -y^3\phi^2n^2/4(n - 1)^2$$

For two plano-convex elements, face to face,)(, this TSC is reduced by 4×.

Figure 21.10 shows the spherical aberration as a function of object distance for convex-plano, equiconvex and plano-convex singlets.

Figure 17.39 shows the effect of splitting a single element into two, three, or four elements, each element with a correspondingly reduced power, when each element is shaped for minimum (undercorrected) spherical aberration. This graph assume that the elements are "thin" (i.e., zero thickness) and that they are in contact—this is physically impossible, but nonetheless the graph illustrates the idea nicely. Note that the second and subsequent elements are meniscus shaped, and with a realistic thickness, a meniscus element has a reduced Petzval field curvature. For an index of 1.5, splitting a lens in two reduces the spherical by a factor of about 5; dividing the lens into three parts reduces it by a factor of about 20. If the lens is split into four parts, the third-order spherical can be reduced to zero. This effect is widely

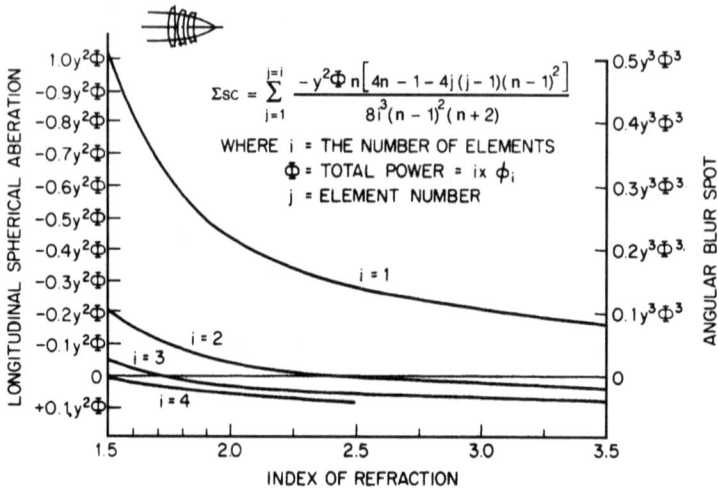

Figure 17.39 The spherical aberration of one, two, three, and four thin lenses, each bent for minimum spherical aberration, as a function of the index of refraction. The number of elements in the set is i. (Object at infinity.)

utilized to improve the image quality of more complex lenses. See, for example, Figs. 16.9, 17.22, and 17.23.

Figure 17.40 is a plot of MTF versus the angular blur β. It is based on geometrical calculation and does not include diffraction effects, so it is reasonably reliable only if the wavefront aberration is large, i.e., more than one or two wavelengths. Here the "cut-off" frequency where the MTF goes to zero occurs at $v\beta = 1.22$. See also Sec. 15.10, Eq. 15.38, and Fig. 15.15.

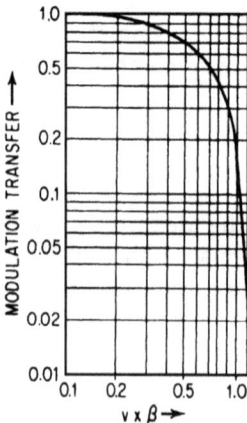

Figure 17.40 The modulation transfer characteristic of a system with an angular blur β (in radians) for a sinusoidal object with a spatial frequency of v cycles per radian. This is a plot of MTF $= 2J_1 (\pi\beta v)/\pi\beta v$ and assumes that the image blur is a uniformly illuminated disk.

Bibliography

See also the references for Chap. 16.

Benford, J., and H. Rosenberger, "Microscope Objectives and Eyepieces," in W. Driscoll (ed.), *Handbook of Optics*, New York, McGraw-Hill, 1978.

Benford, J., "Microscope Objectives," in Kingslake (ed.), *Applied Optics and Optical Engineering*, Vol. 3, New York, Academic, 1965.

Betensky, E., in Shannon and Wyant (eds.), *Applied Optics and Optical Engineering*, Vol. 8, New York, Academic, 1980 (photographic lenses).

Betensky, E., M. Kreitzer, and J. Moskovich, "Camera Lenses" in *Handbook of Optics*, Vol. 2, New York, McGraw-Hill, 1995, Chap. 16.

Cook, G., in Kingslake (ed.), *Applied Optics and Optical Engineering*, Vol. 3, New York, Academic, 1965 (photographic objectives).

Fischer, R. (ed.), *Proc. International Lens Design Conf.*, S.P.I.E., Vol. 237, 1980.

Goldberg, N., "Cameras," in *Handbook of Optics*, Vol. 2, New York, McGraw-Hill, 1995, Chap. 15.

Inoue, S., and R. Oldenboug, "Microscopes," in *Handbook of Optics*, Vol. 2, New York, McGraw-Hill, 1995, Chap. 17.

Johnson, R. B., "Lenses," in *Handbook of Optics*, Vol. 2, New York, McGraw-Hill, 1995, Chap. 1.

Kidger, M., Fundamental Optical Design, SPIE, 2002

Kidger, M., Intermediate Optical Design, SPIE, 2004

Kingslake, R., *Optics in Photography*, S.P.I.E. Press, Bellingham, WA, 1992.

Kingslake, R., *A History of the Photographic Lens*, San Diego, Academic, 1989.

Laikin, M., *Lens Design*, New York, Marcel Dekker, 1991.

Patrick, F., in Kingslake (ed.), *Applied Optics and Optical Engineering*, Vol. 5, New York, Academic, 1965 (military optical instruments).

Riedl, M. J., *Optical Design for Infrared Systems*, S.P.I.E., Vol. TT20, 1995.

Rosin, J., in Kingslake (ed.), *Applied Optics and Optical Engineering*, Vol. 3, New York, Academic, 1965 (eyepieces and magnifiers).

Shannon, R., in Shannon and Wyant (eds.), *Applied Optics and Optical Engineering*, Vol. 8, New York, Academic, 1980 (aspherics).

Smith, W. J., *Modern Lens Design*, New York, McGraw-Hill, 1992.

Smith, W. J. (ed.), *Lens Design*, S.P.I.E., Vol. CR41, 1992.

Smith, W., in W. Driscoll (ed.), *Handbook of Optics*, New York, McGraw-Hill, 1978.

Smith, W., in Wolfe and Zissis (eds.), *The Infrared Handbook*, Washington, Office of Naval Research, 1985.

Taylor, W., and D. Moore (eds.), *Proc. International Lens Design Conf.*, S.P.I.E., Vol. 554, 1985 and *Proc. Int. Optical Design Conf.* of 1990, 1994, 1998, 2002, 2006 (with various edition) SPIE

Exercises

See the note preceding the exercises for Chapter 16.

1 Design a symmetrical eyepiece, using BK7 and SF2 glass, for a 10 × telescope.

2 Design a separable (Lister) 10×, NA 0.25 microscope objective by determining the zero spherical form for each doublet. Analyze the field curvature of the combination and compare with the Petzval projection lens.

3 Design a split-front crown triplet (see Fig. 17.23), using SK4 and SF5 glasses, for a speed of $f/2.8$, a focal length of 125 mm, and a total field of 20°, suitable for a 35-mm slide projector.

Design of Mirror and Catadioptric Systems

18.1 Reflecting Systems

The increasing use of optical systems in the nonvisual regions of the spectrum, i.e., the ultraviolet and infrared regions, has resulted in a corresponding increase in the use of reflecting optics. This is due primarily to the difficulty in procuring completely satisfactory refractive materials for these regions, and secondarily, to the fact that many of the applications permit the use of relatively unsophisticated mirror systems.

The material difficulty is of two kinds. Many applications require the use of a broad spectral band, and a refractive material must transmit well over the full band to be of value. Secondly, chromatic aberration can be difficult to correct over a wide spectral band, and the residual secondary spectrum is sometimes intolerable. A review of Chap. 10 will demonstrate quite clearly the advantages of a reflector in this regard; an ordinary aluminized mirror actually has much better reflectance in the infrared than in the visible and (with special attention) aluminum mirrors suitable for the ultraviolet can be fabricated. Pure reflecting systems are, of course, completely free of chromatic aberration over any desired bandwidth.

18.2 The Spherical Mirror

The simplest reflecting objective is the spherical mirror. For distant objects the spherical mirror has undercorrected spherical aberration, but the aberration is only one-eighth of that of an equivalent glass lens at "minimum bending." The sphere is an especially interesting system when the aperture stop is located at the center of curvature, as shown in Fig. 18.1, because the system is then monocentric, and any line through the center of the stop may be considered to be the optical axis. The image quality is thus practically uniform for any angle of obliquity and the only aberration present is spherical aberration. Coma and astigmatism are zero, and the image surface is a sphere of radius approximately equal to the focal length, centered about the center of curvature. We can approximate the spherical aberration by use of the third-order surface contribution equations. Setting $n = -n' = 1.0$ in Eq. 6.2n, we find that the spherical aberration

$$ \text{SC} = \frac{y^2}{4R} \quad \left[\text{SC} = \frac{(m-1)^2}{4R} y^2 \right] \tag{18.1} $$

where y is the semiaperture, R is the radius, and m is the magnification. The first expression applies for an infinite object distance. The bracketed expression applies to finite conjugates. Using Eq. 15.21 to determine the minimum diameter of the blur spot B, we find that

$$ B = \frac{y^3}{4R^2} \quad \left[B = \frac{(m-1)^2 y^3}{(m+1)4R^2} \right] \tag{18.2} $$

This expression can be converted into the angular blur (in radians) by dividing by the image distance l' (or focal length) to get

$$ \beta = \frac{y^3}{2R^3} \quad \left[\beta = \frac{(m-1)^2 y^3}{(m+1)^2 2R^3} \right] \tag{18.3} $$

By substituting $f = R/2$ and $(f/\#) = f/2y = R/4y =$ relative aperture or NA $= 2y/R$, we obtain the following convenient expression for the

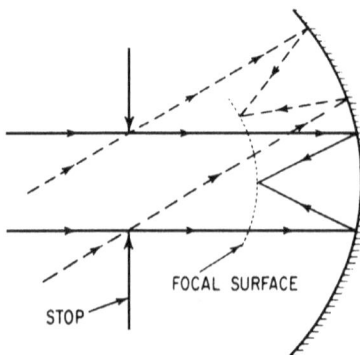

STOP

FOCAL SURFACE

Figure 18.1 A spherical reflector with the stop at its center of curvature forms its image on a concentric spherical focal surface. The image is free of coma and astigmatism when the stop is at this position.

angular blur size of a spherical mirror as a function of its speed (for infinite object distance)

$$\beta = \frac{1}{128(f/\#)^3} = \frac{0.00781}{(f/\#)^3} = \frac{NA^3}{16} \text{ radians} \qquad (18.4)$$

Although this is exact only for the third-order spherical, the expression is quite reliable up to speeds of $f/2$. At $f/1$ the exact ray-traced value of β is 0.0091, at $f/0.75$ it is about 0.024, and at $f/0.5$ it is about 0.13 radians.

When the stop is *not* at the center of curvature, coma and astigmatism are present, and (for an infinite object distance) the third-order contributions are

$$CC^* = \frac{y^2 (R - l_p) u_p}{2R^2} = \frac{(R - l_p) u_p}{32 (f/\#)^2} \qquad (18.5)$$

$$AC^* = \frac{(l_p - R)^2 u_p{}^2}{4R} \qquad (18.6)$$

$$PC = \frac{u_p{}^2 R}{4} = \frac{h^2}{2f} \qquad (18.7)$$

where u_p is the half-field angle and l_p is the mirror-to-stop distance. Note that when l_p is equal to R, CC^* (the sagittal coma) and AC^* (one-half the separation of the S and T fields) are zero. For the case of the stop located at the mirror, we find the minimum angular blur sizes to be
Coma$_s$:

$$\beta = \frac{-u_p}{16 (f/\#)^2} \text{ radians} \qquad (18.8)$$

Compromise focus astigmatism:

$$\beta = \frac{u_p{}^2}{2 (f/\#)} \text{ radians} \qquad (18.9)$$

Equations 18.4, 18.8, and 18.9 provide a very convenient way of estimating the image size for a spherical mirror when combined with the knowledge that (1) coma and astigmatism are zero with the stop at the center of curvature and (2) coma varies linearly (per Eq. 18.6) and astigmatism varies quadratically (per Eq. 18.6) with the distance of the stop from the center of curvature. The sum of the spherical, coma, and astigmatism blur angles gives a fair estimate of the effective size of a point image for a spherical mirror.

18.3   The Paraboloid Reflector

Reflecting surfaces generated by rotation of the conic sections (circle, parabola, hyperbola, and ellipse) share two valuable optical properties. First, a point object located at one focus is imaged at the other focus without spherical aberration. The paraboloid of revolution, Fig. 18.2, described by the equation

$$x = \frac{y^2}{4f} \qquad (18.10)$$

has one focus at f and the other at infinity, and is thus capable of forming perfect (diffraction limited) images of distant *axial* objects. The second characteristic of a conicoid is that if the aperture stop is located at the plane of a focus, as for example in Fig. 18.2a, then the image is free of astigmatism.

However, the paraboloid is not completely free of aberrations; it has both coma and astigmatism. Since it has no spherical aberration, the position of the stop does not change the amount of coma, which is given by Eq. 18.8. The amount of astigmatism *is* modified by the stop position. With the stop at the mirror the astigmatism is given by Eq. 18.9; when the stop is at the focal plane, the astigmatism is zero and the image is located on an approximately spherical surface of radius f, as shown in Fig. 18.2a.

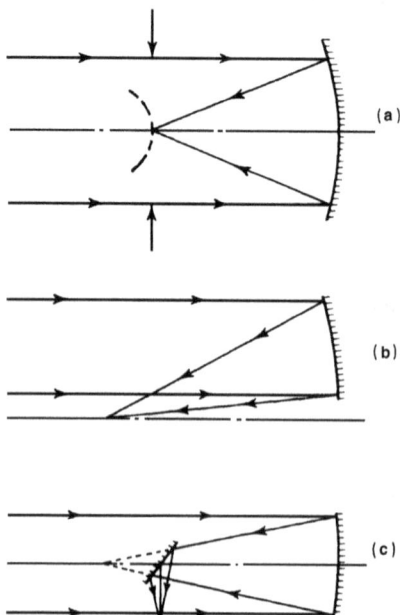

Figure 18.2 (a) A parabolic reflector is free of astigmatism when the stop is at the focus. (b) The Herschel mount for a paraboloid uses an off-axis aperture to keep the focus out of the entering beam. (c) The newtonian mount utilizes a 45° plano reflector to direct the focus to an accessible point outside the main tube of the telescope.

18.4   The Ellipsoid and Hyperboloid

The imaging properties of these conic sections are made use of in the classic Gregorian and Cassegrain telescopic systems, as indicated in Figs. 18.3 and 18.4, respectively.

The primary mirror in each of these is a paraboloid which produces an aberration-free axial image at its focus. The secondary mirror is located so that its first focus coincides with the focus of the paraboloid. Thus the final image is located at the second focus of the secondary mirror and is completely free of spherical aberration. The paraboloid, ellipsoid, and hyperboloid all suffer from coma (compare the magnification produced by the dotted versus the solid lines in Fig. 18.3) and astigmatism, so that the image is aberration-free only exactly on the axis.

It should be apparent that either the Gregorian or Cassegrain objective systems could be made up with almost any arbitrary (within reason) surface of rotation for the primary mirror; some surface then could be found for the secondary mirror which would produce a spherical-free image. This is, in effect, an extra degree of freedom which can be used by the designer to improve the off-axis imagery of these systems.

The *Ritchey-Chretien* objective uses this extra degree of freedom to correct both spherical and coma simultaneously in the Cassegrain configuration. Both mirrors are hyperboloids. The same idea can be applied to the Gregorian or any other two-mirror configuration.

There are many variations on the classic forms. In the Gregorian (Fig. 18.3) two elliptical surfaces plus a weak field lens can be arranged to correct coma and astigmatism as well as spherical. In the Cassegrain (Fig. 18.4) a thick concentric meniscus in the image space can correct the Petzval without changing the spherical or coma. A weak

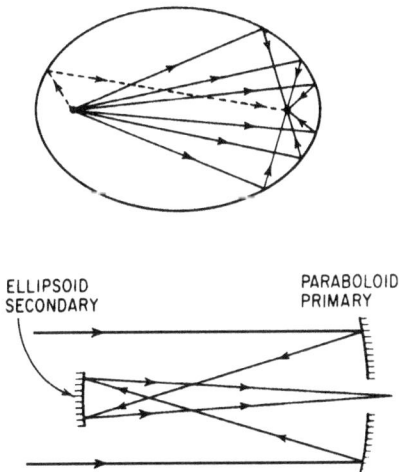

ELLIPSOID
SECONDARY

PARABOLOID
PRIMARY

Figure 18.3 Upper: A point object at one focus of an elliptical reflector is imaged at the other focus without spherical aberration. Lower: The classical Gregorian telescope uses a parabolic primary mirror and an elliptical secondary so that the image is free of spherical.

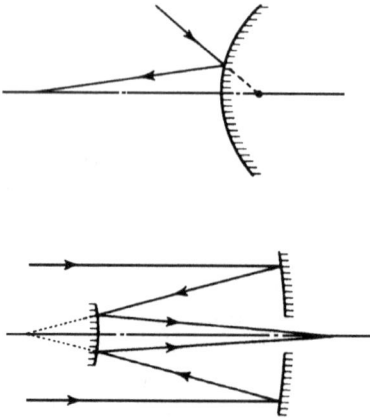

Figure 18.4 Upper: A ray directed toward one focus of a hyperbola is reflected through the other focus. Lower: The classical Cassegrain objective uses a parabolic primary mirror with a hyperboloid secondary. When the primary image is at the focus of the secondary mirror, the final image has no spherical aberration. If the osculating radii of the surfaces are equal, the Petzval field is flat.

negative element near the secondary can correct the coma. A meniscus cemented doublet near the focus can correct the astigmatism and Petzval. The equations for conic sections are in Sec. 18.6.

18.5   Equations for Two-Mirror Systems

The third-order aberration surface contribution equations (Eqs. 6.2) can be used to evaluate the aberrations of a system of two mirrors. The following equations apply to *any* two-mirror system, regardless of configuration. The curvatures of the primary and secondary mirrors are given by

$$C_1 = \frac{(B - F)}{2DF}$$

$$C_2 = \frac{(B + D - F)}{2DB}$$

where F is the effective focal length of the combination, B is the back focus (i.e., the distance from mirror #2 to the focus), and D is the spacing between mirrors (the sign of D is here taken as positive). Note that *any* configuration can be obtained by suitably choosing F, B, and D. The Cassegrain has a positive focal length, the Gregorian a negative one. Both have a focal length which is long compared to D. The Schwarzschild (see Fig. 17.11) configuration results if B is chosen long compared with D.

If we assume an object at infinity and place the stop at the primary mirror, the third-order aberration sums are given by

$$\Sigma TSC =$$

$$\frac{Y^3 \left[F (B-F)^3 + 64D^3 F^4 K_1 + B (F-D-B) (F + D - B)^2 - 64B^4 D^3 K_2 \right]}{8D^3 F^3}$$

$$\Sigma CC =$$
$$\frac{HY^2 \left[2F\,(B-F)^2 + (F-D-B)\,(F+D-B)\,(D-F-B) - 64B^3D^3K_2\right]}{8D^2F^3}$$

$$\Sigma TAC = \frac{H^2Y\left[4BF\,(B-F) + (F-D-B)\,(D-F-B)^2 - 64B^3D^3K_2\right]}{8BDF^3}$$

$$\Sigma TPC = \frac{H^2Y\left[DF - (B-F)^2\right]}{2BDF^2}$$

where Y = the semiaperture of the system
  H = the image height
  B = distance from mirror #2 to image (i.e., the back focal length)
  F = system focal length
  D = spacing (use positive sign)
  ΣTSC = transverse third-order spherical aberration sum
  ΣCC = third-order sagittal coma sum
  ΣTAC = transverse third-order astigmatism sum
  ΣTPC = transverse Petzval curvature sum

and where K_1 and K_2 are the equivalent fourth-order deformation coefficients for the primary and secondary mirrors. For a conic section, K is equal to the conic constant κ (kappa) divided by 8 times the cube of the surface radius. Thus $K = \kappa/8R^3$ and $\kappa = 8KR^3$

We can readily solve for the standard design forms. If both mirrors are independently corrected for spherical aberration, we get the classical *Cassegrain* or *Gregorian,* and

$$K_1 = \frac{(F-B)^3}{64D^3F^3} \quad (\text{KAPPA} = -1)$$

$$K_2 = \frac{(F-D-B)\,(F+D-B)^2}{64B^3D^3}$$

$$\Sigma TSC = 0.0$$

$$\Sigma CC = \frac{HY^2}{4F^2}$$

$$\Sigma TAC = \frac{H^2Y\,(D-F)}{2BF^2}$$

Note that the coma is a function of only the field (H) and the NA; B and D do not enter. *All classic Cassegrains and Gregorians have the same third-order coma.*

For a *Ritchey-Chretien**, we can solve for K_1 and K_2 to get both third-order spherical and coma corrected, and

$$K_1 = \frac{[2BD^2 - (B - F)^3]}{64D^3F^3} = \frac{2BD^2}{(B - F)^3} - 1.0$$

$$K_2 = \frac{[2F(B - F)^2 + (F - D - B)(F + D - B)(D - F - B)]}{64B^3D^3}$$

$$\Sigma TSC = 0.0$$

$$\Sigma CC = 0.0$$

$$\Sigma TAC = \frac{H^2Y(D - 2F)}{4BF^2}$$

Note that these values for K_1 and K_2 will correct the third-order aberrations, but not the higher-order aberrations. It is a simple task to adjust the deformation/conic terms so that the marginal spherical and coma are corrected, but in high speed systems there will be residuals for a conic system. If desired, these residuals can be corrected by varying the sixth- and higher-order deformation terms of a general aspheric surface.

A *Dall-Kirkham system* has an easier to make spherical secondary, and all of the correction is accomplished by the aspheric primary. Thus

$$K_1 = \frac{[F(F - B)^3 - B(F - D - B)(F + D - B)^2]}{64D^3F^4}$$

$$K_2 = 0.0$$

$$\Sigma TSC = 0.0$$

$$\Sigma CC = \frac{HY^2[2F(B-F)^2 + (F-D-B)(F + D-B)(D-F-B)]}{8D^2F^3}$$

$$\Sigma TAC = \frac{H^2Y[4BF(B - F) + (F - D - B)(D - F - B)^2]}{8DBF^3}$$

---

*Chretien, "Rev. d'Optique Theorique et Instrumentale," Jan and Feb 1922.

A sort of *inverse Dall-Kirkham* has a spherical primary and an aspheric secondary:

$$K_1 = 0.0$$

$$K_2 = \frac{[F(B - F)^3 + B(F - D - B)(F + D - B)^2]}{64B^4D^3}$$

$$\Sigma TSC = 0.0$$

$$\Sigma CC = \frac{HY^2[2BD^2 - (B - F)^3]}{8BD^2F^2}$$

$$\Sigma TAC = \frac{H^2Y[(F - B)^3 + 4BD(D - F)]}{8B^2DF^2}$$

These expressions, as mentioned above, are perfectly general, and apply to any and all two-mirror systems. They are of course limited to the third order, but are surprisingly accurate up to a speed of about $f/3$. One can use these results as starting forms for the development of faster or more complex designs, incorporating an aspheric corrector plate or a third mirror to achieve additional correction of, for example, astigmatism. The results of these expressions make excellent starting designs for higher-speed systems.

Note that the conics may appear to violate the principles of image illumination laid down in Chap. 12. For example, a paraboloid can readily be constructed with a diameter more than twice its focal length; a paraboloid with a speed of say $f/0.25$ is quite feasible and will indeed be free of spherical aberration on the axis, whereas in preceding chapters, we may have led the reader to believe that a speed of $f/0.5$ was the largest aperture attainable.

This apparent paradox can be resolved by an examination of Fig. 18.5 which shows an $f/0.25$ parabola. Note that the focal length is equal to f *only* for the axial zone and that for marginal zones the focal length is much larger; for marginal zones the effective focal length of a parabola is given by

$$F = f + x = f + \frac{y^2}{4f} \tag{18.11}$$

The parabola is thus far from an *aplanatic* (spherical- and coma-free) system. For an $f/0.25$ paraboloid the marginal zone focal length is twice that of the paraxial zone and the magnification is correspondingly larger. Thus, if the object has a finite size, the image formed by the

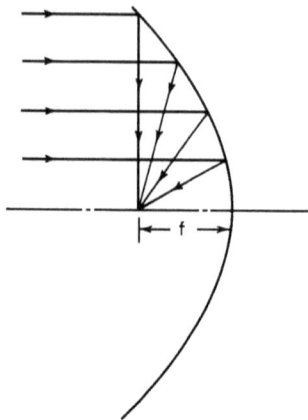

Figure 18.5 Illustrating the extreme variation of focal length with ray height in an *f*/0.25 parabolic reflector.

marginal zones of this mirror will be twice as large as those from the axial zone; this is, of course, nothing but ordinary coma (the "variation of magnification with aperture"). The parabola is thus aberration-free only *exactly* on the axis.

The apparent contradiction of our image illumination principles is thus resolved since we had assumed aplanatic systems in their derivations. From another viewpoint, we can remember that although the parabola forms a perfect image of an infinitesimal (geometrical) point, such a point (being infinitesimal) cannot emit a real amount of energy; the moment one increases the object size to any real dimension, the parabola has a real field, the image becomes comatic, and the energy in the image is spread out over a finite blur spot. This reduces the image illumination to that indicated as the maximum in Chap. 12.

The Cassegrain objective system is used (usually in a modified form) in a great variety of applications because of its compactness and the fact that the second reflection places the image behind the primary mirror where it is readily accessible. It suffers from a very serious drawback when an appreciable field of view is required, in that an extreme amount of baffling is necessary to prevent stray radiation from flooding the image area. Figure 18.6 indicates this difficulty and the type of baffles frequently used to overcome this problem. An exterior "sunshade," which is an extension of the main exterior tube of the scope, is frequently used in addition to the internal baffles.

Because of their uniaxial character, aspheric surfaces are much more difficult to fabricate than ordinary spherical surfaces. A strong paraboloid may cost an order of magnitude more than the equivalent sphere; ellipsoids and hyperboloids are a bit more difficult, and nonconic aspherics are more difficult still. Thus one might well think twice

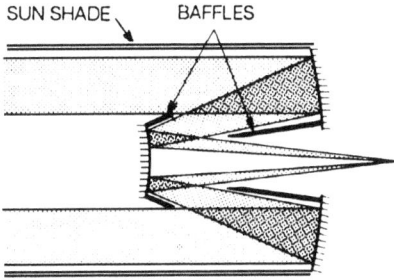

Figure 18.6 Complex conical baffles are necessary in a Cassegrain objective to prevent stray radiation from flooding the image plane.

(or three times) before specifying an aspheric. Often a spherical system can be found which will do nearly as well at a fraction of the cost. This is also true in refracting systems of moderate size where several ordinary spherical elements can be purchased for the cost of a single aspheric. For very large one-of-a-kind systems, however, aspherics are frequently a sound choice. This is because the large systems (e.g., astronomical objectives) are, in the final analysis, handmade, and the aspheric surface adds only a little to the optician's task.

Computer-controlled single-point diamond machining has become a practical technique for fabricating aspheric surfaces. While this is especially true for infrared optics, aspheric surfaces which owe their feasibility to diamond turning are showing up in many commercial applications such as high-level photographic optics. Extremely stable and precise machine tools (e.g., lathes, mills) can produce surfaces with turning marks which are small enough to allow their use in high-quality optical systems. A limitation on diamond turning is that there is only a small number of materials which can be diamond turned. Included are germanium, silicon, aluminum, copper, nickel, zinc sulfide, selenide, and plastics. Note that glass and ferrous metals are not included in this list. However, glass molding techniques have reached a quality level that permits the molding of aspheric surfaces which are useable in diffraction-limited systems. For example, both molded glass and plastic aspheric lenses are made for laser disk objectives. The precision molds for these processes are made on computer-controlled equipment, and in some cases they are also diamond turned.

18.6   Conic Section through the Origin

The conic surfaces are often used in optical systems. They have two interesting optical properties: If the object point is located at one focal point of a conic, the image point is located at the other focal point, and there is no spherical aberration (but there is coma). The second property is that if the aperture stop is located in the plane of a focal point, there is no astigmatism.

There are three commonly used conic constants used to define conic surfaces. In the equations below r is the radius (at the axis), c is the curvature ($c = 1/r$), and p is a conic constant. The constant kappa [$\kappa = (p - 1) = (-e^2)$] is frequently used in optical design. The eccentricity $e = \sqrt{-\kappa} = \sqrt{(1 - p)}$ is rarely used in optical design. The values of the constants for the conic sections are as follows:

Ellipse (A)*	$p > 1$	$\kappa > 0$	$e^2 < 0$
Circle	$p = 1$	$\kappa = 0$	$e = 0$
Ellipse (B)**	$1 > p > 0$	$0 > \kappa > -1$	$0 < e^2 < +1$
Parabola	$p = 0$	$\kappa = -1$	$e = 1$
Hyperbola	$p < 0$	$\kappa < -1$	$e^2 > +1$

*the focal points are not on the x-axis
**the focal points are on the x-axis

Equations for a conic section which passes through the origin:

$$y^2 - 2rz + pz^2 = 0$$

$$Z = \frac{r \pm \sqrt{r^2 - py^2}}{p} = \frac{cy^2}{1 + \sqrt{1 - pc^2y^2}}$$

$$Z = \frac{y^2}{2r} + \frac{1}{2^2 2!}\frac{py^4}{r^3} + \frac{1 \cdot 3}{2^3 3!}\frac{p^2 y^6}{r^5} + \frac{1 \cdot 3 \cdot 5}{2^4 4!}\frac{p^3 y^8}{r^7} + \frac{1 \cdot 3 \cdot 5 \cdot 7}{2^5 5!}\frac{p^4 y^{10}}{r^9} + \cdots$$

In optics the conic surfaces are generated by rotating the conic section about the z-axis. The focal points for the rotated ($p>1$) ellipse (A) are not on the axis, so the imagery between these foci is not stigmatic, and a simple test during fabrication is not available as it is with the ($1 > p > 0$) ellipse (B) where both foci are on these axis and there is stigmatic imagery between the foci. The former, ellipse (A), is sometimes called an oblate spheroid (or ellipsoid) [oblate = pressed or squashed]; the latter, ellipse (B), is referred to as a prolate spheroid (or ellipsoid) (prolate = stretched). There is some confusion about these terms in the optical literature. The surface generated by a parabola is a paraboloid, that generated by a circle is a sphere (spheroid), and that generated by a hyperbola is a hyperboloid.

Distance to foci:

$$\frac{r}{p}\left(1 \pm \sqrt{1 - p}\right)$$

Magnification between foci:

$$-\left[\frac{1 + \sqrt{1 - p}}{1 - \sqrt{1 - p}}\right]$$

Intersects axis at:

$$Z = 0, \frac{2r}{p}$$

Distance between conic and a circle of the same vertex radius r (i.e., the departure from a sphere):

$$\Delta Z = \frac{(p - 1) y^4}{2^2 2! \ r^3} + \frac{1 \cdot 3 \ (p^2 - 1) y^6}{2^3 3! \ r^5} + \frac{1 \cdot 3 \cdot 5 \ (p^3 - 1) y^8}{2^4 4! \ r^7} + \cdots$$

Angle between the normal to the conic and the z-axis:

$$\phi = \tan^{-1}\left[\frac{-y}{(r - pZ)}\right]$$

$$\sin \phi = \frac{-y}{[y^2 + (r - pZ)^2]^{1/2}}$$

Radius of curvature:

Meridional: $$R_t = \frac{R_s^3}{r^2} = \frac{[y^2 + (r - pZ)^2]^{3/2}}{r^2}$$

Sagittal (distance to axis along the surface normal):

$$R_s = [y^2 + (r - pZ)^2]^{1/2}$$

18.7 The Schmidt System

The Schmidt objective (Fig. 18.7) can be viewed as an attempt to combine the wide uniform image field of the stop-at-the-center sphere with the "perfect" imagery of the paraboloid. In the Schmidt, the reflector is a sphere and the spherical aberration is corrected by a thin refracting aspheric plate at the center of curvature. Thus the concentric character of the sphere is preserved in great measure, while the spherical aberration is completely eliminated (at least for one wavelength).

The aberrations remaining are chromatic variation of spherical aberration and certain higher-order forms of astigmatism or oblique spherical which result from the fact that the off-axis ray bundles do not strike the corrector at the same angle as do the on-axis bundles. The action of a given zone of the corrector is analogous to that of a thin

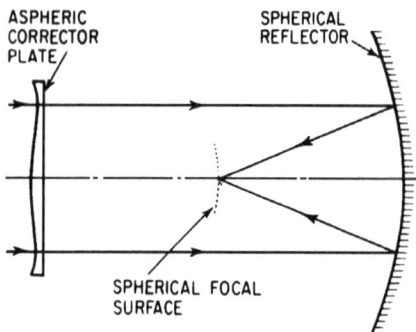

Figure 18.7 The Schmidt system consists of a spherical reflector with an aspheric corrector plate at its center of curvature. The aspheric surface in the $f/1$ system shown here is greatly exaggerated.

refracting prism. For the non-oblique rays, the prism is near minimum deviation; as the angle of incidence changes, the deviation of the "prism" is increased, introducing overcorrected spherical. Since the action is different in the tangential plane than in the sagittal plane, astigmatism results. This combination is oblique spherical aberration. The meridional angular blur of a Schmidt system is well approximated by the expression

$$\beta = \frac{u_p{}^2}{48\,(f/\#)^3} \text{ radians} \tag{18.12a}$$

The axial spherochromatic blur is approximately

$$\beta = \frac{1}{256V(f/\#)^3} \text{ radians} \tag{18.12b}$$

See Fig. 19.52 for a finished Schmidt design.

There are obviously an infinite number of aspheric surfaces which may be used on the corrector plate. If the focus is maintained at the paraxial focus of the mirror, the *paraxial* power of the corrector is zero and it takes the form of a weak concave surface. The best forms have the shape indicated in Fig. 18.7, with a convex paraxial region and the minimum thickness at the 0.866 or 0.707 zone, depending on whether it is desired to minimize spherochromatic aberration or to minimize the material to be ground away in fabrication. The performance of the Schmidt can be improved slightly by (1) incompletely correcting the axial spherical to compensate for the off-axis overcorrection, (2) "bending" the corrector slightly, (3) reducing the spacing, (4) using a slightly aspheric primary to reduce the load on, and thus the overcorrection introduced by, the corrector. Further improvements have been made by using more than one corrector and by using an achromatized corrector.

A near-optimal corrector plate has a surface shape given by the equation

$$Z = 0.5Cy^2 + Ky^4 + Ly^6$$

where

$$C = \frac{3}{128\,(n-1)\,f\,(f/\#)^2}$$

$$K = \frac{\left[1 - \dfrac{3}{64\,(f/\#)^2}\right]^2}{32\,(1-n)\,f^3}$$

$$L = \frac{1}{85.8\,(1-n)\,f^5}$$

and f is the focal length, $f/\#$ is the speed or f-number, and n is the index of the corrector plate.

The aspheric corrector of the Schmidt is usually easier to fabricate than is the aspheric surface of the paraboloid reflector. This is because the index difference across the glass corrector surface is about 0.5 compared to the effective index difference of 2.0 at the reflecting surface of the paraboloid, making it only one-fourth as sensitive to fabrication errors.

An aspheric corrector plate of this type can be added to most optical systems. One must remember that an aspheric surface placed at the aperture stop (as in the Schmidt system) will affect *only* the spherical aberration and that the aspheric must be placed well away from the stop if it is to be used to correct coma or astigmatism. An aspheric plate can be added to any of the two-mirror systems described in previous sections; if both mirrors are aspheric, the addition of the corrector plate provides enough degrees of freedom to correct spherical, coma, and astigmatism. Corrector plates have been used in the entrance beam or in the image space. An example is the "Schmidt Cassegrain," where both mirrors of the Cassegrain configuration are simple spheres. The aspheric corrector plate is the front window of the system and is often used to support the secondary mirror. This is an economical and commercially successful system.

18.8   The Mangin Mirror

The Mangin mirror is perhaps the simplest of the catadioptric (i.e., combined reflecting and refracting) systems. It consists of a second-surface spherical mirror with the power of the first surface chosen to correct the spherical aberration of the reflecting surface. Figure 18.8

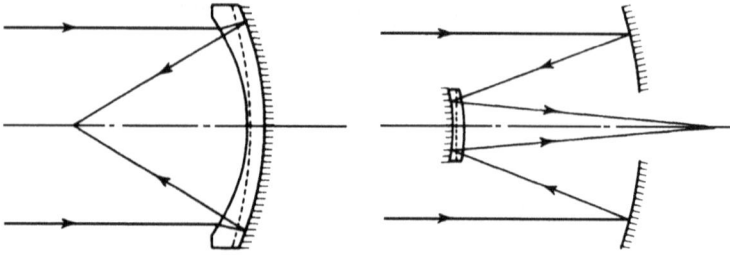

Figure 18.8 In the Mangin mirror (left) the spherical aberration of the second surface reflector is corrected by the refracting first surface. In the right-hand sketch, the spherical is corrected by a Mangin-type secondary. The dotted lines indicate the manner in which color correction can be achieved. In a doublet Mangin, glass choice can be used as a design freedom to correct coma.

shows a Mangin mirror. The design of a Mangin is straightforward. One radius is chosen arbitrarily (a value about 1.6 times the desired focal length is suitable for the reflector surface) and the other radius is varied systematically until the spherical aberration is corrected. For an index of 1.5 this radius equals the focal length. The correction is exact for only one zone, however, and an undercorrected zonal residual remains. The size of the angular blur spot resulting from the zonal spherical can be approximated (for apertures smaller than about $f/1.0$) by the empirical expression

$$\beta = \frac{10^{-3}}{4\,(f/\#)^4}\ \text{radians} \tag{18.13}$$

Note that this is the minimum-diameter blur and that the "hard-core" blur diameter is smaller, as discussed in Chap. 15. At larger apertures, the angular blur predicted by Eq. 18.13 is too small; for example, at $f/0.7$ the blur is about 0.002 radians, twice as large as that predicted by Eq. 18.13.

Since the Mangin is roughly equivalent to an achromatic reflector plus a pair of simple negative lenses, the system has a very large over-corrected chromatic aberration. This can be corrected by making an achromatic doublet out of the refracting element. For the simple Mangin, the chromatic angular blur is approximated by

$$\beta = \frac{1}{6V\,(f/\#)}\ \text{radians} \tag{18.14}$$

where V is the Abbe V-value of the material used. Note that this is only about one-third of the chromatic of a simple lens. [$\beta = \frac{1}{2}V(f/\#)$]

The coma blur of the Mangin primary mirror is approximately one-half of that given by Eq. 18.8. Since the spherical aberration is corrected, little change in the coma results from a shift of the stop position.

The Mangin principle may be applied to the secondary mirror of a system as well as to the primary. The right-hand sketch of Fig. 18.8 shows a Cassegrain type of system in which the secondary is an achromatic Mangin mirror. Such a system is relatively economical and light in weight, since all surfaces are spheres and only the small secondary needs to be made of high-quality optical material. The power of a thin second-surface reflecting element is given by

$$\phi = 2C_1 (n - 1) - 2C_2 n$$

The Mangin mirror is often used as an element of a more complex system. For example, the primary or secondary of a system may be a Mangin; as such, it serves to correct aberrations without adding significantly to the weight of the system and often effectively replaces an expensive aspheric surface. See for example Figs. 19.33 and 19.34.

18.9   The Bouwers (Maksutov) System

The Bouwers (or Maksutov) system may be considered a logical extension of the Mangin mirror principle in which the correcting lens is separated from the mirror to allow two additional degrees of freedom, producing a great improvement in the image quality of the system.

A popular version of this device is the Bouwers concentric system, shown in Fig. 18.9. In this system, all surfaces are made concentric to the aperture stop, which (as we have noted in the case of the simple spherical mirror) results in a system with uniform image quality over the entire field of view. This is an exceedingly simple system to design, since

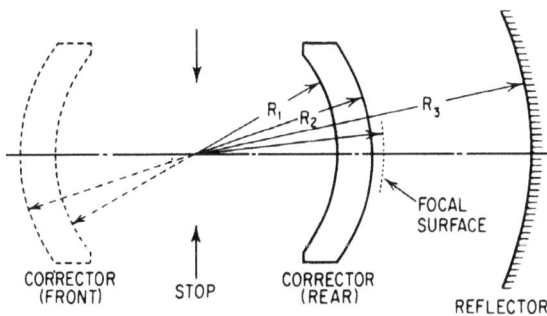

Figure 18.9   In the Bouwers concentric catadioptric system, all the surfaces are concentric about the aperture. The "front" and "rear" versions of the corrector are identical and produce identical correction. The rear system is more compact, but the front system can be better corrected, since it can utilize a greater corrector thickness without interference with the focal surface. Occasionally, correctors in both locations are utilized simultaneously.

there are only three degrees of freedom, namely, the three curvatures. One chooses R_1 to set the scale of the lens (a value of R_1 equal to about 85 percent of the intended focal length is appropriate) and R_2 to provide an appropriate thickness for the corrector, and then determines the value of R_3 for which the marginal spherical is zero. Because of the monocentric construction, coma and astigmatism are zero, and the image is located on a spherical surface which is also concentric to the stop and the radius of which equals the focal length of the system. Thus only a few rays need be traced to completely determine the correction of the system.

One of the interesting features of this system is that the concentric corrector element may be inserted anywhere in the system (as long as it remains concentric) and it will produce *exactly* the same image correction. Two equivalent positions for the corrector are shown in Fig. 18.9. A third position is in the convergent beam, between the mirror and the image.

If we accept the curved focal plane, the only aberrations of the Bouwers concentric system are residual zonal spherical aberration and longitudinal (axial) chromatic aberration. In general, as the corrector thickness is increased, the zonal is reduced and the chromatic is increased.

The concentric system described above is used for applications requiring a wide field of view. When the field requirements permit, the zonal spherical or the chromatic may be reduced by departing from the concentric mode of construction, although this is, of course, accomplished at the expense of the coma and astigmatism correction.

If one of the thick-lens equations (Eq. 3.21 or 3.22) is differentiated with respect to the index, the result can be set equal to zero and the equation solved for the shape of an element the power or image distance of which does not vary with a change in index (or a change in wavelength). This is an achromatic singlet. It takes the shape of a thick meniscus element, and this can be used as an achromatic corrector, just as in Fig. 18.9. This is the basis of the Maksutov system.

Another means of effecting chromatic correction is shown in Fig. 18.10a, in which the corrector meniscus is made achromatic. Note that concentricity is destroyed by this technique, although if the crown and flint elements are made of materials with the same index but different *V*-values (e.g., 617:549, and 617:366), the concentricity can be preserved for the wavelength at which the indices match, and only the chromatic correction will vary with obliquity.

A very powerful system results if the concentric Bouwers system is combined with a Schmidt-type aspheric corrector plate, as shown in Fig. 18.10b. Since the aspheric plate needs only correct the small zonal residual of the concentric system, its effects are relatively weak and the variation of effects with obliquity are correspondingly small.

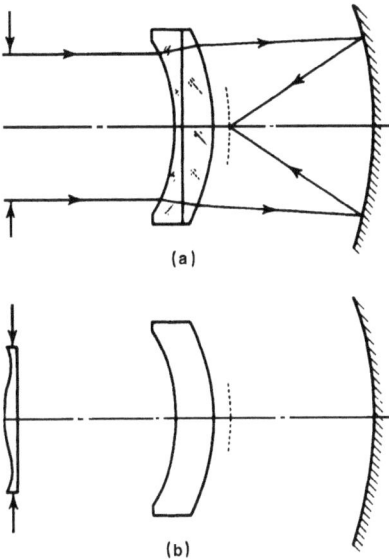

Figure 18.10 (a) An achromatized meniscus corrector. (b) An aspheric corrector plate at the stop removes the residual zonal spherical aberration of the concentric system.

The Baker-Nunn satellite tracking cameras are based on this principle, although their construction is more elaborate, using double-meniscus correctors and three (achromatized) aspheric correctors at the stop.

The basic Bouwers-Maksutov meniscus corrector principle has been utilized in a multitude of forms. A few of the possible Cassegrain embodiments of the principle are shown in Fig. 18.11. The reader can probably devise an equal number in a few minutes. An arrangement similar to that shown in Fig. 18.11c is frequently used in homing missile guidance systems. The corrector makes a reasonably aerodynamic window, or dome, and although the system is not concentric, the primary and secondary can be gimballed as a unit about the center of curvature of the dome so that the "axial" correction is maintained as the direction of sight is varied.

There are a tremendous number of variations of the catadioptric principle. Refractive correctors in almost every conceivable form have been combined with mirrors. Positive field lenses have been used to flatten the overcorrected Petzval surface of the basic concave reflector, coma-correcting field elements have been used with paraboloids, and multiple thin nonmeniscus correctors have been used with spheres, to name just a few of the variations on the device. The basic strength of this general system is, of course, the relatively small aberration inherent in a spherical reflector; the corrector's task is to remove the faults without losing the virtues.

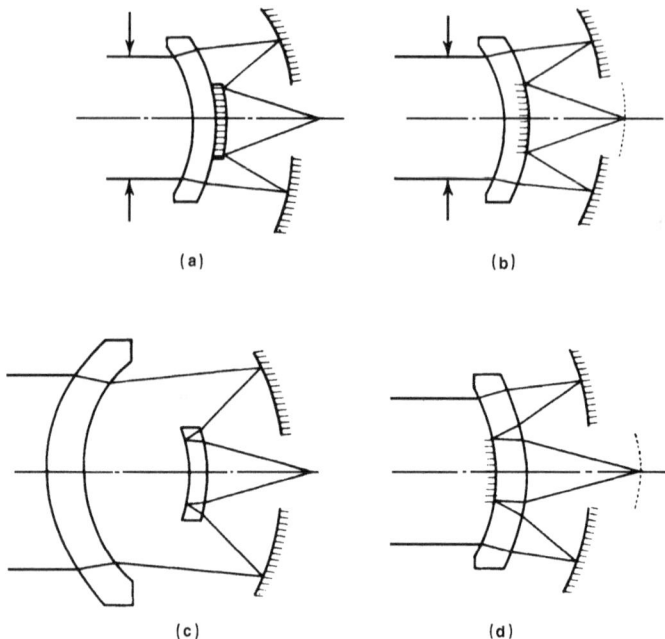

Figure 18.11 Four of the many possible Cassegrain versions of the meniscus corrector catadioptric system. Systems (b) and (d) are often used in compact commercial telescopes.

Two or more closely spaced thin-corrector elements the total power of which is effectively zero can be shaped to correct the aberrations of a spherical mirror. If the glass is the same for all of the corrector elements, then the combination will have little or no chromatic, primary or secondary. Additional examples of catadioptric systems are shown in Figs. 19.32, 19.39, and 19.34.

Note that the L. Jones reference in the bibliography is an excellent and complete survey of the field of reflective and catadioptric objectives. It contains prescriptions, diagrams and interesting, informative notes for each system.

18.10   The Rapid Estimation of Blur Sizes for Simple Optical Systems

It is frequently useful to be able to estimate the size of the aberration blur produced by an optical system without going to the trouble of making a raytrace analysis. In preliminary engineering work or the preparation of technical proposals, where time is limited, the following material (which is based largely on third-order aberration analysis or empirical studies) can be of value.

The aberrations are expressed in terms of the angular size β (in radians) of the blur spot which they produce; β may be converted to B, the linear diameter of the blur, by multiplying by the system focal length. In this section the object will be assumed to be at infinity.

Where the blur size for more than one aberration is given, the sum of all the aberration blurs will yield a conservative (i.e., large) estimate of the total blur.

Where the blur is due to chromatic aberration, the blur angle given encompasses the total energy in the image of a point. Occasionally it is of value to know that 75 to 90 percent of the energy is contained in a blur one-half as large, and 40 to 60 percent of the energy is contained in a blur one-quarter as large as that given by the equations. In the visible, the chromatic blur is usually reduced by a factor of about 40 by achromatizing the system.

The blurs given for spherical aberration are the minimum-diameter blur sizes; these values are the most useful for work with detectors. For visual or photographic work, a "hard-core" focus, as discussed in Chap. 15, is preferable, and the blurs given here should be modified accordingly.

Note that with the exception of Eqs. 18.15 and 18.16, all the blurs are based on geometrical considerations. It is, therefore, wise to evaluate Eq. 18.15 or 18.16 first to be certain that the geometrical blurs are not smaller than the diffraction pattern before basing further effort on the geometrical results.

More complete discussions of the individual systems may be found in the preceding section.

Diffraction-limited systems

The diameter of the first dark ring of the Airy pattern is given by

$$\beta = \frac{2.44\lambda}{D} \text{ radians} \tag{18.15}$$

$$B = 2.44\lambda \ (f/\#) = \frac{1.22\lambda}{NA} \tag{18.16}$$

where λ is the wavelength, D is the clear aperture of the system, $(f/\#)$ = f/D is the relative aperture, and f is the focal length. The "effective" diameter of the blur (for modulation transfer considerations) is about one-half the above.

Spherical mirror

Spherical aberration: $\beta = \dfrac{0.0078}{(f/\#)^3}$ radians   (18.17)

Sagittal coma: $\beta = \dfrac{(l_p - R)\, U_p}{16R\,(f/\#)^2}$ radians   (18.18a)

for a conic section $\beta = \dfrac{(pl_p - R)U_p}{16R\,(f/\#)^2}$ (18.18b)

Astigmatism: $\beta = \dfrac{(l_p - R)^2 U_p^2}{2R^2\,(f/\#)}$ radians   (18.19a)

for a conic section $\beta = \dfrac{(pl_p^2 - 2l_pR + R^2)U_p^2}{2R^2\,(f/\#)}$ (18.19b)

where l_p is the mirror-to-stop distance, R is the mirror radius, $(l_p - R)$ is the center-to-stop distance, and U_p is the half-field angle in radians. The focal plane of a spherical mirror is on a spherical surface concentric to the mirror when the stop is at the center of curvature. See Sec. 18.6 for the conic constant p.

Paraboloidal mirror

Spherical aberration: $\beta = 0$ (18.20)

Sagittal coma: $\beta = \dfrac{U_p}{16\,(f/\#)^2}$ radians   (18.21)

Astigmatism: $\beta = \dfrac{(l_p + f)\, U_p^2}{2f\,(f/\#)}$ radians   (18.22)

where the symbols have been defined above.

Schmidt system (neutral zone at 0.866)

Spherical aberration: $\beta = 0$ (18.23)

Higher-order aberrations: $\beta = \dfrac{U_p^2}{48\,(f/\#)^3}$ radians   (18.24)

Spherochromatic: $\beta = \dfrac{1}{256V\,(f/\#)^3}$ (18.24a)

Mangin mirror   (Stop at the mirror):

Zonal spherical:
$$\beta = \frac{10^{-3}}{4\,(f/\#)^4} \text{ radians} \qquad (18.25)$$

Chromatic aberration:
$$\beta = \frac{1}{6V\,(f/\#)} \text{ radians} \qquad (18.26)$$

($3\times$ better than a thin lens)

Sagittal coma:
$$\beta = \frac{U_p}{32\,(f/\#)^2} \text{ radians} \qquad (18.27)$$

Astigmatism:
$$\beta = \frac{U_p^{\,2}}{2\,(f/\#)} \text{ radians} \qquad (18.28)$$

Simple thin lens   (Minimum spherical shape)

Spherical aberration:
$$\beta = \frac{K}{(f/\#)^3} \text{ radians}$$

$$= \frac{n(4n-1)}{(f/\#)^3 128(n-1)^2\,(n+2)} \qquad (18.29)$$

$$\frac{C_1}{(C_1 - C_2)} = \frac{n(2n+1)}{2(n+2)} = 0.857$$

$K = 0.067$	for $n = 1.5$	
$= 0.027$	for $n = 2.0$	$= 1.25$
$= 0.0129$	for $n = 3.0$	$= 2.1$
$= 0.0103$	for $n = 3.5$	$= 2.54$
$= 0.0087$	for $n = 4.0$	$= 3.0$

Chromatic aberration:
$$\beta = \frac{1}{2V\,(f/\#)} \text{ radians} \qquad (18.30)$$

Sagittal coma:
$$\beta = \frac{U_p}{16\,(n+2)\,(f/\#)^2} \text{ radians} \qquad (18.31)$$

Astigmatism:
$$\beta = \frac{U_p^{\,2}}{2\,(f/\#)} \text{ radians (at compromise focus)} \qquad (18.32)$$

where n is the index of refraction, V is the reciprocal relative dispersion, and the stop is at the lens.

Concentric Bouwers. The expressions for monochromatic aberrations are empirical and are derived from the performance graphs and tables given by Bouwers and by Lauroesch and Wing (see references).

Rear concentric (solid line in Fig. 18.9). The maximum corrector thickness of this form must be limited to keep the image from falling inside the corrector. With the thickest possible corrector:
Zonal spherical:

$$\beta \approx \frac{4 \times 10^{-4}}{(f/\#)^{5.5}} \text{ radians} \tag{18.33}$$

General concentric

Zonal spherical:

$$\beta \approx \frac{10^{-4}}{\left(\dfrac{t}{f} + 0.06\right)(f/\#)^5} \text{ radians} \tag{18.34}$$

Chromatic aberration:

$$\beta \approx \frac{tf\,\Delta n}{2n^2 R_1 R_2 \,(f/\#)} \text{ radians} \tag{18.35}$$

or very approximately:

$$\beta \approx 0.6\,\frac{t}{f}\,\frac{\Delta n}{n^2\,(f/\#)} \text{ radians} \tag{18.36}$$

Corrected concentric

Higher-order aberrations:

$$\beta \approx \frac{9.75\,(U_p + 7.2U_p^3)\cdot 10^{-5}}{(f/\#)^{6.5}} \text{ radians} \tag{18.37}$$

where t is the corrector plate thickness, f is the system focal length, Δn is the dispersion of the corrector material, n is the index of the corrector, and R_1 and R_2 are the radii of the corrector. These expressions apply for corrector index values in the 1.5 to 1.6 range and for relative apertures to the order of $f/1.0$ or $f/2.0$. For speeds faster than $f/1.0$, the monochromatic blur angles are larger than above (e.g., about 20 percent larger at $f/0.7$). The use of a high-index corrector ($n > 2$) will reduce the monochromatic blur somewhat at high speeds.

Bibliography

Bouwers, W., *Achievements in Optics*, New York, Elsevier, 1950.
Dimitroff and J. Baker, *Telescopes and Accessories*, London, Blakiston, 1945.
Fischer, R. (ed.), *Proc. International Lens Design Conf.*, S.P.I.E., Vol. 237, 1980.
Jones, L., "Reflective and Catadioptric Objectives," in *Handbook of Optics*, Vol. 2, New York, McGraw-Hill, 1995, Chap. 18.
Kidger, M. *Fundamental Optical Design*, SPIE, 2002.

Kidger, M. Intermediate Optical Design, SPIE, 2004.
Korsch, D., *Reflective Optics,* New York, Academic Press, 1991.
Lauroesch, T., and C. Wing, *J. Opt. Soc. Am.,* Vol. 49, 1959, p. 410 (Bouwers systems).
Maksutov, D., *J. Opt. Soc. Am.,* Vol. 34, 1944, p. 270.
Riedl, M. J., *Optical Design for Infrared Systems,* S.P.I.E., Vol. TT20, 1995.
Rutten, H., and M. van Venrooij, *Telescopic Optics,* Richmond, VA, Willmann-Bell, 1988.
Schroeder, D., *Astronomical Optics,* San Diego, Academic, 1987.
Shannon, R., in Shannon and Wyant (eds.), *Applied Optics and Optical Engineering,* Vol. 8, New York, Academic, 1980 (aspherics).
Smith, W. J., *Modern Lens Design,* New York, McGraw-Hill, 1992.
Smith, W. J. (ed.), *Lens Design,* S.P.I.E., Vol. CR41, 1992.
Smith, W., in W. Driscoll (ed.), *Handbook of Optics,* New York, McGraw-Hill, 1978.
Smith, W., in Wolfe and Zissis (eds.), *The Infrared Handbook,* Washington, Office of Naval Research, 1985.
Taylor, W., and D. Moore (eds.), *Proc. International Lens Design Conf.,* S.P.I.E., Vol. 554, 1985, also Proc. Intr. Optical Design Conf. 1990, 1994, 1998, 2002, 2006. SPIE and OSA (with various edition).

Exercises

See the note preceding the exercises for Chap. 16.

1 Design a $20\times$ ($f = 8$ mm) two-mirror reflecting microscope objective and determine an appropriate combination of aperture and field over which the aberrations will not exceed the Rayleigh limit.

2 For an aperture of $f/2$ and a half-field of 0.1 radians, determine the relative angular-blur-spot size of the various systems listed in Sec. 18.9. Where stop position is critical, consider (a) the best position, and (b) the most compact arrangement. Assume a V of 100 for the refracting materials.

3 Design an $f/1$ Bouwers concentric system. Achromatize the design (choose a crown and flint with matching indices) and analyze the off-axis chromatic and chromatic variation of the aberrations.

Chapter

19

Selected Lens Designs, Analyzed and Annotated

19.1   Introduction

This chapter consists entirely of the annotated ray analysis and design data for a number of additional lenses which have not appeared in the preceding pages of this text. Most of these designs were selected from Smith, *Modern Lens Design,* 2nd ed, New York, McGraw-Hill, 2005.

All of these designs are shown at a nominal focal length of 100 mm, regardless of the focal length at which they are likely to be used. While this makes the comparison of designs easy, when evaluating the performance one must be aware that, for example, a microscope objective may be used at a focal length which is one or two orders of magnitude smaller than 100 mm, and a telescope objective may well be used at a focal length an order of magnitude larger.

The intent of this chapter is twofold: (1) to provide examples and illustrations of the many design features and the techniques which are used in real lens designs. This is obviously for the benefit of those who are interested in learning more about the craft of lens design; and (2) to provide feasible starting points for those who wish to select or adapt a design configuration for their own particular application.

19.2   Lens Data Tables

The lens data tables are quite straightforward, presenting radius, axial thickness, material name, index of refraction, Abbe V-value, and the semi-aperture (sa) for each surface. The sign convention for the radius is: positive if the center of curvature is to the right of the surface, and negative if it is to the left. A blank in the radius column indicates a plane surface (i.e., an infinite radius). The system data given with the lens table is as follows:

EFL = Effective focal length
BFL = Back focal length
 NA = Numerical aperture (f-Number)
 GIH = Gaussian image height (half field in degrees)
PTZ/F = (Petzval radius)/EFL
 VL = Vertex length
 OD = Object distance

The materials have been selected to match as closely as possible the materials given with the original design data. Most glasses are from the Schott catalog; close equivalents are available from other glass manufacturers. However, one should be aware that in recent years many glass types have been dropped by the glass manufacturers in favor of new formulations which are more environmentally acceptable. With a few exceptions, most of the listed glasses have current equivalents which are satisfactory substitutes. [Obviously if one is attempting to reduce or the correct secondary spectrum, one must always look beyond the six-digit (nnnvvv) glass code and check the partial dispersion as given in a "precision" melt sheet.]

19.3 Raytrace Figures*

The six plots on the left side of the raytrace figures show the ray intercepts plotted against the relative height of the ray at the entrance pupil. For example, in Fig. 1.1 the left-most plots are for a fan of rays in the Y-Z (tangential or meridional) plane, and the adjacent half-plot is for a fan in the X-Z (sagittal) plane. (The left half of the sagittal plot is omitted because the sagittal plot is point symmetric about the origin.) Three field angles are shown, on axis, ±0.7°, and ±1.0° (in this example). The ray intercept plots are plotted for three colors: The cross (+) is for the middle wavelength (588 nm in this case), the triangle is for the short wavelength (486 nm), and the square is for the long wavelength (656 nm). The vertical scale is shown as ±0.02 mm. (Note that the scale for each figure is adjusted to accommodate the amount of aberration of the lens.)

In the upper right row of figures, the left-most shows the field curvature of the para-principal rays. The crosses (+) are for the sagittal rays and the x's are for the tangential rays. The horizontal scale here

---

*Figure 6.2 and Sec. 6.5 provide a similar explanation of the ray aberration plots which are used in this book, plus an annotated sample plot.

is shown as ±0.10 mm. The central plot is the longitudinal spherical aberration in three colors. Note that this plot represents exactly the same data as the on-axis ray intercept plot in the lower left corner of the figure. The scale is ±0.10 mm. The right-most of these three plots is the longitudinal shift of the paraxial focus as a function of wavelength. The scale is ±1.0 mm.

The upper of the two lower central plots is percentage distortion versus. field angle. The vertical scale is ±0.01 percent. The lower plot shows the lateral color of the principal ray. The triangle plot is the height difference for the short wavelength (486 nm) minus the middle wavelength (588 nm). The square plot is the difference between the long (656 nm) and the middle wavelengths. The vertical scale is ± 0.005 mm. The distance between the two plots is the red-to-blue lateral color.

I have chosen to use the ray aberration representation of the performance of the lens designs in this chapter for several reasons: Aberration plots show clearly how well a design has been corrected. They make numerical comparison relatively straightforward. Many of the designs listed in patents are (deliberately or otherwise) not representative of the best correction of which the design form is capable. Given the ray aberration plots, a designer can easily estimate the level of correction that can be achieved in a redesign. This is done by applying third-order changes (to the aberration plots), under the assumption that third-order aberration changes are easily effected by modest changes in the prescription, and that the fifth- and higher-order aberrations are stable and unlikely to change. If the focal length is changed, the aberrations (with the exception of percent distortion) will simply scale with the focal length. If the NA or f/# is changed, third-order theory can provide a fair estimate of how the aberrations will change.

19.4   A Note Re the Modulation Transfer Function

It is tempting to present an MTF versus frequency plot for each lens design. Such a plot represents all the information necessary to evaluate the performance, and it is excellent for comparing two lenses which are evaluated under identical conditions for the same application. Modern software makes an MTF plot only a mouse click away, and the reader of this volume can readily get the MTF for these lens designs at focal lengths and under conditions which are appropriate to his application. Unfortunately, an MTF plot provides no information about which aberrations are adversely affecting the performance, or how one should go about improving the design. And unless a lens is effectively perfect, a single MTF plot at an arbitrarily chosen focus may well give a very misleading evaluation of the lens. An MTF evaluation is correct *only* for the focus at which the MTF has been calculated. To be valid, an MTF analysis (and the choice of the "best" focus) *must* take into account the following:

the application,

the relative importance of the axial image quality vs. the off-axis image quality,

the relative importance of resolution vs. contrast,

the focal length at which the lens is to be used,

the relative aperture setting,

and several other factors.

Note well that an MTF plot *cannot* be scaled when the the lens is scaled (except in the highly unlikely coincidence that the wavelength is going to be scaled by the same factor). A through-focus MTF plot at a well chosen spatial frequency is a valuable aid *during the design process*. Sadly, a more complete plot, incorporating several frequencies and several field positions, although possibly presenting enough data to choose the best focus for a particular application, quickly becomes too complex and cluttered to be useful.

19.5   Index to the Lenses

Table 19.1 indicates the index of lenses described in Chap. 19.

TABLE 19.1   Index of Lenses

Fig.#	Title	Speed	Field
19.1	Airspaced Triplet telescope objective	$f/2.8$	±1°
19.2	Apochromatic doublet tel. obj.	$f/7.0$	±1.0°
19.3	Apochromatic cemented triplet	$f/7.0$	±1.0°
19.4	Airspaced apochromatic triplet	$f/7.0$	±1.0°
19.5	Apochromatic triplet	$f/7.0$	+1.0°
19.6	Doublet magnifier	$f/2.0$	±17.4°
19.7	Four element eyepiece	$f/3.3$	±25.2°
19.8	Symmetrical eyepiece	$f/5.6$	±25.2°
19.9	Erfle eyepiece	$f/5.0$	±35°
19.10	Cooke triplet	$f/4.5$	±25.2°
19.11	Triplet	$f/2.5$	±16.2°
19.12	Broadband triplet	$f/8.0$	±14°
19.13	Reversed telephoto	$f/2.8$	±37°
19.14	Fisheye lens	$f/8.0$	±85.4°
19.15	Telephoto camera lens	$f/5.0$	±6.0°
19.16	Telephoto camera lens	$f/4.0$	±6.0°
19.17	Goerz Dagor	$f/8.0$	±26.6°
19.18	Dogmar enlarger	$f/4.5$	±26°
19.19	Tessar, reversed doublet	$f/3.0$	±28°
19.20	Reversed Tessar	$f/2.8$	±25.2°
19.21	Heliar	$f/3.5$	±25.2°

19.22	Five element Tessar	$f/2.7$	±16.2°
19.23	Petzval projection modification	$f/1.6$	±9.1°
19.24	Six element Petzval	$f/1.4$	±6.8°
19.25	Six element Petzval	$f/1.25$	±12.4°
19.26	R-Biotar radiographic lens	$f/0.9$	±6.8°
19.27	Split-front triplet	$f/1.5$	±15.1°
19.28	Ernostar	$f/1.4$	±15.1°
19.29	Sonnar	$f/2.0$	±19.8°
19.30	Sonnar	$f/1.2$	±18°
19.31	Microscope objective	$f/0.53$	±3.2°
19.32	Catadioptric Cassegrain	$f/1.5$	±1.25°
19.33	Catadioptric Cassegrain	$f/8.0$	±1.76°
19.34	Catadioptric Cassegrain	$f/1.2$	±5.1°
19.35	Double Gauss	$f/1.25$	±12.4°
19.36	Double Gauss	$f/2.0$	±22.3°
19.37	Double Gauss 7-elem.	$f/1.4$	±23°
19.38	Double Gauss 7-elem.	$f/1.4$	±23.8°
19.39	Double Gauss 8-elem.	$f/1.1$	±15.1°
19.40	Angulon 6-elem.	$f/5.6$	±37.6°
19.41	Angulon 8-elem.	$f/4.7$	±45°
19.42	Cooke triplet w/field corrector	$f/4.0$	±31.5°
19.43	Projection TV lens	$f/0.9$	±26°
19.44	Infrared triplet	$f/0.75$	±6°
19.45	Infrared anastigmat	$f/0.55$	±10°
19.46	Infrared 12x Cassegrain telescope	–	±1.6°
19.47	F-Theta scanning lens	$f/5.0$	±15°
19.48	Laser disk lens	$f/0.9$	±0.6°
19.49	Laser collimator doublet	$f/2.5$	±1.0°
19.50	Internal focusing telephoto	$f/5.6$	±3°
19.51	Petzval portrait lens	$f/3.3$	±17.2°
19.52	Schmidt system	$f/2.0$	±5°
19.53	Narrow angle Cooke triplet	$f/3.5$	±4°
19.54	Hologon	$f/8.0$	±55°
19.55	Sonnar	$f/1.4$	±21.8°
19.56	Sonnar	$f/2.9$	±24°
19.57	Microscope objective	0.92 NA	±11°
19.58	Microscope objective	0.57 NA	±2.1°
19.59	Schwarzschild system	$f/1.0$	±3°
19.60	Schwarzschild system, optimized	$f/1.0$	±3°
19.61	Flat field Schwarzschild, TOA	$f/1.0$	±3°
19.62	Flat field Schwarzschild, optimized	$f/1.25$	±3°

19.6   The Lenses

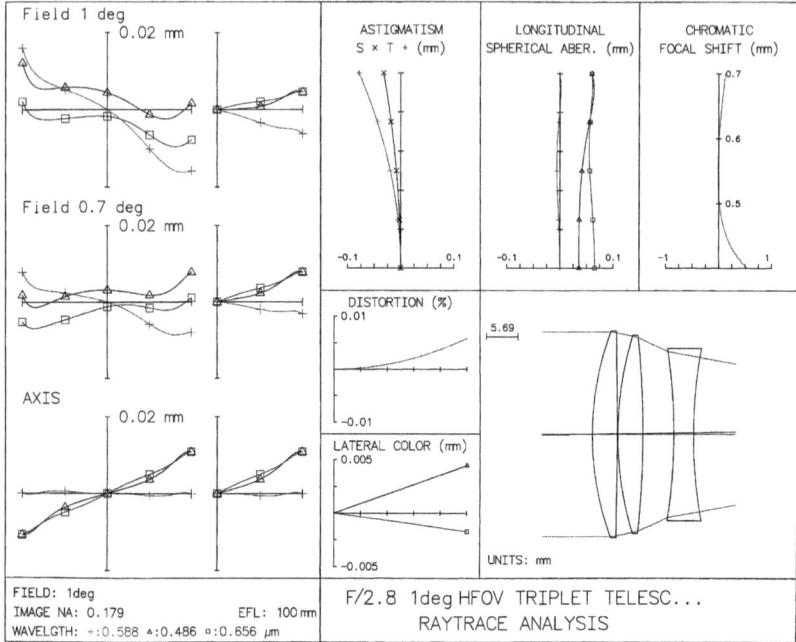

FIELD: 1deg		
IMAGE NA: 0.179	EFL: 100 mm	
WAVELGTH: +:0.588 ▲:0.486 □:0.656 µm		

F/2.8 1deg HFOV TRIPLET TELESC...
RAYTRACE ANALYSIS

F/2.8 Airspaced Triplet Telescope Objective. At F/2.8, this lens is much too fast for a telescope objective, but it serves as an excellent example of how the higher order aberrations may be controlled. Splitting the crown element in two reduces the undercorrected zonal spherical aberration. The large airspace between the crowns and the flint controls the spherochromatism, and also reduces the zonal spherical. The result is a lens so well corrected for spherical and spherochromatism that (on axis) it is is effectively perfect, except for the very obvious secondary spectrum. This problem can be corrected by the use of glasses with unusual partial dispersions (at the expense of increased aberrations and/or the higher cost of a difficult-to-work glass).

The off-axis plots show the effects of the unsymmetrical format of the lens and the relatively wide separation of the positive crowns and the negative flints. Note the higher order coma, and especially the lateral and longitudinal chromatic aberrations in these plots. A symmetrical + – + construction rather than this + + – arrangement would reduce these problems.

This lens requires tight tolerances on the centering and mounting of the elements. A form which cements one crown to the flint [e.g., (+ –) +, + (+ –), or + (– +)] is much less sensitive to misalignment and is easier to fabricate.

F/2.8 1° HFOV Triplet Telescope Objective

radius	thickness	mat'l	index	V-no	sa		
						EFL	= 100
50.098	4.500	BK7	1.517	64.2	18.0	BFL	= 75.13
−983.420	0.100	air			18.0	NA	= −0.1788 (F/2.8)
56.671	4.500	BK7	1.517	64.2	17.3	GIH	= 1.75
−171.150	5.571	air			17.3	PTZ/F	= −1.749
−97.339	3.500	SF1	1.717	29.5	15.0	VL	= 18.17
81.454	75.132	air			0.0	OD	infinite conjugate

Figure 19.1

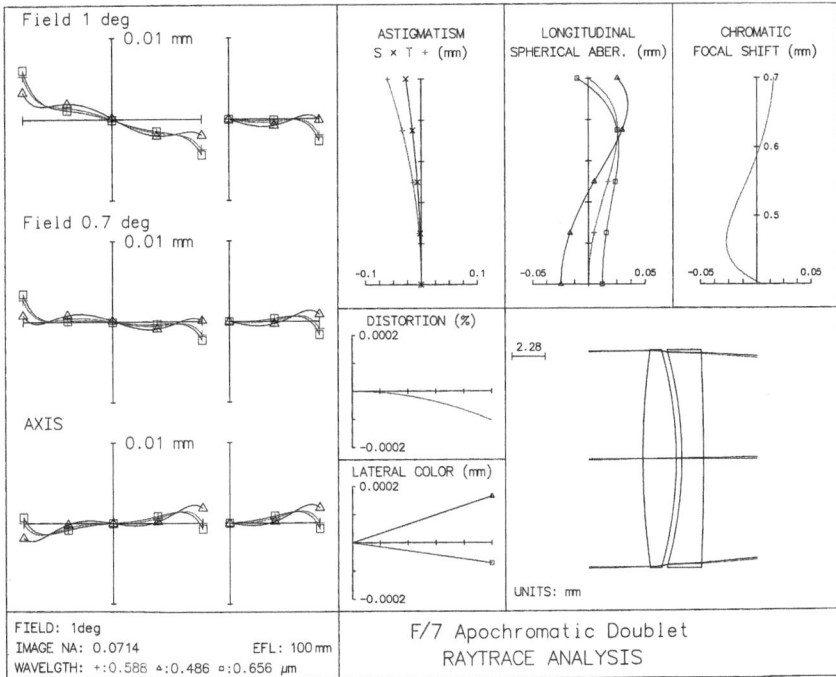

Field 1 deg 0.01 mm	ASTIGMATISM S × T + (mm)	LONGITUDINAL SPHERICAL ABER. (mm)	CHROMATIC FOCAL SHIFT (mm)

FIELD: 1deg
IMAGE NA: 0.0714   EFL: 100 mm
WAVELGTH: +:0.588 ▲:0.486 □:0.656 µm

F/7 Apochromatic Doublet
RAYTRACE ANALYSIS

F/7.0 Apochromatic Doublet Telescope Objective. The glasses in this doublet have the same partial dispersion which allows the secondary spectrum to be corrected to exactly zero at the 0.7 ray. However the spherochromatic is not zero, nor is the zonal spherical, which is overcorrected. Note that the inner airspaced radii are identical. The crown glass, FK51, is close to calcium fluoride, and shares many of its economic, environmental, and durability difficulties, as well as its dispersion.

F/7 Apochromatic Doublet

radius	space	mat'l		sa
52.520	2.30	FK51	487845	7.2
−25.731	0.3826			7.2
−25.731	1.50	KzFSN2	558542	7.2
−166.165	96.85			7.2

EFL = 100.0
BFL = 96.85
NA   = −0.0714 (F/7.0)
GIH = 1.745 (1.0°)
PTZ/F = −1.39
VL   = 4.183

Figure 19.2

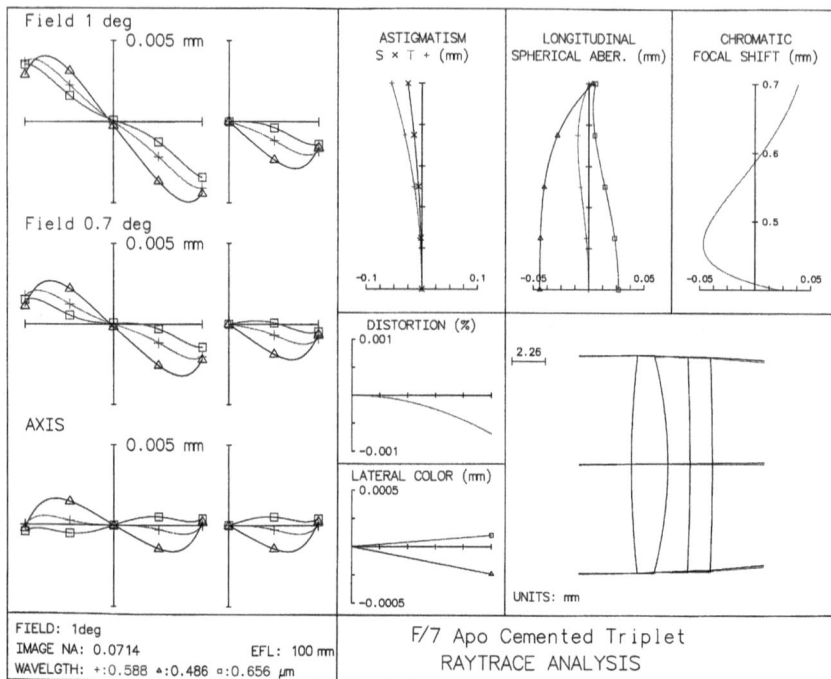

Field 1 deg 0.005 mm	ASTIGMATISM S × T + (mm)	LONGITUDINAL SPHERICAL ABER. (mm)	CHROMATIC FOCAL SHIFT (mm)

FIELD: 1deg
IMAGE NA: 0.0714   EFL: 100 mm
WAVELGTH: +:0.588 ▲:0.486 □:0.656 μm

F/7 Apo Cemented Triplet
RAYTRACE ANALYSIS

F/7.0 Apochromatic Cemented Triplet. This lens is corrected for spherical, coma, and chromatic aberrations. Its worst on-axis aberration is ordinary spherochromatic aberration. Airspacing the lens could improve matters. PK51A and KzFSN2 are unusual partial dispersion glasses.

F/7 Apochromatic Triplet

radius	space	mat'l		sa
61.29	2.50	PK51A	529770	7.20
−27.81	1.50	KzFSN2	558542	7.20
−153.5	1.50	SFL57	847236	7.20
−193.06	97.39			7.20

EFL   = 100.0
BFL   = 97.29
NA   = −0.0714 (F/7.0)
GIH   = 1.745 (1.0°)
PTZ/F = −1.446
VL   = 5.50

Figure 19.3

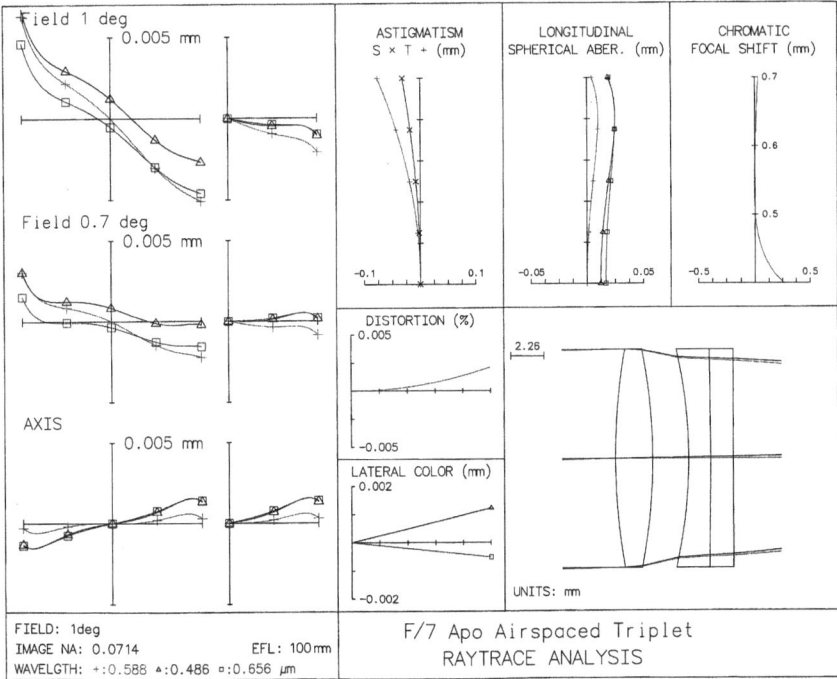

FIELD: 1deg	
IMAGE NA: 0.0714	EFL: 100mm
WAVELGTH: +:0.588 ▲:0.486 □:0.656 µm	

F/7 Apo Airspaced Triplet
RAYTRACE ANALYSIS

F/7.0 Airspaced Apochromatic Triplet. This is the cemented triplet of Fig. 19.3 with an airspace which is used to correct the spherochromatic which afflicted the cemented version. Note the lateral chromatic resulting from the unsymmetrical construction.

F/7.0 Airspaced Apochromatic Triplet

radius	space	mat'l		sa
40.27	2.50	PK51A	529770	7.20
−36.32	2.46	air		7.20
−31.18	1.50	KzFSN2	558542	7.20
−326.92	1.50	SFL57	847236	7.20
2,961.4	89.24			

EFL   = 100.01
BFL   = 89.24
NA   = 0.0714 (F/7.0)
GIH   = 1.746 (1.0°)
PTZ/F = −1.625
VL   = 7.96

Figure 19.4

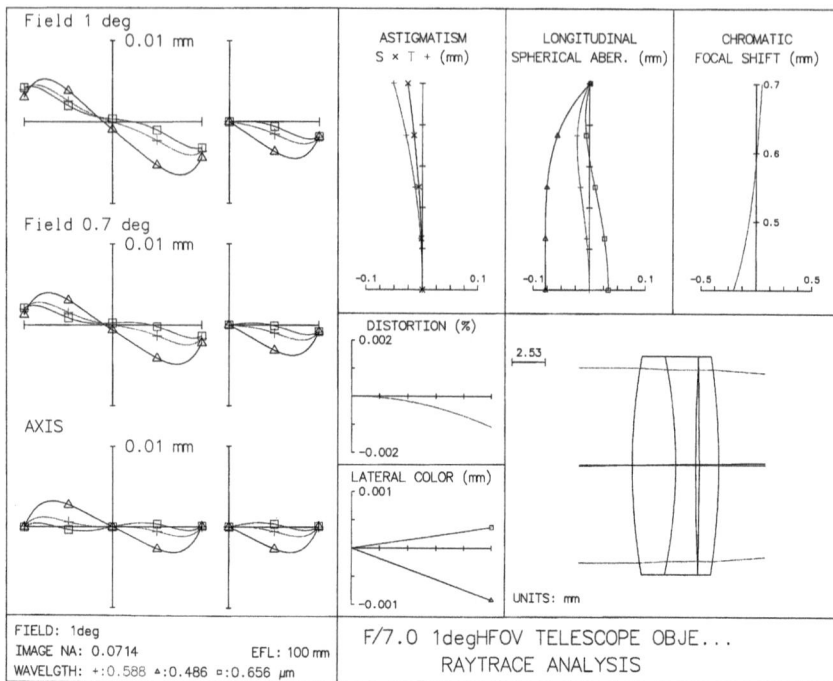

F/7.0 Apochromatic Triplet Telescope Objective. This lens is corrected for spherical, coma, and chromatic aberrations. The glass types are those with unusual partial dispersions. The weak, positive third element serves to adjust or "trim" the partial dispersion of the second element so that the combination of the second and third elements has exactly the same effective partial dispersion as that of the first element. Note that the airspace is necessary to fully correct both spherical and coma. The secondary spectrum is corrected at the margin, but the performance is limited by the spherochromatism, which has the normal sign with overcorrected spherical aberration in the blue and undercorrected in the red. Presumably this could be ameliorated by airspacing the first and second elements (which might also allow the second and third elements to be cemented). A doublet using the glass types of the two stronger elements can be as good or better than the triplet.

F/7.01° HFOV Telescope Objective

radius	thickness	mat'l	index	V-no	sa
44.144	3.300	PK51	1.529	77.0	8.0
−39.524	1.500	LSF18	1.913	32.4	8.0
158.460	0.278	air			8.0
−418.801	1.500	SF57	1.847	23.8	8.0
−54.845	97.720	air			0.0

EFL = 100
BFL = 97.72
NA   = −0.0714 (F/7.0)
GIH = 1.75
PTZ/F = −1.142
VL   = 6.58
OD   infinite conjugate

Figure 19.5

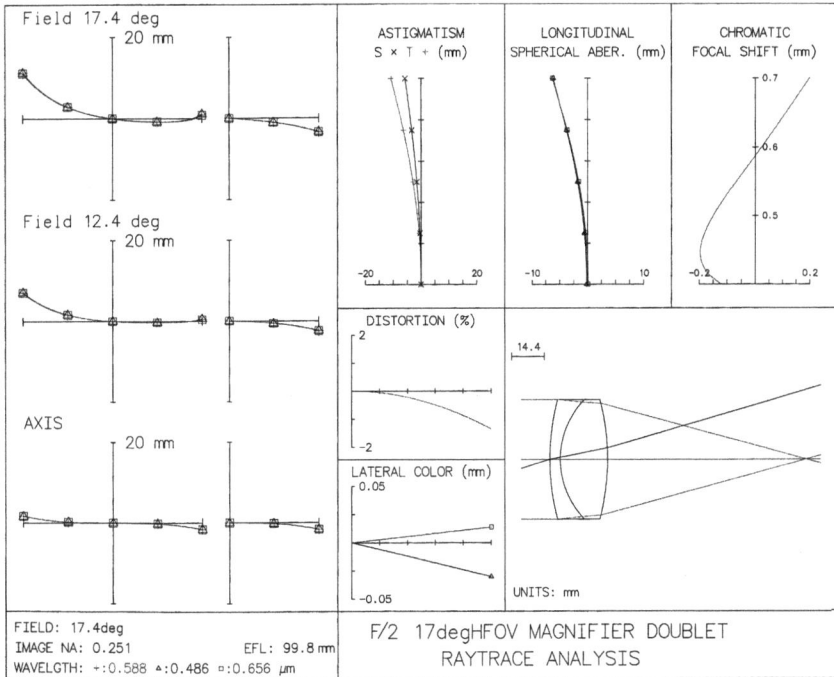

Field 17.4 deg 20 mm	ASTIGMATISM S × T + (mm)
Field 12.4 deg 20 mm	LONGITUDINAL SPHERICAL ABER. (mm)
AXIS 20 mm	CHROMATIC FOCAL SHIFT (mm)

DISTORTION (%)

14.4

LATERAL COLOR (mm)

UNITS: mm

FIELD: 17.4deg
IMAGE NA: 0.251   EFL: 99.8 mm
WAVELGTH: +:0.588 ▲:0.486 □:0.656 μm

F/2 17degHFOV MAGNIFIER DOUBLET
RAYTRACE ANALYSIS

F/2.0 Doublet Magnifier. A magnifier differs from an eyepiece in that, for a magnifier the aperture/pupil is at the user's eye, which is movable, in contrast to an eyepiece, which has a fixed exit pupil that determines the location of the eye. Thus the correction of a magnifier should extend over a large aperture. In use, the aperture stop of a magnifier is determined by the pupil of the eye, so that only a small portion of the full aperture ray intercept plot is applicable. The astigmatism, lateral color, and distortion are affected by the eye location. This lens is quite good as a general purpose magnifier or as a slide viewer lens.

F/2 17.4° HFOV Magnifier Doublet

radius	thickness	mat'l	index	V-no	sa
96.960	4.440	SF2	1.648	33.8	25.0
35.100	20.370	BAK1	1.572	57.5	25.0
−96.960	92.046	air			25.0

EFL = 99.76
BFL = 92.05
NA   = −0.2549 (F/2.00)
GIH = 31.32 (HFOV = 17.43)
PTZ/F = −1.485
VL   = 24.81
OD   infinite conjugate

Figure 19.6

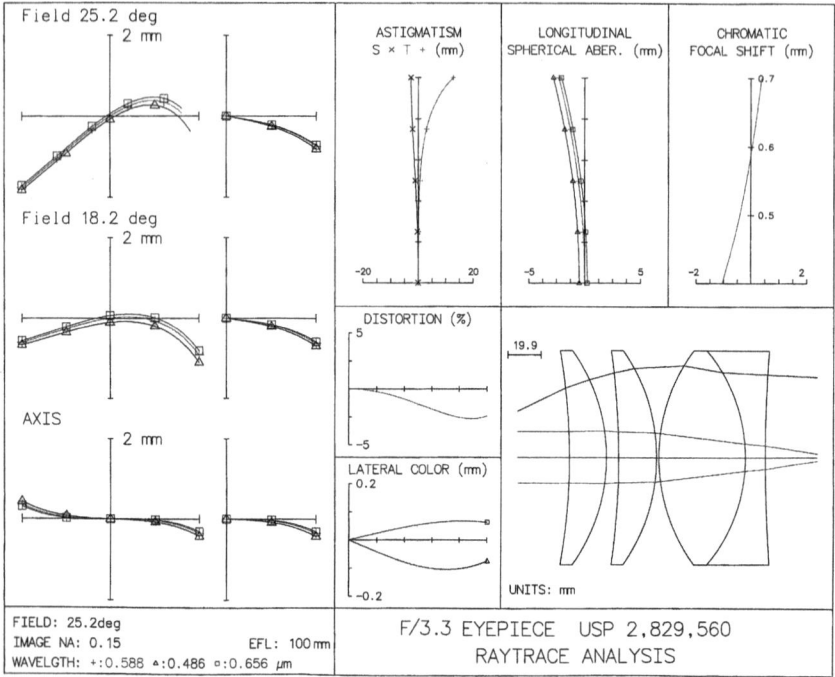

F/3.3 Four Element Eyepiece (USP 2, 829, 560). This excellent eyepiece is often used as a 10× microscope eyepiece. The design is robust enough that many commercially advantageous variations such as identical meniscus elements, an equiconvex doublet crown, and a plano-concave flint are often found (with moderate adjustments of the glass types used).

F/3.3 Eyepiece USP 2,829,560

radius	thickness	mat'l	index	V-no	sa
−352.361	21.900	BK7	1.517	64.2	62.0
−105.274	7.280	air			62.0
−440.723	22.500	BK7	1.517	64.2	62.0
−107.043	1.360	air			62.0
102.491	52.100	BK7	1.517	64.2	62.0
−93.493	11.800	SF61	1.751	27.5	62.0
794.281	47.485	air			62.0

EFL = 100.1
BFL = 47.48
NA   = −0.1508 (F/3.3)
GIH = 47.03 (HFOV = 25.17)
PTZ/F = −1.552
VL   = 116.94
OD   infinite conjugate

Figure 19.7

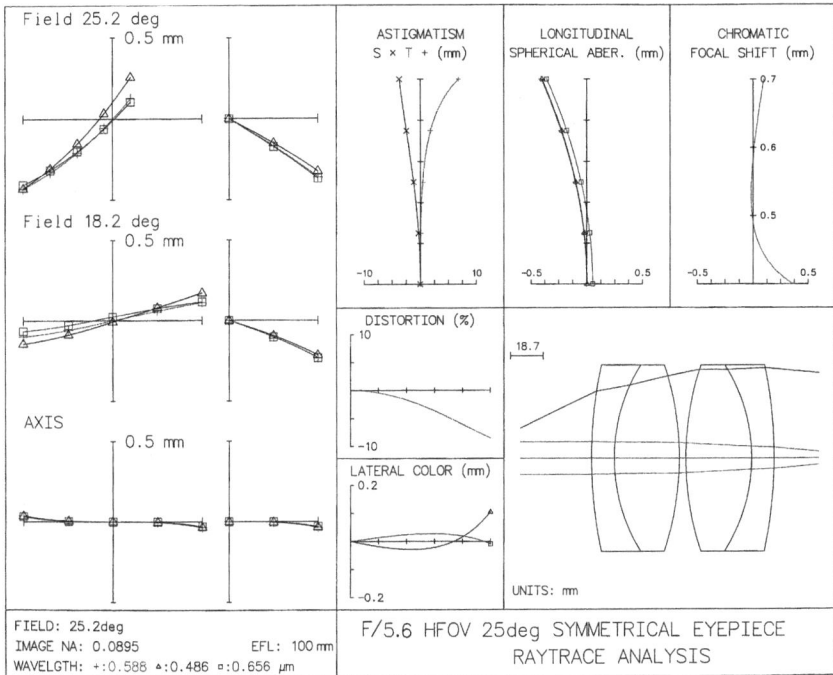

FIELD: 25.2deg
IMAGE NA: 0.0895
WAVELGTH: +:0.588 ▲:0.486 □:0.656 μm
EFL: 100 mm

F/5.6 HFOV 25deg SYMMETRICAL EYEPIECE
RAYTRACE ANALYSIS

F/5.6 Symmetrical, or Plössl, Eyepiece.

This is an excellent, economical general purpose eyepiece and magnifier. The simple design is so robust that almost any pair of scrap achromats will make a reasonable eyepiece or a good magnifier. A variety of forms that include equiconvex crowns, plano-concave flints, and unsymmetrical doublets have been used. One patented design uses four different glasses in plano doublets with equiconvex crowns. A more common glass choice than the SK crown shown here uses an ordinary crown (K or BK) and a dense flint (F or SF), with only a modest reduction in performance.

F/5.6 HFOV 25° Symmetrical Eyepiece

radius	thickness	mat'l	index	V-no	sa
236.748	12.694	SF61	1.751	27.5	50.8
93.577	36.813	SK1	1.610	56.7	50.8
−155.314	3.808	air			50.8
155.314	36.813	SK1	1.610	56.7	50.8
−93.557	12.694	SF61	1.751	27.5	50.8
−236.748	65.443	air			50.8

EFL = 100
BFL = 65.44
NA   = −0.0895 (F/5.6)
GIH = 47.00 (HFOV = 25.17)
PTZ/F = −1.345
VL   = 102.82
OD   infinite conjugate

Figure 19.8

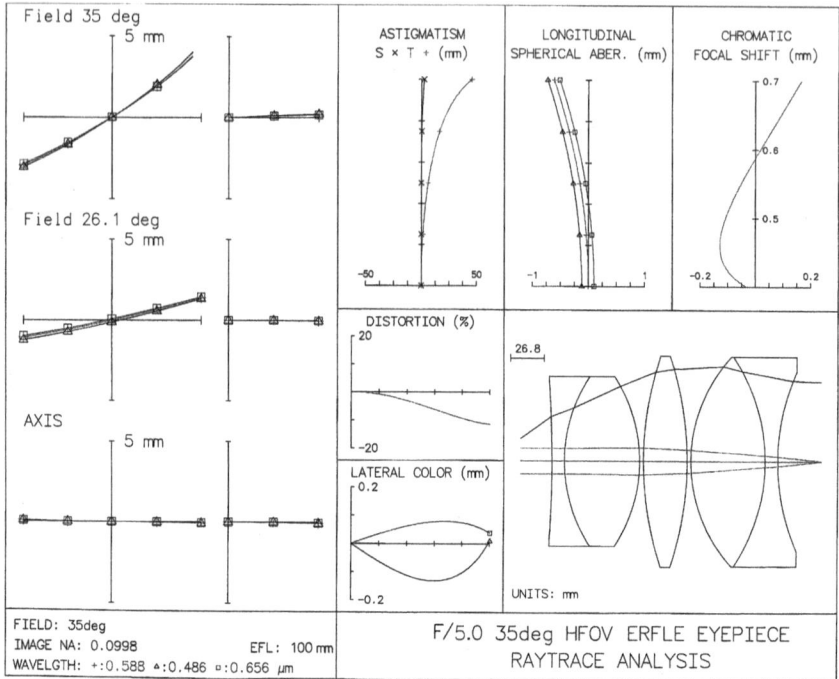

FIELD: 35deg
IMAGE NA: 0.0998   EFL: 100 mm
WAVELGTH: +:0.588 ▲:0.486 □:0.656 µm

F/5.0 35deg HFOV ERFLE EYEPIECE
RAYTRACE ANALYSIS

F/5.0 Wide-Angle Erfle Eyepiece. The Erfle format is the basis of most wide-angle (60° to 70°) eyepieces. It features a long eye relief and a flat field. These benefits result from the use of high index crowns and (especially) from the concave surface close to the focal plane. This acts both as a field flattener and as a negative field lens which lengthens the eye relief. Many variations of this format have been used, including a doublet center lens, two central singlets, triplet cemented components, airspaced doublets, lanthanum crowns, etc. Wide-angle eyepieces are difficult to use because the pupil of the eye shifts laterally when rotating to view at large off-axis angles and moves outside the exit pupil. Spherical aberration of the pupil is especially important in a wide-angle eyepiece.

F/5.0 35° HFOV Erfle Eyepiece

radius	thickness	mat'l	index	V-no	sa
−1000.000	10.000	SF19	1.667	33.0	66.0
117.650	60.000	SK11	1.564	60.8	66.0
−119.130	3.000	air			66.0
253.230	35.000	BK7	1.517	64.2	82.0
−253.230	3.000	air			82.0
118.130	60.000	SK11	1.564	60.8	81.5
−142.860	10.000	SF19	1.667	33.0	81.5
166.670	37.862	air			70.0

EFL = 100.2
BFL = 37.86
NA = −0.0999 (F/5.0)
GIH = 70.13 (HFOV = 34.99)
PTZ/F = −1.865
VL = 181.00
OD   infinite conjugate

Figure 19.9

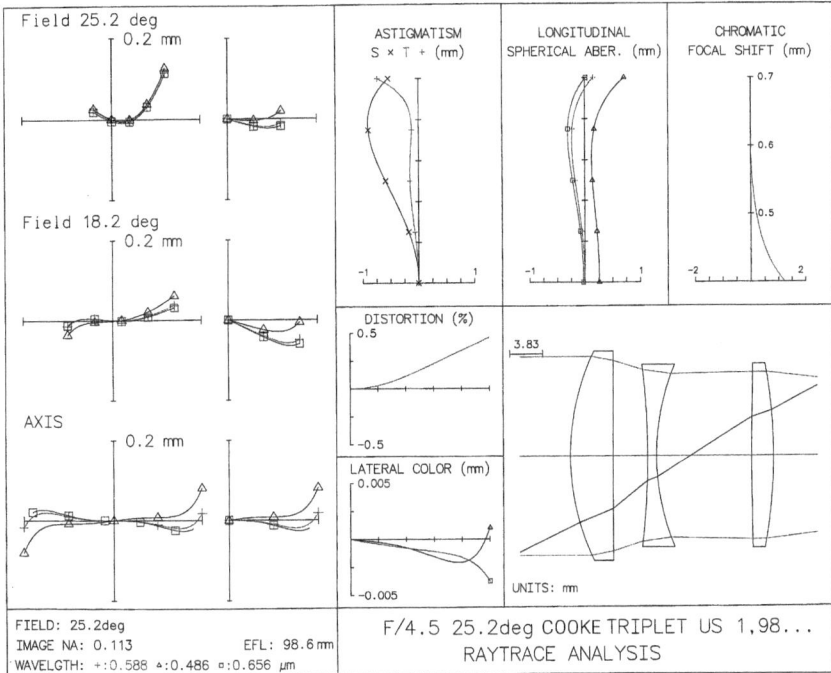

FIELD: 25.2deg	F/4.5 25.2deg COOKE TRIPLET US 1,98...	
IMAGE NA: 0.113 EFL: 98.6 mm	RAYTRACE ANALYSIS	
WAVELGTH: +:0.588 ▲:0.486 □:0.656 μm		

F/4.5 Cooke Triplet Camera Lens. This is a fairly typical F/4.5 camera triplet using lanthanum glass crowns. Note the unsymmetrical arrangement of the lens which becomes favorable in the triplet form as the speed is reduced and the field angle is increased. Additional Cooke triplets are presented in Fig. 16.13 (F/2.7, ±20°), Fig. 16.14 (F/3.0, ±23°), and Fig. 16.15 (F/6.3, ±27°).

F/4.5 25.2° Triplet US 1,987,878/1935 Schneider

radius	thickness	mat'l	index	V-no	sa
26.160	4.916	LAK12	1.678	55.2	11.7
1201.700	3.988	air			11.7
−83.460	1.038	SF2	1.648	33.8	10.2
25.670	4.000	air			10.2
	6.925	air			9.2
302.610	2.567	LAK22	1.651	55.9	10.3
−54.790	81.433	air			10.3

EFL = 98.56
BFL = 81.43
NA   = −0.1127 (F/4.4)
GIH = 46.33 (HFOV = 25.17)
PTZ/F = −2.831
VL   = 23.43
OD   infinite conjugate

Figure 19.10

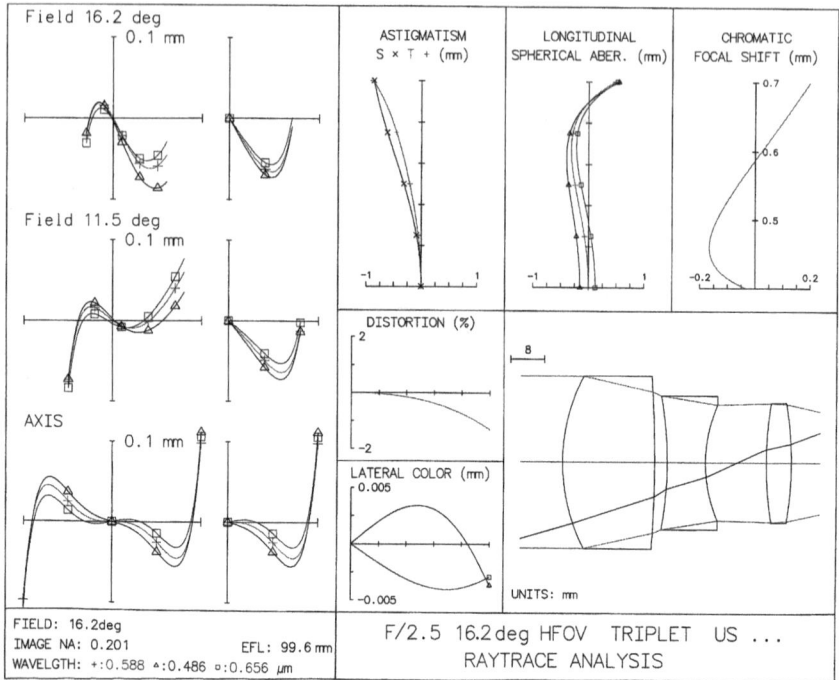

Field 16.2 deg

ASTIGMATISM
S × T + (mm)

LONGITUDINAL
SPHERICAL ABER. (mm)

CHROMATIC
FOCAL SHIFT (mm)

Field 11.5 deg

DISTORTION (%)

AXIS

LATERAL COLOR (mm)

UNITS: mm

FIELD: 16.2deg
IMAGE NA: 0.201   EFL: 99.6 mm
WAVELGTH: +:0.588 △:0.486 □:0.656 μm

F/2.5 16.2 deg HFOV   TRIPLET   US ...
RAYTRACE ANALYSIS

F/2.5 Triplet. This high speed triplet uses lanthanum crowns and two extremely thick front elements to cover a modest 32° field of view. The nicely balanced aberration correction of this lens is an impressive bit of design. At a very short focal length the thick elements will seem more reasonable.

F/2.5 16.2° HFOV Triplet (US 2,720,816)

radius	thickness	mat'l	index	V-no	sa
42.200	22.070	LAK13	1.694	53.3	20.0
−283.530	3.000	air			20.0
−84.930	9.170	SF5	1.673	32.2	15.6
33.230	7.670	air			13.5
	7.000	air			13.6
84.930	6.000	LAK13	1.694	53.3	14.0
−84.930	65.476	air			14.0

EFL = 99.58
BFL = 65.48
NA   = −0.2001 (F/2.5)
GIH = 28.88 (HFOV = 16.17)
PTZ/F = −2.52
VL   = 54.91
OD   infinite conjugate

Figure 19.11

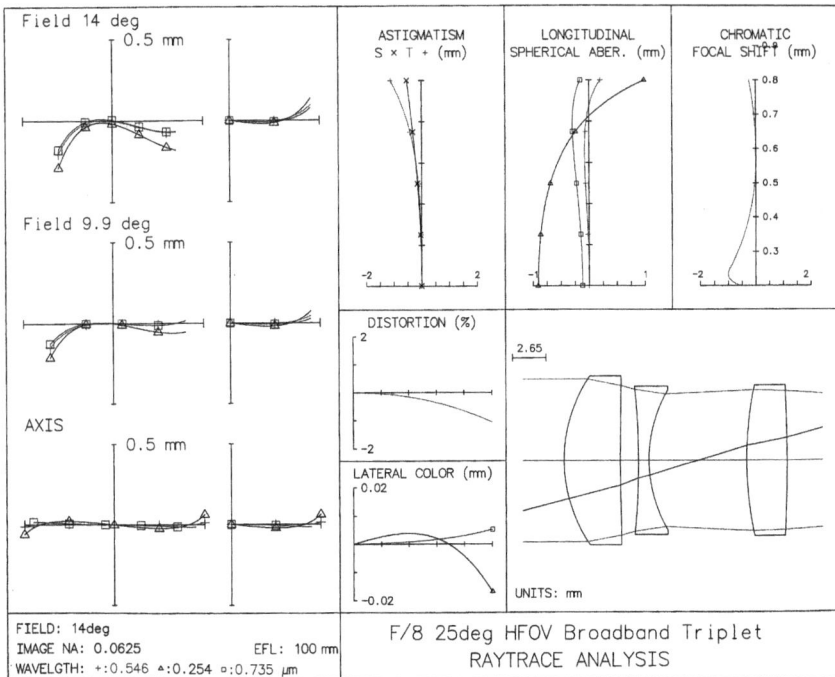

Field 14 deg	ASTIGMATISM S × T + (mm) / LONGITUDINAL SPHERICAL ABER. (mm) / CHROMATIC FOCAL SHIFT (mm)

FIELD: 14deg
IMAGE NA: 0.0625   EFL: 100 mm
WAVELGTH: +:0.546 ▲:0.254 □:0.735 μm

F/8 25deg HFOV Broadband Triplet
RAYTRACE ANALYSIS

F/8.0 Broadband Triplet. This lens operates over a large spectral range, from 254 to 735 nm. To get reasonable transmission in the UV end of this range calcium fluoride ($n = 1.435$, $V = 12.5$) is used for the crown elements and fused quartz ($n = 1.460$, $V = 9.0$) is used as the flint. Note the large spherochromatic and the reversed secondary spectrum. The low index materials limit the speed and field that can be satisfactorily utilized.

F/8.0 Broadband Triplet

radius	space	mat'l		sa
11.35	4.41	CaF_2	435125	6.50
196.93	1.46	air		6.50
−53.80	0.83	SiO_2	460090	5.70
10.32	4.00	air		5.70
stop	3.78	air		5.23
27.40	3.24	CaF_2	435125	5.80
−116.37	85.20			

EFL = 100.10
BFL = 85.20
NA   = 0.0625 (F/8.0)
GIH = 24.96 (14.0°)
PTZ/F = −4.096
VL   = 7.96

Figure 19.12

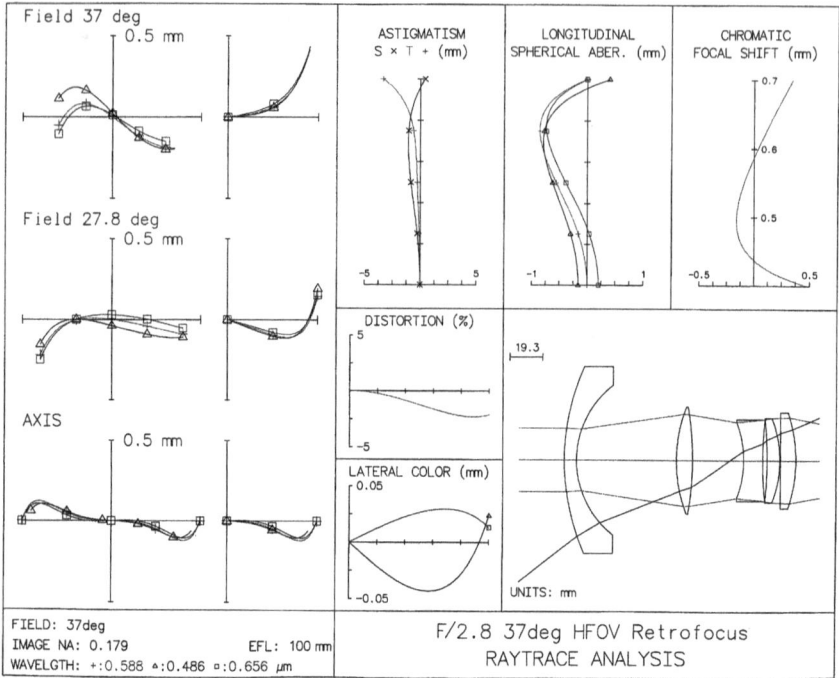

FIELD: 37deg	
IMAGE NA: 0.179 EFL: 100 mm	F/2.8 37deg HFOV Retrofocus
WAVELGTH: +:0.588 ▲:0.486 ▫:0.656 μm	RAYTRACE ANALYSIS

F/2.8 Reversed Telephoto. This relatively simple retrofocus covers a wide (74°) field at a speed of *f*/2.8. It has a back focus which is 30% longer than the focal length. Note that the design has been nicely balanced so that the customary achromatic front negative doublet is not necessary in order to correct the lateral and axial color simultaneously.

F/2.8 MORI USP 4,203,653

radius	thickness	mat'l	index	V-no	sa
126.010	6.967	LAK03	1.670	51.6	52.4
52.254	57.332	air			42.0
82.117	9.356	BASF5	1.603	42.5	30.1
−168.816	29.960	air			29.8
−56.038	11.148	SF56	1.785	26.1	21.4
226.239	2.588	air			23.0
−208.842	7.266	TAF3	1.804	46.5	23.0
−64.382	0.398	air			23.8
−1740.099	8.659	LAK18	1.729	54.7	25.5
−83.502	129.306	air			27.0
	0.011	air			80.0

EFL = 100
BFL = 129.3
NA   = −0.1785 (F/2.8)
GIH = 75.37 (HFOV = 37.00)
PTZ/F = −4.703
VL   = 133.67
OD   infinite conjugate

Figure 19.13

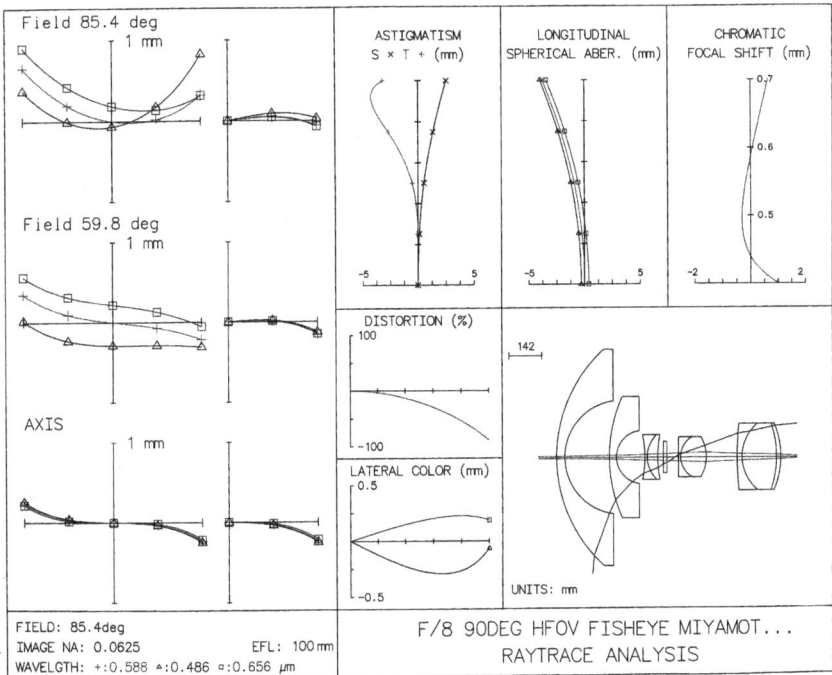

FIELD: 85.4deg	
IMAGE NA: 0.0625	EFL: 100 mm
WAVELGTH: +:0.588 ▲:0.486 □:0.656 μm	

F/8 90DEG HFOV FISHEYE MIYAMOT...
RAYTRACE ANALYSIS

F/8 "Fish eye" Lens. The fish eye lens covers a field of 180° or more by taking advantage of the heavily overcorrected spherical aberration of the powerful negative meniscus front elements, which strongly deviate the principal ray. This spherical aberration of the pupil causes the entrance pupil to move forward, off the axis, and to tilt at wide angles of view. This produces the extreme barrel distortion which is necessary to form an image of this large field of view on a plane surface. (Without distortion the image of a 180° field would be infinite.)

F/8 90° HFOV Fisheye Miyamoto Josa 1964

radius	thickness	mat'l	index	V-no	sa		
599.383	35.030	BK7	1.517	64.2	448.4		
235.825	190.161	air			234.0		
605.513	30.025	FK5	1.487	70.4	251.8		
111.094	120.102	air			110.1		
−452.384	10.008	FK5	1.487	70.4	93.5		
127.733	45.038	SF56	1.785	26.1	93.5		
462.892	25.021	air			93.5		
	15.013	K3	1.518	59.0	65.4		
	36.281	air			65.5		
	13.762	air			15.8	EFL	= 100
38507.649	10.008	SF56	1.785	26.1	84.1	BFL	= 150.1
95.081	110.093	LAF2	1.744	44.7	84.1	NA	= −0.0626 (F/8.0)
−162.638	130.110	air			84.1	GIH	= 133.60
1376.167	20.017	SF56	1.785	26.1	139.0		(HFOV = 85.40)
177.275	150.127	BSF52	1.702	41.0	139.0	PTZ/F	= 49.15
−400.339	18.766	BASF6	1.668	41.9	139.0	VL	= 959.56
−337.536	150.119	air			139.0	OD	infinite conjugate

Figure 19.14

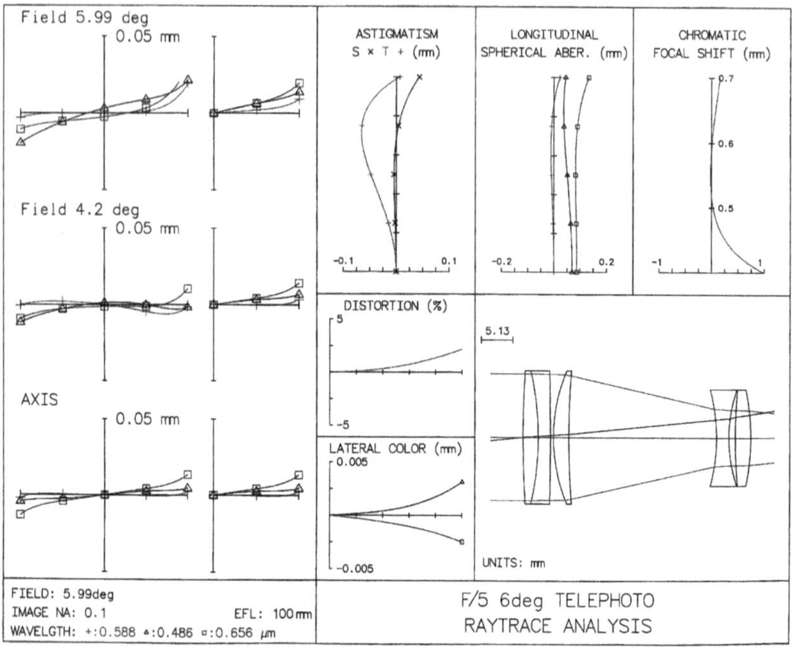

FIELD: 5.99deg
IMAGE NA: 0.1   EFL: 100 mm
WAVELGTH: +:0.588 ▲:0.486 □:0.656 μm

F/5 6deg TELEPHOTO
RAYTRACE ANALYSIS

F/5.0 Telephoto Camera Lens. This is a simple and fairly typical telephoto, with an 80% telephoto ratio, (i.e., the front lens to image distance is 20% less than the focal length). In a telephoto the longitudinal magnification of the rear negative component enlarges the aberrations of the front component, so that the front component must be well corrected. The front configuration shown here, with a cemented doublet plus a singlet, is less sensitive to misalignment than a front of three airspaced elements would be. This design would make an excellent 200 mm lens for a 35 mm camera, and with higher index glass could perform well at a higher speed. If the telephoto ratio is reduced in order to get a more compact lens, the component powers become larger, as do the aberrations, and a more complex design is necessary to maintain the image quality.

F/5 6° Telephoto

radius	thickness	mat'l	index	V-no	sa
149.035	2.500	SK4	1.613	58.6	10.5
−46.003	2.000	SF14	1.762	26.5	10.5
−477.921	0.500	air			10.5
26.522	2.500	SK4	1.613	58.6	10.5
132.322	24.060	air			10.5
−28.605	2.000	SK4	1.613	58.6	7.6
22.989	1.050	air			7.6
82.834	2.500	F5	1.603	38.0	7.6
−36.911	42.897	air			7.6

EFL = 100
BFL = 42.9
NA   = −0.1000 (F/5.0)
GIH = 10.50 (HFOV = 5.99)
PTZ/F = 7.68
VL   = 37.11
OD   infinite conjugate

Figure 19.15

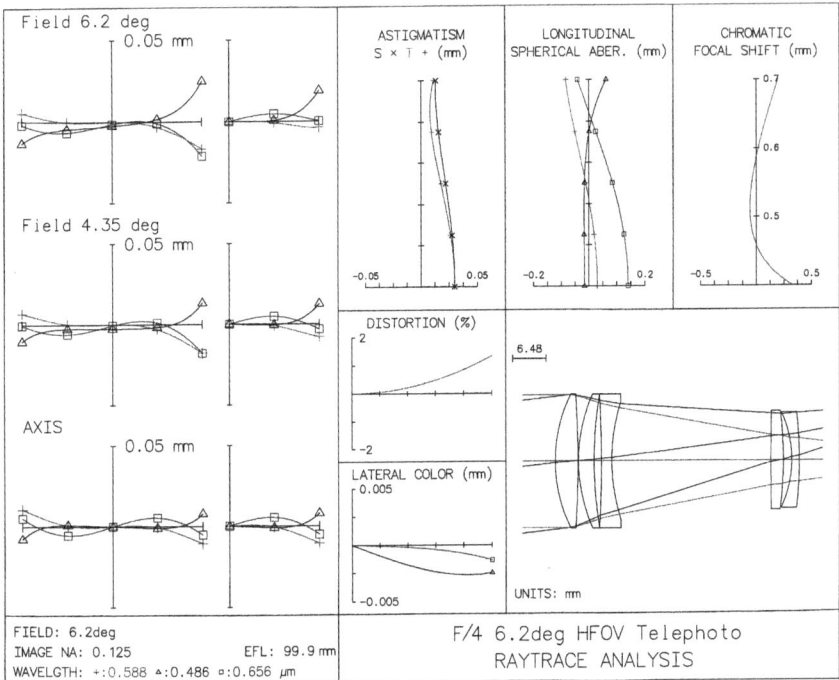

| | | | ASTIGMATISM S × T + (mm) | LONGITUDINAL SPHERICAL ABER. (mm) | CHROMATIC FOCAL SHIFT (mm) |

FIELD: 6.2deg
IMAGE NA: 0.125   EFL: 99.9 mm
WAVELGTH: +:0.588 ▲:0.486 □:0.656 μm

F/4 6.2deg HFOV Telephoto
RAYTRACE ANALYSIS

F/4.0 Telephoto Camera Lens. Designed as a 200 mm telephoto for a 35 mm camera, this five element lens achieves a higher speed than the previous example (Fig. 19.15) by airspacing all of the elements and modifying the glass choice. The telephoto ratio is 0.80.

F/4.0 Telephoto Camera Lens

radius	space	mat'l		sa
27.03	4.50	BK10	498670	12.6
−176.93	0.10	air		12.6
30.66	3.00	BK10	498670	12.6
76.46	1.40	air		12.6
−212.41	2.00	SF5	673322	12.6
36.22	30.84	air		11.0
506.55	2.50	SF57	847238	9.3
−67.74	1.66	air		9.3
−20.57	1.50	LaK8	713538	9.3
−78.32	32.46			

EFL   = 99.90
BFL   = 32.46
NA   = 0.125 (F/4.0)
GIH   = 10.86 (6.2°)
PTZ/F = −23.6
VL   = 47.5
OD   = infinity

Figure 19.16

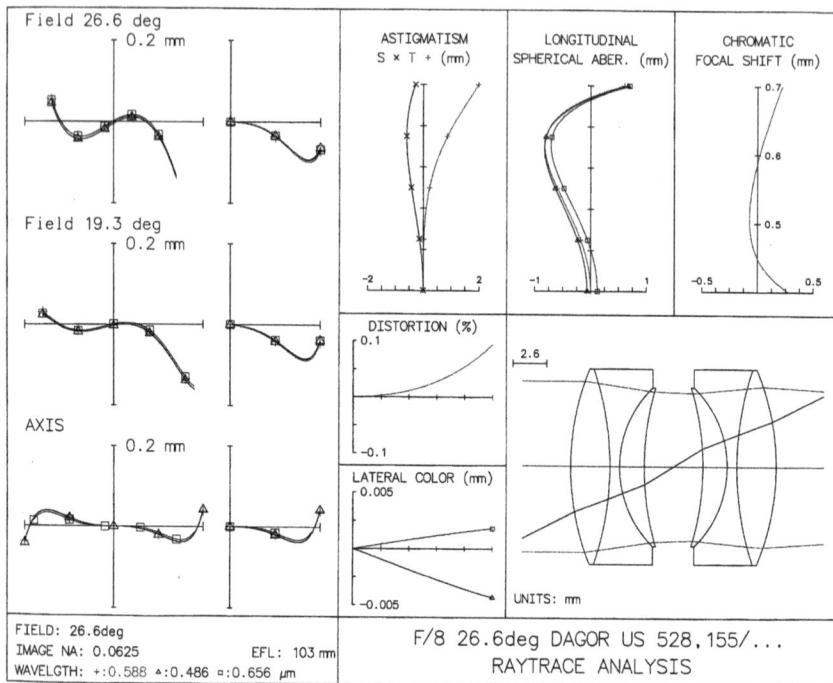

Field 26.6 deg

ASTIGMATISM
S × T + (mm)

LONGITUDINAL
SPHERICAL ABER. (mm)

CHROMATIC
FOCAL SHIFT (mm)

Field 19.3 deg

AXIS

DISTORTION (%)

LATERAL COLOR (mm)

UNITS: mm

FIELD: 26.6deg
IMAGE NA: 0.0625   EFL: 103 mm
WAVELGTH: +:0.588 ▲:0.486 □:0.656 μm

F/8 26.6deg DAGOR US 528,155/...
RAYTRACE ANALYSIS

F/8.0 Goerz Dagor USP 528, 155/1894. The classic Dagor combines the front and rear doublets of the Protar into a cemented triplet. The outer crowns are high index, which combined with a portion of the lower index inner flints constitute a *new achromat*. This combination has a flatter Petzval field than the *old achromat*, but the cemented interface contributes undercorrected spherical. The balance of the triplet is an old achromat, with the flint index higher than the crown, so that its cemented surface provides the overcorrection necessary to balance the spherical. The symmetrical configuration takes care of the coma, distortion, and lateral color. There were many symmetrical triplets (as well as quadruplets and quintuplets) which with various arrangements utilized the principles of the Dagor.

In a further development the inner crown was split off. Then the airspace could be used to correct the spherical and the low index crown could be replaced with a higher index glass. (See Fig. 17.15.) This split doubled the number of air-glass interfaces, but the improvement in image quality was enough to compensate for the lower transmission and increased reflections.

F/8 26.6° Goerz Dagor (US 528, 155/1894)

radius	thickness	mat'l	index	V-no	sa		
19.100	3.056	SK6	1.614	56.4	7.4		
−22.635	0.764	BALF3	1.571	52.9	7.4	EFL	= 103.3
8.272	1.910	K4	1.519	57.4	6.0	BFL	= 96.27
20.453	2.292	air			6.0	NA	= −0.0622 (F/8.0)
	2.292	air			5.6	GIH	= 51.67
−20.453	1.910	K4	1.519	57.4	6.0		(HFOV = 26.57)
−8.272	0.764	BALF3	1.571	52.9	6.0	PTZ/F	= −3.706
22.635	3.056	SK6	1.614	56.4	7.4	VL	= 16.04
−19.100	96.267	air			7.4	OD	infinite conjugate

Figure 19.17

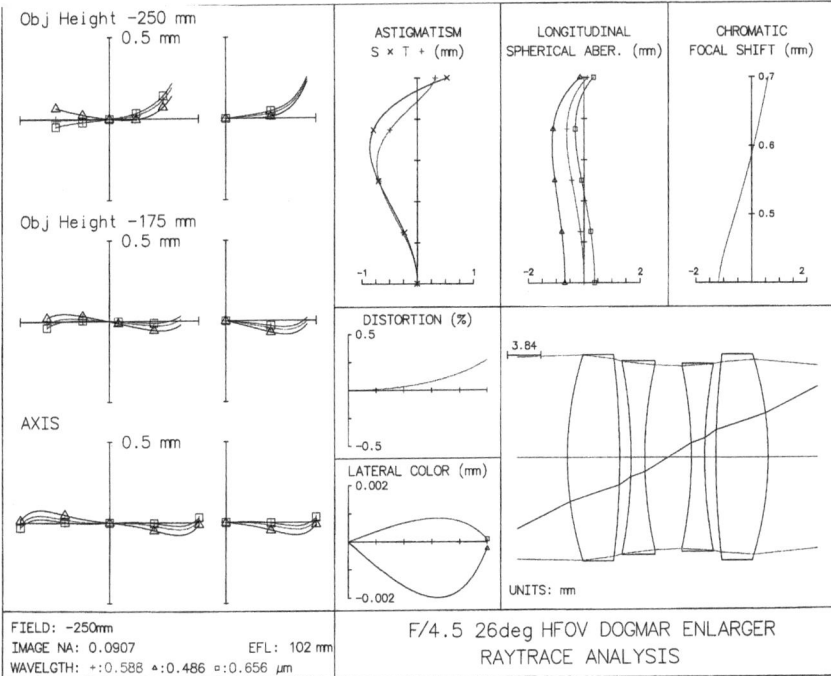

Obj Height -250 mm
0.5 mm

Obj Height -175 mm
0.5 mm

AXIS
0.5 mm

ASTIGMATISM
S × T + (mm)

LONGITUDINAL
SPHERICAL ABER. (mm)

CHROMATIC
FOCAL SHIFT (mm)
0.7
0.6
0.5

DISTORTION (%)
0.5
-0.5

LATERAL COLOR (mm)
0.002
-0.002

3.84

UNITS: mm

FIELD: -250mm
IMAGE NA: 0.0907   EFL: 102 mm
WAVELGTH: +:0.588 ▲:0.486 □:0.656 μm

F/4.5 26deg HFOV DOGMAR ENLARGER
RAYTRACE ANALYSIS

F/4.5 Dogmar Enlarger Lens. The Dogmar, while it looks like a split-flint Cooke triplet, was originally conceived as a symmetrical pair of triplet components, where the center element of each triplet is an air lens. The Dogmar is an excellent general purpose lens. While originally fully symmetrical, it is usually made slightly unsymmetrical to improve the correction, especially when it is used as a camera lens. Because the design is relatively insensitive to conjugate change it makes an excellent and economical enlarger lens. With glasses chosen to reduce the secondary spectrum, the Dogmar form is often used in commercial process lenses. Another Dogmar design is shown in fig. 17.24.

F/4.5 26° HFOV Dogmar Enlarger

radius	thickness	mat'l	index	V-no	sa
37.210	6.000	LAF2	1.744	44.7	11.5
-96.590	1.280	air			11.5
-56.740	1.550	F2	1.620	36.4	10.8
46.770	2.690	air			10.4
	2.690	air			10.3
-47.880	1.550	F2	1.620	36.4	10.2
54.560	1.280	air			10.5
96.590	6.000	LAF2	1.744	44.7	11.5
-37.210	118.116	air			11.5

EFL   = 102.4
BFL   = 118.1
NA   = -0.0906 (F/5.5)
GIH   = 62.84
PTZ/F = -5.337
VL   = 23.04
OD   = 500.00 (MAG = -0.251)

Figure 19.18

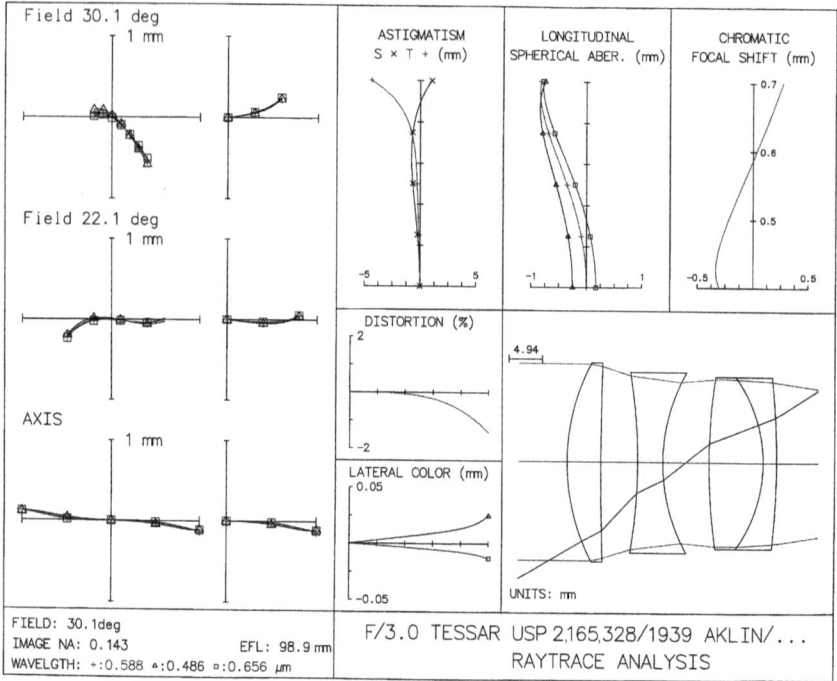

Field 30.1 deg

ASTIGMATISM
S × T + (mm)

LONGITUDINAL
SPHERICAL ABER. (mm)

CHROMATIC
FOCAL SHIFT (mm)

Field 22.1 deg

DISTORTION (%)

AXIS

LATERAL COLOR (mm)

UNITS: mm

FIELD: 30.1deg
IMAGE NA: 0.143   EFL: 98.9 mm
WAVELGTH: +:0.588 ▲:0.486 □:0.656 μm

F/3.0 TESSAR USP 2,165,328/1939 AKLIN/...
RAYTRACE ANALYSIS

F/3.0 Tessar with Reversed Rear Doublet. The powerful Tessar design form has been modified many times and in many ways. Here we see the rear doublet reversed. This orientation is said to be better when higher index glasses are utilized in the design. An ordinary Tessar design is shown in fig. 17.19.

F/3 Aklin USP 2,165,328

radius	thickness	mat'l	index	V-no	sa
30.322	5.054	SK16	1.620	60.3	16.8
390.086	5.579	air			16.8
−78.533	3.760	LF7	1.575	41.5	16.0
26.178	4.320	air			14.0
	2.634	air			14.0
82.072	8.076	SK52	1.639	55.5	14.8
−21.128	2.021	KF9	1.523	51.5	14.8
−114.906	81.484	air			16.0

EFL   = 100
BFL   = 81.48
NA   = −0.1682 (F/3.0)
GIH   = 53.17 (HFOV = 28.00)
PTZ/F = −3.373
VL   = 31.44
OD   infinite conjugate

Figure 19.19

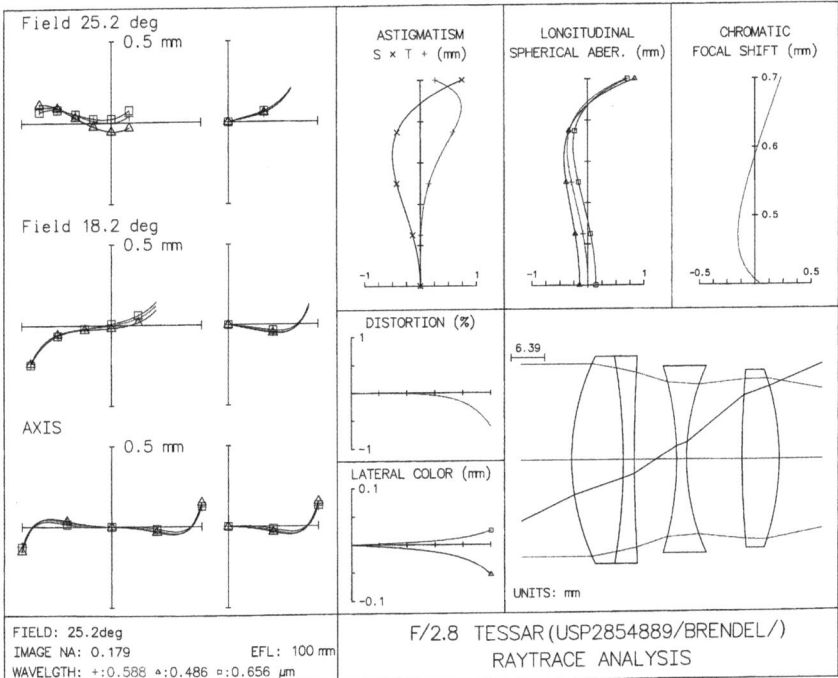

FIELD: 25.2deg
IMAGE NA: 0.179   EFL: 100 mm
WAVELGTH: +:0.588 ▵:0.486 ▫:0.656 μm

F/2.8 TESSAR (USP2854889/BRENDEL/)
RAYTRACE ANALYSIS

F/2.8 Reversed Tessar. This reversed Tessar achieves good performance with high index lanthanum crowns and higher index glass all around.

F/2.8 Tessar USP2854889/Brendel/

radius	thickness	mat'l	index	V-no	sa
42.970	9.800	LAK9	1.691	54.7	19.2
−115.330	2.100	LLF7	1.549	45.4	19.2
306.840	4.160	air			19.2
	4.000	air			15.0
−59.060	1.870	SF7	1.640	34.6	17.3
40.930	10.640	air			17.3
183.920	7.050	LAK9	1.691	54.7	16.5
−48.910	79.831	air			16.5

EFL = 100
BFL = 79.83
NA   = −0.1774 (F/2.8)
GIH = 47.01 (HFOV = 25.17)
PTZ/F = −3.025
VL   = 39.62
OD   infinite conjugate

Figure 19.20

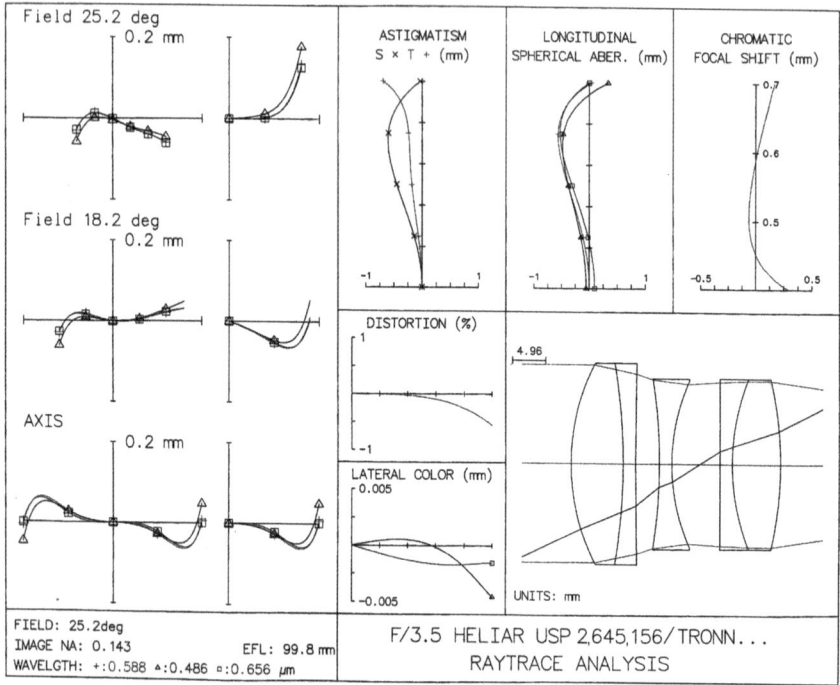

FIELD: 25.2deg
IMAGE NA: 0.143   EFL: 99.8 mm
WAVELGTH: +:0.588 △:0.486 □:0.656 µm

F/3.5 HELIAR USP 2,645,156/ TRONN...
RAYTRACE ANALYSIS

F/3.5 Heliar USP 2,645, 156. The five element Heliar is an improvement over the Tessar, because its added element allows a more nearly symmetrical construction, which reduces the high orders of coma, distortion, and lateral color. As with the Dogmar and the Tessar, the Heliar makes a excellent general purpose lens, and it has been used successfully for camera lenses, enlargers, aerial survey, process cameras. Just as the Tessar is better than the Cooke triplet, the Heliar is better than the Tessar.

F/3.5 Heliar USP 2,645,156/ Tronnier/

radius	thickness	mat'l	index	V-no	sa
30.810	7.700	LAKN7	1.652	58.5	14.5
−89.350	1.850	F5	1.603	38.0	14.5
580.380	3.520	air			14.5
−80.630	1.850	BAF9	1.643	48.0	12.3
28.340	4.180	air			12.0
	3.000	air			11.6
	1.850	LF5	1.581	40.9	12.3
32.190	7.270	LAK13	1.694	53.3	12.3
−52.990	81.857	air			12.3

EFL   = 99.81
BFL   = 81.86
NA   = −0.1428 (F/3.5)
GIH   = 46.91 (HFOV = 25.17)
PTZ/F = −3.682
VL   = 31.22
OD   infinite conjugate

Figure 19.21

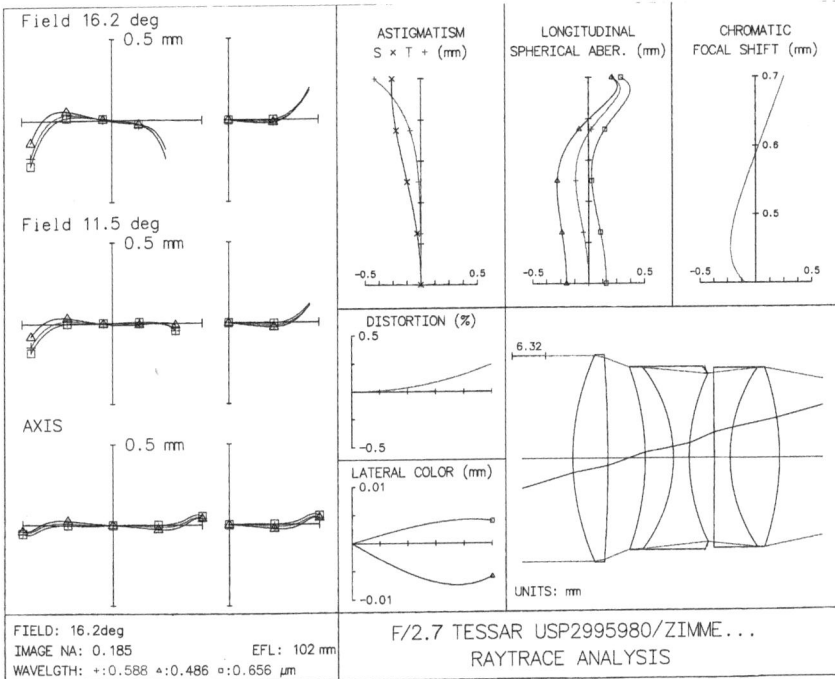

FIELD: 16.2deg		
IMAGE NA: 0.185	EFL: 102 mm	F/2.7 TESSAR USP2995980/ZIMME...
WAVELGTH: +:0.588 ▵:0.486 ▫:0.656 μm		RAYTRACE ANALYSIS

F/2.7 Tessar with a Doublet Center Component. This variation on the the Tessar form combines the advantages of the doublet rear component with a Merte' surface in the center component. The Merte' surface is low power, strongly curved convergent surface. Because of the very small index break across the surface, the spherical aberration contribution for rays near the axis is quite small. However the rays at the margin of the aperture make very large angles of incidence at this surface and the resulting undercorrected high order spherical contribution at this ray height is very large. The result is that the negative 7th order spherical effectively reduces the spherical. This is a good lens, but it is rarely used, probably because a Merte' surface is critical and costly to fabricate. The six element Hektor is shown in fig. 17.21 at F/1.8, ±18.3°.

F/2.7 Tessar USP 2995980/ Zimmerman/

radius	thickness	mat'l	index	V-no	sa
44.650	6.700	LAK9	1.691	54.7	19.0
−267.950	4.000	air			19.0
	3.000	air			16.7
−49.040	5.400	SF4	1 755	27.6	16.8
−26.710	3.000	SF7	1.640	34.6	16.8
34.870	4.800	air			16.8
−1326.300	3.000	F1	1.626	35.7	16.5
29.160	9.270	LAFN2	1.744	44.8	16.5
−49.930	84.745	air			16.5

EFL = 101.9
BFL = 84.74
NA   = −0.1847 (F/2.7)
GIH = 29.55 (HFOV = 16.17)
PTZ/F = −5.145
VL   = 39.17
OD   infinite conjugate

Figure 19.22

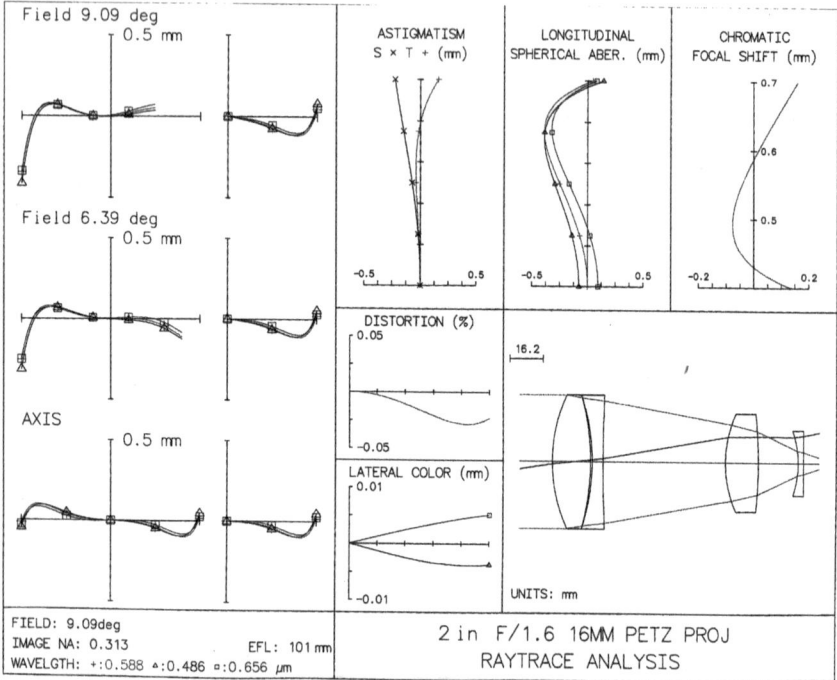

FIELD: 9.09deg	
IMAGE NA: 0.313	EFL: 101 mm
WAVELGTH: +:0.588 ▲:0.486 □:0.656 μm	

2 in F/1.6 16MM PETZ PROJ
RAYTRACE ANALYSIS

F/1.6 Petzval Projection Modification. This lens reduces the unfortunate inward field curvature of the regular four-element Petzval lens by using dense barium crowns and splitting the rear doublet with such a wide spacing that the rear flint acts as a field flattener. The front doublet is airspaced to correct the spherical. This results in a somewhat larger zonal spherical aberration and some undercorrected fifth-order linear (y^4h) coma. It is worth noting that all the projection lens field flatteners are made of flint glass. An ordinary Petzval projection lens with ordinary glasses is shown in fig. 17.26 at F/1.6, ± 6.8°. A Petzval lens with a field flattener in shown in fig. 17.27.

2 in F/1.6 16 mm Petz Proj

radius	thickness	mat'l	index	V-no	sa
73.962	18.550	DBC1	1.611	58.8	31.6
−114.427	0.776	air			31.6
−99.183	5.300	EDF3	1.720	29.3	31.6
660.831	59.678	air			31.6
55.173	15.900	DBC1	1.611	58.8	23.1
−228.329	19.769	air			23.1
−44.891	2.650	EDF1	1.720	29.3	15.4
2130.600	15.724	air			15.4

EFL = 100.7
BFL = 15.72
NA = −0.3139 (F/1.60)
GIH = 16.12 (HFOV = 9.09)
PTZ/F = −3.838
VL = 122.62
OD infinite conjugate

Figure 19.23

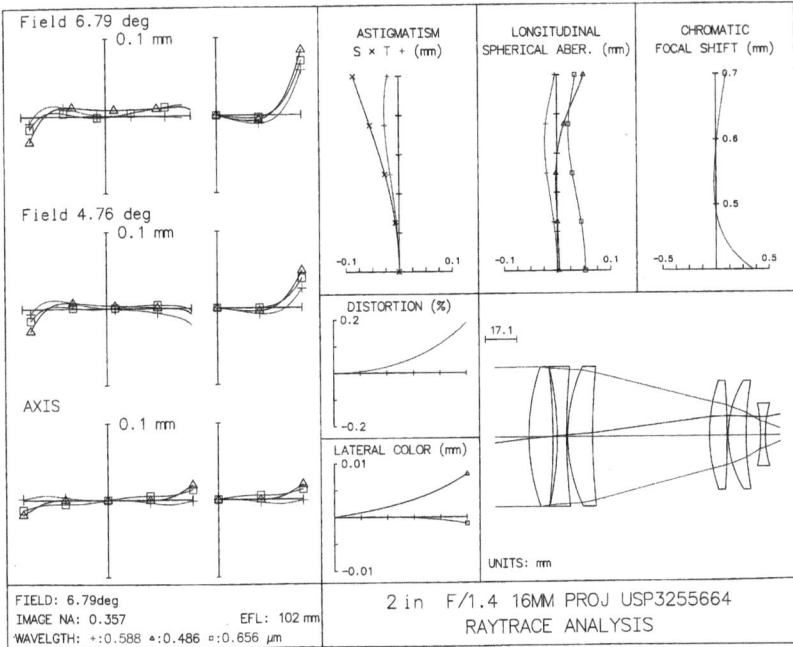

Field 6.79 deg	ASTIGMATISM S × T + (mm)	LONGITUDINAL SPHERICAL ABER. (mm)
		CHROMATIC FOCAL SHIFT (mm)

FIELD: 6.79deg
IMAGE NA: 0.357   EFL: 102 mm
WAVELGTH: +:0.588 ▲:0.486 □:0.656 μm

2 in F/1.4 16MM PROJ USP3255664
RAYTRACE ANALYSIS

F/1.4 Six Element Petzval Projection Lens Modification. This lens was originally designed as a modification of Fig. 19.23 by splitting both crowns in two. This made an excellent lens; the performance was limited by fifth-order linear coma produced by the strong negative air lens between the two front elements (which was used to correct the spherical). The lens shown here is a subsequent redesign done with an automatic optimization program (with the glass types as additional degrees of freedom). This eliminated most of the high order coma. The limiting aberrations in this design are sagittal oblique spherical and secondary spectrum.

2 in F/1.4 16 mm Proj USP3255664 1966 Chg'd Glass

radius	thickness	mat'l	index	V-no	sa
108.061	13.000	LAK21	1.640	60.1	36.5
−345.695	2.315	air			36.5
−159.702	5.000	SF10	1.728	28.4	36.5
361.952	0.600	air			36.5
86.990	13.200	PSK52	1.603	65.4	36.5
360.984	63.402	air			36.5
89.189	9.200	PSK53	1.620	63.5	28.5
−419.257	0.600	air			28.5
53.078	9.600	PSK53	1.620	63.5	28.5
152.163	8.027	air			28.5
−78.525	2.600	SF5	1.673	32.2	16.5
55.142	17.322	air			16.5

EFL   = 102.1
BFL   = 17.32
NA   = −0.3574 (F/1.40)
GIH   = 12.15 (HFOV = 6.79)
PTZ/F = −5.738
VL   = 127.54
OD   infinite conjugate

Figure 19.24

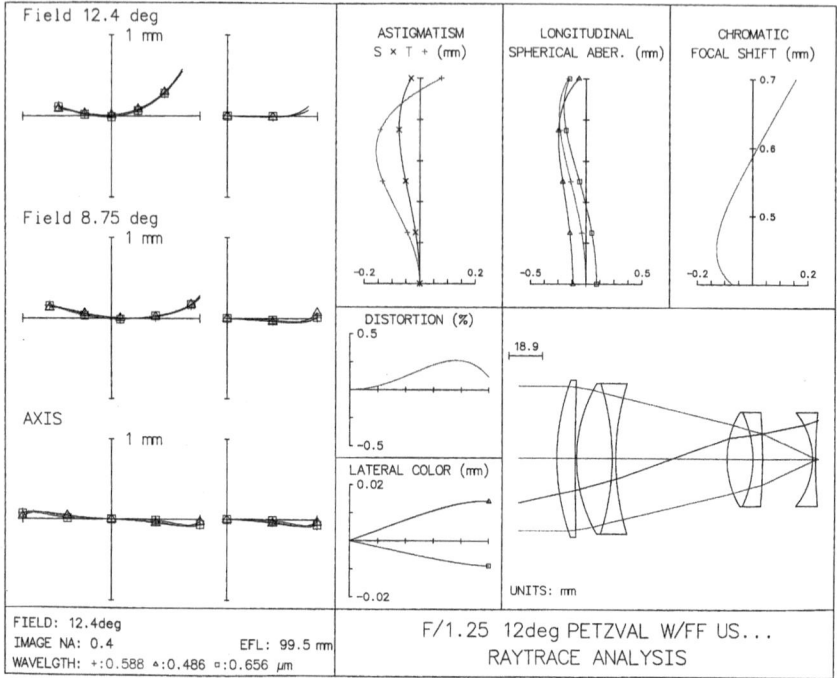

FIELD: 12.4deg	
IMAGE NA: 0.4	EFL: 99.5 mm
WAVELGTH: +:0.588 ▲:0.486 □:0.656 μm	

F/1.25 12deg PETZVAL W/FF US...
RAYTRACE ANALYSIS

F/1.25 Six Element Petzval Modification. This is an "ordinary glass" Petzval with a split-front crown and a field flattener element. If this design is reoptimized to remove the coma, it can be a superbly corrected high speed lens. It is presented here to show the effect of back focal length on the axial aberrations in a field flattener lens. A longer back focus would require higher powered components and they would produce larger aberration residuals. The almost-zero back focus of this lens is what makes possible the relatively high level of aberration correction. Of course there are only a very few applications which could tolerate such an extremely short working distance.

F/1.25 12 deg Petzval W/FF USP2649021/ Angenieux/

radius	thickness	mat'l	index	V-no	sa
121.110	10.380	BK7	1.517	64.2	43.2
1600.000	0.860	air			43.2
81.320	19.900	BK7	1.517	64.2	41.4
−138.430	2.080	F1	1.626	35.7	41.4
138.430	32.150	air			41.4
	31.000	air			26.9
49.310	14.710	BK7	1.517	64.2	26.0
−60.560	5.540	F1	1.626	35.7	26.0
−332.230	28.300	air			26.0
−40.660	1.730	SF17	1.650	33.7	26.0
288.100	1.665	air			26.0

EFL   = 99.5
BFL   = 1.665
NA   = −0.4008 (F/1.25)
GIH   = 21.89 (HFOV = 12.41)
PTZ/F = 337.3
VL   = 146.65
OD   infinite conjugate

Figure 19.25

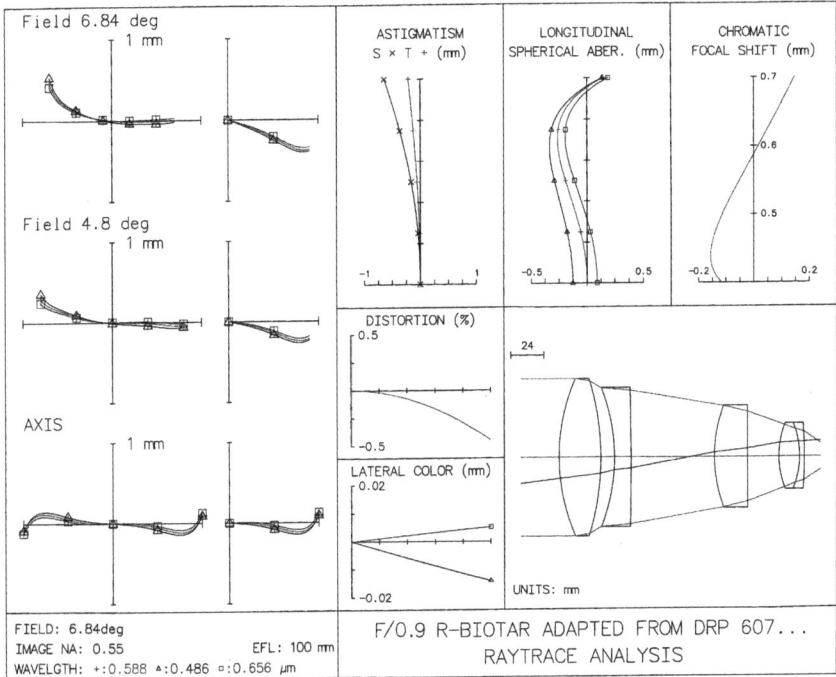

Field 6.84 deg	ASTIGMATISM S × T + (mm)
Field 4.8 deg	LONGITUDINAL SPHERICAL ABER. (mm)
AXIS	CHROMATIC FOCAL SHIFT (mm)
	DISTORTION (%)
	LATERAL COLOR (mm)

FIELD: 6.84deg
IMAGE NA: 0.55   EFL: 100 mm
WAVELGTH: +:0.588 ▲:0.486 □:0.656 μm

F/0.9 R-BIOTAR ADAPTED FROM DRP 607...
RAYTRACE ANALYSIS

UNITS: mm

F/0.9 R-Biotar Radiographic Objective. The R-Biotar, an extremely high speed Petzval modification, achieves good performance with a relatively simple construction. The high order aberrations are controlled by very careful spacing and power distribution. The airspace in the front doublet is used to reduce the spherochromatism and zonal spherical; this requires accurate fabrication. Kodak produced an F/1.08 mm movie projection lens by adding a field flattener to a higher index version of this design.

F/0.9 R-Biotar Adapted from DRP 607631/1932

radius	thickness	mat'l	index	V-no	sa
135.100	30.000	SKN18	1.639	55.4	55.0
−183.000	9.800	air			55.0
−129.000	11.700	SF4	1.755	27.6	48.6
−1813.800	43.800	air			48.6
	16.600	air			39.3
97.700	23.800	FK3	1.465	65.8	35.8
	22.600	air			35.8
60.200	15.100	SSK2	1.622	53.2	23.2
−59.800	3.400	SF4	1.755	27.6	23.2
−369.200	25.677	air			23.2

```
EFL   = 99.97
BFL   = 25.68
NA    = −0.5483 (F/0.91)
GIH   = 12.00 (HFOV = 6.84)
PTZ/F = −0.8335
VL    = 176.80
OD    infinite conjugate
```

Figure 19.26

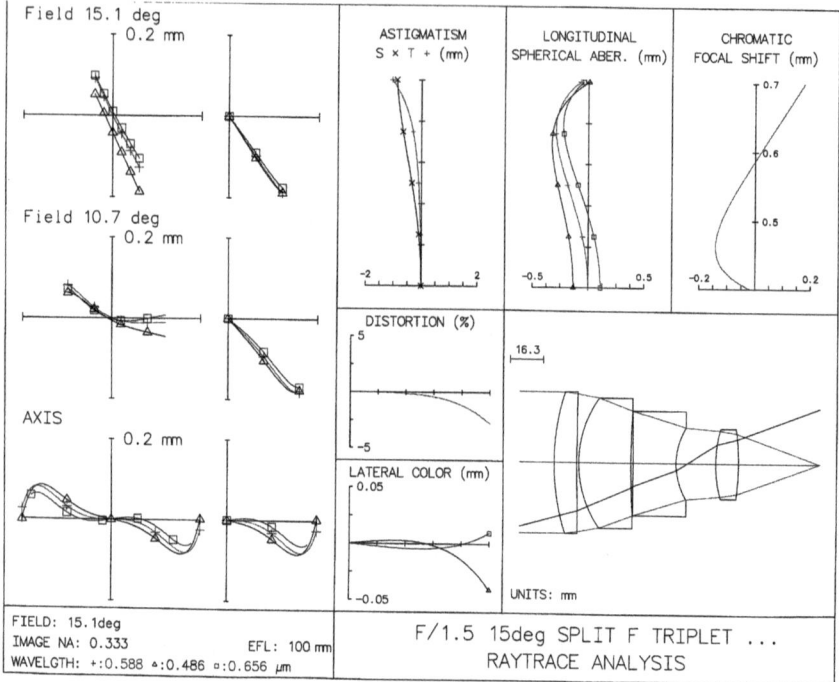

Field 15.1 deg

ASTIGMATISM S × T + (mm)

LONGITUDINAL SPHERICAL ABER. (mm)

CHROMATIC FOCAL SHIFT (mm)

Field 10.7 deg

AXIS

DISTORTION (%)

16.3

LATERAL COLOR (mm)

UNITS: mm

FIELD: 15.1deg
IMAGE NA: 0.333   EFL: 100 mm
WAVELGTH: +:0.588 ▲:0.486 □:0.656 μm

F/1.5 15deg SPLIT F TRIPLET ...
RAYTRACE ANALYSIS

F/1.5 Split-Front Triplet with Thick Elements. The thin element split-front triplet is an excellent design form. If, during the design process for a faster lens, the element thicknesses are allowed to vary, the second and third elements expand to fill the front (second) airspace. The three front elements then closely resemble the front components of the double Gauss lens (Figs. 19.35, 19.36, 19.37) and serve the same functions in maintaining the performance of the design. This design could be considered a relative of the Sonnar and Ernostar forms. An ordinary split-front crown triplet is shown in fig, 17.23 and a split rear in fig. 17.22.

F/1.5 15° Split-F Triplet (US 2,310,502)

radius	thickness	mat'l	index	V-no	sa
107.000	11.000	SK4	1.613	58.6	33.4
1000.000	1.000	air			32.7
52.630	25.200	SK4	1.613	58.6	30.8
350.900	1.100	air			24.9
−2632.000	21.000	SF63	1.748	27.7	24.8
31.940	5.500	air			17.0
	14.100	air			16.9
58.140	11.000	BAF9	1.643	48.0	16.5
−125.300	39.693	air			16.5

EFL   = 100.4
BFL   = 39.69
NA   = −0.3331 (F/1.50)
GIH   = 27.10 (HFOV = 15.11)
PTZ/F = −1.782
VL   = 89.90
OD   infinite conjugate

Figure 19.27

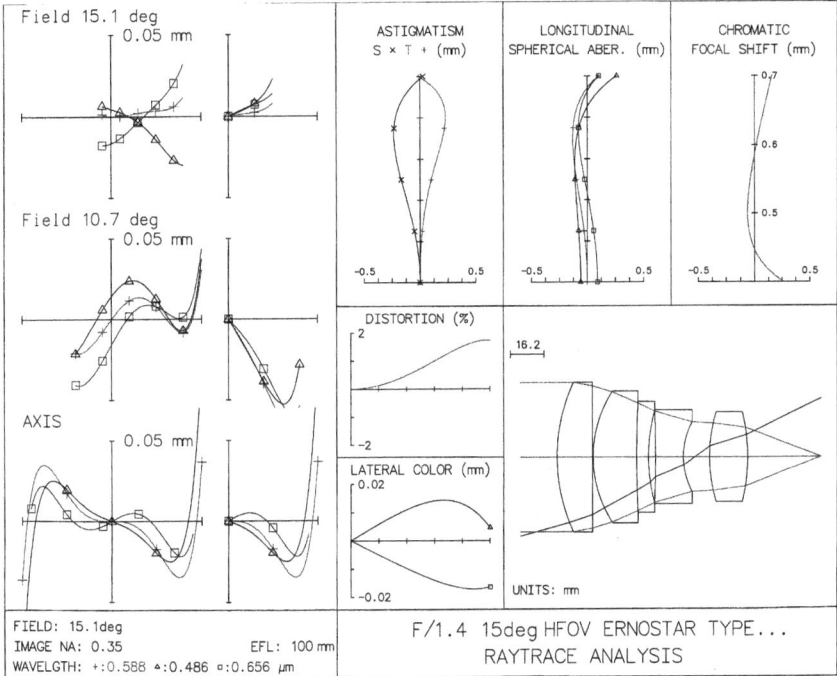

Field 15.1 deg
0.05 mm

ASTIGMATISM
S × T + (mm)

LONGITUDINAL
SPHERICAL ABER. (mm)

CHROMATIC
FOCAL SHIFT (mm)

Field 10.7 deg
0.05 mm

AXIS
0.05 mm

DISTORTION (%)

16.2

LATERAL COLOR (mm)

UNITS: mm

FIELD: 15.1deg
IMAGE NA: 0.35
WAVELGTH: +:0.588 ▲:0.486 ▫:0.656 μm   EFL: 100 mm

F/1.4 15deg HFOV ERNOSTAR TYPE...
RAYTRACE ANALYSIS

F/1.4 Ernostar. In this modification of the split-front triplet, the meniscus second element is made an achromatic doublet. Some high index glass and a modest (30°) field of view allow reasonable performance at a high speed for this relatively simple design form.

F/1.4 15° HFOV Ernostar Type from US 3,024,697

radius	thickness	mat'l	index	V-no	sa
81.100	17.100	SK16	1.620	60.3	35.0
	0.320	air			35.0
56.100	20.720	LAK8	1.713	53.8	31.0
479.000	5.830	SF7	1.640	34.6	26.0
77.800	3.550	air			22.0
−645.000	13.680	SF11	1.785	25.8	22.0
33.150	5.740	air			16.4
	7.030	air			16.3
72.000	18.130	LAF2	1.744	44.7	21.0
−87.440	35.745	air			21.0

EFL = 100
BFL = 35.74
NA   = −0.3502 (F/1.43)
GIH = 27.00 (HFOV = 15.11)
PTZ/F = −2.548
VL   = 92.08
OD   Infinite conjugate

Figure 19.28

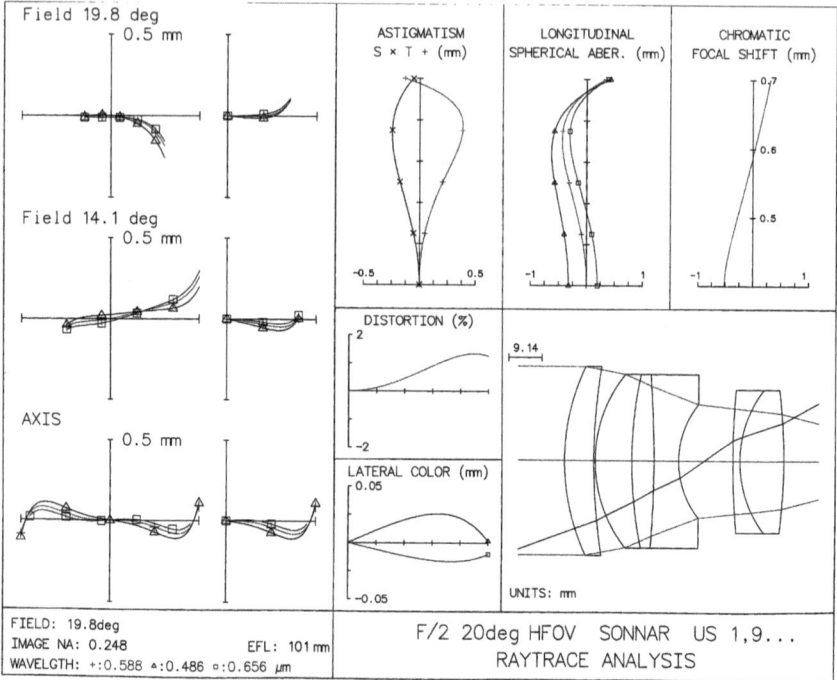

Field 19.8 deg

Field 14.1 deg

AXIS

ASTIGMATISM S × T + (mm)

LONGITUDINAL SPHERICAL ABER. (mm)

CHROMATIC FOCAL SHIFT (mm)

DISTORTION (%)

LATERAL COLOR (mm)

UNITS: mm

FIELD: 19.8deg
IMAGE NA: 0.248   EFL: 101 mm
WAVELGTH: +:0.588 ▲:0.486 □:0.656 μm

F/2 20deg HFOV   SONNAR   US 1,9...
RAYTRACE ANALYSIS

F/2.0 Sonnar. A direct descendent from the split-front triplet, this lens fills the front (second) airspace of the split triplet with low index glass, achieving apochromatic color correction. Replacing the rear element with a Tessar type rear doublet, this design manages a 40° field of view.

F/2 20° HFOV Sonnar US 1,998,704

radius	thickness	mat'l	index	V-no	sa
57.000	8.000	SK16	1.620	60.3	25.0
146.300	0.400	air			25.0
36.200	10.000	BAF53	1.670	47.1	23.0
110.000	6.000	FK3	1.465	65.8	23.0
−300.000	6.800	SF8	1.689	31.2	23.0
23.700	7.000	air			15.1
	8.000	air			14.9
200.000	2.000	BAK4	1.569	56.1	19.0
30.700	12.000	BAF53	1.670	47.1	19.0
−152.640	48.771	air			19.0

EFL   = 100.8
BFL   = 48.77
NA   = −0.2475 (F/2.0)
GIH   = 36.29 (HFOV = 19.80)
PTZ/F = −3.794
VL   = 60.20
OD   infinite conjugate

Figure 19.29

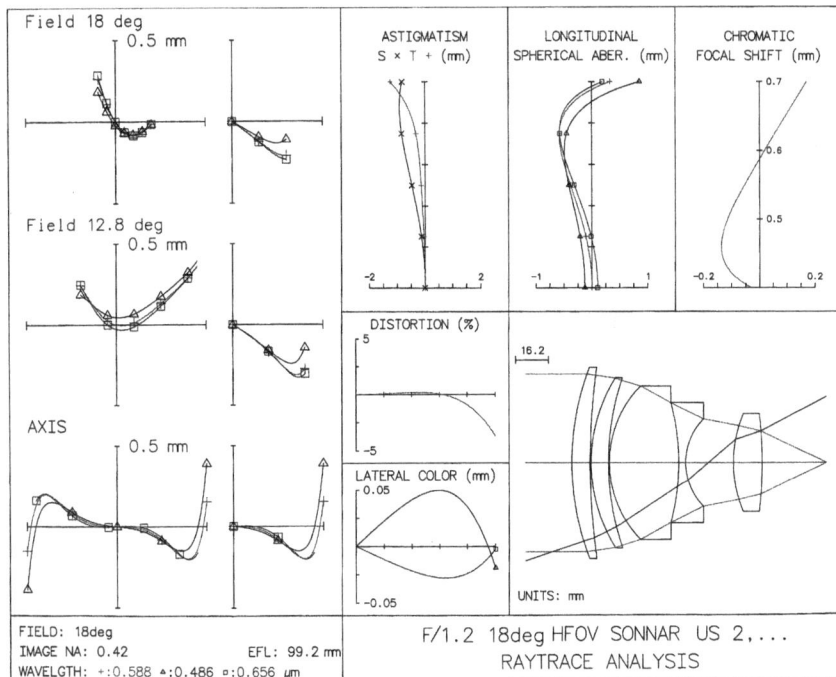

Field 18 deg ⊤ 0.5 mm	ASTIGMATISM S × T + (mm)	LONGITUDINAL SPHERICAL ABER. (mm)	CHROMATIC FOCAL SHIFT (mm)
Field 12.8 deg ⊤ 0.5 mm			
AXIS ⊤ 0.5 mm	DISTORTION (%) LATERAL COLOR (mm)		UNITS: mm

FIELD: 18deg
IMAGE NA: 0.42   EFL: 99.2 mm
WAVELGTH: +:0.588 ▲:0.486 □:0.656 μm

F/1.2 18deg HFOV SONNAR US 2,...
RAYTRACE ANALYSIS

F/1.2 Sonnar. Another variation on the split-front triplet, this Sonnar covers a 36° field at a really high F/1.2 speed. The Ernostars and the Sonnars were designed by the same man; they vied with the double Gauss for the role of the standard lens for the (then) new 35 mm camera.

F/1.2 18° HFOV Sonnar US 2,012,822

radius	thickness	mat'l	index	V-no	sa
121.480	8.810	SK4	1.613	58.6	45.0
310.680	0.500	air			45.0
73.270	8.380	SK4	1.613	58.6	40.0
118.550	0.500	air			40.0
50.420	33.700	SK7	1.607	59.5	36.1
−105.160	3.390	SF52	1.689	30.6	28.2
29.030	10.940	air			20.6
	12.000	air			20.1
59.390	13.720	SK4	1.613	58.6	23.0
−190.620	31.406	air			23.0

EFL = 99.2
BFL = 31.41
NA   = −0.4185 (F/1.19)
GIH = 32.24 (HFOV = 18.00)
PTZ/F = −1.837
VL   = 91.94
OD   infinite conjugate

Figure 19.30

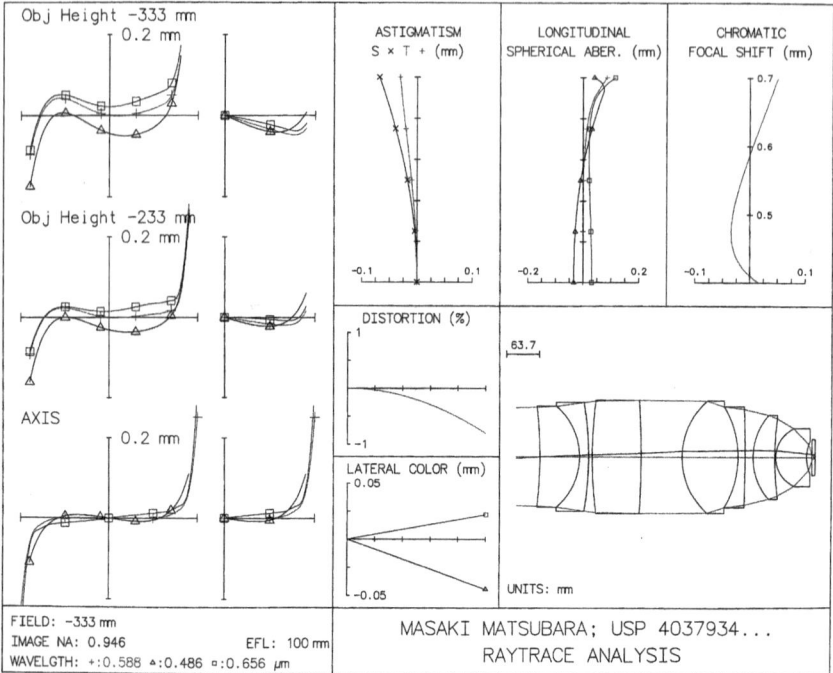

NA 0.95 Microscope Objective. Modern microscope objectives have much flatter fields than the classic forms, which are based on the Petzval projection lens. The flatter field is obtained in several ways: strongly meniscus thick doublets similar to those in the double Gauss; thick singlet meniscus elements; and a field flattening concave surface close to the focal plane. This lens is of the latter type, with the plane surface of the hyperhemisphere of the classical "aplanatic front" changed to a concave surface. Being close to the focus, it flattens the Petzval field without greatly reducing the power. Note that FK51 (487845) glass and calcium fluoride (454949) are used in positive elements to produce an apochromatic correction. At a focal length of one or two millimeters, the aberrations scale to very small values.

95NA 60× Microscope Objective #1 Masaki Matsubara; USP 4037934

radius	thickness	mat'l	index	V-no	sa
−753.114	76.280	FK51	1.487	84.5	92.0
−121.010	17.373	PCD4	1.618	63.4	94.0
−577.791	6.889	air			101.3
808.826	93.153	PCD4	1.618	63.4	103.7
−1635.724	77.878	air			105.3
139.381	107.531	CAF	1.434	94.9	104.3
−175.224	13.878	SF3	1.740	28.3	94.0
−2129.653	15.576	air			89.7
116.217	41.635	CAF	1.434	94.9	78.3
571.301	1.697	air			71.7
59.007	58.907	LAF28	1.773	49.6	54.0
70.289	10.667	air			30.0
	6.000	K3	1.518	59.0	33.3
	0.002	air			33.3

EFL = 100
BFL = −0.001501
NA = −0.9472 (F/0.53)
GIH = 5.56
PTZ/F = −1.853
VL = 527.46
OD = 5834.39
(MAG = −0.016)

Figure 19.31

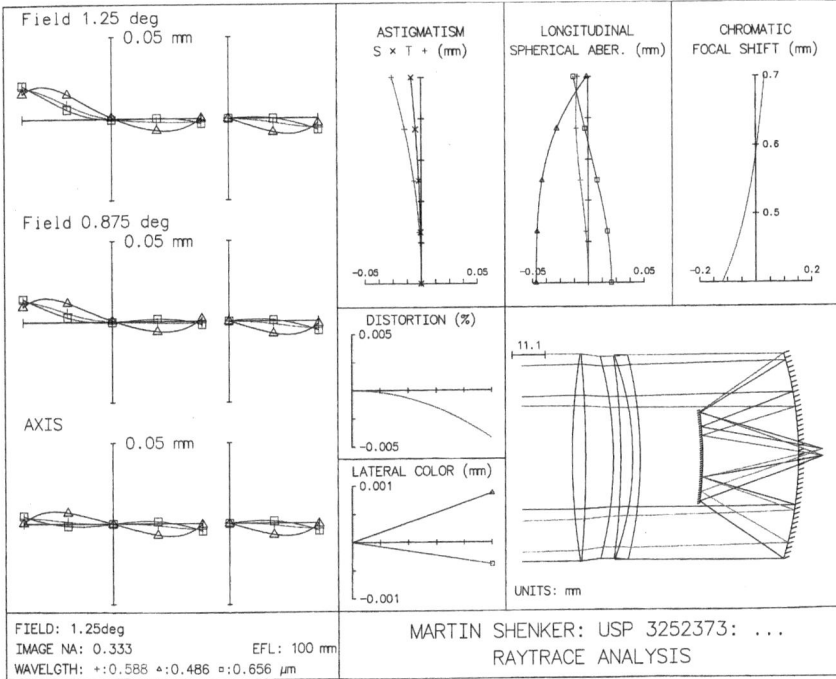

```
Field 1.25 deg
          0.05 mm
```
ASTIGMATISM S × T + (mm)

LONGITUDINAL SPHERICAL ABER. (mm)

CHROMATIC FOCAL SHIFT (mm)

```
Field 0.875 deg
          0.05 mm
```

```
AXIS
          0.05 mm
```

DISTORTION (%)

LATERAL COLOR (mm)

UNITS: mm

FIELD: 1.25deg
IMAGE NA: 0.333   EFL: 100 mm
WAVELGTH: +:0.588 ▲:0.486 □:0.656 μm

MARTIN SHENKER: USP 3252373: ...
RAYTRACE ANALYSIS

F/1.5 Catadioptric Cassegrain Lens. Thin refracting correctors located at the aperture stop are often used to correct the monochromatic aberrations of spherical surfaced mirror systems. If two or more thin, closely spaced, elements are all made of the same glass and have a total power approximating zero, it is possible to have a corrector component which produces neither primary nor secondary chromatic aberration. In this high speed Cassegrain, the three element corrector achieves an apochromatic correction for the marginal rays, but a significant amount of spherochromatic remains as the performance limiter. Note that the secondary mirror will obscure the central half of the aperture.

F/1.5 Catadioptric Telephoto #2 Martin Shenker; USP 3252373

radius	thickness	mat'l	index	V-no	sa	
212.834	4.463	UBK7	1.517	64.3	33.3	EFL = 100
390.476	9.174	air			33.3	BFL = 39.93
−125.482	2.480	UBK7	1.517	64.3	32.5	NA = −0.3328
−231.298	3.967	air			32.5	(F/1.50)
−91.834	2.480	UBK7	1.517	64.3	32.5	GIH = 2.18
−133.883	20.400	air			32.9	PTZ/F = −151.1
	32.047	air			33.1	VL = 43.35
−111.690	31.661	mirror			33.2	OD infinite conjugate
−111.690	39.925	mirror			15.0	

Figure 19.32

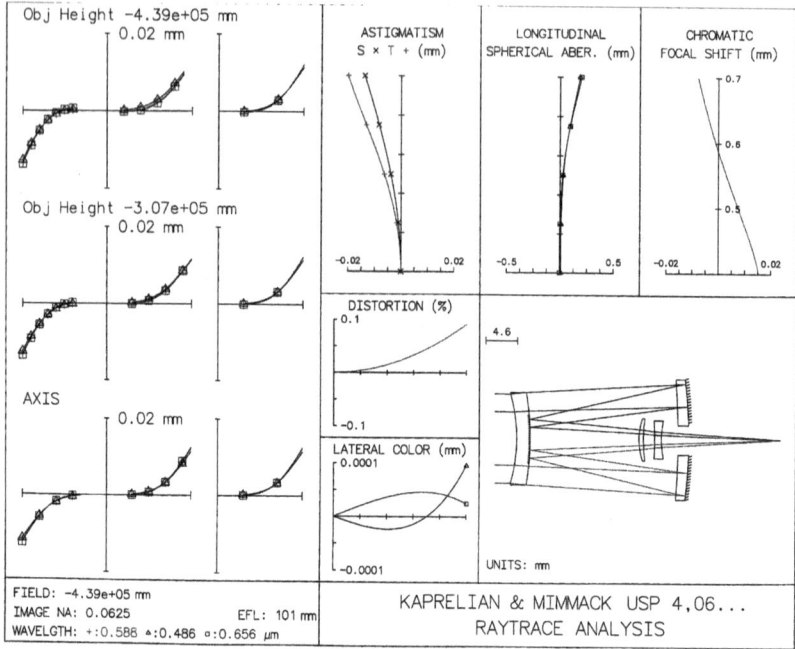

F/8.0 Catadioptric Cassegrain Lens. This design illustrates several very effective features often used in catadioptric designs: The thick meniscus corrector (a la Bouwers/Maksutov) is a spherical aberration corrector. Its second surface also serves as the secondary mirror, which negates the need for a spider mount. This surface is also aspheric and, as the secondary, can be used to affect other aberrations in addition to spherical (to which an aspheric at the aperture stop is limited to correcting). The primary is a second surface mirror on a meniscus element which is positive powered, and thus is not a classical Mangin mirror. The two elements in the convergent imaging cone are field correctors, especially effective with coma, astigmatism, and Petzval curvature. Astute glass choice and power distribution have produced a lens free of all forms of axial chromatic. A minor adjustment should be able to remove the tiny amount of over-corrected spherical.

F/8.0 Kaprelian & Mimmack USP 4,061,420

radius	thickness	mat'l	index	V-no	sa		
−22.500	1.929	BK7	1.517	64.2	6.3		
−37.943	21.394	air			6.5		
ad	−2.053E-06						
ae	3.815E-09						
af	−6.336E-11						
ag	8.243E-14						
−107.429	1.571	BK7	1.517	64.2	8.4		
−66.714	1.571	BK7	1.517	64.2	8.5		
−107.429	21.386	mirror			8.2		
−37.943	15.686	mirror			3.5		
ad	−2.053E-06						
ae	3.815E-09					EFL	= 100.6
af	−6.336E-11					BFL	= 17.37
ag	8.243E-14					NA	= −0.0621 (F/8.0)
9.179	0.643	KF9	1.523	51.5	3.0	GIH	= 3.09
20.029	1.657	air			2.9	PTZ/F	= 4.5
−32.614	0.957	KF9	1.523	51.5	2.7	VL	= 20.88
12.271	17.374	air			2.6	OD	infinite conjugate

Figure 19.33

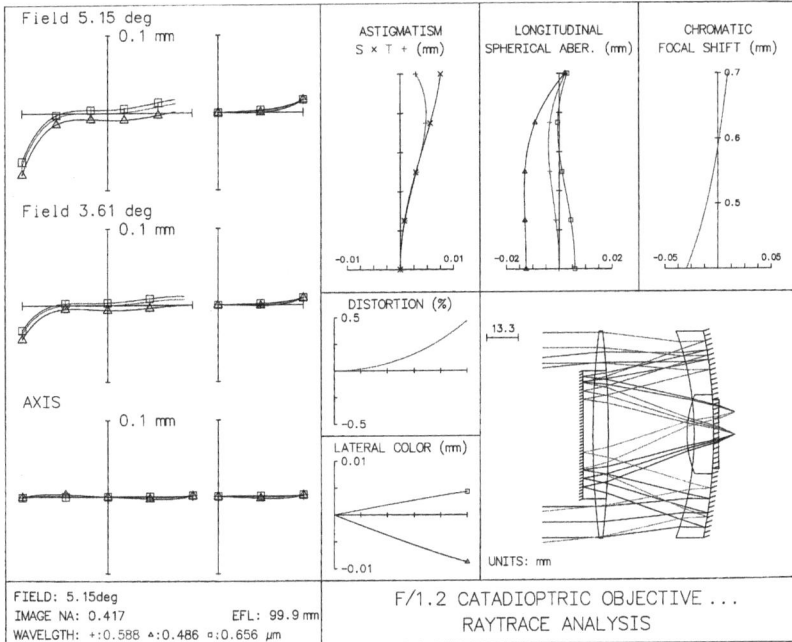

FIELD: 5.15deg		
IMAGE NA: 0.417	EFL: 99.9 mm	
WAVELGTH: +:0.588 ▲:0.486 □:0.656 µm		

F/1.2 CATADIOPTRIC OBJECTIVE ...
RAYTRACE ANALYSIS

F/1.2 Catadioptric Cassegrain Lens. This catadioptric is unusual in several respects. The corrector lens is a positive element rather than the more conventional weak negative meniscus lens. This has the advantage of reducing the necessary diameter of the primary mirror, which here is a classical Mangin mirror. Note that all the glass is BK7 (517642) except the very last element which is SF10 (728264). The secondary is a second surface mirror on an almost flat surface of what is effectively a positive element. The field corrector is a meniscus doublet. At the margin the chromatic correction is apochromatic, but there is some negligible spherochromatic. Lateral color is significant, as is some fifth-order coma.

F/1.2, EFL = 99.9 Catadioptric Objective USP 4,547,045

radius	thickness	mat'l	index	V-no	sa		
340.785	6.500	BK7	1.517	64.2	42.0	EFL	= 99.86
−375.235	36.000	air			42.0	BFL	= 6.783
−120.616	8.000	BK7	1.517	64.2	42.0	NA	= −0.4166
−215.820	8.000	BK7	1.517	64.2	42.0		(F/1.20)
−120.616	36.000	mirror			42.0	GIH	= 9.00
−375.235	6.500	BK7	1.517	64.2	26.0	PTZ/F	= 33.85
340.785	4.000	BK7	1.517	64.2	26.0	VL	= 52.50
−1316.482	4.000	BK7	1.517	64.2	26.0	OD	infinite conjugate
340.785	6.500	BK7	1.517	64.2	26.0		
−375.235	33.000	mirror			26.0		
41.443	3.000	BK7	1.517	64.2	16.0		
−120.616	8.000	BK7	1.517	64.2	16.0		
−215.820	2.000	SF10	1.728	28.4	14.0		
379.752	6.783	air			14.0		

Figure 19.34

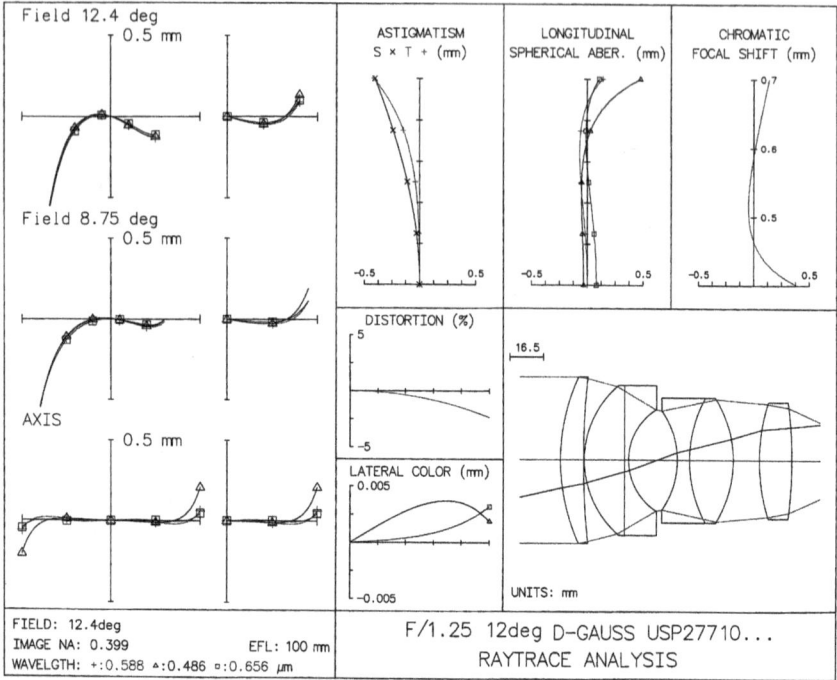

FIELD: 12.4deg
IMAGE NA: 0.399   EFL: 100 mm
WAVELGTH: +:0.588 ▲:0.486 □:0.656 μm

F/1.25 12deg D-GAUSS USP27710...
RAYTRACE ANALYSIS

F/1.25 Double Gauss Lens. This is a fairly basic double Gauss which uses high index lanthanum crowns to achieve the high speed of F/1.25. The large spacing of the last element is frequently utilized to reduce the sagittal oblique spherical aberration. The plano fourth surface is an economical choice. Note that this and all the other double Gauss designs in this chapter use a high index flint in the front element. Since the chromatic of the front element affects the relative height of the red and blue rays on the subsequent components, it can be used to correct the spherochromatism. It is also a way of getting a high index positive element without paying the high price for a lanthanum glass element. Another basic Double Gauss is shown in fig. 17.16.

F/1.25 12° D-Gauss USP2771006/ Werfeli

radius	thickness	mat'l	index	V-no	sa		
93.320	11.320	LAF3	1.717	48.0	40.0		
358.290	0.400	air			40.0		
46.320	20.000	BAF9	1.643	48.0	36.0		
	2.000	LF2	1.589	40.9	36.0	EFL	= 100.1
28.680	14.000	air			24.5	BFL	= 56.42
	10.000	air			24.3	NA	= −0.3992 (F/1.25)
−41.320	6.000	SF14	1.762	26.5	24.0	GIH	= 22.03
60.800	22.000	LAF2	1.744	44.7	30.0		(HFOV = 12.41)
−55.000	13.000	air			30.0	PTZ/F	= −3.801
90.200	16.000	LAF3	1.717	48.0	28.0	VL	= 114.72
−212.580	56.424	air			28.0	OD	infinite conjugate

Figure 19.35

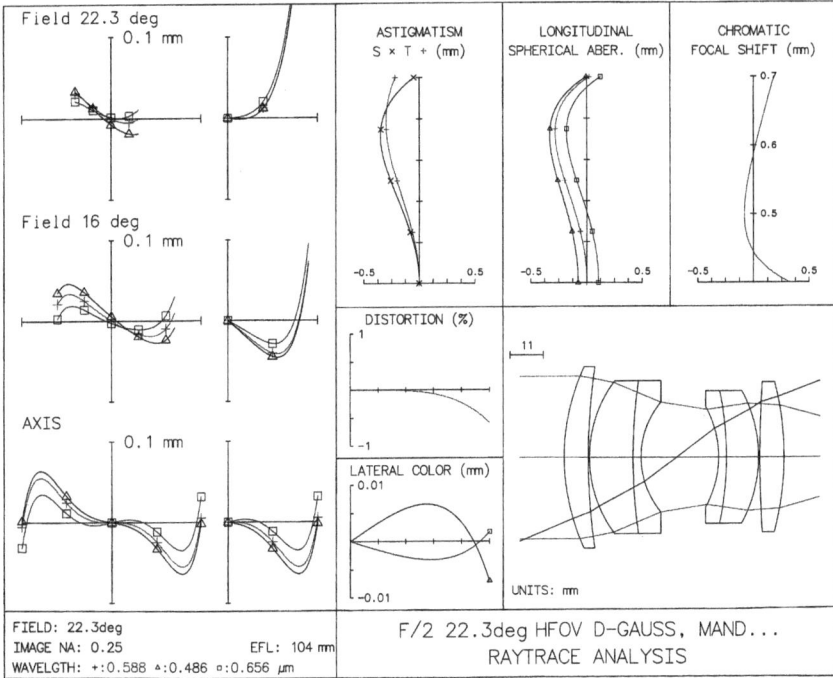

<table>
<tr><td>FIELD: 22.3deg</td><td></td></tr>
<tr><td>IMAGE NA: 0.25</td><td>EFL: 104 mm</td></tr>
<tr><td>WAVELGTH: +:0.588 ▲:0.486 □:0.656 μm</td><td></td></tr>
</table>

F/2 22.3deg HFOV D-GAUSS, MAND...
RAYTRACE ANALYSIS

F/2.0 Double Gauss Lens. This excellent design of an F/2.0 objective for a 35 mm camera was the ultimate result (and best) from an extensive design study reported by Mandler at the 1980 International Lens Design Conference. With a back focal length of only about 60 percent of the focal length, it is not suitable as a standard single lens reflex objective, requiring a clearance of 38 to 40 mm.

F/2 22.3° HFOV D-Gauss, Mandler SPIE V237 1980

radius	thickness	mat'l	index	V-no	sa		
67.080	8.000	LAF23	1.689	49.7	29.0		
191.260	0.400	air			29.0		
39.860	14.380	LAFN2	1.744	44.8	24.4		
171.680	2.600	SF1	1.717	29.5	24.4	EFL	= 103.9
27.080	11.840	air			17.6	BFL	= 62.18
	13.820	air			16.8	NA	= −0.2507 (F/2.00)
−32.200	2.600	SF13	1.741	27.6	16.1	GIH	= 42.61
−99.480	10.460	LAFN2	1.744	44.8	21.2		(HFOV = 22.29)
−43.960	0.400	air			21.2	PTZ/F	= 5.749
371.440	8.000	LAF21	1.788	47.5	24.0	VL	= 72.50
−91.040	62.176	air			24.0	OD	infinite conjugate

Figure 19.36

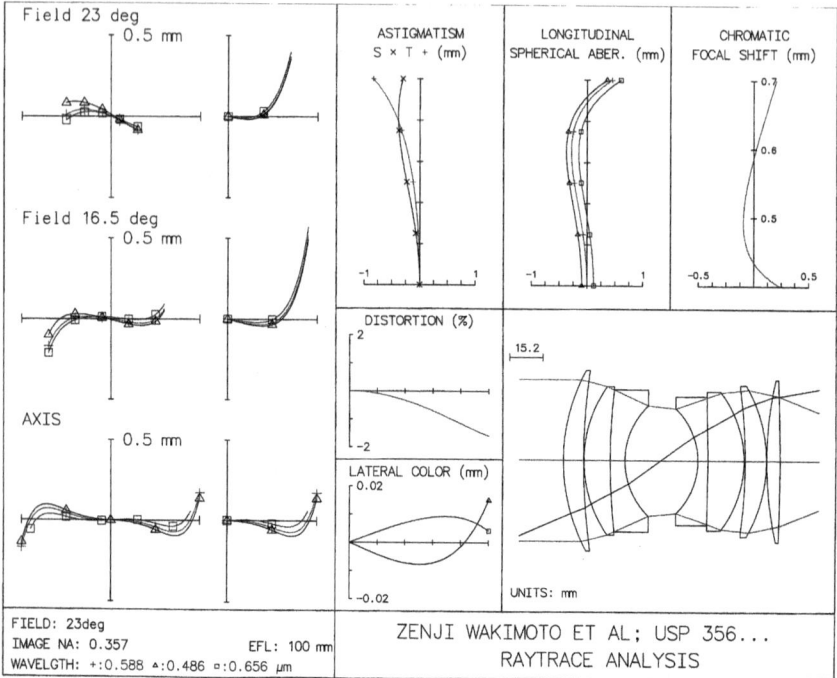

F/1.4 Double Gauss with Split Rear Crown. Splitting the rear crown element of the basic 6-element lens is probably the most effective of any of the basic modifications to the Double Gauss design. It allows a high speed of F/1.4, and has been widely used as a better-than-standard 35 mm camera lens.

F/1.4 46° Zenji Wakimoto ET AL; USP 3560079; Camera Lens #2

radius	thickness	mat'l	index	V-no	sa
81.400	9.300	LAF15	1.749	35.0	40.0
265.120	0.190	air			40.0
56.200	12.020	LAF20	1.744	44.9	32.9
271.320	6.590	SF15	1.699	30.1	31.5
33.410	16.330	air			24.6
	17.000	air			23.9
−32.270	4.650	FD1	1.717	29.5	23.3
−857.270	16.470	LAC12	1.678	55.5	28.5
−49.530	0.580	air			30.8
−232.560	9.690	LAC8	1.713	53.9	32.0
−66.100	0.190	air			33.5
158.910	6.200	LAC8	1.713	53.9	36.0
−890.280	74.557	air			36.0

EFL   = 100.1
BFL   = 74.56
NA   = −0.3563 (F/1.40)
GIH   = 42.50 (HFOV = 23.00)
PTZ/F = −5.88
VL   = 99.21
OD   infinite conjugate

Figure 19.37

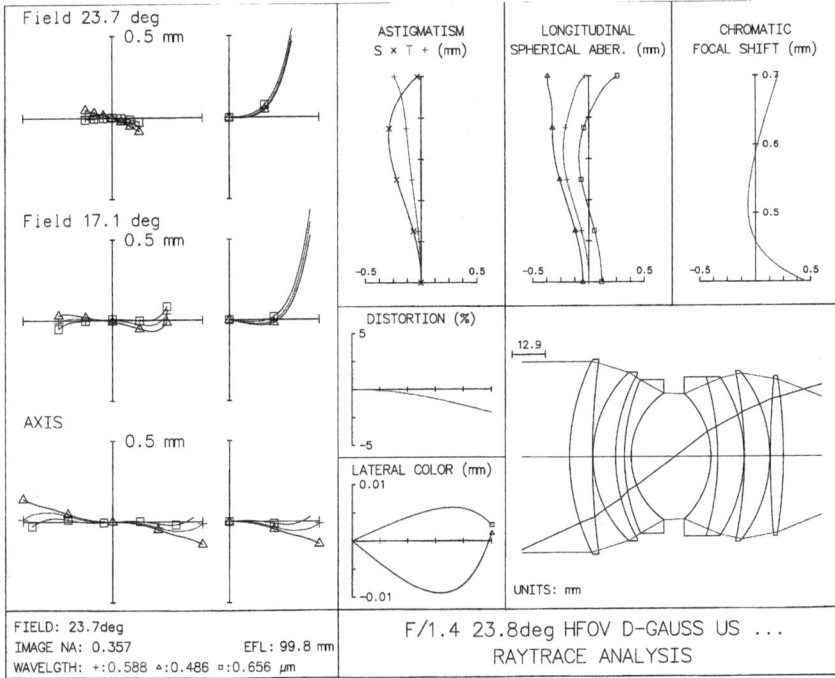

FIELD: 23.7deg	F/1.4 23.8deg HFOV D-GAUSS US ...
IMAGE NA: 0.357 EFL: 99.8 mm	RAYTRACE ANALYSIS
WAVELGTH: +:0.588 ▲:0.486 □:0.656 μm	

F/1.4 Split-Front Doublet 7-Element Double Gauss.

F/1.4 Split-Front Doublet 7-Element Double Gauss. Recent Double Gauss camera lenses have advantageously air-spaced the front doublet of the split-rear crown version of the Double Gauss. Many of the newer designs for camera lenses follow this configuration.

F/1.4 23.8° HFOV D-Gauss US 3,851,953/1974 Nakagawa

radius	thickness	mat'l	index	V-no	sa
78.186	8.862	LASF5	1.835	42.7	36.4
259.846	0.232	air			36.4
42.498	8.784	LASF3	1.808	40.6	31.5
60.180	3.436	air			30.8
69.416	2.606	SF56	1.785	26.1	28.7
28.654	16.888	air			23.7
	14.000	air			23.6
−31.582	2.646	SF56	1.785	26.1	23.5
−99.808	9.982	LASF3	1.808	40.6	29.7
−52.702	0.192	air			29.7
−150.994	9.692	LAK8	1.713	53.8	31.9
−50.290	0.232	air			31.9
338.224	5.174	LAK8	1.713	53.8	30.0
−186.406	73.796	air			30.0

EFL = 99.77
BFL = 73.8
NA = −0.3531 (F/1.40)
GIH = 43.90 (HFOV = 23.75)
PTZ = −6.057
VL = 82.73
OD infinite conjugate

Figure 19.38

F/1.1 Split-Front and Rear-Crown Double Gauss. This 8-element version of the Double Gauss achieves an extremely high speed at a modest field by splitting both of the outer crowns. This particular prescription can be corrected for chromatic by substituting BaSF6 (668–419) glass for the SF5 (673–322) glass in the front crown.

F/1.1 15° D-Gauss USP 2701982/ Angenieux/

radius	thickness	mat'l	index	V-no	sa		
164.120	10.990	SF5	1.673	32.2	54.0		
559.280	0.230	air			54.0		
100.120	11.450	BAF10	1.670	47.1	51.0		
213.540	0.230	air			51.0		
58.040	22.950	LAK9	1.691	54.7	41.0		
2551.000	2.580	SF5	1.673	32.2	41.0		
32.390	15.660	air			27.0		
	15.000	air			25.5	EFL	= 99.93
−40.420	2.740	SF15	1.699	30.1	25.0	BFL	= 55.74
192.980	27.920	SK16	1.620	60.3	36.0	NA	= −0.4521 (F/1.10)
−55.530	0.230	air			36.0	GIH	= 26.98
192.980	7.980	LAK9	1.691	54.7	35.0		(HFOV = 15.11)
−225.280	0.230	air			35.0	PTZ/F	= −2.979
175.100	8.480	LAK9	1.691	54.7	35.0	VL	= 126.67
−203.540	55.742	air			35.0	OD	infinite conjugate

Figure 19.39

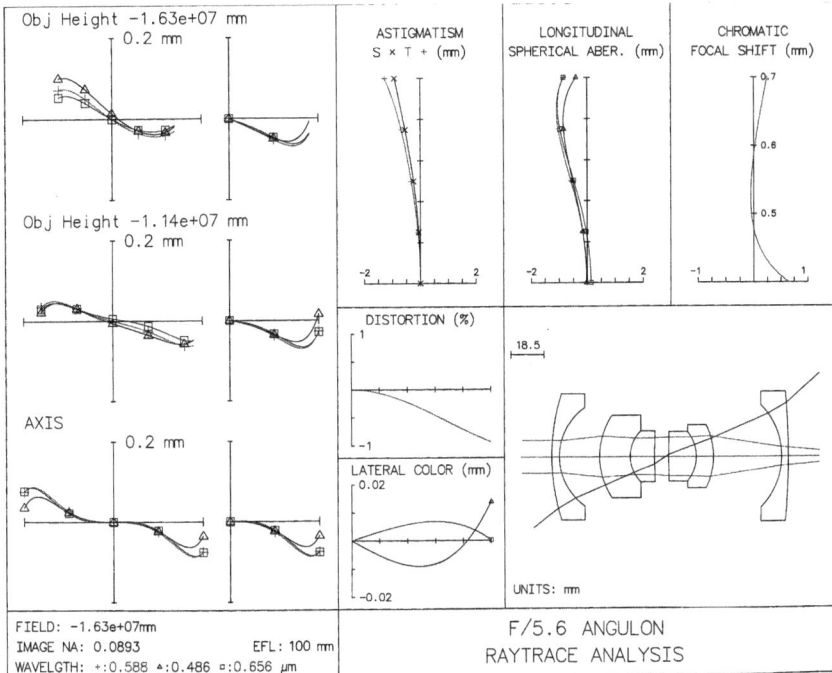

Obj Height -1.63e+07 mm			
0.2 mm	ASTIGMATISM S × T + (mm)	LONGITUDINAL SPHERICAL ABER. (mm)	CHROMATIC FOCAL SHIFT (mm)

FIELD: -1.63e+07mm
IMAGE NA: 0.0893   EFL: 100 mm
WAVELGTH: +:0.588 ▲:0.486 ▫:0.656 μm

F/5.6 ANGULON
RAYTRACE ANALYSIS

F/5.6 Angulon-Type Wide Angle Lens. This type of lens replaced the previous Metrogon and Topogon style of wide angle lenses because the aberrations of the strong negative outer elements enlarged the off-axis pupil size, and improved the off-axis illumination significantly. One might consider this to be a sort of inside-out Cooke triplet; the widely spaced negative outer elements in low ray-height locations flatten the Petzval field. Note that they also serve to reduce the angular field that the inner components must handle.

F/5.6 Angulon

radius	space	mat'l		sa
110.52	4.14	FK5	487704	34.0
32.048	22.30			26.6
39.402	16.88	LaFN3	717480	22.3
19.453	13.27	K10	501564	13.8
157.295	6.26			10.6
STOP	1.70			10.3
251.613	15.61	SK16	620603	10.6
−20.718	9.58	SF8	689312	13.4
−37.760	38.12			17.0
−37.760	3.82	FK5	487704	27.6
−369.543	53.70			35.0

EFL = 100.
BFL = 53.7
NA   = −0.0901 (F/5.6)
GIH = 76.884 (HFOV = 37.6°)
PTZ/F = −43.0
VL   = 140.17

Figure 19.40

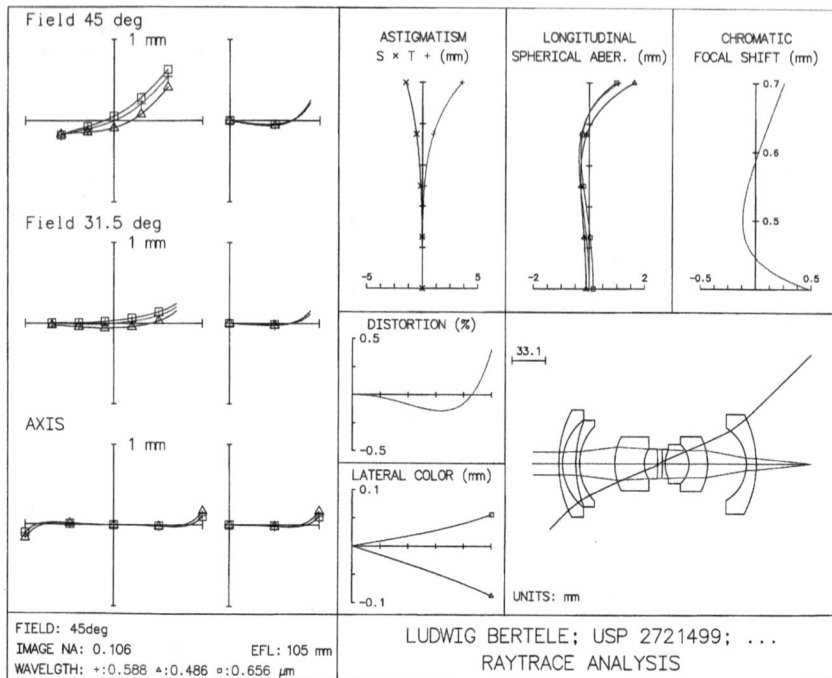

FIELD: 45deg
IMAGE NA: 0.106   EFL: 105 mm
WAVELGTH: +:0.588 ▲:0.486 □:0.656 μm

LUDWIG BERTELE; USP 2721499; ...
RAYTRACE ANALYSIS

F/4.7 Angulon-Type Wide Angle Lens. The addition of two more elements to the previous design allows an increase in speed from F/5.6 to F/4.7 and an increase in the angular field from 75° to 90°. In another variation on this theme, the rear negative was split in two and the central components were both triplets.

F/4.5 90° Field Ex. #2 Ludwig Bertele; USP 2721499

radius	thickness	mat'l	index	V-no	sa		
109.140	3.700	PK1	1.504	66.8	51.5		
52.630	13.200	air			42.0		
110.250	3.700	FK5	1.487	70.2	41.6		
50.720	35.000	air			36.0		
56.240	29.300	LAC10	1.720	50.3	26.0		
25.370	13.300	BACD7	1.607	59.5	14.4		
−194.920	1.700	air			14.3	EFL	= 104.6
	3.000	air			14.1	BFL	= 53.86
−252.700	2.800	BAK1	1.572	57.5	14.0	NA	= −0.1052 (F/4.7)
30.590	23.700	SSK2	1.622	53.2	13.8	GIH	= 104.59
−25.510	18.200	FD20	1.720	29.3	19.0		(HFOV = 45.00)
−57.380	39.000	air			26.5	PTZ/F	= −52.14
−40.150	9.800	SBC2	1.642	58.1	35.0	VL	= 196.40
−102.640	53.863	air			48.0	OD	infinite conjugate

Figure 19.41

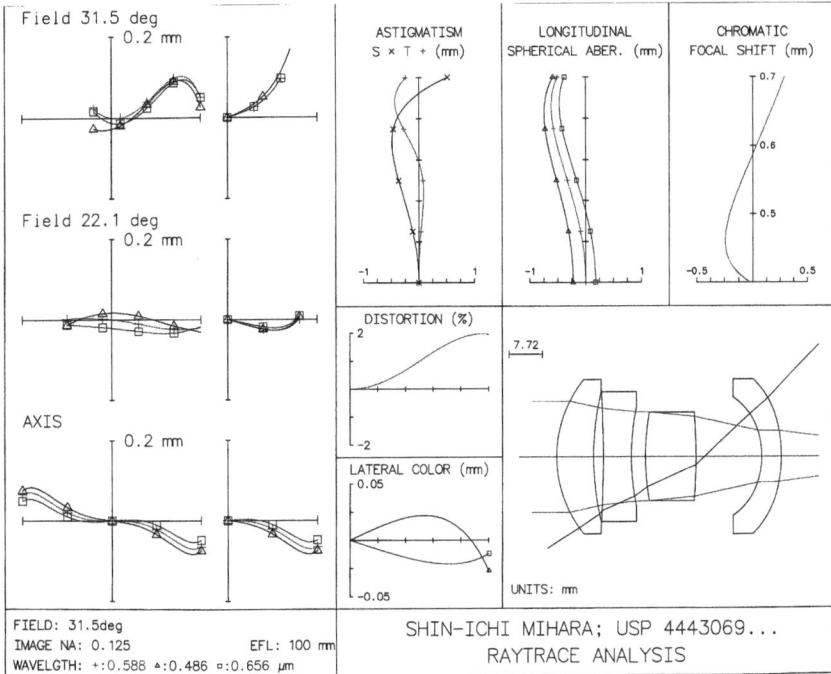

FIELD: 31.5deg
IMAGE NA: 0.125   EFL: 100 mm
WAVELGTH: +:0.588 ▲:0.486 □:0.656 μm

SHIN-ICHI MIHARA; USP 4443069...
RAYTRACE ANALYSIS

F/4.0 Triplet Anastigmat with a Field Corrector. This external aperture stop Cooke (?) triplet has a double aspheric field lens which is both a negative field flattener and a field corrector. This design might be appropriate for an inexpensive "point-and-shoot" camera.

F/4 63° Camera Lens #5 Shin-Ichi Mihara; USP 4443069

radius	thickness	mat'l	index	V-no	sa
27.131	8.540	TACB	1.729	54.7	17.4
61.089	2.286	air			14.6
−253.240	6.487	SF03	1.847	23.9	14.6
54.314	3.200	air			11.0
54.188	11.827	FD19	1.667	33.1	10.0
−110.136	2.000	air			9.1
	13.091	air			8.5
−19.281	4.670	FF2	1.533	45.9	13.3
ad		−2.423E-05			
ae		−2.615E-08			
af		−3.123E-11			
ag		−1.339E-12			
−32.458	45.808	air			17.4
ad		−1.514E-05			
ae		2.269E-09			
af		−4.269E-11			
ag		2.142E-14			

EFL = 100
BFL = 45.81
NA   = −0.1259 (F/4.0)
GIH = 61.28
       (HFOV = 31.50)
PTZ/F = −4.862
VL   = 52.10
OD   infinite conjugate

Figure 19.42

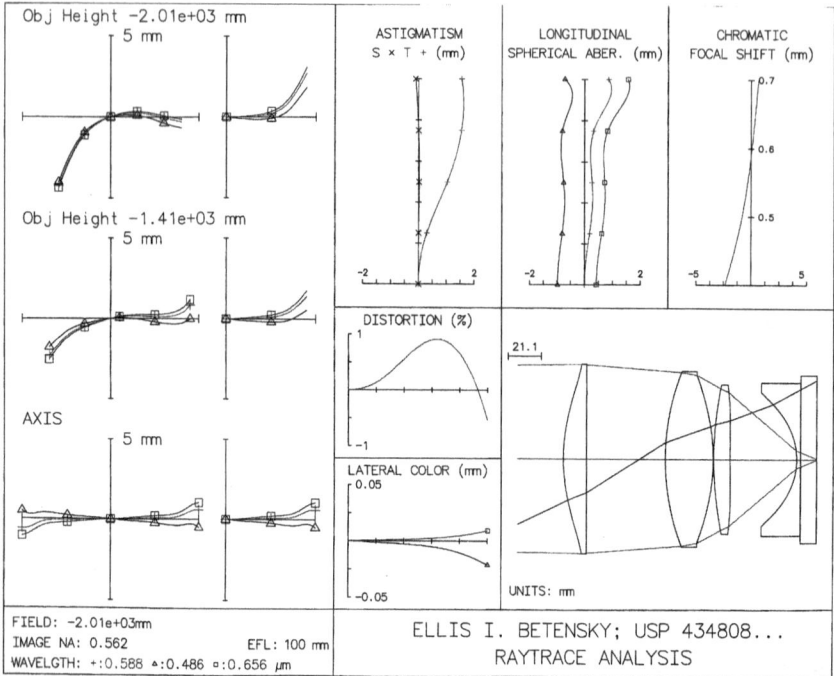

Obj Height −2.01e+03 mm	ASTIGMATISM S × T + (mm)	LONGITUDINAL SPHERICAL ABER. (mm)	CHROMATIC FOCAL SHIFT (mm)

FIELD: −2.01e+03mm
IMAGE NA: 0.562   EFL: 100 mm
WAVELGTH: +:0.588 ▲:0.486 □:0.656 μm

ELLIS I. BETENSKY; USP 434808...
RAYTRACE ANALYSIS

F/0.9 Projection TV Lens. This type of lens is basically a strong positive element plus a strong field flattener. The aberrations are corrected by aspheric surfaces. Additional weak elements are used for correction. In this example all the elements are plastic and two are aspheric. Plastic lenses have a big problem in that their focus position is extremely sensitive to temperature change. A *basic* form of this type of lens has a weak front aspheric corrector, a spherical surfaced glass lens which has almost all of the positive power, and an aspheric field flattener which, as in this design, is in optical contact with the CRT tube. The glass element is used to control the thermal defocusing problem.

116.1 MM F/1 CRT Proj Lens # 2 Ellis I. Betensky; USP 4348081

radius	thickness	mat'l	index	V-no	sa
107.980	14.772	ACRYL	1.490	57.9	58.3
kappa	1.326				
ad	−4.681E-07				
ae	−8.327E-12				
af	−1.795E-14				
ag	−4.346E-19				
25294.500	49.837	air			58.3
143.896	30.256	ACRYL	1.490	57.9	54.1
−118.000	0.380	air			51.8
kappa	−5.183				

radius	thickness	mat'l	index	V-no	sa
ad		−8.768E-08			
ae		−3.073E-12			
af		1.679E-15			
ag		4.462E-19			
214.441	10.630	ACRYL	1.490	57.9	46.0
1802.141	42.244	air			44.9
kappa		0.010			
ad		−3.744E-07			
ae		−6.689E-11			
af		6.865E-15			
ag		3.108E-18			
43.843	2.301	ACRYL	1.490	57.9	41.4
kappa		−5.599			
ad		−8.039E-06			
ae		4.937E-09			
af		−2.235E-12			
ag		5.099E-16			
	0.081	air			47.2
	10.354	K5	1.522	59.5	51.8
	0.255	air			51.8

EFL   = 100
BFL   = −0.2546
NA   = −0.5620 (F/0.89)
GIH   = 48.82
PTZ/F = −5.116
VL   = 160.85
OD   = 4256.60 (MAG = −0.024)

Figure 19.43

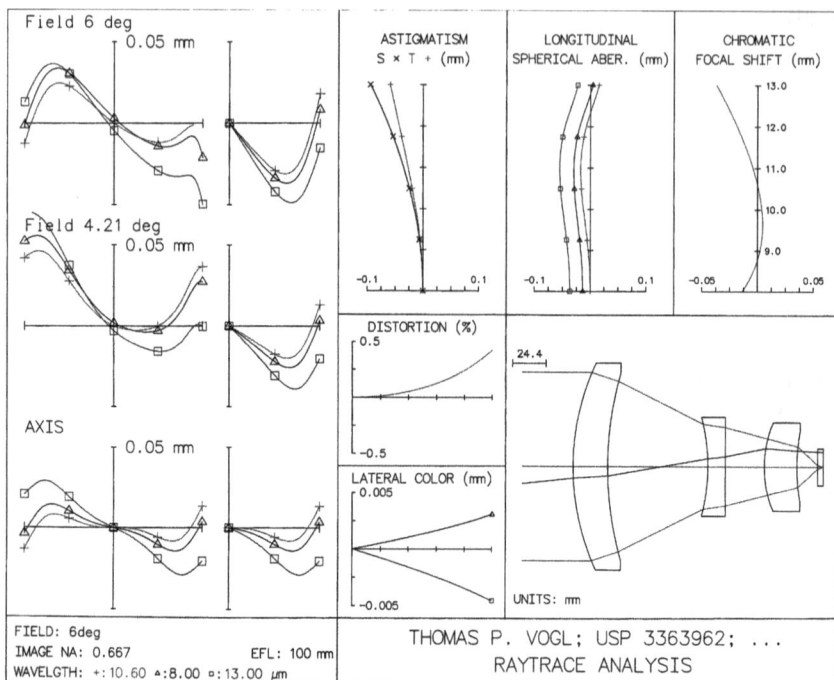

Figure raytrace analysis panel contents:

Field 6 deg — 0.05 mm

ASTIGMATISM S × T + (mm)

LONGITUDINAL SPHERICAL ABER. (mm)

CHROMATIC FOCAL SHIFT (mm)

Field 4.21 deg — 0.05 mm

AXIS — 0.05 mm

DISTORTION (%)

LATERAL COLOR (mm)

UNITS: mm

FIELD: 6deg
IMAGE NA: 0.667   EFL: 100 mm
WAVELGTH: +:10.60 ▲:8.00 □:13.00 μm

THOMAS P. VOGL; USP 3363962; ...
RAYTRACE ANALYSIS

F/0.75, 12° FOV Infrared Triplet. One might regard this as an infrared Cooke triplet. The positive elements are germanium, which has an extremely high V-value in the 8–12 μm wavelength region, and the central negative element is cesium iodide (CsI). The high index of germanium (4.0) and the relatively low index (1.74) of cesium iodide, as well as the + − + triplet configuration, flatten the field. The element shapes differ from the usual Cooke triplet shapes because the extremely high index materials not only have reduced aberration contributions, but their aberration maxima and minima which occur at bendings (shapes) differ significantly from those for lower index materials. For example, at an index of 4.0, the shape for minimum spherical is strongly meniscus (as in the front element of this lens and Fig. 17.38) In contrast, the shape for minimum spherical with an index of 1.5 is biconvex.

3 in F/.75 Far Infrared Lens Ex. 2 Thomas P. Vogl; USP 3363962;

radius	thickness	mat'l	index	V-no	sa
167.262	25.488	GERMA	4.003	779.6	73.2
233.147	73.172	air			66.3
−129.923	12.523	CSI	1.739	180.6	35.0
−22862.688	28.626	air			34.3
80.063	23.582	GERMA	4.003	779.6	31.5
137.090	15.448	air			25.1
	4.167	ZNSIR	2.192	17.0	13.1
	0.005	air			13.1

EFL = 100
BFL = −0.004675
NA   = −0.6640 (F/0.75)
GIH = 10.51 (HFOV = 6.00)
PTZ/F = −5.183
VL   = 183.01
OD   infinite conjugate

Figure 19.44

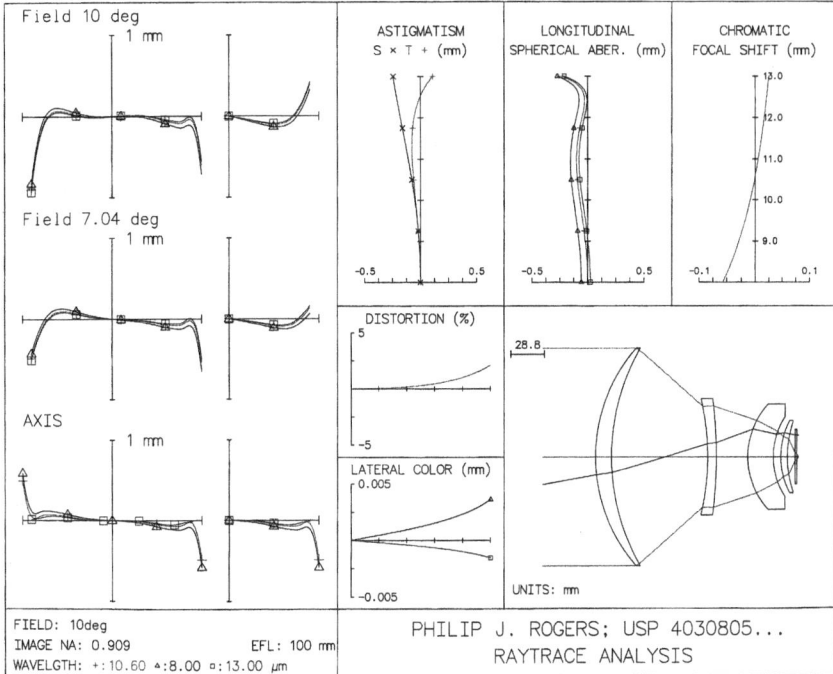

FIELD: 10deg	
IMAGE NA: 0.909 EFL: 100 mm	PHILIP J. ROGERS; USP 4030805...
WAVELGTH: +:10.60 ▲:8.00 □:13.00 μm	RAYTRACE ANALYSIS

F/0.55 20° FOV All Ge Infrared Anastigmat. Not only is a speed of F/0.55 remarkable in and of itself, but a lens of this speed is an exceedingly difficult lens design to get started. At extreme high speeds the elements are powerful and the strongly deviated rays are often aberrated as well; as a result the rays often completely miss the subsequent surfaces. Here the high index of the germanium is a mixed blessing. A high index means flatter curves and lower aberration contributions, but at the germanium index of 4.0, the critical angle is only 14°, and getting a ray out of an element may be prevented by total internal reflection (TIR). The spacings and the high order undercorrected spherical contributions of the fourth element in this design have balanced the spherical with an undercorrected seventh order. Note also that the low dispersion of the germanium in this spectral region, as well as the clever design have kept the chromatic small relative to the other aberrations, in spite of the all-germanium construction.

F/.55 IR Lens; Ex. 3 Philip J. Rogers; USP 4030805;

radius	thickness	mat'l	index	V-no	sa		
136.368	13.715	GERMA	4.003	779.6	91.0		
181.438	45.720	air			91.0		
	37.387	air			70.4		
−201.977	7.481	GERMA	4.003	779.6	47.9	EFL	= 99.98
−366.057	26.423	air			48.7	BFL	= 1.593
63.474	22.560	GERMA	4.003	779.6	44.2	NA	= −0.9055 (F/0.55)
58.048	6.840	air			33.5	GIH	= 17.63
62.196	5.837	GERMA	4.003	779.6	30.5		(HFOV = 10.00)
87.739	6.796	air			28.9	PTZ/F	= −4.706
	1.473	GERMA	4.003	779.6	23.0	VL	= −174.23
	1.593	air			23.0	OD	infinite conjugate

Figure 19.45

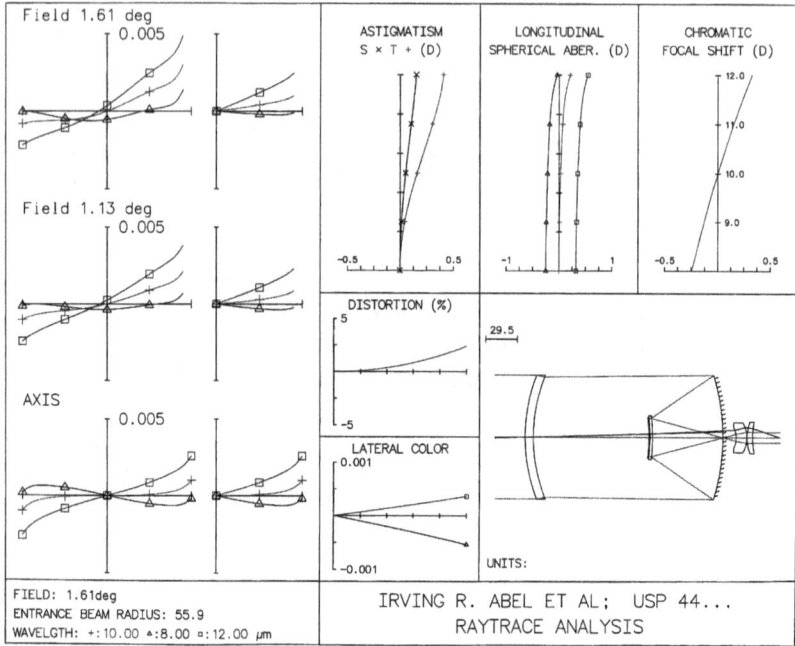

FIELD: 1.61deg
ENTRANCE BEAM RADIUS: 55.9
WAVELGTH: +:10.00 ▲:8.00 □:12.00 μm

IRVING R. ABEL ET AL; USP 44...
RAYTRACE ANALYSIS

Twelve Power Compact Cassegrain Infrared Telescope. The Cassegrain configuration produces a short objective system. The meniscus zinc sulfide Maksutov/Bouwers type corrector helps with the spherical and keeps the outside environment at bay. The rest of the elements are germanium, including the second surface "Mangin" style secondary mirror. The positive "eyepiece" produces a real, accessible exit pupil, which can be the locus of a scanner or of a cold stop for a cooled detector in a dewar. This eyepiece configuration of dual meniscus elements is a common one in IR telescopes, and is useful in managing distortion and spherical aberration of the pupil. There are three aspheric surfaces.

11.9× Compact Infrared Telescope Irving R. Abel et al; USP 4411499

radius	thickness	mat'l	index	V-no	sa		
152.400	7.620	ZNSIR	2.200	23.0	56.0		
144.780	121.900	air			56.0		
	54.500	air			56.0		
−182.016	64.897	mirror			56.2		
kappa	−0.751						
−171.653	2.540	GERMA	4.003	869.1	19.6		
−139.065	2.540	GERMA	4.003	869.1	19.0		
kappa	−4.700						
ad	1.350E-07						
ae	−1.380E-10						
af	6.731E-14						
−171.653	64.897	mirror			19.6		
	12.698	air			0.0		
−20.777	10.185	GERMA	4.003	869.1	9.6	EFL	= −1.175e + 04
−26.594	0.254	air			13.6	BFL	= 1026
39.218	5.080	GERMA	4.003	869.1	14.3	NA	= 0.0058 (F/105.3)
kappa	−0.356					GIH	= −330.78
59.436	26.416	air			13.3	PTZ/F	= −0.5364
Afo Un-	0.000	air			0.0	VL	= 212.24
supported						OD	infinite conjugate

Figure 19.46

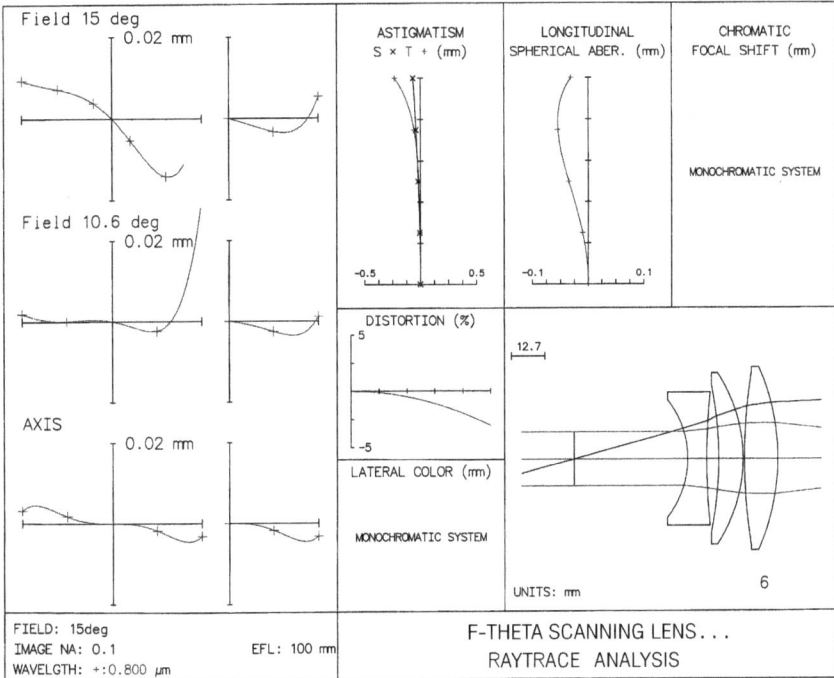

Field 15 deg	ASTIGMATISM	LONGITUDINAL	CHROMATIC
0.02 mm	S × T + (mm)	SPHERICAL ABER. (mm)	FOCAL SHIFT (mm)

MONOCHROMATIC SYSTEM

Field 10.6 deg
0.02 mm

-0.5   0.5   -0.1   0.1

DISTORTION (%)
5

12.7

AXIS
0.02 mm

-5
LATERAL COLOR (mm)

MONOCHROMATIC SYSTEM

UNITS: mm   6

FIELD: 15deg
IMAGE NA: 0.1   EFL: 100 mm
WAVELGTH: +:0.800 μm

F-THETA SCANNING LENS...
RAYTRACE ANALYSIS

F/5.0 F-Theta Scanning Lens. This design is an obvious configuration for this application. A collimated laser beam is deflected by a scanning mirror at the external entrance pupil (which is the aperture stop for the system). If a "perfect" lens were used, the location of the scanned image spot would be given by $H = F \tan \theta$. The scan lens has barrel distortion in an amount such that $H = F \cdot \theta$, so that the scanning spot has a velocity which is constant across the field, producing a uniform exposure. For a monochromatic laser system, the design need not be achromatic, and we can use inexpensive glasses, i.e., a low-index crown for the negative element and a high-index flint for the positive elements to help correct the Petzval field curvature. Of course, the system is a hyperchromat.

F/5, EFL = 55, H°= 14.31, FOV = 30 F–Theta Scanning

radius	thickness	mat'l	index	V-no	sa		
	43.550	air			10.0	EFL	= 100
-33.679	7.349	BK7	1.511	21.7		BFL	= 129.7
227.078	4.536	air			24.6	NA	= -0.1002 (F/5.0)
-137.219	9.073	SF11	1.765	27.6		GIH	= 26.79
-57.486	0.544	air			31.7		(HFOV = 15.00)
207.716	12.702	SF11	1.765	31.9		PTZ/F	= -31.85
-80.622	129.740	air			33.8	VL	= 77.76
						OD	infinite conjugate

Figure 19.47

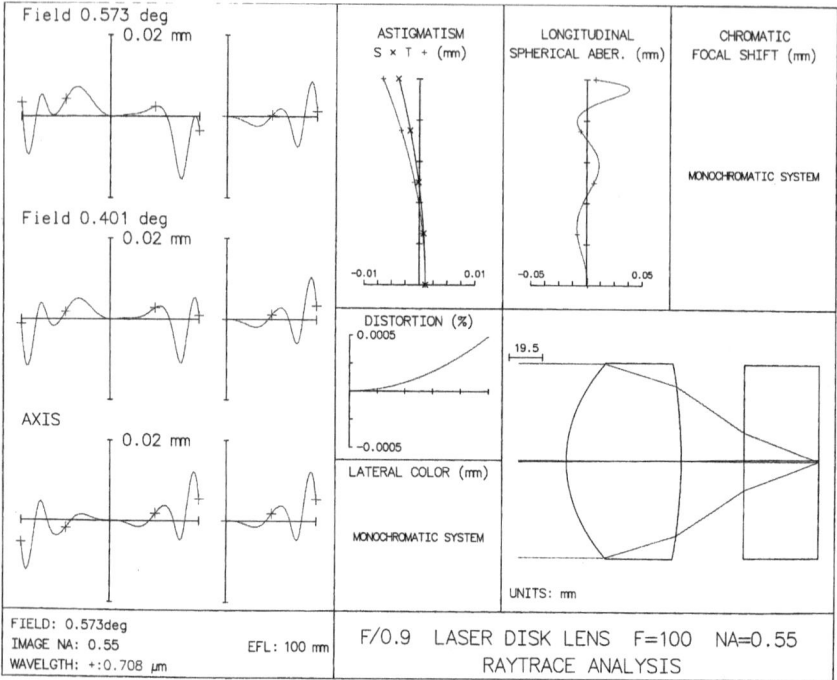

Field 0.573 deg	ASTIGMATISM S × T + (mm)	LONGITUDINAL SPHERICAL ABER. (mm)
		CHROMATIC FOCAL SHIFT (mm)

Field 0.573 deg 0.02 mm

Field 0.401 deg 0.02 mm

AXIS 0.02 mm

ASTIGMATISM S × T + (mm)

LONGITUDINAL SPHERICAL ABER. (mm)

CHROMATIC FOCAL SHIFT (mm)

MONOCHROMATIC SYSTEM

-0.01   0.01   -0.05   0.05

DISTORTION (%)
0.0005

19.5

-0.0005

LATERAL COLOR (mm)

MONOCHROMATIC SYSTEM

UNITS: mm

FIELD: 0.573deg
IMAGE NA: 0.55
WAVELGTH: +:0.708 μm

EFL: 100 mm

F/0.9 LASER DISK LENS F=100 NA=0.55
RAYTRACE ANALYSIS

F/0.9 Laser Disk Lens. The ability to mold precision aspheric surfaces economically in either glass or plastic has been widely utilized in singlet lenses for laser disk reading and writing objectives (as well as in many other systems). Obviously the thickness of the material covering the written surface must be an integral part of the design, especially at high speeds. The large axial thickness of the lens is useful and necessary to reduce the negative, inward curving astigmatism which is inherent in thin positive singlet lenses. With two aspheric surfaces we can get an almost perfect correction over a small field. These lenses are used at focal lengths of a few millimeters, which reduce the already small residual aberrations to completely negligible amounts. The economy of a molded plastic element may be offset by the defocus caused by temperature changes.

F/0.9 Laser Disk Lens F = 100 NA = 0.55

radius	thickness	mat'l	index	V-no	sa		
71.519	67.467	BAF5	1.601		0.0		
kappa		−0.379					
ae		−2.918E-11				EFL	= 100
af		4.403E-15				BFL	= 0
ag		−1.489E-18				NA	= −0.5500 (F/0.91)
−244.305	36.667	air		55.5		GIH	= 1.00
kappa		−73.482				PTZ/F	= −1.473
ae		−7.716E-11				VL	= 148.13
af		3.097E-14				OD	infinite conjugate
ag		−6.207E-18					
	44.000	CARBO	1.577		55.0		

Figure 19.48

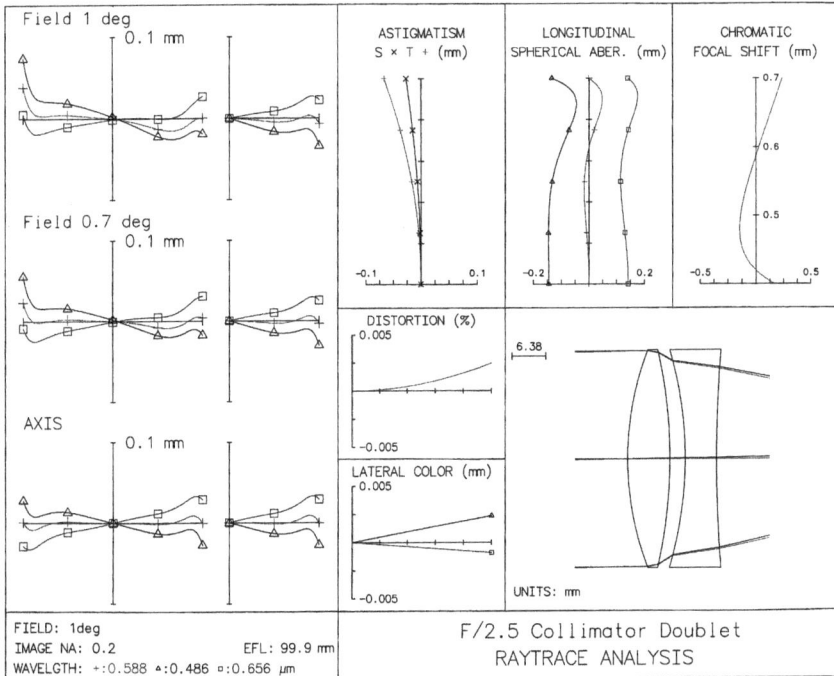

| Field 1 deg | | |
| ASTIGMATISM S × T + (mm) | LONGITUDINAL SPHERICAL ABER. (mm) | CHROMATIC FOCAL SHIFT (mm) |

FIELD: 1deg
IMAGE NA: 0.2   EFL: 99.9 mm
WAVELGTH: +:0.588 ▲:0.486 □:0.656 μm

F/2.5 Collimator Doublet
RAYTRACE ANALYSIS

F/2.5 Laser Collimator Doublet. This high speed collimator is basically a telescope objective which has been shaped and air-spaced to correct the spherical, coma, and zonal spherical, as described in Chap. 16. Since the laser is monochromatic, no color correction is needed and both elements can be made from high-index flint glasses. This design differs from prior art (Fig. 14.46 in the 3rd edition of *M. O. E.*) in that it has been designed to have no spherochromatism. Thus it is well corrected, and can be used at, different wavelengths; obviously it has axial chromatic and must be refocused when the wavelength is changed.

F/2.5 Laser Collimator Doublet

radius	space	mat'l		sa
55.381	8.0	LASFN30	800903	20.2
−89.402	2.978			20.2
−68.677	6.00	SF6	799517	20.2
243.105	82.46			20.2

EFL = 99.94
BFL = 82.46
NA = 0.200 (F/2.5)
GIH = 1.744 (HFOV = 1.0°)
VL = 16.978

Figure 19.49

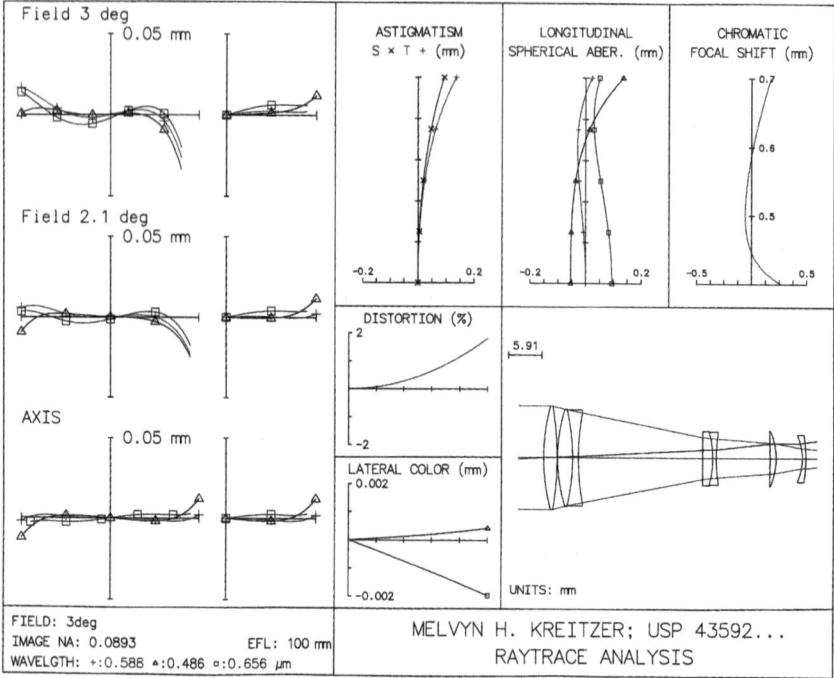

| Field 3 deg | ASTIGMATISM S × T + (mm) | LONGITUDINAL SPHERICAL ABER. (mm) | CHROMATIC FOCAL SHIFT (mm) |

FIELD: 3deg
IMAGE NA: 0.0893   EFL: 100 mm
WAVELGTH: +:0.588 ▲:0.486 □:0.656 μm

MELVYN H. KREITZER; USP 43592...
RAYTRACE ANALYSIS

F/5.6 Internal Focusing Telephoto Lens. This telephoto lens is focused by shifting the inner doublet, while the other components are fixed in place. This has two advantages (1) It is not necessary to shift the entire lens to focus (2) Telephoto lenses are sensitive to changes in object distance. Here the aberration correction can be maintained when the lens is focused close up. The telephoto ratio is a severe 0.66, producing a much shorter lens (and a more difficult lens to design) than the more normal ratio of 0.80. This lens might be a 400 mm lens for a 35 mm camera.

390 mm F/5.6 6° Tel M. Kreitzer; USP 4359272

radius	thickness	mat'l	index	V-no	sa
33.072	2.386	C3	1.518	59.0	8.9
−53.387	0.077	air			8.9
27.825	2.657	C3	1.518	59.0	8.4
−35.934	1.025	LAF7	1.749	35.0	8.3
40.900	22.084	air			7.8
	1.794	FD110	1.785	25.7	4.7
−16.775	0.641	TAFD5	1.835	43.0	4.6
27.153	9.607	air			4.5
−120.757	1.035	CF6	1.517	52.2	4.8
−12.105	4.705	air			4.8
−9.386	0.641	TAF1	1.773	49.6	4.0
−24.331	18.960	air			4.1

EFL = 100
BFL = 18.96
NA   = −0.0892 (F/5.6)
GIH = 5.24
PTZ/F = 2.097
VL   = 46.65
OD   infinite conjugate

Figure 19.50

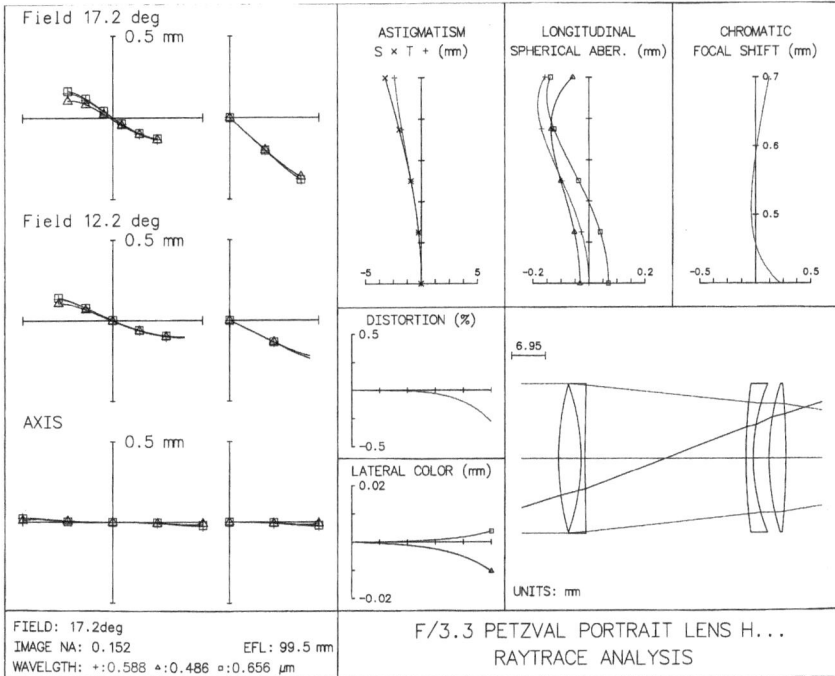

FIELD: 17.2deg
IMAGE NA: 0.152   EFL: 99.5 mm
WAVELGTH: +:0.588 ▲:0.486 □:0.656 μm

F/3.3 PETZVAL PORTRAIT LENS H...
RAYTRACE ANALYSIS

F/3.3 Petzval Portrait Lens. This lens was designed by Joseph Petzval in 1840 for the Daguerreotype camera (with the aid of two corporals and eight bombardiers who were "skilled in computing"). This team produced two designs in about 6 months. This lens was extremely fast for its day. The landscape lens previously used had a speed of about F/15, almost 20 times slower. Ironically, Petzval's design is noted for its large curvature of (the Petzval) field, but, modified as the Petzval projection lens, it is now widely used when a high speed, well corrected lens is needed to cover a narrow field.

F/3.3 HFOV = 17° Petzval Portrait Lens

radius	thickness	mat'l	index	V-no	sa
55.900	4.700	K3	1.518	59.0	15.1
−43.700	0.800	LF7	1.575	41.5	15.1
460.400	16.800	air			15.1
	16.800	air			13.1
110.600	1.500	LF7	1.575	41.5	15.0
38.900	3.300	air			15.0
48.000	3.600	BK7	1.517	64.2	15.0
−157.800	70.731	air			15.0

EFL = 99.52
BFL = 70.73
NA   = −0.1517 (F/3.3)
GIH = 30.85 (HFOV = 17.22)
PTZ/F = −1.268
VL   = 47.50
OD   infinite conjugate

Figure 19.51

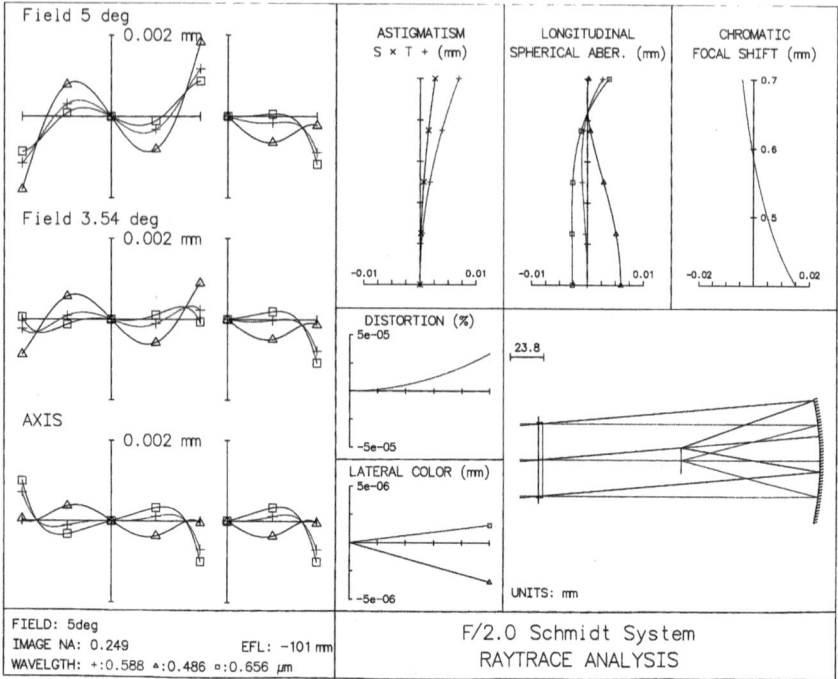

F/2.0 Schmidt System. The Schmidt system/camera/telescope is based on two principles. (1) A spherical mirror with the aperture stop located at the center of curvature is free of coma and astigmatism. (2) An aspheric surface located at the aperture stop affects only spherical aberration; the other Seidel aberrations are unaffected. The fourth-order (and sixth) deformation terms in the aspheric corrector correct the spherical aberration, but the spherical contribution of the corrector varies with its index, producing spherochromatism. The convex radius on the corrector introduces just enough undercorrected chromatic aberration to balance the spherochromatism. The performance is limited by fifth-order oblique spherical, which is apparent in the off-axis ray intercept plots. A wider field is possible, but the oblique spherical varies as $y^3 h^2$, so the image quality continues to degrade (unless the speed is reduced).

F/2.0 ±5° Schmidt System

radius	space	mat'l		sa
9433.1	3.00	BK7	517642	26
AD = −5.9584e-08		AE = −1.5984e-12		
plano	198.04	air		26
−200.00	−99.45	air		43
−100.22	(image surface)			8.8

EFL = 100.56
BFL = 99.45
NA = 0.25 (F/2.0)
GIH = 8.8 (±5°)
PTZ/F = −1.00
VL = 201

Figure 19.52

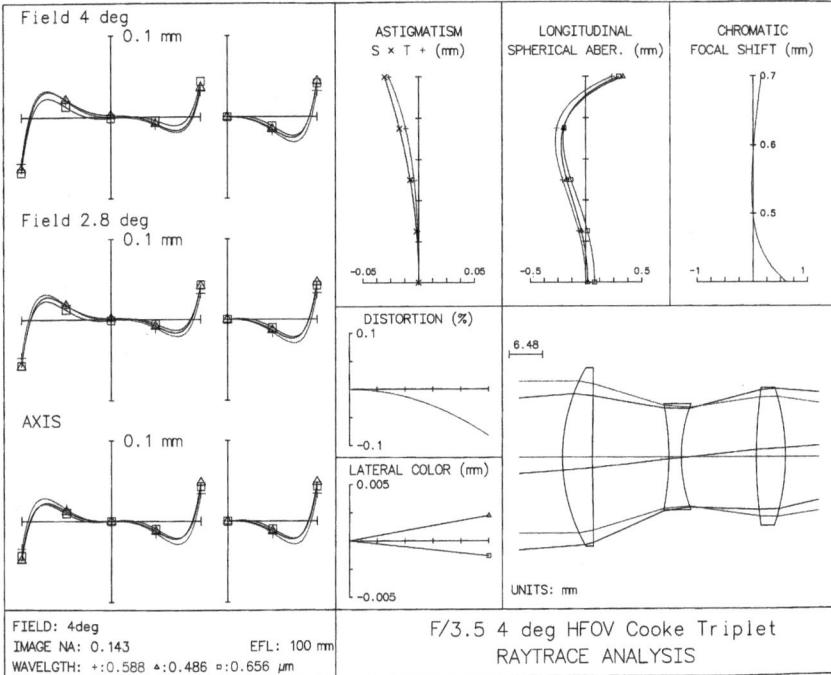

Field 4 deg	ASTIGMATISM S × T + (mm)
	LONGITUDINAL SPHERICAL ABER. (mm)
	CHROMATIC FOCAL SHIFT (mm)

FIELD: 4deg
IMAGE NA: 0.143   EFL: 100 mm
WAVELGTH: +:0.588 ▲:0.486 □:0.656 μm

F/3.5 4 deg HFOV Cooke Triplet
RAYTRACE ANALYSIS

UNITS: mm

F/3.5 Narrow Angle Cooke Triplet. A narrow angle anastigmat is a difficult lens to get out of an automatic lens design program. The program usually wants to turn out a lens with near perfect correction on the axis and very poor imagery only slightly off axis—in other words it wants to design a telescope objective. One way to get out of this trap is to start with a triplet with a normal field angle and reduce the field in small increments, optimizing at each step along the way.

F/3.5 ±4° Cooke Triplet

radius	space	mat'l		sa
33.14	5.83	SSKN5	659509	16.8
623.0	14.69	air		16.8
−52.10	2.47	SF63	748277	10.0
26.90	1.98	air		10.0
stop	12.59	air		9.37
90.43	5.83	SSKN5	659509	13.0
−39.76	76.96			13.0

EFL = 100.00
BFL = 76.96
NA   = 0.143 (F/3.5)
GIH = 6.993 (4.0°)
PTZ/F = −6.29
VL   = 43.39

Figure 19.53

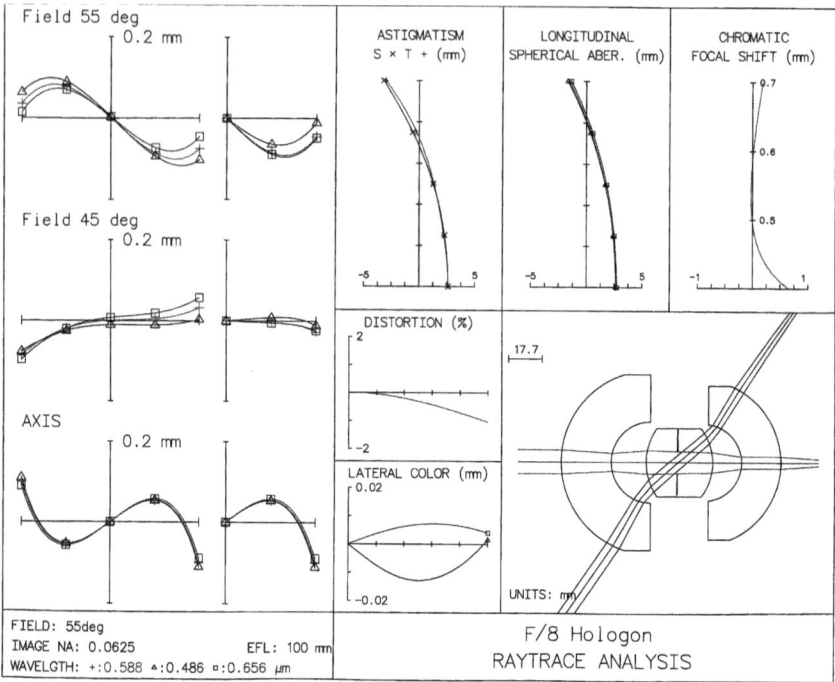

F/8 Hologon
RAYTRACE ANALYSIS

FIELD: 55deg
IMAGE NA: 0.0625   EFL: 100 mm
WAVELGTH: +:0.588 ▲:0.486 □:0.656 µm

F/8.0 Hologon. The Hologon can be regarded as an inverse Cooke triplet, with the crown in the center and the flints as the outer lenses. Its surprisingly good wide angle quality results from its near monocentric construction. The thick elements are necessary to correct the aberrations. The Hologon may be regarded as a descendant of the monocentric water filled F/30 Sutton Panoramic ball lens (1859). In the forties Baker designed an unsymmetrical but monocentric solid ball lens. Both of these lenses had concentric spherical image surfaces, whereas the Hologon is designed to be used with a flat film.

F/8.0 ±55° Hologon

radius	space	mat'l		sa
47.41	26.42	SF4	755274	45.3
20.98	18.63	air		21.0
29.72	17.20	BK7	517642	17.7
stop	18.23	BK7	517642	5.67
−28.38	15.41	air		17.7
−19.78	21.13	SF4	755274	19.7
−42.52	(51.27–2.62)			39.1

EFL   = 100.00
BFL   = 51.27
NA   = 0.0625 (F/8.0)
GIH   = 142.81 ((55()
PTZ/F = (24.6
VL   = 117.02

Figure 19.54

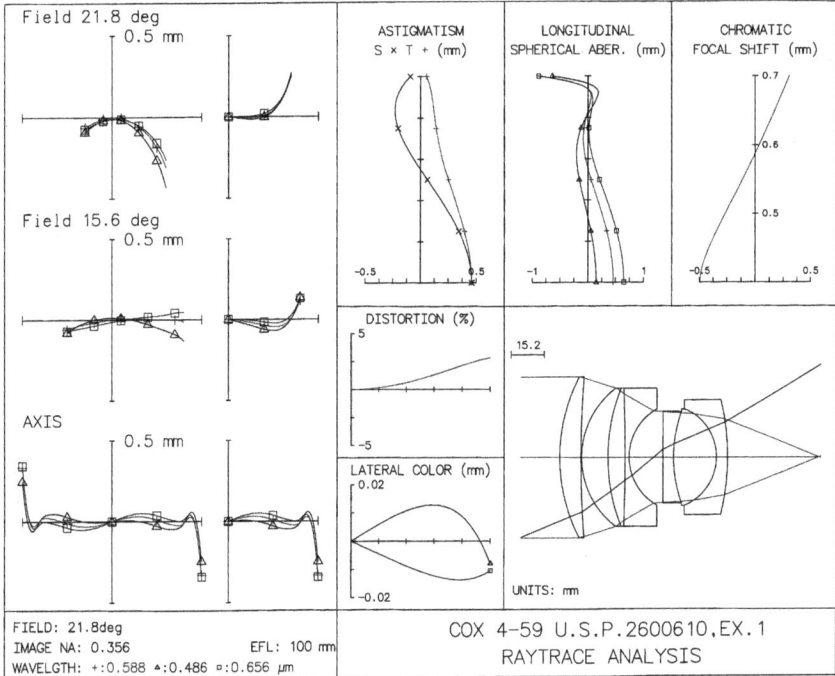

FIELD: 21.8deg	
IMAGE NA: 0.356	EFL: 100 mm
WAVELGTH: +:0.588 ▲:0.486 □:0.656 μm	

COX 4-59 U.S.P.2600610,EX.1
RAYTRACE ANALYSIS

F/1.4 Sonnar. This high speed 7-element version of the Sonnar is one of the many variations on the split-front triplet scheme, designed by Bertele under the names Ernostar and Sonnar for Ernemann and Zeiss in the twenties and thirties. Bertele was very skilled at controlling the high-order aberrations with cemented surfaces of different powers and index breaks. Note the Merte' surface at surface #9.

F/1.4 ±21.8° Sonnar

radius	space	mat'l		sa
75.61	9.65		693430	35.8
450.47	0.30	air		35.8
37.57	12.09	LAKN14	697554	30.5
80.10	7.87	FK5	487704	30.5
−817.22	1.83	SF18	722292	30.5
24.44	11.07	air		21.3
stop	4.06	air		20.08
plane	4.88	K10	501564	20.3
58.38	19.81	BaFN11	667484	20.3
−22.76	5.08	SK2	607566	21.3
−104.42	(43.12–0.46)			25.4

EFL = 100.00
BFL = 43.12
NA   = 0.356 (F/1.4)
GIH = 40.0 (±218°)
PTZ/F = −3.49
VL   = 76.64

Figure 19.55

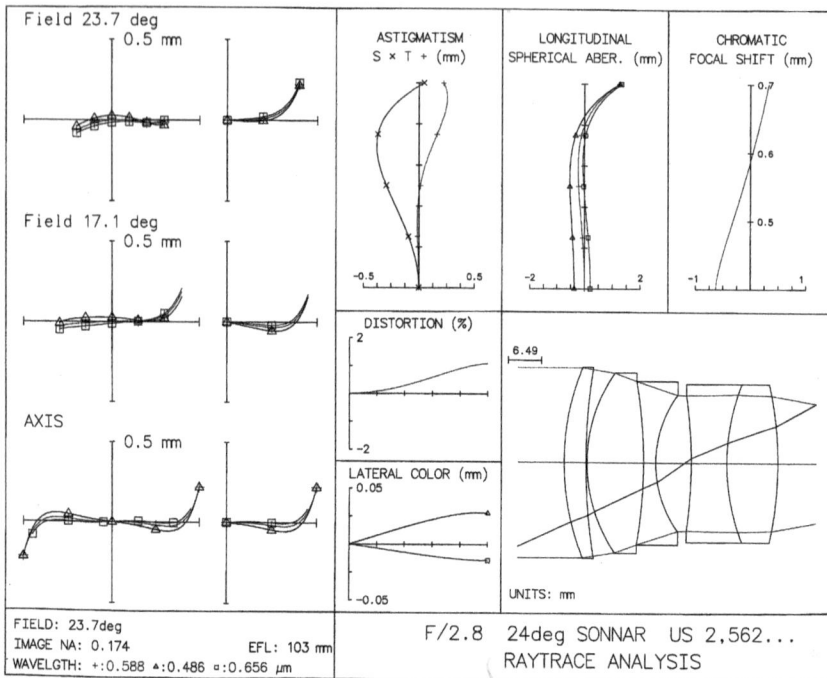

```
Field 23.7 deg
     0.5 mm

Field 17.1 deg
     0.5 mm

AXIS
     0.5 mm
```

ASTIGMATISM
S × T + (mm)

LONGITUDINAL
SPHERICAL ABER. (mm)

CHROMATIC
FOCAL SHIFT (mm)

DISTORTION (%)

LATERAL COLOR (mm)

UNITS: mm

FIELD: 23.7deg
IMAGE NA: 0.174   EFL: 103 mm
WAVELGTH: +:0.588 ▲:0.486 □:0.656 μm

F/2.8   24deg SONNAR   US 2,562...
RAYTRACE ANALYSIS

F/2.9 Five Element Sonnar. This is one of many designs which are derived from the split-front crown Cooke triplet, and were used as objective lenses for early 35 mm cameras. The first three elements resemble the first three elements of the Double Gauss lens.

F/2.8 24° Sonnar US 2,562,012

radius	thickness	mat'l	index	V-no	sa
46.000	4.300	SK9	1.614	55.2	18.0
110.000	0.210	air			18.0
29.300	11.000	BAF51	1.652	44.9	17.1
−92.170	2.360	LAFN7	1.750	34.9	15.5
21.680	5.650	air			13.0
	1.300	air			12.9
−90.000	7.000	K10	1.501	56.4	13.0
39.500	10.000	LAK9	1.691	54.7	15.0
−70.000	71.265	air			15.0

EFL = 103.2
BFL = 71.27
NA   = −0.1718 (F/2.9)
GIH = 45.43 (HFOV = 23.75)
PTZ/F = −4.319
VL   = 41.82
OD   infinite conjugate

Figure 19.56

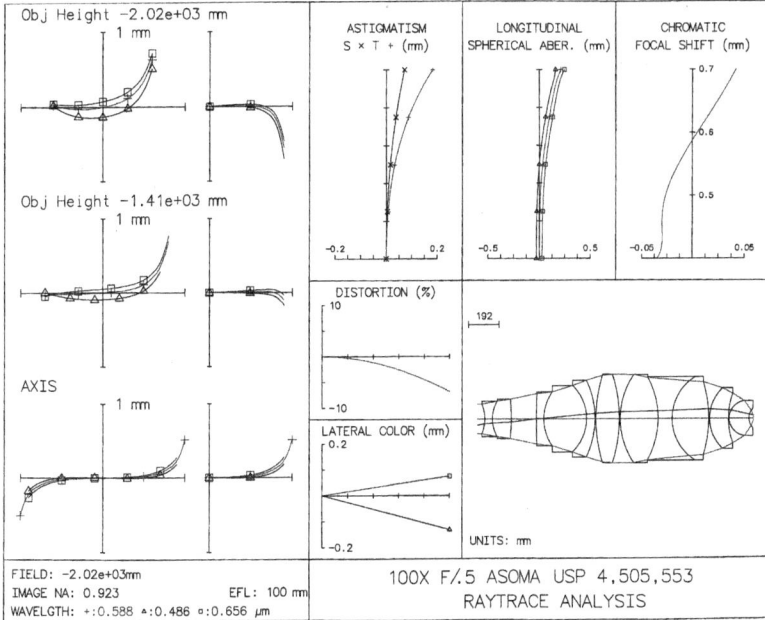

| Obj Height −2.02e+03 mm | ASTIGMATISM S × T + (mm) | LONGITUDINAL SPHERICAL ABER. (mm) | CHROMATIC FOCAL SHIFT (mm) |

DISTORTION (%)

LATERAL COLOR (mm)

192

UNITS: mm

FIELD: −2.02e+03mm
IMAGE NA: 0.923   EFL: 100 mm
WAVELGTH: +:0.588 ▲:0.486 □:0.656 μm

100X F/.5 ASOMA USP 4,505,553
RAYTRACE ANALYSIS

100 × 0.92 NA Microscope Objective. This flat field microscope objective achieves a very flat Petzval surface using two devices. 1) The construction is basically a reverse telephoto, with a negative doublet followed by a nine element positive component. 2) The other Petzval corrector is the concave surface on the "front" element adjacent to the focal plane, which acts as a field flattener. Note the extensive use of calcium fluoride to correct the secondary spectrum. In evaluating the aberrations, remember that in use the focal length (and the aberrations) will be about 1.5% of what we show here.

100 × 0.92 NA Microscope Objective

radius	space	mat'l		sa
object	9852.	air		
−236.24	48.37	H NBFD5	762403	90.5
210.18	117.51	H FDS9	847238	107.9
−1904.8	156.67	air		119.7
4406.6	47.00	O LAL14	697555	153.5
277.80	270.1	CaF2	434954	163.9
−277.85	52.88	O BPM4	613438	202.2
−529.38	5.88	air		233.6
875.81	145.76	CaF2	434954	263.2
−875.66	5.88	air		269.2
473.37	239.13	CaF2	434954	268.7
−502.26	89.65	KzFSN5	654396	250.9
295.15	227.92	O PHM51	617628	254.7
−767.46	5.88	air		223.2
257.02	115.07	BK10	498670	185.6
452.04	4.59	air		152.9
107.34	124.73	SK14	603606	106.9
83.53	33.53			60.0

EFL = 100.44
BFL = 32.5
NA   = −0.923 (F/0.54)
GIH = 20.10 (±11.4°)
PTZ/F = + 59.3
VL   = 1645.

Figure 19.57

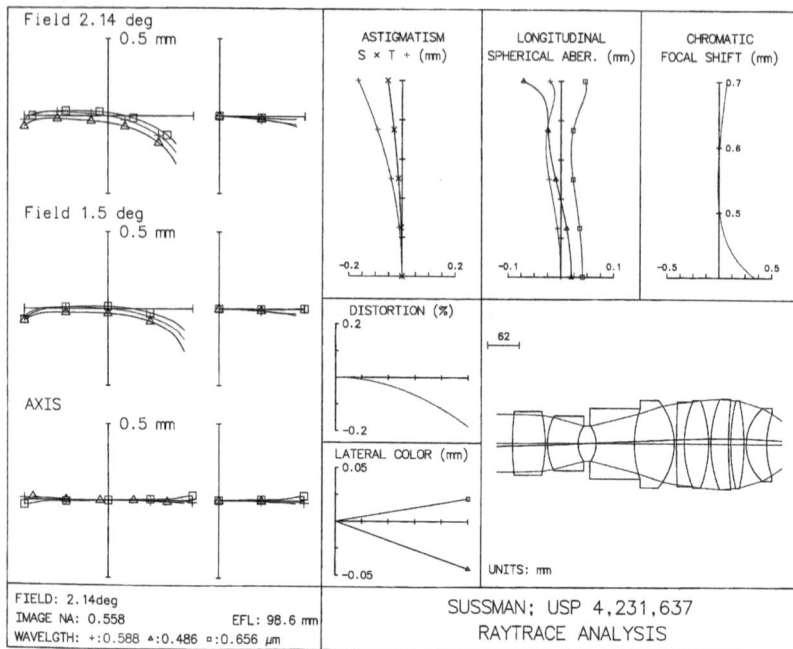

Field 2.14 deg 0.5 mm	ASTIGMATISM S × T + (mm)	LONGITUDINAL SPHERICAL ABER. (mm)	CHROMATIC FOCAL SHIFT (mm)

FIELD: 2.14deg
IMAGE NA: 0.558   EFL: 98.6 mm
WAVELGTH: +:0.588 ▲:0.486 □:0.656 μm

SUSSMAN; USP 4,231,637
RAYTRACE ANALYSIS

0.56 NA Microscope Objective. This microscope objective is corrected for use with its image at infinity. What is effectively a telescope is used to view the image, and allows the insertion of plates and tilted beam splitters between the objective and the telescope. Since that space is collimated, these things do not introduce aberrations. The features of the design which help to control the Petzval are elements two and three, which are thick meniscus, working much like the inner doublets of the double Gauss lens. The other Petzval correcting feature is the last element which is also a thick meniscus. Note the use of CaF2, FK51 and KzFS4, which are materials with unusual partial dispersions which reduce the secondary spectrum.

Sussman; USP 4,231,637

radius	thickness	mat'l	index	V-no	sa		
553.260	64.900	FK51	1.487	84.5	60.6		
−247.644	4.400	air			57.2		
115.162	59.400	LLF2	1.541	47.2	52.1		
57.131	17.600	air			34.0		
	17.600	air			33.6		
−57.646	74.800	SF5	1.673	32.2	36.0		
196.614	77.000	FK51	1.487	84.5	67.0		
−129.243	4.400	air			83.0		
2062.370	15.400	KZFS4	1.613	44.3	77.5		
203.781	48.400	CAF	1.434	94.9	80.5	EFL	= 98.58
−224.003	4.400	air			83.2	BFL	= 96.19
219.864	35.200	CAF	1.434	94.9	86.0	NA	= −0.5658 (F/0.90)
793.300	4.400	air			84.3	GIH	= 3.68
349.260	26.400	FK51	1.487	84.5	83.7	PTZ/F	= 44.28
−401.950	4.400	air			82.7	VL	= 498.30
91.992	39.600	SK11	1.564	60.8	70.0	OD	infinite conjugate
176.000	96.189	air			59.0		

Figure 19.58

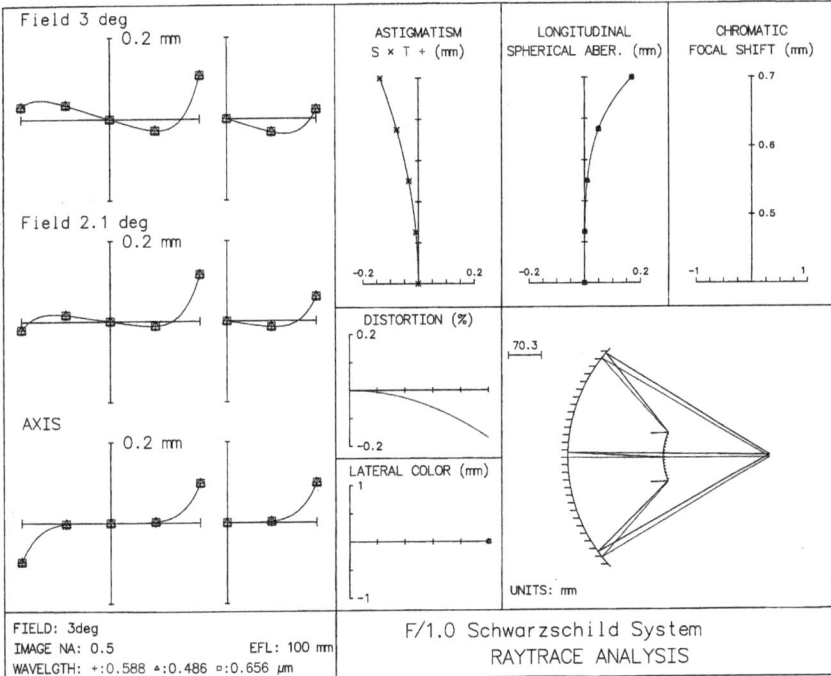

Field 3 deg 0.2 mm	ASTIGMATISM S × T + (mm)
Field 2.1 deg 0.2 mm	DISTORTION (%)
AXIS 0.2 mm	LATERAL COLOR (mm)

FIELD: 3deg
IMAGE NA: 0.5   EFL: 100 mm
WAVELGTH: +:0.588 ▲:0.486 □:0.656 μm

F/1.0 Schwarzschild System
RAYTRACE ANALYSIS

F/1.0 Schwarzschild System. This is the third-order corrected version of the all-sphere Schwarzschild objective. The equations to which this example is designed are in Sec. 17.2 Eqs. 17.6 to 17.11 (personal communication, Max Riedl). The system is monocentric, so coma and astigmatism are zero. The image surface is a sphere, concentric, and concave to the incoming light. The third-order spherical is also zero, but, as can be seen from the ray intercept plots, at F/1.0 the high-order spherical aberration is quite large. Since it is basically fifth-order, at a slower speed it might well be acceptable.

F/1.0, ±3° Schwarzschild System

radius	space	mat'l	sa
123.61	−200.02	mirror	50
323.62	423.62	mirror	222.3

EFL $-$ 100
BFL = 423.62
NA   = 0.50 (F/1.0)
GIH = 5.24 (±3°)
PTZ/F = −1.0
VL   = 423.6

Figure 19.59

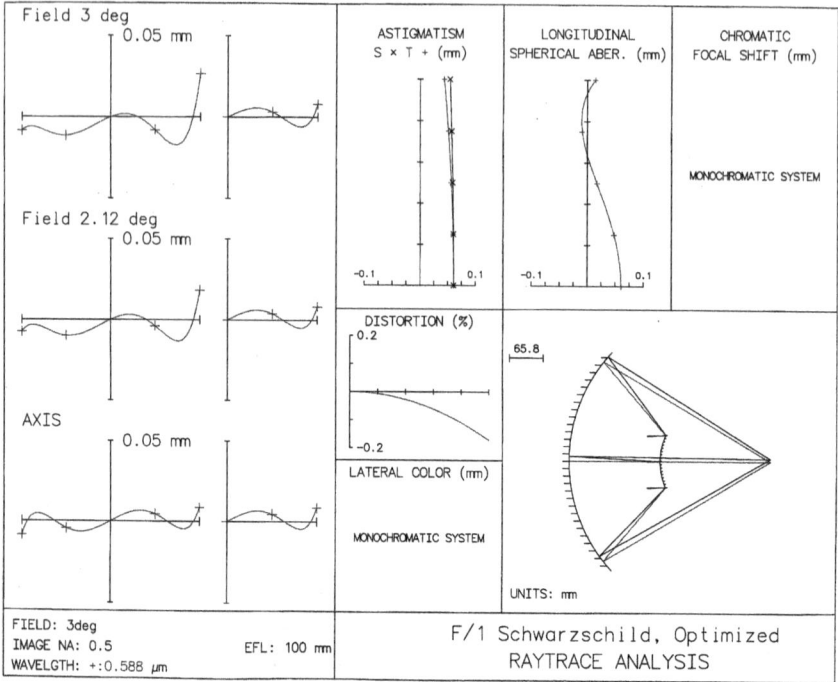

| Field 3 deg | ASTIGMATISM S × T + (mm) | LONGITUDINAL SPHERICAL ABER. (mm) | CHROMATIC FOCAL SHIFT (mm) |

FIELD: 3deg
IMAGE NA: 0.5
WAVELGTH: +:0.588 μm
EFL: 100 mm

F/1 Schwarzschild, Optimized
RAYTRACE ANALYSIS

F/1.0 Schwarzschild System, Optimized. The radii and spacing of the third-order Schwarzschild objective are optimized in this example, and a curved focal surface is added to compensate for the curved field of the basis system. The surfaces are spherical and there is an apparent residual of fifth-order spherical and coma, which could presumably be eliminated with a sixth-order deformation on one or both surfaces.

F/1.0, ±3° Schwarzschild System, Optimized

radius	space	mat'l	sa
120.12	−178.59	mirror	50
298.21	397.35	mirror	208
−101.54			
(image surf.)			

EFL = 100.00
BFL = 397.35
NA   = 0.50 (F/1.0)
GIH = 5.24 (±3°)
PTZ/F = −1.006
VL   = 397.35

Figure 19.60

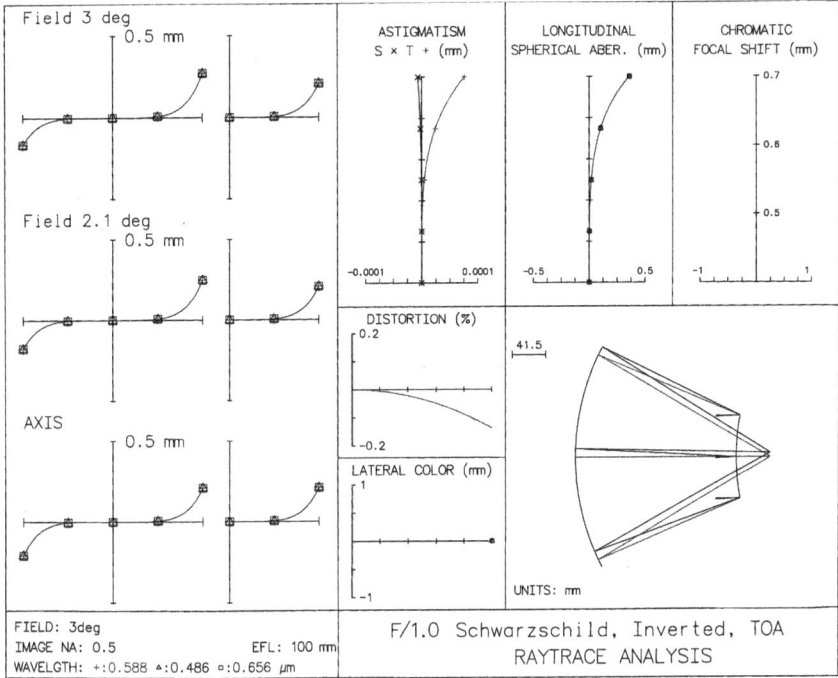

FIELD: 3deg	
IMAGE NA: 0.5 EFL: 100 mm	F/1.0 Schwarzschild, Inverted, TOA
WAVELGTH: +:0.588 ▲:0.486 □:0.656 μm	RAYTRACE ANALYSIS

F/1.0 Flat Field Schwarzschild, TOA. With two conic surfaces this third-order design (personal communication w/Max Riedl) is free of spherical, coma, and has a flat Petzval surface. As can be seen in the ray intercept plots, the higher-order spherical and astigmatism are severe, and a slower speed would be appropriate if the design were not optimized.

F/1.0 ±3° Flat Field Third-Order Schwarzschild

radius	space	mat'l		sa
282.84	−199.99	mirror	50	cc = 5.8284
282.84	241.42	mirror	131	cc = 0.17157

EFL = 100.00
BFL = 241.42
NA   = 0.5 (F/1.0)
GIH = 5.24 (±3°)
PTZ/F = infinity
VL   = 241.42

Figure 19.61

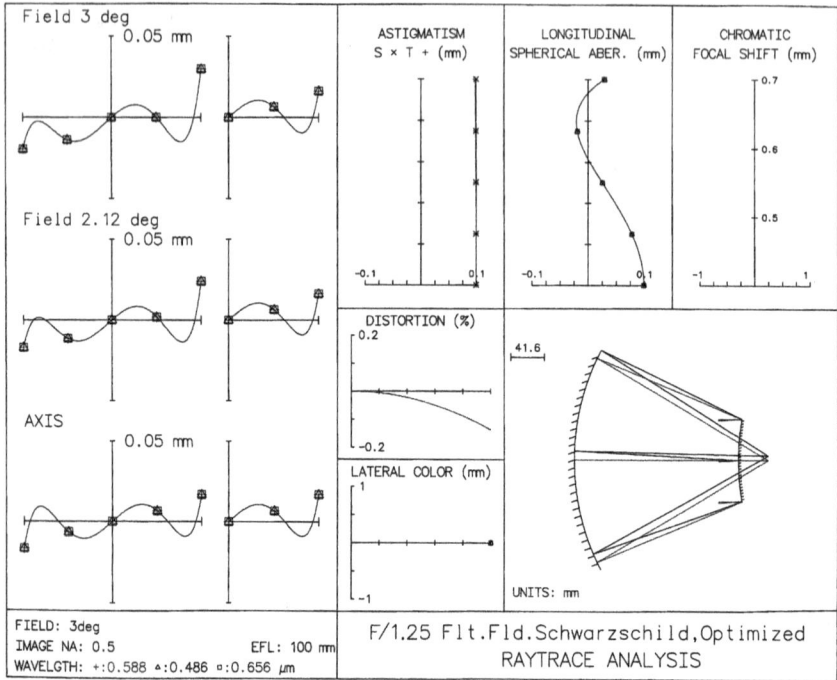

F/1.25 Flat Field Schwarzschild, Optimized. This design was optimized from the preceding third-order version in Fig. 19.61; the radii, the spacing, and the conic constants have been adjusted to balance the fifth-order spherical. As with the optimized concentric Schwarzschild, the fifth-order spherical is the dominant residual.

F/1.0, ±3° Flat Field Schwarzschild, Optimized

radius	space	mat'l		sa
287.85	−204.37	mirror	50	cc = 5.97173
285.58	242.00*	mirror	131	cc = 0.17577
	*defocus −0.1			

EFL = 100.00
BFL = 242.00
NA   = 0.50 (F/1.00)
GIH = 5.24 (±3°)
PTZ/F = ±181.
VL   = 242.0

Figure 19.62

Bibliography

Betensky, E., M. Kreitzer, and J. Moskovich, "Camera Lenses," in *OSA Handbook of Optics,* Vol. 2, New York, McGraw-Hill, 1995, Chap. 16.

Jones, L., "Reflective and Catadioptric Objectives," in *OSA Handbook of Optics,* Vol. 2, New York, McGraw-Hill, 1995, Chap. 18.

Smith, W. J., *Modern Lens Design*, 2d ed., New York, McGraw-Hill, 2005.

20

The Practice of Optical Engineering

This chapter will briefly survey the factors involved in reducing an optical system to practice. A short description of the optical manufacturing process will be followed by a discussion of the specification and tolerancing of optics for the shop. The mounting of optical elements will be considered next, and the chapter will be concluded with a section on optical laboratory measurement techniques.

20.1 Optical Manufacture

Materials. The starting point for quantity production of optics is most frequently a rough molded glass blank or pressing. This is made by heating a weighed chunk of glass to a plastic state and pressing it to the desired shape in a metal mold. The blank is made larger than the finished element to allow for the material which will be removed in processing; the amount removed must (at a minimum) be sufficient to clean up the outer layers of the blank which are of low quality and may contain flaws or the powdery fireclay used in molding. Typically a lens blank will be about 3-mm thicker than the finished lens and 2-mm larger in diameter. A prism blank will be large enough to allow removal of about 2 mm on each surface. These allowances vary with the size of the piece and are less for a clean blank. When the blank is of an expensive material, such as silicon or one of the more exotic glasses, the blanking allowances are held to the absolute minimum to conserve material.

Although most blanks are single, a cluster form is frequently economical for small elements. A cluster may consist of five or ten blanks connected by a thin web which is ground off to free the individual blanks. If molded blanks are unobtainable, either because of the small quantity involved or the type of material, a rough blank may be prepared by chipping or sawing a suitable shape from stock material.

Rough blanks can be checked fairly satisfactorily for the presence of strain (which results from poor annealing of the glass) by the use of a polariscope. An accurate check of the index requires that a plano surface be polished on a sample piece; however, if a batch of blanks is known to have been made from a single melt or run of glass, only one or two of the blanks need be checked, because the index within a melt is quite consistent. Since the final annealing process raises the index, the presence of strain is frequently accompanied by a low index value.

When the shape of a blank is such that there are large variations in thickness from center to edge, it is difficult to get a uniform anneal. A variation of index within the blank may result. This is especially true for certain of the exotic optical glasses which are difficult to anneal. Glass in slab form is easier to anneal uniformly and is thus more homogeneous; it is often required for especially critical lenses for this reason.

Rough shaping. The preliminary shaping of an element is often accomplished by using diamond-charged grinding wheels. In the case of spherical surfaces, the process involved is *generating*. The blank is rotated in a vacuum chuck and is ground by a rotating annular diamond wheel the axis of which is at an angle to the chuck axis, as indicated in Fig. 20.1. The geometry of this arrangement is such that a sphere is generated; the radius is determined by the angle between the two axes

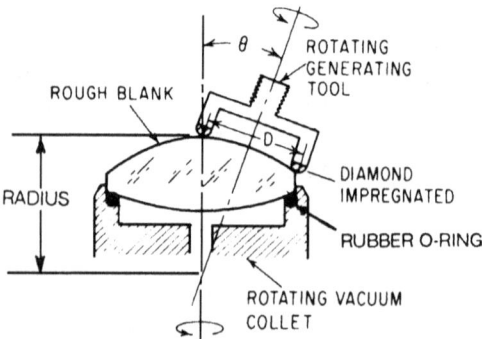

Figure 20.1 Schematic diagram of the generating process. The annular diamond tool and the glass blank are both rotated. Since their axes intersect at an angle (θ), the surface of the blank is generated to a sphere of radius $R = D/2 \sin \theta$.

and by the effective diameter of the diamond tool (which will usually overhang the edge of the lens). The thickness is, of course, governed by how far the work is advanced into the tool. Flat work can be roughed out in a similar manner, with the two axes parallel. Rectangular shapes can be formed by milling, again using diamond tools.

Blocking. It is customary to process optical elements in multiples by fastening or blocking a suitable number on a common support. There are two primary reasons for this: The obvious reason is economy, in that several elements are processed simultaneously; the less apparent reason is that a better surface results when the processing is averaged over the larger area represented by a number of pieces.

The elements are fastened to the blocking tool with pitch, although various compounds of waxes and rosins are also used for special purposes. Pitch has the useful property of adhering tenaciously to almost anything which is hot and not sticking to cold surfaces. The pitch bond is readily broken by chilling the pitch to a brittle state and shocking it with a brisk but light tap. Typically the elements are fastened to the blocker by pitch buttons which are molded to the back of the elements (suitably warmed); the buttons are then stuck to the heated blocker, as indicated in Fig. 20.2. (The surfaces of the elements are maintained in alignment by placing the buttoned elements into a lay-in tool of the proper radius and then pressing the heated blocker into contact with the pitch buttons.)

The cost of processing an element is obviously closely related to the number of elements which can be blocked on a tool. There is no

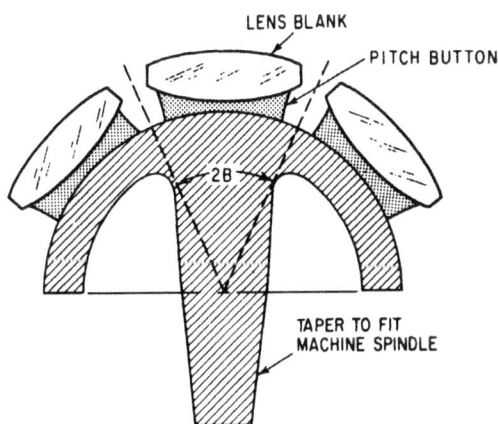

Figure 20.2 Section of a blocking tool with blanks fastened in place with buttons of blocking pitch. The maximum number of lenses that can be blocked on a tool is determined by the angle B (see Eq. 20.2).

simple way to determine this number exactly; however, the following expressions (which are "limiting-case" expressions, modified to fit the actual values) are accurate to within about one element per tool.

$$\text{No. per tool} = \frac{3}{4}\left(\frac{D_t}{d}\right)^2 - \frac{1}{2} \qquad (20.1)$$

rounded downward to the nearest integer, where D_t is the diameter of the blocking tool and d is the effective diameter of the piece (and should include an allowance for clearance between the elements).

For spherical surfaces:

$$\text{No. per tool} = \frac{6R^2}{d^2}\left[\frac{\text{SH}}{R}\right] - \frac{1}{2} \qquad (20.2a)$$

where R is the surface radius, d is the lens diameter (including a clearance allowance), and SH is the sagittal height of the tool. For a tool which subtends 180°, SH $= R$ and this reduces to

$$\text{No. per tool} = \frac{6R^2}{d^2} - \frac{1}{2} = \frac{1.5}{(\sin B)^2} - \frac{1}{2} \qquad (20.2b)$$

rounded downward to the nearest integer, where B is the half-angle subtended by the lens diameter (plus spacing allowance) from the center of curvature of the surface, as indicated in Fig. 20.2.

Where there are only a few lenses per tool, Table 20.1 is convenient and more accurate for 180° tools.

Grinding. The surface of the element is further refined by a series of grinding operations, performed with loose abrasive in a water slurry and cast iron grinding tools. If the elements have not been generated, the grinding process begins with a coarse, fast-cutting emery. Otherwise,

TABLE 20.1   The Number of Blanks Which can be Blocked on a 180°
Tool is Determined by the Angle Subtended by the Blank Diameter from
the Center of Curvature of the Radius as Shown in Fig. 20.2.

No. per tool	Maximum d/D_t	Maximum sin B	$2B$
2	0.500	0.707	90°
3	0.462	0.655	81.79°
4	0.412	0.577	70.53°
5	0.372	0.507	60.89°
6	—	0.500	60°
7	0.332	0.447	53.13°
8	0.301	0.398	46.91°
9	0.276	0.383	45°
10	—	0.369	43.24°
11	0.253	0.358	41.88°
12	0.243	0.346	40.24°

it begins with a medium grade and proceeds to a very fine grade which imparts a smooth velvety surface to the glass.

The grinding (and polishing) of a spherical surface is accomplished to a high degree of precision with relatively crude equipment by taking advantage of a unique property of a spherical surface, namely, that a concave sphere and a convex sphere of the same radius will contact each other intimately regardless of their relative orientations. Thus, if two mating surfaces which are approximately spherical are contacted (with abrasive between them) and randomly moved with respect to each other, the general tendency is for both surfaces to wear away at their high spots and to approach a true spherical surface as they wear. (For a detailed analytical treatment of the subject of relative wear in optical processing, the reader is strongly urged to consult the reference by Deve, listed at the end of this chapter.)

Usually the convex piece (either blocker or grinding tool) is mounted in a power-driven spindle and the concave piece is placed on top as shown in Fig. 20.3. The upper tool is constrained only by a ball pin-and-socket arrangement and is free to rotate as driven by its sliding contact with the lower piece; it tends to assume the same angular rate of rotation as the lower piece. The pin is oscillated back and forth so that the relationship between the two tools is continuously varied. By adjusting the offset and amplitude of the motion of the pin, the optician can modify the pattern

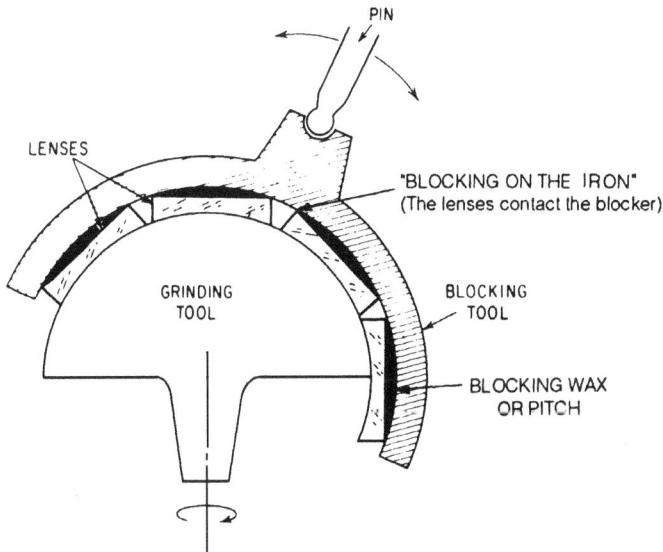

Figure 20.3   In grinding (or polishing), a semirandom scrubbing action is set up by the rotation of the lower (convex) tool about its axis and the back-and-forth oscillation of the upper (concave) tool. Note that the upper tool is free to rotate about the ball end of the driving pin and takes on a rotation induced by the lower tool.

of wear on the glass and thus effect minute corrections to the value and uniformity of the radius generated by the process.

Each successively finer grade of emery is used until the grinding pits left by the preceding operation are ground out. The abrasives used include garnet, carborundum, sapphire, and diamond powder.

Polishing. The mechanics of the polishing process are quite analogous to the grinding process. However, the polishing tool is lined with a layer of pitch and the polishing compound is a slurry of water and rouge (iron oxide) or cerium oxide. The polishing pitch will cold flow and thus take on the shape of the work in a very short time.

The polishing process is a peculiar one that is still incompletely understood. It appears that the surface of the glass is hydrolyzed by the polishing slurry and the resulting gel layer is scraped away by the particles of polishing compound embedded in the polishing pitch. This analysis explains many of the phenomena associated with polishing, such as scratches and cracks which are "flowed" shut by polishing, but which later open up when heated or exposed to the atmosphere. But when one considers that historically, polishing tools have been made from materials as diverse as felt, lead, taffeta, leather, wood, copper, and cork, and that polishing compounds other than rouge have been successfully used, and that many optical materials (e.g., silicon, germanium, aluminum, nickel, and crystals) have a different chemistry than glass, it would seem that a variety of polishing mechanisms is quite likely. Some polishing agents are actually etchants of the material that they polish; some materials can be polished dry.

Polishing is continued until the surface is free of any grinding pits or scratches. The accuracy of the radius is checked by the use of a test plate (or test glass). This is a very precisely made master gage which has been polished to an exact radius and which is a true sphere to within a tiny fraction of a wavelength. The test plate is placed in contact with the work, and the difference in shape is determined by the appearance of the interference fringes (Newton's rings) formed between the two. The relative curvatures of the two surfaces can be determined by noting whether the gage contacts the work at the edge or the center. If the number of rings is counted, the difference between the two radii can be closely approximated from the formula

$$\Delta R \approx N\lambda \left(\frac{2R}{d} \right)^2 \tag{20.3a}$$

$$N = \frac{\Delta R D^2}{4\lambda R^2} = \frac{\Delta C D^2}{4\lambda} \tag{20.3b}$$

where ΔR is the radius difference, N is the number of fringes, λ is the wavelength of the illumination, R is the radius of the test plate, C is the curvature ($1/R$), and d is the diameter over which the measurement is made. One fringe indicates a change of one-half wavelength in the spacing between the two surfaces. A noncircular fringe pattern is an indication of an aspheric surface. An elliptical fringe indicates a toroidal surface.

Small corrections are made either by adjustment of the stroke of the polishing machine or by scraping away portions of the polishing tool so that the wear is concentrated on the portion of the work which is too high.

Centering. After both surfaces of an element are polished, the lens is centered. This is done by grinding the rim of the lens so that the mechanical axis (defined by the ground edge of the lens) coincides with the optical axis, which is the line between the centers of curvature of the two surfaces. In *visual* centering the element is fastened (with wax or pitch) to an accurately trued tubular tool on a rotating spindle. When the lens is pressed on the tool, the surface against the tool is automatically aligned with the tool and hence with the axis of rotation. While the pitch is still soft, the operator slides the lens laterally until the outer surface also runs true. If the lens is rotated slowly, any decentration of either surface is detectable as a movement of the reflected image (of a nearby target) formed by that surface, as indicated in Fig. 20.4. For high-precision work, the images may be viewed with a telescope or microscope to increase the operator's sensitivity to the image motion. The periphery of the lens is then ground to the desired diameter with a diamond-charged wheel. Bevels or protective chamfers are usually ground at this time.

For economical production of moderately precise optics, a mechanical centering process is used. In this method, called "cup" or "bell" centering, the lens element is gripped between two accurately trued tubular tools. The pressure of the tools causes the lens to slip sideways until the distance between the tools is at a minimum, thus centering

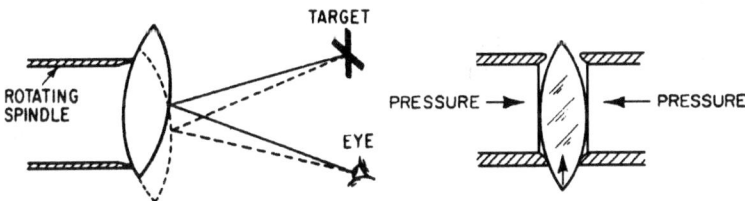

Figure 20.4 Left: In visual centering the lens is shifted laterally until no motion of the image of a target reflected from the lens surface can be detected as the lens is rotated. Right: In mechanical centering the lens is pressed between hollow cylinders. It slides laterally until its axis coincides with the common axis of the two tools.

the lens. A lubricant such as STP can improve the centering. The lens is then rotated against a diamond wheel to grind the diameter to size.

The manufacture of the lens is completed by low-reflection coating the surfaces as required and by cementing, if the element is part of a compound component; these processes are outlined in Chap. 11.

Modifications of the standard processing techniques are sometimes required for unusual materials. Brittle materials (e.g., calcium fluoride) must be treated gently, especially in grinding. A finer, softer abrasive is required; sometimes soap is added to the abrasive and soft brass grinding tools are used in place of cast iron. At the other extreme, sapphire (Al_2O_3) cannot be processed with ordinary materials because of its extreme hardness, and diamond powder is used for both grinding and polishing.

Materials which are subject to attack by the grinding or polishing slurry are sometimes processed using a saturated solution of the optical material in the liquid of the slurry. For example, if a glass is attacked by water, one could make up the slurry with water in which a powder of the glass has been boiled or soaked for several days. Alternately a slurry of kerosene or oil sometimes works well. Other liquids which have been used in slurries include ethylene glycol, glycerol, and triacetate.

High-speed processing.   For optics where the surface accuracy requirements are not high, the processes described above can be materially accelerated. Ordinary grinding usually takes tens of minutes. Polishing may take from 1 or 2 hours up to 8 or 10 hours in difficult cases. These operations can be speeded up by increasing both the speed of the spindle rotation and the pressure between work and tool. Tool wear and deformation are then a problem, so tools which are very resistant to change are used. Grinding is accomplished using tools faced with pellets or pads of diamond particles sintered in a metal matrix; loose abrasive is not used. This is called pellet grinding or pel-grinding. Polishing is done with a metal (typically aluminum) tool faced with a thin (0.01 to 0.02 in) layer of plastic (e.g., polyurethane). Processing times are to the order of minutes; a surface may be generated, ground, and polished in 5 or 10 minutes. Since the tools are not compliant, it is necessary that the radius from the generator have an exact relationship to the radius of the diamond pellet grinding tool, and that the ground radius match (to within a few fringes, à la Eq. 20.3) the radius required by the hard plastic polishing tool. This process is widely used for sunglasses, filters, inexpensive camera lenses, eyepieces and the like. The surface geometry tends to be marginal as regards accuracy of figure, and the fixed-abrasive grinding does cause some subsurface fracturing, but the process is fast and economical. The tooling required and the fine-tuning adjustments of the steps of the process limit its application to large-quantity production.

Other techniques.   There are several other processes which can be used to fabricate optics. While they tend to be more suitable for low precision work, some have been developed to the point that they are used to fabricate diffraction limited optics.

Sagging is usually a relatively low precision approach. A polished plane parallel plate is placed over a mold (often aspheric) and the glass is heated until it sags into the mold. The surface which does not contact the mold is the one that is used. Large mirrors with deep sags can be made this way. In many cases the contact surface is ground and polished to make a transmitting element. Schmidt corrector plates have been made this way.

Molding ranges in quality from surfaces suitable for condensers to molded glass or plastic elements which are diffraction limited, such as laser disk reading lenses. Many camera lenses incorporate molded plastic elements. Almost all inexpensive or disposable cameras have plastic lenses. Molded plastic (and glass) aspheric elements are widely used in lenses with quality levels ranging from high speed TV projection lenses to top of the line zoom and camera lenses. Small elements are especially suited for molding.

Replication is another version of molding. A negative master is made and the substrate is machined or ground and polished to a form very close to the desired shape. The master is coated with a parting agent and any required thin film interference coatings. The master and substrate are pressed together with a few drops of low shrinkage epoxy between them. When the epoxy is set the master is removed and the substrate has an epoxy surface defined by the shape of the master. The epoxy layer is about 0.001 in thick to avoid shrinkage problems. The substrate can be glass, Pyrex, fused quartz, or a very well stabilized metal such as aluminum. Most useful as a mirror fabrication system, replication can produce aspheric surface integral with structural members, and on surfaces which would be impossible to access for regular grinding and polishing. Flat mirrors on aluminum substrates up to about 18 in in size, and aspherics 8 in in diameter on 0.06 in thick aluminum backings have been made.

Nonspherical surfaces.   Aspherics, cylinders, and toroids do not share the universality of the spherical surface, and their manufacture is difficult. While a sphere is readily generated by a random grinding and polishing (because any line through the center is an axis), optical aspherics have only one axis of symmetry. Thus the simple principle of random scrubbing which generates a sphere must be replaced by other means. An ordinary spherical optical surface is a true sphere to within a few millionths of an inch. For aspherics this precision can only be obtained by a combination of exacting measurement and skilled "hand correction" or its equivalent.

Cylindrical surfaces of moderate radius can be generated by working the piece between centers (i.e., on a lathe). However, any irregularity in the process tends to produce grooves or rings in the surface. This can be counteracted by increasing the rate of working *along* the axis relative to the rate of rotation *about* the axis. It is difficult to avoid a small amount of taper (i.e., a conical surface) in working cylinders. Large-radius cylinders are difficult to swing between centers and are usually handled with an x-y rocking mechanism which constrains the axes of work and tool to parallelism so as to avoid a saddle surface.

Aspherics of rotation, such as paraboloids, ellipsoids, and the like, can be made in modest production quantities if the precision required of the surface is of a relatively low order, as, for example, in an eyepiece. The usual technique is to use a cam-guided grinding rig (with a diamond wheel) to generate the surface as precisely as possible. The problem is then to fine-grind and polish the surface without destroying its basic shape. The difficulty is that any random motion which works the surface uniformly tends to change the surface contour toward a spherical form. Extremely flexible tools which can follow the surface contour are required; however, their very flexibility tends to defeat their purpose, which is to smooth or average out small local irregularities left in the surface by the generating process. Pneumatic (i.e., air-filled, elastic) or spongy tools have proved quite successful for this purpose.

Where precise aspherics are required, "hand" or "differential" correction is practically a necessity. The surface is ground and polished as accurately as possible and is then measured. The measurement technique must be precise enough to detect and quantify the errors. For high-quality work, this means that the measurements must be able to indicate surface distortions of a fraction of a wavelength. The Foucault knife edge test and the Ronchi grating tests are widely used for this purpose; these tests can usually be applied directly to the aspheric surface, although there are many aspheric applications (e.g., the Schmidt corrector plate) where the test must be applied to the complete system to determine the errors in the aspheric.

When the surface is close to the required figure, it can be tested with an interferometer, just as a spherical surface on a lens is tested with a test plate (which is of course a simple interferometer). However, for a nonspherical surface some sort of arrangement is necessary to reshape the wave front reflected from the aspheric so that it matches the reference wave front of the interferometer. For a conic surface, auxiliary mirrors can be arranged so that the conic is imaging from focal point to focal point, and a perfect conic will then produce a perfectly spherical wave front. A more generally applicable approach is the use of a *null lens,* which is designed and very carefully constructed to distort the reflected wave front into an exactly spherical shape. For a paraboloid tested at its center of curvature, the null lens can be as simple as one or two plano-convex lenses whose undercorrected spherical cancels the overcorrection of the paraboloid. For general aspherics, the null lens may need to be quite complex.

When the surface errors have been measured and located, the surface is corrected by polishing away the areas which are too high. This can be accomplished (with a full-size polisher and a very short stroke) by scraping away those areas of the polisher which correspond to the low areas of the surface. In making a paraboloid of low aperture, such as used in a small astronomical telescope, the surface is close enough to a sphere that the correction can often be effected simply by modifying the stroke of the polisher. However, for large work and for difficult aspherics, it is usually better to use small or ring (annular) tools and to wear down the high zones by a direct attack. A certain amount of delicacy and finesse is required for this approach; if the process is continued for a minute or so longer than required, the result is a depressed ring which then requires that the entire balance of the surface be worn down to match this new low point.

A few companies have developed equipment which more or less automates this process. In one technique, a *computer-controlled polisher* uses a small polishing tool (or a tool consisting of three small tools which are driven to spin about their centroid) which is directed to dwell on the regions of the work which are high and need to be polished down. The location and dwell time are determined from interferograms of the surface, plus a knowledge of the wear pattern which the polishing tool produces. The use of a small, driven polisher means that the device is not limited to polishing annular zones on the work, and thus unsymmetrical surface errors can be efficiently corrected.

Another computer-controlled process is called *magnetorheologic* polishing. Here the polishing slurry includes a magnetic iron compound. The slurry is moved past the rotating lens, and at the lens a magnetic field causes the slurry to become stiff. This produces a localized polishing (or wearing) action on the surface. By rocking, spinning, and advancing the lens into the moving slurry under computer control, the surface can be locally polished to achieve the desired surface figure. Again, an unsymmetrical figure error can be corrected by synchronizing the localized polishing action with the position of the lens.

Single-point diamond turning. Extremely accurate, numerically controlled lathes and milling machines are now available which are capable of generating both the finish and the precise geometry required for an optical surface. The cutting tool used is a single-crystal diamond, and the optic is machined as in a lathe or as with a fly-cutter in a mill. A single-point machining operation leaves tool marks—the finished surface is scalloped, and in some respects resembles a diffraction grating. This is one limitation of the process, and finished surfaces are often lightly "postpolished" to smooth out the turning marks. The more severe limitation is that only a few materials are suitable for machining, and unfortunately, optical glass is not one of them. However, several useful materials are turnable, including copper, nickel, aluminum, silicon, germanium, zinc selenide and sulfide, and, of course, plastics. Thus mirrors and infrared optics can be fabricated this way. Infrared optics do not require the same level of precision as do visual-wavelength optics, simply because a quarter-wave is almost 20 times larger at 10 μm than in the visible wavelengths. With this process an aspheric surface is just about as easy to make as a spherical surface. It has found significant acceptance in the infrared and military applications.

20.2   Optical Specifications and Tolerances

Many otherwise fully competent optical workers come to grief when it is necessary for them to send their designs to the shop for fabrication. The two most common difficulties are underspecification, in the sense of incompletely describing what is required, and overspecification, wherein tolerances are established which are much more severe than necessary.

Optical manufacture is an unusual process. If enough time and money are available, almost any degree of precision (that can be measured) can be attained. Thus, specifications must be determined on a dual basis: (1) the limits which are determined by the performance requirements of the system, and (2) the expenditure of time and money which is justified by the application. Note well that optical tolerances which represent an equal level of difficulty to maintain may vary widely in magnitude. For example, it is not difficult to control the sphericity of a surface to one-tenth of a micrometer; the comparable (in terms of difficulty) tolerance for thickness is about 100 μm (0.1 mm), three orders of magnitude larger. For this reason it is rare to find "box" tolerances in optical work; each dimension, or at least each class of dimension, is individually toleranced.

Every essential characteristic of an optical part should be spelled out in a clear and unambiguous way. Optical shops are accustomed to this, and if a specification is incomplete, either time must be wasted in questioning the specification to determine what the requirements are, or the shop must arbitrarily establish a tolerance. Either procedure is undesirable.

The following paragraphs are an attempt to provide a general guide to the specification of optics. The discussion will include the basis for the establishment of tolerances, the conventional methods of specifying desired characteristics, and an indication of what tolerances a typical shop may be expected to deliver.

The intelligent choice of specifications and tolerances for optical fabrications is an extremely profitable endeavor. The guiding philosophy in establishing tolerances should be to allow as large a tolerance as the requirement for satisfactory performance of the optical system will permit. Designs should be established with the aim of minimizing the effect produced by production variations of dimensions. Frequently, simple changes in mounting arrangements can be made which will materially reduce fabrication costs without detriment to the performance of a system. One should also be certain that the tightly specified dimensions of a system are the truly critical dimensions, so that time and money are not wasted in adhering to meaningless demands for accuracy.

Surface quality. The two major characteristics of an optical surface are its quality and its accuracy. *Accuracy* refers to the dimensional characteristics of a surface, i.e., the value and uniformity of the radius. *Quality* refers to the finish of the surface, and includes such defects as pits, scratches, incomplete or "gray" polish, stains, and the like. Quality is usually extended to similar defects within the element, such as bubbles or inclusions. In general (with the exception of incomplete polish which is almost never acceptable) these factors are merely cosmetic or "beauty defects" and may be treated as such. The percentage of light absorbed or scattered by such defects is usually a completely negligible fraction of the total radiation passing through the system. However, if the surface is in or near a focal plane, then the size of the defect must be considered relative to the size of the detail it may obscure in the image. Also, if a system is *especially* sensitive to stray radiation, such defects may assume a functional importance. In any case, one may evaluate the effect of a defect by comparing its area with that of the system clear aperture at the surface in question.

The standards of military specification MIL-O-13830 (now formally obsolete) and ISO 10110 are widely utilized in industry. The surface quality is specified by a number such as 80–50, in which the first two digits relate to the *apparent* width of a tolerable scratch and the second two indicate the diameter of a permissible dig, pit, or bubble in hundredths of a millimeter. Thus, a surface specification of 80–50 would permit a scratch of an *apparent* width which matched (by visual comparison) a #80 standard scratch and a pit of 0.5-mm diameter. The total length of all scratches and the number of pits are also limited by the specifications. In practice, the size of a defect is judged by a visual comparison with a set of graded standard defects. Digs and pits can, of course, be readily measured with a microscope; unfortunately the apparent width of a scratch is not directly related to its physical size, and this portion of the specification is not as well founded as one might desire. However, the concept of a visual comparison with a standard is a good and efficient one.

McLeod and Sherwood, who originated this method of specifying surface quality, in their 1945 article describing it, said that the number of a scratch was equal to the measured width in microns (micrometers) of a scratch made by a certain technique. Recently the government has used a relationship which indicates that the width in micrometers is only one-tenth of the scratch number. There is a widespread (and not unreasonable) suspicion that the widths of the standard scratches (which are maintained on physical pieces of glass) have become smaller in the decades since the 1940s (when the system originated).

Surface qualities of 80–50 or coarser (i.e., larger) are relatively easily fabricated. Qualities of 60–40 and 40–30 command a small premium in cost. Surfaces with quality specifications of 40–20, 20–10, 10–5, or similar combinations require extremely careful processing, and the more critical are considerably more expensive to fabricate. Such specifications are usually reserved for field lenses, reticle blanks, or laser optics.

Surface accuracy.   Surface accuracy is usually specified in terms of the wavelength of light from a sodium lamp (0.0005893 mm) or HeNe laser (0.0006328 mm). It is determined by an interferometric comparison of the surface with a test plate gage, by counting the number of (Newton's) rings or "fringes" and examining the regularity of the rings. As previously mentioned, the space between the surface of the work and the test plate changes one-half wavelength for each fringe. The accuracy of the fit between work and gage is described in terms of the number of fringes seen when the gage is placed in contact with the work.

Test plates are made truly flat or truly spherical to an accuracy of a small fraction of a fringe. Spherical test plates, however, have radii which are known to an accuracy only as good as the optical-mechanical means which are used to measure them. Thus the radius of a test plate is frequently known only to an accuracy of about one part in a thousand or one part in ten thousand. Further, test plates are expensive (several hundred dollars per set) and are available as "stock tooling" only in discrete steps. Thus it frequently pays to enquire what radii the optical shop has as standard tooling.

The usual shop specification for surface accuracy is thus with respect to a *specific* test plate, and it takes the form of requiring that the piece must fit the gage within a certain number of rings and must be spherical (or flat in the case of plane surfaces) within a number of rings. A fit of from five to ten rings, with a sphericity (or "regularity") of from one-half to one ring is not a difficult tolerance. Fits of from one to three rings with correspondingly better regularity can be achieved in large-scale production at a very modest increase in cost. Note that an irregularity of a small fraction of a ring is difficult to detect when the fit is poor. Thus, little is saved by specifying a ten-ring fit and a quarter-ring sphericity, since the fit must be considerably better than ten rings to be certain that the irregularity is less than one-quarter ring. The usual ratio is to have a fit of no worse than four or five times the maximum allowable irregularity. The change in radius due to a poor fit is frequently negligible in effect. For example, the radius difference between two (approximately) 50-mm radii at a 30-mm diameter which corresponds to five rings is (by Eq. 20.3) only about 33 μm.

The surface figure can be measured easily with an interferometer. While it is more difficult to control radius value with an interferometer than with a test plate, the interferometer is far superior when it comes to testing for sphericity or regularity. This is because the effective radius of the comparison wave front can be adjusted to match that of the surface under test, and also because the viewpoint of the interferometer is always normal to the surface and is thus not subject to the obliquity errors which afflict test plate readings.

If possible, one should avoid specifying accurate surfaces on pieces the thickness-to-diameter ratio of which is low. Such elements tend to spring and warp in processing, and extreme precaution is necessary to hold an accurate surface figure. A common rule of thumb is to make the axial thickness at least one-tenth of the diameter for negative elements; where there is a good edge thickness, one-twentieth or one-thirtieth of the diameter is sometimes acceptable. For extremely precise work, especially on plane surfaces, the optician might prefer a thickness of one-fifth to one-third of the diameter.

The performance effects of errors in radius values (i.e., departures from the nominal design radii) are usually not too severe. In fact, it is the practice of some purchasers of optics *not* to indicate a tolerance on the specified radii, but to specify final performance in terms of focal length and resolution. It is usually possible for a well-tooled optical shop to select judiciously (from its tooling list) nearby radii which produce a result equivalent to the nominal design. If tolerances are specified on radius values, one should bear in mind the fact that most effects produced by a radius variation are not proportional to ΔR, but to ΔC (or $\Delta R/R^2$). To take a simple example, we can differentiate the thin-lens focal-length equation

$$\phi = \frac{1}{f} = (n-1)(C_1 - C_2) = (n-1)\left(\frac{1}{R_1} - \frac{1}{R_2}\right)$$

with respect to the first surface to get the following:

$$d\phi = (n - 1)\, dC_1$$

$$df = f^2\, (n - 1)\, dC_1 = f^2\, (n - 1)\, \frac{dR_1}{R_1^{\,2}}$$

In a more complex system, the change in focal length resulting from a change in the ith curvature is approximated by

$$df \approx \left(\frac{y_i}{y_1}\right) f^2\, (n'_i - n_i)\, dc_i$$

$$df \approx \left(\frac{y_i}{y_1}\right) f^2\, (n'_i - n_i)\, \frac{dR_i}{R_i^{\,2}}$$

The point is that if a uniform tolerance is to be established for all radii in a system, the uniform tolerance should be on curvature, *not* on radius. Therefore, radius tolerances should be proportional to the square of the radius. For example, given a lens with a radius of 1 in on one side and a radius of 10 in on the other, if we vary the 1-in radius by 0.001 in, the effect on the focal length is the same as a change of 0.100 in on the 10-in *radius*. If the second surface had a radius of 100 in, then the equivalent radius change would be about 10 in.

The preceding is, of course, based on focal-length considerations only. With regard to aberrations, it is difficult to generalize, since one surface of a system may be very effective in changing a given aberration while another may be totally ineffective. The relative sensitivity is determined by the heights of the axial and principal rays at the surface, the index break across the surface, and the angles of incidence at the surface. A good estimate of the effect that any tolerance has on the aberrations of a system can be determined by use of the third-order surface contribution equations of Chap. 6.

Surface flatness is often specified in *millidiopters*, as is surface astigmatism. This can be translated into interference fringes by the following equation:

$$N = 10^3 y^2\, \phi/(n - 1)\lambda$$

where N is the number of fringes, y is the semi-diameter in mm, λ is the wavelength in μm, ϕ is the surface power in millidiopters, and n is the index of refraction. For a first surface mirror $(n - 1)$ equals 2, and for a second surface mirror $(n - 1)$ equals $2n$.

The effect of surface irregularity is more readily determined. Consider the case where the Newton's rings are not circular; this is an indication of axial astigmatism, since the power in one meridian is stronger than in the other. Here it is convenient to call on the Rayleigh quarter-wave criterion. The OPD produced by a "bump" of height H on a surface is equal to $H(n' - n)$, or, expressing it in terms of interference rings (remembering that each fringe or ring represents one-half wavelength change in surface contour),

$$\text{OPD} = \tfrac{1}{2} \, (\#\text{FR}) \, (n' - n) \text{ wavelengths}$$

where (#FR) is the number of fringes of irregularity.

Thus, to stay within the Rayleigh criterion, the total OPD, summed over the whole system, should not exceed one-fourth wavelength; this is expressed by the following inequality:

$$\sum (\#\text{FR}) \, (n' - n) < 0.5$$

Thus, a single element of index 1.5 could have one-half fringe of astigmatism (or any other surface irregularity) on both surfaces before the Rayleigh criterion was exceeded (assuming that the nominal correction was perfect and that the irregularities were additive).

Note that the expressions above do not take into account the fact that the system will probably be refocused to minimize the effects of any surface irregularity. See the discussion of OPD and spherical aberration in Sec. 15.3, for example. For astigmatism, refocusing reduces the OPD by a factor of 2.

When a spherical surface is tilted a departure from flatness will introduce aberrations, predominately astigmatism. At 45° a reflecting surface will introduce OPD = 1.26 N waves, where N is the number of circular fringes. For a refracting surface the OPD is a quarter of this. In general, a tilted surface (at best focus) will produce an astigmatism OPD of:

$$\text{OPD} = (n/n')^2 \, u_P^2 \, N(n' - n) \text{ waves}$$

A spherical surface has a slope equal to $\lambda N/y$ radians, and the ray deviation is $(n - 1)$ times the slope.

Thickness. The effects of thickness and spacing variation on the performance of a system are readily analyzed, either by raytracing or by a third-order aberration analysis. The importance of thickness variation differs greatly from system to system. In the negative doublets of a Biotar (double-Gauss) objective, the thickness is extremely critical, especially as regards spherical aberration; for this reason the crown and flint elements are usually selected so that their *combined* thickness is very close to the design nominal. At the other extreme, the thickness variation of a plano-convex eyepiece element may be almost totally ignored, since it ordinarily has little or no effect on anything.

In general, thicknesses and spacings may be expected to be critical where the slope of the marginal axial ray is large. Anastigmats in general, and meniscus anastigmats in particular, are prone to this sensitivity. High-speed lenses, large-NA microscope objectives, and the like are usually sensitive.

Unfortunately the thickness of an optical element is not as readily controlled as some of the other characteristics. In production procedures where many elements are processed on the same block, the maintenance of a uniform nominal thickness requires precise blocking and tooling. The grinding operation, while precise enough in terms of radius, is difficult to control in terms of its extent. For close thickness control, the generating operation must be accurate and each subsequent grinding stage must be exactly timed so that the proper finish, radius, and thickness are arrived at simultaneously.

A reasonable thickness tolerance for precise work is ±0.1 mm (±0.004 in). This can cause a shop some difficulty on certain lens shapes and on larger lenses; where a relaxation is possible, a tolerance of ±0.15 or ±0.2 mm is more economical. It is possible to hold ±0.05 mm in large-scale production by taking care throughout the fabrication procedure. The rejection rate at this tolerance can become disastrous if the smallest mischance occurs. Of course it is possible, by handworking and selection, to produce pieces to any desired tolerance level; the author has seen ±0.01 mm held in moderate production quantities (although at rather immoderate cost).

When the quantity of lenses to be produced is large enough, the benefits of spot blockers may warrant the initial tooling cost. The spot blocker is a metal blocking tool with machined seats into which the elements are cemented. The tool is designed with a specific element in mind, and takes into account the exact diameter, thickness and the radius of the side which is cemented into the seat, as well as the radius which is to be processed. When the lenses are ground true, the elements are all the proper thickness. The spot blocker allows better control of the thickness than pitch button blocking (illustrated in Fig. 20.2). Note that "blocking on the iron" as shown in Fig. 20.3 is equivalent to spot blocking.

Centering.   The tolerances in centering are (1) on the diameter of the piece, and (2) on the accuracy of the centering of the optical axis with the mechanical axis. If the piece is to be centered (i.e., as a separate operation), the diameter can be held to a tolerance of plus nothing, minus 0.03 mm by ordinary techniques, and this is the standard tolerance in most shops. A small economy is effected by a more liberal tolerance. Tighter tolerances are possible, but are not often necessary for ordinary work.

The concentricity of an element is most conveniently specified by its *deviation*. This is the angle by which an element deviates an axial ray of light directed toward the mechanical center of the lens. The deviation angle is an especially useful measure of decentration, since the deviation of a group of elements is simply the (vector) sum of the deviations of the individual elements. Figure 20.5 is an exaggerated sketch of a decentered element. The optical and mechanical axes are shown separated by an amount Δ (the decentration). Since a ray parallel to the optical axis must pass through the focal point, the angular deviation δ in radians of the ray aimed along the mechanical axis is given by the decentration divided by the focal length.

$$\delta = \frac{\Delta}{f} \text{ radians} \qquad (20.4)$$

Note that a decentered element may be regarded as a centered element plus a thin wedge of glass. The angle of the wedge W is given by the difference between the maximum and minimum edge thicknesses divided by the diameter of the element

$$W = \frac{E_{max} - E_{min}}{d} \text{ radians} \qquad (20.5)$$

Since the deviation of a thin prism is given by $D = (n - 1)A$, we can similarly relate the wedge angle of an element to its deviation by

$$\delta = (n - 1) \, W \text{ radians} \qquad (20.6)$$

If an element is centered on a high-production mechanical (clamping) centering machine, the limit on the accuracy of the concentricity obtained is determined by the residual difference in edge thickness which the cylindrical clamping tools cannot "squeeze out." On most machines, this is to the order of 0.0005 in when residual tooling and spindle errors are also taken into account. Thus the residual wedge angle for a lens with a diameter d is given by

$$W = \frac{0.0005 \text{ in}}{d}$$

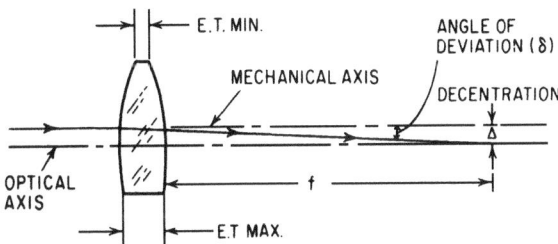

Figure 20.5 Showing the relationships between the optical and mechanical axes, and the decentration and angle of deviation in a decentered lens.

and the resulting deviation is

$$\delta = \frac{0.0005 \text{ in } (n-1)}{d}$$

Thus, for ordinary lenses ($n = 1.5$ to 1.6) a reasonable estimate of the deviation is given by

$$\delta \approx \frac{1}{d} \text{ minutes} \approx \frac{1.67(n-1)}{d} \text{ minutes} \tag{20.7}$$

where d is in inches, and the centering is done mechanically.

If the centering is accomplished visually (as indicated in the left-hand sketch of Fig. 20.4) then the ability of the eye to detect motion is the limiting factor. If we assume that the eye can detect an angular motion of 6 or 7×10^{-5} radians, then the deviation will be approximately

$$\delta = (n-1)\left(\frac{1}{R} \pm 0.06\right) \pm \text{(contact and spindle errors)} \tag{20.8}$$

where δ is in minutes and R is the radius of curvature of the outer surface in inches.*

The term $(n-1)/R$ is from the visually undetected "wobble" of the outer radius and the $0.06 (n-1)$ term is due to the tilt in the tool which the eye could not detect in the truing of the tool (this is tested by pressing a flat glass plate against the rotating tool and observing any motion in the reflected image). The eye can, of course, be aided by means of a telescope or microscope which will further reduce the amount of decentration which can be detected by a factor equal to the magnification.

Occasionally lenses are not put through a separate centering operation. When this is the case, the concentricity of the finished lens is determined by the wedge angle which is left in by the grinding operations. If the blocking tooling is carefully worked out, it is possible to produce elements with a wedge (i.e., the difference between the edge thickness of opposite edges) to the order of 0.1 or 0.2 mm. Centering is often omitted on inexpensive camera lenses, condensers, magnifiers, or almost any single element of a simple optical system. Simple elements made from rounded circles of window glass are often left uncentered.

Prism dimensions and angles. The linear dimensions of prisms can be held to tolerances approximating those of an ordinary machined part,

---

*Equation 20.8 assumes that the image reflected from the outer radius is viewed at 10 in. This is obviously impossible if R is a convex surface with a radius longer than 20 in, and it is impractical if R is a long concave radius. Thus for $|R| > 20$ in, one should substitute 0.05 for the $1/R$ term.

although the fabrication requirements of a prism are more difficult because of the finish and accuracy requirements of an optical surface. Thus tolerances of 0.1 or 0.2 mm are usually reasonable and tighter tolerances are possible.

Prism angles can be held to within 5 or 10 minutes of nominal by the use of reasonably good blocking forms. Indeed it is possible, although exceedingly difficult, to make angles accurate to a few percent of these tolerances if one takes exquisite pains with the design, fabrication, correction, and use of the blocking tools. Usually angles which must be held to tolerances of a few seconds (such as roof angles) are "hand corrected." Such angles are checked with an autocollimator, either by comparison with a standard or by using the internal reflections to make the piece a retrodirector. Angles of 90° and 45° (among others) can be self-checked in this way since their internal reflections form constant-deviation systems of 180° deviation (as discussed in Chap. 7).

Prism size tolerances are usually based on the necessity to limit the image displacement errors (lateral or longitudinal) which they produce. Angular tolerances are usually established to control angular deviation errors. One can usually find one or two angles in a prism system which are more critical than the others; these can be tightly controlled and the other angles allowed to vary. For example, with respect to the deviation of a pentaprism, an angular error in the 45° angle between the reflecting faces is six times as critical as an error in the 90° angle between the entrance and exit faces, and the other two angles have no effect on the deviation. On occasion, prism tolerances are based on aberration effects. Since a prism is equivalent to a plane parallel plate and introduces overcorrected spherical and chromatic; an increase in prism thickness in a nominally corrected system will overcorrect these aberrations. Some prism angle errors are equivalent to the introduction of thin-wedge prisms into the system. The angular spectral dispersion of a thin wedge is $(n - 1)W/V$ (where W is the wedge angle and V is the Abbe V-value of the glass) and the resultant axial lateral color may limit the allowable angular tolerances.

Materials. The characteristics of the refractive materials used in optical work which are of primary concern are index, dispersion, and transmission. For ordinary optical glass procured from a reputable source, visual transmission is rarely a problem. Occasionally, where a thick piece of dense glass is used in a critical application, transmission limits or color must be specified. Similarly, the dispersion, or V-value, is seldom a problem, except in special cases. For apochromatic systems where the partial dispersion ratio is exceedingly critical, very special precautions are required.

The index of refraction is usually of prime concern in optical glass. As indicated in Chap. 10, the standard index tolerance is ±0.0005 or ±0.001, depending on the glass type. The glass supplier can hold the index more closely than this by selection or by extra care in the processing; either increases the cost somewhat. In practice the glass supplier will ordinarily use up only a fraction of this tolerance, since the index within a single melt or batch of glass is remarkably consistent. Thus, within a single lot of glass the index may vary only one in the fourth place. However, bear in mind that this variation *may* be centered about a value which is 0.0005 or 0.0016 from the nominal index. It is sometimes economical to accept the standard tolerance and to adjust a design to compensate for the variation of a lot of glass in cases where the index is critical.

Transmission and spectral characteristics are often poorly specified. For filters and coatings, ambiguity can usually be avoided by specifying spectral reflection (or transmission) *graphically,* i.e., by indicating the area of the reflection (or transmission) versus wavelength plot within which the characteristics of the part must lie. One should also indicate whether or not the spectral characteristics outside the specified region are of importance. For example, in a bandpass filter, it is important to indicate how far into the long- and short-wavelength regions the blocking action of the filter must extend.

Figure 20.6 is a table of typical tolerances and may be used as a guide. Bear in mind, however, that the values given are *typical* and that there are many special cases that this sort of tabulation cannot cover and that the "tolerance profile" of a shop will likely differ somewhat than the Table.

Relative cost factor.   For the classic or *normal* optical production methods, the cost of an element varies roughly as the following expressions:

$$\text{Quantity cost factor} = 1.07 + 2.26\,Q^{-0.42}$$

$$\text{Quality cost factor} = \frac{1.5}{\sqrt[20]{S \cdot D \cdot P \cdot R \cdot T}}$$

where Q = quantity of pieces to be made
  S = scratch number
  D = dig number
  P = surface power tolerance in fringes
  R = surface regularity tolerance in fringes
  T = thickness tolerance in millimeters

These equations are useful to determine the effect on cost of changing the specifications.

Surface Quality	Diameter, mm	Deviation (concentricity), min	Thickness, mm	Radius	Regularity (asphericity)	Linear Dimension, mm	Angles
Low cost	120-80	± 0.2	> 10	± 0.5	Gage	Gage	± 0.5	Degrees
Commercial	80-50	± 0.07	3–10	± 0.25	10 Fr	3 Fr	± 0.25	± 15'
Precision	60-40	± 0.02	1–3	± 0.1	5 Fr	1 Fr	± 0.1	± 5'–10'
Extraprecise	60-40	± 0.01	< 1	± 0.05	1 Fr	⅕ Fr	As req'd.	Seconds
Plastic	80-50		1	± 0.02	10 Fr	5 Fr	0.02	minutes

Figure 20.6 Tabulation of typical optical fabrication tolerances.

Additive tolerances. In analyzing an optical system to determine the tolerances to be applied to specific dimensions, one can readily calculate the partials of the system characteristics with respect to the dimensions under consideration. Thus, one obtains the value of the partial derivative of the focal length (for example) with respect to each thickness, spacing, curvature, and index; likewise for the other characteristics, which may include back focus, magnification, field coverage, as well as the aberrations or wave-front deformations. Then each dimensional tolerance, multiplied by the appropriate derivative, indicates the contribution of that tolerance to the variation of the characteristic. Now if it were necessary to be *absolutely* certain that (for example) the focal length did not vary more than a certain amount, one would be forced to establish the parameter tolerances so that the sum of the absolute values of the derivative-tolerance products did not exceed the allowable variance. Although this "worst-case" approach is occasionally necessary, one can usually allow much larger tolerances by taking advantage of the laws of probability and statistical combination.

As a simple example, let us consider a stack of disks, each 0.1 in thick. We will assume that each disk is made to a tolerance of ±0.005 in and that the probability of the thickness of the disk being any given value between 0.095 and 0.105 in is the same as the probability of its being any other value in this range. This situation is represented by the rectangular frequency distribution curve of Fig. 20.7a. Thus, for example, there is 1 in 10 chance that any given disk will have a thickness between 0.095 and 0.096 in. Now if we stack two disks, we know that it is *possible* for their combined thickness to range from 0.190 to 0.210 in. However, the *probability* of the combination having either of these extreme thickness values is quite low. Since the probability of either of the disks having a thickness between 0.095 and 0.096 is 1 in 10, if we randomly select two disks, the probability of *both* falling in this range is one-tenth of one-tenth, or 1 in 100. Thus, the probability of a pair of disks having a thickness between 0.190 and 0.192 is 1 in 100; similarly for a combined thickness of 0.208 to 0.210 in. The probability of a combined thickness of 0.190 to 0.191 (or 0.209 to 0.210) is much less; 1 in 400.

The frequency distribution curve representing this situation is shown in Fig. 20.7b as a triangular distribution. Figure 20.7c shows frequency distribution curves for 1-, 2-, 4-, 8-, and 16-element assemblies. These curves have been normalized so that the area under each is the same and the extreme variations have been equalized. The important point here is that the probability of an assembly taking on an extreme value is tremendously reduced when the number of elements making up the assembly is increased. For example, in a stack of 16 disks with a nominal total thickness of 1.6 in and a possible total variation in thickness of ±0.080 in, the probability of a random stack having a

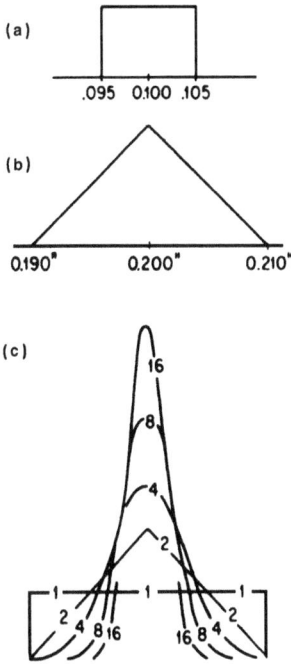

(a)

.095 0.100 .105

(b)

0.190"   0.200"   0.210"

(c)

16
8
4
2
1   1   1
2   4 8 16   16 8 4 2

Figure 20.7 Showing the manner in which additive tolerances combine in assembly. Plot *A* shows a uniform probability in a dimension of a single piece. When two such pieces are combined, the resulting frequency distribution is shown in *B*. Normalized curves for assemblies of 1, 2, 4, 8, and 16 pieces are shown in *C*.

thickness less than 1.568 in or more than 1.632 in (i.e., 1.600 in ±0.032 in) is less than 1 in 100.

The importance of this in setting tolerances is immediately apparent. In the stacked-disks example, if the range of thicknesses represented by 1.568 to 1.632 in for 16 disks were the greatest variation that could be tolerated, we could be absolutely sure of meeting this requirement *only* by tolerancing each individual disk at ±0.002 in. However, if we were willing to accept a rejection rate of 1 percent in large-scale production, we could set the thickness tolerance at ±0.005 in. If the cost of the pieces made to the tighter tolerance exceeded the cost of the pieces made to the looser tolerance by as little as 1 percent (plus one sixteen-hundredth of the assembly, processing, and final inspection costs), the looser tolerance would result in a less costly product.

In a frequency distribution curve such as those shown in Fig. 20.7 the area under the curve between two abscissa values represents the (relative) number of pieces which will fall between the two abscissa values. Thus the probability of a characteristic falling between two values is the area under the curve between the two abscissas divided by the total area under the curve.

The "peaking-up" characteristic of multiple assemblies can also be represented by the two plots shown in Fig. 20.8. The graph on the left

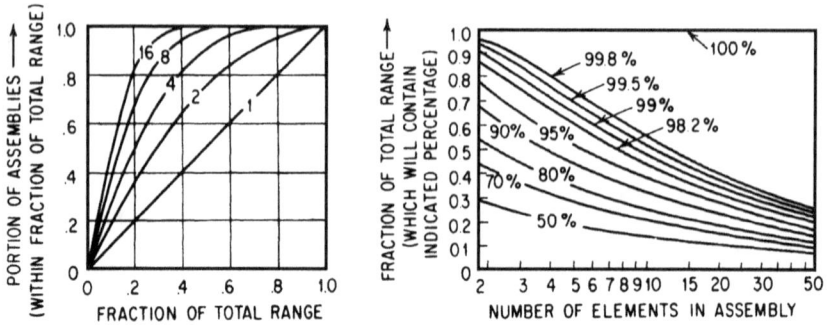

Figure 20.8 Probability distributions of additive tolerances in multiple assemblies. See text for details.

shows the percentage of assemblies which fall within a given central fraction of the total tolerance range as a function of that fraction. The number of elements per assembly is indicated on each curve. These curves were derived from Fig. 20.7c. The graph on the right in Fig. 20.8 is simply another way of presenting the same data. If one were interested in an assembly of 10 elements, the intersection of the abscissa corresponding to 10 and the appropriate curve would indicate that all but 0.2 percent (using the 99.8 percent curve) of the assemblies would fall within 0.55 of the total tolerance range represented by the sum of all 10 tolerances, and that over one-half of the assemblies (using the 50 percent curve) would fall within 0.15 of the total possible range.

The preceding discussion has been based upon the unlikely assumptions that (1) each individual piece had a rectangular frequency distribution, and (2) each tolerance was equal in effect. This is rarely true in practice. The frequency distribution will, of course, depend on the techniques and controls used in fabricating the part, and the tolerance sizes may represent the partial derivative tolerance products from such diverse sources as tolerances on index, thickness, spacing, and curvature. Note, however, that in Fig. 20.7c the progression of curves may be started at any point. If, for example, the production methods produce a triangular distribution (such as that shown for an assembly of two elements), then the curve marked 4 (for "four elements") will be the frequency distribution for two elements (of triangular distribution) and so on.

Note also that as more and more elements are included in the assembly, the curve becomes a closer and closer approximation to the normal distribution curve which is so useful in statistical analysis (except that the tolerance-type curves do not go to infinity as do normal curves). One useful property of the normal curve for an additive assembly is that its "peakedness" is proportional to the square root of the number of elements in assembly. Thus if 99 percent of the individual pieces are expected to fall within some given range, then for an assembly of 16 elements, 99 percent would be expected to fall within $\sqrt{1/16}$, or one-quarter of the total range. A brief examination will indicate that even the rectangular distribution assumed for Figs. 20.7 and 20.8 tends to follow this rule when there are more than a few elements in the assembly.

A rule of thumb frequently used to establish tolerances may be represented as follows:

$$T = D \sqrt{\sum_{i=1}^{n} t_i^2} \qquad (20.9)$$

This is frequently referred to as the RSS rule, shorthand for the square Root of the Sum of the Squares. What the RSS rule means is this: If some percentage (say 99 percent) of the part tolerances produces effects less than t (and varies according to a normal, or gaussian, distribution), then the same percentage (i.e., 99 percent in our example) of the assemblies will show a total tolerance effect less than T.

For most cases, the *customary* value of T (given by Eq. 20.9 with $D = 1$) is usually about 40 or 50 percent high in practice, depending on how the dimensional errors are distributed. If the values of t_i are *only* either plus or minus t (i.e., an end point distribution with nothing in between), then D in Eq. 20.9 is equal to 1.0. If all of the tolerances have a rectangular (uniform) distribution as in Fig. 20.7a, then $D = 0.58$. If the distribution is a 2σ truncated gaussian, then D should be about 0.43. Obviously, if the distribution is more peaked up than this gaussian, then D should be even smaller.

The probability of the combined tolerance effects exceeding some maximum acceptable value (S) is given in the following table, where S/T is the ratio of the maximum acceptable (S) to the RSS value (T)—with an appropriate value for D, and F is the statistically expected fraction of the assemblies which will exceed the acceptable value.

S/T	F
0.67	50%
0.8	42%
1.0	32%
1.5	13%
2.0	5%
2.5	1%

While this section may seem to be a far cry from optical engineering, consider that a simple Cooke triplet has the dimensions (as given above) which affect its focal length and aberrations: six curvatures, three thicknesses, two spacings, three indices, and three *V*-values. These total fourteen for monochromatic characteristics and seventeen for chromatic aberrations. Such a system is eminently qualified for statistical treatment. *Note that the validity of this approach does not depend on a large production quantity; it depends on a random combination of a certain number of tolerance effects.*

There are two obvious features of the RSS rule which are well worth noting. One is the square-root effect: If you have *n* tolerance effects of a size ±*x,* then the RSS rule says that a random combination will produce an effect equal to ±*x* times the square root of *n*. For example, given 16 tolerance effects of ±1 mm, we should expect a variation of only ±4 mm, not ±16 mm. The other feature is that the larger effects dominate the combination. As an example, consider a case with nine tolerance effects of ±1 mm and one tolerance effect of ±10 mm. If we use the RSS rule on this, we get an expected variation equal to the square root of 109, or ±10.44 mm. Compare this with the fact that the single ±10-mm tolerance has an RSS of ±10 mm. The addition of the nine ±1-mm tolerances changed the expected variation by only 4.4 percent.

One possible way to establish a tolerance budget using this principle is as follows:

1. Calculate the partial derivatives of the aberrations with respect to the fabrication tolerances (radius, asphericity, thickness and spacing, index, homogeneity, surface tilt, etc.). Express the aberrations as OPDs (wave-front deformation).

2. Select a preliminary tolerance budget. Figure 20.6 can be used as a guide to appropriate tolerance values.

3. Multiply the individual tolerances by the partial derivatives calculated in step 1.

4. Compute RSS for all the aberrations for each individual tolerance. This will indicate the relative sensitivity of each tolerance.

5. Compute RSS for all of the effects calculated in step 4 combined.

6. Find the OPD_{DES} of the nominal design, either by direct calculation or from the design MTF, the Strehl, or whatever measure is convenient. Determine the OPD_{SPEC}, which is the maxium OPD allowed by the performance specifications. Since the design aberrations and the tolerance effect aberrations are random, they can be RSS-ed, and we can solve that relationship for the tolerance OPD_{TOL}. Thus the maximum allowed OPD_{TOL} due to tolerances is given by:

$$OPD_{TOL} = (OPD^2_{SPEC} - OPD^2_{DES})^{1/2}$$

7. Adjust the tolerance budget so that the result of step 6 is equal to the required performance. Since the larger effects dominate the

RSS, if you are tightening the tolerances (as is quite likely on the first go-round), you should tighten the most sensitive ones (and possibly loosen the least sensitive). Note that there is no economic gain if you loosen tolerances beyond the level at which costs or prices cease to go down. Conversely, one should be sure that the tolerances are not tightened beyond a level at which fabrication becomes impossible—since cost rises asymptotically toward infinity as this level is approached.

8. After one or two adjustments (steps 2 through 7) the tolerance budget should converge to one which is reasonable economically and which will produce an acceptable product.

If the tolerances necessary to get an acceptable performance are too tight to be fabricated economically, there are several ways which are commonly used to ease the situation:

1. A *test plate fit* is a redesign of the system using the measured values of the radii of existing test plates. This eliminates the radius tolerance (except for the variations due to the test glass "fit" in the shop, and any error in the measurement of the radius).

2. A *melt fit* can effectively eliminate the effects of index and dispersion variation. Again, this is a redesign, using the measured index of the actual piece of glass to be used, instead of the catalog values.

3. A *thickness fit* uses the measured thicknesses of the actual fabricated elements to be assembled; this amounts to an adjustment of the airspaces during the assembly process.

4. *Inverse Design* Measure the aberrations of the assembly, including the aberrations resulting from tilts and misalignments. Use the optimization program to determine a prescription (including tilts and decenters) which has the same aberrations as the assembly. Then modify the assembly by introducing errors which are the opposite of the changes which were found to produce the existing aberrations. This scheme is useful in adjusting multi-mirror assemblies.

The redesigns called for in all four "fitting" operations above, while hardly trivial, are not major undertakings when an automatic lens design program is used.

While the above may tend to induce a desirable relaxation in tolerances, one or two words of caution are in order. As previously mentioned, the index of refraction distribution within a melt or lot of glass may or may not be centered about the nominal value. When it is centered about a nonnominal value, the preceding analysis is valid only with respect to the central value, not the nominal value. Further, in some optical shops, there is a tendency to make lens elements to the high side of the thickness tolerance; this allows scratched surfaces to be reprocessed and will, of course, upset the theoretical probabilities. Another tendency is for polishers to try for a "hollow" test glass fit, i.e., one in which there is a convex air lens between the test plate and the work. This is done because a block of lenses which is polished "over" is difficult to bring back. Surprisingly, these nonnormal distributions have very little effect on Eq. 20.9 (if there are enough elements in the assembly).

Thus, the situation is seen to be a complex one, but nonetheless one in which a little careful thought in relaxing tolerances to the greatest allowable extent can pay handsome dividends. For those who wish to avoid the labor of a detailed analysis, the use of Eq. 20.9, or even the assumption that the tolerance buildup will not exceed one-half or one-third of the possible maximum variation, are fairly safe procedures in assemblies of more than a few elements. Above all, when cost is important, one should try to establish tolerances which are readily held by normal shop practices.

See Sec. 20.6 for an example of tolerance budgeting.

20.3   Optical Mounting Techniques

General.   In optical systems, just as in precise mechanical devices, it is best to observe the basic principles of kinematics. A body in space has six degrees of freedom (or ways in which it may move). These are translation along the three rectangular coordinate axes and rotation about these three axes. A body is fully constrained when each of these possible movements is *singly* prevented from occurring. If one of these motions is inhibited by more than one mechanism, then the body is overconstrained and one of two conditions occurs; either all but one of the (multiple) constraints are ineffective or the body (and/or the constraint) is deformed by the multiple constraint.

The laboratory mount indicated in Fig. 20.9 is a classical example of a kinematic mount. Here it is desired to uniquely locate the upper piece with respect to the lower plate. At *A* the ball-ended rod fits into a conical depression in the plate. This (in combination with gravity or a springlike pressure at *D*) constrains the piece from any lateral translations. The *V*-groove at *B* eliminates two rotations, that about a vertical axis at *A* and that about the axis *AC*. The contact between the ball end and the plate at *C* eliminates the final rotation (about axis *AB*). Note that there are no extra constraints and that there are no critical tolerances. The distances *AB, BC,* and *CA* can vary widely without introducing any binding effects. There is one unique position which will be taken by the piece; the piece may be removed and replaced and will always assume exactly the same position.

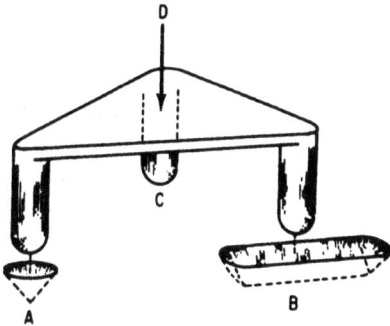

Figure 20.9 An example of a kinematic locating fixture. The three ball-ended legs of the stool rest in a conical hole at *A,* a *V*-groove (aligned with *A*) at *B,* and on a flat surface at *C.*

A perfectly kinematic system is frequently undesirable in practice and semikinematic methods are often used. These substitute small-area contacts for the point and line contacts of a pure kinematic mount. This is necessary for two reasons. Materials are often not rigid enough to withstand point contact without deformation, and the wear on a point contact soon reduces it to an area contact in any case.

Thus, in the design of an instrument, optical or otherwise, it is best to start by defining the degrees of freedom to be allowed and the degrees of constraint to be imposed. These can be outlined first by geometrical points and axes and then reduced to practical pads, bearings, and the like. This sort of approach results in a thorough and clear understanding of the effects of manufacturing tolerances on the function of the device and often indicates relatively inexpensive and simple methods by which a high order of precision can be maintained.

Blackening. Light which is reflected or scattered from the mounting structure or from the element edges will reduce the image contrast (and MTF) and will occasionally produce ghost images if the reflecting surface is smooth. The ground edges of lens elements can be blackened with ink to reduce scattered light without increasing the diameter. Black paint works well, but the increase in diameter caused by the paint thickness must be taken into account. Brass mounting components can be blackened by immersing the clean, hot piece in a solution of two parts (by volume) of cupric carbonate, three parts ammonium hydroxide, and six parts distilled water. Aluminum parts can be black anodized. The inner surfaces of mounts can be "scored" with a threading tool, or sandblasted to roughen the surface. A flat black paint will reduce the reflection; *Floquil* brand flat black model locomotive paint works almost as well as the specialized paints.

Lens mounts. Optical lens elements are almost always mounted in a close-fitting sleeve. A number of methods are used to retain the element in the mount; several are sketched in Fig. 20.10. In sketches (a) and (b) the lenses are retained by spring rings. In the left-hand mount (a), the spring catches in a V-groove, and if the mount is properly executed, the spring wire (which in its free state assumes a larger diameter) presses against the face of the element and the outer face of the groove. The lens is thus under a light pressure. The flat spring retainer (b) is less satisfactory, since the retainer will readily slip out unless the spring is strong or has sharp edges which bite into the mount. Other methods suitable for retaining low-precision elements include staking or upsetting ears of metal from the cell which clamp a thin metal washer over the lens element. Condenser systems are often mounted between three rods which are grooved as indicated in Fig. 20.10c. This provides a loose mount which leaves the condenser elements free to expand with the heat from the projection lamp without being constricted by the mount; it also allows cooling air to circulate freely. Both points are especially important in the mounting of a heat absorbing filter.

Where precision is required, the cell is fitted rather closely to the lens. For good-quality optics the lens diameter may be toleranced $+0.000$, -0.001 in and the inside cell diameter toleranced $+0.001$, -0.000 in with 0.001- or 0.0005-in clearance between the nominal diameters. For small lenses which demand high precision, these tolerances can be halved, at the expense of some difficulty in production. Large-diameter optics are usually specified to somewhat looser tolerances. Lenses are most commonly retained by a threaded lock ring, as indicated

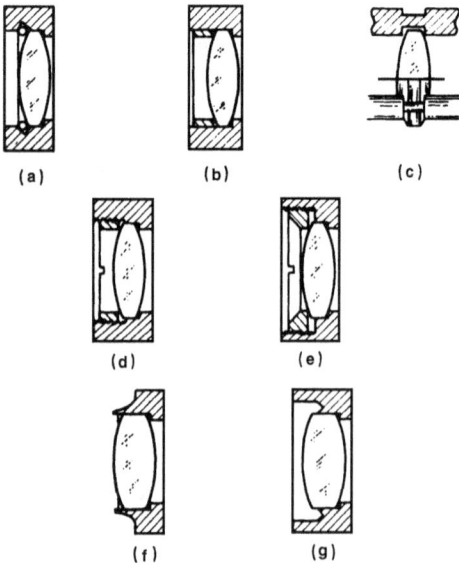

(a)   (b)   (c)

(d)   (e)

Figure 20.10 Several methods of retaining optical elements. (a) Wire spring ring in a V-groove; (b) flat spring ring; (c) three grooved rods at 120°; (d) and (e) threaded lock ring; (f) spinning shoulder, before burnishing and (dotted) after; (g) cemented in place with trough for cement overflow.

(f)   (g)

in Fig. 20.10d or e. Sometimes the lock ring has an unthreaded pilot the diameter of which is the same as the lens in order to be certain that the lens will ride on the bored seat and not on the threads. A separate spacer may be substituted for the pilot. The fit of the threaded parts should be loose so that the lens takes its orientation from the seat and shoulder, rather than from the threaded lock ring which frequently cocks.

A lens may be spun into the mount, as shown in Fig. 20.10f. In this method the mount is made with a thin spinning shoulder which protrudes past the edge of the lens (which is preferably beveled). This spinning shoulder is a few thousandths of an inch thick at the outside edge and has an included angle of 10 or 20°. The lens is inserted and the thin lip is turned over, usually by rotating the cell while the lip is bent over. Care and skill are required, but there are a number of advantages to this technique. The pressure of the spinning shoulder tends to center the lens in the mount. In assemblies requiring extreme precision, the seat can be bored to fit the lens diameter and the lens can be spun in place without removing the piece from the lathe; the result is concentricity of an order which is difficult to duplicate by any other means.

Another technique which results in both economy and precision is to cement the lens into its seat. The cement has a modest centering action, and with a good plastic cement the lens is securely retained. Care should be taken to provide an overflow groove (Fig. 20.10g) so that excess cement is kept away from the surface of the lens.

For optics which must withstand a difficult thermal and/or vibration environment, a useful form of mount is achieved by making the inside diameter of the cell oversize and cementing the element in place with a compliant, elastomeric RTV type of cement. The lens is trued in the mount before it is cemented in place. This technique is especially useful for large-diameter elements where the thermal expansion difference between the element and the mount is a serious problem; the layer of RTV between the element and the cell is made thick enough to take up the expansion difference.

In an assembly where several lenses and spacers are retained by a single lock ring, care must be taken that the thickness tolerances on the lenses and spacers are not allowed to build up to a point where the outside lens (1) extends beyond its seat and is not constrained by the seat diameter, or (2) is down into the mount so far that the lock ring cannot seat down on it. Another point to watch is that the mouth of a long inside-diameter bore is frequently bell-shaped, and a lens located near the mouth may have several thousandths of an inch more lateral (diametral) freedom than intended. In critical assemblies, it frequently pays to locate the lens well inside the mouth of the bore.

When elements of different diameters are to be mounted together, the mount can be designed so that the lens seats can all be bored in one operation. This not only tends to reduce the cost of the mount but eliminates a possible source of decentration of each element with respect to the others, which can occur when the lens seats are bored in two or more separate operations.

The microscope style of element mounting shown in Fig. 20.11 illustrates a number of valuable devices. The lens seat and the outside support diameter of each cell can be turned in the same operation; indeed, in a critical system, the optical element may be spun in place without removing the piece from the lathe. (Cementing the lenses in place can be substituted for spinning.) All the cells are seated in the same bore of the main mount and they are isolated from the lock-ring threads (not shown) by a long spacer. All these techniques contribute to maintaining the exquisite concentricity necessary in a first-class microscope objective.

In mounting any type of optical element, it is important to avoid any warping or twisting. In the case of lens elements (which are in effect clamped between a shoulder and a lock ring, or their equivalents), this is not too difficult, since the pressure points are opposite each other and result in compression of the lens. More care is necessary in mounting mirrors and prisms, however, since it is quite easy to make the mistake of restraining a mirror in such a way that its surface is warped out of shape. One way to avoid this is to be sure that for each point at which pressure is exerted, there is a pad directly opposite so that no twisting moment is introduced.

Figure 20.12 serves as an indication of how few constraints are necessary to kinematically define the location of a piece. This illustration might apply to a piece of cubical shape. The three points in the XZ plane define a plane on which the lower face of the piece rests; these points take up one translational and two rotational degrees of freedom. The two points in the YZ plane take up one translation and one rotational freedom. Note that if there were three nonaligned points in this plane, they would then define an angle between the XZ and YZ faces of the piece; if the piece had a different angle, then there would be *two* ways in which the piece could be seated. The single point in the XY plane eliminates the last remaining of the six available degrees of

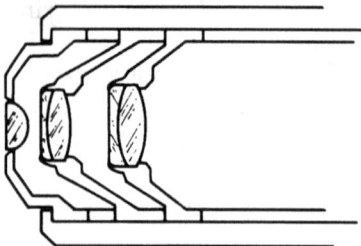

Figure 20.11 Mounting detail, microscope objective.

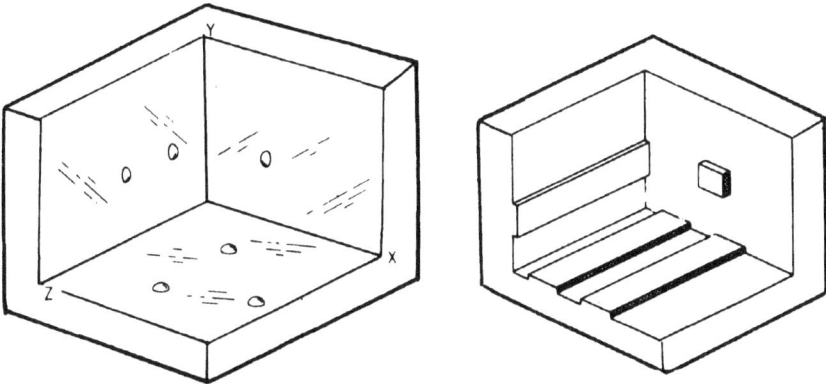

Figure 20.12 Kinematic and semikinematic position defining mount for a rectangular piece.

freedom. A flexible pressure on the near corner of the piece will now uniquely locate the piece in this mount.

The sketch on the right illustrates one way of putting this type of mount into practice. The points are replaced by pads or rails. As shown, the two rails in the XZ plane must be carefully machined in the same operation to assure that they are exactly coplanar: this is not difficult, but if it were, the substitution of a short pad for one rail would eliminate any difficulty on this score.

Prisms and mirrors are usually clamped or bonded to their mounts. In clamp mounts the pressure is usually exerted by a screw on a metal pressure pad. A piece of cork or compressible composition material is placed between the glass and the metal pad to distribute the pressure evenly over the glass; this prevents the pressure from being exerted at a single point. There are a number of excellent cements available for bonding glass pieces to metal mounts. Some care is necessary in designing the mount when bonding a thin mirror, since the cement may warp the mirror (toward the shape of the mount) if the cemented area is large.

20.4   Optical Laboratory Practice

The lens bench.   An optical bench or lens bench consists, in essence, of a collimator which produces an infinitely distant image of a test target, a device for holding the optical system under test, a microscope for the examination of the image formed by the system, and a means for supporting these components. Each of the components may take various forms, depending on the usage for which it is primarily designed.

The collimator consists of a well-corrected objective and an illuminated target at the focus of the objective. For visual work, the objective is usually a well-corrected achromat; for infrared work, a paraboloidal mirror is used, usually in an "off-axis" or Herschel configuration. The target may be a simple pinhole (for star tests or energy distribution studies), a resolution target, or a calibrated scale if a "focal" collimator is desired. The collimator provides a target at infinity.

The lens holder can range in complexity from a simple platform with wax to stick the lens in place, to a T-bar nodal slide which generates a flat image surface. The microscope is usually equipped with at least one micrometer slide, and frequently with two or three orthogonal slides so that accurate measurements may be made.

In subsequent paragraphs, we will discuss some of the applications of the lens bench and will describe the components of the bench more fully in the context of their applications.

The measurement of focal length.   There are two basic lens bench techniques for the routine measurement of effective focal length: the nodal slide method and the focal collimator. Both schemes are sketched in Fig. 20.13.

The *nodal slide* is a pivoted lens holder equipped with a slide which allows the lens to be shifted axially (i.e., longitudinally) with respect to the pivotal axis. Thus, by moving the lens forward or backward, the lens can be made to rotate about any desired point. Now note that, if the lens is pivoted about its second nodal point (as indicated in Fig. 20.13), the ray emerging from this point (which by definition emerges from the system parallel to its incoming direction) will coincide with the bench axis (through the nodal point). Thus there will be no lateral motion of the image when the lens is rotated about the second nodal point. Once the nodal point has been located in this manner, the

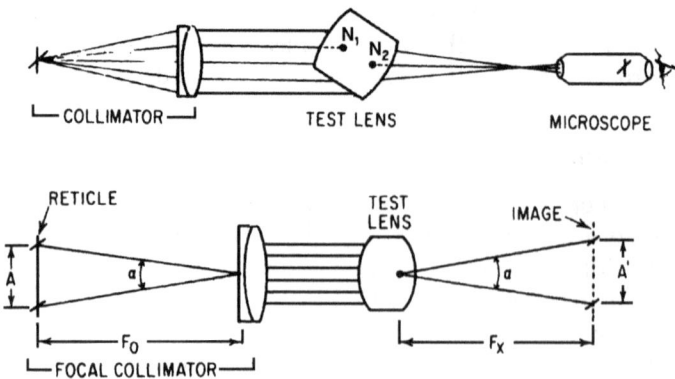

Figure 20.13 Illustrating the nodal slide (upper) and the focal collimator (lower) methods of measuring focal length on the optical bench.

lens is then realigned with the collimator axis and the location of the focal point is determined. Since the nodal points and principal points are coincident when a lens is in air, the distance from the nodal point to the focal point is the effective focal length.

This technique is basic and applicable to a wide variety of systems. Its limitations are primarily in the location of the nodal point. The operation of swinging the lens, shifting its position, swinging again, and so on, is tedious, and since it is discontinuous, it is difficult to make an exact setting. If the axis of the test lens is not accurately centered over the axis of rotation of the nodal slide, there will be no position at which the image stands still. Lastly, the measurement of the distance from the axis of rotation to the position of the aerial image is subject to error unless the equipment is carefully calibrated.

A *focal collimator* consists of an objective with a calibrated reticle at its focal point. The focal length of the objective and the size of the reticle must be accurately known. The test lens is set up and the size of the image formed by the lens is accurately measured with the measuring microscope. From Fig. 20.13 it is apparent that the focal length of the test lens is given by

$$F_x = A' \left(\frac{F_0}{A} \right) \tag{20.10}$$

where A' is the measured size of the image, A is the size of the reticle, and F_0 is the focal length of the collimator objective. Note that the focal collimator may be used to measure negative focal lengths as well as positive; one simply uses a microscope objective with a working distance longer than the (negative) back focus of the lens under test.

It is apparent from Eq. 20.10 that any inaccuracies in the values of A', A, or F_0 are reflected directly in the resultant value of the focal length. Further, any error in setting the longitudinal position of the measuring microscope at the focus will be reflected in F_x. Note that both the nodal slide and focal collimator methods assume that the test lens is free of distortion. If an appreciable amount of distortion exists, the measurements must be made over a small angle; this, of course, will limit the accuracy possible.

In setting up a focal collimator, it is necessary to determine the collimator constant (F_0/A) to as high a degree of accuracy as possible. The value of A, the reticle spacing, can be readily measured with a measuring microscope. The focal length of the collimating lens can be determined to a high degree of accuracy by a finite conjugate version of the focal collimator technique. An accurate scale (or glass plate with a pair of lines) is set up 20 to 50 ft from the collimator lens, as shown in Fig. 20.14. The measuring microscope is used to measure the size of the image of the target accurately, and the distance from object to image is

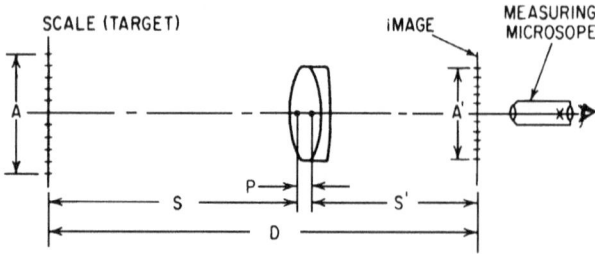

Figure 20.14   Setup for basic measurement of focal length.

measured. The value of p, the distance between the principal points is estimated, either from the design data of the lens or by assuming it to be about one-third the lens (glass) thickness. (As long as p is small compared to D, the error introduced by an inaccurate value of p is small.) Now since $D = s + s' + p$ and $A{:}s = A'{:}s'$, s and s' can be determined and substituted (with due regard for the sign convention) into

$$\frac{1}{s'} = \frac{1}{f} + \frac{1}{s} \tag{20.11}$$

and the value of the effective focal length determined. The necessity of estimating a value for p can be eliminated, if desired, by measuring the front focal length and applying the newtonian equation for magnification (Eq. 2.6) or, alternatively, by measuring front and back focal lengths (as outlined in the next paragraph), determining $p = \text{ffl} + \text{bfl} + t - 2f$, and repeating the original calculation; after a few iterations the calculation will converge to the exact p and f.

Collimation and the measurement of front and back focal lengths.   A basic method of locating the focal points is by autocollimation. As indicated in Fig. 20.15, an illuminated target is placed near the focus of the lens under test and a plane mirror is placed in front of the lens so as to reflect the light back into the lens. When the reflected image is focused on a screen in the same plane as the target, both screen and target lie in the focal plane. For accurate work an autocollimating microscope, shown in Fig. 20.16, produces excellent results. The lamp and condenser

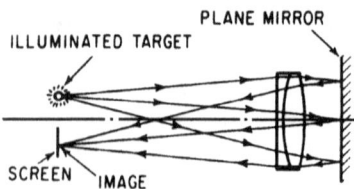

Figure 20.15   Autocollimation as a method of locating the focal points. When the object and reflected image are in the same plane (the focal plane), the system is autocollimated.

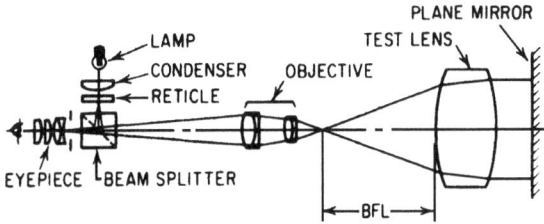

Figure 20.16 The autocollimating microscope is used to measure back focal length by focusing first on the surface of the test lens and then on the autocollimated image at the focal point.

illuminate the reticle, which may consist of clear lines scribed through an aluminized mirror. The reticle is then imaged at the focus of the microscope objective. The eyepiece of the microscope is positioned so that its focal plane is exactly conjugate with the reticle. Thus when the microscope is focused on the focal plane of the test lens, the reticle image is autocollimated by the test lens-plane mirror combination and is seen in sharp focus at the eyepiece. The microscope is then moved in to focus on the rear surface of the test lens; the distance traveled by the microscope is equal to the back focus of the lens.

The lens bench collimator itself may be adjusted for exact collimation using this technique. When the collimator reticle and the reflected image of the microscope reticle are simultaneously in focus, then the collimator is in exact adjustment. Note that the mirror must be a precise plano surface if accurate results are expected.

For routine measurements of back focus the bench collimator is substituted for the plane mirror, and if no autocollimating microscope is available, a little powder or a grease pencil mark on the rear surface of the test lens can be used as an aid in focusing on the lens surface.

In the absence of many of the usual laboratory trappings, it is still possible to make reasonably accurate determinations of focal lengths and focal points. A lens may be collimated simply enough by focusing it on a distant object. The error in collimation can be determined by the newtonian equation $x' = -f^2/x$, where x is the object distance less one focal length and x' is the error in the determination of the focal position. A set of distant targets, such as building edges, smokestacks, and the like, the angular separations of which are accurately known can often be substituted for a focal collimator in determining focal lengths.

Measurement of telescopic power.   The power of a telescopic system can be measured in three different ways. If the focal lengths of the objective and eyepiece (including any erectors) can be measured, their quotient equals the magnification. The ratio of the diameters of the entrance and exit pupils will also yield the magnifying power. Occasionally the multiplicity of stops in a telescope will introduce some confusion as to whether the pupils measured are indeed conjugates; in this case the image of a transparent scale laid across the objective can be measured at (or near) the exit pupil to determine the ratio. When the field of view is sharply defined, the magnification can be determined by taking the ratio of the tangents of the half-field angles at the eyepiece and the objective. Note that the almost inevitable distortion in telescopic eyepieces will usually cause this measurement of power to differ from measurements made by focal lengths or pupil diameters. One should ascertain that the telescope is in afocal adjustment before measuring the power. One way of doing this is to use a low-power (3 to 5×) auxiliary telescope (or dioptometer) previously focused for infinity at the eyepiece; this reduces the effect of visual accommodation when the focus is adjusted.

The measurement of aberrations.   In most instances the aberrations of a test lens can be readily measured on the lens bench by simulating a raytrace. For the measurement of spherical or chromatic aberration, a series of masks, each with a pair of small (to the order of a millimeter in diameter) holes, is useful. As indicated in Fig. 20.17, such a mask, centered over the test lens, simulates the passage of two "rays." When the image is examined with a microscope, a double image of the target is seen, except when the microscope is focused at the intersection of the two rays. By measuring the relative longitudinal position of the ray intersections for masks of various hole spacings, the spherical aberration can be determined. If the measurements are made in red and blue light, the data will yield the chromatic and spherochromatic aberration of the lens.

Figure 20.18 indicates how a similar three-hole mask can be used to measure the tangential coma of a test lens. A multiple-hole mask can also be used to measure and plot a ray intercept curve, if desired. The technique for measurement of field curvature is indicated in Fig. 20.19. The bench collimator is equipped with a reticle consisting of horizontal

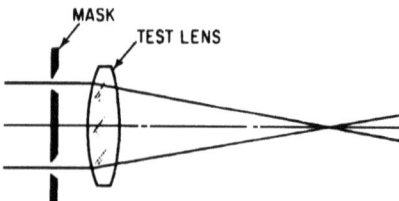

Figure 20.17   A two-hole mask can be used to locate the focus of a particular zone of a lens to determine the spherical aberration.

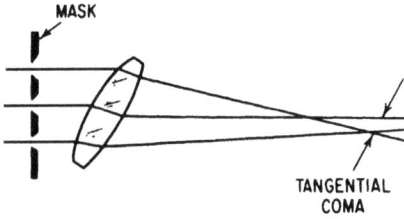

Figure 20.18 A three-hole mask can be used to measure the coma of a test lens.

and vertical lines. The focal length of the test lens is measured. The lens is then adjusted so that its second nodal point is over the center of rotation of the nodal slide and the position of the focal point (with the lens axis parallel to the bench axis) is noted. The lens is then rotated through some angle θ. From Fig. 20.19 it is apparent that the intersection of the (flat) focal plane of the lens with the bench axis will shift away from the lens by an amount equal to

$$\text{efl}\left(\frac{1}{\cos \theta} - 1\right)$$

as the lens is pivoted through an angle θ. The bench microscope is used to measure D, the amount by which the focus shifts along the axis. Two measurements are necessary, one for the sagittal focus and one for the tangential focus; this is the reason for the orthogonal line pattern of the reticle. Now the departure (along the *bench* axis) of the image surface from a flat plane is equal to

$$D - \text{efl}\left(\frac{1}{\cos \theta} - 1\right)$$

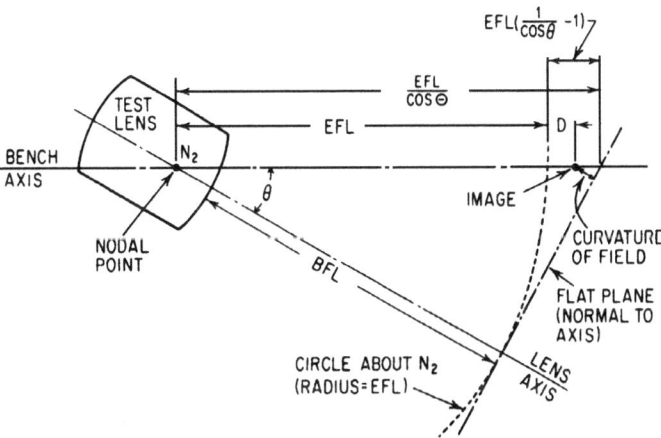

Figure 20.19 Geometry of the measurement of field curvature using the lens bench nodal slide.

and the curvature of field (parallel to the *lens* axis) is given by

$$ x = \cos \theta \left[D - \text{efl} \left(\frac{1}{\cos \theta} - 1 \right) \right] $$

Much of the numerical work in determining the field curvature by this method can be eliminated by the use of a T-bar attachment to the nodal slide. The cross bar of the T acts as a guide for the bench microscope, causing it to focus on the flat field position as the lens is pivoted. Thus one may measure the value of $x/\cos \theta$ directly; the use of the T-bar eliminates several sources of potential errors inherent in the method described above, although it does complicate the construction of the nodal slide. Measure at $\pm\theta$ to detect a tilted field.

Distortion is a difficult aberration to measure. The nodal slide may be used. The lens is adjusted on the slide so that no lateral image shift is produced by a *small* rotation of the lens. Then as the lens is pivoted through larger angles, any lateral displacement of the image is a measure of the distortion. An alternate method is to use the lens to project a rectilinear target and to measure the sag or curvature of the lines in the image, or to measure the magnification of targets of several different angular sizes. The difficulty with *any* method of measuring distortion is that one invariably winds up basing the work on measurements of magnification (or whatever) vanishingly close to the axis, and the accuracy of such small measurements is usually quite low.

The star test. If the object imaged by a lens is effectively a "point," i.e., if its nominal image size is smaller than the Airy disk, then the image will be a very close approximation to the diffraction pattern. A microscopic examination of such a "star" image can indicate a great deal about the lens to the experienced observer. One should be sure that the microscope NA is larger than that of the lens being tested. On the axis, the star image of a perfectly symmetrical (about the axis) system obviously must be a symmetrical pattern. Therefore, any asymmetry in the on-axis pattern is an indication of a lack of symmetry in the system. A flared or coma-shaped pattern on axis generally indicates a decentered or tilted element in the system. If the axial pattern is cruciform or shows indications of a dual focus, the cause may be axial astigmatism due either to a toroidal surface, a tilted or decentered element, or an index inhomogeneity.

The axial pattern may also be used to determine the state of correction of spherical and chromatic aberration. The outer rings in the diffraction pattern of a well-corrected lens are relatively inconspicuous, and the pattern, when defocused, looks the same both inside and outside the best focus point. In the presence of undercorrected spherical, the pattern will show rings inside the focus and will be blurred outside the focus; the reverse is true of overcorrected spherical. When the spherical aberration is a zonal residual, the ring pattern tends to be heavier and more pronounced than that from simple under- or overcorrected spherical.

In the case of undercorrected chromatic, the pattern inside the focus will have a blue center and a red or orange outer flare. As the microscope focus is moved away from the lens, the center of the pattern may turn green, yellow, orange, and will finally become red with a blue halo. The reverse sequence will result from overcorrected chromatic. A chromatically "corrected" lens with a residual secondary spectrum usually shows a pattern with a characteristic yellow-green (apple green) center surrounded by a blue or purple halo.

Off-axis star patterns are subject to a much wider range of variations. The classical comet-shaped coma pattern is easily recognized, as is the cross- or onion-shaped pattern due to astigmatism. However, it is rare to find a system with a "pure" pattern off-axis, and it is much more common to encounter a complex mixture of all the aberrations, which are difficult, if not impossible, to sort out.

The star test is a very useful *diagnostic* tool requiring only minimal equipment, and, in skilled hands, it can be highly effective. The novice should be warned, however, that reliable judgments of relative quality are difficult, and a considerable amount of experience is necessary before one can safely depend on a star check for even simple comparative evaluations. It should *not* be used for quality control acceptance tests. Again, be certain that the microscope NA is larger than that of the system under test.

The Foucault test. The Foucault, or knife-edge, test is performed by moving a knife (or razor-blade) edge laterally into the image of a small point (or line) source. The eye, or a camera, is placed immediately behind the knife, and the exit pupil of the system is observed. The arrangement of the Foucault test is shown in Fig. 20.20. If the lens is perfect and the knife is slightly ahead of the focus, a straight shadow will move across the exit pupil in the same direction as the knife. When the knife is behind the focus, the direction of the shadow movement is the reverse of the knife direction. When the knife passes exactly through the focus, the entire pupil (of a perfect lens) is seen to darken uniformly.

The same type of analysis can be applied to *zones* of the pupil. If an entire zone or ring of the pupil darkens suddenly and uniformly as the knife is advanced into the beam, then the knife is cutting the axis at the focus of that particular zone. This is the basis of most of the quantitative measurements made with the Foucault test. The technique generally

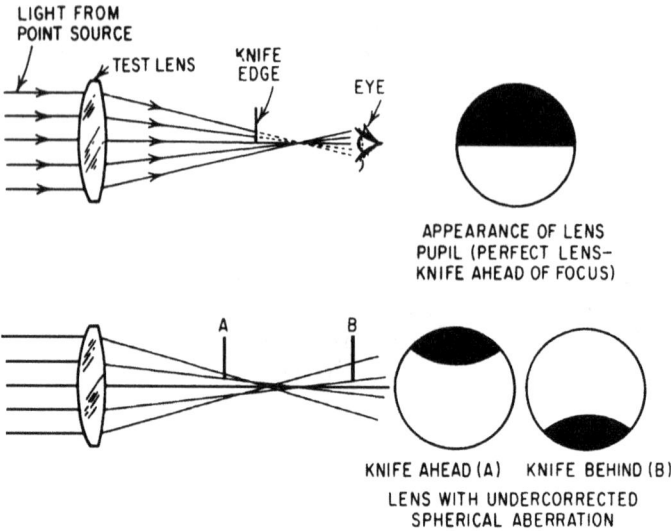

Figure 20.20 The Foucault knife-edge test. Upper: On a perfect lens the knife shadow has a straight edge. Lower: The shadow has a curved edge in the presence of spherical aberration. When the knife cuts through the focus, the pupil (or the zone of the focus) darkens uniformly.

used is to place a mask over the lens with two symmetrically located apertures to define the zone to be measured. The knife is shifted longitudinally until it cuts off the light through both apertures simultaneously. It is then located at the focus for the zone defined by the mask. The process is repeated for other zones, and the measured positions of the knife are compared with the desired positions.

This test is extremely useful in the manufacture of large concave mirrors, which can be tested either at their focus or at their center of curvature. For the center-of-curvature test, the source is a pinhole closely adjacent to the knife (Fig. 20.21), and a minimum of space and equipment is required. Obviously if the mirror is a sphere, all zones will have the same focus, and a perfect sphere will darken uniformly as the knife passes through the focus. When the surface to be tested is an aspheric,

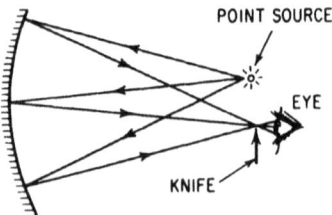

Figure 20.21 The knife-edge test applied to a concave mirror by placing both knife and source at the center of curvature.

the desired foci for the various zones are computed from the design data and the measurements are compared with the calculated values. It is a relatively simple matter to convert these focus differences into errors in the surface contour; in this way the optician can determine which zones of the lens or mirror require further polishing to lower the surface.

If the aspheric surface equation is expressed in the form

$$x = f(y)$$

then the equation of the normal to the surface at point (x_1, y_1) is

$$y = y_1 + f(y_1) f'(y_1) - xf'(y_1)$$

[where $f'(y) = dx/dy$], and the intersection of the normal with the (optical) axis is then

$$x_0 = x_1 + \frac{y_1}{f'(y_1)}$$

As an example, for a paraboloid represented by

$$x = \frac{y^2}{4f}$$

$$f'(y) = \frac{dx}{dy} = \frac{y}{2f}$$

and the axial intersection of the normal through the point (x_1, y_1) is

$$x_0 = x_1 + \frac{y_1}{(y_1/2f)} = x_1 + 2f = \frac{y_1^2}{4f} + 2f$$

This last equation gives the longitudinal position at which the knife edge should uniformly darken a ring of semidiameter y_1, when a parabola is tested at the center of curvature (as in Fig. 20.21) and knife and source are simultaneously moved along the axis.

In practice, the knife edge is adjusted longitudinally until the central zone of the mirror darkens uniformly. The distance from the knife to mirror is then equal to $2f$. Then a series of measurements is made using masks with half-spacings of y_1, y_2, y_3, etc., each measurement yielding an error e_1, e_2, e_3, etc., where e is the longitudinal distance from the "desired" position for the knife to the actual position.

These data may be readily converted into the difference between the actual slope of the surface and the desired slope by reference to Fig. 20.22. When e is small, we can (to a very good approximation) write for our parabolic example

$$\frac{A}{e} = \frac{y}{\sqrt{4f^2 + y^2}}$$

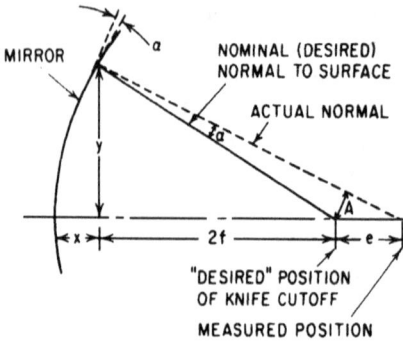

Figure 20.22 Geometry of knife-edge test used to determine the surface contour of a concave (paraboloidal) mirror.

where the term in the right-hand denominator is the distance from the surface to the axis taken along the normal. Now the angle α between the actual normal and the desired normal is equal to

$$\alpha = \frac{A}{\sqrt{4f^2 + y^2}}$$

and substituting for A from the previous expression, we get

$$\alpha = \frac{ye}{4f^2 + y^2}$$

Note that α is also the amount by which the slope of the surface is in error; we can determine the actual departure of the surface from its desired shape by reference to Fig. 20.23. Taking the surface error at the axis as zero, the departure from the desired curve at y_1 is given by

$$d_1 = \frac{-y_1\alpha_1}{2}$$

At y_2 it is

$$d_2 = d_1 - \frac{1}{2}(y_2 - y_1)(\alpha_1 + \alpha_2)$$

At y_3 it is

$$d_3 = d_2 - \frac{1}{2}(y_3 - y_2)(\alpha_2 + \alpha_3)$$

In general we can write

$$d_n = \frac{1}{2}\sum_{i=1}^{i=n}(y_{i-1} - y_i)(\alpha_{i-1} + \alpha_i)$$

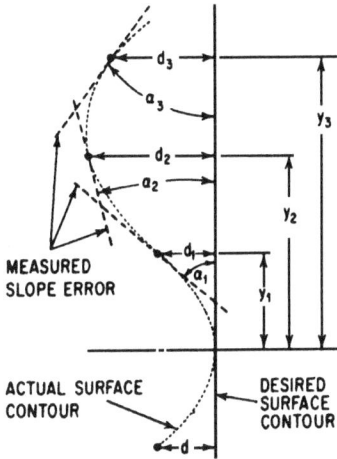

Figure 20.23 Conversion of measured errors of surface slope (α) into the departure (d) of the actual surface from the desired surface.

where y_0 and α_0 are assumed zero, and the sign of d is positive if the actual surface is above (to the right in Figs. 20.22 and 20.23) the desired surface.

The method outlined above can be readily applied to any concave aspheric. Since it checks the aspheric only at discrete intervals, it must, of course, be supplemented with an overall knife-edge check to be certain that the surface contour is smooth and free from ridges or grooves. The testing of convex surfaces is more difficult; they are usually checked in conjunction with another mirror chosen so that the combination has an accessible "center focus." The computation of the normal is more involved in this case, but the principles involved are exactly the same.

The Schlieren test. The Schlieren test is actually a modification of the Foucault test in which the knife blade is replaced by a small pinhole. Thus any ray which misses the pinhole causes a darkened region in the aperture of the optical system. The Schlieren test is especially useful in detecting small variations in index of refraction, either in the optical system or in the medium (air) surrounding it. In wind-tunnel applications, the tunnel is set up between a collimating optical system and a matching system which focuses the image on the pinhole. When the test is recorded photographically, it is possible to derive quantitative data on the airflow from density measurements on the film.

Resolution tests. Resolution is usually measured by examining the image of a pattern of alternating bright and dark lines or bars. Conventionally, the bright and dark bars are of equal width. A target consisting of several sets of bar patterns of graded spacing is used, and the finest pattern in which the bars can be distinguished (and in which the number of bars in the image is equal to the number in the object) is taken as the limiting resolution of the system under test.

The resolution patterns in use vary in two details of (relatively minor) significance: the number of lines or bars per pattern and the length of the lines relative to their width. The most common practice is to use three bars (and two spaces) per pattern, with a length of five, or more, bar widths. The USAF 1951 target is of this type and the patterns are graded in frequency with a ratio of the sixth root of 2 between patterns. The National Bureau of Standards Circular No. 533 includes both high- (25:1) and low- (1.6:1) contrast three-bar patterns which are approximately 1-in long and range in frequency from about one-third line per millimeter to about three lines per millimeter in steps of the fourth root of 2. A number of transparent (on film or glass) targets are commercially available; these are, for the most part, based on the USAF target.

Figure 20.24 shows two types of resolution test targets. The USAF 1951 target is probably the most widely used and accepted resolution target. The radial target is interesting since it nicely demonstrates the 180° phase shift of the optical transfer function. This produces the "spurious resolution" which is illustrated in Fig. 20.24c. See also Figs. 15.14 and 15.15.

The number of line pairs per millimeter in the USAF 1951 target is given in Table 20.2. The "groups" differ by a factor of 2. There are six sets of three-bar targets in each group; each set differs by the sixth root of 2.

In evaluating the resolution of a system it is important to adopt a rational criterion for deciding when a pattern is "resolved." The following is *strongly* recommended: A pattern is resolved when the lines can be discerned, and when all coarser (lower-frequency) patterns also meet this requirement. This implicitly requires that the number of lines in the image be the same as in the target, and also rules out spurious resolution. Do *not* allow any consideration of "sharpness," "definition," "crispness," "clearly resolved," "contrast," or the like to enter the evaluation;

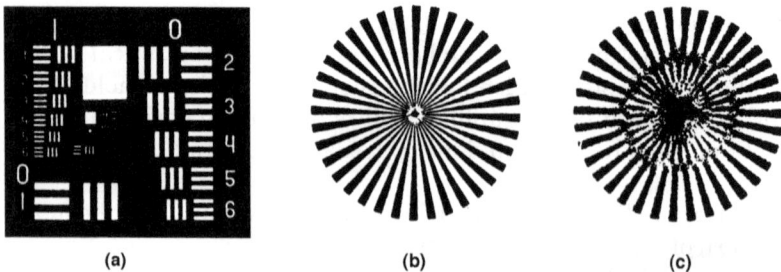

(a)   (b)   (c)

Figure 20.24 (a) USAF1951 resolution chart; (b) Siemens star resolution chart; (c) defocusing a well-corrected lens can cause a 180° phase shift which reverses the contrast of the pattern, causing areas which should be dark to be light and vice versa.

TABLE 20.2   Number of Lines/mm in USAF Resolving Power Test Target 1951

				Group Number					
Element	−2	−1	0	1	2	3	4	5	6	7
1	0.250	0.500	1.00	2.00	4.00	8.00	16.00	32.0	64.0	128.0
2	0.280	0.561	1.12	2.24	4.49	8.98	17.95	36.0	71.8	144.0
3	0.315	0.630	1.26	2.52	5.04	10.10	20.16	40.3	80.6	161.0
4	0.353	0.707	1.41	2.83	5.66	11.30	22.62	45.3	90.5	181.0
5	0.397	0.793	1.59	3.17	6.35	12.70	25.39	50.8	102.0	203.0
6	0.445	0.891	1.78	3.56	7.13	14.30	28.50	57.0	114.0	228.0

these are all subjective and involve individual interpretation. They lead to interminable arguments. The only consideration that should be used is "Can you discern the lines?"

The resolution of a photographic system is tested by photographing a suitable target and examining the film under a microscope. In order to obtain optimum results, the photographic processes must be carried out with extreme care, especially with regard to the selection of the best focus, exposure and development, and the elimination of any vibration in the system. If the microscope used in examination of the test film has a power approximately equal to the number of lines per millimeter in the pattern, the visual image will have a frequency equal to one line per millimeter and will be easy to view.

Objective lenses can be tested on an optical bench with a resolution target in the collimator. For lenses with an appreciable angular coverage, an accurate T-bar nodal slide is practically a necessity if reliable off-axis results are to be obtained. Projection of a resolution target is a very convenient means of checking the resolution of lenses designed to cover areas less than a few inches in size. Care must be taken to ensure that the illumination system of the projector completely fills the aperture of the lens under test; otherwise, the results may be misleading. In all resolution tests, the alignment of the lens axis perpendicular to the target and film planes is a critical factor. The resolution of telescopic systems can be checked by visual observation of a suitably distant or collimated target. Since the limiting resolution of a telescope is frequently (by design) close to the limiting resolution of the eye, a common practice is to view the image through a low-power auxiliary telescope. Such a telescope serves a dual purpose in that it reduces the effect of the observer's visual acuity on the measurement and also reduces the effect that involuntary accommodation (focusing) can have.

The classical criterion for resolution, namely, the ability of a system to separate two point sources of equal intensity, is seldom used (except in astronomy). This is largely because a test using line objects is much easier to make.

Measurement of the modulation transfer function. The measurement of the MTF (frequency response) is, in principle, quite straightforward. The basic elements of the equipment are shown in Fig. 20.25. The test pattern is one in which the brightness varies as a sinusoidal function of one dimension. Such a target is not an easy thing to prepare; fortunately the errors introduced by a target which is not truly sinusoidal are unimportant for most purposes. Some instruments utilize "square-wave" targets. The target pattern is imaged by the test lens on a narrow slit the direction of which is exactly parallel to the target pattern. The light passing through the slit is measured by a photodetector.

As the target or *the slit* is shifted laterally, the amount of light falling on the detector will vary, and the image modulation is given by

$$M_i = \frac{max - min}{max + min}$$

where max and min represent the maximum and minimum illumination on the photodetector. The object modulation M_0 is similarly derived from the maximum and minimum brightness levels of the target. The MTF (or frequency response, or sine-wave response, or contrast transfer) is then the ratio $M_i : M_0$.

A provision is usually made to vary the spatial frequency of the target pattern so that the response may be plotted against frequency. The target portion of the system may be as simple as a set of interchangeable targets which are slowly traversed by hand, or it may be a fully

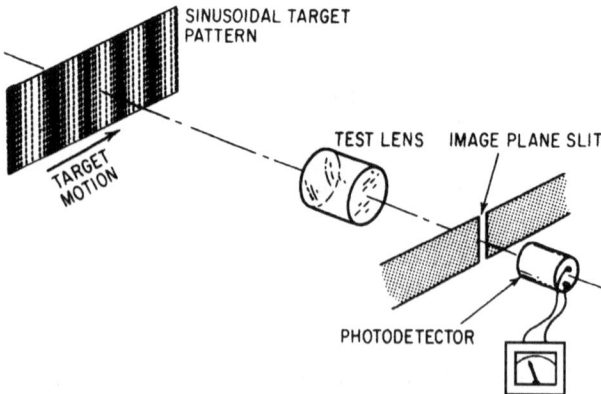

Figure 20.25 The basic elements of modulation transfer (frequency response) measurement equipment. The motion of the target scans its image across the narrow slit, where the maximum and minimum illumination levels are measured. By using targets of different spatial frequency, a plot of the modulation transfer function (vs. frequency) can be obtained.

automatic device which translates the target and scans a range of frequencies simultaneously.

The image-plane slit is almost never just a slit, since the manufacture of a slit of the required narrow dimensions can be fairly difficult. Instead, the image is magnified by a first-class microscope objective; this allows the use of a wider slit.

Obviously any real slit width will have some effect on the measurements, and a slit as narrow as the sensitivity of the photodetector will allow should be used. The effect of the slit width on the response may be readily calculated, since it simply represents a line spread function of rectangular cross section, and the data can be adjusted accordingly where necessary. The response of a slit of width w at a frequency v is $\sin(\pi v w)/(\pi v w)$. The measured MTF can be divided by this to correct for the effect of the finite width of the slit.

The source of illumination and the spectral response of the photodetector must, of course, be matched to the application for which the system under test is to be used. Otherwise, serious errors in measurement will result from the unwanted radiation outside the spectral band for which the system has been designed. Usually a set of filters can be found which will provide the proper response.

Another technique which is much more widely used than that described above is based on a *knife-edge scan*. A knife edge is passed through the image of a point (or slit) and the light passing by the edge is measured. If the measured light I is plotted against the lateral position of the knife edge y, the slope of the curve (dI/dy) is exactly equal to the line spread function of the lens. The MTF can be calculated from the line spread function using the methods outlined in Sec. 15.8. Most commercial MTF equipment is set up so that the knife-edge scan data are read directly into a computer which processes the data to calculate the MTF at whatever frequency is desired. Note that this technique does not require a sinusoidal target, nor does it require a separate target for each frequency. As in any MTF measurement, the spectral distribution of the source and the response of the light-measurement sensor must match that of the application.

The wave-front shape as measured by an interferometer can also be used to determine the MTF. The fringe pattern is scanned to digitize the data, and it is computer-processed to calculate the MTF at any desired frequency as in the knife-edge scan. This is entirely adequate for mirror systems or systems which operate at the laser wavelength. For systems which utilize a finite-width spectral band or a different wavelength, the results are not correct.

The analysis of "unknown" optics.   It is frequently necessary to determine the constructional parameters of an existing optical system. An example might be the analysis of a sample system to determine the reason for its failure to perform to the designer's expectations. Another example might be the analysis of an existing lens so that its design data can be used as the starting point for a new design. Yet another reason might involve patent coverage. For the most part this amounts to the measurement of the radii, thicknesses, spacings, and indices of the system components.

Since the measurements to be made are frequently of a precision barely adequate for the purpose, it is best to provide as many interdependent checks on the process as possible. Thus the first steps should include accurate measurements of effective, back, and front focal lengths, as well as the aberrations, so that when all of the measured system data are collected, a calculation of the complete (measured) system can provide a final comparison check on the overall accuracy of the analysis.

The thicknesses and spacings of a system are readily measured. For small systems a micrometer (equipped with ball tips for concave surfaces) is sufficient. A depth gage or an oversize plunger caliper (Nonius gage) is useful for larger systems. If a dimension can be deduced from two different measurements (as a check), the extra time involved is usually a worthwhile investment.

The radius of an optical surface can be measured in many ways. The simplest is probably by use of a thin templet, or "brass gage," cut to a known radius and pressed into contact with the surface. Differences between the gage and glass of a few ten-thousandths of an inch can easily be detected this way, but such a gage is not useful unless it very nearly matches the surface.

The classical instrument for radius measurement is the spherometer, the basic principles of which are outlined in Fig. 20.26. The spherometer measures the sagittal height of the surface over a known diameter; the radius is determined from the formula

$$R = \frac{Y^2 + S^2}{2S} \pm r$$

where Y is the semidiameter of the spherometer ring, S is the measured sagittal height, and r is the radius of the ball if the spherometer legs are ball-tipped. If the surface is convex, r is negative. This is a good approximation if $r \ll R$. Since the sagittal height is a rather small dimension and thus subject to relatively large measurement errors, the accuracy of a spherometer leaves something to be desired even when extreme precautions are taken. One of the best ways to use a spherometer is as a comparison device, by measuring both the unknown radius and a (nearly equal) carefully calibrated standard radius (e.g., a test glass or a ball).

The diopter gage, or lens measure, or Geneva lens gage is a handy tool which can provide a quick approximate measure of the surface

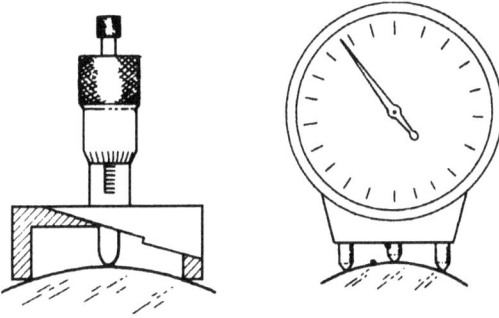

Figure 20.26 Left: Simple ring spherometer determines the radius of a surface through a measurement of the sagittal height. Right: The diopter gage or lens measure is a spherometer calibrated to read surface curvature in diopters.

curvature. As shown in Fig. 20.26, it consists of a dial gage with its plunger between two fixed points. The dial of a diopter gage is calibrated in diopters; the readings may be converted to radii by the formula

$$R = \frac{525}{D} \text{ m}$$

where the 525 is the constant representing $1000\,(n-1)$ for an "average" opthalmic glass. The accuracy of a typical diopter gage is to the order of 0.1 diopter. The lens gage can be calibrated by measuring several test plates of accurately known radius.

Probably the best way to measure a concave radius is by use of an autocollimating microscope. The microscope is first focused on the surface and is then focused at the center of curvature (where the microscope reticle image is imaged back on itself by reflection from the surface). The distance traveled by the microscope between these two positions is equal to the radius. The *precision* of this method can be to the order of micrometers; the *accuracy* is obviously dependent on the accuracy of the measurement method used. If the microscope used is of fairly high power (say 150× with NA = 0.3), the quality of the reflected image at the center of curvature is an excellent indication of the sphericity of the surface. Convex surfaces can be measured in this way provided that the working distance of the microscope objective is longer than the radius. A series of long-focal-length objectives is useful in this regard, although the precision of the method drops as the NA of the objective is lowered (long-focal-length objectives usually have a small NA) due to the increased depth of focus. If a precise determination of a long convex radius is necessary, a mating concave surface can be made so that it fits perfectly (as tested by interference rings) and the measurement is made on the concave glass. Master test plates are measured by this technique. Note that Eq. 20.3 can be used to calculate small radius differences from interference fringe readings.

If a separate piece of glass from which the lens under analysis was made is available, the measurement of its index can be made with considerable precision. The minimum deviation of a test prism may be measured on a laboratory spectrometer, and the prism equations of Chap. 7 used to find the index. Alternatively, a Pulfrich refractometer measurement can be made. Either method will readily yield the index value accurate to the fourth decimal place. When one is constrained to measure the lens element itself, without destroying it, the problem is more difficult. A crude determination of the index can be made for normal glasses (i.e., not the newer "light" glasses) by measuring the density of the element. A plot of the catalog values of the index against density is then used to determine (very approximately) the corresponding index. The classic relationship between index (n) and density (D) is very approximately $n = (11 + D)/9$.

A somewhat more general method is to measure the axial thickness of the element and then to measure the apparent optical thickness by focusing a measuring autoreflecting microscope first on one surface and then the other. A simple paraxial calculation, taking into account the refractive properties of the surface radius through which the second surface is viewed, will yield a value for the index. Depending on the thickness of the element, the index value achieved will probably be almost completely unreliable in the third place, due to the large relative inaccuracy in the measurement of the apparent thickness and to the spherical aberration introduced by the thickness of the glass.

If one measures the radii carefully and makes a good determination of the paraxial focal length of the element, the thick-lens formula for focal length can be solved to determine the index of refraction. Although this method requires skilled laboratory technique, it is capable of producing results which are accurate to one or two digits in the third place. Note that if care is not taken to eliminate the effects of spherical aberration from the focal-length measurement, the resulting index value will tend to err on the high side. Another nondestructive technique involves immersion in index-matching liquids, then measuring the index of the liquid.

20.5   Tolerance Budget Example

This section will describe the process of establishing a tolerance budget. In the interest of brevity we use a relatively simple case, and also make some simplifying assumptions. Our intent is to demonstrate the process as simply as possible.

TABLE 20.3   14 mm NA 0.42 Laser Recording Lens

	Radius	Space	Mat'l	Clear Aperture
0	object	76.539		
1	+50.366	2.80	SF11	11.65
2	−39.045	0.4353	air	11.62 (edge contact)
3	−19.836	2.00	SF11	11.62
4	−34.36	0.20	air	11.90
5	+17.42	2.65	SF11	11.81
6	+79.15	11.84	air	11.22
7	+7.08	2.24	SF11	5.24
8	+15.665	3.182	air	4.13
9	plane	2.032	acrylic	
10	plane			

We will use a four element 14 mm $f/1.2$ laser disk lens, working at a wavelength of 0.82 μm. The application for this lens will control focus, shifting the lens to maintain best focus. Thus, defocus and curvature of field are not a problem, although astigmatism and coma are. The performance requirement is that the Strehl ratio must be 75 percent or better over a total field of 0.7 mm. A Strehl of 75 percent corresponds to an OPD of 0.082 waves RMS or 0.288 waves (peak to valley), using $S = (1 - 2\pi^2\omega^2)^2$, where ω is the RMS OPD, and RMS equals (P-V)/3.5.

The nominal design, shown in Table 20.3 and Fig. 20.27, has an OPD of 0.04 waves on axis and 0.23 waves at the edge of the field. Using 0.23 waves, we find that if the combined fabrication tolerances produce an OPD of 0.173 waves, 0.23 and 0.173 will RSS (Eq. 20.9) to yield an OPD of 0.288 waves.

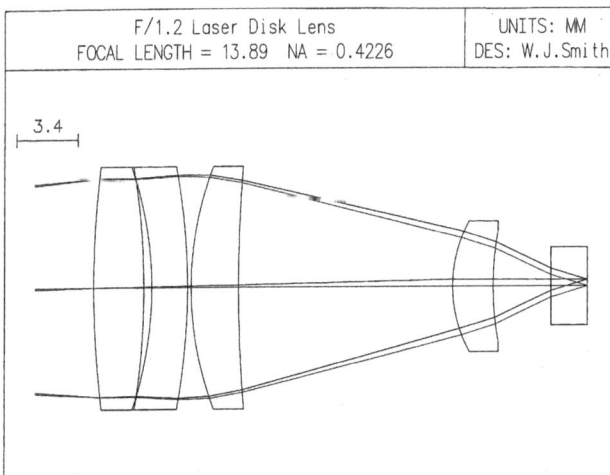

```
F/1.2 Laser Disk Lens              UNITS: MM
FOCAL LENGTH = 13.89  NA = 0.4226  DES: W.J.Smith
```

Figure 20.27   Lens used in example of tolerance budgeting.

The "change table" is calculated by changing each construction parameter of the lens, one at a time, by a small amount (to the order of several times the expected tolerance values) and calculating the change produced in the aberrations of interest. For this example we are interested in the changes produced in the transverse spherical aberration (TA), the tangential coma (COMA$_T$), and the astigmatism (ASTIG). The change sizes used in creating Table 20.4 were:

1. a surface radius change corresponding to 10 fringes on a test plate (see Eq. 20.3),
2. a spacing change of 0.2 mm,
3. an index change of 0.001, and
4. a surface tilt of 0.001 radians (3.4 minutes).

TABLE 20.4   The Aberration Changes in Wavelengths of (P-V) OPD Produced by Small Parameter Changes

	TA	COMAT	ASTIG.	RSS	RSS OF CLASS
R1	+.014λ	+.007λ	+.003λ	.016λ	
R2	−.005	−.020	−.002	.021	
R3	−.051	+.027	−.005	.058	Radius
R4	+.017	−.021	−.005	.027	.101λ
R5	−.027	−.010	+.002	.029	
R6	+.028	−.006	−.003	.029	
R7	−.013	+.004	+.003	.014	
R8	+.057	+.017	−.005	.060	
T1	−.001λ	+.003λ	+.002λ	.004λ	
T2	−.020	+.029	+.000	.035	Thickness
T3	−.037	−.004	+.003	.037	.091λ
T4	+.017	−.021	+.002	.027	
T5	+.021	−.029	+.005	.036	
T6	+.037	−.044	+.008	.059	
T7	−.008	−.009	+.002	.012	
N1	+.007λ	.000λ	−.004λ	.008λ	
N3	+.002	.000	−.003	.004	Index
N5	−.005	.000	.000	.005	.011λ
N7	−.004	.000	.000	.004	
TR1	---	+.043λ	+.009λ	.044λ	
TR2	---	−.069	+.010	.070	
TR3	---	+.179	−.015	.180	
TR4	---	−.093	+.009	.093	Tilt
TR5	---	+.101	+.013	.102	.277λ
TR6	---	−.106	+.006	.106	
TR7	---	+.024	−.014	.028	
TR8	---	−.080	+.010	.081	
RSS TOTAL	.110λ	.286λ	.034λ	.308λ	.308λ

Most computer programs have a feature which will readily produce a change table.

In addition to the aberrations tabulated earlier, we should include the effects of surface irregularity or asphericity, which will produce a (P-V) OPD of:

$$\text{OPD} = {}^1\!/_2\,(\#\text{FR})(n' - n)(\lambda_1/\lambda_2) \text{ wavelengths}$$

Inserting appropriate numbers, we get:

$$\text{OPD} = 0.5 \times 1 \times (1.746 - 1)(0.59/0.82) = 0.275 \text{ wavelengths, per fringe}$$

Where (#FR) is the number of fringes, $(n' - n)$ is the index break across the surface, λ_1 is the test wavelength, and λ_2 is the wavelength of use. Since we have eight surfaces, one fringe of irregularity on each surface (when RSS-ed) produces 0.778 waves of (P-V) OPD.

As a preliminary trial budget we can simply use as tolerances the values used in making the change table. As indicated along the bottom and the right edge of the table, RSS-ing all of the effects yields a value of 0.308 waves of (P-V) OPD. If we now RSS the irregularity OPD of 0.778 waves with the tolerance 0.308 waves, we get 0.837 waves.

This exceeds the 0.288 waves which we calculated would yield a 75 percent Strehl by a factor of 2.9×. We could get a budget which met the specifications by dividing all of the assumed tolerances by three. This yields the following budget:

1. ±three rings test plate fit,

2. ±0.07 mm thickness,

3. ±0.0003 index, and

4. ±0.3 fringe of irregularity.

An inspection of the change table indicates that the sensitivity of the tolerances varies widely, ranging from the total insensitivity of coma to the index changes, to significant effects from the radius and thickness changes and to the strong contributions from tilts and irregularities. We have previously noted that the larger items dominate the RSS process, so that a rational approach might be to reduce the tolerance on the most sensitive dimensions, and possibly increase the tolerance on those dimensions which are relatively insensitive.

There are a few practical considerations which affect the budget adjustment process. One is that the test plate fit and the irregularity should have no more than a four- or five-to-one relationship. This is because it is difficult to determine the size of an irregularity if the test plate fit shows too many fringes. Another consideration is that most optical shops have a customary profile of tolerances. Thus if we specify

TABLE 20.5   The Tolerance Budget

Tolerance		(P-V) RSS OPD
Radius test plate fit	: 1 ring	.010λ
Surface irregularity	: 0.2 ring	.156
Index variation	: .001	.011
Surface tilt	: .0002 radians	.055
Thickness tolerance	T1: 0.10	.002
	T2: 0.02	.004
	T3: 0.05	.009
	T4: 0.04	.005
	T5: 0.05	.009
	T6: 0.03	.009
	T7: 0.07	.004
	RSS	.167λ
	Design edge of field OPD	.230
	RSS Total	.284λ
	Strehl ratio	75.7%

a tolerance which is larger than that which is customary, there is little or no saving to be had; the piece will be made to the shop's customary tolerances. The same is true for optical glass. In the other direction however, a tighter tolerance than the shop profile will result in a higher cost, and too tight a tolerance may push the cost toward infinity.

If we consider the RSS column in the change table, this indicates the relative sensitivity of each dimension. This can be used as an indicator for which dimension needs a tight tolerance and which can tolerate a loose one. Following this line, we get the following budget, for which the RSS OPD is 0.167 waves, just slightly better than the 0.173 wave required to achieve the specified 75 percent Strehl ratio.

See Sec. 20.3 for the probabilities connected with the RSS sums. In the interest of simplicity we have considered each thickness independently and neglected the effect of an element thickness on the adjacent airspace. If the surface irregularity is astigmatism (which is likely in production) then its orientation is a random factor and the combined OPD will be about a third less. If the irregularity is a "gull-wing" pattern, the combination will be less random and the tolerance should be smaller.

Bibliography

Baird, K., and G. Hanes, in Kingslake (ed.), *Applied Optics and Optical Engineering*, Vol. 4, New York, Academic, 1967 (interferometers).

DG-G-451, Flat and Corrugated Glass.

Deve, C., *Optical Workshop Principles*, London, Hilger, 1945.

Habell, K., and A. Cox, *Engineering Optics*, London, Pitman, 1948.

Hopkins, R., in Shannon and Wyant (eds.), *Applied Optics and Optical Engineering*, Vol. 8, New York, Academic, 1980 (lens mounting).

Ingalls, G., *Amateur Telescope Making*, books 1, 2, and 3, *Scientific American*, 1935, 1937, 1953.

JAN-P-246 Slide Projectors.

Karow, H. Fabrication Methods for Precision Optics, Wiley 1993.

Malacara, D., "Optical Testing," in *Handbook of Optics,* Vol. 2, New York, McGraw-Hill, 1995, Chap. 30.

Malacara, D., and Z. Malacara, "Optical Metrology," in *Handbook of Optics,* Vol. 2, New York, McGraw-Hill, 1995, Chap. 29.

MIL-A-003920 Thermosetting Optical Cement.

MIL-C-48497 Scratch and Dig for Opaque Coatings.

MIL-C-675 Antireflection Coatings.

MIL-G-1366 Aerial Photography Window Glass.

MIL-G-16592 Plate Glass.

MIL-L-19427 Anamorphic Projection Lenses.

MIL-M-13508 Front Surface Aluminized Mirrors.

MIL-O-13830 Scratch and Dig Specifications.

MIL-O-16898 Packaging Optical Elements.

MIL-P-47160 Optical Black Paint.

MIL-P-49 16-mm Projectors.

MIL-R-6771 Glass Reflectors, Gunsight.

MIL-STD-1241 Optical Terms and Definitions.

MIL-STD-150 Photographic Lenses.

MIL-STD-34 Drawings for Optical Elements and Systems.

MIL-STD-810 Interference Filters.

McLeod and Sherwood, *J. Opt. Soc. Am.,* Vol. 35, 1945, pp. 136–138 (origin of the scratch and dig standards).

Offner, A., *Applied Optics,* Vol. 2, 1963, pp. 153–155 (null lens for parabola).

Parks, R., "Optical Fabrication," in *Handbook of Optics,* Vol. 1, New York, McGraw-Hill, 1995, Chap. 40.

Parks, R., in Shannon and Wyant (eds.), *Applied Optics and Optical Engineering,* Vol. 10, San Diego, Academic, 1987 (fabrication).

Photonics Buyers Guide, Optical Industry Directory, annually, Laurin Publishing Co., Pittsfield, Mass.

Rhorer and Evans, "Fabrication of Optics by Diamond Turning," in *Handbook of Optics,* Vol. 1, New York, McGraw-Hill, 1995, Chap. 41.

Sanger, G., in Shannon and Wyant (eds.), *Applied Optics and Optical Engineering,* Vol. 10, San Diego, Academic, 1987 (fabrication, diamond turning).

Scott, R., in Kingslake (ed.), *Applied Optics and Optical Engineering,* Vol. 3, New York, Academic, 1965 (optical manufacturing).

Shannon, R. R., "Optical Specifications," in *Handbook of Optics,* Vol. 1, New York, McGraw-Hill, 1995, Chap. 35.

Shannon, R. R., "Tolerancing Techniques," in *Handbook of Optics,* Vol. 1, New York, McGraw-Hill, 1995, Chap. 36.

Shannon, R., in Kingslake (ed.), *Applied Optics and Optical Engineering,* Vol. 3, New York, Academic, 1965 (testing).

Shannon, R., in Shannon and Wyant (eds.), *Applied Optics and Optical Engineering,* Vol. 8, San Diego, Academic, 1980 (aspherics).

Strong, J., *Procedures in Experimental Physics,* Englewood Cliffs, N.J., Prentice-Hall, 1938.

Strong, J., *Procedures in Applied Optics,* New York, Dekker, 1989.

The Optical Industry Directory, Pittsfield, Mass., Photonics Spectra (published annually),

Twyman, F., *Prism and Lens Making,* London, Hilger, 1988.

Yoder, P. R., *Mounting Lenses in Optical Systems,* S.P.I.E., Vol. TT21, 1995.

Yoder, P. R., "Mounting Optical Components," in *Handbook of Optics,* Vol. 1, New York, McGraw-Hill, 1995, Chap. 37.

Yoder, P., *Opto-Mechanical System Design,* New York, Dekker, 1986.

Young, A., in Kingslake (ed.), *Applied Optics and Optical Engineering,* Vol. 4, New York, Academic, 1967 (optical shop instruments).

Zschommler, W., *Precision Optical Glassworking,* New York, Macmillan/S.P.I.E., 1984.

Getting the Most Out of "Stock" Lenses*

21.1 Introduction

This chapter is intended as a guide and an aid to the use of "catalog" or "stock" lenses, that is, lenses that can be bought from one of the many suppliers to the trade (or perhaps lenses of unknown origin which you may find in the back of a desk drawer or in a dusty cabinet in your laboratory). The ideas behind this topic are that (1) there is often a "best" way to utilize a lens for a particular application, (2) various types of lenses have different capabilities as to their coverage and speed, and (3) there are a number of simple ways to measure and test lenses, ways which do not necessarily require a lot of costly equipment.

21.2 Stock Lenses

The benefits of utilizing stock lenses (as opposed to having optics custom-made) are obvious to most of us. Cost is probably the first advantage that comes to mind. Although stock optics are priced at retail, with the substantial mark-up that this implies, this cost is significantly offset by the fact that stock optics are made in larger quantities than when one or two sets are made to order. The second big advantage is time. The optics are, almost by definition, usually in stock and available for immediate delivery.

---

*This chapter is adapted from Smith, *Practical Optical System Layout*, New York, McGraw-Hill, 1997.

Of course, there are drawbacks. The most obvious is that the stock lens has not been specifically designed for the application for which you are using it and cannot be expected to represent the ultimate in performance. Despite this, many systems assembled from stock lenses have turned out to be entirely satisfactory for their intended applications. Many stock lens systems have also been at best only crude prototypes and eventually have had to be replaced by custom designed and fabricated optics. This is not to imply that they were without value; much can be learned from a proof of concept or a rough prototype. Our aim here is to make our stock lens system as good as the available stock lenses will allow.

Another drawback with catalog optics is the need to select the optics from a limited list of available diameters and focal lengths, although if one has an extensive set of suppliers' catalogs, this may be less of a problem than it would appear at first glance.

Many vendors now include the nominal prescriptions for some of their optics in their catalogs, and many optical software providers include these prescriptions in their database. When this data is available, the performance of the stock lens system can be evaluated (using the software) without having to make a mock-up. Unfortunately, however, many lenses are sold without construction data. In some cases (particularly for more complex lenses such as anastigmats, microscope optics, and camera lenses) construction data are regarded as proprietary, and the prescriptions are not released even to large-volume OEM customers. In other cases the vendor simply may not know the prescription data; this is usually the case where the optics are salvage, scrap, or overruns.

An often overlooked problem with "stock" optics is the very real possibility of a limited supply. This is obviously to be expected in the case of salvaged or overrun lenses, where the total supply is often limited to stock on hand. Since vendors occasionally share the same inventory on a cooperative basis, it is wise to be sure that you are not counting the same lot of lenses twice when you check on the available quantity of the optics you plan to use. Another possibility is that the vendor may decide to drop your lens from the line or even retire from the business. If you go into production on an instrument which incorporates stock lenses, it is nightmare time when you discover that there are not any more.

An inexpensive insurance policy against this problem is to squirrel away a few sets of optics, so that when the ax falls you still have a reasonable chance of survival. A skilled optical engineer with good laboratory equipment can measure the radii, thickness, spacings, and index of a sample lens. Even if the measurements are not too accurate, an optical design program can optimize or touch up the measured data to suit your purposes. Then you can have some more fabricated. Even if the vendor publishes the construction data of a lens, it is still wise to hold on to a couple of samples; the published data may or may not be accurate.

In the case of salvage optics, you have probably ordered from a catalog listing of diameter and focal length. You should be aware of a couple of factors. The listings are probably based on measurements made on a sample lens, and the listing probably gives the focal length and diameter to the nearest millimeter. There may be more than one lens in the vendor's stock which fits the diameter and focal length description to this accuracy. The next time you order the same catalog item you may get a different lens, and this lens may perform differently.

21.3   Some Simple Measurements

This section is written for the benefit of the reader who does not have access to the usual optical measurement and laboratory equipment. A lens bench with collimator and a measuring microscope are essential for really accurate measurements of the imaging (i. e., gaussian) properties of a lens. However, there are a number of simple ways that an approximate measurement of focal length and back focal length can be made.

The measurement of the *back focal length,* or bfl (see Sec, 2.2 and Fig. 2.1), is relatively easy. When using a lens bench, a target at infinity is provided by the collimator; the location of the focal point is determined by focusing the lens bench microscope alternately on the focal point and on the last surface of the lens. A distant object (a tree, a building, a telephone pole) makes a reasonable substitute for a collimator. To get a rough measurement, one simply focuses the target on a light-colored wall or a piece of typing paper as shown in Fig. 21.1 and measures the lens-to-image distance with a scale. If the lens has

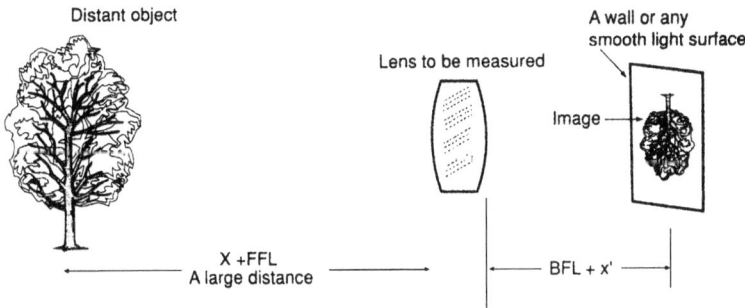

Figure 21.1   The back focal length (bfl) of a lens can be measured using a distant object (instead of a collimator) and measuring the lens-to-image distance with a scale or ruler. If the object is not effectively at infinity, the newtonian focus shift $(x' = -f^2/x)$ is subtracted from the measurement.

spherical aberration (and most simple lenses do) your measurement will come up a bit shorter than the paraxial back focus. If there is enough light, the spherical aberration can be minimized by reducing the lens aperture with a mask.

Using an object which is not collimated (i.e., not located at infinity) will introduce a small error in your measurement. The error in locating the focal point is indicated by Newton's equation (Eq. 2.3) as $x' = -f^2/x$, where x' is the error, f is the focal length, and x is the distance to your target. For example, if you measure the back focus of a 2-in-focal-length lens using a target down the hall which is only 50 ft (600 in) away, the error will be $x' = 2^2/600 = 0.007$ in; this is much less than the error in your crude measurement of the back focus is likely to be. Even in cases where this error is significant, you can always calculate the error and subtract it from the measured BFL to improve the validity of the result.

One problem associated with this technique is that stray light falling on the image will make it difficult to see and focus. A cardboard screen with a hole for the lens is one way around this problem; using a light bulb as a target in a darkened room is another. Figure 21.2 shows a matchbox slipped over the scale and slid into best focus as a handy tool for making the measurement.

The measurement of the *effective focal length* (efl) is considerably more difficult, because it involves the location of the principal points. Figure 3.7 shows the principal point locations for simple lenses of various shapes. A fair estimate can be made for single elements and most

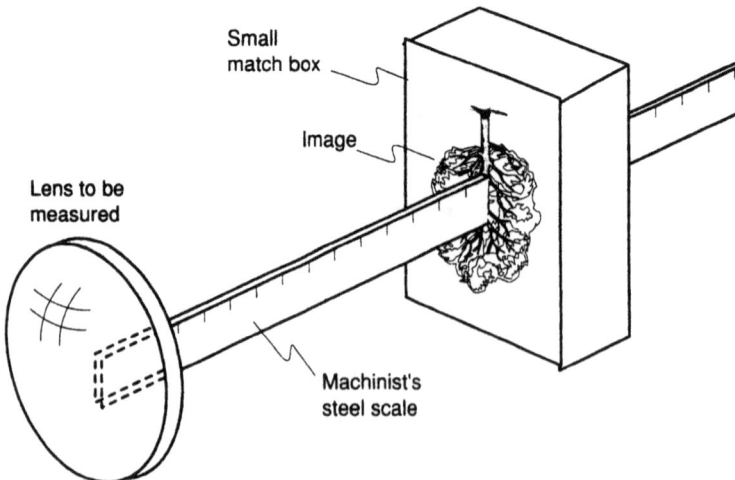

Figure 21.2 A matchbox-sized box (or a folded piece of pasteboard) slipped over a machinist's steel scale makes a convenient way to measure the back focal length of a lens.

cemented doublets by assuming that the space between the principal points is approximately one-third of the axial thickness of the lens. [A somewhat better estimate for singlets is $t(n - 1)/n$.] For plano-convex forms, one principal point is always located at the curved surface; for equiconvex lenses the points are evenly spaced within the lens. As illustrated in Fig. 21.3, adding a suitable fraction of the axial thickness to the measured bfl will then get an estimate for the efl. For an anastigmat, such as a Cooke triplet or a Tessar, the principal point locations are more difficult to estimate. Adding one-half to two-thirds of the vertex length of the lens to the bfl is about the best estimate one can make. In a complex lens the principal points are often almost coincident, and occasionally reversed.

To get a better value for the focal length one must measure a magnification. Magnification is simply the ratio of the image size to the object size. An illuminated object of known size is imaged and the image size is measured; admittedly, doing this accurately may be easier said than done, but it can be done. The setup is shown schematically in Fig. 21.4. The object-to-image distance (often called the total track length) is measured. The estimated spacing between the principal points is subtracted from the track length. This adjusted length is scaled to get s and s'. Dividing the adjusted track length T by $(m + 1)$,

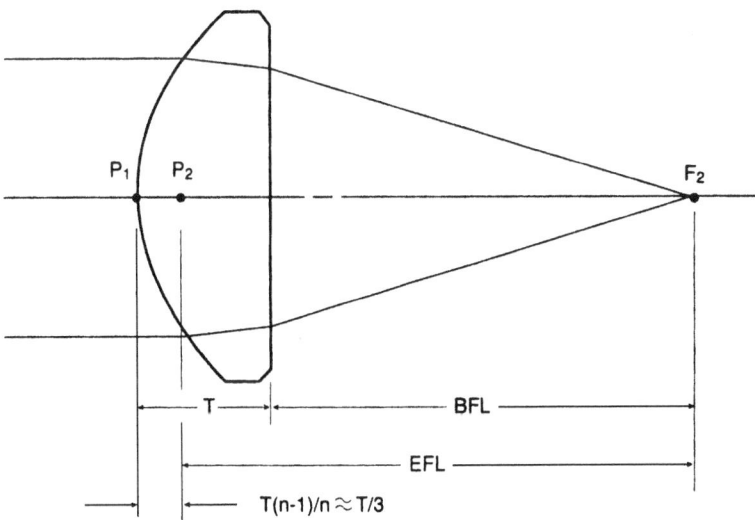

Figure 21.3 One way of estimating the effective focal length (efl) is to add a suitable fraction of the element thickness to the measured back focal length (bfl). For a plano-convex element, the convex side should face the distant target to minimize spherical aberration. If the lens is reversed, the measured bfl will equal the efl, but the spherical aberration will be much larger and will affect the measurement.

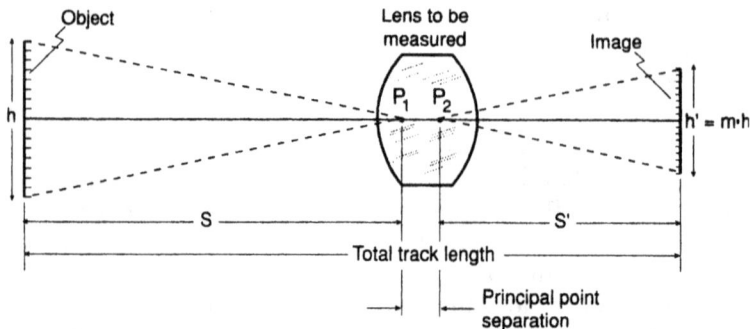

Figure 21.4 Setup for measuring effective focal length (efl) by measuring the magnification at finite conjugates: The magnification (here taken as the ratio of measured image size to object size—positive for an inverted image) and the track length are measured. Then the gaussian conjugates can be found by dividing the total track length (minus the principal point separation) by the magnification plus 1 to get S. Then $S' = mS$, and the Gauss equation can be solved for the focal length.

where m is the ratio of the image to object size will get s; then $(T - s)$ equals s'. (Note that we use a positive sign for the magnification m in this case.) Now we solve the Gauss equation (Eq. 2.4) to get the focal length.

If this process is carried out accurately for several different object-to-image distances, it is possible to eliminate the estimation of the principal point separation by making a simultaneous solution for the exact value.

Sample calculation

The object is a back-illuminated transparent scale, 15-in long, and the lens forms an image which is measured at 3.5 in. The magnification is thus $m = 3.5/15 = 0.2333$. The object-to-image distance is 40 in, and the lens is 1 in thick. Assuming that the principal points are separated by one-third the lens thickness, or 0.333, the adjusted track length is 39.667 in. We get the Gauss object distance by dividing the track length by $(m + 1)$, or 1.2333, which gives us $s = 39.667/1.2333 = 32.162$ in and $s' = 39.667 - 32.162 = 7.5045$ in. Substituting s and s' into Eq. 2.4, and solving for f gives us the focal length. (Note that our sign convention requires s to be negative.)

$$\frac{1}{s'} = \frac{1}{s} + \frac{1}{f}$$

$$\frac{1}{7.504} = \frac{1}{-32.162} + \frac{1}{f}$$

$$f = 6.0847 \text{ in}$$

Yet another way to measure efl involves the use of a distant target the *angular* size of which is known. The size of the image of the target is measured. Then the focal length of the lens is equal to half the measured image size divided by the tangent of the half angle which the object subtends. If the object is not at infinity, the Newton correction of f^2/x can be applied (as in the discussion of bfl above). A distant building with vents, chimneys, elevator towers, etc., on its roof line can be measured with a theodolite through a convenient window to serve as a target of this type. Alternatively, the measurement of a lens of accurately known focal length can be used to determine the angle.

21.4   System Mock-up and Test

A simple way of testing an optical system is to make a mock-up by fastening the elements to a ruler (or other convenient straightedge) with one of the available waxes which are commonly used for this purpose. This material is basically beeswax, formulated so that, when warmed in the hand, it is soft, pliable, and adherent to most things, but it becomes relatively firm at room temperature. This wax is available in stick form as optician's "red wax" or in an unpigmented tan color. Universal Photonics, Inc., of Hicksville, NY, carries "Red (or White) Sticky Wax" in stick or bulk form, and Central Scientific has a tacky wax packaged in 1- or 2-lb cans. The optics are conveniently stuck on the edge of the scale as indicated in Fig. 21.5, and a fair impression of

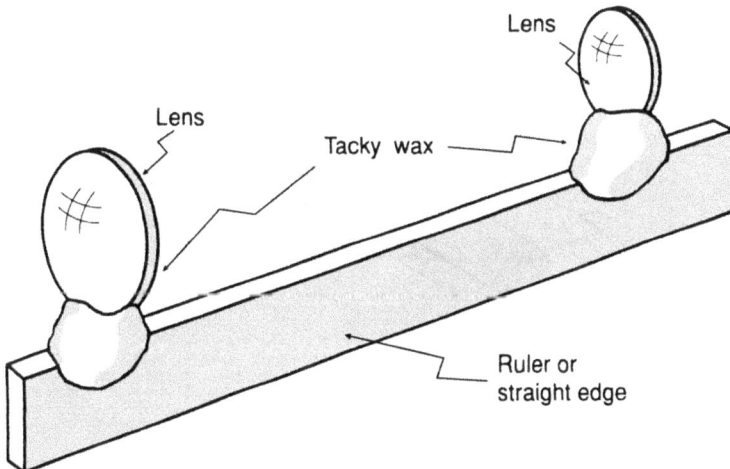

Figure 21.5   A straightedge and some tacky wax make a convenient way to mockup an optical system. This is especially handy for trying out visual systems such as telescopes or microscopes.

the performance can be obtained by viewing through the optics or projecting the image on a suitable screen.

A somewhat more elaborate lens bench can be made inexpensively
from a cold-rolled hexagonal bar as shown in Fig. 21.6. Two bars are
fastened parallel to each other, spaced apart so that short carrier sections can slide along the length of the bars. Each carrier section is
drilled to accept a vertical rod, which is adjustable for height and fixed
by tightening a set-screw. The lens can be waxed on the end of the rod,
or a short length of angle iron can be threaded on the end of the rod to
serve as a V block to hold the optics. This bench allows easy and rapid
adjustment of the spacings and alignments of the system components.

In making a mock-up of a system, the alignment of the optics with
respect to the optical axis can be extremely important. Establish an
axial center point on the object, and add the downstream components
one at a time, making sure that each image they form is well centered
on the axis. Visually sighting over (or through) the optics is often
helpful, as is the use of an HeNe laser beam as an alignment aid. Be
especially careful in the adjustment of mirrors and prisms; many people make the mistake of underestimating how serious the effect of a

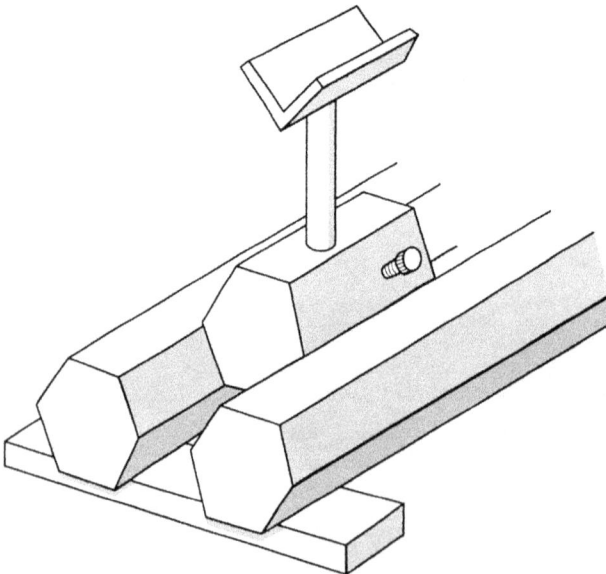

Figure 21.6 An inexpensive lens bench can be made from hexagonal steel bars. Two bars are fastened together as shown here so
the short sections of bar will slide between them. The optics can
be waxed on the sliders, or angle iron V troughs threaded on rods
which are adjustable for height alignment can be used to support
the optics.

misaligned reflecting component can be. If cylindrical elements are included in the system, the orientation of the cylinder axes is critical, especially if there are orthogonal cylinders or if the object is a slit.

The simplest method of performance testing is by the visual evaluation of resolution. As indicated earlier, resolution is not the be-all, end-all of image evaluation, but it is easy and quick to test. A bar target similar to that shown in Fig. 21.7 is readily obtainable; one can be "homemade" with black drafting tape and white paper, or a copy of USAF 1951 can be purchased for a few dollars (or Fig. 21.7 can be xeroxed and used). Another type of test target can be made by scratching fine lines through the aluminized surface of a first-surface mirror, and illuminating it from the rear. The image of a pinhole target can be examined to analyze both aberrations and alignment problems (which are indicated by a nonsymmetrical blur spot on the axis).

For testing eyepieces, magnifiers, or similar devices, where the appearance of an extended field is important, a piece of graph paper makes an excellent test target. One can readily evaluate the image distortion and curvature of field, as well as the effect of changing the eye position, with this sort of target.

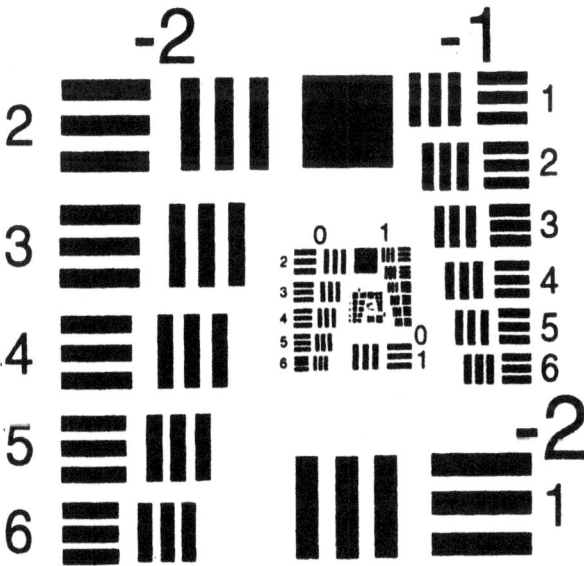

Figure 21.7 A three-bar resolution target. Each pattern differs from the next by a factor equal to the sixth root of 2 (1.1225), and thus the groups of six differ by a factor of 2. Targets of this type are commercially available on film or as metal evaporated on glass.

An often overlooked factor in developing an optical system is the deleterious effect of stray light. This is light from outside the field of view which is reflected or scattered from some part of the assembly (typically the housing of the optics) into the field of view, where it can either reduce the image contrast or produce ghost images. This can come as a great surprise when the first complete model of the system is made, because often the system mock-up is mechanically quite different from the final product, and the stray light may not be present in the mock-up. There are two ways that stray light can be handled. In a system which has an internal pupil, a glare stop, as shown in Fig 9.5 and 13.8, is both invaluable and effective. Stops can be placed at every internal pupil and also at every internal image plane. The other technique is simply blackening the offending (reflecting) member. Sometimes it is difficult to know where the problem is. An effective method of locating the source is to simply look into the optics from the location in the image where the stray light is showing up. In other words, put your eye there and look back into the optics. This view is most sensitive if the eye is placed in a location where the image should be dark, e.g., outside the field of view. Then you can see the (image of) structure from which the light is reflected. Another place to look is at the exit pupil, which can be examined with a magnifying glass. The image of the inside of the optical instrument is typically focused near the exit pupil. Once you have located the culprit, a flat black paint (such as Floquil flat black model locomotive paint, available from your local hobby shop) will usually fix things. Another very effective material is black flocked paper, which can be glued on the reflecting surface. Flocked black paper is a good absorber; it can be obtained from Edmund Scientific in Barrington, NJ.

21.5   Aberration Considerations

In addition to aperture and field, most aberrations vary with the shape of the lens element. For example, if the object is distant, spherical aberration is a minimum for a particular shape. For glass lenses this shape is approximately plano-convex, with the convex surface facing the distant object. The variation of spherical aberration with lens shape, as a function of index, is shown in Fig.17.38.

For distant objects, biconvex and meniscus (one side convex, the other concave) lenses have more aberration. But for very high index lenses, such as silicon ($n = 3.5$) or germanium ($n = 4.0$), the minimum spherical shape is meniscus with the convex side facing the distant object as indicated in Fig. 17.38. But if we consider a different object location, the shape for minimum spherical changes. For example, at one-to-one imagery, the equiconvex shape has the least spherical. Thus for any

given element in an optical system, one orientation may be much preferred to the other. Figure 21.10 shows the variation of the angular
spherical aberration blur with object distance for three different lens
shapes.

The position of the aperture stop, relative to the lens shape, can have
a big effect on the off-axis aberrations (coma, astigmatism, distortion,

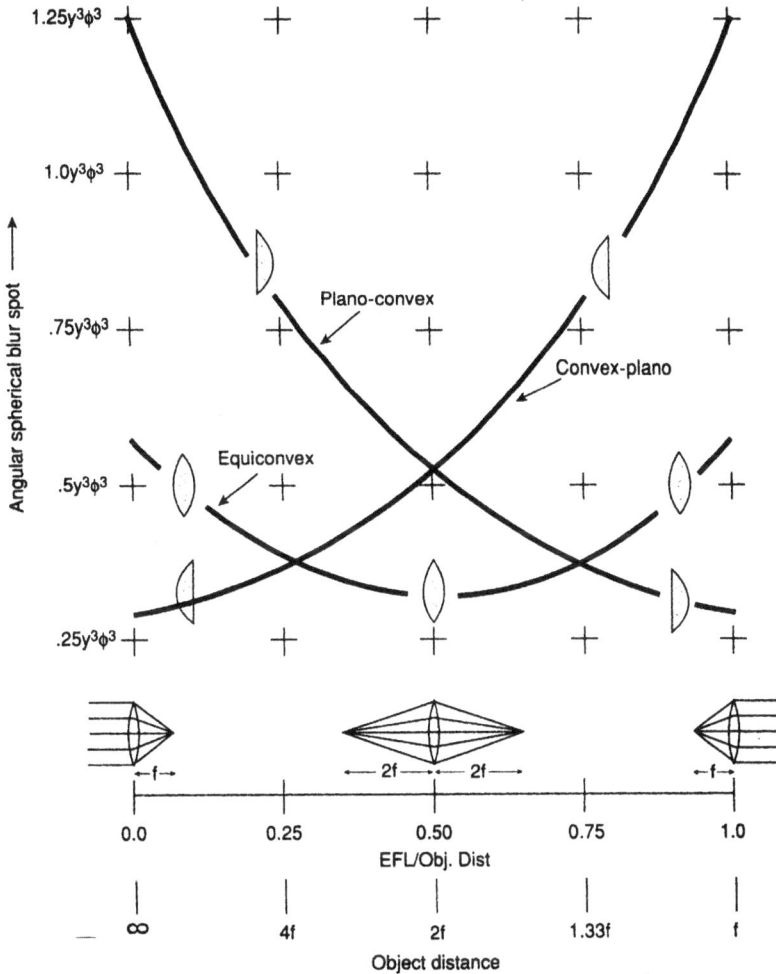

Figure 21.10 Spherical aberration varies as a function of the object distance. The
graph plots this variation for three element shapes for a lens with an index of
$n = 1.80$. Note that for the plano-convex shape the minimum spherical occurs
when the curved side faces the longer conjugate, whereas for the equiconvex
shape the minimum is at 1:1 magnification. The size of the blur spot can be found
by multiplying the angular blur by the image distance. ϕ is the element power
and y is the semiaperture.

and lateral color). Often, reversing a lens will make a significant difference in the system performance at the edge of the field, because it changes its orientation with respect to the stop.

In general, the higher the power of a lens, the more aberration it will introduce to the system. So a good way to reduce the aberrations is to substitute two low-power elements for a single high-power element as shown in Fig. 21.11. When the elements are properly shaped to take advantage of this "split," the spherical aberration can be reduced by a factor of 5 or so. A split is most effective against spherical aberration. Another technique for reducing spherical is to spread the "work" equally, where work is the amount that the lens bends the marginal light ray. A quick look at Eq. 4.1 indicates that "work" is simply $y\phi$, the product of ray height and lens power.

Petzval field curvature is a function of lens power and index; it is not affected by shape or object distance, as are the other aberrations. In essence, it amounts (roughly) to the sum of all the positive power in the system minus the sum of all the negative power. Looked at this way, it is apparent that the usual field curvature problem that we encounter when using stock optics is due to having only positive elements in a system; this is why the field curvature is almost always curved inward, toward the lens. Flat-field lenses correct this by introducing negative power where it does not have much effect on the ray slope, namely, where the ray height is low. Anastigmat lenses get their correction

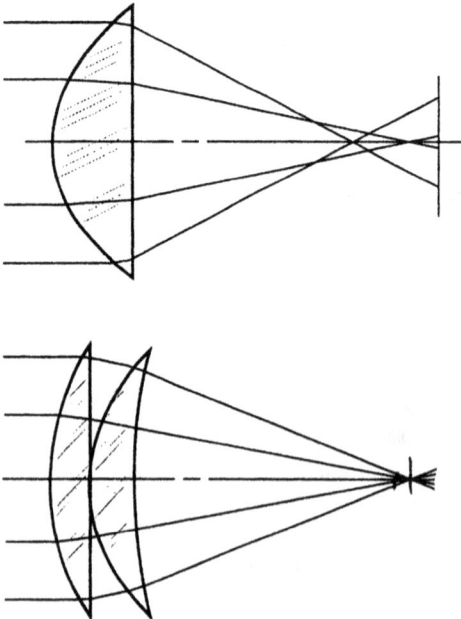

Figure 21.11 The spherical aberration of an element can be reduced by a factor of 5 or more by splitting it into two elements, each shaped to minimize its spherical aberration.

(a)

(b)

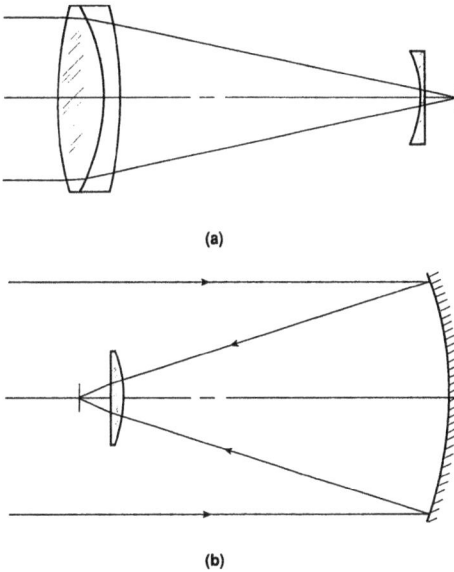

Figure 21.12 A field flattener is a lens placed close to the image where it has little effect except on the Petzval field curvature. A negative lens, as shown in (a), will flatten an inward-curving field such as that produced by positive components. In (b), the concave mirror has a backward-curving field which is corrected by a positive field-flattener lens.

by using high-index glass and separating positive and negative power in order to lower the ray height on the negative elements relative to the positive elements. In a "stock" layout we are usually stuck with lots of positive lenses, but sometimes there is one thing that we can do, and that is the introduction of what is called a *field flattener*. This is usually a negative element placed at or near a focal plane (where the ray height is very low and the lens has little effect on the other aberrations or on the image size) as shown in Fig. 21.12a. A field flattener can have a very salubrious effect on a system with too much inward field curvature. Note that the reverse is also true; a positive field lens will increase the inward field curvature. Note also that while a positive converging lens has inward field curvature, a converging (concave) mirror has backward field curvature and needs a positive field flattener as shown in Fig. 21.12b

21.6 How to Use a Singlet (Single Element)

When using "stock" lenses our choice of elements is quite limited. Indeed, the usual choice that optical catalogs offer us is between a plano-convex (or nearly so) lens and a biconvex (which is probably equiconvex) lens, as indicated in Fig. 21.13. In the following discussions, a biconvex which has one surface *much* more strongly curved than the other can be regarded as a plano-convex. If both surfaces are similarly shaped, the lens can be treated as equiconvex (although in this case there may be a preferred orientation).

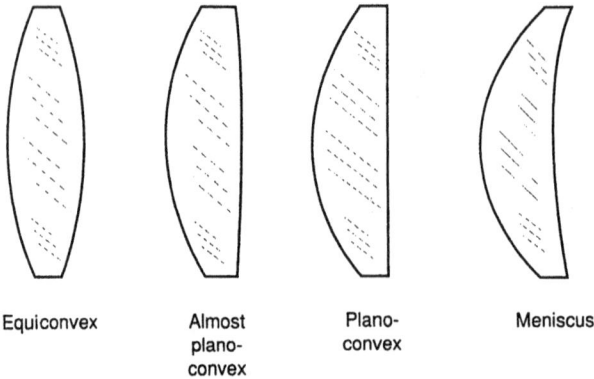

| Equiconvex | Almost plano-convex | Plano-convex | Meniscus |

Figure 21.13 The forms of single-lens elements which are commonly available from catalogs.

We can start our considerations by assuming that we have a system which requires a well-corrected image over a small field of view. What this means is that we will want to minimize the spherical aberration in the image and that we will not worry too much about the off-axis image. We consider the three cases diagrammed in Fig. 21.14.

For a *telescope objective type,* that is, a system with the object a long distance away, say more than 5 or 10 focal lengths, we choose a plano-convex lens and orient it with the curved surface toward the distant object.

For a *microscope objective type*, that is, a lens which will magnify the object by five or more times, we again choose a plano-convex but face the plano side toward the object.

For a *relay lens type,* that is, a lens with a magnification between $(-)5\times$ and $(-)0.2\times$, a biconvex lens is the choice. If the lens is not equiconvex, orient the more strongly curved surface (i.e., the shorter radius) toward the longer conjugate.

For a wider field of view, we must be more concerned with the off-axis aberrations. In this case the location of the aperture stop can be critical, since its position will affect coma, astigmatism, and field curvature. In general, we must sacrifice the image quality at the axis in order to get better performance at the edge of the field. Usually a plano-convex or a meniscus (one side convex, the other concave) is the best bet, with the aperture stop on the plano (or concave, if meniscus) side of the lens, as shown in Fig. 21.15. Note that the lens wants to sort of "wrap around" the stop. This is the reason that most camera lenses have an external shape which is almost like a sphere with the stop in the center. When there is no separate stop and the object is some distance away, a plano-convex lens with the plano toward the object often

(a) "Telescope objective type"

(b) "Microscope objective type"

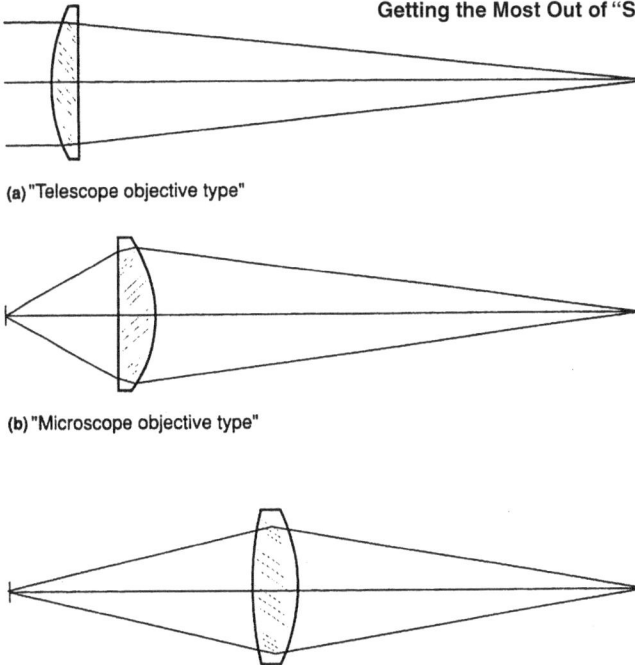

Figure 21.14 For applications with small fields of view there are three common cases: (a) the "telescope objective" type, where the object is a long distance to the left; a plano-convex lens with the convex side facing the distant object minimizes the spherical aberration; (b) the "microscope objective" type, where the image is distant, and the convex side of a plano-convex lens faces the image to minimize spherical; and (c) the "relay lens" type where neither conjugate is greatly longer than the other; a biconvex lens is best, with the more strongly curved surface facing the longer conjugate.

works well (because the coma in the image produces a sort of field flattening effect).

When a singlet is used as a *magnifying glass* and held close to the eye as in Fig. 21.16a, a plano-convex lens with the plano side toward the eye works best. Here the pupil of the eye acts as the aperture stop, and the lens is "wrapped around" it. This usage is the same as found in head-mounted displays (HMD) and is also much like that in a telescope eyepiece. However, if the lens is a foot or two from the eye, as shown in Fig. 21.16b or as in a tabletop slide viewer or in a head-up display (HUD), the plano side should face away from the eye. This is because the image of the eye formed by the lens is a pupil of the system, and, with the lens well away from the eye, this image is on the far side of the lens. We want the plano side to face the stop/pupil, so that the lens wraps around the pupil. For a general-purpose magnifier which is used both near to and far from the eye, an equiconvex shape is probably the best compromise, although the two-lens magnifier as described later is much better.

Stop

Stop

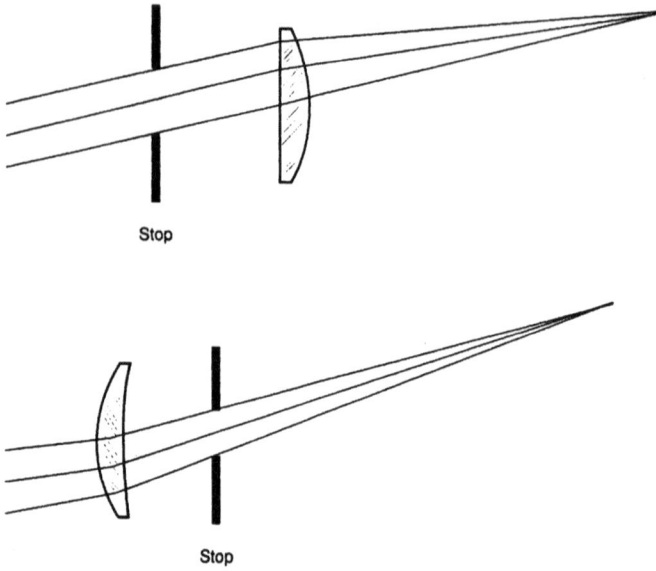

Figure 21.15 For applications where a wider field of view is covered, the lens is oriented with the field aberrations (coma and astigmatism) as the prime concern. An aperture stop, spaced away from the lens on the "concave" or plano side (as shown in this figure), can have a favorable effect on the off-axis imagery.

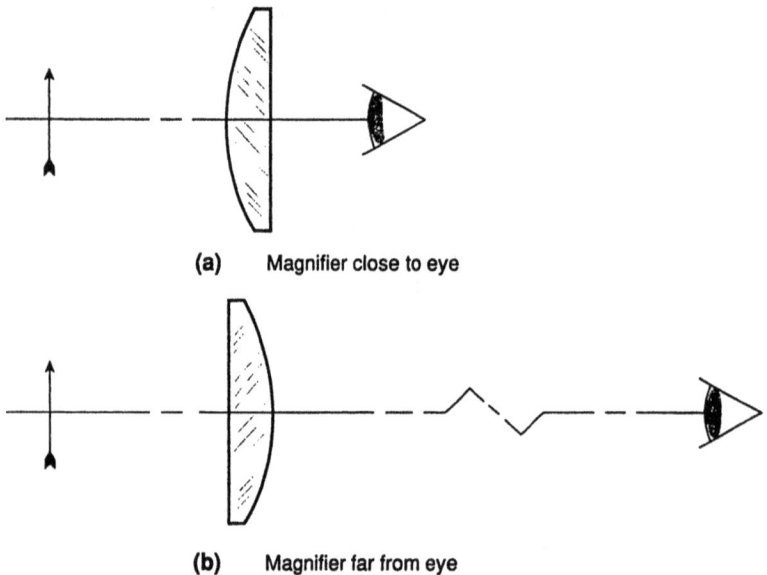

(a)   Magnifier close to eye

(b)   Magnifier far from eye

Figure 21.16 When a plano-convex lens is used as a magnifier, the best orientation depends on the location of the eye, which acts as the aperture stop. When close to the eye, the plano side should face the eye; this orientation minimizes distortion, coma, and astigmatism. When far from the eye, the convex side should face the eye.

Note that these comments can also be applied to a plano-convex cemented achromatic doublet.

21.7   How to Use a Cemented Doublet

Most stock cemented doublets are designed to be corrected for chromatic and spherical aberration, and probably coma as well, when used with an object at infinity. In other words, they are effectively telescope objectives and are designed to cover a small field of view. As illustrated in Fig. 21.17, the external form is usually biconvex, with one surface much more strongly curved than the other, i.e., close to a plano-convex shape. Just as with the plano-convex singlet, the more strongly curved surface should face the distant object. If the doublet is used at finite conjugates, the strong side should face the longer conjugate.

If neither of the exterior surfaces is more significantly curved than the other, the odds are that the lens was not designed for use with an infinite conjugate. It may be corrected for use at finite conjugates, or, what is more likely, it may have been part of a more complex assembly. Here, some experimentation is in order. Try both orientations and observe the performance. Again, as with the singlet, it is highly probable that the stronger surface will want to face the longer conjugate.

A meniscus-shaped doublet is rarely found as a "stock" lens; such a doublet is most likely either surplus or salvage, and its shape results from the design of which it was originally part. Although a (thick) meniscus is very useful as a lens design tool (to flatten the field), such a lens will probably not be too useful in your system mock-up; it might work out as part of an eyepiece, with the concave side adjacent to either the eye or the field stop.

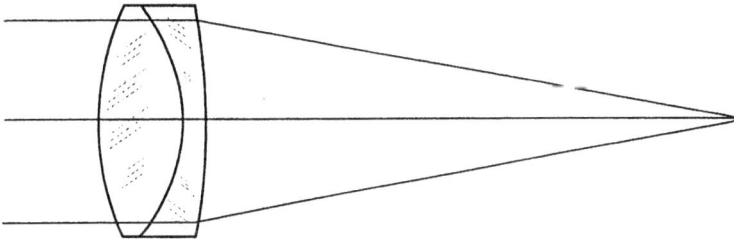

Figure 21.17 Most "stock" achromatic doublets are designed as telescope objectives and are corrected for chromatic and spherical aberration as well as coma with the object at infinity. The more strongly curved surface should face the more distant conjugate.

21.8 Combinations of Stock Lenses

Often the use of two lenses instead of one can make a big improvement in system performance. The following paragraphs discuss a number of possibilities.

High-speed (or large NA) applications.   The usual problem in fast systems is spherical aberration. Using two lenses instead of one, with each shaped to minimize the spherical aberration, can alleviate the situation. The optimum division of power is equal; both elements have the same focal length, and the sum of their powers equals the power of the single element which they are replacing. For a distant object, the first element should be plano-convex, with the convex side facing the object. Ideally, the second element should be meniscus, with the convex side facing the first element as sketched in Fig. 21.18a. But since

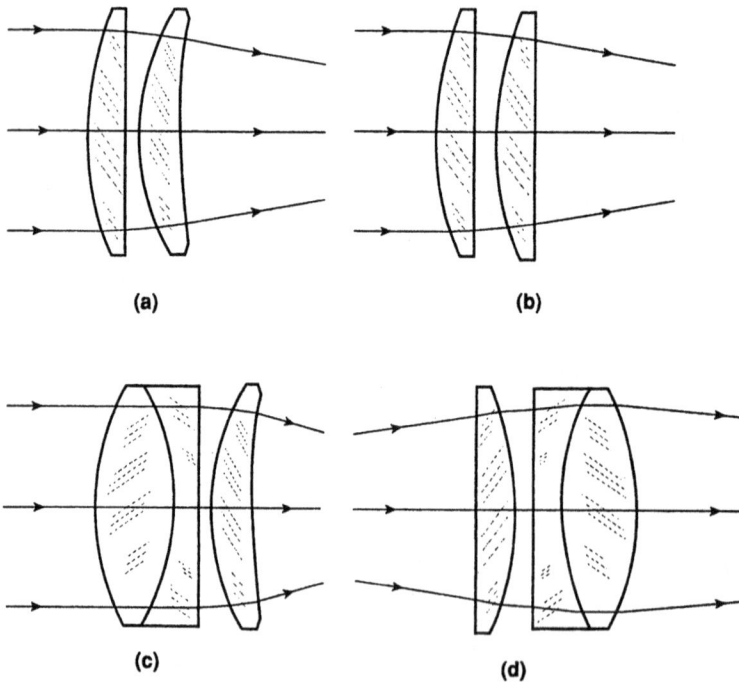

(a)   (b)

(c)   (d)

Figure 21.18   When lenses are used at high speed (large NA or small *f*-number), spherical aberration is the usual problem. It can be reduced by using two elements instead of one. The first should be oriented to minimize spherical for the object location and the second shaped for *its* object location. For distant objects the best arrangements are shown in (a), (b), and (c). The first element is plano-convex and the best shape for the second is meniscus; if the second is also a plano-convex it should be oriented as in (b). If one element is a doublet, it should face the longer conjugate as indicated in (c) and (d).

meniscus stock elements are hard to come by, the usual stock lens arrangement is another plano-convex with its convex side also facing the object, as in Fig. 21.18b. If one plano-convex singlet is stronger than the other, place it in the convergent beam. If one of the lenses is a doublet, it should probably be the one facing the distant object, followed by the singlet as in Fig. 21.18c. If both are doublets, put their strongly curved surfaces toward the object. If the application is microscope like, then of course the arrangement is reversed as shown in Fig. 21.18d.

When the system is to work at finite conjugates, for example, at one-to-one or at a small magnification, then the best arrangement is usually with the convex surfaces facing each other (provided that the angular field is small). Figure 21.19 shows pairs of singlets and doublets working at one-to-one. This arrangement allows each half of the combination to work close to its design configuration, i.e., with the object at infinity, and if the angular field is not large, the space may be varied to get a desired track length. If the magnification is not one-to-one, using different focal lengths (the ratio of which equals the magnification) can be beneficial.

A *projection condenser* is usually two or three elements, shaped and arranged to minimize spherical aberration. There is often an aspheric surface. With two elements, the more strongly curved surfaces face each other; if they are not the same power, the stronger (shorter focal length) faces the lamp. With three elements, the one nearest the lamp is often meniscus with the concave surface facing the lamp. The other two are often plano-convex, with their curved sides facing. If the elements are spherical-surfaced, they should each be approximately the

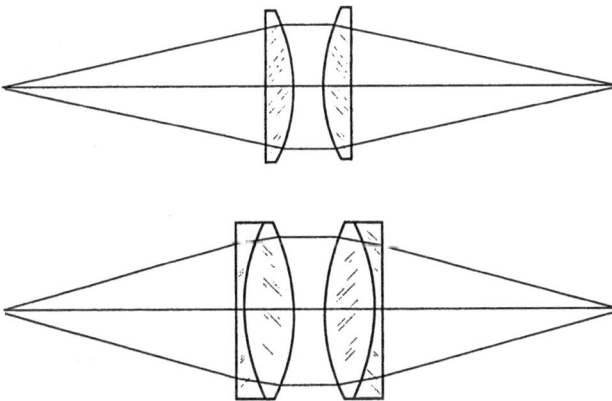

Figure 21.19 For small fields and magnifications which are close to 1:1, the lenses should be oriented facing each other as shown in order to minimize the spherical. At 1:1 the light is collimated between the lenses; at other magnifications it is nearly so.

same power. If one is aspheric, it is often stronger than the others and is the one next to the lamp.

Eyepieces and magnifiers. Very good magnifiers can be made from two plano-convex elements with the curved sides facing each other as shown in Fig. 21.20a. This arrangement works well, either close to the eye or at arm's length. As a telescope eyepiece, the spacing between them is often increased to about 50 or 75 percent of the singlet focal length, so that one element acts as a field lens, as in Fig. 21.20b; this increased spacing also reduces the lateral color and helps with coma and astigmatism. (If the elements have different focal lengths, the lens near the eye should have the shorter focal length.) This is the classical "Ramsden" eyepiece. If the eyelens is a doublet, it is the "Kellner" eyepiece shown in Fig. 21.20c; usually the flatter side of the doublet faces the eye. Some versions of this popular binocular eyepiece are closely spaced, and some are used in a reversed orientation. A few trials with a graph paper target and your eye at the exit pupil location will tell you which arrangement suits your stock lenses the best.

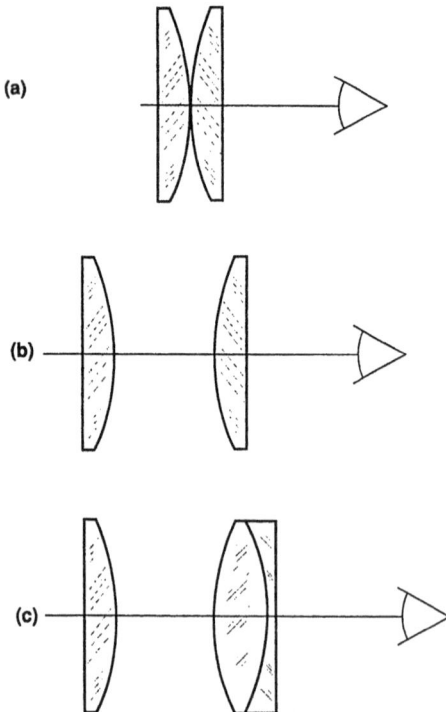

Figure 21.20 (a) Two plano-convex elements, convex to convex, make a good magnifier which works well both near the eye and at a distance. (b) For use as a telescope eyepiece, the spacing is increased to reduce the coma and lateral color (and to allow the left-hand lens to act as a field lens). (c) The Kellner eyepiece uses a doublet as the eyelens to further correct the lateral color. This eyepiece is often found in ordinary prism binoculars.

Two achromats work even better. With the strong curves of two identical achromats facing each other as in Fig. 21.21a, this makes one of the best general-purpose magnifiers and eyepieces. This is the "Plössl" or "symmetrical" eyepiece, justly popular for its high quality, low cost, versatility, and long eye relief. Depending on exactly what the shape of your doublet is and what your eye relief is, you may want to reverse the orientation of one or the other (but not both) of the doublets as indicated in Fig. 21.21b and c.

Wide-field combinations. Let us face it right up front. It is *very* difficult to put together stock elements so that they perform well over a wide field of view. Usually your best bet is to obtain a corrected assembly such as a triplet anastigmat or a camera lens. But there are a few things we can do to optimize the situation when we do not have a suitable anastigmat available.

As mentioned earlier, to obtain a wide-field coverage we often must sacrifice the image quality in the center of the field. We have two basic

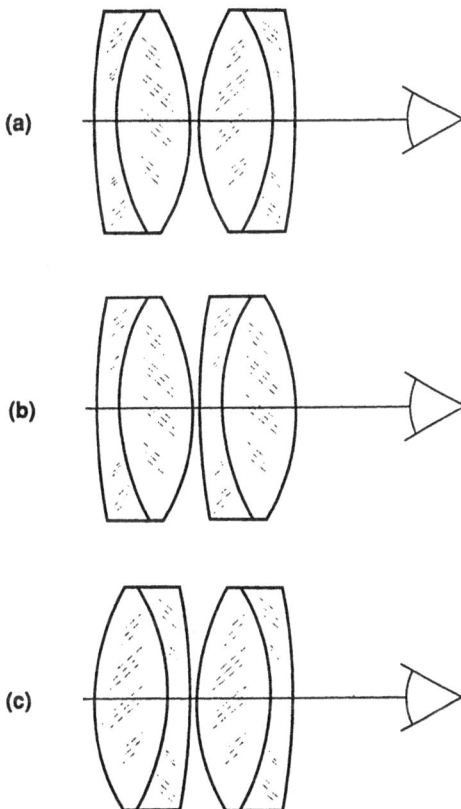

(a)

(b)

(c)

Figure 21.21 (a) Two doublets, crown to crown, make an excellent eyepiece and also an excellent magnifier. This is the symmetrical, or Plössl, eyepiece. (b) and (c) Depending on the shape of the doublets and the eye relief of the telescope, one of these alternate orientations may work well as an eyepiece.

tools which we can use to improve the image quality at the edge of the field. One is the placement of the aperture stop, and the other is the *symmetrical principle.* If a system is symmetrical about the stop (in a left-to-right sense as shown in Fig. 21.22), then the system is free of coma, distortion, and lateral color. Strictly speaking, the system must work at unit magnification to be fully symmetrical, but much of the benefit of symmetry is obtained even if the object is at infinity. Of course, symmetry works whether we are doing wide or narrow fields of view. But whereas we orient the elements "strong-side-facing" as in Fig. 21.22a to get the least spherical aberration in a narrow-field application, we usually want the strong surfaces facing outward for wider fields of view as in Fig. 21.22b. Plano-convex or meniscus elements are the shapes of choice for this. The elements are spaced a modest, but significant, distance from the aperture stop, which is midway between them. The spacing is significant because it affects the astigmatism; there is an optimum spacing which yields the best compromise between the amount of astigmatism and the flatness of the field.

Relay systems. For a relay system which requires some given magnification, consider using two achromats, such that the ratio of their focal lengths equals the desired magnification, and the sum of their focal

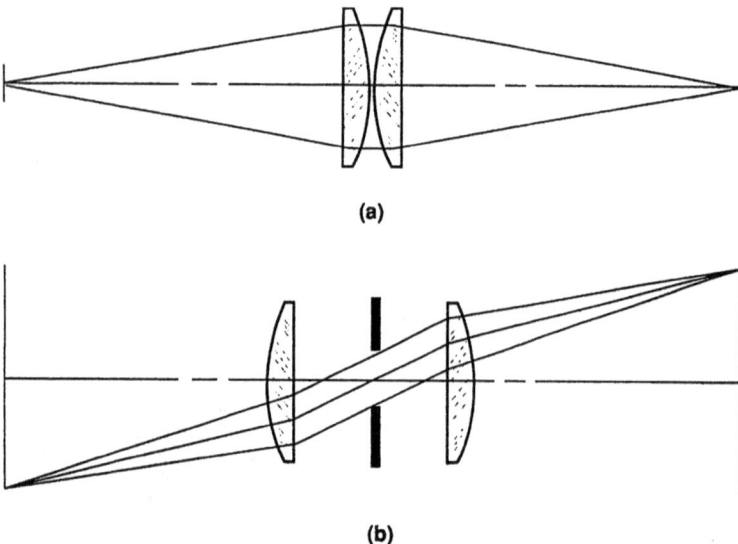

(a)

(b)

Figure 21.22 Left to right (or mirror) symmetry will automatically eliminate coma, distortion, and lateral color. With two plano-convex elements, the orientation shown in (a) would be best for a small field, but for a wider field, the orientation in (b) will usually work better.

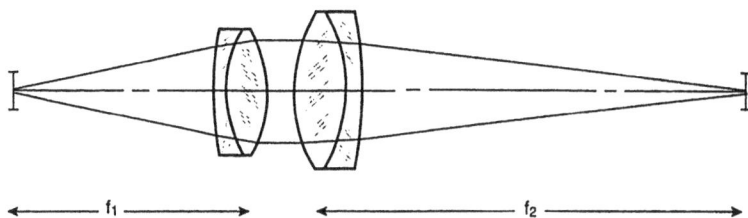

Figure 21.23 A well-corrected narrow-field relay can be made from two achromatic doublets by choosing their focal lengths so that their ratio equals the desired magnification $m = -f_2/f_1$. When this is done, the light between the lenses is collimated and each lens works at its design conjugates (assuming the lenses were designed for an infinitely distant object).

lengths is approximately equal to the desired object-to-image distance, as shown in Fig. 21.23. The rays in the space between the lenses will be collimated, and the spacing between them will not be a critical dimension. Note that a 45° tilted-plate beam splitter can be used in a collimated beam without introducing astigmatism. If the achromats are corrected for an infinite object distance, the relay image will also be corrected.

The two-achromat relay can produce an excellent image over a small field. A wider-field system can be made from two photographic lenses, again used face to face, with collimated light between them, as shown in Fig. 21.24. Since photo lenses are longer than the achromatic doublets we discussed in the preceding paragraph, one must be aware of vignetting. Most photographic objectives vignette when used at full aperture, often by as much as 50 percent. For an oblique beam (tilting upward as it goes left to right) the beam is clipped at the bottom by the aperture of the left lens, and clipped at the top by the right-hand lens. When two camera lenses are used face to face, their vignetting characteristics are usually such that the combination has much worse vignetting than either lens alone. Thus this sort of relay is usually

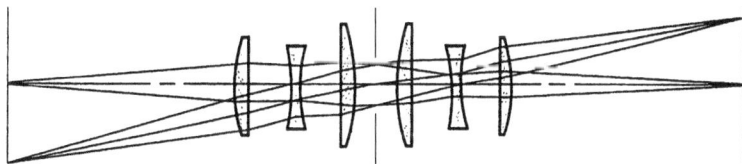

Figure 21.24 When a wider field than two doublets (as shown in Fig. 21.23) can cover is needed, two camera lenses can be used, face to face, to make a high-quality relay system. If the relay is to have magnification, the focal lengths of the lenses should be chosen so that their ratio equals the magnification. If an iris diaphragm is to be used, it should be located between the lenses (unless the field is small). Note that with some lenses vignetting may be a problem.

limited by vignetting to a smaller field than one might expect. Note also that if an iris diaphragm is to be used, it should be *between* the lenses, rather than using the iris of one of the lenses (unless the field of view is quite small). This arrangement of readily available stock camera or enlarging lenses makes an excellent, well-corrected finite conjugate imaging system.

The above technique of using photo lenses has the virtue of using them as they were designed to be used, namely, with one conjugate at infinity. Most photo lenses retain their image quality down to object distances of about 25 times their focal lengths, more or less, depending on the design type. But at close distances the image quality deteriorates. In general, high-speed lenses tend to be quite sensitive to object distance. Slower (i.e., low NA, large *f*-number) lenses can be used successfully over a wider range of conjugate distances.

A *close-up attachment* is simply a weak positive lens placed in front of a camera lens. If, as shown in Fig. 21.25 the focal length of the attachment lens is approximately equal to the object distance, then the object is collimated (imaged at infinity) and the camera lens sees the object as if it were at infinity. The attachment lens is ideally a meniscus, with the concave side facing the camera lens (so that it wraps around the stop), although a plano-convex form is often quite acceptable. If the field is quite narrow, the reverse orientation of the lens *might* be better. Note that the use of a close-up attachment is equivalent to combining two positive lenses to get a lens with a shorter focal length. You can also use a weak negative focal length attachment to increase the focal length of a camera lens which is too short for your application.

Beam expander. A laser beam expander is simply a telescope used "backward" to increase the diameter and to reduce the divergence of the laser beam. The galilean form of telescope is the most frequently used

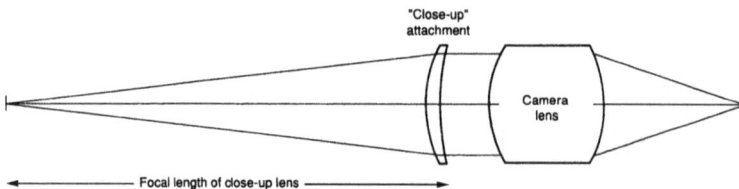

Figure 21.25 Many camera lenses lose image quality when the object is close. A "close-up attachment" is simply a weak positive element the focal length of which approximates the object distance, so that the light is collimated for the camera lens. The attachment is usually a meniscus lens the shape of which is a compromise between minimum spherical aberration and minimum coma and astigmatism.

because it can be executed with simple elements and has no internal focus point (which might induce atmospheric breakdown with a high-power laser). The Kepler telescope can also be used, and its internal focal point can provide a spatial filter capability, but it is more difficult to correct the Kepler because both components are positive, converging lenses.

Since the laser light is monochromatic and the beam angle is small, we are mostly concerned with correcting spherical aberration. The objective (positive) component of the galilean scope is the big contributor of spherical aberration, so it is important that it be shaped to minimize spherical. If the expander is to be made from two simple elements as diagrammed in Fig. 21.26a, the negative element must contribute enough overcorrected spherical to balance that from the objective lens. Thus our "stock lens" choice is often a plano-convex element for the objective and a plano-concave element for the negative, with both lenses oriented so that their plano sides face the laser. (A meniscus form for the negative has more overcorrected spherical and might produce a better correction.) For higher-power beam expanders, a well-corrected doublet objective is necessary; it should be combined with a planoconcave element with its concave side facing the laser as shown in Fig. 21.26b to minimize its overcorrection of the spherical aberration.

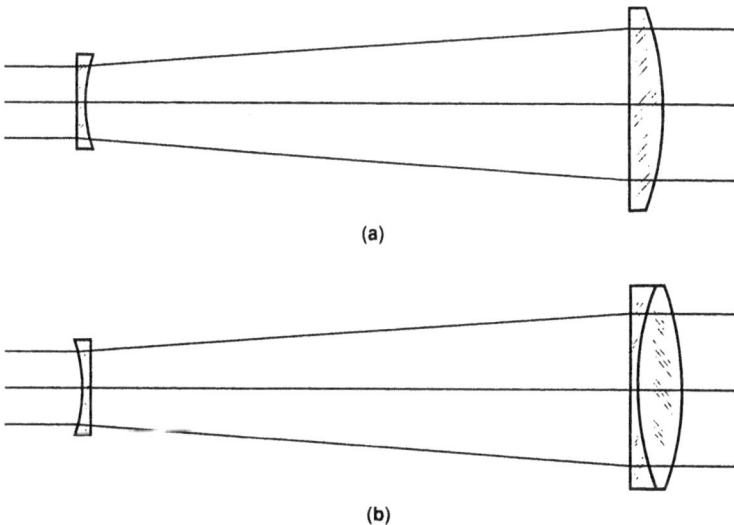

(a)

(b)

Figure 21.26   A low-power laser beam expander can be made from a plano-convex "eyelens" and a plano-convex "objective," with both plano sides facing the laser. For higher powers an achromatic doublet is used as the objective to reduce the spherical aberration, and the plano-concave negative element is reversed.

21.9 Sources

Sources of stock lenses

The following are some of the companies which stock lenses. Most have catalogs. Some have their catalogs on disk; many of these include the lens prescriptions (radii, thickness, index), and a few have free computer programs which can be used to calculate the performance of their lenses.

Ealing Electro-Optics, Inc.
89 Doug Brown Way
Holliston, MA 01746
Tel: 508/429–8370;
Fax: 508/429–7893;
http://www.ealing.com

Edmund Scientific
101 East Gloucester Pike
Barrington, NJ 08007
Tel: 609/573–6852;
Fax: 609/573–6233;
John_Stack@edsci.com

Fresnel Optics, Inc.
1300 Mt. Read Blvd.
Rochester, NY 14606
Tel: 716/647–1140;
Fax: 716/254–4940

Germanow-Simon Corp., Plastic
Optics Div.
408 St. Paul St.
Rochester, NY 14605–1734
Tel: 800/252–5335;
Fax: 716/232–2314;
gs optics@aol.com

Janos Technology Inc.
HCR#33, Box 25, Route 35
Townshend, VT 05353–7702
Tel: 802/365–7714;
Fax: 802/365–4596;
optics@sover.net

JML Optical Industries, Inc.
690 Portland Ave., Rochester, NY
14621–5196

Tel: 716/342–9482;
Fax: 716/342–6125;
marty@jmlopt.com
http://www.jmlopt.com

Melles Griot, Inc.
19 Midstate Drive, Ste. 200
Auburn, MA 01501
Tel: 508/832–3282;
Fax: 508/832–0390;
76245,2764@compuserve.com

Newport Corporation
1791 Deere Ave., Irvine, CA 92714
Tel: 714/253–1469;
Fax: 714/253–1650;
pgriffith@newport.com

Optics for Research
P.O. Box 82, Caldwell,
NJ 07006–0082
Tel: 201/228–4480;
Fax: 201/228–0915;
dwilson@ofr.com

Optometrics USA, Inc.
Nemco Way, Stony Brook Ind. Park
Ayer, MA 01432
Tel: 508/772–1700;
Fax: 508/772–0017;
opto@optometrics.com

OptoSigma Corp.
2001 Deere Ave.
Santa Ana, CA 92705
Tel: 714/851–5881;
Fax: 714/851–5058;
optosigm@ix.netcom.com

Oriel Instruments
250 Long Beach Blvd.,
P.O. Box 872
Stratford, CT 06497–0872
Tel: 203/377–8282;
Fax: 203/378–2457;
res_sales@oriel.com

Reynard Corporation
1020 Calle Sombra
San Clemente, CA 92673
Tel: 714/366–8866;
Fax: 714/498–9528

Rodenstock Precision Optics, Inc.
4845 Colt Road, Rockford, IL
61109–2611

Rolyn Optics
706 Arrowgrand Circle, Covina, CA
91722–9959

Tel: 818/915–5707;
Fax: 818/915–1379

Spectral Systems
35 Corporate Park Drive
Hopewell Junction, NY 12533
Tel: 914/896–2200;
Fax: 914/896–2203

Spindler & Hoyer Inc.
459 Fortune Blvd.
Milford, MA 01757
Tel: 508/478–6200;
800/334–5678;
Fax: 508/478–5980

Their catalog on disk includes an "Optical Design Program for WINDOWS."

Directories

Several directories are available which can help in locating sources of optical things. Probably the most complete is the *Photonics Buyer's Guide,* published by Laurin Publishing Co., Inc., Berkshire Common, P.O. Box 4949, Pittsfield, MA 01202–4949, Tel: 413/499–0514, Fax: 413/442–3180, email: Photonics@laurin.com. This is the lead volume of a four-volume set; it lists optical products by category, giving sources for each type of product. A second volume, the *Photonics Corporate Guide,* lists the names, addresses, etc., of the source companies. *Laser Focus World* magazine and *Lasers & Optronics* magazine also publish optical buyers' guides which are distributed to subscribers.

Raytracing and Aberration Calculation

A.1 Introduction

The importance of a knowledge and understanding of raytracing techniques has been significantly reduced by the near ubiquitous use of personal computers for the design and analysis of optical systems. It is now more important to understand how to effectively *use* an optical software program (such as OSLO, ZEMAX, or CodeV) than it is to understand exactly *how* the raytracing is done. Nonetheless, it is occasionally of value to access the raytracing equations. To that end we present here the equations for raytracing meridional rays, i.e., those which lie in the y-z plane, for the benefit of those who may need to calculate one or two rays using a pocket calculator. These equations are designed for this usage, and are not suited for automatic computer use. We also present equations which are designed for computer use (i.e., which do not "blow up" unless the ray can not be calculated (a) because it never intersects the surface, or (b) because it encounters total internal reflection (TIR) at the surface. The equations for tracing paraxial rays are given in Chap. 3, Eqs. 3.16 and 3.17 (as are a set of equations suitable for tracing meridional rays using an electro-mechanical desk calculator, Eqs. 3.1 through 3.7).

This appendix also contains Coddington's equations, which trace paraxial type rays about a meridional principal ray (sometimes called para-principal rays), and allow the calculation of sagittal and tangential field curvature. Several aspects of the calculation of the other specific aberrations are also discussed here.

A.2 Meridional Rays

Meridional rays are those rays which are coplanar with the optical axis of the system. The plane in which both ray and axis lie is called the meridional plane, and, in an axially symmetrical system, a meridional ray remains in this plane as it passes through the system. The two-dimensional nature of the meridional ray makes it relatively easy to trace. Although a great amount of information about an optical system can be obtained by tracing a few meridional rays plus a Coddington trace or two (Sec. A.6), given the speed of the modern computer, meridional rays are usually traced as a special case of a skew or general raytrace. However, if rays are to be traced with an electronic pocket calculator, then meridional rays are the obvious choice. The formulas in this section are designed to take advantage of the trigonometric capabilities of this type of calculator (See Fig. A.1).

Opening: 1. Given Q and $\sin U$ at the first surface.

or 2. $$Q = -L \sin U \tag{A.1a}$$

or 3. $$Q = H \cos U - s \sin U \tag{A.1b}$$

Refraction:

$$\sin I = Qc + \sin U \tag{A.1c}$$

$$\sin I' = \frac{n \sin I}{n'} \tag{A.1d}$$

$$U' = U - I + I' \tag{A.1e}$$

$$Q' = \frac{Q (\cos U' + \cos I')}{(\cos U + \cos I)} \tag{A.1f}$$

Transfer:

$$Q_{j+1} = Q'_j + t \sin U'_j \tag{A.1g}$$

$$U_{j+1} = U'_j \tag{A.1h}$$

Closing:

$$L'_k = \frac{-Q'_k}{\sin U'_k} \tag{A.1i}$$

or

$$H' = \frac{Q'_k + s'_k \sin U'_k}{\cos U'_k} \tag{A.1j}$$

Miscellaneous:

$$y = \frac{Q\,[1 + \cos\,(I-U)]}{(\cos U + \cos I)} = \frac{Q'\,[1 + \cos\,(I-U)]}{(\cos U' + \cos I')} = \frac{\sin\,(I-U)}{c} \qquad \text{(A.1k)}$$

$$z = \frac{Q\,\sin\,(I-U)}{(\cos U + \cos I)} = \frac{1-\cos\,(I-U)}{c} \qquad \text{(A.1l)}$$

$$D_{1\,to\,2} = \frac{t - z_1 + z_2}{\cos U'_1} \qquad \text{(A.1m)}$$

The symbols used are, for the most part, the same as those defined in Sec. 3.3, capitalized to differentiate them from the lowercase paraxial symbols. Symbols new to this section are

Q	the distance from the vertex of the surface to the incident ray, perpendicular to the ray; positive if upward.
Q'	the distance from the surface vertex to the refracted ray, perpendicular to the ray.
I	the angle of incidence at the surface; positive if the ray must be rotated clockwise to reach the surface normal (i.e., the radius).
I'	the angle of refraction.
z	the longitudinal coordinate (abscissa) of the intersection of the ray with the surface; positive if the intersection is to the right of the vertex.
$D_{1\,to\,2}$	the distance along the ray between surface 1 and surface 2.

The physical meanings of the symbols are indicated in Fig. A.1.

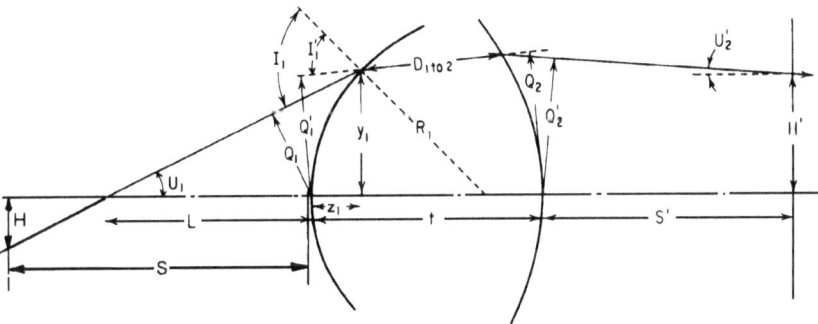

Figure A.1 Diagram illustrating the symbols used in the meridional raytracing equations.

A.3   Skew Rays: Spherical Surfaces

A skew ray is a perfectly general ray; however, the application of the term "skew" is usually restricted to rays which are not meridional rays. A skew ray must be defined in three coordinates x, y, and z, instead of just z and y as in the case of meridional rays. Until the advent of the electronic computer, skew rays were rarely traced because of the lengthy computation involved. Since a skew ray takes only a bit longer to trace on an electronic computer than a meridional ray, the reverse situation is now common, and meridional rays are usually traced as special cases of general rays. The general raytracing equations given below are slightly modified from those presented by D. Feder in the *Journal of the Optical Society of America*, vol. 41, 1951, pp. 630–636.

The ray is defined by the coordinates x, y, and z of its intersection point with a surface, and by its direction cosines, X, Y, and Z. (Symbols L, M, N are often used instead of X, Y, Z.) The origin of the coordinate system is at the vertex of each surface. Figure A.2 shows the meanings of these terms. Note that if x and X are both zero, the ray is a meridional ray and direction cosine Y equals sin U. The direction cosines are the projections, on the coordinate axes, of a unit-length vector along the ray. The direction cosines may be visualized as the length, height, and width of a rectangular solid or box which has a diagonal equal to one (1.0). (Note that the *optical* direction cosine is simply the direction cosine as defined above, multiplied by the index of refraction.)

The computation is opened by determining the values for x, y, z, X, Y, and Z with respect to an arbitrarily chosen reference surface, which may be plane (the usual choice) or spherical. Convenient choices for the

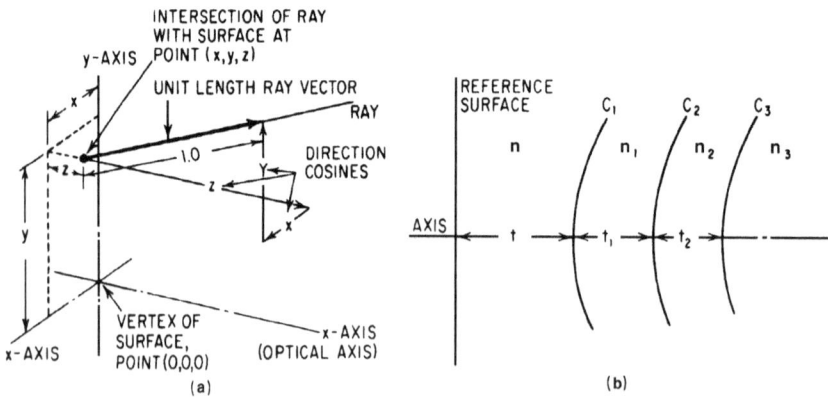

Figure A.2 Symbols used in skew raytracing Eqs. A.2a through A.2o. (a) The physical meanings of the spatial coordinates (x, y, z) of the ray intersection with the surface and of the ray direction cosines, X, Y, and Z. (b) Illustrating the system of subscript notation. Note that the nomenclature L, M, N is often used for X, Y, Z.

location of the reference surface are at the object (which allows the easy use of a curved object surface, if appropriate), at the vertex of the first surface, or at the entrance pupil. Note that Eq. A.2a is simply the equation of a sphere (and thus assures that the ray origin point lies in the reference surface), and that Eq. A.2b assures that the square of the unit vector along the ray is equal to 1.0.

Opening (at the reference surface):

$$c\,(x^2 + y^2 + z^2) - 2z = 0 \tag{A.2a}$$

$$X^2 + Y^2 + Z^2 = 1.0 \tag{A.2b}$$

Transfer to the first (or next) surface:

$$e = tZ - (xX + yY + zZ) \tag{A.2c}$$

$$M_{1z} = z + eZ - t \tag{A.2d}$$

$$M_1^{\,2} = x^2 + y^2 + z^2 - e^2 + t^2 - 2tz \tag{A.2e}$$

$$E_1 = \sqrt{Z^2 - c_1(c_1 M_1^{\,2} - 2M_{1z})} \tag{A.2f}$$

$$L = e + \frac{(c_1 M_1^{\,2} - 2M_{1z})}{Z + E_1} \tag{A.2g}$$

$$z_1 = z + LZ - t \tag{A.2h}$$

$$y_1 = y + LY \tag{A.2i}$$

$$x_1 = x + LX \tag{A.2j}$$

Refraction:

$$E'_1 = \sqrt{1 - \left(\frac{n}{n_1}\right)^2 (1 - E_1^{\,2})} \tag{A.2k}$$

$$g_1 = E'_1 - \frac{n}{n_1} E_1 \tag{A.2l}$$

$$Z_1 = \frac{n}{n_1} Z - g_1 c_1 z_1 + g_1 \tag{A.2m}$$

$$Y_1 = \frac{n}{n_1} Y - g_1 c_1 y_1 \tag{A.2n}$$

$$X_1 = \frac{n}{n_1} X - g_1 c_1 x_1 \tag{A.2o}$$

Terms without subscript refer to the reference surface and the following space. Terms subscripted with 1 refer to the first surface and the following space.

The symbols have the following meanings:

x,y,z	The spatial coordinates of the ray intersection with the reference surface.
x_1,y_1,z_1	The spatial coordinates of the ray intersection with surface #1.
M_1	The distance (vector) from the vertex of surface #1 to the ray, perpendicular to the ray.
M_{1z}	The z component of M_1.
E_1	The cosine of the angle of incidence at surface #1.
L	The distance along the ray from the reference surface (x, y, z) to surface #1 (x_1, y_1, z_1). L_j is the distance from surface j to $j+1$.
E'_1	The cosine of the angle of refraction (I') at surface #1.
X, Y, Z	The direction cosines of the ray in the space between the reference surface and surface #1 (before refraction).
X_1, Y_1, Z_1	The direction cosines after refraction by surface #1.
c	The curvature (reciprocal radius $= 1/R$) of the reference surface.
c_1	The curvature of surface #1.
n	The index between the reference surface and surface #1.
n'	The index following surface #1.
t	The axial spacing between the reference surface and surface #1.

Notice that the choice of the positive value for the square root in Eq. A.2f selects that intersection of the ray with the surface which is nearer the surface vertex. Also, if the argument under the radical in Eq. A.2f is negative, it indicates that the ray misses (never intersects) the spherical surface. If the argument under the radical in Eq. A.2k is negative, it indicates that the angle of incidence exceeds the critical angle; the ray is thus subject to total internal reflection (TIR) and cannot pass through the surface.

The calculation is opened by inserting c, two of the coordinates (x, y, z), and two of the direction cosines (X, Y, Z) into Eqs. A.2a and b and solving for the third coordinate and the third direction cosine. Then the intersection of the ray with the first surface (x_1, y_1, z_1) is determined from Eqs. A.2c through A.2j. Next, the ray direction cosines after refraction at surface #1 (X_1, Y_1, Z_1) are found from Eqs. A.2k through A.2o. This completes the raytrace through the first surface; at this point Eqs. A.2a and A.2b (with unit subscripts) may be used to check the accuracy of the computation.

To transfer to the second surface, the subscripts of Eqs. A.2c through A.2j are advanced by one, and x_2, y_2, and z_2 are determined. Similarly, the direction cosines after refraction (X_2, Y_2, Z_2) at surface #2 are found by Eqs. A.2k through A.2o with the subscripts incremented.

This process is repeated until the intersection of the ray with the final surface of the system, which is usually the image plane, has been determined. This completes the calculation.

Note that any ray which intersects the axis is a meridional ray; thus it is only necessary to trace skew rays from off-axis object points. Further, there is no loss of generality in assuming that the object point lies in the $y-z$ plane of the coordinate system (because we assume a system with axial symmetry). Therefore, any skew ray can be started with x equal to zero. When this is done, it is apparent that the two halves of the optical system, in front of, and behind the $y-z$ plane are mirror images of each other and that any ray X_k, Y_k, Z_k passing through x_k, y_k, z_k has a mirror image $(-X_k)$, Y_k, Z_k passing through $(-x_k)$, y_k, z_k in the other half of the system. For this reason, it is only necessary to trace skew rays through one-half of the system aperture; rays through the other half are represented by the same data with the signs of x and X reversed.

A.4   Skew Rays: Aspheric Surfaces

For raytracing purposes, an aspheric surface of rotation is conveniently represented by an equation of the form

$$z = f(x, y) = \frac{cs^2}{[1 + \sqrt{1 - c^2 s^2}]} + A_2 s^2 + A_4 s^4 + \cdots + A_j s^j \qquad \text{(A.3a)}$$

where z is the longitudinal coordinate (abscissa) of a point on the surface which is a distance s from the z axis. Using the same coordinate system as Sec. A.3, the radial distance s is related to coordinates y and x by

$$s^2 = y^2 + x^2 \qquad \text{(A.3b)}$$

As shown in Fig. A.3, the first term of the right-hand side of Eq. A.3a is the equation for a spherical surface of radius $R = 1/c$. The subsequent terms represent deformations to the spherical surface, with A_2, A_4, etc., as the constants of the second, fourth, etc., power deformation terms. Since any number of deformation terms may be included, Eq. A.3a is quite flexible and can represent some rather extreme aspherics. Note that Eq. A.3a is redundant in that the second-order deformation term ($A_2 s^2$) is not necessary to specify the surface, since it can be implicitly included in the curvature c. The importance of the inclusion of this term is that otherwise a large value of c (i.e., a short radius) could be required to describe the surface, and rays which would actually intersect the aspheric surface might not intersect the reference sphere. If necessary, the reference sphere may be a plane.

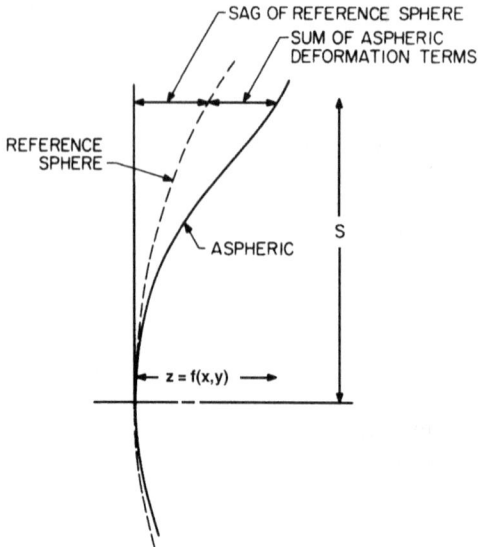

Figure A.3 Showing the significance of Eq. A.3a, which defines
an aspheric surface by a deformation from a reference spherical surface. The z coordinate of a
point on the surface is the sum of
the z coordinate of the reference
sphere and the sum of all the
deformation terms.

Note that the equation of a conic section, with conic constant k, is
given by:

$$Z = \frac{cs^2}{[1 + \sqrt{1 - (k + 1)\, c^2 s^2}\,]}$$

Aspheric surfaces which are *conic sections* (paraboloid, ellipsoid,
hyperboloid) also can be represented by a power series; see Chap. 18
for further details.

The difficulty in tracing a ray through an aspheric surface lies in
determining the point of intersection of the ray with the aspheric,
since this cannot be determined directly. In the method given here,
this is accomplished by a series of approximations, which are continued until the error in the approximation is negligible.

The first step is to compute x_0, y_0, and z_0, the intersection coordinates
of the ray with the spherical surface (of curvature c) which is usually
a fair approximation to the aspheric surface. This is done with Eqs. Ac
through Aj of the preceding section.

Then the z coordinate of the aspheric ($\bar{z}_0$) corresponding to this distance from the axis is found by substituting $s_0^2 = y_0^2 + x_0^2$ into the
equation for the aspheric (A.3a)

$$\bar{z}_0 = f(y_0, x_0) \tag{A.3c}$$

Then compute

$$l_0 = \sqrt{1 - c^2 s_0^2} \tag{A.3d}$$

$$m_0 = -y_0[c + l_0(2A_2 + 4A_4 s_0^2 + \cdots + jA_j s_0^{(j-2)})] \tag{A.3e}$$

$$n_0 = -x_0[c + l_0(2A_2 + 4A_4 s_0^2 + \cdots + jA_j s_0^{(j-2)})] \tag{A.3f}$$

$$G_0 = \frac{l_0(\bar{z}_0 - z_0)}{(Zl_0 + Ym_0 + Xn_0)} \tag{A.3g}$$

where X, Y, and Z are the direction cosines of the incident ray.

Now an improved approximation to the intersection coordinates is given by

$$x_1 = G_0 X + x_0 \tag{A.3h}$$

$$y_1 = G_0 Y + y_0 \tag{A.3i}$$

$$z_1 = G_0 Z + z_0 \tag{A.3j}$$

The process is sketched in Fig. A.4.

The approximation process is now repeated (from Eqs. A.3c to A.3j) until the error is negligible, i.e., until (after k times through the process)

$$z_k = \bar{z}_k \tag{A.3k}$$

to within sufficient accuracy for the purposes of the computation.

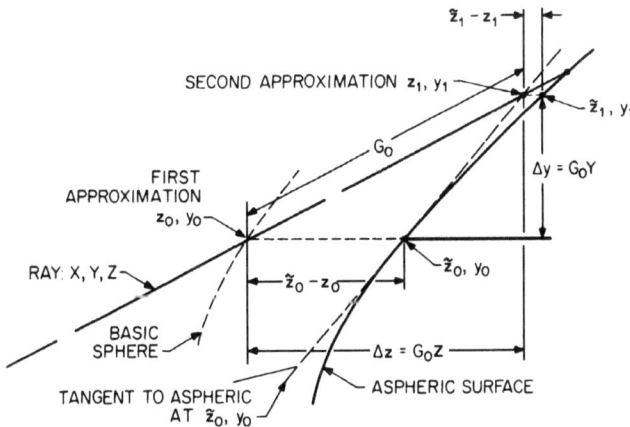

Figure A.4 Determination of the ray intersection with an aspheric surface. The intersection is found by a convergent series of approximations. Shown here are the relationships involved in finding the first approximation after the intersection with the basic reference sphere has been determined.

The refraction at the surface is carried through with the following equations:

$$P^2 = l_k^{\ 2} + m_k^{\ 2} + n_k^{\ 2} \tag{A.3l}$$

$$F = Zl_k + Ym_k + Xn_k \tag{A.3m}$$

$$F' = \sqrt{P^2\left(1 - \frac{n^2}{n_1^{\ 2}}\right) + \frac{n^2}{n_1^{\ 2}}F^2} \tag{A.3n}$$

$$g = \frac{1}{P^2}\left(F' - \frac{n}{n_1}F\right) \tag{A.3o}$$

$$Z_1 = \frac{n}{n_1}Z + gl_k \tag{A.3p}$$

$$Y_1 = \frac{n}{n_1}Y + gm_k \tag{A.3q}$$

$$X_1 = \frac{n}{n_1}X + gn_k \tag{A.3r}$$

This completes the trace through the aspheric. The spatial intersection coordinates are x_k, y_k, and z_k, and the new direction cosines are X_1, Y_1, and Z_1.

In Eqs. A.3d through A.3G, l, m, n are P times the direction cosine of the surface normal, and in Eqs. A.3e and A.3f the bracketed [] term is the approximate curvature at s.

In Eqs. A.3m and A.3n, $F = P \cos l$ and $F' = P \cos l'$.

The optical path equals $n \left(L + \sum\limits_{i=1}^{k} G_i\right)$

A.5   Coddington's Equations

The tangential and sagittal curvature of field can be determined by a process which is equivalent to tracing paraxial rays along a principal ray, instead of along the axis. In Chap. 5 it was pointed out that the slope of the ray intercept plot was equal to Z_t, the tangential field curvature. This slope *could* be determined by tracing two closely spaced meridional rays and computing

$$Z_t = \frac{H'_1 - H'_2}{\tan U'_2 - \tan U'_1} = \frac{-\Delta H'}{\Delta \tan U'}$$

and a similar process using close sagittal (skew) rays would yield Z_s, the sagittal field curvature.*

---

*Note that despite the currently almost universal use of z to represent the optical axis, it is still common usage to symbolize field curvature as x_t and x_s.

Coddington's equations are equivalent to tracing a pair of infinitely close rays, and the formulation has a marked similarity to the paraxial raytracing equations. However, object and image distances as well as surface-to-surface spacings are measured along the principal ray instead of along the axis, and the surface power is modified for the obliquity of the ray.

Figure A.5 shows a principal ray passing through a surface with sagittal and tangential ray fans originating at an object point and converging to their foci. The distance along the ray from the surface to the focus is symbolized by s and t for the object distance and by s' and t' for the image distance. The sign convention is as usual; if the focus or object point is to the left of the surface, the distance is negative; to the right, positive. In Fig. A.5, s and t are negative, s' and t' are positive.

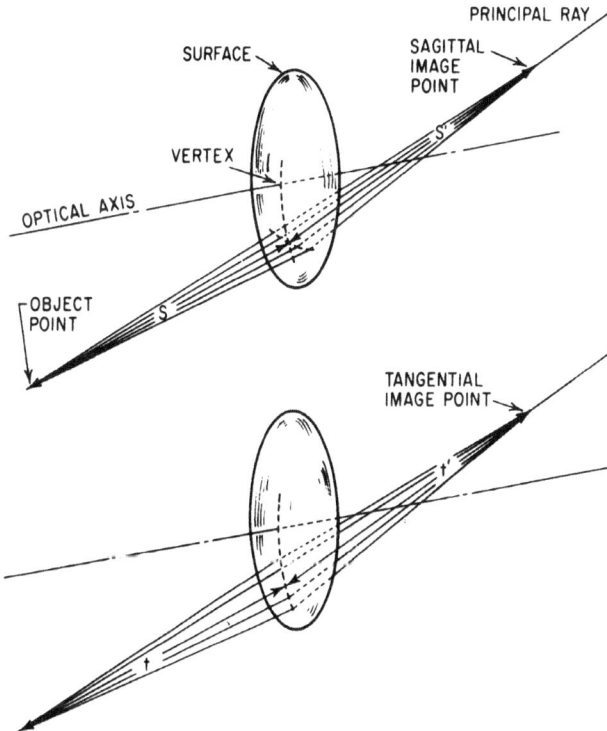

Figure A.5 A principal ray passing through a surface with sagittal and tangential ray fans originating at an object point and converging to their foci.

The computation is carried out by tracing the principal ray through the system using the meridional formulas of Sec. A.2, determining the *oblique power* for each surface by

$$\phi = c\,(n' \cos I' - n \cos I) \qquad \text{(A.4a)}$$

and determining the distance (D) from surface to surface along the ray by Eq. A.1m. The initial values of s and t are determined (Eq. A.1m is often useful in this regard) and then the focal distances are determined by solving the following equations for s' and t'.

$$\frac{n'}{s'} = \frac{n}{s} + \phi \qquad \text{(sagittal)} \qquad \text{(A.4b)}$$

$$\frac{n' \cos^2 I'}{t'} = \frac{n \cos^2 I}{t} + \phi \qquad \text{(tangential)} \qquad \text{(A.4c)}$$

The values of s and t for the next surface are given by

$$s_2 = s'_1 - D \qquad \text{(A.4d)}$$
$$t_2 = t'_1 - D \qquad \text{(A.4e)}$$

where D is the value given by Eq. A.1m.

The calculation is repeated for each surface of the system; the final values of s' and t' represent the distances along the ray from the last surface to the final foci. The final curvature of field (with respect to a reference plane an axial distance l' from the last surface) can be found from

$$z_s = s' \cos U' + z - l' \qquad \text{(A.4f)}$$
$$z_t = t' \cos U' + z - l' \qquad \text{(A.4g)}$$

where z is determined for the last surface by Eq. A.1l.

A.6   Aberration Determination

This section will briefly indicate the computational procedures involved in determining the numerical values of the various aberrations discussed in Chap. 5. Since this discussion will be somewhat condensed, the reader may wish to review Chap. 5 at this point.

Some of these procedures can be used to gain a reasonably complete analysis of the performance of a system by tracing only a few rays. Such techniques were quite popular before the advent of the personal computer.

We will assume that the paraxial focal distance l' (from the vertex of the last surface of the system to the paraxial image) has been determined. It is also useful to predetermine the size and location of the entrance pupil.

Spherical aberration

Trace a marginal meridional ray from the axial intercept of the object (through the edge of the entrance pupil of the system) and determine its final axial intercept L' and/or its intersection height H' in the paraxial focal plane. Then the longitudinal spherical aberration (LA′) is given by

$$LA' = L' - l' \tag{A.5a}$$

and the transverse spherical aberration (TA′) is given by

$$TA' = H' = -(LA') \tan U' \tag{A.5b}$$

The spherical aberration is overcorrected if the sign of $n'LA'$ is positive and undercorrected if the sign is negative.

The zonal spherical aberration is determined by tracing a second ray through the 0.707 zone (i.e., a ray which strikes the entrance pupil at a distance from the axis equal to 0.707 times the distance for the marginal ray). The zonal aberration is found from Eqs. A.5a and b. Rays may also be traced through other zones of the aperture if a more complete description of the axial correction of the system is required. The customary choice of the $0.707 = \sqrt{0.5}$ zone for zonal rays derives from the fact that, for most systems, the longitudinal spherical can be approximated by

$$LA' = aY^2 + bY^4 \tag{A.5c}$$

where Y is the ray height and a and b are constants. Thus, if the marginal spherical, at a ray height of Y_m, is corrected to zero, the maximum longitudinal zonal aberration occurs at

$$Y = \sqrt{\frac{Y_m^{\,2}}{2}} = 0.707Y_m$$

The maximum transverse spherical TA′ occurs at

$$Y = \sqrt{0.6Y_m^{\,2}} = 0.775Y_m$$

Since spherical is a function of ray height, it is common to use a subscript to identify the ray, as in LAm or LAz.

Coma

Three meridional rays are traced from an off-axis object point: a principal ray through the center of the entrance pupil and upper and lower rim rays through the upper and lower edges of the pupil. The final intersection heights of these rays with the paraxial focal plane are determined. Then the tangential coma is given by

$$\text{Coma}_T = H'_A + H'_p + \frac{(H'_A - H'_B)(\tan U'_A - \tan U'_p)}{(\tan U'_B - \tan U'_A)} \qquad \text{(A.5d)}$$

For most lenses, where the ray slope U' is a smooth uniform function of the ray position in the pupil, the following simplified equation is sufficiently accurate. This can be evaluated when examining a ray intercept plot by connecting the ends of the plot with a straight line and noting the sag of the principal ray from the line.

$$\text{Coma}_T = \frac{H'_A + H'_B}{2} - H'_p$$

where H'_p is the intercept for the principal ray and H'_A and H'_B are the intercepts of the rim rays.

Ordinarily, sagittal coma is very nearly equal to one-third of the tangential coma (especially near the axis). Sagittal coma can be determined by tracing a skew ray through the entrance pupil at $y = 0$, $x =$ the radius of the pupil. Then the displacement of the y intersection coordinate in the image plane from H'_p gives the sagittal coma (note that in this instance the image plane should be the plane of intersection of the upper and lower rim rays, i.e., where $H'_A = H'_B$).

The variation of coma with field angle (or image height) can be determined by repeating the process for another object height. The variation of coma with aperture is found by tracing zonal oblique rays.

OSC

The offense against the (Abbe) sine condition (OSC) is an indication of the amount of coma present in regions near the optical axis. It is determined by tracing a paraxial and a marginal ray from the axial object point and substituting their data into

$$\text{OSC} = \frac{\sin U}{u} \cdot \frac{u'}{\sin U'} \cdot \frac{(l' - l'_p)}{(L' - l'_p)} - 1 \qquad \text{(A.5e)}$$

where u and u' are the initial and final slopes of the paraxial ray, U and U' are the initial and final slopes for the marginal ray, l' and L' are the final intercept lengths of the paraxial and marginal rays, and l'_p is the final intercept of the principal ray (thus l'_p is the distance from the last surface to the exit pupil). If the object is at infinity, the initial y and Q are substituted for u and $\sin U$ in Eq. A.5e.

For regions near the axis

$$\text{Coma}_s = H' \, (\text{OSC})$$
$$\text{Coma}_t = 3H' \, (\text{OSC})$$

(A.5f)

Figure A.6 shows (a) third-order coma balancing fifth-order (linear) coma and (b) third-order coma balancing fifth-order (elliptical) coma. Note that the slope (or tilt) of the plot in Fig. A.6b usuals the OSC, demonstrating that a small OSC is a necessary but not sufficient condition for coma convection.

Distortion

Distortion is found by tracing a meridional principal ray from an offaxis object point through the center of the entrance pupil and determining its intersection height H'_p in the paraxial focal plane. A paraxial principal ray may be traced from the same object point to determine the paraxial

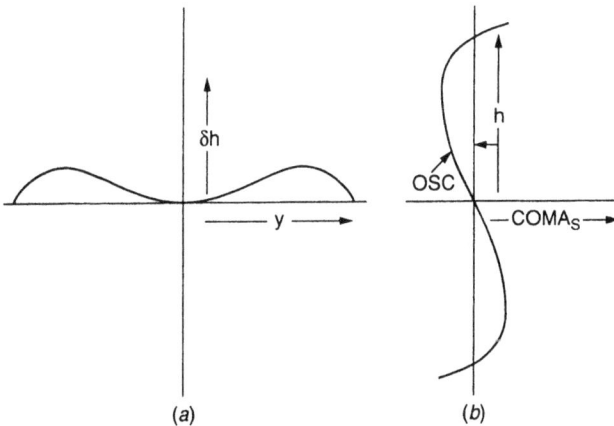

(a)   (b)

Figure A.6 (a) A ray intercept plot showing undercorrected fifth-order linear coma (which varies as hy^4) balanced by overcorrected third-order coma (which varies as hy^2). (b) A plot of coma versus field height showing overcorrected fifth-order elliptical coma (which varies as h^3y^2) balanced by undercorrected third-order coma (which varies as hy^2). Note that OSC = (sagittal coma)/h, as h approaches zero.

image height h', or the optical invariant Inv may be used as indicated in Chap. 4, Eq. 4.16. Then

$$\text{Distortion} = H'_p - h' \qquad\qquad \text{(A.5g)}$$

Distortion is frequently expressed as a percentage of the image height, thus:

$$\text{Percent distortion} = \frac{H'_p - h'}{h'} \times 100 \qquad\qquad \text{(A.5h)}$$

The variation of distortion with image height or field angle is found by repeating the process for several object heights.

Astigmatism and curvature of field

Trace a principal ray from an off-axis object point through the center of the entrance pupil. Then trace close sagittal and tangential rays by Coddington's equations (Sec. A.5) and determine the final z'_s and z'_t with respect to the paraxial image plane; z'_s and z'_t are then the sagittal and tangential curvature of field for this image point.

Alternatively, a meridional ray from the object point passing through the system close to the principal ray can be traced. Then

$$Z_t = \frac{H'_p - H'}{\tan U' - \tan U'_p} \qquad\qquad \text{(A.5i)}$$

will provide a close approximation to z'_t, since Z'_t approaches z'_t, as the two rays approach each other. A similar procedure with a close skew ray will yield Z'_s.

Since the variation of field curvature with image height is usually of interest, z'_s and z'_t may be determined for additional object heights or field angles and plotted against obliquity.

Note that it is common to refer to the field curvature (z_s and z_t) as x_s and x_t, in conformance with earlier usage when the optical axis was denoted as the x axis.

Chromatic aberration—axial (or longitudinal)

Paraxial longitudinal chromatic aberration is found by determining the paraxial image points for the longest and shortest wavelengths in the spectral bandpass of the system. This is done by determining l' using the indices of refraction associated with one wavelength and then with the other. For visual systems, the long wavelength is usually taken as C-light (λ = 0.6563 μm hydrogen line) and the short wavelengths as F-light (λ = 0.4861 μm hydrogen line). The longitudinal chromatic aberration is then

$$\text{LchA}' = l'_F - l'_C \tag{A.5j}$$

The transverse measure of axial chromatic can be found from

$$\text{TAch} = -\text{LchA} \tan U'_K$$

or by calculating the height of the rays in the mid-wavelength focal surface and

$$\text{TAch} = h'_F - h'_C$$

The chromatic aberrations for other zones of the aperture are found by tracing meridional rays from the axial object point for each wavelength and substituting the final axial intercepts into Eq. A.5j.

The secondary spectrum is found by tracing axial rays in at least three wavelengths—long, middle, and short—and plotting their axial intercepts against wavelength. A numerical value for the secondary spectrum is strictly valid only when the long- and short-wavelength images are united at a common focus, so that

$$l'_F = l'_C$$

then

$$SS' = l'_d - l'_F = l'_d - l'_C \tag{A.5k}$$

where the subscripts C, d, and F indicate long, middle, and short wavelengths. For visual work, C, F, and d represent the C and F lines of hydrogen and the helium d line at 0.5876 μm.

The *spherochromatism* (chromatic variation of spherical aberration) is found by determining the spherical aberration at various wavelengths. Thus, for visual work the spherochromatism would be the spherical in F light minus the spherical in C light.

Chromatic aberration—lateral

Lateral chromatic aberration, or chromatic difference of magnification, is determined by tracing a principal ray from an off-axis object point through the center of the entrance pupil in both long and short wavelengths and finding the final intersection heights with the focal plane. Then

$$\text{TchA} = H'_F - H'_C \tag{A.5l}$$

for visual work. Alternatively, the paraxial lateral color can be found by tracing paraxial "principal" rays in two colors and substituting h'_F and h'_C into Eq. A.5l. The chromatic difference of magnification is given by

$$CDM = TchA/h'$$

Lateral chromatic aberration should not be confused with the transverse expression for axial (longitudinal) chromatic aberration, which is given by

$$TAch = H'_F - H'_C = -(LchA) \tan U' \qquad (A.5m)$$

where the data are derived from rays traced from an object point *on the optical axis.*

Optical path difference (wave-front aberration)

Recalling (from Chap. 1) that a wave front which forms a "perfect" image is spherical in shape and is centered about the image point, it is apparent that the aberration of an image formed by an optical system can be expressed in terms of the departure of the wave front from an ideal spherical wave front. The velocity of light in a medium of index n is given by $c/n,$ where c is the speed of light in vacuum, and the time required for a point on a wave front to travel a distance D through the medium is nD/c. Thus, if a number of rays from an object point are traced through an optical system, and the distances along each ray from surface to surface are computed (by Eqs. A.1m or A.2g), including the distance from object point to the first surface, then the points for which $\Sigma nD/c$, or ΣnD, are equal, are points through which the wave front passes at the same instant. A smooth surface through these points is the locus of the wave front.

The departure of the wave front from the ideal reference sphere is equal to the difference in the optical paths to the reference sphere. Thus the wave-front aberration or *optical path difference* (OPD) can be found by tracing rays from the object to the surface of a reference sphere centered on the image point and determining

$$OPD = (\Sigma nD)_A - (\Sigma nD)_B \qquad (A.5n)$$

Note that the choice of the reference image point location will have a great effect on the size of the OPD, since a shift of the reference point is equivalent to focusing (in the longitudinal direction) or to scanning the image plane for the point image (when shifting the reference point laterally).

OPD is usually measured in wavelengths, or fractions thereof. For example, the Rayleigh criterion may be expressed as follows: An image will be "sensibly" perfect if there exists not more than one-quarter wavelength difference in optical path over the wave front with reference to a sphere centered at the selected image point. The numerical precision required to obtain significant results in an OPD calculation is higher than that required for ordinary raytracing. The OPD is customarily determined with respect to a spherical surface (centered about the reference point) with a radius equal to the distance from the exit pupil to the reference point. Some programs do the equivalent of using a reference sphere of infinite radius to simplify the calculation. If the wave-front aberration is small, this gives a result that is almost exact.

B

Some Standard Dimensions

Vidicon, CCD & CMOS Formats
(diagonal in parentheses)

name	format
1/4 inch	2.4 × 3.2 mm (4.0 mm)
1/4 inch	2.7 × 3.6 mm (4.5 mm)
1/3.6 inch	3.0 × 4.0 mm (5.0 mm)
1/3.2 inch	3.42 × 4.54 mm (5.68 mm)
1/3 inch	3.6 × 4.8 mm (6.0 mm)
1/2.7 inch	3.98 × 5.27 mm (6.60 mm)
1/2 inch	4.8 × 6.4 mm (8.0 mm)
1/2 inch CCD	4.2 × 6.4 mm (7.66 mm)
1/1.8 inch	5.32 × 7.18 mm (8.93 mm)
2/3 inch	6.6 × 8.8 mm (11.0 mm)
1 inch	9.6 × 12.8 mm (16.0 mm)
Pulmicon	12.7 × 17.1 mm (21.2 mm)
4/3 inch	13.5 × 18.0 mm (22.5 mm)

Flange Focus Distances

C-Mount	17.53 mm
Olympus	46.2 mm
Nikon	46.5 mm
Panavision	51.56 mm
Arriflex	52.0 mm
T-Mount	55.1 mm
BNCR	61.47 mm

Standard Microscope Cover Glass

Thickness: t = 0.18 mm (0.16 to 0.19 mm)
index: n_d = 1.523 ±0.005
v-value: V = 56 ±2

Film Formats (diagonal in parentheses)

Still Cameras	
35 mm camera "double frame"	24 × 36 mm (43.27 mm)
35 mm camera "single frame"	18 × 24 mm (30.00 mm)
	also 17 × 24 (29.41) and 18 × 23 (29.21)
APS camera	16.7 × 30.2 mm (34.51 mm)
110 Camera	13 × 17 mm (21.4 mm)
Double 16 camera	10 × 14 mm (17.20 mm)
Minox camera	8 × 11 mm (13.6 mm)
120/620 square	56 × 56 mm (79.2 mm)
6. × 7. cm	56 × 69.5 mm (89.3 mm)
120/620 rect.	56 × 82.5 mm (99.7 mm)
116	63.5 × 107.95 mm (125.24 mm)
4 × 5 inch	101.6 × 127.0 mm (162.64 mm)
5 × 7	127.0 × 177.8 mm (218.5 mm)
8 × 10	203.2 × 254.0 mm (325.3 mm)
11 × 14	279.4 × 355.6 mm (452.2 mm)

Motion Picture Cameras	
8 mm	3.51 × 4.80 mm (5.95 mm)
Super-8	4.22 × 5.77 mm (7.15 mm)
16 mm	7.42 × 10.22 mm (12.63 mm)
35 mm silent	19.05 × 25.37 mm (31.73 mm)
35 mm sound	16.00 × 22.00 mm (27.20 mm)
Anamorphic	18.7 × 22.0 mm (28.87 mm)
70 mm	22.1 × 48.6 mm (53.39 mm)

Glossary

The following is a glossary of terms common to the field of lens and optical system design. An optical practitioner who is new to this field might well profit from a thorough perusal of this list, since the jargon of optical engineering may differ from that of physics or even that of basic optics. These definitions actually go a long way toward establishing a basic understanding of the art. Note that the meanings of some terms are determined by the context of their use.

Abbe *V*-number The reciprocal relative dispersion of an optical material. For visual work $V = (n_d - 1)/(n_F - n_c)$, where d, F, and C indicate the Fraunhofer wavelengths: 0.5876, 0.4861, and 0.6563 μm, respectively. Often called V-value or v-value, or nu-number.

Abbe sine condition (OSC) A condition where the magnification is constant across the aperture, and coma near the axis is zero. Paraxial magnification (m = u/u′) and trigonometric magnification (M = sin U/sin U′) should be equal, so that m/M = (u′ sin U)/(u sin U′) = 1.0. A departure from this equality is called the offense against the sine condition, or OSC. Near the axis, OSC = (sagittal coma)/h'.

aberration An image defect whereby all rays from a point source do not converge to a point image at the desired location. An aberrated wave front departs from a perfect sphere centered on the desired image point. The primary aberrations are: spherical, coma, astigmatism, field curvature, distortion, axial chromatic, and lateral color. Relative to the ideal paraxial image, the aberration of a ray may be measured as a transverse displacement, a longitudinal displacement, an angular deviation, or a wavefront deformation.

achromat An optical system free of primary chromatic aberration. Usually defined as a system where two different wavelengths (e.g., C- and F-light) are brought to a focus at the same location. Usually accomplished by the use of materials of differing V-values.

afocal system An optical system which forms an image of an infinitely distant object at infinity, i.e., a system where both input and output beams are collimated. A telescope is an afocal system, as is a beam expander. An afocal system can also form a real image of an object at a finite distance.

air path, equivalent   The distance in air which is optically equivalent to a distance in a medium. Equal to $\Sigma(D/n)$.

Airy disk   The central bright patch of the diffraction pattern which is formed as the image of a point source. The disk size is defined by the diameter of the first dark ring of the pattern, equal to $1.21\lambda/NA$ (or, very approximately, equal to the *f*-number in microns). Usually implies a perfect or near-perfect lens with a circular aperture.

anamorphic   A system having a different magnification or focal length in each of two mutually perpendicular meridians. It may be composed of prisms or lenses which have cylindrical or toroidal surfaces.

anastigmat   Strictly, without astigmatism. The term is usually applied to a lens system where an effort has been made to flatten the field and reduce the astigmatism. An anastigmat lens often has a *node* with zero astigmatism at some field angle.

angle of incidence, refraction, reflection   The angle between an incident, refracted, or reflected ray and the normal to the surface erected at the point where the ray intersects the surface.

aperture stop   That feature of an optical system which most severely limits the diameter of the axial beam which can pass through the system. The feature is usually the clear aperture of a lens element or a mechanical aperture, such as the iris diameter in a camera lens. The chief or principal ray crosses the axis at, and passes through the center of, the aperture stop. In many compound optical systems (e.g., a telescope or microscope) the aperture stop is located at the objective lens. Note that for off-axis object points the beam size may be limited (vignetted) by more than one physical feature of the system.

aplanat(ic)   A lens or surface which is free of both spherical aberration and coma.

apochromat   A lens in which three colors of light are brought to a common focus. Usually requires the use of materials with unusual partial dispersions. (Pronounced APO-chro-mat, *not* a-POCH-ro-mat.) Formerly implied chromatic correction for three wavelengths and spherical aberration correction for two wavelengths.

apodization   Where the transmission of a system is varied over the aperture in order to modify the diffraction pattern. Originally intended to eliminate the rings (feet = pod) around the Airy disk in the diffraction pattern. Similar effects can result from a variation of the beam intensity distribution, e.g., as in a gaussian laser beam.

apparent field of view   The angular field of view as seen by the eye through a telescope. The apparent field equals the real (object) field times the telescope magnification.

aspheric surface   A su1rface which departs from a true spherical shape. The conic section surfaces (paraboloid, ellipsoid, and hyperboloid) are aspherics, as are more general aspheric surfaces. Aspheric surfaces are often used to correct aberrations.

astigmatism   An aberration which causes a fan of rays in one meridian to focus at a different location than a fan in the orthogonal meridian. May be caused by oblique imagery or by a toroidal surface. Primary astigmatism varies as the square of the field angle.

axial astigmatism, coma   A tilted surface, or one which is not axially symmetric, will produce aberrations on the optical axis which are only found off the axis for systems which are axially symmetrical.

axial chromatic   An aberration which causes light of different wavelengths to be focused at different distances from the lens.

axial object point   An object point which is on the optical axis.

axial ray   A ray from the axial intercept of the object to the edge or margin of the entrance pupil. A marginal ray.

axis, optical   The common axis of rotational symmetry for an optical system. For a spherically surfaced lens element, the line connecting the centers of curvature of the surfaces.

back focal length (bfl)   The distance from the vertex of the last surface of a system to the second focal point.

baffle   Opaque shielding to reduce or eliminate stray light. Often a thin annular piece around the inner surface of a lens mounting structure, in order to prevent the reflection of light which grazes the inside wall of the mount.

Barlow lens   A negative lens placed behind the primary (telescope) objective to increase its focal length and the telescope magnification.

beam expander   An afocal system (usually galilean) used to increase the beam diameter and reduce the beam divergence, usually of a laser beam.

bending   The lens design process of changing the shape of an element while maintaining its power at a constant value.

binary optics   Diffractive optics where the ideal smooth (kinoform) surface is approximated by a stepped or staircase surface.

blocker   A tool to support or carry lens elements in the grinding and polishing operations. Classically an iron tool to which the elements are fastened with pitch. A spot blocker is a metal tool with accurately machined recesses in which the elements are cemented.

boresight error   A misalignment of the optical axis with respect to the aiming axis, or a misalignment of two optical axes.

buried surface   A cemented interface separating two materials with very nearly identical refractive indices, but with different dispersions. Used (now rarely) by some designers during the late stages of the design process to correct chromatic aberration without affecting the other aberrations.

cardinal points   The Gauss points. The first and second focal points and principal points. The nodal points are often also considered cardinal points.

Cassegrain   A two-mirror objective with a concave primary mirror (classically a paraboloid) and a convex secondary mirror (a hyperboloid), corrected for spherical aberration.

catadioptric and catoptric   Optical systems consisting of only mirrors (catoptric), or of mirrors and refracting surfaces (catadioptric). A purely refracting system is called *dioptric*.

CDM (chromatic difference of magnification)   See chromatic aberration and lateral chromatic.

chief or principal ray   There are several definitions, depending on the use to which the concept is put:

> The oblique ray which passes through the center of the aperture stop of an optical system.
>
> The central ray of a vignetted oblique beam.
>
> The oblique ray aimed toward the center of the entrance pupil.

chromatic aberration   An aberration which results from the dispersion of the materials used in an optical system. See axial chromatic and lateral chromatic.

clear aperture   The diameter of the transmitting (or, if a mirror, the reflecting) portion of a surface, lens, or system.

coating, low reflection   Very thin (fraction of a wavelength) layer(s) of optical materials which reduce the reflection from a lens surface. May be a single quarter-wave thick layer of low index material, e.g., MgF_2, or multiple layers of different materials.

coherent   Light equivalent to that from a true point source (spatial coherence) and/or monochromatic light (temporal coherence.) Many laser beams are relatively coherent.

cold finger   The chilled support for a detector.

cold stop   A cooled aperture within a vacuum Dewar in an infrared system, which is (ideally) the aperture stop or a pupil of the system. Its purpose is to prevent the detector from seeing anything but the optics and the imaged scene, especially the (warmer) interior of the system.

collective surface   A surface which refracts light rays toward the optical axis. A convergent surface. A positive surface. A surface where $(n' - n)/r$ is positive.

collimated light   A beam of light wherein all the rays originating at a point are parallel to each other, and wherein all wavefronts are plane. Light from a point source at infinity is collimated. A "collimated beam" from a source which is not a true point expands as it travels; its divergence angle equals the source size divided by the focal length of the optical system.

coma   An off-axis aberration where annular zones of the aperture have different magnifications. The resulting image of a point looks like a comet. Axial coma is produced by tilted surfaces. See also Abbe sine condition and OSC.

component   One or more lens elements (usually in a group) which are treated as a unit.

concave surface   A hollow curved surface, i.e., one which is lower in the center, sunken. The inner surface of a hollow sphere.

condenser   The lens or component in a projection system which collects light from a source and directs it through the aperture of the projection lens. A form of field lens. In Koehler illumination the condenser images the source into the pupil of the projection lens in order to produce uniform illumination on the screen.

conjugate   Object and image are conjugates, as are the distances associated with them.

Conrady $(D - d)\delta n$ A technique used to control or correct chromatic aberration by making the chromatic variation of the optical path of the marginal ray $(\Sigma D\delta n)$ equal to that of the axial or principal ray $(\Sigma d\delta n)$, where δn is the dispersion.

converging lens or surface   One which bends rays toward the optical axis. A positive lens or surface.

convex surface   A surface which is higher at the center than at the edge, outward curving, and bulging. The outer surface of a sphere.

Cooke triplet   A triplet anastigmat with two outer positive crown elements and an inner negative flint element, all spaced apart.

coordinate system   The optical axis is the z coordinate, the y coordinate is vertical, and the x coordinate is normal to the meridional plane. A right-handed system. The coordinate origin is usually located at the vertex of the surface.

cosine fourth   The illumination in the image plane of a nominal optical system varies approximately as the fourth power of the cosine of the angle of obliquity of the chief ray. Assumes no distortion of the image or of the pupil, and a small NA.

cosmetic defects   Scratches, digs, pits, inclusions, etc., which (in most cases) do not, affect the function of an optical system.

critical angle   See TIR (total internal reflection). Arcsin n'/n.

critical illumination   Where the light source is focused on the subject.

CRT Cathode ray tube.

crown   An optical shop term for a convex shaped element.

crown glass   A low dispersion glass. A glass with a V-value of more than 50 (for an index >1.6) or more than 55 (for an index <1.6)

curvature   The reciprocal of the surface radius. ($c = 1/r$) A measure of the departure from a flat surface.

curvature of field   The departure of the image surface from a desired flat surface. Measured longitudinally, or as the Petzval radius, ρ.

cut-off frequency   The spatial frequency beyond which an optical system cannot transmit information. Equal to $2NA/\lambda$ or $1/\lambda(f/\#)$. In the visible spectral region the cut-off frequency is $1800/(f/\#)$ line pairs per mm.

$(D - d)\delta n$ See Conrady $(D - d)\delta n$.

damped least squares (DLS)   A modification of the least squares solution technique (which see), where the squares of the weighted parameter changes are added to the merit function (as a penalty, to prevent nonlinear relationships from producing big changes, leading to extreme or impossible solutions).

decentered element   An element whose principal points are displaced from the optical axis.

decentered surface   A surface the center of curvature or axis of symmetry of which is not on the optical axis.

degree of freedom   A constructional parameter which may be varied in the design process.

depth of focus (or field) The longitudinal shift of the image sensor (or of the object) which produces an image degradation which is acceptable for the application. Depth of focus and depth of field are related by the longitudinal magnification. In photography the criterion is often based on the size of the defocus blur spot compared with the smallest perceptible or the smallest recordable blur (e.g., film grain or pixel size). Another criterion is based on the Rayleigh quarter-wave limit for OPD; this allows a depth of focus equal to $\pm 2\lambda(f/\#)^2$, or $\pm\lambda/2NA^2$.

dewar An insulated, evacuated container for a cooled, long-wavelength infrared detector with a transmitting entrance window, a cold stop and a cold finger.

dialyte A lens consisting of (two) separated positive and negative elements or components.

dichroic filter A filter with two separate transmission (or reflection) bands.

diffraction The cause of the spreading or divergence of a wave front which occurs when it encounters an obstruction such as an aperture or an opaque edge. The Airy disk and the associated rings are caused by the diffraction resulting from the aperture of the optical system.

diffraction limited Strictly, when system performance is limited solely by diffraction. Often colloquially applied to a system with a Strehl ratio of 0.8 or more, or an OPD of one quarter wave or less.

diffractive optics Optics where the effects are produced by diffraction (as opposed to refraction). A diffractive lens surface is a Fresnel surface modulo 2π [i.e., step height $= \lambda/(n' - n)$].

diopter A measure of the power of a lens or surface. Equal to the reciprocal of the focal length of a lens (in meters). For a surface, $\varphi = (n' - n)/r$. Distances (as reciprocals) may also be expressed in diopters.

dioptric A system of refracting surfaces. (Catoptric is a system of reflecting surfaces, and catadioptric is a system with both refracting and reflecting surfaces.)

dispersion The change of index with wavelength. For visual work it is usually taken as the index difference between the red and blue Fraunhofer hydrogen lines C (656.3 nm) and F (486.1 nm), thus ($n_F - n_C$). This is total or principal dispersion. See also partial dispersion.

dispersive surface A surface which refracts light rays away from the optical axis. A diverging surface. A negative surface. A surface where $(n' - n)/r$ is negative.

distortion   An aberration in which the magnification varies over the field of view. It is called pincushion or positive if the magnification increases toward the edge of the field, and barrel or negative if it decreases. Note that distortion reverses sign if the object and image are interchanged.

diverging lens or surface   One which bends light rays away from the optical axis. A negative lens or surface. A lens with a negative focal length. A surface where $(n' - n)/r$ is negative.

doublet lens   Either (1) a closely spaced or cemented pair of elements, one positive and one negative, or (2) two separated components with a stop between them.

effective focal length (fel)   See focal length.

element   A lens which is a single piece of glass (or a mirror).

empty magnification   Magnification (in a telescope or microscope) which is larger than the magnification at which the diffraction limited resolution matches the resolution of the eye. An increase in magnification which does not serve to increase the information content of the image, although the increased magnification may make the image easier to recognize, locate or process, and also reduce the visual strain on the user. For a visual telescope, empty magnification begins when the power exceeds approximately MP = 11D, where D is the diameter of the objective in inches. For a microscope it begins when the magnifying power exceeds approximately MP = 225NA. (The constants in these relationships depend on the resolution assumed for the eye.)

endoscope   A miniature periscope (which see) usually for medical use.

entrance or exit pupil   The image of the aperture stop which is seen from object or image space. All light rays passing through the system must enter through the entrance pupil and exit through the exit pupil. The principal ray passes through the center of both pupils.

entrance window   The image of the field stop in object space.

erect image   An image in which top and bottom, and left and right correspond to the orientation of the original object.

erector lens   A lens which relays the image and re-inverts an inverted image to produce an erect final image.

etendue or **throughput**   The product of the light beam area and the solid angle of the beam, or the product of the pupil area and the solid angle of the field of view, or the detector area times the lens speed. It is constant through the system (demonstrating the conservation of brightness-luminance-radiance). Related to the square of the Lagrange invariant. See invariant.

evanescent wave   In total internal reflection (TIR), the electric vector actually penetrates a very small distance (to the order of a wavelength) into the lower index medium. TIR may be frustrated if the evanescent wave is captured by another material very close to the surface.

exit pupil   The image of the aperture stop as seen from image space. All light rays passing through the optical system must emerge through the exit pupil. In a visual system, the eye must be placed at the exit pupil to see the full field of view.

exit window   The image of the field stop in image space.

eye box   The area or volume within which the eye must be placed and specifications must be met. Usually specified for a system without an (imaged) exit pupil.

eye relief   The relief or clearance distance between the exit pupil (which is the usual location for the eye) and the last surface of a visual optical system such as a telescope or microscope. Sometimes a speci-fied eye location, as in eye box.

fiber optics   A very small diameter, flexible cylinder of glass or plastic which transmits light by total internal reflection. Used singly to trans-mit information by modulation of the light, or in coherent bundles to transmit images, or in random bundles for illumination.

field curvature   The departure of the image surface from a plane, when the image is formed as a curved surface due to astigmatism and/or Petzval aberrations. See curvature of field.

field flattener   A lens (usually negative) placed close to the image to flatten the Petzval curvature without greatly affecting the image size or the other aberrations. In some mirror systems it is a positive lens.

field lens   A lens placed at or near an internal image of an optical device (e.g., telescope, microscope, or endoscope) in order to converge the oblique beams so that they pass through the clear apertures of the following components (e.g., the eye lens of a telescope). The lens is usu-ally positive with the aim of widening the field of view. Occasionally a negative field lens is used to flatten the Petzval field and/or to increase the eye relief; This requires a larger diameter eye lens.

field stop   The aperture (or feature) which limits or defines the part of the object which is imaged. Usually located at an image plane.

field of view   That part of the object which is included in the final image. May be expressed as an angle or as a linear dimension. Abbreviated FOV. The "real" FOV of a telescope is the angular field in object space; The "apparent" FOV is the corresponding field in image space.

fifth-order aberration   See third-order aberration.

first order   A term applied to paraxial calculations and characteristics.

fish-eye lens   A lens with a field of view of 180° or more. A reversed telephoto type with a large amount of barrel distortion.

flint   An optical shop term for a concave shaped element.

flint glass   An optical glass with a V-value less than 50 (for index >1.6), or less than 55 (for index <1.6). Named for the broken flints added to the melt in making fine glass for tableware.

floating lens   An element or component which moves independently of the balance of the system, usually to maintain good aberration correction while focusing, or as a focusing device.

f-number   The "speed" or relative aperture of a lens system. The ratio of the effective focal length to the diameter of the entrance pupil. A measure of the illuminating capability of a lens. Usually written f/n where n is the f-number, e.g., $f/6.3$; or 1:6.3, or f:6.3. Sometimes abbreviated as $f/\#$ or f/no. For an aplanatic system with the object at infinity, the f-number equals 0.5/NA. A "fast" lens with a small f-number can record an image in a short period of time, and thus produces a non-blurred image of a rapidly moving object. The f-numbers now in common use for iris diaphragms are usually taken from the series: 1, 1.4, 2, 2.8, 4, 5.6, 8, 11, 16, 22, 32, although the largest aperture is often not from this series. See also working f-number

focal length   The effective (or equivalent) focal length is the distance from the second principal point to the second focal point. Often abbreviated efl or simply f. It is the limiting value of $f = h'/\tan \theta$ as h' and θ approach zero, where h' is the image height and θ is the angle subtended by an infinitely distant object.

focal point   The image of an infinitely distant axial point source object. The second or back focal point is the image of a point which is to the left of the lens, and the first or front focal point is the image of a point to the right.

focus (noun)   The (usually longitudinal) location of the sharpest image.

focus (verb)   The act of changing the relative positions of elements and sensor in order to get a sharp image.

focus shift   The change of best focus position as the size of the aperture is changed. Usually due to spherical aberration. Also the image shift due to thermal or other environmental change.

Fraunhofer lines   A series of dark lines in the solar spectrum, e.g., C, d, and F lines. A listing of lines commonly used in optical design follows (wavelengths in nanometers):

Nd laser	1060.0 nm	D	589.29 ave
t	1013.98	d	587.56
s	852.11	e	546.07
A'	768.19	F	486.13
r	706.52	F'	479.99
C	656.28	g	435.83
C'	643.85	h	404.66
HeNe laser	632.8	i	365.01

Fresnel surface   A surface wherein annular zones are stepped and the surface of each zone has a curvature and slope corresponding to that of an ordinary surface. Often used in condensing, signaling and illuminating systems. Allows crude imagery with a thin, light element. A diffractive lens is one with a Fresnel surface where the step height is $\lambda/(n-1)$.

front focal length (ffl)   The distance from the vertex of the first surface of a lens system to the first focal point.

fringes   Dark bands caused by interference, as seen in an interferometer or with a test plate.

frustrated TIR   When a material in the lower index medium is placed very close to the TIR surface, the reflection can be frustrated or modified.

f-theta lens   A lens with barrel distortion, the image height of which is given by $h' = f\theta$ rather than $h' = f\tan\theta$. Used in scanner systems to get a uniform exposure across the scan.

gauss objective   An airspaced achromatic telescope objective where both crown and flint elements are meniscus and convex to the distant object.

ghost image   An image formed by (undesired) multiple reflections from the lens surfaces. Often due to the sun or other bright object outside of the field.

global optimum   See optimum, global.

Gregorian   A two-mirror objective with a concave primary mirror (classically a paraboloid) and a concave secondary mirror (an ellipsoid), corrected for spherical aberration, which provides an erect image.

grinding tool   A cast iron (typ. mehanite) tool used with an abrasive slurry to grind optical elements to a spherical shape. May also be a tool consisting of many diamond impregnated metal "buttons" bonded to a support structure.

H–tan U plot   A ray intercept plot where the intercept height of a ray is plotted against the tangent of the ray slope angle.

hyperfocal distance   The distance at which a lens is focused so that the image is within the depth of focus for object distances from infinity to one half of the hyperfocal distance.

image   A pictorial representation of an object, formed by the distribution of light at the focus of an optical system.

incoherent   Light from an extended (i.e., nonpoint) source and/or non-monochromatic light.

index   See refractive index.

invariant   In general, an expression which has the same value everywhere in an optical system. The Lagrange or optical invariant is equal to the product of the index, the image (or object) height, and the half convergence (or divergence) angle of the axial beam. Often expressed as $hnu = h'n'u'$ or $m = h'/h = nu/n'u'$. At a general surface the invariant is given by $(y_p nu - ynu_p) = (y_p n'u' - yn'u_p')$. Throughput or etendue is the three dimensional version of the invariant, and can be regarded as the product of the object (or image) area times the solid angle of collection (or illumination); alternately the pupil area times the solid angle of the field of view. Another paraxial invariant (used in stop shift theory) is $\delta y_p/y$, where δy_p is the change of the principal ray height produced at a surface by the shift of the stop, and y is the axial ray height at the surface.

inverted image   This terminology is neither definitive nor consistently used. This term may be used to describe an image which is inverted both top to bottom and left to right (as in the real image formed by an ordinary lens), or alternately, an image which is inverted in only one meridian (as for example, by reflection from a plane surface).

iris diaphragm   A mechanically adjustable aperture formed by thin pivoting arcuate leaves.

keystone distortion   A distortion which produces a trapezoidal image of a rectangular object, resulting from nonparallel object and image planes.

Koehler illumination   Uniform illumination produced by imaging the light source in the pupil of the projection lens.

landscape lens   A meniscus singlet with a separate aperture stop on its concave side.

Lagrange invariant   See invariant.

lateral chromatic (color)   The variation of image height or magnification with wavelength. The chromatic difference of magnification, $CDM = \delta h'/h'$.

least squares   An optimization method which minimizes the sum of the squares of the operands in the merit-defect-error function, based on the assumption that the operands are linearly related to the variable parameters. Usually it will locate the nearest local optimum in parameter space. See also damped least squares.

lens hood   A (usually) cylindrical shield which extends into object space to prevent extraneous stray light from entering the lens. A sun-shade.

light pipe   A polished rod or cylinder which transmits light by total internal reflection. Can be used to "scramble" or homogenize illumination.

line spread function   The distribution of illumination across the width of the image of a line.

longitudinal   Having to do with distance measured along, or parallel to, the optical axis.

Lyot stop   A glare stop placed at a pupil (an image of the aperture stop).

macro lens   A lens corrected for, or adjustable for use with near-by objects.

magnification, angular   Telescope magnification. The angle subtended by the image divided by the angle subtended by the object. Usually the ratio of the tangents of the half-field angles

magnification: lateral, linear, or transverse   The ratio of image height to object height, measured normal to the axis.

magnification, longitudinal   The ratio of the longitudinal motion of the image to that of the object, or the ratio of the longitudinal length (thickness) of the image to that of the object. For small motions or lengths, it is equal (in the limit) to the square of the lateral magnification, and is thus always positive, so that the object and image always move in the same direction.

magnification, microscopic   In a microscope or magnifier, the ratio of the angle subtended by the image to that subtended by the object, when the object is viewed at a conventional distance of 10 in, which is assumed to be the distance of most distinct vision. Thus, if the image is at infinity, $MP = 10 \text{ in}/F$, where F is the focal length of the magnifier or microscope in inches.

marginal ray   The ray (usually from an axial object point) which passes through the edge or margin of the lens aperture. An axial ray.

Marechal criterion   A Strehl ratio of 80 percent or more indicates a "sensibly perfect" image. Corresponds roughly with the Rayleigh quarter-wave criterion.

medial image   In astronomical circles, a compromise focus position midway between the sagittal and tangential images, presumed to be the best focus.

member   In a photographic lens the front member consists of those elements before the aperture stop or iris diaphragm, and the rear member consists of those following the stop.

meridional plane   Any plane which includes the optical axis. Also called the tangential plane.

meridional ray   A ray which lies in the meridional plane; a ray which intersects the optical axis. A tangential ray.

merit function   A collection of weighted terms (or targets, or operands) which are squared and summed. The terms may represent aberrations, physical dimensions, magnifications, image locations, or any characteristic which can be calculated; very often an operand is the difference between a calculated value and the desired value. Since each entry represents an undesired characteristic, the merit function might better be called a defect function or an error function. The purpose of the merit function is to represent, with a single number, the worth of an optical system. The smaller the merit function, the better the system.

microscope, compound   A two component system where the objective lens forms a magnified image of a small object which is further magnified by the eyepiece.

microscope, simple   A magnifying glass.

modulation   The contrast in an object or image the luminance or illuminance of which varies sinusoidally. Defined as $M = $ (max. $-$ min.)/ (max. $+$ min.), where max. and min. are the maximum and minimum levels of luminance or illuminance.

monocentric   A system in which all surfaces have a common center of curvature.

MTF (Modulation Transfer Function)   The ratio of the image modulation (or contrast) to that of the object, expressed as a function of the spatial frequency, where the object modulation is a sinusoidal variation of brightness/luminance/radiance and the image modulation is a sinusoidal variation of the illuminance/irradiance. MTF $= (M_{image})/(M_{object})$. MTF is the real part of the complex optical transfer function (OTF), in which the imaginary part is the phase transfer function (PTF). It was originally known as the sine-wave response, the frequency response and the contrast transfer function.

NA See numerical aperture.

narcissus The image of a cold detector formed by reflection from a surface of the optical system, which when (nearly) in focus produces a dark central spot in the image.

new achromat A cemented achromat where the positive crown element has a higher index than the negative flint element. The convergent cemented surface cannot correct the spherical aberration, but the Petzval sum can be smaller than an "old" achromat or an equivalent singlet.

nodal points Two axial points such that an oblique ray aimed toward the first nodal point emerges from the lens parallel to its original direction, and appears to come from the second nodal point. If the system is immersed in the same medium on both sides (e.g., air), the nodal points and the principal points are coincident.

node In an anastigmat lens, the image height at which the astigmatism is zero, i.e., where the sagittal and tangential fields cross.

null lens An optical system designed to convert the nominal design wave front of a system under test into a perfectly spherical wave front, for easy testing by an interferometer.

numerical aperture, NA The numerical aperture, NA $= n \sin U$, where n is the index of the medium in which the image is formed, and U is the half angle of the imaging cone. For an infinitely distant object, NA $= 1/(2\ f\text{-number})$, and f-number $= 1/(2\text{NA})$.

Nyquist frequency The spatial frequency resolution limit imposed by the size of the pixels in a digital sensor (e.g., CCD). The Nyquist frequency is equal to the reciprocal of twice the pixel spacing d, or freq $= 1/(2d)$. In practice, $1/(2.8d)$ is a realistic spatial frequency limit.

object That which is being imaged by the optical system.

objective lens In a camera, telescope, microscope, or other optical system, the lens which is closest to the object.

oblique beam or ray A beam or ray which originates from an off-axis object point.

off-axis   Regions which are not at, or close to, the optical axis.

offense against the sine condition (OSC)   See Abbe sine condition.

old achromat   A cemented achromat where the positive crown element has a lower index than the negative flint element. The dispersive cemented surface works to correct the spherical aberration, but the Petzval sum is worse than that of a singlet.

operand   The name commonly used for an entry or target in the merit function. It usually defines a system characteristic which is desired to be minimized or controlled.

ophthalmologist   An MD who specializes in the physiology of the eye.

optical axis   The common axis of symmetry of a lens or optical system. See axis, optical.

optical glass   An amorphous, clear, highly transmissive material, made with accurately controlled index of refraction and chromatic dispersion.

optical path difference (OPD)   Wave-front aberration. The departure of the actual wave front from an ideal spherical wavefront. The difference between the optical paths, $\Sigma(nD)$, of two rays measured from their common point of origin in the object to their intersection with a reference sphere centered on the ideal object point. Usually measured in wavelengths, or fractions thereof.

optical path length   The index times the path distance along a ray. $OP = \Sigma(nD)$. Related to the transit time of light along a ray through a system.

optical transfer function (OTF)   The complex function of spatial frequency used to describe the imagery of an optical system. It consists of the real part (the MTF, or Modulation Transfer Function), and the imaginary part (the PTF, or Phase Transfer Function).

optician   One who fabricates (grinds, polishes, etc.) optical elements.

optimum, global   The best possible form for an optical system, i.e., that with the smallest possible merit function. In practice, it is unknowable for even a modestly complex system.

optimum, local   A solution to the design problem which represents a minimum of the merit/defect/error function, such that any small parameter change (or any combination of small parameter changes) will produce an increase in the merit function. An optical system may have many local optima.

optometrist   One who measures the eye and prescribes corrective spectacles. An O.D.

OSC The offense against the sine condition. See Abbe sine condition.

parameter See variable.

paraxial A region where all angles are treated as infinitesimals, so that $\phi = \sin\phi = \tan\phi$, and the equations for raytracing are simple linear (i.e., nontrigonometric) expressions. "A thin thread-like region about the optical axis." Paraxial equations describe the imagery of perfect aberration-free optical systems. Also called first order or gaussian.

paraxial ray A raytraced according to the paraxial rules, where the ray heights and angles are infinitesimals. The linearity of the paraxial raytracing equations allows the use of fictitious, real, finite values for height and slope.

parfocal Lenses which are adjusted so that when in focus, the distance from the image (or object) to a mounting shoulder is the same for all lenses in the set.

partial dispersion The difference in refractive index for two wavelengths, expressed as a fraction of the total dispersion, e.g., $P_{F,d} = (n_F - n_d)/(n_F - n_C)$.

periscope An optical system used to image a relatively wide field of view through a long, narrow space. It usually consists of an objective, followed by alternating field and relay lenses, and is terminated by an eyepiece or camera. A medical endoscope is a miniature periscope. A coherent fiber bundle or a radial index gradient rod may substitute for the field lens-relay lens system. Alternate definition: A rhomboid prism or a pair of parallel mirrors used to displace, but not deviate, the line of sight, e.g., a child's toy periscope.

periscopic lens A camera lens consisting of two meniscus singlets, spaced apart, and oriented with concave surfaces facing, with an aperture stop midway between.

Petzval curvature or sum A measure of the basic field curvature of a lens. The Petzval sum is equal to $\Sigma(n' - n)/nn'r$. The radius of curvature (ρ) of the Petzval surface equals (minus) the reciprocal of the Petzval sum. For a thin element, $\rho = -n/\varphi = -nf$. The Petzval surface is the locus of the image when there is no astigmatism.

pixel A picture element. In a digital sensor, the image is sensed (and recorded) by a tiled array of very small detector elements or pixels.

point source In geometrical optics a point source is a true geometric point, with dimensions of zero by zero. In radiometry it often means "sufficiently small."

point spread function   The distribution of illumination in the image of a point.

polishing tool   A metal tool lined with pitch, felt or polyurethane, used with a polishing compound such as rouge or cerium oxide to polish or shine a finely ground surface.

power   The power of a lens is the reciprocal of its effective focal length. The power of a surface is equal to $(n' - n)/r$. If the dimensions are in meters, the unit of power is the diopter. A positive power converging lens, or surface bends rays toward the axis; a negative power, diverging lens bends rays away from the axis. Power may also refer to the angular magnifying power of a telescope, microscope, or magnifier.

principal plane   The hypothetical surface in a lens at which it appears that an incoming paraxial ray, parallel to the optical axis, is bent. If the incoming and the emerging rays are extended until they intersect, the intersection point is on the principal plane. The second principal plane is defined by rays from the left, and the first by rays from the right. For real trigonometric rays (i.e., not paraxial) this surface is curved, approximating a sphere centered on the object or image point. The surface is a plane only in the infinitesimal paraxial region. See cardinal points and focal length.

principal point   The point at which the principal plane (or surface) intersects the optical axis.

principal ray   See chief ray. Infrequently, an oblique ray directed toward the first principal point.

pupil   Any image of the aperture stop. See also entrance pupil and exit pupil.

rapid rectilinear   A symmetrical camera lens consisting of two identical meniscus achromatic doublets, oriented with concave surfaces facing, and the crown elements facing.

ray   See axial ray, chief ray, marginal ray, meridional ray, oblique ray, paraxial ray, principal ray, rim ray, sagittal ray, trigonometric ray, exact ray, tangential ray, zonal ray.

ray intercept plot   A plot of the intercept locations of a fan of rays versus the relative position of the ray in the lens aperture or pupil. Usually plotted for a fan of tangential (meridional) rays or a fan of sagittal rays. The plotted location is usually the difference between the location of the ray and that of a reference ray, such as the axis or the principal ray. For tangential rays the y coordinate is plotted; for sagittal rays, the x coordinate. The relative pupil position of the ray is sometimes given as the tangent of the ray slope, hence the term "$H - \tan U$ plot."

Rayleigh criterion: image quality   If the wave front forming the image is perfectly spherical to within a quarter of a wavelength, Rayleigh felt that the image would be sensibly perfect. A system meeting the quarter-wave criterion is often (incorrectly) called "diffraction limited." See also Marechal.

Rayleigh criterion: resolution   Two points are assumed to be resolvable by a perfect optical system if their images are separated by $0.61\lambda/NA$, or if the object points have an angular separation of $1.22\lambda/D$ radians where D is the entrance pupil diameter (or $5.5/D$ seconds of arc, where D is in inches, and visual wavelengths are assumed). The Sparrow and Dawes resolution criteria are approximately 20 percent smaller.

real field of view   The angular field of view in object space.

real image   An image which can be formed on a screen (as opposed to a virtual image, which is located inside the optical system and cannot be directly accessed).

refraction   The bending or directional change of a light ray upon passing from one medium to another. Refraction follows Snell's Law, $n \sin l = n' \sin l'$, where n and n' are the indices of refraction of the media and l and l' are the angles of incidence and refraction.

refractive index   The ratio of the velocity of light in vacuum (or, commonly, in air) to its velocity in the medium being characterized ($n = c/v$). Related to the bending of a light ray at a surface, as in Snell's law: $n \sin l = n' \sin l'$.

relative aperture   The ratio of the focal length of a lens to its clear aperture. May be given as a ratio, 1:6.3, or as a fraction, $f/6.3$. In this example 6.3 is called the f-number. With an object at infinity, f-number = $0.5/NA$.

relay lens   A lens used to transfer an image longitudinally from one location to another. It reimages and erects a real internal image, as in a periscope or terrestrial telescope.

resolution   The ability of an optical system to image small details. The smallest separation of lines in an image. Often specified as lines per millimeter or as line pairs per millimeter or as cycles per millimeter. See also cut-off frequency, Rayleigh criterion, modulation transfer function.

reversed telephoto or retrofocus   A lens system whose back focus is longer that its effective focal length. It consists of a negative front component followed by a positive rear component.

reverted   Inverted in one meridian. Use this term with care.

rim ray   A ray through the edge or rim of the aperture. The upper and lower rim rays are oblique meridional rays through the top and bottom of the clear aperture.

Risley prisms   A matched pair of wedges which, when counter-rotated, linearly deviate a light beam.

Ritchey-Chretien   A Cassegrain system with both mirrors hyperbolic, shaped to correct both spherical and coma.

RMS   The acronym for root mean square, the square root of the average of the squares of a set of numbers. The rms spot size and the rms OPD are often used as operands in a merit function.

RSS   The acronym for the square root of the sum of the squares. Used to assess the combined effect of multiple random fabrication tolerances.

sagittal plane or ray   A plane normal to the meridional/tangential plane and which includes the principal ray. Usually defined in object space. A ray which lies in the sagittal plane in object space.

Scheimpflug condition   Defines the tilted plane on which the image of a tilted object is formed.

Schwarzschild   A two-mirror objective with a convex primary mirror and a large concave secondary mirror.

secondary spectrum   The residual chromatic aberration when the primary chromatic aberration has been corrected. For example, if red and blue light have been brought to a common focus, the distance from that focus to the yellow-green focus is the secondary spectrum.

Seidel aberrations   The third-order aberrations, which are: spherical, coma, astigmatism, Petzval, and distortion. Also called the primary aberrations.

semi-coherent   Light from a very small but finite sized source and/or light with a very small spectral bandwidth. Also the illumination produced when the image of the light source only partially fills the pupil of the lens, as in microscopy or microlithography.

simulated annealing   A controlled, random search, design process which allows the lens design to temporarily become worse in order to find a new optimum.

skew ray   A general ray, not limited to the tangential/meridional plane.

sky lens   A very wide angle (ca 180 deg) lens designed to photograph the entire hemisphere of the sky. Usually consists of an extremely strong, large negative front lens component spaced away from a rear positive component. A "fish eye" lens.

Snell's law   The change in direction of a ray crossing the boundary between two media, is governed by Snell's law, which is: $n \sin l = n' \sin l'$, where n and n' are the refractive indices of the two media, and l and l' are the angles of incidence and refraction (the angle between the ray and the surface normal).

solve   A capability of an optical design program which algebraically solves for a construction parameter (such as a curvature or a spacing) which will produce a desired paraxial ray slope or intersection height. The ray chosen is usually the axial marginal ray or the principal/chief ray. An "angle solve" or a "height solve."

speed   See f-number.

spherical aberration   The difference between the focus location of rays through the center of a lens aperture (i.e., paraxial rays) and those through the margin (or other parts) of the aperture.

spherochromatism   The variation of spherical aberration with wavelength, or the variation of axial chromatic with ray height.

spot diagram   A plot of the intersection points of rays (from an object point) in the image surface, where each intersection point is represented by a spot. If the rays are uniformly distributed in the aperture, the spot diagram is a representation of the illumination distribution in the image (when diffraction effects are neglected). Usually several hundred rays are plotted, often in several colors, to make a single diagram.

stigmatic   Perfect imagery; all rays from an object point pass through the same image point.

stop   See aperture stop or field stop.

Strehl ratio   The ratio of the peak intensity of the point spread function for an aberrated lens to the peak for an aberration-free lens. A Strehl ratio of 80 percent (called the Marechal criterion) corresponds to the quarter-wave Rayleigh criterion (exactly for defocusing, approximately for the other aberrations). The Strehl ratio has an excellent correlation with other image quality metrics for well corrected systems.

sun shade   See lens hood.

symmetrical lens   Most lenses are rotationally, axially symmetrical. Another type of symmetry is "front-to-back" or "mirror" symmetry where the elements before the aperture stop are the same as those which follow it. Front-to-back symmetry eliminates coma, distortion, and lateral color aberrations.

tangential plane; tangential ray   See meridional.

telecentric   A system with the entrance and/or exit pupil located at infinity, so that the associated principal ray is parallel with the axis. Used to prevent a change in image size when the system is slightly defocused, as in metrology or microlithography.

telephoto lens   A lens the length from the first surface of which to the focal point is shorter than its effective focal length. The ratio of the two is called the telephoto ratio which, for a true telephoto, is less than one. The lens consists of a positive front component followed by a negative rear component. The name is sometimes incorrectly applied to an ordinary lens of long focal length.

telescope   An afocal system which produces a magnified image of a distant object. A Keplerian or astronomical telescope comprises two positive components producing an inverted image. A Galilean or "Dutch" telescope has a positive objective and a negative eyepiece, and produces an erect image. A terrestrial telescope includes an erecting relay lens component.

test plate or test glass   A precisely made spherical gage of glass (typically Pyrex or fused quartz) which is placed in contact with a surface to be tested. Interference fringes indicate the distance between the gage and the work. Noncircular fringes indicate a nonspherical surface. A "full fit" means the gage and work contact in the center; a "hollow fit" indicates an edge contact.

thin lens   A *concept* which is useful in preliminary optical system layout. It assumes that the optical components have zero axial thickness, so that the principal points and the lens are coincident.

third order   Aberrations which (in transverse measure) vary as the combined exponents of the aperture (y) and field (h) equaling three, i.e., y^3, y^2h, yh^2, yh^2, and h^3. The corresponding five (third order) Seidel aberrations are: spherical, coma, astigmatism, Petzval, and distortion. In the fifth-order aberrations the exponents add up to five (y^5, y^4h, yh^4, yh^4, and h^5), and there are two additional aberrations–elliptical coma (y^2h^3) and oblique spherical aberration (y^3h^2).

throughput or etendue   See also invariant. The invariant product of pupil area and the solid angle field of view, or the product of beam diameter and the solid angle of the convergence cone. Related to the square of the Lagrange invariant.

TIR (total internal reflection) In passing from a medium of higher index to a lower index, a ray is totally reflected back into the higher index medium if the angle of incidence exceeds the critical angle, which equals arcsin (n'/n).

T-number An equivalent f-number which includes the effect of transmission losses.

$$\text{T-number} = (f\text{-number})/\sqrt{\text{transmission}}$$

Total dispersion The index difference between the wavelengths which define the spectral band of interest. For visual work it is often taken as the index difference between F-light and C-light.

track length; total track (TT) The object to image distance.

transverse Measured in a direction normal to the optical axis.

trigonometric ray A ray the path of which is traced according to Snell's law (which see) as opposed to a paraxial ray. Also called an exact ray.

triplet anastigmat A three-element lens with two positive outer crown elements and a central negative flint element, all spaced apart. A Cooke triplet.

triplet, cemented Three elements cemented together; often as a magnifier such as a Hastings or a Coddington triplet.

tube length In a microscope, the distance between the focal points of the objective and eyepiece. Often standardized at 160 mm or 215 mm.

tunnel diagram An "unfolded" diagram of a prism system where each reflection is unfolded so that a ray path through the prism can be drawn as a straight line. See, for example, Figs. 7.16, 7.17, 7.19, 7.38, and 7.39.

variable A construction element (such as a surface curvature, a surface spacing, a material index or dispersion, a surface asphericity, or the like) which may be varied for the purpose of improving a system. A variable parameter.

varifocal lens A lens assembly whose spacings can be changed to change the focal length. A zoom lens is a varifocal lens which also maintains focus as the focal length is changed.

vertex In an axially symmetric system, the vertex of a surface is the point where the surface intersects the optical axis.

vertex length The axial distance from the vertex of the first optical surface of a system to the vertex of the last.

vignetting The mechanical clipping or obscuration of the edges of oblique beams by the apertures of elements which are spaced away from the stop. It reduces the off-axis illumination (in addition to the cosine-fourth reduction). Often introduced to reduce manufacturing cost and/or to eliminate aberrated portions of the beam. In visual or photographic systems, as much as 50 percent vignetting is not uncommon.

virtual image An image formed within or before the optical system and which therefore cannot be accessed or focused on a screen, as opposed to a "real" image which can.

visible light Light to which the human eye is sensitive and which can be perceived. Usually considered to include wavelengths from 380 nm to 780 nm, at which wavelengths the photopic response is less than 0.0001 of the peak photopic response at 555 nm.

V-**value** See Abbe *V*-number.

wave front A surface wherein all points have the same optical path distance (Σnd) from the object point, i.e., where the light has the same phase.

wave front aberration The departure of the wave front from a perfectly spherical surface. See optical path difference (OPD).

working *f*-number Describing the convergence of the imaging cone when the object is at a finite distance; equal to 0.5/NA. As opposed to the conventional "infinity" *f*-number. This is not an established convention; caution is advised.

zonal aberration The aberration of rays in the mid-zones (e.g., 0.7) of the aperture or field, usually when the aberrations of the central and marginal rays are corrected.

zonal ray Usually a ray the height in the aperture of which is 0.707 of the height of the marginal ray.

zoom lens A lens the focal length of which is changed by changing the component spacings and which also maintains focus while zooming.

Index

Abbe prism, 143, 143*f*
Abbe sine condition (OSC), 378, 421, 709
Abbe *V*-number, 115, 210, 709
  for glass, 211
aberration, 498–500. *See also* angular
  aberration; astigmatism; blur;
  chromatic aberration; coma;
  distortion; offense against Abbe sine
  condition; optical path difference;
  primary aberration; residual
  aberration; spherical aberration;
  transverse aberration
  aperture and, 88–89, 89*f*
  in biconvex lenses, 668
  correction of, 91–94
  definition of, 709
  field and, 88–89, 89*f*
  lens power and, 670
  lens shape and, 87*f*, 88*f*
  measurement of, 638–640
  OPD and, 76–77
  P-V OPD and, 654*f*, 655
  ray intercept curves for, 97*f*
  ray-tracing and, 698–705
  RL and, 379
  in stock lenses, 668–671
  tolerances for, 373–378
aberration polynomial
  astigmatism, 79
  coma and, 77–79
  distortion and, 81–83
  field curvature and, 79
  OPD and, 373
  primary aberration and, 72–83
  spherical aberration and, 74–77
absorption
  in optical materials, 205–210
  of ultraviolet, 208
absorption filters, 224–227
accommodation, 158, 169*f*

achromat(s)
  airspaced, 428–429
  AMTIR and, 427
  calcium fluoride and, 427
  definition of, 709
  diffractive singlet, 433
  doublet, 675*f*
    athermalization in, 429–430
    chromatic aberration in, 675
    coma in, 423*f*, 675
    spherical aberration in, 423*f*, 424*f*,
      426*f*, 675
    SS of, 427
    in telescopes, 91*f*, 419–421, 419*f*, 465
  germanium and, 427
  prisms, 126–128, 127*f*
  telescope objectives, 91*f*, 419*f*
    design forms, 421–430
    thin lens, 417–421
  triplet, 427–428
    Schott glass for, 428
acid resistance, 452
additive tolerances, 622–628
  frequency distribution curves in,
    622–623, 623*f*
aerial image modulation (AIM), 386
afocal attachments, 493–494
afocal systems, 291, 709
AIM. *See* aerial image modulation
air
  index of refraction of, 4
  optical system in, 26
airspaced achromats, 428–429
airspaced anastigmats, 479–487
airspaced apochromatic triplet, 537*f*
airspaced triplet telescope objective, 534*f*
Airy disk, 190, 261, 710
  illumination in, 192*f*
  Strehl ratio and, 375
alignment telescope, 465*f*

www.ingramcontent.com/pod-product-compliance
Lightning Source LLC
Chambersburg PA
CBHW060415220326
41598CB00021BA/2184